D.L. Domingue • C.T. Russell
Editors

The MESSENGER Mission to Mercury

Foreword by D.L. Domingue and C.T. Russell

Previously published in *Space Science Reviews* Volume 131, Issues 1–4, 2007

Springer

D.L. Domingue
The Johns Hopkins University
Applied Physics Laboratory
Laurel, MD, USA

C.T. Russell
Institute of Geophysics & Planetary Physics
University of California
Los Angeles, CA, USA

Cover illustration: Artist's rendition of the MESSENGER spacecraft orbiting Mercury.

Library of Congress Control Number: 2007941871

ISBN-978-0-387-77211-0 e-ISBN-978-0-387-77214-1

Printed on acid-free paper.

9 8 7 6 5 4 3 2 1

springer.com

Contents

Space Sci Rev (2007) 131: 1–2
DOI 10.1007/s11214-007-9275-2

Foreword

D.L. Domingue · C.T. Russell

Published online: 21 September 2007

Fifteenth and sixteenth century explorers conquered the oceans of this world with wooden sailing ships, reaching every corner of the globe by relying on the precarious nature of the winds and their strength of will. These were long, difficult journeys oftentimes in harsh environments. Success meant great rewards both financially, by opening new trade routes, and scientifically, by making discoveries that still benefit us today. Twentieth and twenty-first century explorers now sail the vast emptiness of space, making new discoveries amongst the stars their ancestors used for navigation. These ventures are difficult, and they are just as costly as they were to the coffers of seafaring nations five centuries ago. Yet we still pursue them, driven to expanding the boundaries of our world and trusting that these voyages will bring home scientific riches, not least of which is a new and deeper understanding of our planetary ancestral roots.

Ancient explorers would return home with wondrous tales and artifacts from exotic ports of call. Our spacecraft make ports of call at the planets themselves, returning tales and wonders in the information and data they send home. Like ancient Mariners before it, the MESSENGER spacecraft braves its own set of harsh environments to visit such ports of call as Venus and Mercury, the innermost and most forbidding of the terrestrial planets. As a second-generation explorer of this region, MESSENGER does not simply pass by its ultimate target, Mercury, but it establishes a long-term presence in orbit, perhaps paving the way for more ambitious settlement later.

This volume describes the MESSENGER mission to Mercury and our present understanding of this exotic, alien land beginning with an overview of the mission by the principal investigator (S.C. Solomon et al.). It is followed by articles on the geology (J. Head, III et al.), surface geochemistry (W. Boynton et al.), surface and interior geophysical properties (M.T. Zuber et al.), the magnetosphere (J.A. Slavin et al.), and the atmosphere

D.L. Domingue
The Johns Hopkins University Applied Physics Laboratory, Laurel, MD 20723, USA

C.T. Russell (✉)
University of California, Los Angeles, CA 90095, USA
e-mail: ctrussell@igpp.ucla.edu

(D. Domingue et al.). The mission to Mercury is no less intriguing than the target. The spacecraft has to operate in extremely harsh thermal and solar environments and the navigation of the interplanetary trade winds involves as much art as science (J. Leary et al. and J. McAdams et al., respectively). While the brains and brawn of such missions are in the spacecraft, the heart and soul reside within the payload. The payload is comprehensive, as befits the multifaceted nature of Mercury and its environment. The dual imaging system (S. Hawkins et al.) will return images of the surface never before seen by a spacecraft. The gamma ray and neutron spectrometer (J. Goldsten et al.) along with the X-ray spectrometer (C. Schlemm et al.) will provide the first information about the elemental chemistry of the Mercurian surface. The MESSENGER magnetometer (B. Anderson et al.) will map the magnetosphere only glimpsed by Mariner 10. The laser altimeter has been designed (J. Cavanaugh et al.) to provide topographic information that will be used to help unravel the mysteries of Mercury's surface evolution. The atmosphere and surface composition spectrometer (W. McClintock and M. Lankton) will provide the first in situ measurements of the atmosphere and the first high spatial mineral maps of the surface. The energetic particle and plasma spectrometer (G. Andrews et al.) will provide insight into the space environment and the intricate connections between solar particles, magnetosphere, atmosphere, and surface properties. And, as is traditional, the last science system to be described is the radio system (D. Srinivasan et al.) that provides the gravity science needed to understand the evolution of the planet's interior. The operation of this mission (M. Holdridge and A. Calloway) is a complex balancing of subsystem operations and constraints that guide the spacecraft through the harsh environment to its final destination and goal. Science operations (H. Winters et al.) describes how the glorious tales of the journey, captured through the observations and measurements of the spacecraft payload, will be disseminated and retold for generations to come.

The success of this volume is due to many people, but first of all the editors wish to thank the authors who had the difficult job of distilling the thousands of documents and the millions of facts such missions produce into highly readable documents. The editors also benefited from an excellent group of referees who acted as a test readership, refining the manuscripts provided by the authors. These referees included: T. Armstrong, R. Arvidson, W. Baumjohann, M. Bielefeld, D. Blewett, D. Blaney, D. Byrnes, A. Cheng, U. Christensen, T. Cole, A. Dombard, W.C. Feldman, K.H. Glassmeier, J. Green, S. Joy, K. Klaasen, A. Konopliv, J. Longuski, W. Magnes, A. Matsuoka, T. McCoy, L. Nittler, T. Perron, T.H. Prettyman, M. Ravine, G. Schubert. M. Smith, H. Spence, P. Spudis, V.C. Thomas, F. Vilas, J. Witte, and D. Yeomans. The MESSENGER PI, S. Solomon, also provided excellent reviews and helped to mold this issue into a consistent view of the mission. Equally important has been the strong support this project received at Springer and the extra effort expended by Fiona Routley, Randy Cruz, and Harry Blom. At UCLA we were skillfully assisted by Marjorie Sowmendran who acted as the interface between the editors, the authors, and the publishers.

Space Sci Rev (2007) 131: 3–39
DOI 10.1007/s11214-007-9247-6

MESSENGER Mission Overview

**Sean C. Solomon · Ralph L. McNutt, Jr. ·
Robert E. Gold · Deborah L. Domingue**

Received: 9 January 2007 / Accepted: 13 July 2007 / Published online: 5 October 2007

Abstract The MErcury Surface, Space ENvironment, GEochemistry, and Ranging (MESSENGER) spacecraft, launched on August 3, 2004, is nearing the halfway point on its voyage to become the first probe to orbit the planet Mercury. The mission, spacecraft, and payload are designed to answer six fundamental questions regarding the innermost planet: (1) What planetary formational processes led to Mercury's high ratio of metal to silicate? (2) What is the geological history of Mercury? (3) What are the nature and origin of Mercury's magnetic field? (4) What are the structure and state of Mercury's core? (5) What are the radar-reflective materials at Mercury's poles? (6) What are the important volatile species and their sources and sinks near Mercury? The mission has focused to date on commissioning the spacecraft and science payload as well as planning for flyby and orbital operations. The second Venus flyby (June 2007) will complete final rehearsals for the Mercury flyby operations in January and October 2008 and September 2009. Those flybys will provide opportunities to image the hemisphere of the planet not seen by Mariner 10, obtain high-resolution spectral observations with which to map surface mineralogy and assay the exosphere, and carry out an exploration of the magnetic field and energetic particle distribution in the near-Mercury environment. The orbital phase, beginning on March 18, 2011, is a one-year-long, near-polar-orbital observational campaign that will address all mission goals. The orbital phase will complete global imaging, yield detailed surface compositional and topographic data over the northern hemisphere, determine the geometry of Mercury's internal magnetic field and magnetosphere, ascertain the radius and physical state of Mercury's outer core, assess the nature of Mercury's polar deposits, and inventory exospheric neutrals and magnetospheric charged particle species over a range of dynamic conditions. Answering the questions that have guided the MESSENGER mission will expand our understanding of the formation and evolution of the terrestrial planets as a family.

S.C. Solomon (✉)
Department of Terrestrial Magnetism, Carnegie Institution of Washington, Washington, DC 20015, USA
e-mail: scs@dtm.ciw.edu

R.L. McNutt, Jr. · R.E. Gold · D.L. Domingue
The Johns Hopkins University Applied Physics Laboratory, Laurel, MD 20723, USA

Keywords Mercury · MESSENGER · Planet formation · Geological history · Magnetosphere · Exosphere

1 Introduction

Mercury is the least studied of the inner planets. A substantially improved knowledge of the planet Mercury is nonetheless critical to our understanding of how the terrestrial planets formed and evolved. Determining the surface composition of Mercury, a body with a ratio of metal to silicate higher than any other planet or satellite, will provide a unique window on the processes by which planetesimals in the primitive solar nebula accreted to form planets. Documenting the global geological history will elucidate the roles of planet size and solar distance as governors of magmatic and tectonic history for a terrestrial planet. Characterizing the nature of the magnetic field of Mercury and the size and state of Mercury's core will allow us to generalize our understanding of the energetics and lifetimes of magnetic dynamos, as well as core and mantle thermal histories, in solid planets and satellites. Determining the nature of the volatile species in Mercury's polar deposits, atmosphere, and magnetosphere will provide critical insight into volatile inventories, sources, and sinks in the inner solar system.

MESSENGER is a MErcury Surface, Space ENvironment, GEochemistry, and Ranging mission designed to achieve these aims. As part of the Discovery Program of the U.S. National Aeronautics and Space Administration (NASA), the MESSENGER spacecraft will orbit Mercury for one Earth year after completing three flybys of that planet following two flybys of Venus and one of Earth. The Mercury flybys will return significant new data early in the mission, while the orbital phase, guided by the flyby data, will enable a focused scientific investigation of the innermost planet. Answers to key questions about Mercury's high density, crustal composition and structure, volcanic history, core structure, magnetic field generation, polar deposits, atmosphere, overall volatile inventory, and magnetosphere will be provided by an optimized set of seven miniaturized scientific instruments. In this paper we first describe the rationale for and scientific objectives of the MESSENGER mission. We then summarize the mission implementation plan designed to satisfy those objectives. Companion papers in this issue provide detailed descriptions of the MESSENGER spacecraft (Leary et al. 2007) and mission design (McAdams et al. 2007), mission (Holdridge and Calloway 2007) and science operations centers (Winters et al. 2007), payload instruments (Anderson et al. 2007; Andrews et al. 2007; Cavanaugh et al. 2007; Goldsten et al. 2007; Hawkins et al. 2007; McClintock and Lankton 2007; Schlemm et al. 2007), and radio science (Srinivasan et al. 2007), as well as more expansive summaries of the principal scientific issues to be addressed by a Mercury orbiter mission (Boynton et al. 2007; Domingue et al. 2007; Head et al. 2007; Slavin et al. 2007; Zuber et al. 2007).

2 Context for MESSENGER Selection

The selection of MESSENGER as a NASA Discovery Program mission was a decision rooted in a 25-year history of Mercury exploration and strategic planning for improving our understanding of the inner planets.

The only spacecraft to visit Mercury to date was Mariner 10. In the course of three flybys of the planet in 1974 and 1975, Mariner 10 imaged about 45% of Mercury's surface

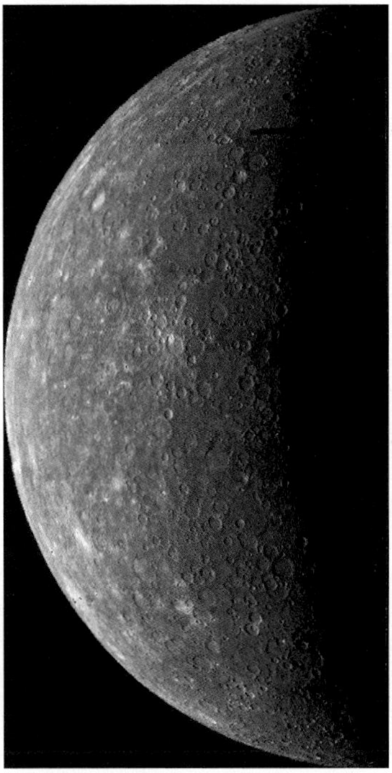

Fig. 1 Mosaic of images of Mercury obtained by the Mariner 10 spacecraft on the incoming portion of its first flyby of Mercury (Robinson et al. 1999)

(Fig. 1) at an average resolution of about 1 km and less than 1% of the surface at better than 500-m resolution (Murray 1975). Mariner 10 discovered the planet's internal magnetic field (Ness et al. 1974, 1975); measured the ultraviolet signatures of H, He, and O in Mercury's tenuous atmosphere (Broadfoot et al. 1974, 1976); documented the time-variable nature of Mercury's magnetosphere (Ogilvie et al. 1974; Simpson et al. 1974); and determined some of the physical characteristics of Mercury's surface materials (Chase et al. 1974).

Immediately following the Mariner 10 mission, a Mercury orbiter was widely recognized as the obvious next step in the exploration of the planet (COMPLEX 1978). Further, the primary objectives of such an orbiter mission were defined: "to determine the chemical composition of the planet's surface on both a global and regional scale, to determine the structure and state of the planet's interior, and to extend the coverage and improve the resolution of orbital imaging" (COMPLEX 1978). In the late 1970s, however, it was thought that the change in spacecraft velocity required for orbit insertion around Mercury was too large for conventional propulsion systems, and this belief colored the priority placed on further exploration of the innermost planet (COMPLEX 1978).

In the mid-1980s, about a decade after the end of the Mariner 10 mission, multiple gravity-assist trajectories were discovered that could achieve Mercury orbit insertion with chemical propulsion systems (Yen 1985, 1989). This finding stimulated detailed studies of Mercury orbiter missions in Europe and the United States between the mid-1980s and early 1990s (Neukum et al. 1985; Belcher et al. 1991). During the same time interval there were important discoveries made by ground-based astronomy, including the Na and K components of Mercury's atmosphere (Potter and Morgan 1985, 1986) and the radar-reflective deposits at Mercury's north and south poles (Harmon and Slade 1992; Slade et al. 1992). A re-examination of the primary objectives of a Mercury orbiter mission during that period affirmed those defined earlier and added "that characterization of

Mercury's magnetic field be [an additional] primary objective for exploration of that planet" (COMPLEX 1990).

In the early 1990s, after re-examining its approach to planetary exploration, NASA initiated the Discovery Program, intended to foster more frequent launches of less costly, more focused missions selected on the basis of rigorous scientific and technical competition. Mercury was the target of a number of early unsuccessful proposals to the Discovery Program for flyby and orbiter missions (Nelson et al. 1994; Spudis et al. 1994; Clark et al. 1999). The MESSENGER concept was initially proposed to the NASA Discovery Program in 1996, and after multiple rounds of evaluation (McNutt et al. 2006) the mission was selected for flight in July 1999.

In parallel with the selection, development, and launch of MESSENGER, the European Space Agency (ESA) and the Institute of Space and Astronautical Science (ISAS) of the Japan Aerospace Exploration Agency (JAXA) have approved and are currently developing the BepiColombo mission to send two spacecraft into Mercury orbit (Grard et al. 2000; Anselmi and Scoon 2001). BepiColombo was selected by ESA as its fifth cornerstone mission in 2000, and ISAS announced its intent to collaborate on the project that same year. The two spacecraft, scheduled for launch on a single rocket in 2013, will be in coplanar polar orbits. An ESA-supplied Mercury Planetary Orbiter will emphasize observations of the planet, and an ISAS-supplied Mercury Magnetospheric Orbiter will emphasize observations of the magnetosphere and its interactions with the solar wind. Payload instruments on the two spacecraft were selected in 2004 (Hayakawa et al. 2004; Schulz and Benkhoff 2006).

3 Guiding Science Questions

The MESSENGER mission was designed to address six key scientific questions, the answers to which bear not only on the nature of the planet Mercury but also more generally on the origin and comparative evolution of the terrestrial planets as a class.

3.1 What Planetary Formational Processes Led to the High Ratio of Metal to Silicate in Mercury?

Mercury's uncompressed density (about 5.3 Mg/m^3), the highest of any planet, has long been taken as evidence that iron is the most abundant contributor to the bulk composition. Interior structure models in which a core has fully differentiated from the overlying silicate mantle indicate that the core radius is approximately 75% of the planetary radius and the fractional core mass is about 60% if the core is pure iron (Siegfried and Solomon 1974); still larger values are possible if the core has a light element such as sulfur alloyed with the iron (Harder and Schubert 2001). Such a metallic mass fraction is at least twice that of the Earth (Fig. 2), Venus, or Mars.

Calculations of dynamically plausible scenarios for the accretion of the terrestrial planets permit a wide range of outcomes for Mercury. Given an initial protoplanetary nebular disk of gas and dust, planetesimals accrete to kilometer size in 10^4 years (Weidenschilling and Cuzzi 1993), and runaway growth of planetary embryos of Mercury- to Mars-size accrete by the gravitational accumulation of planetesimals in 10^5 years (Kortenkamp et al. 2000). During runaway growth, Mercury-size bodies can experience substantial migrations of their semimajor axes (Wetherill 1988). Further, each of the terrestrial planets probably formed from material originally occupying a wide range in solar distance, although some correlation

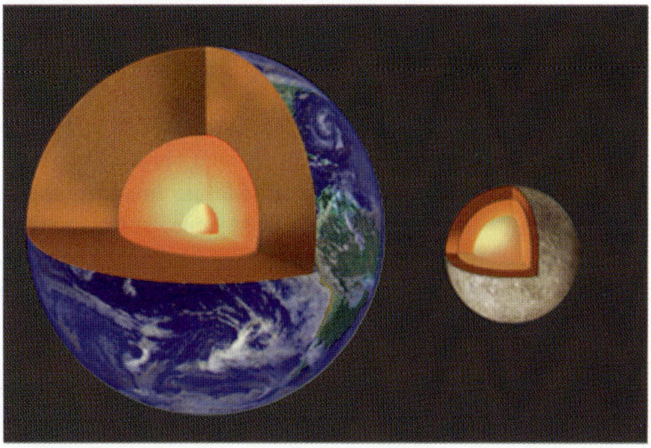

Fig. 2 Schematic cut-away views of the fractional volumes occupied by the central metallic cores of Mercury and Earth. The solid inner core and fluid outer core of the Earth are shown to approximate scale. Mercury's outer core is likely fluid (Margot et al. 2007), but the core radius and the nature of any inner core remain to be determined

is expected between the final heliocentric distance of a planet and those of the planetesimals from which it formed (Wetherill 1988, 1994).

Three explanations for the high metal fraction of Mercury have been put forward. The first invokes differences in the response of iron and silicate particles to aerodynamic drag by nebular gas to achieve fractionation at the onset of planetesimal accretion (Weidenschilling 1978). The second and third explanations invoke processes late in the planetary accretion process, after the Mercury protoplanet had differentiated silicate mantle from metal core. In one, the high metal content of Mercury is attributed to preferential vaporization of silicates by radiation from a hot nebula and removal by a strong solar wind (Cameron 1985; Fegley and Cameron 1987). In the other, selective removal of silicate occurred as a result of a giant impact (Benz et al. 1988; Wetherill 1988, 1994).

These three hypotheses lead to different predictions for the bulk chemistry of the silicate fraction of Mercury (Lewis 1988; Boynton et al. 2007). Under the giant impact hypothesis, the residual silicate material on Mercury would be dominantly of mantle composition. The FeO content would reflect the oxidation state of the material from which the protoplanet accreted, but the loss of much of the original crust would deplete Ca, Al, and alkali metals without enriching refractory elements. The vaporization model, in contrast, predicts strong enrichment of refractory elements and depletion of alkalis and FeO (Fegley and Cameron 1987). Under both of these hypotheses, the present crust should represent primarily the integrated volume of magma produced by partial melting of the relic mantle. Under the aerodynamic sorting proposal (Weidenschilling 1978), the core and silicate portions of Mercury can be prescribed by nebular condensation models, suitably weighted by solar distance, except that the ratio of metal to silicate is much larger (Lewis 1988). This hypothesis permits a thick primordial crust, i.e., one produced by crystal-liquid fractionation of a silicate magma ocean. Determining the bulk chemistry of the silicate portion of Mercury thus offers an opportunity to discern those processes operating during the formation of the inner solar system that had the greatest influence on producing the distinct compositions of the inner planets.

Present information on the chemistry and mineralogy of the surface of Mercury, however, is too limited to distinguish clearly among the competing hypotheses. Ground-based reflectance spectra at visible and near-infrared wavelengths do not show a consistent absorption feature near 1 μm diagnostic of Fe^{2+} (Vilas 1985; Warell et al. 2006), limiting the average FeO content to be less than about 3–4 weight percent (Blewett et al. 1997). Very reduced compositions comparable to enstatite achondrite meteorites with less than 0.1% FeO are compatible with Mercury's reflectance, although a generally red spectral slope is thought to be the result of nanophase iron metal, altered by space weathering from silicates originally containing a few percent FeO (Burbine et al. 2002). Earth-based mid-infrared observations show emission features consistent with the presence of both calcic plagioclase feldspar containing some sodium and very-low-FeO pyroxene; variations in spectral features with Mercury longitude indicate that surface mineralogical composition is spatially heterogeneous (Sprague et al. 2002). Mature lunar highland anorthosite soils are regarded as good general spectral analogues to Mercury surface materials (Blewett et al. 2002).

On the basis of the low FeO content of Mercury's surface materials inferred from Earth-based spectra and Mariner 10 color images, surface units interpreted as volcanic in origin are thought to average no more than about 3% FeO by weight (Robinson and Taylor 2001). On the grounds that the solid/liquid partition coefficient for FeO during partial melting of mantle material is near unity, the mantle FeO abundance has been inferred to be comparable (Robinson and Taylor 2001). This deduction, together with a general increase in bulk silicate FeO content with solar distance for the terrestrial planets and the eucrite parent body, has been taken to suggest both that the inner solar nebula displayed a radial gradient in FeO and that Mercury was assembled dominantly from planetesimals that formed at solar distances similar to that of Mercury at present (Robinson and Taylor 2001).

Substantial progress on understanding the composition of Mercury must await remote sensing by an orbiting spacecraft (Boynton et al. 2007). Also important to an assessment of bulk composition and formation hypotheses would be an estimate of the thickness of Mercury's crust. Variations in crustal thickness can be estimated by a combined analysis of gravity and topography measurements (Zuber et al. 2007). Moreover, an upper bound on mean crustal thickness can be obtained from isostatically compensated long-wavelength topographic variations, on the grounds that the temperature at the base of the crust cannot have been so high that variations in crustal thickness were removed by viscous flow on timescales shorter than the age of the crust (Nimmo 2002).

3.2 What Is the Geological History of Mercury?

A generalized geological history of Mercury has been developed from Mariner 10 images (Head et al. 2007). The 45% of Mercury's surface imaged by Mariner 10 can be divided into four major terrains (Spudis and Guest 1988). Heavily cratered regions have an impact crater density suggesting that this terrain records the period of heavy bombardment that ended about 3.8 Ga on the Moon (Neukum et al. 2001). Intercrater plains, the most extensive terrain type, were emplaced over a range of ages during the period of heavy bombardment. Hilly and lineated terrain occurs antipodal to the Caloris basin—at 1,300 km in diameter the largest and youngest (Neukum et al. 2001) well-preserved impact structure on Mercury—and is thought to have originated at the time of the Caloris impact by the focusing of impact-generated shock and seismic waves. Smooth plains, cover 40% of the area imaged by Mariner 10. Smooth plains are the youngest terrain type and are mostly associated with large impact basins. They are in a stratigraphic position similar to that of the lunar maria. On the basis of the areal density of impact craters on the portion of Mercury's surface imaged by Mariner 10, as well as the scaling of cratering flux from the Moon to Mercury, smooth plains emplacement may have ended earlier on Mercury than did mare volcanism on the Moon (Neukum et al. 2001).

The role of volcanism in Mercury's geological history, however, is uncertain. Both volcanic and impact ejecta emplacement mechanisms have been suggested for the intercrater and smooth plains, and the issue remains unresolved because no diagnostic morphological features capable of distinguishing between the two possibilities are clearly visible at the

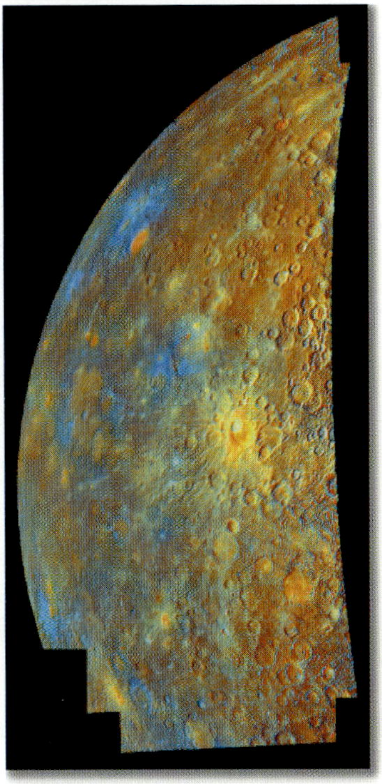

Fig. 3 Enhanced color composite showing portions of the incoming hemisphere of Mercury during the first Mariner 10 encounter (Robinson and Lucey 1997). The red component is the inverse of the opaque index (increasing redness indicates decreasing opaque mineralogy), the green component is the iron-maturity parameter, and blue shows the relative visible color. Smooth plains units (*center left*) display distinct colors and embaying boundaries consistent with material emplaced as a fluid flow. Both characteristics support the hypothesis that the plains are volcanic in origin. Other color variations have been interpreted as evidence for pyroclastic material, differences in composition between impact-excavated material and its surroundings, and differences in soil maturity (Robinson and Lucey 1997)

typical resolution of Mariner 10 images (Milkovich et al. 2002). Ground-based infrared and millimeter observations of Mercury have been interpreted as indicating a generally basalt-free surface and thus a magmatic history dominated either by intrusions or by eruptions of only low-FeO (FeO plus TiO_2 less than 6% by weight) lavas (Jeanloz et al. 1995). Recalibration of Mariner 10 color images and reprojection using color parameters sensitive to iron content, soil maturity, and opaque mineral abundances (Robinson and Lucey 1997) indicate that geological units are distinguishable on the basis of color (Fig. 3). In particular, the correlation of color boundaries with lobate boundaries of smooth plains previously mapped from Mariner 10 images supports the inference that the plains units are volcanic deposits compositionally distinct from underlying older crustal material (Robinson and Lucey 1997).

Mercury's tectonic history is unlike that of any other terrestrial planet. The most prominent tectonic features on the surface are lobate scarps, 20 to 500 km in length and hundreds of meters to several kilometers in height (Watters et al. 1998). On the basis of their asymmetric cross sections, rounded crests, sinuous but generally linear to arcuate planforms, and transection relationships with craters, the scarps (Fig. 4) are interpreted to be the surface expression of major thrust faults (Strom et al. 1975). Because the scarps are more or less evenly distributed over the well-imaged portion of the surface and display a broad range of azimuthal trends, they are thought to be the result of global contraction of the planet. From the lengths and heights of the scarps, and from simple geometric fault models or fault length-displacement relationships, the inferred 0.05–0.10% average contractional strain if extrapolated to the full surface area of the planet would be equivalent to a decrease of 1–2 km in planetary radius (Strom et al. 1975; Watters et al. 1998). Scarp development postdated the intercrater plains, on the grounds

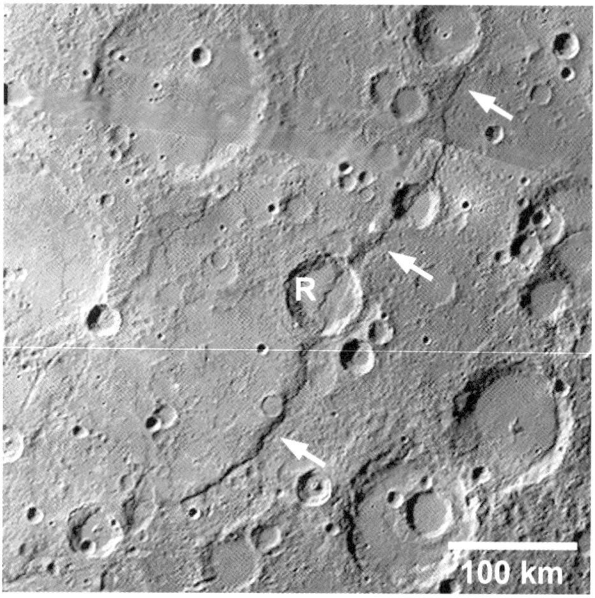

Fig. 4 Mariner 10 image mosaic of Discovery Rupes, the longest known lobate scarp on Mercury (Strom et al. 1975). The scarp is 550 km long and displays 1 km or more of topographic relief (Watters et al. 1998). *Arrows* denote the approximate direction of underthrusting of the crustal block on the right beneath the block to the left. The crater Rameau (*R*), transected by the scarp, is 60 km in diameter. Image courtesy M.S. Robinson

that no scarps are embayed by such plains material, and extended until after emplacement of smooth plains units (Strom et al. 1975).

This estimate of global contraction poses a potentially strong constraint on models for cooling of Mercury's interior. Thermal history calculations that incorporate parameterized core and mantle convection as well as the generation and upward transport of mantle partial melt (Hauck et al. 2004) indicate that models consistent with 0.05–0.10% surface contraction since the end of heavy bombardment are limited to those with a mantle rheology appropriate to anhydrous conditions, modest concentrations of heat-producing elements, and a significant fraction of a light alloying element (e.g., S) in the core to limit inner core solidification. A further constraint on thermal models may come from estimates of the depth of faulting that accompanied scarp formation. Modeling of topographic profiles across several of Mercury's longest known scarps yields inferred depths of faulting of 30–40 km, and from an estimate of the temperature limiting brittle behavior a thermal gradient may be derived (Watters et al. 2002; Nimmo and Watters 2004), although the age appropriate to that estimate and the degree to which it is representative of the global average gradient at that time are not known.

Recent ground-based imaging has yielded information on the hemisphere of Mercury not viewed by Mariner 10. Optical to near-infrared images of the sunlit portion of Mercury have been made by several groups using short-exposure, high-definition techniques (Baumgardner et al. 2000; Dantowitz et al. 2000; Warell and Limaye 2001; Ksanfomality et al. 2005; Warell and Valegård 2006; Ksanfomality and Sprague 2007). Resolution of the best such images approaches 200 km, and both bright and dark features appear in common locations on those portions of the surface imaged with independent methods (Mendillo et al. 2001). Dark features are thought to be plains (Mendillo et al. 2001), and a majority of the bright features are likely to be young rayed craters, which have comparable densities on Mercury's two hemispheres (Warell and Limaye 2001). A large basin comparable in diameter to Caloris has been identified at about 8°N, 80°E (Ksanfomality et al. 2005). Radar images at a resolution as good as 1.5–3 km have been obtained of a number of radar-bright features on the side of Mercury not imaged by Mariner 10 (Harmon 1997, 2002; Harmon et al. 2007). At the highest resolution these features appear to be of impact origin (Harmon 2002; Harmon et al. 2007), including one previously speculated to be a volcanic construct on the basis of earlier radar images of coarser resolution (Harmon 1997).

To make a substantial improvement in our knowledge of the full geological history of Mercury, global multicolor imaging of the surface from an orbiting spacecraft is required. Average resolution should be significantly better than that typical of Mariner 10 images, and a capability for targeted high-resolution imaging is desirable. Topographic information would aid in landform identification and could be obtained from an altimeter, stereo photogrammetry (Cook and Robinson 2000), or a combination of the two methods.

3.3 What Are the Nature and Origin of Mercury's Magnetic Field?

Mercury's intrinsic magnetic field, discovered by Mariner 10 (Ness et al. 1976), has a dipole component nearly orthogonal to Mercury's orbital plane and a moment near 300 nT-R_M^3, where R_M is Mercury's mean radius (Connerney and Ness 1988). The origin of this field, however, is not understood (Stevenson 2003). Mercury's magnetic field cannot be externally induced on the grounds that the measured planetary field is far greater in magnitude than the interplanetary field (Connerney and Ness 1988). The dipole field could be a remanent or fossil field acquired during lithospheric cooling in the presence of an internal or external field (Srnka 1976; Stephenson 1976), or it could be the product of a modern core dynamo (Schubert et al. 1988; Stevenson 2003). Permanent magnetization from an external source has been discounted on the grounds that a thick shell of coherently magnetized material is needed to match the observed dipole moment, and the lithosphere of Mercury would not have been able to cool and thicken sufficiently in the time interval during which strong solar or nebular fields were present (Schubert et al. 1988). Permanent magnetization from an internal source has been questioned on the grounds that a high specific magnetization of the shell and a characteristic interval between field reversals much longer than on Earth are both required (Schubert et al. 1988).

The hypothesis that Mercury's internal field is remanent received renewed attention after the discovery of strongly magnetized regions in the crust of Mars (Acuña et al. 1999). Mars may not be a good analogue to Mercury in all respects, because the potential magnetic carriers on Mars are iron-rich oxides (Kletetschka et al. 2000) and, as discussed earlier, Mercury's crust appears to be very low in Fe^{2+}. The possibility remains, however, that Mercury's crust may contain sufficient metallic iron or iron sulfides (Sprague et al. 1995) to display magnetic thermoremanence and crustal fields detectable from orbit.

A fresh look at the idea that crustal remanence may give rise to the dipolar field has come from a consideration of the strong variation of solar heating with latitude and longitude on Mercury (Aharonson et al. 2004). Because Mercury's obliquity is small, equatorial regions are heated by the Sun to a greater degree than polar regions. Further, Mercury's eccentric orbit and 3 : 2 spin–orbit resonance result in two equatorial "hot poles" that view the Sun at zenith when Mercury is at perihelion (and two equatorial "cold poles" midway between them). Despite a theorem that a uniform spherical shell magnetized by an internal field displays no external field after the internal field has been removed (Runcorn 1975), a result that is not strictly correct when the magnetizing effect of the crustal field is included (Lesur and Jackson 2000), the thickness of Mercury's crust that is below the Curie temperature of a given magnetic carrier varies spatially (Aharonson et al. 2004). As a result, there is a strong dipolar contribution to the external field that would be produced by a crust magnetized by a past internal field, the predicted dipole moment (Aharonson et al. 2004) is within the range of estimates for Mercury (Connerney and Ness 1988), and the predicted ratio of quadrupole to dipole terms (Aharonson et al. 2004) is testable with spacecraft measurements.

A challenge to the hypothesis that Mercury's magnetic field is the product of a hydromagnetic dynamo in a liquid, metallic outer core is that the field is comparatively weak. At a dipole moment three orders of magnitude less than Earth's (Connerney and Ness 1988), Mercury's field is difficult to reconcile with the common expectation for dynamos that Lorentz and Coriolis forces in the outer core are comparable in magnitude (Stevenson 2003), a condition known as magnetostrophic balance. Explanations for the weak external field involving a dynamo otherwise broadly similar to Earth's include thin-shell (Stanley et al. 2005) and thick-shell (Heimpel et al. 2005) dynamos for which a comparatively strong toroidal field maintains magnetostrophic balance and a dynamo that operates only deep in a fluid outer core beneath an electrically conductive but stable layer of liquid metal (Christensen 2006). For the first class of models, strong radial magnetic flux patches outside the cylinder aligned with the spin axis and tangent to the inner core should be found at different latitudes for the thin-shell and thick-shell models (Zuber et al. 2007), and for the latter model the multipolar expansion of external field strength is predicted to have little energy beyond the quadrupole term (Christensen 2006), so there are clear tests of these models that can be made from orbital magnetic field measurements.

A hydromagnetic dynamo as an explanation for Mercury's field (Schubert et al. 1988; Stevenson 2003) requires both that a substantial fraction of Mercury's core is presently fluid and that there are sufficient sustained sources of heat or chemical buoyancy within the core to drive the convective motions needed to maintain a dynamo. Because it is not known that either requirement is met in Mercury, and because of Mercury's weak field strength, more exotic dynamo models have also been considered. If the fluid outer core is sufficiently thin and the core–mantle boundary is distorted by mantle convective patterns, thermoelectric currents might be driven by temperature differences at the top of the core (Stevenson 1987; Giampieri and Balogh 2002). A thermoelectric dynamo is likely to produce a field richer in shorter wavelength harmonics than an Earth-like dynamo (Stevenson 1987), and these harmonics may correlate with those for the gravity field (Giampieri and Balogh 2002), so testing for such a dynamo should be possible from orbital measurements.

The presence of significant heat production within the core would expand the range of conditions under which a modern core dynamo would be expected. New laboratory experiments have reopened the question of whether a significant fraction of potassium in a differentiating terrestrial planet may partition into a liquid metal phase at high pressures (Murthy et al. 2003). Although potassium is not expected to be abundant on Mercury on the basis of several of the cosmochemical hypotheses for the planet's high metal fraction, potassium derived from surface materials is present in the atmosphere and even a small fraction of ^{40}K in the core could have a pronounced impact on the history of core cooling and the energy available to maintain a core dynamo. Tidal dissipation in the outer core may be important for maintaining a fluid state, but uncertainties in Mercury's internal structure prevent a definitive assessment (Bills 2002).

As a result of Mercury's small dipole moment, the planet's magnetosphere (Fig. 5) is among the smallest in the solar system and stands off the solar wind only 1,000–2,000 km above the surface (Slavin et al. 2007). Although the magnetosphere shares many features with that of Earth, because of its small size the timescales for wave propagation and convective transport are much shorter at Mercury, and the proximity to the Sun renders the driving forces more intense. Strong variations in magnetic field and energetic particle characteristics observed by Mariner 10 have been interpreted as evidence of magnetic substorms and magnetic reconnection in the tail (Siscoe and Christopher 1975;

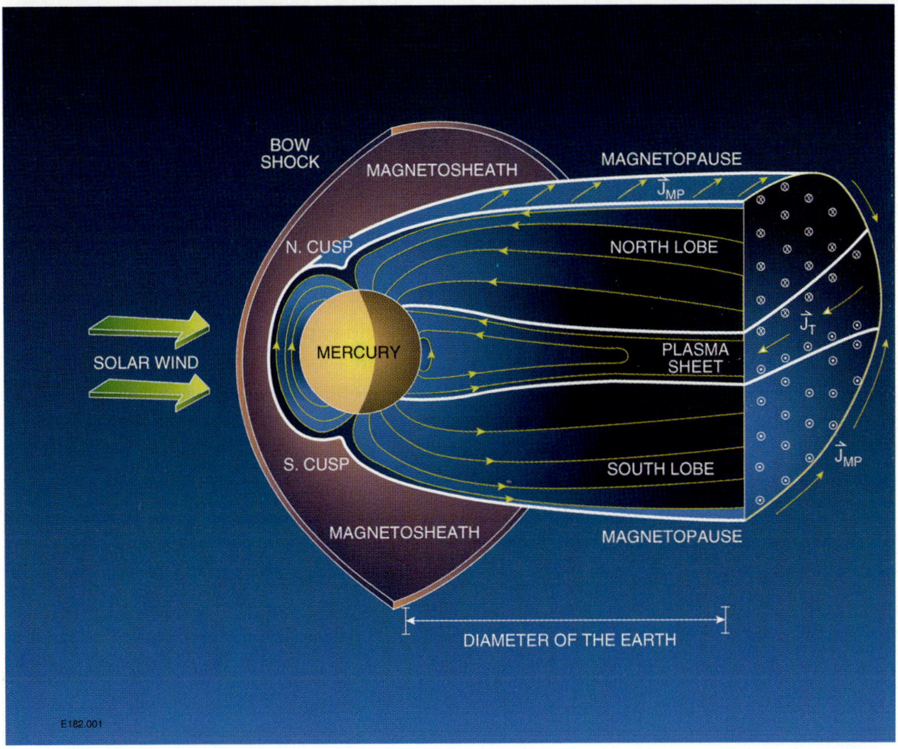

Fig. 5 A simplified, schematic view of Mercury's magnetic field and magnetosphere consistent with Mariner 10 observations and scaling of analogous features from the Earth's magnetosphere. Not depicted are the expected intense temporal variations in magnetospheric characteristics and dynamics and the consequent strong interactions among the solar wind, magnetosphere, exosphere, surface regolith, and planetary interior. From Slavin (2004)

Baker et al. 1986; Eraker and Simpson 1986; Christon 1987). The absence of a significant conducting ionosphere at Mercury, however, implies that the associated current systems close in Mercury's regolith (Janhunen and Kallio 2004) or through a process of pick-up ion formation (Cheng et al. 1987). Magnetic reconnection at the dayside magnetopause may erode the subsolar magnetosphere and allow solar wind ions to impact the planetary surface, but induced currents in Mercury's interior may act to resist magnetospheric compression (Hood and Schubert 1979). All of these factors are expected to lead to complex interactions among the solar wind, magnetosphere, exosphere, regolith, and interior (Slavin et al. 2007).

Determining the geometry of Mercury's intrinsic magnetic field and the structure of Mercury's magnetosphere will elucidate all of these issues. A challenge to the determination of the internal field, however, is that external sources can dominate the total measured field, as was the situation for Mariner 10 (Ness et al. 1976). Errors from external fields were such that the uncertainty in Mercury's dipole moment derived from Mariner 10 data is a factor of 2, and higher order terms are linearly dependent (Connerney and Ness 1988). Simulations of field recovery from orbital observations to be made by MESSENGER (Korth et al. 2004), however, indicate that the effects of the dynamics of the solar wind and Mercury's magnetosphere can be substantially reduced and important aspects of the internal field determined.

3.4 What Are the Structure and State of Mercury's Core?

An observation that can demonstrate the existence and determine the radius of a liquid outer core on Mercury (Fig. 2) is the measurement of the amplitude of Mercury's forced physical libration (Peale 1988). The physical libration of the mantle (manifested as an annual variation in the spin rate about the mean value) is the result of the periodically reversing torque on the planet as Mercury rotates relative to the Sun. The amplitude of this libration ϕ_0 is approximately equal to $(B - A)/C_m$, where A and B are the two equatorial principal moments of inertia of the planet and C_m is the polar moment of inertia of the solid outer part of the planet (Peale 1988). The moment differences also appear in expressions for the second-degree coefficients of the planetary gravity field expanded in spherical harmonics. The latter relations, the libration amplitude, and an expression resulting from Mercury's resonant state and relating the planet's small but non-zero obliquity to moment differences and other orbital parameters together yield C_m/C, where C is the polar moment of inertia of the planet (Peale 1988). The quantity C_m/C is unity for a completely solid planet and about 0.5 if Mercury has a fluid outer core (Peale 1988).

Two conditions on the above relationship for ϕ_0 are that the fluid outer core does not follow the 88-day physical libration of the mantle and that the core does follow the mantle on the timescale of the 250,000-year precession of the spin axis (Peale 1988). These constraints lead to bounds on the viscosity of outer core material, under the assumption that coupling between the outer core and solid mantle is viscous in nature, but the bounds are so broad as to be readily satisfied. Alternative core–mantle coupling mechanisms, including pressure forces on irregularities in the core–mantle boundary, gravitational torques between the mantle and an axially asymmetric solid inner core, and magnetic coupling between the electrically conductive outer core and a conducting layer at the base of the mantle, do not violate either of the required conditions (Peale et al. 2002; Zuber et al. 2007).

Of the four quantities needed to determine whether Mercury has a fluid outer core, two of them—the second-degree coefficients in the planet's gravitational field—can be determined only by tracking a spacecraft near the planet (Anderson et al. 1987). Two means for determining the remaining two quantities—the obliquity and the forced libration amplitude—from a single orbiting spacecraft have been proposed. One makes use of imaging from a spacecraft with precise pointing knowledge (Wu et al. 1997), while the other involves repeated sampling of the global topography and gravity fields (Smith et al. 2001). The MESSENGER mission will use the latter approach (Zuber et al. 2007). Mercury's obliquity and libration amplitude can also be determined from Earth-based radar observations, using either multiple images of features on Mercury viewed with a common geometry but at differing times (Slade et al. 2001) or correlations of the speckle pattern in radar images of the planet obtained at two widely separated antennas (Holin 2002). Observations made with the latter method indicate that $C_m/C < 1$ at 95% confidence (Margot et al. 2007), a result strongly indicative of a molten outer core. Improved estimates of C_m/C as well as the determination of C require a more precise determination of the planetary gravity field from tracking an orbiting spacecraft.

3.5 What Are the Radar-Reflective Materials at Mercury's Poles?

The discovery in 1991 of radar-bright regions near Mercury's poles and the similarity of the radar reflectivity and polarization characteristics of these regions to those of icy satellites and the south residual polar cap of Mars led to the proposal that these areas host deposits of

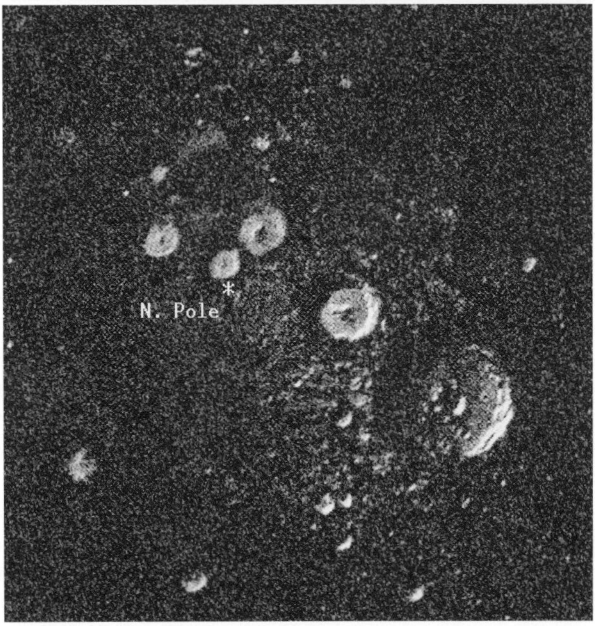

Fig. 6 Radar image of the north polar region of Mercury, obtained by the Arecibo Observatory in July 1999 (Harmon et al. 2001). The radar illumination direction is from the upper left, and the resolution is 1.5 km. Mercury polar deposits are the radar-bright regions within crater floors

surface or near-surface water ice (Harmon and Slade 1992; Slade et al. 1992). Subsequent radar imaging at improved resolution (Fig. 6) has confirmed that the radar-bright deposits are confined to the floors of near-polar impact craters (Harmon et al. 2001). Because of the small obliquity of the planet, sufficiently deep craters are permanently shadowed and are predicted to be at temperatures at which water ice is stable for billions of years (Paige et al. 1992). Such water ice is not likely to represent exposed portions of larger subsurface polar caps, on the grounds that polar craters display depth-to-diameter ratios similar to those of equatorial craters, contrary to the terrain softening expected in areas of subsurface ice (Barlow et al. 1999). While a contribution from interior outgassing cannot be excluded, impact volatilization of cometary and meteoritic material followed by random-walk transport of water molecules to polar craters can provide sufficient polar ice to match the characteristics of the deposits (Moses et al. 1999).

The highest-resolution images of polar deposits show that they extend more than 10° in latitude from the pole and that for larger craters farther from the pole the radar-bright material is concentrated on the side of the crater floor farthest from the pole (Harmon et al. 2001). Both of these characteristics are consistent with thermal models for water ice insulated by burial beneath a layer of regolith tens of centimeters thick (Vasavada et al. 1999), although the detection of radar-bright features in craters as small as 10 km in diameter and the observation that some radar-bright deposits within about 30° of longitude from the equatorial "cold poles" extend up to 18° southward from the pole pose difficulties for current thermal models (Harmon et al. 2001).

Two alternative explanations of the radar-bright polar deposits of Mercury have been suggested. One is that the polar deposits are composed of elemental sulfur rather than water ice, on the grounds that sulfur would be stable in polar cold traps and the presence of sulfides in the regolith can account for a high disk-averaged index of refraction and low microwave opacity of surface materials (Sprague et al. 1995). The second alternative hypothesis is that the permanently shadowed portions of polar craters are radar bright not because of trapped volatiles but because of either unusual surface roughness (Weidenschilling 1998) or low dielectric loss (Starukhina 2001) of near-surface silicates at extremely cold temperatures. This second suggestion can be tested by carrying out impact experiments with very cold silicate targets (Weidenschilling 1998) or measuring dielectric losses of silicates at appropriate temperatures and frequencies (Starukhina 2001), while the first proposal can potentially be tested by measurements from an orbiting spacecraft.

Determining the nature of the polar deposits from Mercury orbit will pose a challenge because the deposits will occupy a comparatively small fraction of the viewing area for most remote sensing instruments (Boynton et al. 2007) and because any polar volatiles may be buried beneath a thin layer of regolith (Vasavada et al. 1999). The most promising measurements include searches of the polar atmosphere with an ultraviolet spectrometer for the signature of excess OH or S (Killen et al. 1997) and neutron spectrometer observations of the polar surface to seek evidence for near-surface hydrogen (Feldman et al. 1997).

3.6 What Are the Important Volatile Species and Their Sources and Sinks on and near Mercury?

Mercury's atmosphere is a surface-bounded exosphere whose composition and behavior are controlled by interactions with the magnetosphere and the surface (Domingue et al. 2007). The exosphere is known to contain six elements (H, He, O, Na, K, Ca). The Mariner 10 airglow spectrometer detected H, He, and O (Broadfoot et al. 1974, 1976), while ground-based spectroscopic observations led to the discovery of Na (Potter and Morgan 1985), K (Potter and Morgan 1986), and Ca (Bida et al. 2000). The exosphere is not stable on timescales comparable to the age of the planet (Hunten et al. 1988), so there must be sources for each of the constituents. H and He are likely to be dominated by solar wind ions neutralized by recombination at the surface, but the other species are likely derived from impact vaporization of micrometeoroids hitting Mercury's surface or directly from Mercury surface materials (Domingue et al. 2007).

Proposed source processes for supplying exospheric species from Mercury's crust include diffusion from the interior, evaporation, sputtering by photons and energetic ions, chemical sputtering by protons, and meteoritic infall and vaporization (Killen et al. 1999). That several of these processes play some role is suggested by the strong variations in exospheric characteristics observed as functions of local time, solar distance, and level of solar activity (Potter et al. 1999; Killen et al. 2001; Hunten and Sprague 2002) as well as by correlations between atmospheric Na and K enhancements and surface features (Sprague et al. 1998). Simulations of Mercury's Na exosphere and its temporal variation in which most of the above source processes are incorporated have shown that evaporation exerts a strong control on the variation of surface Na with time of day and latitude (Leblanc and Johnson 2003). These simulations provide good matches to measurements of changes in the Na exosphere with solar distance and time of day (Sprague et al. 1997) and observations (Potter et al. 2002b) of Mercury's sodium tail (Fig. 7).

The presence of the volatile elements Na and K in Mercury's exosphere poses a potential challenge for the hypotheses advanced to account for Mercury's high ratio of metal to silicate. Whether Mercury is metal rich because of mechanical segregation between metal and silicate grains in the hot, inner solar nebula (Weidenschilling 1978) or because of extensive volatilization or impact removal of the outer portions of a differentiated planet (Cameron 1985; Fegley and Cameron 1987; Benz et al. 1988; Wetherill 1988), the planetary crustal concentrations of volatile elements should be very low. For several of the proposed sources of exospheric Na and K, surface abundances ranging from a few tenths of a percent to a few

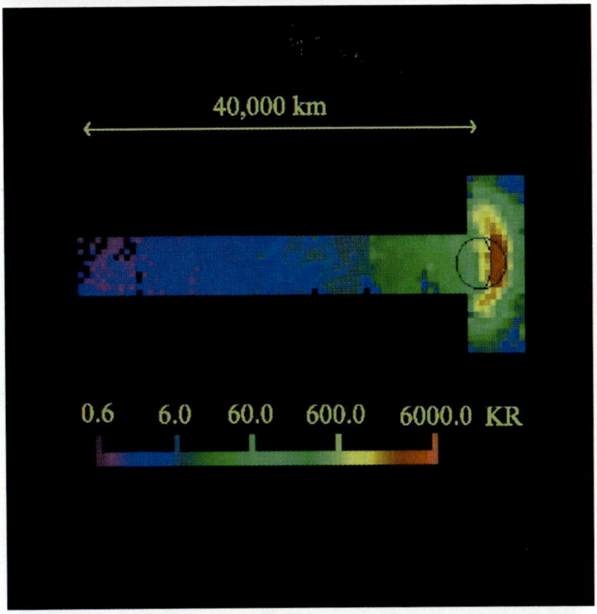

Fig. 7 Composite image of the sodium D2 emission line in the vicinity of Mercury obtained at the McMath-Pierce Solar Telescope at the National Solar Observatory on May 26, 2001 (Potter et al. 2002a). The Na tail is in the anti-sunward direction, and south is at the top. The *color scale* for intensity (in kiloRayleighs) is logarithmic

percent by weight are commonly required (Killen et al. 2001). Simulations of variations in the exospheric Na abundance, however, can match all observations with a supply of fresh Na no greater than that predicted by meteoritic impact volatilization (Leblanc and Johnson 2003).

A spacecraft in orbit about Mercury will provide a range of opportunities for elucidating further the nature of the exosphere. Limb scans conducted with an ultraviolet–visible spectrometer can monitor variations in the major exospheric constituents and search for new species. Surface sources of exospheric materials can be mapped with gamma-ray, X-ray, and neutron spectrometers. Measurement of energetic and thermal plasma ions will detect solar-wind pick-up ions that originated as exospheric neutral atoms.

4 Mission Science and Measurement Objectives

The six guiding science questions lead naturally to six science objectives for the MESSENGER mission, which in turn lead to corresponding sets of measurement objectives to be accomplished by the spacecraft (Fig. 8).

Addressing the origin of Mercury's anomalously high ratio of metal to silicate leads to the scientific objective to map globally the major element chemistry and mineralogy of the planet's surface. To differentiate among the leading formation hypotheses for Mercury, the elements mapped should include both volatile (e.g., K) and refractory (e.g., Ca, Al) species. Spectral measurements from visible to near-infrared wavelengths at spatial resolutions of several kilometers or better are needed to search for absorption features diagnostic of mineralogy. The global maps should at least regionally be at a resolution sufficient to distinguish the compositions of the principal geological units and to determine whether the composition of material excavated from depth and ejected by young impact craters differs from that of surrounding surface materials (cf. Blewett et al. 2007). MESSENGER will obtain major-

Guiding Questions	Science Objectives	Measurement Objectives
What planetary formational processes led to the high ratio of metal to silicate in Mercury?	Map the elemental and mineralogical composition of Mercury's surface	Surface elemental abundances: GRNS and XRS Spectral measurements of surface: MASCS (VIRS)
What is the geological history of Mercury?	Image globally the surface at a resolution of hundreds of meters or better	Global imaging in color: MDIS (WAC) Targeted high-resolution imaging: MDIS (NAC) Global stereo: MDIS Spectral measurements of geological units: MASCS (VIRS) Northern hemisphere topography: MLA
What are the nature and origin of Mercury's magnetic field?	Determine the structure of the planet's magnetic field	Mapping of the internal field: MAG Magnetospheric structure: MAG, EPPS
What are the structure and state of Mercury's core?	Measure the libration amplitude and gravitational field structure	Gravity field, global topography, obliquity, libration amplitude: MLA, RS
What are the radar reflective materials at Mercury's poles?	Determine the composition of the radar-reflective materials at Mercury's poles	Composition of polar deposits: GRNS Polar exosphere: MASCS (UVVS) Polar ionized species: EPPS Altimetry of polar craters: MLA
What are the important volatile species and their sources and sinks on and near Mercury?	Characterize exosphere neutrals and accelerated magnetosphere ions	Neutral species in exosphere: MASCS (UVVS) Ionized species in magnetosphere: EPPS Solar wind pick-up ions: EPPS Elemental abundances of surface sources: GRNS, XRS

Fig. 8 The guiding questions, science objectives, and measurement objectives for the MESSENGER mission. Each question will be answered by observations from two or more elements of the MESSENGER payload, and the observations from each instrument will address multiple questions

element maps of Mercury's surface at 10% relative uncertainty or better at the 1,000-km scale and determine local composition and mineralogy at the ~20-km scale.

Assessing the geological history of Mercury leads to the scientific objective to image globally the planetary surface at a horizontal resolution of hundreds of meters or better coupled with spectral measurements of major geologic units at visible and near-infrared wavelengths. Viewing geometry for imaging should be optimized to discern geological features over a range of scales. High-resolution imaging and the determination of topographic profiles across key geological features from altimetry or stereo will aid in the interpretation of surface geological processes. MESSENGER will obtain a global image mosaic (monochrome) with at least 90% coverage at 250-m average resolution or better, image at least 80% of the planet stereoscopically, obtain a global multi-spectral map at 2 km/pixel average resolution or better, and map the topography of the northern hemisphere at a 1.5-m average height resolution.

Addressing the nature and origin of Mercury's internal magnetic field leads to a requirement to make measurements of the vector magnetic field both near the planet and throughout the planet's magnetosphere. Repeated measurements from orbit are needed to separate internal from external contributions to the field. Measurement of the distributions of energetic particles and plasma boundaries will be critical in the interpretation of magnetospheric structure and dynamics and their relationship to the internal field and solar wind conditions. MESSENGER will obtain a multipole model of Mercury's internal magnetic field resolved through quadrupole terms with an uncertainty of less than ~20% in the dipole magnitude and direction.

Determining the size of Mercury's core and whether its outer core is liquid or solid requires the measurement of Mercury's obliquity, the amplitude of Mercury's physical libration, and the magnitude of the second-degree coefficients in the harmonic expansion of Mercury's gravitational field. These quantities can be measured by repeated altimetric measurements of Mercury's long-wavelength shape and by the determination of Mercury's gravitational field from ranging and range-rate measurements from an orbiting spacecraft. MESSENGER will provide a global gravity field to spherical harmonic degree and order 16 and determine the ratio of the polar moment of inertia of the solid outer shell of the planet to the polar moment of inertia of the entire planet (C_m/C) to ~20% or better.

Determining the nature of Mercury's polar deposits is a challenging goal for a spacecraft in an orbit that does not feature a low-altitude periapsis over one of the poles, but several measurements are promising. Ultraviolet spectrometry of Mercury's near-limb region can reveal whether species diagnostic of candidate polar deposit materials (e.g., OH, S) are present at excess levels in the polar exosphere. Gamma-ray and neutron spectrometry, for sufficiently strong signals, could detect an enhancement of near-surface H in the floors of polar craters. Imaging and altimetry of high-latitude craters can confirm which areas are in permanent shadow and strengthen thermal models for polar regions. By use of all of these methods, MESSENGER aims to identify the principal component the polar deposits at Mercury's north pole.

Determining the volatile budget on Mercury and the sources and sinks for dynamic variations in the exosphere leads to measurement requirements for the identification of all major neutral species in the exosphere and all major charged species in the magnetosphere. The former can be accomplished by ultraviolet and visible wavelength spectrometry of the exosphere with sufficient spectral resolution to detect and identify emission lines diagnostic of known and possible species. The latter can be carried out by in situ analysis of the energies and compositions of charged particles within and in the vicinity of Mercury's magnetosphere. Measurements of surface composition will illuminate the question of the extent to which surface materials act as sources for the exosphere, and measurements of magnetosphere-solar wind interactions will inform questions on the sources and sinks of magnetospheric and exospheric species. MESSENGER will obtain altitude profiles at 25-km resolution of the major neutral exospheric species and characterize the energy distributions of major ion species, both as functions of local time, Mercury heliocentric distance, and solar activity.

5 Payload Overview

The measurement objectives for MESSENGER (Fig. 8) are met by a payload consisting of seven instruments plus radio science. The instruments (Fig. 9) include the Mercury Dual Imaging System (MDIS), the Gamma-Ray and Neutron Spectrometer (GRNS), the X-Ray Spectrometer (XRS), the Magnetometer (MAG), the Mercury Laser Altimeter (MLA), the Mercury Atmospheric and Surface Composition Spectrometer (MASCS), and the Energetic Particle and Plasma Spectrometer (EPPS). The instruments communicate to the spacecraft through fully redundant Data Processing Units (DPUs). The mass and power usage of each instrument are listed in Table 1. A brief summary of each of the seven instruments is given below. This summary updates an overview of the payload published early in the design stage of the project (Gold et al. 2001). Detailed descriptions of each instrument can be found in companion papers in this volume (Anderson et al. 2007; Andrews et al. 2007; Cavanaugh et al. 2007; Goldsten et al. 2007; Hawkins et al. 2007; McClintock and Lankton 2007; Schlemm et al. 2007). The MESSENGER radio science (RS) capabilities and objectives are described in another companion paper (Srinivasan et al. 2007).

5.1 MDIS

The MDIS instrument (Hawkins et al. 2007) includes both a wide-angle camera (WAC) and a narrow-angle camera (NAC) with an onboard pixel summing capability. That combi-

Table 1 Some characteristics of MESSENGER instruments

Instrument	Mass[a] (kg)	Power[b] (W)
MDIS	8.0	7.6
GRNS	13.1	22.3
XRS	3.4	6.9
MAG	4.4	4.2
MLA	7.4	16.4
MASCS	3.1	6.7
EPPS	3.1	7.8
DPUs	3.1	12.3
Miscellaneous[c]	1.7	
Total	47.2	84.4

[a] Mass includes mounting hardware and captive thermal control components. The mass for MDIS includes the calibration target. The MAG mass includes the boom

[b] Nominal average power consumption per orbit; actual values will vary with instrument operational mode and spacecraft position in orbit

[c] Includes purge system, payload harnesses, and magnetic shielding for the spacecraft reaction wheels

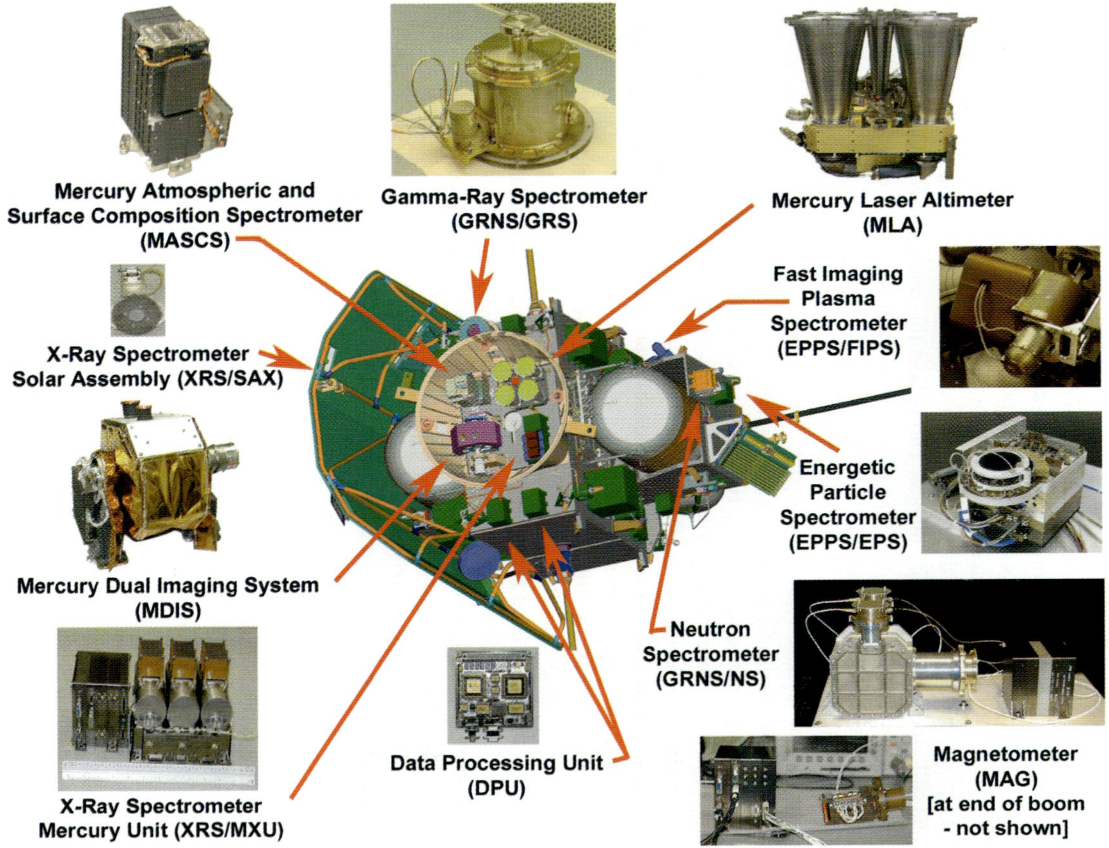

Fig. 9 MESSENGER payload instruments and their locations on the spacecraft. The Magnetometer is mounted at the end of a 3.6-m boom (not shown) that extends in the anti-sunward direction. The Solar Assembly for X-rays (SAX) is mounted on the Sun-facing side of the spacecraft's sunshade (Leary et al. 2007)

nation of features was chosen to provide images of a nearly uniform horizontal resolution throughout MESSENGER's elliptical orbit while minimizing downlink requirements. Because of the geometry of the orbit and limitations on off-Sun pointing by the spacecraft, the WAC and NAC are mounted on opposite sides of a pivoting platform to provide for optical navigation and planetary mapping during the Mercury flybys. MDIS is the only MESSENGER instrument with a pointing capability independent of the spacecraft attitude. The MDIS pivot can point from 50° toward the Sun to 40° anti-sunward centered on nadir, where it is co-aligned with the other optical instruments, all of which are mounted on the spacecraft lower deck (Fig. 9). The pivot platform drive has a redundant-winding stepper motor system and a resolver to measure the platform rotation to a precision <75 µrad.

The thermal design for MDIS faced the challenge that the instrument must work in cold space and yet be able to point at the >700-K sub-solar region of Mercury for extended periods and still produce high-quality images. Throughout this range of environmental conditions, the charge-coupled device (CCD) camera heads are maintained between −10 and −40°C to minimize their dark noise. The MDIS thermal protection system includes high-heat-capacity beryllium radiators, diode heat pipes to shut off thermal conduction when viewing the hot planet, phase-change "wax packs" to limit temperatures during hot periods, and flexible thermal links to tie these elements together.

The WAC is a refractive design with a 10.5° field of view (FOV) and a 12-position filter wheel to provide full-color mapping. The NAC is an off-axis reflective design with a 1.5° FOV and a single band-limiting filter. The passband is a compromise between limiting the light at Mercury to keep the exposure times reasonable and providing high throughput for stellar imaging required for optical navigation.

The CCD camera heads use highly integrated, low-mass electronics with 12-bit intensity resolution. The CCD detectors are 1,024 × 1,024 pixel frame-transfer devices with electronic shuttering. There is no mechanical shutter. There are both manual and automatic exposure controls, and the exposure range is from 1 ms to ∼10 s. The cameras can be commanded to perform on-chip summing of 2 × 2 pixels for 512 × 512 pixel images as required. The imager hardware can also compress the images from 12-bit to 8-bit quantization with a variety of look-up tables. Images are sent directly to the spacecraft solid-state recorder. They are later read back into the main spacecraft processor for additional image compression as commanded on an image-by-image basis.

5.2 GRNS

The GRNS instrument (Goldsten et al. 2007) includes two sensors, a Gamma-Ray Spectrometer (GRS) and a Neutron Spectrometer (NS). The GRS is a cryocooled, high-purity germanium detector with an active shield and measures elemental abundances of O, Si, S, Fe, H, K, Th, and U. Because it was not practical to mount the GRS on a long boom in the Mercury thermal environment, the signal-to-background ratio was maximized by choice of detector. Developing an actively cooled detector to operate at <90 K in the >700 K environment at Mercury was a significant design challenge. The GRS sensor has a 50 × 50 mm cylindrical detector with a Stirling-cycle cooler and an active scintillator shield of boron-loaded plastic. A triple-layer thermal shield surrounds the germanium detector to minimize heat leaks. The boron-loaded plastic scintillator shield is viewed by a large photomultiplier tube (PMT). The anti-coincidence shield removes the cosmic-ray background and softer component of the spacecraft gamma-ray background. The boron loading in the shield also responds directly to slow neutrons and thereby supplements the NS data. The GRS electronics use a novel signal processing design that achieves linearity and stability that nearly equal the performance of a full digital signal processing system with a minimal amount of radiation-hardened electronics.

The NS part of the GRNS is particularly sensitive to the presence of H but may also provide information on Fe content. The NS sensor has two lithium glass scintillators on the ends separated by a beveled cube of neutron-absorbing, borated plastic scintillator. The glass scintillator plates are loaded with lithium enriched in ^{6}Li to detect thermal and epithermal neutrons. Because the MESSENGER orbital velocity is about 3 km/s, the difference in counts in the ram and wake directions greatly enhances the discrimination of thermal and epithermal neutrons. The borated-plastic central scintillator counts epithermal neutrons from all directions and measures the energy depositions of fast neutrons. All three scintillators are viewed by individual PMTs.

5.3 XRS

XRS is an improved version of the Near Earth Asteroid Rendezvous (NEAR) Shoemaker X-ray spectrometer to measure the atomic surface abundances of Mg, Al, Si, Ca, Ti, and Fe by solar-induced X-ray fluorescence (Schlemm et al. 2007). Three gas proportional counters measure low-energy X-rays from the planet, and a Si-PIN detector mounted on the spacecraft sunshade (Fig. 9) views the solar X-ray input. The detectors cover the energy range from 1 to 10 keV. XRS proportional counters have a 12° FOV, provided by a high-throughput, Cu–Be honeycomb collimator. A matched filter technique is used to separate the lower energy X-ray lines (Al, Mg, and Si). The proportional counter tubes are improved from the NEAR Shoemaker design by the addition of anticoincidence wires surrounding most of the tube, a low-emission carbon liner in the sensitive volume, and field-equalizing structures at the ends of the tube to prevent the charge build-up that was seen on that spacecraft. The planet-viewing portion of the instrument, the Mercury X-ray Unit (MXU) is mounted on the lower spacecraft deck (Fig. 9). The XRS solar monitor consists of a small (0.03 mm^2 aperture) detector protected by a pair of thin Be foils. The outer foil reaches >500°C and is the hottest component on the spacecraft, while the detector, just 4 cm away, sits at -45°C.

5.4 MAG

MAG is a three-axis, ring-core, fluxgate magnetometer of the same basic design as that flown on many planetary missions (Anderson et al. 2007). The MAG sensor head is mounted on a lightweight, 3.6-m carbon-fiber boom extending in the anti-sunward direction. Because the sensor can protrude from the shadow of the spacecraft when the spacecraft is pointed near its allowable off-Sun limits, the sensor has its own sunshade. The MAG detector samples the field at a 20-Hz rate, and hardware anti-aliasing filters plus software digital filters provide selectable readout intervals from 0.05 s to 100 s. Readout intervals greater than 1 s generate a sample of the 0.5-Hz filtered signal at the time of the readout. MAG data are output with 16-bit quantization, which eliminates the need for range switching during orbital operations at the ±1530-nT full-scale range. Auto-ranging is provided at the less sensitive range, $\pm51{,}300$ nT full scale, in the event that large crustal fields are present. Spacecraft-induced stray fields were minimized during subsystem development and fabrication. The reaction wheels and a few propulsion system valves were provided with shielding and compensation magnets, respectively, as needed to meet requirements on background magnetic fields.

5.5 MLA

MLA includes a diode-pumped, Q-switched, Cr:Nd:YAG laser transmitter operating at 1,064 nm wavelength and four receiver telescopes with sapphire lenses (Cavanaugh et al. 2007). MLA is mounted on the spacecraft lower deck (Fig. 9), along with the other optical instruments. A silicon avalanche photodiode (APD) and a time-interval unit, based on an application-specific integrated circuit (ASIC) chip, measure altitudes to 30-cm precision or better and ranges up to 1,200 km. Because of MESSENGER's elliptical orbit at Mercury, MLA will operate for about 30 minutes around the periapsis of each orbit. The laser transmits pulses at 8 Hz through a beam expander with a heat-absorbing sapphire window. The four 115-mm-diameter receiver telescopes comprise a multi-aperture receiver, which collects the laser return pulses from Mercury and passes them via four optical fibers through an optical bandpass filter to reject the solar background before going to the silicon APD detector.

5.6 MASCS

The MASCS instrument combines an exospheric and a surface-viewing instrument in a single package (McClintock and Lankton 2007). A moving-grating Ultraviolet-Visible Spectrometer (UVVS) will observe emissions from the Mercury exosphere during limb scans, and a Visible-Infrared Spectrograph (VIRS) will observe the planetary surface. The two spectrometers are contained in the same package and fed by a single front-end telescope. The Cassegrain telescope feeds the UVVS Ebert-Fastie spectrometer directly. Its moving diffraction grating design is optimized for measuring the very weak emissions of the exosphere with excellent signal-to-noise ratio. UVVS spans the spectral range from 115 to 600 nm with three photon-counting PMT detectors. When scanning the limb, it has 25-km altitude resolution and an average spectral resolution of 1 nm. VIRS is fed by a fused-silica fiber-optic bundle from the focal plane of the front-end telescope. A holographic diffraction grating images onto two semiconductor line-array detectors. A dichroic beam splitter separates the visible (300–1,025 nm) and infrared (0.95–1.45 μm) spectra. The 512-element visible detector is silicon, and the 256-element infrared detector is made of InGaAs. MASCS requires no active cooling. The instrument is mounted on the lower spacecraft deck (Fig. 9).

5.7 EPPS

The EPPS (Andrews et al. 2007) instrument consists of an Energetic Particle Spectrometer (EPS) and a Fast Imaging Plasma Spectrometer (FIPS). FIPS measures thermal and low-energy ions with a unique electrostatic analyzer and a time-of-flight (TOF) spectrometer section. The FIPS analyzer is sensitive to ions entering over nearly a full hemisphere, with energy per charge (E/q) up to >15 keV/q. Particles of a given E/q and polar angle pass through the dome-shaped electrostatic deflection system and into the position-sensing TOF telescope. The ions are then post-accelerated by a fixed voltage before passing through a very thin ($\sim$1 μg/cm^2) carbon foil. Secondary electrons from the foil are measured with a position-sensitive detector that reads out the initial incidence angle. Mass per charge of an ion is measured by the E/q (set by the deflection voltage) and the TOF. The deflection voltage is stepped to cover the full E/q range in about one minute. The EPS sensor measures the TOF and residual energy of ions from 10 keV/nucleon to $\sim$3 MeV and electrons to 400 keV. Time-of-flight is measured from secondary electrons as the ions pass through two foils, while total energy is measured by a 24-pixel silicon detector array. The FOV, 160° by 12°, is divided into six segments of 25° each. The EPPS common electronics process all of the TOF, energy, and position signals from both EPS and FIPS. EPS is mounted on the rear deck of the spacecraft, whereas FIPS is mounted on the side of the spacecraft (Fig. 9), where it can observe the plasma over a wide range of pitch angles.

5.8 RS

The radio frequency (RF) telecommunications system used to conduct radio science (RS) as well as communicate with the MESSENGER spacecraft (Srinivasan et al. 2007) includes two opposite-facing, high-gain phased-array antennas, two fanbeam medium-gain antennas, and four low-gain antennas. The RF signals are transmitted and received at X-band frequencies (7.2 GHz uplink, 8.4 GHz downlink) by the NASA Deep Space Network (DSN). Precise observations of the spacecraft's Doppler velocity and range assist in navigating the spacecraft and will be inverted to determine the planet's gravitational field, provide improvements to the planet's orbital ephemeris, and sharpen knowledge of the planet's rotation state, including obliquity and forced physical libration. The times of occultation of the spacecraft RF signal by the planet will be used to determine local values of Mercury's radius, of particular importance for Mercury's southern hemisphere, most of which will be out of range of the MLA instrument.

5.9 Complementarity of Instruments

As illustrated in Fig. 8, each of the mission science objectives will be addressed by at least two elements of the MESSENGER payload (including Radio Science). Mercury's elemental surface composition will be mapped by GRNS and XRS, which are complementary in their elemental sensitivity and the depth of near-surface material contributing to detected signals; mineralogical information will be obtained from the VIRS sensor on MASCS and, with much less spectral resolution, the color imaging that will be carried out by the WAC on MDIS. Mercury's geological history will primarily be derived from mosaics of MDIS images, in color and in high-resolution monochrome, but the interpretation of unit definition will be aided by spectral reflectance measurements by MASCS and the interpretation of geological features will be enhanced by information on topography measured by MLA and obtained from MDIS with stereogrammetry. Mercury's magnetic field will be mapped by MAG, while plasma and energetic particle characteristics measured by EPPS will help to define the principal magnetospheric boundaries consistent with internal field structure. The key parameters necessary to determine Mercury's core radius and the nature of the outer core can be derived independently from MLA and RS observations. The composition of polar deposits will be addressed by GRNS, MLA observations will address the topographic cold trap hypotheses, and MASCS and EPPS observations will address whether the polar regions have enhancements in neutral or ionized species that may be derived from polar deposit material. The processes governing the exosphere will be variously addressed by the UVVS sensor on MASCS, the EPPS measurements, and the chemical observations of potential surface source regions by GRNS and XRS.

Just as each science objective is met with data from multiple payload elements, each instrument addresses two or more of the guiding science questions. This dual complementarity provides for important crosschecks between sets of observations and ensures that mission science requirements can be met even in the case of problems with one of the payload instruments.

6 Spacecraft Overview

The requirements on the MESSENGER spacecraft (Santo et al. 2001) flowed directly from the science requirements (Solomon et al. 2001) and mission design (McAdams et al. 2007). The Delta II 7925H-9.5 launch vehicle was the largest available to a Discovery-class mission. This vehicle provided 1,107 kg of lift mass to achieve the necessary heliocentric orbit. This fact, coupled with the complex trajectory requiring that 599 kg (54%) of the spacecraft launch mass be propellant, limited the spacecraft dry mass—a challenging constraint for designing a fully redundant spacecraft with MESSENGER's functionality. A schematic view of the MESSENGER spacecraft, described in greater detail in a companion paper (Leary et al. 2007), is shown in Fig. 10, and an image of the spacecraft in the process of being mated to the launch vehicle is shown in Fig. 11.

The MESSENGER spacecraft structure, primarily lightweight composite material, was integrated at the outset of design with a dual-mode propulsion system. The propulsion system features state-of-the-art lightweight fuel tanks and can provide 2,250 m/s velocity change (ΔV) capability. A ceramic-cloth sunshade eliminates most of the solar input throughout the cruise and orbital phases of the mission. The spacecraft is three-axis stabilized and momentum biased to ensure Sun pointing while allowing instrument viewing by rotation about the spacecraft–Sun line. Power is provided by two specially designed 2.6-m^2 solar arrays consisting of two-thirds mirrors and one-third solar cells for thermal management. Generally passive thermal management techniques have been used on the rest of the spacecraft to minimize the required power while protecting the spacecraft from the harsh

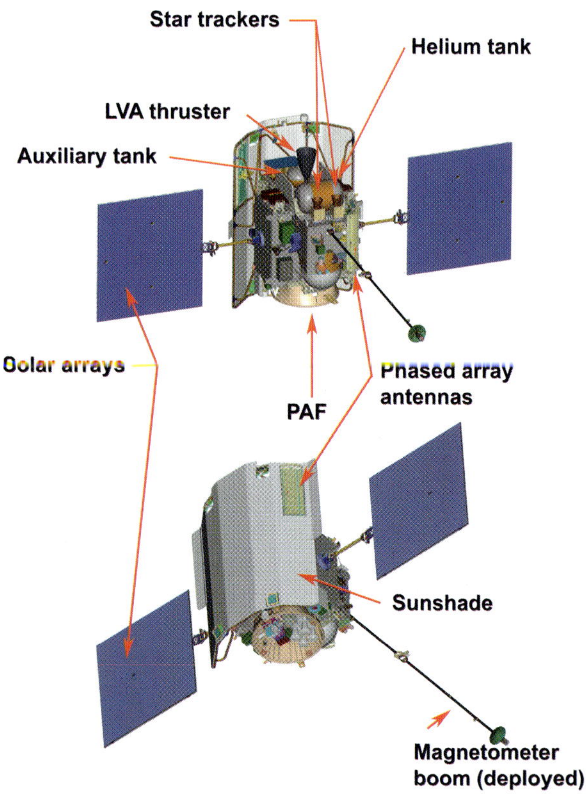

Fig. 10 Schematic view of the MESSENGER spacecraft from two perspectives. The identified tanks and the large velocity adjustment (LVA) thruster are part of the propulsion system. The payload attach fitting (PAF) mated the spacecraft to the third stage of the launch vehicle at the time MESSENGER was launched and now encloses four of the payload instruments

Fig. 11 The MESSENGER spacecraft on July 14, 2004, after it was attached to the payload assist module of the Delta II third stage at Astrotech Space Operations in Titusville, FL. The flat, reflective panels are the solar arrays stowed in their launch positions; solar cells are the dark strips between the optical solar reflectors (mirrors) that permit high-temperature operation. The gold reflective material is thermal blanket. A portion of the stowed magnetometer boom is visible between the solar arrays

environment near the Mercury dayside. A single redundant processor performs all nominal spacecraft functions, while two other processors monitor spacecraft health and safety. Telecommunications are provided by redundant transponders, solid-state power amplifiers, and a diverse antenna suite that includes two phased-array antennas, the first electronically steered antennas designed for use in deep space.

Because the spacecraft is solar powered (except for a battery needed for eclipses), power generation increases as the spacecraft moves sunward. Early in the mission the spacecraft was oriented with the sunshade pointed away from the Sun at solar distances greater than 0.85 AU, allowing a substantial reduction in needed heater power. Peak power demand occurs during science operations in orbit about Mercury. During the orbital phase, there are eclipses of varying lengths, and for the longest eclipses (>35 minutes) science operations are constrained by limits established to the permitted depth of discharge of the spacecraft battery.

7 Mission Timeline

MESSENGER was launched successfully by a Delta II 7925H-9.5 rocket on August 3, 2004 (Fig. 12). A summary of major mission milestones from launch to orbit insertion is given

Table 2 Key events in the MESSENGER mission

Event	Date	UTC
Launch	3 August 2004	06:15:56.5
Earth flyby	2 August 2005	19:13:08.4
DSM-1	12 December 2005	11:30:00.0
Venus flyby 1	24 October 2006	08:33:59.9
Venus flyby 2	5 June 2007	23:10:10.9
DSM-2	17 October 2007	22:30:00.0
Mercury flyby 1	14 January 2008	18:37:08.8
DSM-3	17 March 2008	19:00:00.0
Mercury flyby 2	6 October 2008	11:39:07.9
DSM-4	6 December 2008	19:00:00.0
Mercury flyby 3	29 September 2009	23:59:47.4
DSM-5	29 November 2009	19:00:00.0
MOI	18 March 2011	07:30:00.0

Times of key events are based on the full-mission reference trajectory database as of January 10, 2007. Times shown for each Deep Space Maneuver (DSM) and for Mercury Orbit Insertion (MOI) correspond to the start times of these propulsive maneuvers. Final times of future events (Venus flyby 2 and later) will differ somewhat from the values shown

in Table 2. The cruise phase of the mission is 6.6 years in duration and includes six planetary flybys—one of Earth, two of Venus, and three of Mercury—as well as a number of propulsive corrections to the trajectory (Fig. 13). At the spacecraft's fourth encounter with Mercury, orbit insertion is accomplished on March 18, 2011. A full description of the design of the MESSENGER mission and how the principal elements of mission design flowed from the science requirements is given in a companion paper (McAdams et al. 2007).

The Earth flyby was accomplished successfully on August 2, 2005, with a closest approach distance of 2,348 km over central Mongolia (McAdams et al. 2007). The event provided important calibration opportunities for four MESSENGER instruments. Prior to closest approach, MDIS acquired images of the Moon for radiometric calibration. Images of Earth (Fig. 14) were acquired with 11 filters of the MDIS wide-angle camera to test optical navigation sequences that will be used to target later planetary flybys, and a movie was assembled from 358 sets of MDIS images taken in three filters every four minutes over a 24-hour period after closest approach. MASCS obtained spectral observations of the Moon that permitted absolute radiometric calibration of UVVS and VIRS as well as intercomparison with MDIS, and MASCS observed Earth's hydrogen corona in the month following closest approach. MESSENGER also measured the magnetic field and charged particle characteristics within Earth's magnetosphere and across major magnetospheric boundaries. About two months prior to the Earth flyby, MESSENGER's MLA instrument set a distance record (24 Gm) for two-way laser transmission and detection in space (Smith et al. 2006).

The first of the two Venus flybys, which occurred on October 24, 2006, and achieved a closest approach distance of 2,987 km, increased the spacecraft's orbit inclination and reduced the orbit period. No scientific observations were made during that flyby, however, because direct communication with the spacecraft was precluded by the fact that Venus and Earth were on opposite sides of the Sun.

The second Venus flyby on June 5, 2007, will lower the spacecraft perihelion distance sufficiently to permit the subsequent three flybys of Mercury. Closest approach for the sec-

Fig. 12 Launch of the MESSENGER spacecraft on August 3, 2004. The Delta II 7925H-9.5 rocket was launched from Cape Canaveral Air Force Station Space Launch Complex 17B, Florida, at 06:15:56.5 UTC

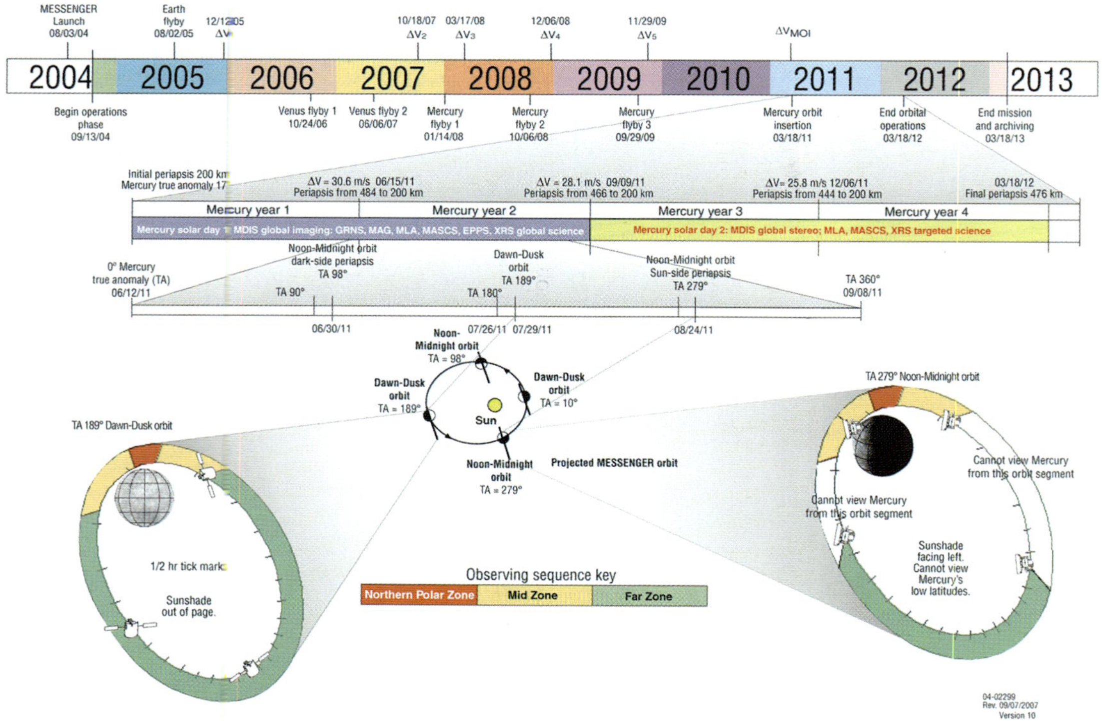

Fig. 13 The MESSENGER mission timeline. The top line shows all significant events from mission launch through end of mission and data archiving. ΔV_1 through ΔV_5 are deep-space propulsive maneuvers, and ΔV_{MOI} is the propulsive burn at Mercury orbit insertion (McAdams et al. 2007). The second line provides further details for the orbital phase of the mission. The third line expands on one Mercury year of observations, from perihelion to perihelion (0° Mercury true anomaly, or TA). The accompanying figure at the center shows the progression of the orbit in local time. Observing strategies are detailed for a dawn-dusk (terminator) orbit (TA = 189°) at lower left and a noon-midnight orbit (TA = 279°) at lower right. The divisions of these orbits by observing sequence are keyed to different data acquisition combinations for the payload instruments (Solomon et al. 2001)

Fig. 14 MDIS WAC image of Earth taken on August 2, 2005, shortly before closest approach during the Earth flyby. Portions of North, Central, and South America are visible

ond Venus flyby is targeted at 313 km altitude over 12°S, 165°E, near the boundary between the lowlands plains of Rusalka Planitia and the rifted uplands of Aphrodite Terra. All of the MESSENGER instruments will be trained on Venus during that flyby. MDIS will image the nightside in near-infrared bands, and color and higher-resolution monochrome mosaics will be made of both the approaching and departing hemispheres. The UVVS sensor will make profiles of atmospheric species on the dayside and nightside as well as observations of the exospheric tail on departure. The VIRS sensor will observe the planet near closest approach to sense cloud chemical properties and search for near-infrared returns from the surface. The laser altimeter will serve as a passive 1,064-nm radiometer and will attempt to measure the range to one or more cloud decks for several minutes near closest approach.

The European Space Agency's Venus Express mission (Svedhem et al. 2005), now in an elliptical polar orbit about Venus, should still be operational in June 2007. The MESSENGER flyby will therefore permit the simultaneous observation of Venus from two independent spacecraft, a situation of particular value for characterization of the particle and field environment at Venus. MESSENGER's EPPS will observe charged particle acceleration at the Venus bowshock and elsewhere. The Magnetometer will provide measurements of the upstream interplanetary magnetic field (IMF), bowshock signatures, and pickup ion waves as a reference for EPPS and Venus Express observations. The encounter will enable two-point measurements of IMF penetration into the Venus ionosphere, primary plasma boundaries, and the near-tail region.

The three flybys of Mercury, in January and October 2008 and September 2009, will provide important new scientific observations of Mercury in advance of the orbital phase of the mission. MDIS will carry out an extensive campaign of imaging during each approach and departure (Solomon et al. 2001), and the geometry of the flybys (McAdams et al. 2007) are such that much of the surface unseen by Mariner 10 will have been imaged by the end of the second flyby (Hawkins et al. 2007). Each flyby will pass within 200 km of Mercury's surface, permitting measurements of the magnetic field and charged particle environment at closer distances from the planet than achieved by Mariner 10 (Connerney and Ness 1988).

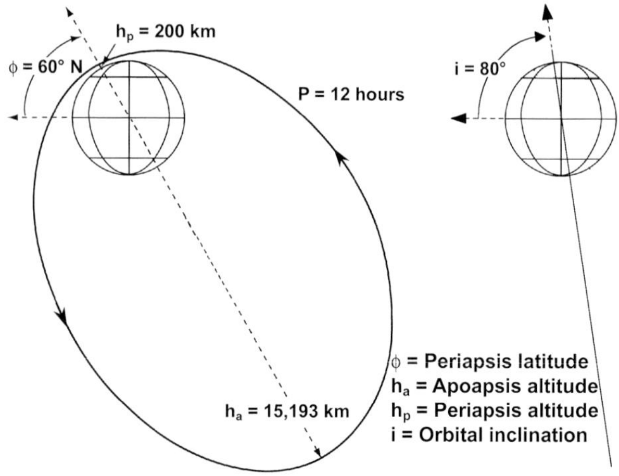

Fig. 15 MESSENGER's nominal orbit around Mercury. Parameters of the orbit were determined by balancing science objectives against propulsion and trajectory constraints and the design of the spacecraft thermal and power systems

The UVVS system on the MASCS instrument will carry out surveys of exospheric species and map the species in Mercury's magnetotail, and VIRS will conduct detailed mapping of dayside surface reflectance at visible and near-infrared wavelengths in search of mineralogical absorption features. The MLA will range to the surface near nightside closest approaches, and the GRNS and XRS instruments will collect early baseline measurements of the Mercury environment.

Within a few days of orbit insertion, the spacecraft will be in its mapping orbit, which has an 80° inclination to Mercury's equator, an initial 200-km minimum altitude over 60°N latitude, and a 12-hour orbit period (Fig. 15). As a result of solar torques, the periapsis latitude drifts northward and the minimum altitude progressively increases. Once per 88-day Mercury year the spacecraft will execute orbit correction maneuvers to return the minimum altitude to 200 km (McAdams et al. 2007). Otherwise propulsive events will be minimized to permit the recovery of Mercury's gravity field from ranging and Doppler velocity measurements (Srinivasan et al. 2007). The orbital phase of the mission is scheduled for one Earth year, or slightly longer than two Mercury solar days (Fig. 13). At the end of the nominal mission the periapsis latitude will be 72°N. Approximately one year after the last propulsive adjustment to its orbit, the spacecraft will impact Mercury's surface.

While in Mercury orbit, observations are staged by altitude and time of day so as to maximize scientific return among all scientific instruments (Fig. 13), subject to restrictions on spacecraft attitude set by the need to maintain sunshade pointing within small angular deviations in yaw and pitch of the sunward direction (Leary et al. 2007). MDIS will build on the flyby imaging to create global color and monochrome image mosaics during the first six months of the orbital phase; a global monochrome base mosaic will be obtained at 250-m/pixel or better average spatial sampling, low emission angle, and moderate incidence angle, and a global color mosaic will be obtained at a resolution of 2 km/pixel or better. Emphasis during the second six months will shift to targeted, high-resolution imaging (up to ~ 20 m/pixel resolution) with the NAC and repeated mapping at a different viewing geometry to carry out global stereogrammetry (Hawkins et al. 2007). GRNS and XRS will build up observations that will yield global maps of elemental composition at resolutions that will vary with latitude, species, and (for XRS) the intensity of the solar X-ray flux (Goldsten et al. 2007; Schlemm et al. 2007). MAG will measure the vector magnetic field over six Mercury sidereal days (each 58.65 Earth days) under a range of solar distances and conditions, which should permit separation of internal and external fields sufficient to resolve Mercury's quadrupole magnetic moment (Korth et al. 2004) and shorter-wavelength features near periapsis latitudes (Anderson et al. 2007). MLA will measure the topography of the northern hemisphere over four Mercury years (Cavanaugh et al. 2007). RS will extend topographic information to the southern hemisphere by occultation measurements of planet radius, and the planet's obliquity and the amplitude of the physical libration will be determined independently from the topography and gravity field (Srinivasan et al. 2007). The VIRS component of the MASCS instrument will produce global maps of surface reflectance from which mineralogy and its variation with geological unit can be inferred, and the UVVS component of the MASCS instrument will produce global maps of exospheric species abundances versus altitude and their temporal variations over four Mercury years and a range of solar activity (McClintock and Lankton 2007). EPPS will sample the plasma and energetic particle population in the solar wind, at major magnetospheric boundaries, and throughout the environment of Mercury at a range of solar distances and levels of solar activity (Andrews et al. 2007).

An additional important constraint on payload observing sequences is imposed by a rate of data downlink from the spacecraft to the DSN that varies strongly with time during the mission orbital phase (Fig. 16). The strategy to deal with such a variable data return is to store most data on the spacecraft solid-state recorder during periods when Mercury is far from Earth and to downlink combinations of stored data and newly acquired data during periods when Mercury is closest to Earth. A data prioritization scheme will assist in managing the downlink process. Under fairly conservative assumptions (downlinking to one 34-m

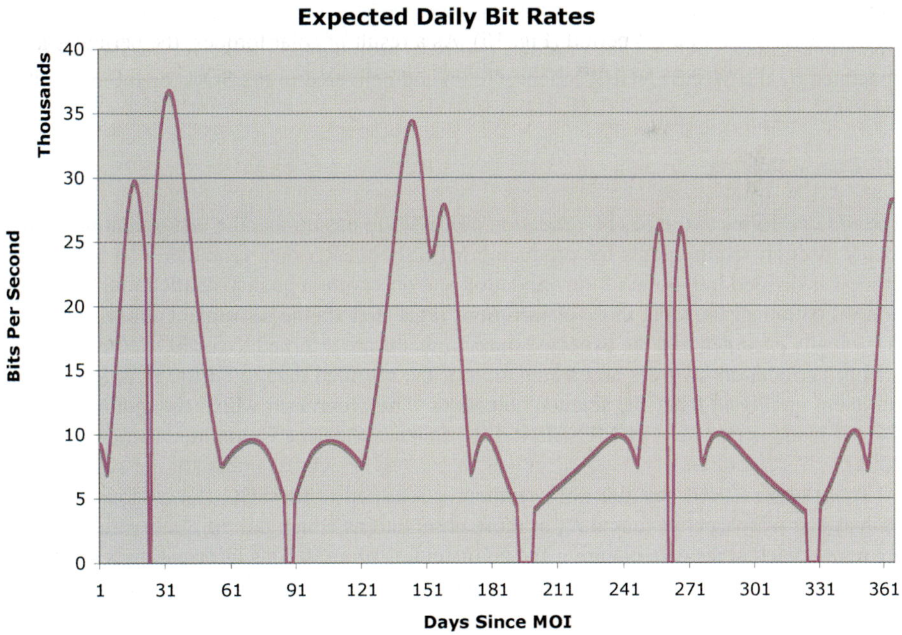

Fig. 16 Downlink data rates per day during the mission orbital phase. Peaks in the curve correspond to times near Mercury inferior conjunction; longer segments with zero data rate correspond to Mercury superior conjunction, and shorter segments correspond to times when Mercury passes between the Sun and Earth. This profile is based on the assumption that downlinked data will be received for 6.5 hours each day by one of the DSN 34-m antennas

DSN station for 6.5 hours per mission day) a total of more than 100 Gb of data will be returned during the mission orbital phase.

The orbital observation strategy is based on a combination of position along the orbit (northern polar zone, mid zone, and far zone as defined in Fig. 13) and a balance between available downlink and solid-state recorder resources. The exploratory nature of this mission requires built-in flexibility in the planning strategy in order to take maximum advantage of what is learned during the flybys and the early part of the orbital phase. Many of the instruments operate in conjunction with each other in observational campaigns that are defined by science objectives but are constrained by limits on data volumes. Margin and multiple opportunities for high-priority observations are therefore incorporated into the strategy where possible.

8 Data Products and Archiving

On the basis of its guiding science questions and measurement objectives, the MESSENGER project has defined a set of data products that will be produced primarily by the MESSENGER Science Team and archived with NASA's Planetary Data System (PDS). These data products and the schedule for delivering them to the PDS are defined in a formal MESSENGER Data Management and Science Analysis Plan (DMSAP) and are discussed in more detail in a companion article (Winters et al. 2007).

Planning and acquisition of science measurements are handled by MESSENGER's Science Planning Group (SPG). The SPG is responsible for ensuring that the data acquisition plan includes all observations needed to meet the mission's measurement objectives. These measurements are made available to the MESSENGER Science Team through the Science Operations Center (SOC). Data products that have been tagged for delivery to the PDS are generated by the MESSENGER Science Team and delivered to the SOC for submission to the PDS.

8.1 Data Validation

The SPG performs two types of validation processes to ensure that the instrument measurements meet all requirements for producing MESSENGER's data products. The validation process is divided into observation validation and observation quality verification. Observation validation ensures that those observations requested via the instrument command loads are actually executed and the expected measurements are returned to the SOC. Observation quality verification involves an examination of the returned data to ensure that they are of sufficient quality to meet the science objectives. The criteria on which the quality assessment is made is provided by MESSENGER's Science Steering Committee (Solomon et al. 2001).

Those observations that are not executed or returned to the SOC (for various reasons, such as loss of spacecraft function), or those observations which fail the quality assessment, are rescheduled in the data planning and commanding process. This information is conveyed to the Science Team via four discipline groups (Solomon et al. 2001) and the SPG. Both validation processes ensure that the data products produced by the MESSENGER team meet the mission's science objectives.

The data acquisition is monitored weekly by the SPG, and the progress toward meeting mission objectives is constantly assessed and reported to the Science Steering Committee. Coverage maps for each instrument's data set are generated daily to assess the mission's science objectives and to validate the data acquisition process.

8.2 Data Products

The data products produced by the MESSENGER mission are divided into two broad categories: raw data or Experimental Data Records (EDRs) and higher-level data products or Reduced Data Records (RDRs). The EDRs are formatted raw instrument data produced by the SOC directly from the spacecraft telemetry for use by the Science Team. The EDRs are used by the SOC and Science Team to produce the RDRs.

The RDRs have been divided into three groups: Calibrated Data Records (CDRs), Derived Data Products (DDPs), and Derived Analysis Products (DAPs). CDRs generally consist of EDR data that have been transformed into physical units. This transformation is done by either Science Team members or the SOC via algorithms provided by the Science Team. DDPs and DAPs are higher-level products produced by the Science Team and delivered to the SOC for submission to the PDS. These higher-level products may be constructed from observations made by more than one instrument. A list of the DDPs and DAPs that the MESSENGER project will be archiving to the PDS may be found in a companion paper (Winters et al. 2007).

8.3 Archiving Plan

The MESSENGER project is working closely with the PDS to facilitate the data archiving process, and toward that end a Data Archive Working Group (DAWG) was established early in the project (Solomon et al. 2001). Through this group the EDR and RDR data formats have been defined and described in instrument software interface specification (SIS) documents. These documents have been reviewed and approved by both the MESSENGER project and the PDS. These baseline efforts permit the archiving process to be streamlined, portions of the process to be automated, and the full delivery schedule for MESSENGER's data products (Winters et al. 2007) to be met.

During the MESSENGER mission there are several designated deliveries of data to the PDS (Winters et al. 2007), each associated with a mission milestone. The first four deliveries are, respectively, six months following the second Venus flyby (EDR data only) and six months following each of the Mercury flybys (EDRs and either calibration documentation or CDRs). Deliveries of orbital data (EDRs and CDRs) are scheduled at six-month intervals following orbit insertion. High-level RDR products (DDPs and DAPs) will be delivered to the PDS one year after the end of the mission, providing the Science Team adequate time to produce these products with the full MESSENGER data set.

9 Conclusions

The MESSENGER mission to Mercury will provide important new information on the formation and evolution of the inner planets. We will have obtained the first global views of Mercury's geology, exosphere, magnetic field, and magnetosphere. We will have ascertained the state and size of Mercury's core, fractionally the largest among the terrestrial planets. We will have learned about the nature of Mercury's polar deposits and what that nature implies for the sources of and storage mechanisms for near-surface volatiles. We will have carried out the first chemical remote sensing of Mercury's surface and from that information obtained new constraints on the planetary processes that led to Mercury's high ratio of metal to silicate. This new information will fuel a new understanding of planetary formation, the early history of the inner solar system, the origin of planetary magnetism, and modes of solar wind-magnetosphere interaction.

It is noteworthy that the MESSENGER mission is a product of NASA's Discovery Program, under which mission concepts are constrained at the outset by cost, schedule, and launch vehicle. Those constraints contributed to the extended duration of the mission cruise phase and limited the number of potential payload instruments. MESSENGER is nonetheless ambitious in its scientific scope for a Discovery mission, a tribute to the fact that scientific requirements guided the development of spacecraft (Leary et al. 2007) and mission design (McAdams et al. 2007) at every stage in the project, from initial concept through all design trades and testing. Those same science requirements now frame decisions made regularly in mission operations (Holdridge and Calloway 2007).

During and following the MESSENGER mission, the MESSENGER team will be working in close communication with the team now developing the BepiColombo mission, which involves the launch in 2013 of two spacecraft that will be inserted into Mercury orbit in 2019. Such communication is intended to ensure that the scientific return will be optimized from both missions (McNutt et al. 2004).

Acknowledgements The MESSENGER mission is supported by the NASA Discovery Program under contracts NASW-00002 to the Carnegie Institution of Washington and NAS5-97271 to The Johns Hopkins University Applied Physics Laboratory. We thank John Harmon, Andrew Potter, Mark Robinson, and James Slavin for permission to use figures; and we are grateful to Brian Anderson, John Cavanaugh, John Goldsten, Edward Hawkins, George Ho, James Leary, William McClintock, James McAdams, Charles Schlemm, Dipak Srinivasan, Richard Starr, Xiaoli Sun, Thomas Zurbuchen, and two anonymous reviewers for providing helpful comments on an earlier draft.

References

M.H. Acuña et al., Science **284**, 790–793 (1999)
O. Aharonson, M.T. Zuber, S.C. Solomon, Earth Planet. Sci. Lett. **218**, 261–268 (2004)
B.J. Anderson et al., Space Sci. Rev. (2007, this issue). doi:10.1007/s11214-007-9246-7
J.D. Anderson, G. Colombo, P.B. Esposito, E.L. Lau, G.B. Trager, Icarus **71**, 337–349 (1987)
G.B. Andrews et al., Space Sci. Rev. (2007, this issue). doi:10.1007/s11214-007-9272-5
A. Anselmi, G.E.N. Scoon, Planet. Space Sci. **49**, 1409–1420 (2001)
D.N. Baker, J.A. Simpson, J.H. Eraker, J. Geophys. Res. **91**, 8742–8748 (1986)
N.G. Barlow, R.A. Allen, F. Vilas, Icarus **141**, 194–204 (1999)
J. Baumgardner, M. Mendillo, J.K. Wilson, Astron. J. **119**, 2458–2464 (2000)
J.W. Belcher et al., Technical Memorandum 4255, NASA, Washington, DC, 1991, 132 pp
W. Benz, W.L. Slattery, A.G.W. Cameron, Icarus **74**, 516–528 (1988)
T.A. Bida, R.M. Killen, T.H. Morgan, Nature **404**, 159–161 (2000)
B.G. Bills, Lunar Planet. Sci. **33** (2002), abstract 1599
D.T. Blewett, P.G. Lucey, B.R. Hawke, G.G. Ling, M.S. Robinson, Icarus **129**, 217–231 (1997)
D.T. Blewett, B.R. Hawke, P.G. Lucey, Meteorit. Planet. Sci. **37**, 1245–1254 (2002)
D.T. Blewett, B.R. Hawke, P.G. Lucey, M.S. Robinson, J. Geophys. Res. **112**, E02005 (2007). doi:10.1029/2006JE002713
W.V. Boynton et al., Space Sci. Rev. (2007). doi:10.1007/s11214-007-9258-3
A.L. Broadfoot, S. Kumar, M.J.S. Belton, M.B. McElroy, Science **185**, 166–169 (1974)
A.L. Broadfoot, D.E. Shemansky, S. Kumar, Geophys. Res. Lett. **3**, 577–580 (1976)
T.H. Burbine, T.J. McCoy, L.R. Nittler, G.K. Benedix, E.A. Cloutis, T.L. Dickinson, Meteorit. Planet. Sci. **37**, 1233–1244 (2002)
A.G.W. Cameron, Icarus **64**, 285–294 (1985)
J.F. Cavanaugh et al., Space Sci. Rev. (2007, this issue). doi:10.1007/s11214-007-9273-4
S.C. Chase, E.D. Miner, D. Morrison, G. Münch, G. Neugebauer, M. Schroeder, Science **185**, 142–145 (1974)
A.F. Cheng, R.E. Johnson, S.M. Krimigis, L.J. Lanzerotti, Icarus **71**, 430–440 (1987)
U.R. Christensen, Nature **444**, 1056–1058 (2006)
S.P. Christon, Icarus **71**, 448–471 (1987)
P.E. Clark, S. Curtis, B. Giles, G. Marr, C. Eyerman, D. Winterhalter, Lunar Planet. Sci. **30** (1999), abstract 1036

COMPLEX, *Strategy for Exploration of the Inner Planets: 1977–1987* (Committee on Lunar and Planetary Exploration, National Academy of Sciences, Washington, 1978), 97 pp
COMPLEX, *1990 Update to Strategy for Exploration of the Inner Planets* (Committee on Lunar and Planetary Exploration, National Academy Press, Washington, 1990), 47 pp
J.E.P. Connerney, N.F. Ness, in *Mercury*, ed. by F. Vilas, C.R. Chapman, M.S. Matthews (University of Arizona Press, Tucson, 1988), pp. 494–513
A.C. Cook, M.S. Robinson, J. Geophys. Res. **105**, 9429–9443 (2000)
R.F. Dantowitz, S.W. Teare, M.J. Kozubal, Astron. J. **119**, 2455–2457 (2000)
D.L. Domingue et al., Space Sci. Rev. (2007, this issue). doi:10.1007/s11214-007-9260-9
J.H. Eraker, J.A. Simpson, J. Geophys. Res. **91**, 9973–9993 (1986)
B. Fegley Jr., A.G.W. Cameron, Earth Planet. Sci. Lett. **82**, 207–222 (1987)
W.C. Feldman, B.L. Barraclough, C.J. Hansen, A.L. Sprague, J. Geophys. Res. **102**, 25565–25574 (1997)
G. Giampieri, A. Balogh, Planet. Space Sci. **50**, 757–762 (2002)
R.E. Gold et al., Planet. Space Sci **49**, 1467–1479 (2001)
J.O. Goldsten et al., Space Sci. Rev. (2007, this issue). doi:10.1007/s11214-007-9262-7
R. Grard, M. Novara, G. Scoon, ESA Bull. **103**, 11–19 (2000)
H. Harder, G. Schubert, Icarus **151**, 118–122 (2001)
J.K. Harmon, Adv. Space Res. **19**, 1487–1496 (1997)
J.K. Harmon, Lunar Planet. Sci. **33** (2002), abstract 1858
J.K. Harmon, M.A. Slade, Science **258**, 640–642 (1992)
J.K. Harmon, P.J. Perillat, M.A. Slade, Icarus **149**, 1–15 (2001)
J.K. Harmon, M.A. Slade, B.J. Butler, J.W. Head, III, M.S. Rice, D.B. Campbell, Icarus **187**, 374–405 (2007)
S.A. Hauck, II, A.J. Dombard, R.J. Phillips, S.C. Solomon, Earth Planet. Sci. Lett. **222**, 713–728 (2004)
S.E. Hawkins, III et al., Space Sci. Rev. (2007, this issue). doi:10.1007/s11214-007-9266-3
H. Hayakawa, Y. Kasaba, H. Yamakawa, H. Ogawa, T. Mukai, Adv. Space Res. **33**, 2142–2146 (2004)
J.W. Head et al., Space Sci. Rev. (2007, this issue). doi:10.1007/s11214-007-9263-6
M.H. Heimpel, J.M. Aurnou, F.M. Al-Shamali, N. Gomez Perez, Earth Planet. Sci. Lett. **236**, 542–557 (2005)
M.E. Holdridge, A.B. Calloway, Space Sci. Rev. (2007, this issue). doi:10.1007/s11214-007-9261-8
I.V. Holin, Lunar Planet. Sci. **33** (2002), abstract 1387
L.L. Hood, G. Schubert, J. Geophys. Res. **84**, 2641–2647 (1979)
D.M. Hunten, A.L. Sprague, Meteorit. Planet. Sci. **37**, 1191–1195 (2002)
D.M. Hunten, T.H. Morgan, D.E. Shemansky, in *Mercury*, ed. by F. Vilas, C.R. Chapman, M.S. Matthews (University of Arizona Press, Tucson, 1988), pp. 562–612
P. Janhunen, E. Kallio, Ann. Geophys. **22**, 1829–1830 (2004)
R. Jeanloz, D.L. Mitchell, A.L. Sprague, I. de Pater, Science **268**, 1455–1457 (1995)
R.M. Killen, J. Benkhoff, T.H. Morgan, Icarus **125**, 195–211 (1997)
R.M. Killen, A. Potter, A. Fitzsimmons, T.H. Morgan, Planet. Space Sci. **47**, 1449–1458 (1999)
R.M. Killen et al., J. Geophys. Res. **106**, 20509–20525 (2001)
G. Kletetschka, P. Wasilewski, P.T. Taylor, Meteorit. Planet. Sci. **35**, 895–899 (2000)
S.J. Kortenkamp, E. Kokubo, S.J. Weidenschilling, in *Origin of the Earth and Moon*, ed. by R.M. Canup, K. Righter (University of Arizona Press, Tucson, 2000), pp. 85–100
H. Korth et al., Planet. Space Sci. **52**, 733–746 (2004)
L. Ksanfomality, A.L. Sprague, Icarus **188**, 271–287 (2007)
L. Ksanfomality, G. Papamastorakis, N. Thomas, Planet. Space Sci. **53**, 849–859 (2005)
J.C. Leary et al., Space Sci. Rev. (2007, this issue). doi:10.1007/s11214-007-9269-0
F. Leblanc, R.E. Johnson, Icarus **164**, 261–281 (2003)
V. Lesur, A. Jackson, Geophys. J. Int. **140**, 453–459 (2000)
J.S. Lewis, in *Mercury*, ed. by F. Vilas, C.R. Chapman, M.S. Matthews (University of Arizona Press, Tucson, 1988), pp. 651–669
J.L. Margot, S.J. Peale, R.F. Jurgens, M.A. Slade, I.V. Holin, Science **316**, 710–714 (2007)
J.V. McAdams, R.W. Farquhar, A.H. Taylor, B.G. Williams, Space Sci. Rev. (2007, this issue). doi:10.1007/s11214-007-9162-x
W.E. McClintock, M.R. Lankton, Space Sci. Rev. (2007, this issue). doi:10.1007/s11214-007-9264-5
R.L. McNutt Jr., S.C. Solomon, R. Grard, M. Novara, T. Mukai, Adv. Space Res. **33**, 2126–2132 (2004)
R.L. McNutt Jr., S.C. Solomon, R.E. Gold, J.C. Leary, the MESSENGER Team, Adv. Space Res. **38**, 564–571 (2006)
M. Mendillo, J. Warell, S.S. Limaye, J. Baumgardner, A. Sprague, J.K. Wilson, Planet. Space Sci. **49**, 1501–1505 (2001)
S.M. Milkovich, J.W. Head, L. Wilson, Meteorit. Planet. Sci. **37**, 1209–1222 (2002)
J.I. Moses, K. Rawlins, K. Zahnle, L. Dones, Icarus **137**, 197–221 (1999)
B.C. Murray, J. Geophys. Res. **80**, 2342–2344 (1975)

V.R. Murthy, W. van Westrenen, Y. Fei, Nature **423**, 163–165 (2003)
R.M. Nelson, L.J. Horn, J.R. Weiss, W.D. Smythe, Lunar Planet. Sci. **25**, 985–986 (1994)
N.F. Ness, K.W. Behannon, R.P. Lepping, Y.C. Whang, K.H. Schatten, Science **185**, 151–160 (1974)
N.F. Ness, K.W. Behannon, R.P. Lepping, Y.C. Whang, J. Geophys. Res. **80**, 2708–2716 (1975)
N.F. Ness, K.W. Behannon, R.P. Lepping, Y.C. Whang, Icarus **28**, 479–488 (1976)
G. Neukum et al., Mercury polar orbiter, A proposal to the European Space Agency in response to a call for new mission proposals issued on 10 July 1985, Deutsche Forschungs- und Versuchsanstalt für Luft- und Raumfahrt, Wessling, Germany, 1985, 33 pp
G. Neukum, J. Oberst, H. Hoffmann, R. Wagner, B.A. Ivanov, Planet. Space Sci. **49**, 1507–1521 (2001)
F. Nimmo, Geophys. Res. Lett. **29**, 1063 (2002). doi:10.1029/2001GL013883
F. Nimmo, T.R. Watters, Geophys. Res. Lett. **31**, L02701 (2004). doi:10.1029/2003GL018847
K.W. Ogilvie et al., Science **185**, 145–151 (1974). doi:10.1029/2003GL018847
D.A. Paige, S.E. Wood, A.R. Vasavada, Science **258**, 643–646 (1992)
S.J. Peale, in *Mercury*, ed. by F. Vilas, C.R. Chapman, M.S. Matthews (University of Arizona Press, Tucson, 1988), pp. 461–493
S.J. Peale, R.J. Phillips, S.C. Solomon, D.E. Smith, M.T. Zuber, Meteorit. Planet. Sci. **37**, 1269–1283 (2002)
A.E. Potter, T.H. Morgan, Science **229**, 651–653 (1985)
A.E. Potter, T.H. Morgan, Icarus **67**, 336–340 (1986)
A.E. Potter, R.M. Killen, T.H. Morgan, Planet. Space Sci. **47**, 1441–1448 (1999)
A.E. Potter, R.M. Killen, T.H. Morgan, Meteorit. Planet. Sci. **37**, 1165–1172 (2002a)
A.E. Potter, C.M. Anderson, R.M. Killen, T.H. Morgan, J. Geophys. Res. **107**, 5040 (2002b). doi:10.1029/2000JE001493
M.S. Robinson, P.G. Lucey, Science **275**, 197–198 (1997)
M.S. Robinson, G.J. Taylor, Meteorit. Planet. Sci. **36**, 841–847 (2001)
M.S. Robinson, M.E. Davies, T.R. Colvin, K. Edwards, J. Geophys. Res. **104**, 30847–30852 (1999)
S.K. Runcorn, Nature **253**, 701–703 (1975)
A.G. Santo et al., Planet. Space Sci. **49**, 1481–1500 (2001)
C.E. Schlemm II, et al., Space Sci. Rev. (2007, this issue). doi:10.1007/s11214-007-9248-5
G. Schubert, M.N. Ross, D.J. Stevenson, T. Spohn, in *Mercury*, ed. by F. Vilas, C.R. Chapman, M.S. Matthews (University of Arizona Press, Tucson, 1988), pp. 429–460
R. Schulz, J. Benkhoff, Adv. Space Res. **38**, 572–577 (2006)
R.W. Siegfried, II, S.C. Solomon, Icarus **23**, 192–205 (1974)
J.A. Simpson, J.H. Eraker, J.E. Lamport, P.H. Walpole, Science **185**, 160–166 (1974)
G. Siscoe, L. Christopher, Geophys. Res. Lett. **2**, 158–160 (1975)
M.A. Slade, B.J. Butler, D.O. Muhleman, Science **258**, 635–640 (1992)
M.A. Slade, R.F. Jurgens, J.-L. Margot, E.M. Standish, in *Mercury: Space Environment, Surface, and Interior* (Lunar and Planetary Institute, Houston, 2001), pp. 88–89
J.A. Slavin, Adv. Space Res. **33**, 1859–1874 (2004)
J.A. Slavin et al., Space Sci. Rev. (2007, this issue). doi:10.1007/s11214-007-9154-x
D.E. Smith, M.T. Zuber, S.J. Peale, R.J. Phillips, S.C. Solomon, in *Mercury: Space Environment, Surface, and Interior* (Lunar and Planetary Institute, Houston, 2001), pp. 90–91
D.E. Smith et al., Science **311**, 53 (2006)
S.C. Solomon et al., Planet. Space Sci. **49**, 1445–1465 (2001)
A.L. Sprague, D.M. Hunten, K. Lodders, Icarus **118**, 211–215 (1995)
A.L. Sprague et al., Icarus **129**, 506–527 (1997)
A.L. Sprague, W.J. Schmitt, R.E. Hill, Icarus **136**, 60–68 (1998)
A.L. Sprague, J.P. Emery, K.L. Donaldson, R.W. Russell, D.K. Lynch, A.L. Mazuk, Meteorit. Planet. Sci. **37**, 1255–1268 (2002)
P.D. Spudis, J.E. Guest, in *Mercury*, ed. by F. Vilas, C.R. Chapman, M.S. Matthews (University of Arizona Press, Tucson, 1988), pp. 118–164
P.D. Spudis, J.B. Plescia, A.D. Stewart, Lunar Planet. Sci. **25**, 1323–1324 (1994)
D.K. Srinivasan, M.E. Perry, K.B. Fielhauer, D.E. Smith, M.T. Zuber, Space Sci. Rev. (2007, this issue). doi:10.1007/s11214-007-9270-7
L.J. Srnka, Phys. Earth Planet. Inter. **11**, 184–190 (1976)
S. Stanley, J. Bloxham, W.E. Hutchison, M.T. Zuber, Earth Planet. Sci. Lett. **234**, 27–38 (2005)
L.V. Starukhina, J. Geophys. Res. **106**, 14701–14710 (2001)
A. Stephenson, Earth Planet. Sci. Lett. **28**, 454–458 (1976)
D.J. Stevenson, Earth Planet. Sci. Lett. **82**, 114–120 (1987)
D.J. Stevenson, Earth Planet. Sci. Lett. **208**, 1–11 (2003)
R.G. Strom, N.J. Trask, J.E. Guest, J. Geophys. Res. **80**, 2478–2507 (1975)

H. Svedhem, D. Titov, D. McCoy, J. Rodriguez-Canabal, J. Fabrega, Eos, Trans. Am. Geophys. Union **86** (Fall Meeting suppl.) (2005), abstract P23E-01
A.R. Vasavada, D.A. Paige, S.E. Wood, Icarus **141**, 179–193 (1999)
F. Vilas, Icarus **64**, 133–138 (1985)
J. Warell, S.S. Limaye, Planet. Space Sci. **49**, 1531–1552 (2001)
J. Warell, P.-G. Valegård, Astron. Astrophys. **460**, 625–633 (2006)
J. Warell, A.L. Sprague, J.P. Emery, R.W.H. Kozlowski, A. Long, Icarus **35**, 99–111 (2006)
T.R. Watters, M.S. Robinson, A.C. Cook, Geology **26**, 991–994 (1998)
T.R. Watters, R.A. Schultz, M.S. Robinson, A.C. Cook, Geophys. Res. Lett. **29**, 1542 (2002). doi:10.1029/2001GL014308
S.J. Weidenschilling, Icarus **35**, 99–111 (1978)
S.J. Weidenschilling, Lunar Planet. Sci. **29** (1998), abstract 1278
S.J. Weidenschilling, J.N. Cuzzi, in *Protostars and Planets III*, ed. by E.H. Levy, J.I. Lunine (University of Arizona Press, Tucson, 1993), pp. 1031–1060
G.W. Wetherill, in *Mercury*, ed. by F. Vilas, C.R. Chapman, M.S. Matthews (University of Arizona Press, Tucson, 1988), pp. 670–691
G.W. Wetherill, Geochim. Cosmochim. Acta **58**, 4513–4520 (1994)
H.L. Winters et al., Space Sci. Rev. (2007, this issue). doi:10.1007/s11214-007-9257-4
X. Wu, P.L. Bender, S.J. Peale, G.W. Rosborough, M.A. Vincent, Planet. Space Sci. **45**, 15–19 (1997)
C.-W. Yen, Ballistic Mercury orbiter mission via Venus and Mercury gravity assists, AAS/AIAA Astrodynamics Specialist Conference, AIAA 83-346, San Diego, CA (1985)
C.-W. Yen, J. Astronaut. Sci. **37**, 417–432 (1989)
M.T. Zuber et al., Space Sci. Rev. (2007, this issue). doi:10.1007/s11214-007-9265-4

Space Sci Rev (2007) 131: 41–84
DOI 10.1007/s11214-007-9263-6

The Geology of Mercury: The View Prior to the MESSENGER Mission

James W. Head · Clark R. Chapman · Deborah L. Domingue · S. Edward Hawkins, III · William E. McClintock · Scott L. Murchie · Louise M. Prockter · Mark S. Robinson · Robert G. Strom · Thomas R. Watters

Received: 9 January 2007 / Accepted: 10 August 2007 / Published online: 10 October 2007

Abstract Mariner 10 and Earth-based observations have revealed Mercury, the innermost of the terrestrial planetary bodies, to be an exciting laboratory for the study of Solar System geological processes. Mercury is characterized by a lunar-like surface, a global magnetic field, and an interior dominated by an iron core having a radius at least three-quarters of the radius of the planet. The 45% of the surface imaged by Mariner 10 reveals some distinctive differences from the Moon, however, with major contractional fault scarps and huge expanses of moderate-albedo Cayley-like smooth plains of uncertain origin. Our current image coverage of Mercury is comparable to that of telescopic photographs of the Earth's Moon prior to the launch of Sputnik in 1957. We have no photographic images of one-half of the surface, the resolution of the images we do have is generally poor (~1 km), and as with many lunar telescopic photographs, much of the available surface of Mercury is distorted by foreshortening due to viewing geometry, or poorly suited for geological analysis and

J.W. Head (✉)
Department of Geological Sciences, Brown University, Providence, RI 02912, USA
e-mail: James_Head@brown.edu

C.R. Chapman
Southwest Research Institute, 1050 Walnut St., Suite 400, Boulder, CO 80302, USA

D.L. Domingue · S.E. Hawkins, III · S.L. Murchie · L.M. Prockter
The Johns Hopkins University Applied Physics Laboratory, Laurel, MD 20723, USA

W.E. McClintock
Laboratory for Atmospheric and Space Physics, University of Colorado, Boulder, CO 80303, USA

M.S. Robinson
Department of Geological Sciences, Arizona State University, Tempe, AZ 85251, USA

R.G. Strom
Lunar and Planetary Laboratory, University of Arizona, Tucson, AZ 85721, USA

T.R. Watters
Center for Earth and Planetary Studies, National Air and Space Museum, Smithsonian Institution, Washington, DC 20560, USA

impact-crater counting for age determinations because of high-Sun illumination conditions. Currently available topographic information is also very limited. Nonetheless, Mercury is a geological laboratory that represents (1) a planet where the presence of a huge iron core may be due to impact stripping of the crust and upper mantle, or alternatively, where formation of a huge core may have resulted in a residual mantle and crust of potentially unusual composition and structure; (2) a planet with an internal chemical and mechanical structure that provides new insights into planetary thermal history and the relative roles of conduction and convection in planetary heat loss; (3) a one-tectonic-plate planet where constraints on major interior processes can be deduced from the geology of the global tectonic system; (4) a planet where volcanic resurfacing may not have played a significant role in planetary history and internally generated volcanic resurfacing may have ceased at ~3.8 Ga; (5) a planet where impact craters can be used to disentangle the fundamental roles of gravity and mean impactor velocity in determining impact crater morphology and morphometry; (6) an environment where global impact crater counts can test fundamental concepts of the distribution of impactor populations in space and time; (7) an extreme environment in which highly radar-reflective polar deposits, much more extensive than those on the Moon, can be better understood; (8) an extreme environment in which the basic processes of space weathering can be further deduced; and (9) a potential end-member in terrestrial planetary body geological evolution in which the relationships of internal and surface evolution can be clearly assessed from both a tectonic and volcanic point of view. In the half-century since the launch of Sputnik, more than 30 spacecraft have been sent to the Moon, yet only now is a second spacecraft en route to Mercury. The MESSENGER mission will address key questions about the geologic evolution of Mercury; the depth and breadth of the MESSENGER data will permit the confident reconstruction of the geological history and thermal evolution of Mercury using new imaging, topography, chemistry, mineralogy, gravity, magnetic, and environmental data.

Keywords Mercury · MESSENGER · Planets and satellites: general · Mariner 10 · Caloris basin

1 Introduction and Background

In the 47 years between the launch of Sputnik, the first artificial satellite of the Earth, and the launch of the MErcury Surface, Space ENvironment, GEochemistry, and Ranging (MESSENGER) spacecraft to Mercury in 2004, the golden age of Solar System exploration has changed the terrestrial planets from largely astronomically perceived objects to intensely studied geological objects. During this transition, we have come to understand the basic range of processes differentiating planetary interiors, creating planetary crusts, and forming and modifying planetary surfaces. We have also learned how the relative importance of processes has changed with time; the chemical and mineralogic nature of surfaces and crusts; the broad mechanical and chemical structure of planetary interiors; and the relationship of surface geology to internal processes and thermal evolution (e.g., Head 2001a, 2001b). Together with these new insights have come outlines of the major themes in the evolution of the terrestrial planets (e.g., Head and Solomon 1981; Stevenson 2000).

These comprehensive advances and the synthesis of our understanding mask an underlying problem: Our level of knowledge of the terrestrial planets is extremely uneven, and this difference threatens the very core of our emerging understanding. Nothing better illustrates

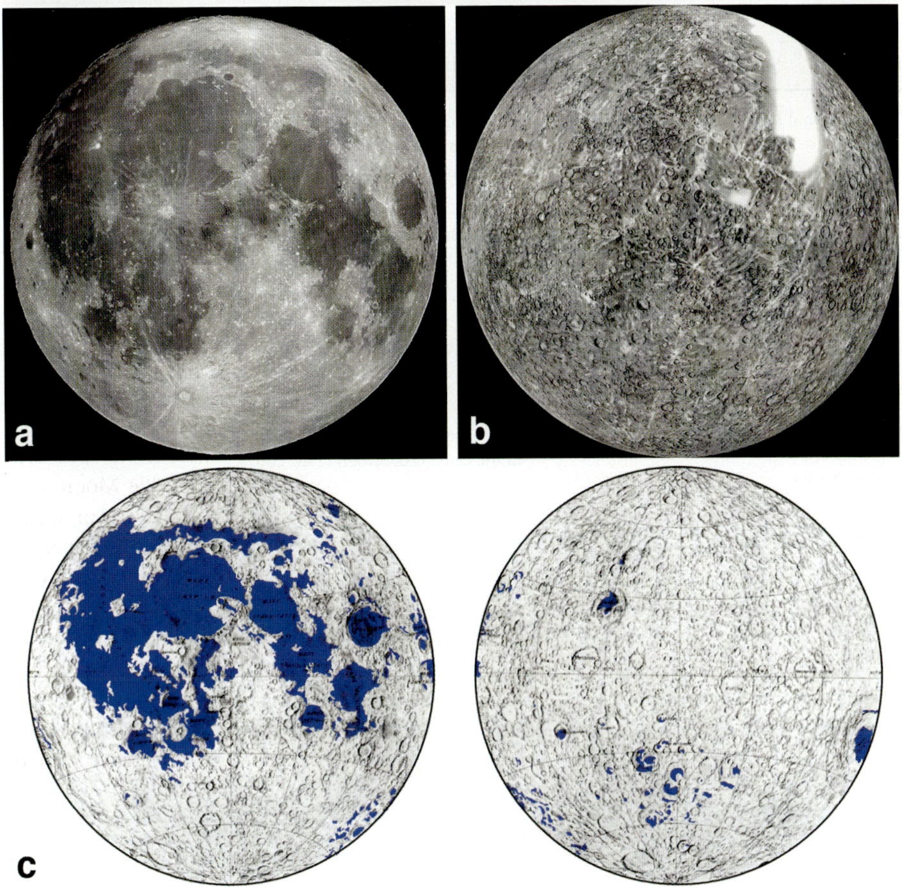

Fig. 1 (**a**) Earth-based telescopic photograph of the Moon typical of the area of the Moon seen prior to the time of the launch of Sputnik in 1957. Lick Observatory photograph. (**b**) Photographic coverage of Mercury from Mariner 10 (launched 1973) available at the time of the launch of the MESSENGER mission to Mercury in 2004, almost a half-century after Sputnik (shaded relief airbrush map; USGS, Flagstaff). (**c**) Map of the Earth's Moon in equal area projection showing the distribution of mare basalts on the nearside and farside. Compare with (**a**) and note the distinctive nearside-farside differences in lunar mare basalt distribution unknown before Luna 3 in 1959, and the general lack of mare deposits on the nearside limbs and southern nearside, a fact underappreciated due to Earth-based telescope viewing geometry (**a**)

this point than our currently poor knowledge of the planet Mercury. Mariner 10 imaged less than one-half of Mercury at a resolution of $\sim$1 km/pixel and even these data are variable in terms of quality due to differences in viewing geometry and solar illumination (Strom 1987). Indeed, our current image data for Mercury are generally comparable in resolution and coverage to our pre-Sputnik, Earth-based telescope photographs of the Moon (Fig. 1). However, the pre-Sputnik Earth-based telescope photographs of the Moon are actually more useful in terms of the range of different illumination conditions available. Recently, radar delay-Doppler mapping has begun to provide data with sufficient spatial resolution to enable some geologic studies of the side of Mercury not seen by Mariner 10 (e.g., Harmon et al. 2007).

Table 1 Instruments on the MESSENGER mission to Mercury (Gold et al. 2001; Santo et al. 2001)

Instrument	Description
Mercury Dual Imaging System (MDIS)	Wide-angle and narrow-angle imagers that map landforms and variations in surface spectra and gather topographic information (Hawkins III et al. 2007)
Gamma-Ray and Neutron Spectrometer (GRNS)	Maps the relative abundances of different elements and helps to determine if there is ice in the polar regions (Goldsten et al. 2007)
X-Ray Spectrometer (XRS)	Detects emitted X-rays to measure the abundances of various elements in the materials of the crust (Schlemm II et al. 2007)
Magnetometer (MAG)	Maps the magnetic field and any regions of magnetized rocks in the crust (Anderson et al. 2007)
Mercury Laser Altimeter (MLA)	Produces highly accurate measurements of topography (Cavanaugh et al. 2007)
Mercury Atmospheric and Surface Composition Spectrometer (MASCS)	Measures the abundances of atmospheric gases and minerals on the surface (McClintock and Lankton 2007)
Energetic Particle and Plasma Spectrometer (EPPS)	Measures the composition, distribution, and energy of charged particles (electrons and various ions) in magnetosphere (Andrews et al. 2007)
Radio Science (RS)	Measures very slight changes in spacecraft velocity to study interior mass distribution, including crustal thickness variations (Srinivasan et al. 2007)

Yet there are striking contradictions brought about by what little information we do have about Mercury. Could a terrestrial (Earth-like) planet form and evolve with no extrusive volcanic activity? Can the internally generated resurfacing of a terrestrial planet conclude at ~3.8 Ga? Can one of the hottest planetary surfaces in the Solar System harbor an inventory of cometary ices? Can a planet containing an iron core proportionally much larger than that of the Earth not show demonstrable surface signs of internal convection? Can we confidently place Mercury in the scheme of geological and thermal evolution without ever having seen more than half of its surface with spacecraft observations? These and other questions formed the basis for the scientific rationale for the MESSENGER mission to Mercury (Solomon 2003). In this contribution, we review our basic current knowledge of the characteristics of the surface of Mercury at several scales, the geological features and processes observed thus far, and how this knowledge relates to its overall geological and thermal evolution. In the course of this review, we identify key unanswered questions, and how future studies and observations, in particular the MESSENGER mission and its instrument complement (Table 1), might address these. We first assess the state of knowledge of the surface from Earth-based remote sensing data, then review the current understanding of the geology of Mercury from Mariner 10 data, and end with a series of outstanding questions that can be addressed by the MESSENGER mission.

2 Remote Sensing and the Nature of the Surface of Mercury

Knowledge of the physical, chemical, mineralogic, and topographic properties of planetary surfaces is critical to understanding geological processes and evolution. Remote observations using instruments designed to characterize the surface at various wavelengths, first with Earth-based telescopes, and then with instruments on flybys and orbiters, have been the traditional manner in which we have learned about planetary surfaces. Two problems are presented by the proximity of Mercury to the Sun, first in making observations of a planet in such close solar proximity, and second the difficulty in placing a spacecraft in orbit around a planet so close to the huge solar gravity well. These factors, coupled with the apparent spectral blandness of Mercury, have resulted in rather limited knowledge of the nature of the optical surface. Here we review current knowledge and outstanding problems that can be addressed with MESSENGER instrument measurements and data.

2.1 Chemistry and Mineralogy

We know very little about the surface composition of Mercury (see detailed discussion in Boynton et al. 2007). Several decades ago it was realized that Mercury has a steeply reddened, quite linear reflectance spectrum throughout the visible and near-infrared (McCord and Clark 1979; Vilas 1988). It is similar to, but even redder than, the reddest lunar spectrum. Debate over the existence of minor spectral features in this spectral range (especially a possible pyroxene band near 0.95 μm) has been resolved in recent years by well-calibrated, higher quality spectra: Mercury's spectrum varies spatially from featureless to one with a shallow but well-resolved pyroxene absorption band (Fig. 2, bottom) (Warell et al. 2006). There are hints of absorption and emission features at longer infrared wavelengths (dominated by thermal emission) (Fig. 2, top), but their reality and the mineralogical implications have been debated (Vilas 1988; Boynton et al. 2007).

As is the case with the Moon, interpretation of such data by comparison with laboratory samples of plausible minerals is complicated by the major role played by space weathering (the modification of the inherent spectral signature of the minerals present by bombardment and modification of the minerals by micrometeorites, solar wind particles, etc.). Because Mercury is closer to the mineral-damaging radiation of the Sun, meteoroid impact velocities are much higher there, and Mercury's greater surface gravity inhibits widespread regolith ejecta dispersal, space weathering is predicted to be even more substantial than on the Moon, and it is likely that Mercury's spectrum is modified by space weathering even more than the lunar spectrum (e.g., Noble and Pieters 2003). Mineral grains at Mercury's optical surface are probably heavily shocked, coated with submicroscopic metallic iron, and otherwise damaged (e.g., Noble and Pieters 2003).

Although exogenous materials space-weather Mercury's surface, they are not expected to contaminate the mineralogical composition of the surface (by addition of exogenous material) to a degree that would generally be recognizable in remote-sensing data. The volumetric contribution of meteoritic material to lunar regolith samples is $\sim 1-2\%$ and there is no reason to expect it to be very different on Mercury. This is primarily because the projectile volume is tiny compared with the volume of planetary surface material that is displaced in a cratering event and cycled through the regolith. In addition, the Moon loses more mass than it gains by impact (Shuvalov and Artemieva 2006) and despite Mercury's higher escape velocity the greater impact velocities probably result in less retention of projectile material on Mercury. Darkening by admixture of fine carbonaceous material is probably overwhelmed by direct space-weathering effects. Small percentages of exogenous material are important

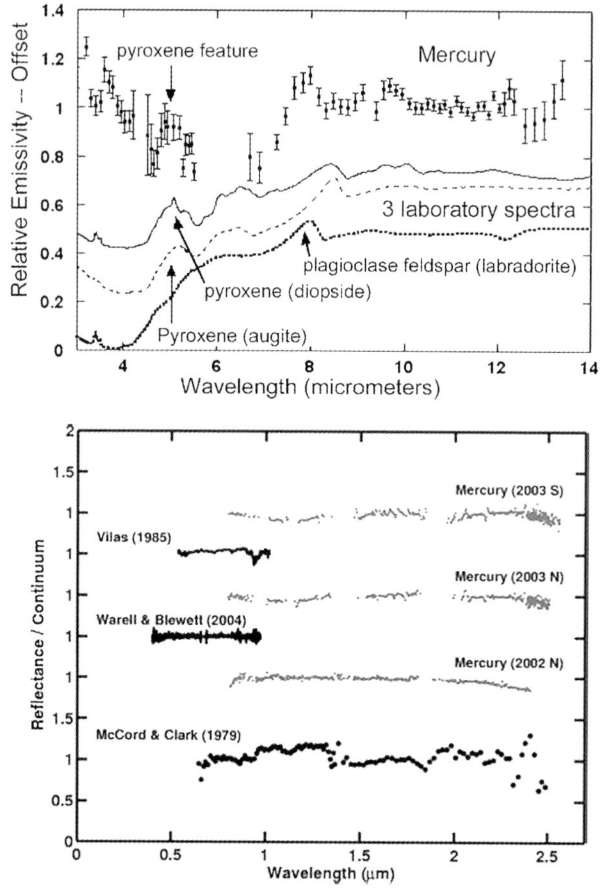

Fig. 2 Spectra of the surface of Mercury. (*Top*) A spectrum for the surface of Mercury in the mid-infrared (Sprague et al. 2002). Comparison with laboratory samples shows a peak near 5 μm that has been attributed to pyroxene and one near 8 μm similar to the spectral characteristics of anorthitic feldspar (Strom and Sprague 2003). (*Bottom*) Infrared Telescope Facility (IRTF) spectra from three different locations on Mercury (*gray*) compared with three other telescopic spectra (*black*). Solar reflectance and thermal emission components for the IRTF spectra have been removed, and each spectrum has been divided by a linear fit to the continuum. All spectra are normalized at 1 μm. The FeO absorption band is seen at 0.8–1.1 μm in the 2003N and 2003S spectra but absent in the 2002N spectrum, indicating lateral variability on the surface. From Warell et al. (2006)

to the degree that they are cold-trapped at the poles or visible in the tenuous atmosphere of Mercury.

A common interpretation of Mercury's nearly featureless spectrum is that its surface is analogous to the lunar anorthositic crust (Tyler et al. 1988; Sprague et al. 1994) (Fig. 2, bottom). But is there evidence for mare-like basalts that might have formed the smooth plains? Recent analyses have revealed the presence of a shallow 0.8–1.3 μm absorption feature centered near 1.1 μm that can be confidently interpreted as a characteristic iron-bearing silicate absorption (Fig. 2, bottom) (Warell et al. 2006), indicating that at least locally, soils may contain up to a few percent FeO. The scale of the observations precludes assignment of these spectra to specific geological units. Analysis of the exosphere of Mercury from ground-based observations has revealed enhanced Na and K emissions (e.g., Sprague et al. 1998) that may be correlated with specific areas on the surface of Mercury, specifically very fresh impact craters.

Ground-based remote sensing has also focused on imaging the parts of Mercury unimaged by Mariner 10 using advanced astronomical techniques (charge-coupled device, or CCD cameras and short exposure times) and modern processing software (combination of multiple images) (Warell and Limaye 2001; Mendillo et al. 2001; Ksanfomality 2004; Ksanfomality et al. 2005). Such efforts have resulted in the interpretation of a very large

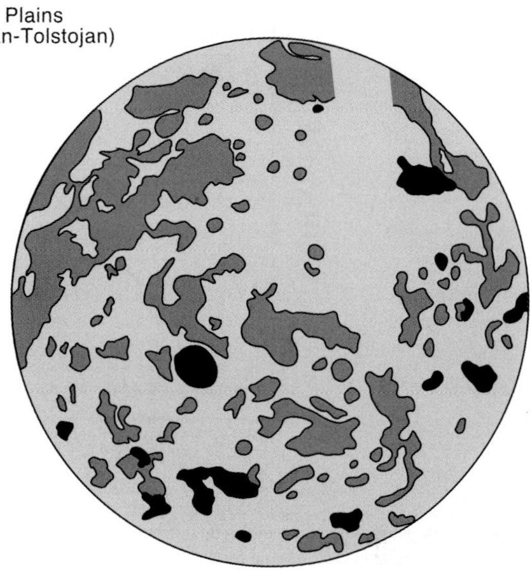

Fig. 3 Distribution of smooth plains on Mercury. Calorian-aged smooth plains are shown in dark gray, and Calorian and/or Tolstojan are shown in black. The remainder is cratered terrain. Together these smooth plains cover about 10.4×10^6 km^2 or 40% of the part of Mercury imaged by Mariner 10. Lambert equal-area projection centered on 0°N, 260°E (100°W), with north to the top. From Spudis and Guest (1988). Copyright, Arizona Board of Regents

impact basin (up to 2,300 km) with a dark central region in the unimaged part of Mercury (Ksanfomality 2004).

No totally self-consistent physical and chemical model for the composition, grain-sizes, and other parameters of Mercury surface soils has yet been devised that is fully compatible with these observations. Until space weathering processes are better understood, it will remain uncertain what firm constraints can be placed on Mercury's surface composition and its variation in relation to geologic units. The results from MESSENGER's numerous instrumental measurements (see also Boynton et al. 2007) (Table 1) will be critical to this understanding.

The Mariner 10 spacecraft carried no instrumentation capable of providing compositionally diagnostic remote-sensing information. The color images taken of Mercury have been reprocessed in recent years, showing slight but real differences in color, which may be correlated with surface morphology (Robinson and Lucey 1997). It is not clear whether variations in titanium content of surface soils might be responsible for the observed variations, as they are for color variations within the lunar maria. Albedo variations may also reflect, in some unknown way, variable composition, but Mercury lacks albedo variations as prominent as those between the highlands and maria of the Moon.

Initial analyses of Manner 10 color images of Mercury led to three major conclusions: crater rays and ejecta blankets are bluer (higher ratio of ultraviolet, or UV, to orange) than average Mercury, color boundaries often do not correspond to photogeologic units, and no low-albedo blue materials are found that are analogous to titanium-rich lunar mare deposits (Hapke et al. 1980; Rava and Hapke 1987). From these early studies it was noted that in a few cases color boundaries might correspond to mapped smooth plains units (Fig. 3); for example, Tolstoj basin (Rava and Hapke 1987) and Petrarch crater (Kiefer and Murray 1987). However, the calibration employed in these earlier studies did not adequately remove vidicon blemishes and radiometric residuals. A recalibration of the Mariner 10 UV (375 nm) and orange (575 nm) images resulted in a significantly increased signal-to-noise ratio (Robinson and Lucey 1997). These improved images were mosaicked and have been interpreted to indicate that color units correspond to previously mapped smooth plains on Mercury, and further that some color units are the re-

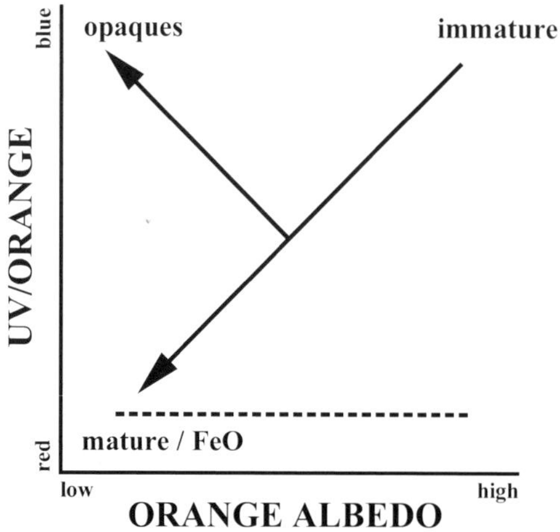

Fig. 4 Trends in the visible color of the lunar surface. The visible color of the lunar surface can be described by two perpendicular trends (opaque mineral concentration and iron-plus-maturity). The addition of ferrous iron to an iron-free silicate material (e.g., anorthosite) reddens the visible slope and lowers the albedo (a translation down the iron-maturity line; upper right to lower left). Color changes in lunar soil during maturation mimic the iron trend. As soils mature they redden (UV brightness/orange brightness) and their albedo decreases (orange brightness); soils translate down the iron-maturity line from upper right to lower left as they age. Adding spectrally neutral opaque minerals, such as ilmenite, results in a color trend that is nearly perpendicular to the iron-maturity line. Opaques lower the albedo but decrease the relative redness (an increase in the UV/orange ratio) of lunar soils. These two trends can be used to map the distribution of opaques (opaque index) and the iron-plus-maturity parameter through a coordinate rotation such that their perpendicular axes become parallel with the X and Y axes of the color-albedo plot (Robinson and Lucey 1997; Lucey et al. 1998); the *dotted line* indicates the position of the iron-maturity line after rotation. Adapted from Robinson and Lucey (1997)

sult of compositional heterogeneities in the crust of Mercury (Robinson and Lucey 1997; Robinson and Taylor 2001).

The newly calibrated Mariner 10 color data were interpreted in terms of the color reflectance paradigm that ferrous iron lowers the albedo and reddens (relative decrease in the UV/visible ratio) soil on the Moon and Mercury (Hapke et al. 1980; Rava and Hapke 1987; Cintala 1992; Lucey et al. 1995, 1998). Soil maturation through exposure to the space environment has a similar effect; soils darken and redden with the addition of submicroscopic iron metal and glass (Fig. 4). In contrast, addition of spectrally neutral opaque minerals (i.e., ilmenite) results in a trend that is nearly perpendicular to that of iron and maturity: Opaque minerals lower the albedo and increase the UV/visible ratio (Hapke et al. 1980; Rava and Hapke 1987; Lucey et al. 1998). For the Moon, the orthogonal effects of opaques and iron-plus-maturity are readily seen by plotting visible color ratio against reflectance (Lucey et al. 1998).

From Mariner 10 UV and orange mosaics a similar plot was constructed for the Mercury observations, and a coordinate rotation resulted in the separation of the two perpendicular trends (opaque mineral abundance from iron-plus-maturity) into two separate images (Robinson and Lucey 1997). The rotated data made possible the construction of two parameter maps: one delineating opaque mineralogy and the other showing variations in iron and maturity (Figs. 5 and 6). The opaque parameter map distinguishes units corresponding to previously mapped smooth plains deposits. The three best examples are the plains associated with Rudaki crater, Tolstoj basin, and Degas crater, each distinguished by their low

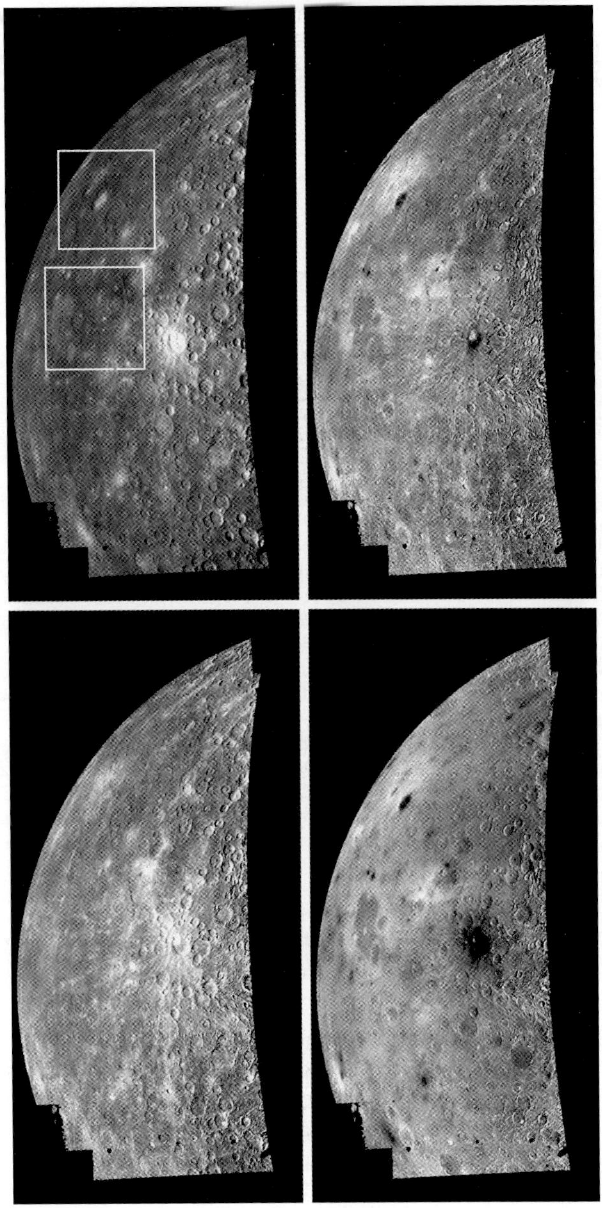

Fig. 5 Essential spectral parameters for the Mariner 10 incoming hemisphere. (*Upper left*) Orange (575 nm) albedo; boxes indicate areas enlarged in Fig. 6 (top is B, bottom is A). (*Upper right*) Relative color (UV/orange); higher tones indicate increasing blueness. (*Lower left*) Parameter 1-iron-maturity parameter; brighter tones indicate decreasing maturity and/or decreasing FeO content. (*Lower right*) Parameter 2-opaque index; brighter tones indicate increasing opaque mineral content. The relatively bright feature in the center right of the albedo image is the Kuiper-Muraski crater complex centered at 12°S, 330°E (30°W). Adapted from Robinson and Lucey (1997)

opaque index relative to their corresponding basement materials (Robinson and Lucey 1997; Robinson et al. 1997, 1998). In all three cases, the basement material is enriched in opaques.

A critical observation is that none of these units show a distinct unit boundary in the iron-plus-maturity image that corresponds to the morphologic plains boundary, leading to the interpretation that the smooth plains have an iron content that differs little from the global average. In the case of the Tolstoj basin (Robinson et al. 1998), a distinct mappable opaque index unit corresponds with the asymmetric NE–SW trending ejecta pattern of the basin, known as the Goya Formation (Schaber and McCauley 1980; Spudis and Guest 1988).

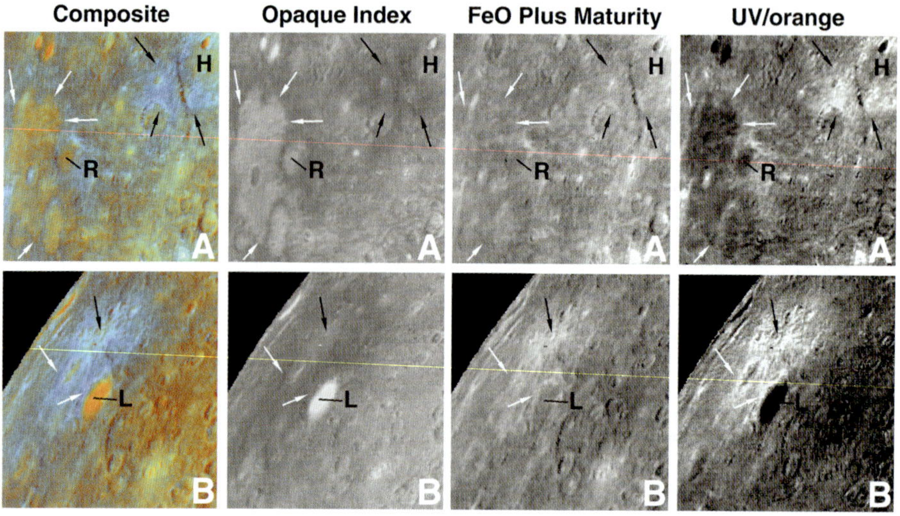

Fig. 6 Color ratio images of portions of Mercury. Enlargement of areas found on the Mariner 10 incoming hemisphere, keying on color units indicative of possible volcanically emplaced materials near the crater Rudaki (*R*; 120 km diameter; *top row*, *A*), Homer (*H*; 320 km diameter; *top row*, *A*), and Lermontov (*L*; 160 km diameter; *bottom row*, *B*). *Red* is formed from the inverse of the opaque index (increasing redness indicates decreasing opaque mineralogy; Fig. 5, lower right), the green component is the iron-maturity parameter (Fig. 5, lower left), and blue shows the relative color (UV/orange ratio; Fig. 5, upper right). The plains unit seen west and south and filling the crater Rudaki exhibits embaying boundaries indicative of material emplaced as a flow, and it has a distinct color signature relative to its surroundings. The blue material on the southwest margin of the crater Homer exhibits diffuse boundaries, is insensitive to local topographic undulations (*black arrows*), and is aligned along a linear segment of a Homer basin ring. A portion of the blue material seen northwest of the crater Lermontov is somewhat concentric to a small impact crater (*black arrow*) and may represent material excavated from below during the impact. However, examination of the iron-maturity parameter and opaque index images (*bottom row*) suggests that the darkest and bluest material (*white arrows*) in the deposit is not associated with an impact ejecta pattern, but rather that the anomalous lighter blue ejecta is composed of the dark material, although less mature and possibly with an admixture of basement material, overlying the darker blue portions of the deposit. Note that the opaque index was inverted relative to that shown in Fig. 5 to enhance contrast in the color composites (upper left and lower left panels). Adapted from Robinson and Lucey (1997)

This stratigraphic relation implies that formation of the Tolstoj basin (~550 km diameter) resulted in excavation of anomalously opaque-rich material from within the crust. The Goya Formation is not a mappable unit in the iron-plus-maturity image, indicating that its FeO content does not differ significantly from the local (and hemispheric) average.

A distinctive unit exhibiting diffuse boundaries (Fig. 6) is found near both Homer and Lermontov craters; examination of the iron-maturity parameter and opaque index images reveals that the darkest and bluest material in this deposit is not associated with an ejecta pattern, leading Robinson and Lucey (1997) to favor a pyroclastic origin (Figs. 6 and 7). The relatively blue color, high opaque index, and low albedo of these materials (for both areas) are consistent with a more mafic material, possibly analogous to a basaltic or gabbroic composition, or simply an addition of opaque minerals. Sprague et al. (1994) reported a tentative identification of basalt-like material in this hemisphere with Earth-based thermal IR measurements, while later microwave measurements were interpreted to indicate a total lack of areally significant basaltic materials on Mercury (Jeanloz et al. 1995). Earth-based spectral measurements have often been unable to resolve a ferrous iron band or to make any unassailable compositional infer-

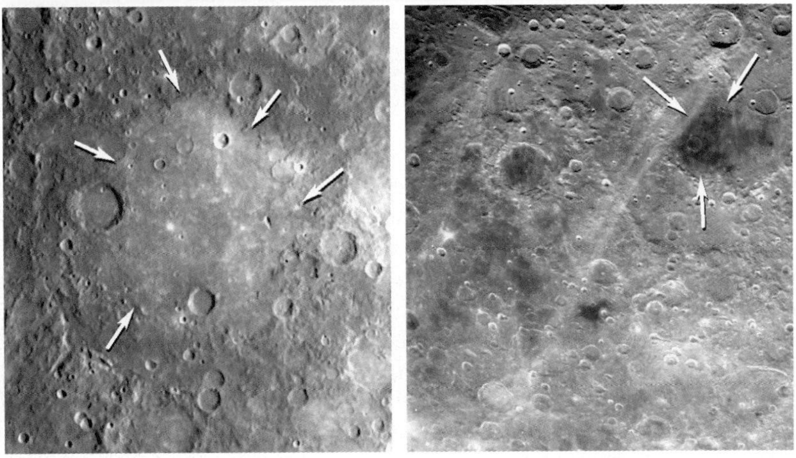

Fig. 7 A visual comparison of plains interpreted by some as flood lavas on Mercury found on the floor of the Tolstoj basin [*left*, 17°S, 196°E (164°W)] with Mare Humboldtianum on the Moon (*right*, 56°N, 280°E). Both data sets were acquired by Mariner 10 with similar resolutions (~1 km per pixel; each image is about 625 km across) and viewing geometries (incidence angle = 65° for Mercury; 55° for the Moon). The most obvious distinguishing characteristic of the lunar mare deposit is its albedo contrast with the underlying highlands (*right*), a contrast not observed for Mercury (*left*). The key morphologic properties used to identify flood lavas on the Moon (other than albedo) are embayment relationships and ponding in topographic lows (usually basins; see *arrows* on both images). For the Moon, classic flow features such as flow fronts and vents are not visible at a scale of 1 km, except in some cases under low-Sun illumination (see Milkovich et al. 2002, and references therein)

ences (Vilas 1988), although a generally anorthositic crust is favored (Blewett et al. 2002; Warell and Blewett 2004). From the data currently available it is not possible to identify basaltic material or of any other rock type; however, the Mariner 10-derived spectral parameters, stratigraphic relations, and morphology are interpreted by numerous workers to be consistent with volcanically emplaced materials (e.g., Spudis and Guest 1988; Robinson and Lucey 1997). The areal extent of these diffuse deposits is small, and thus it is unlikely that current Earth-based observation could detect their presence. Regardless of the mode of emplacement, the materials found around the craters Homer and Lermontov, and the plains units identified earlier (Figs. 5–7), argue that significant compositional units occur within the crust of Mercury and that at least some of them were likely to have been emplaced by volcanic processes.

Thus, Mariner 10 data provide clues to the nature and distribution of spectrally distinctive parts of the crust of Mercury related to processes of crustal differentiation, impact excavation, maturation, plains relationships, and possible pyroclastic volcanism. MESSENGER (Table 1) will provide high-resolution multispectral images of much of the surface of Mercury that, together with the results of high-spectral-resolution data, will permit characterization of the mineralogy of the surface. Together with data on crustal chemistry (e.g., Boynton et al. 2007), MESSENGER will thus provide a more global characterization of the chemistry and mineralogy of the crust, and the documentation of variations in a host of geological environments. For example, analysis of the ejecta deposits and central peaks of craters with a range of diameters can provide essential information on the crustal structure of Mercury, as has been done on the Moon (e.g., Tompkins and Pieters 1999), and examination of the range of mineralogy of the plains can lead to important insight into the origin and source heterogeneity of volcanically emplaced plains, as has been done on the Moon (e.g., Hiesinger et al. 2003).

2.2 Physical Properties: Photometry

The physical properties of the regolith on Mercury (such as porosity, particle size distribution, surface roughness, and particle albedo and structure) can be constrained through the analysis and modeling of photometric observations. To date these photometrically derived properties for Mercury have been studied through the analysis of both telescopic observations (Danjon 1949; de Vaucouleurs 1964; Dollfus and Auriere 1974; Mallama et al. 2002; Warell and Limaye 2001) and Mariner 10 spacecraft measurements (Hapke 1984; Bowell et al. 1989; Robinson and Lucey 1997). There have been several studies of Mercury's photometric properties using Hapke's (1993) model (Veverka et al. 1988; Domingue et al. 1997; Mallama et al. 2002; Warell 2004), especially in comparison to similar studies of the Moon (Veverka et al. 1988; Mallama et al. 2002; Warell 2004). Early modeling of Mercury's photometric phase curve using this model was performed by fitting Danjon's (1949) disk-integrated observations and comparing the resulting fits to disk-resolved measurements taken from Mariner 10 images (Veverka et al. 1988; Domingue et al. 1997). Danjon's data set covers 3° to 123° phase angle, which does not adequately constrain the opposition surge (in terms of detecting any coherent backscatter effects, thus poorly constraining regolith porosity and particle size distribution) or the backscattering regime (phase angles beyond 120°, which constrain surface roughness versus albedo and particle structure). More recent observations by Mallama et al. (2002) extended the phase angle coverage range (2° to 170°), especially in the backscattering regime. In addition, disk-resolved photometric measurements are now available from high-resolution CCD images obtained with the Swedish Vacuum Solar Telescope (SVST; Warell and Limaye 2001). Warell (2004) improved previous modeling efforts by simultaneously fitting both disk-integrated (Mallama et al. 2002) and disk-resolved (Warell and Limaye 2001) observations, using a more comprehensive Hapke model (Hapke 1981, 1984, 1986, 1993, 2002) and a Henyey–Greenstein single particle scattering function (which can be compared with the laboratory studies of particle scattering behavior versus particle structure by McGuire and Hapke 1995).

The photometric studies of Veverka et al. (1988), based on analysis of disk-integrated data from Danjon (1949), found that in comparing the regoliths on the Moon and Mercury, Mercury's regolith was less backscattering, possibly more compact, and similar in surface roughness. Similar modeling by Mallama et al. (2002) of their disk-integrated observations found that, in comparison, the regoliths of these two objects are similar in compaction and particle size distribution, and that the surface of Mercury is smoother. The results from Warell's (2004) simultaneous modeling of the disk-integrated and disk-resolved photometric observations are more in line with the results from Mallama et al. (2002). Compared with the lunar regolith, Warell (2004) showed that Mercury's surface has a slightly lower single scattering albedo, similar porosity, a smoother surface, and a stronger backscattering anisotropy in the single-particle scattering function. The larger range in phase angle coverage of the Mallama et al. (2002) data, modeled by both Mallama et al. (2002) and Warell (2004), provides a better determination of the surface roughness and particle scattering properties.

Porosity determinations based on Hapke's model are strongly coupled to assumptions made about the particle size distribution and the ratio of the radii of largest (r_{largest}) to smallest (r_{smallest}) particle within the regolith. The lunar regolith has been shown to have a grain-size distribution, Y, given by

$$Y = \frac{\sqrt{3}}{\ln\left(r_{\text{largest}}/r_{\text{smallest}}\right)} \tag{1}$$

(Bhattacharya et al. 1975). If this particle size distribution is assumed to hold true for both the Moon and Mercury, then the relationship

$$h = -\left(\frac{3}{8}\right) Y \ln(\rho), \tag{2}$$

where h is the Hapke opposition width parameter and ρ is the porosity, can be used to estimate regolith porosity. Mallama et al. (2002) found an h value of 0.065 for Mercury, whereas Warell's (2004) preferred solutions for Mercury and the Moon gave h values of 0.09 and 0.11, respectively. For $r_{\text{largest}}/r_{\text{smallest}}$ ratio values from 100 to 10,000, the porosity difference between the surface of the Moon and Mercury is $\sim$7%, with Mercury's regolith being slightly more porous (38% porosity with $r_{\text{largest}}/r_{\text{smallest}} = 1{,}000$).

Values for the Hapke surface roughness parameter vary between 20° and 25° (Veverka et al. 1988; Bowell et al. 1989; Domingue et al. 1997) to 8° to 16° (Mallama et al. 2002; Warell 2004). The disk-integrated observations of Mallama et al. (2002) and the disk-resolved observations of Warell and Limaye (2001) support a smoother surface on Mercury. However, the Mariner 10 disk-resolved data better match a surface with the higher, lunar-like roughness values. This discrepancy is most likely due to the variation in roughness across the surface of Mercury and the relative sampling of the surface by the different data sets (Warell 2004).

Analysis of the high-resolution CCD images of Mercury obtained with the SVST shows that there is an inverse relationship between the spectral slope and emission angle (Warell and Limaye 2001). A similar relationship between spectral slope and emission angle is observed for the Moon, but the relationship is more pronounced in the Mercury observations. Warell's (2002) interpretation is that the regolith of Mercury is more backscattering than the lunar regolith. The more backscattering nature of the surface is also seen in Warell's (2004) modeling of the integral phase curve and CCD images. When comparing the single particle scattering characteristics of the modeling solutions of the Moon and Mercury with the laboratory studies of McGuire and Hapke (1995), Warell (2004) found that the particles from both objects are characterized by grains with internal scatterers. The comparisons indicate that in general the regolith grains on Mercury have a higher number of internal scatterers and are more like the lunar mare materials than the lunar highlands. The backscattering nature of the grains on both the Moon and Mercury are commensurate with highly space-weathered, ground-up materials. The MESSENGER mission will provide important new information on the physical properties of the surface of Mercury from imaging observations at different viewing geometries, laser altimeter backscatter properties, albedo characterization of different geological environments, and the reflectance properties of surfaces of different ages.

2.3 Radar Observations

Earth-based radar observations from Arecibo and Goldstone have provided information on surface scattering properties, equatorial topography, deposits in permanently shadowed crater interiors, and preliminary information about the morphology and morphometry of portions of Mercury not observed by Mariner 10 (e.g., Harmon and Campbell 1988; Clark et al. 1988; Harmon and Slade 1992; Anderson et al. 1996; Harmon et al. 1986,

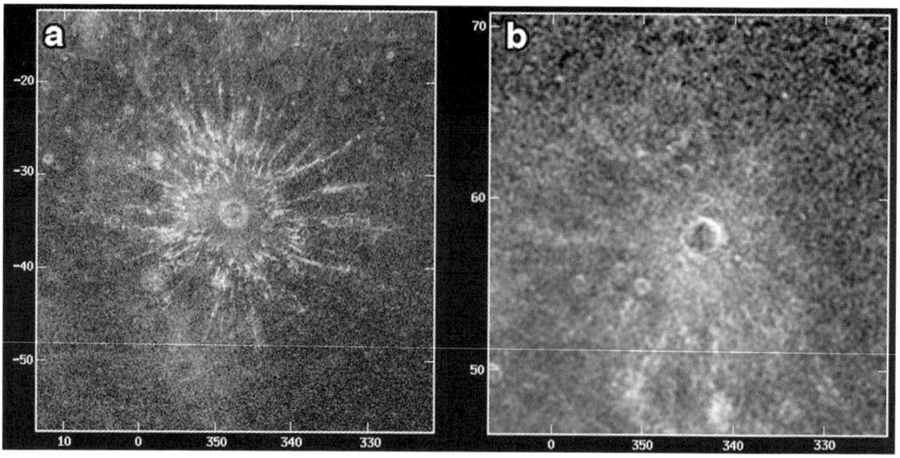

Fig. 8 Arecibo radar images from part of the surface not imaged by Mariner 10. (**a**) Feature "A," an 85-km-diameter crater whose radar ray system may be the most well-developed in the solar system (SC, same as transmitted sense polarization; i.e., same component transmitted and received). (**b**) Feature "B," a 95-km-diameter impact crater with a very bright halo but less distinct ray system. Feature "B," originally considered to be a candidate for a large volcanic edifice, is now clearly seen to be a very bright-haloed impact crater with a less distinct ray system than that of feature "A" (OC, opposite of transmitted sense polarization; i.e., opposite component received from that transmitted). Images from Harmon et al. (2007)

1994, 2001, 2007). Doppler spectrum shape and depolarization data yield information on dielectric properties and surface roughness, complementing the optical data. These data help confirm the presence of a regolith layer and show that the smooth plains are smooth at radar wavelengths (root mean square, or rms, slopes of about 4°). Quantitative data on equatorial topography have been very useful for the analysis of equatorial radius (~2,439.7 km) and shape, the range of altitudes (~7 km, from −2.4 to +4.6 km), and definition of the zero-altitude datum (+0.3 km), the most probable altitude as shown in the peak of the equatorial altimetric histogram (Harmon et al. 1986). Radar altimetry provided high-resolution topographic profiles for major features on Mercury (Harmon et al. 1986) showing a systematic difference in the depths of large craters between Mercury (shallower) and the Moon, and systematic differences between shadow measurements and radar measurements (17% lower) for large crater depths on Mercury. Other profiles documented the steep topography associated with major lobate fault systems (a 3 km drop in 70 km) and the rounded topography associated with arcuate scarps. Radar altimetry of basins and smooth plains shows the usefulness of depth determinations for basin degradation studies and regional topography for revealing large-scale undulations (downbowing) in the smooth plains. Altimetry of portions of Mercury not imaged by Mariner 10 revealed the extension of the circum-Caloris smooth plains into the unimaged hemisphere and suggested the presence of similar cratered terrain and plains there (Harmon et al. 1986).

Harmon et al. (2007) recently presented dual-polarization, delay-Doppler radar images of nonpolar and unimaged regions of Mercury obtained from several years of observations with the upgraded Arecibo S-band (12.6-cm) radar telescope. The images are dominated by radar-bright features associated with fresh impact craters. As previously reported, three of the most prominent crater features are located in the hemisphere not imaged by Mariner 10 and consist of feature "A", a crater 85 km in diameter whose radar ray system may be the most well-developed in the solar system (Fig. 8a), feature "B", a crater 95 km in diameter with a very bright halo but less distinct ray system (Fig. 8b), and feature "C", with rays and secondary craters distributed asymmetrically about a 125-km-diameter source crater. Feature "B", originally considered to be a candidate for a large volcanic edifice (Harmon 1997), is now clearly seen to be a very bright-haloed impact crater with a less distinct ray system than that of feature "A" (compare Figs. 8a and 8b). Two excellent examples of large ejecta/ray systems preserved in an intermediate state of degradation were also described. Although no evidence for volcanic edifices or central sources of lava flows are reported by Harmon et al. (2007) in the unimaged portion of Mercury, diffuse radar albedo variations are seen that have no obvious association with impact ejecta. Some smooth plains regions such as the circum-Caloris plains in Tir, Budh, and Sobkou Planitiae and the interior of Tolstoj basin show high depolarized brightness relative to their surroundings, which is the reverse of the mare/highlands contrast seen in lunar radar images. In contrast, Caloris basin appears dark and featureless in the images. The high depolarized brightness of the smooth plains could be due to (1) compositional differences from the lunar maria (lower iron and titanium content and thus less electrically lossy than mare lavas); (2) rougher small-scale surface texture which, if the plains are volcanic, could be related to differences in lava rheology; (3) a different roughness state due to the relative youth of the surface; and/or (4) a higher dielectric constant (Harmon et al. 2007).

Thus, we anticipate that the MESSENGER mission image and altimeter data will provide important new insight into surface topography in terms of the statistics of crater depths, the documentation of large degraded basins, crater degradation processes, tectonics, plains emplacement, and a determination of the features and stratigraphic relationships necessary to reconstruct the geologic history of Mercury.

3 The Geology of Mercury: General Terrain Types, Stratigraphy, and Geologic Time Scale

Prior to Mariner 10 nothing was known about the geological features and terrain types on Mercury; this situation changed virtually overnight with the first Mariner 10 images and the two subsequent flybys (Murray 1975). Trask and Guest (1975) used traditional photogeologic methods and the Mariner 10 images covering about 45% of the planet to produce the first geologic terrain map of Mercury. They recognized (1) a widespread unit, intercrater plains, (2) heavily cratered plains, (3) the Caloris basin and related deposits, (4) smooth plains, (5) hilly and lineated terrain antipodal to the Caloris basin, and (6) numerous younger craters and their related deposits, drawing attention to the similarities in units and geological history of Mercury and the Moon.

Subsequent more detailed analyses of the images were undertaken in a comprehensive geological mapping program at a scale of 1 : 5 M (e.g., Schaber and McCauley 1980; De Hon et al. 1981; Guest and Greeley 1983; McGill and King 1983; Trask and Dzurisin 1984; Spudis and Prosser 1984; Grolier and Boyce 1984). These geological maps, together with specific studies assessing key geological processes (e.g., Gault et al. 1975; Strom et al. 1975; Pike 1988; Schultz 1988; Strom and Neukum 1988; Melosh and McKinnon 1988; Thomas et al. 1988), provided the basis for our current state of knowledge about the geological history of Mercury (e.g., Murray et al. 1975; Spudis and Guest 1988). A time-stratigraphic system for Mercury (e.g., Spudis 1985) based on the rock-stratigraphic classification constructed during the 1 : 5 M quadrangle mapping and the earlier definition and subdivision of the Caloris Group (McCauley et al. 1981), has facilitated a correlation of geological events over the hemisphere imaged by Mariner 10 (Spudis and Guest 1988, plate 1-6) (Fig. 9) and

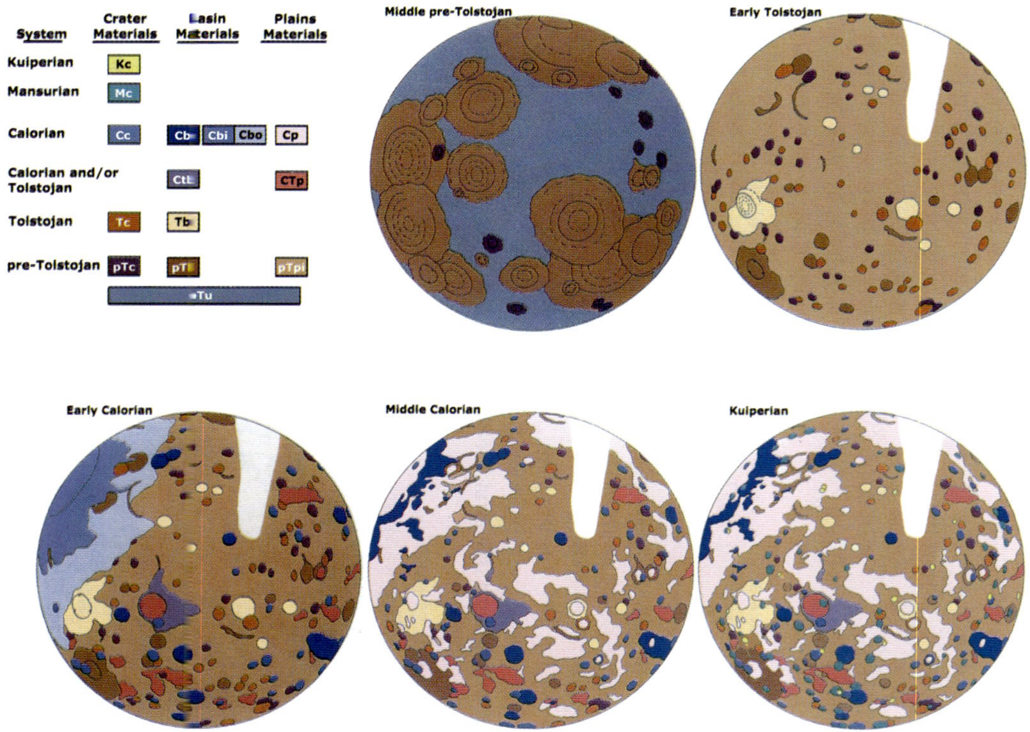

Fig. 9 Stratigraphic subdivisions of Mercury and a sequence of maps portraying the geologic setting at different times in the history of Mercury. After Spudis and Guest (1988). Copyright, Arizona Board of Regents

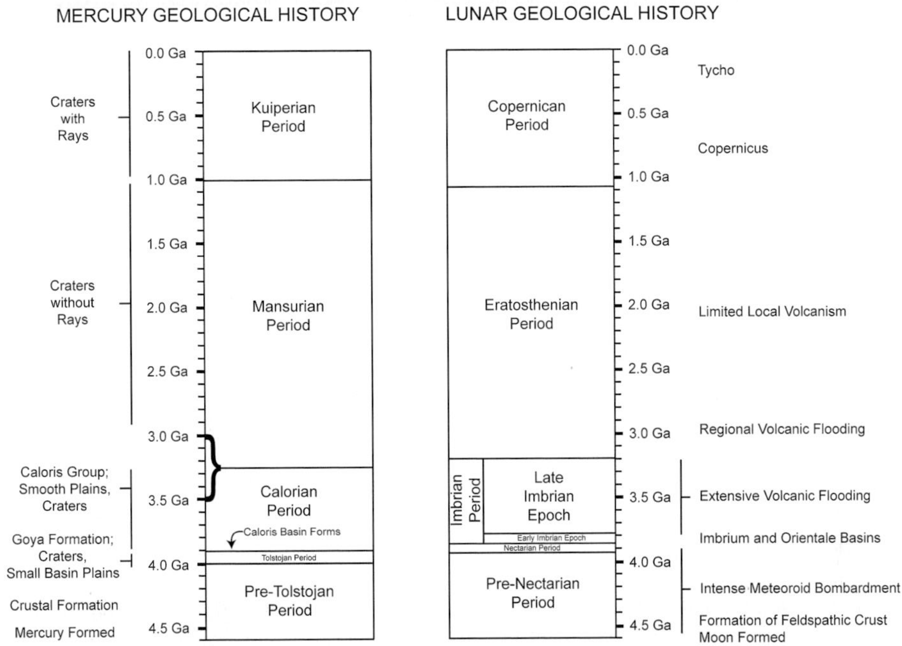

Fig. 10 The geological time scale of Mercury compared with that of the Moon. The absolute ages for Mercury are assumed to be tied to the lunar time scale but are not independently known. From Head (2006)

has permitted the continued comparison of the geological histories of Mercury and other planetary bodies begun soon after Mariner 10 (Murray et al. 1975).

Currently the geological history of Mercury is divided into five time-stratigraphic systems (Spudis and Guest 1988) (Figs. 9 and 10). The oldest predates the Tolstoj basin and consists largely of crater and multiringed basin deposits and extensive intercrater plains emplaced prior to the Tolstoj basin. Assuming that the heavily cratered terrains were produced by the same late heavy bombardment that is interpreted to have occurred on the Moon, this system is thought to predate 4.0 Ga and is approximately analogous to the pre-Nectarian on the Moon. This is also the very important period of crustal formation and early evolution during which time the impact rate was sufficiently high that the surface geological record was largely obliterated, and thus existing crater data are not very informative. For example, on Mars during this time, the crust formed, major crustal magnetic anomalies were emplaced, the fundamental global dichotomy in crustal thickness and topography was formed, and Tharsis, a major manifestation of internal thermal evolution, was emplaced (see Solomon et al. 2005). Despite our lack of knowledge of similar or analogous events in this period of the history of Mercury, MESSENGER and its instrument complement (Table 1) hold promise for detecting crustal magnetic anomalies, determining the origin of the magnetic field and assessing the properties of the outer core (Margot et al. 2007) and its implications for convection (e.g., Solomatov and Reese 2001), establishing the gravity field, determining global shape and topography, characterizing the elemental and mineralogical nature of the crust, establishing the major mode of crustal isostatic compensation (e.g., Zuber et al. 2007), and correlating all of these with the global geological context and history.

The base of the next youngest period, the Tolstojan System, is defined by the Tolstoj basin-forming event, and consists of Tolstoj and other crater and basin deposits as well as plains materials. This is assumed to date from ~4.0 to 3.9 Ga and is equivalent to the Nectarian on the Moon. The base of the next overlying unit, the Calorian System, is defined by the Caloris basin-forming event (evidently the last major basin formed on Mercury, Fig. 10) and consists of Caloris basin deposits, smaller crater deposits and widespread smooth plains deposits. It is thought to extend from about 3.9 Ga to perhaps as young as 3.0–3.5 Ga and is analogous to the Imbrian Period on the Moon. The next youngest Mansurian Period is defined by the crater Mansur and consists of impact crater deposits that contain no bright rays (analogous to the Eratosthenian Period on the Moon). There is no evidence for regional volcanic or tectonic activity during this time in the portion of Mercury imaged by Mariner 10. The Mansurian is thought to span the period from the end of the Calorian to about 1 Ga, although the absolute chronology has not been determined to better than a factor of a few. The youngest Kuiperian Period is defined by the bright-rayed crater Kuiper; deposits consist of impact craters still maintaining their bright rays, and there is no evidence for any regional volcanic or tectonic activity. The Kuiperian extends from ~1.0 Ga to the present and is analogous to the Copernican Period on the Moon.

These five systems define a context for the occurrence of other geological activity (Fig. 10). Widespread contractional deformation during the Calorian Period, after the formation of the Caloris basin and the emplacement of Calorian smooth plains, resulted in the lobate scarp thrust faults and wrinkle ridges in the imaged hemisphere. This suggests that the compressional stresses that formed these tectonic landforms peaked after the end of the period of heavy bombardment (Watters et al. 2004). Long-wavelength folds may also have formed in the period of global contraction (e.g., Hauck et al. 2004). Smaller wrinkle ridges formed on the smooth plains, and their emplacement and deformation have been dated as later than the Caloris impact event but closely associated with Calorian time. Spudis and Guest (1988) marshaled evidence in favor of a volcanic origin for the smooth plains on Mercury, citing (1) their planet-wide distribution (Figs. 3 and 9), (2) their total volume well in excess of what could be explained by impact ejecta, and (3) crater density data that indicated that major expanses of circum-Caloris smooth plains substantially postdate Caloris and all other major basins (see their Table III). On the basis of these data, they concluded that although the evidence is indirect, it is compelling enough to conclude that Mercury underwent large-scale volcanic resurfacing subsequent to the Caloris basin-forming impact. The extent and duration of the Calorian Period, and thus of the emplacement of the smooth plains and their deformation by wrinkle ridges, is unknown. By analogy with the lunar maria, Spudis and Guest (1988) estimated its duration to be from about 3.9 Ga to perhaps as young as 3.0–3.5 Ga, but others have estimated that the duration is much shorter (e.g., Strom and Neukum 1988; Neukum et al. 2001; Strom et al. 2005). In the next section, we address the important question of impact cratering rates and the absolute time scale. It is clear, however, that the MESSENGER mission (Table 1) will obtain a significantly better understanding of the geological history of Mercury through acquisition of data showing the geology of the other half of its surface, data to obtain better crater size frequency distributions for age determinations, and topography to study geological and stratigraphic relationships.

4 Geological Processes on Mercury: Impact Cratering and Basin Formation

The Mariner 10 images offered the opportunity to study the impact cratering process in a planetary environment similar to the Moon in some ways (lack of an atmosphere and its effects during crater formation and modification), similar to Mars in others (gravity), and different from both in terms of mean impact velocity. The morphology and morphometry of impact craters can provide significant insight into the physics of the cratering processes, as Mercury is a unique locale for calibrating the effects of impact velocity and gravity on a volatile-depleted silicate crust (e.g., Schultz 1988; Pike 1988). Thus, Mercury was viewed as a laboratory for the assessment of these variables and the Mariner 10 data as the first results. Analyses were undertaken to characterize the morphology and morphometry of fresh and degraded craters, and to assess the size-frequency distribution of impact craters to estimate ages of regional geological units defined by geological relationships and thus contribute to reconstruction of the geological history.

As with the Moon, Mars, Venus, and Earth, the morphologic complexity of impact craters (Fig. 11) was observed to increase systematically with diameter (Pike 1988). Key morphologic parameters were determined to be size dependent (e.g., depth, rim height, rim, floor and peak diameter; presence of bowl shape, flat floor, central peak, scalloped walls, wall terraces, etc.). Lunar-like classes of fresh craters were defined and ranged in increasing diameter from simple, to complex, to protobasin, to multiringed basin. The data permitted the relatively precise determination of transitions in depth/diameter relationships between the crater classes. The diameter of the transition from simple to complex craters on Mercury ($\sim$10.3 km) provided a comparison with that of the Moon, Mars, and Earth, and confirmed a strong inverse relationship with surface gravity and approach velocity. The new data showed that impact craters on Mercury and the Moon differed significantly in some other size-dependent aspects of crater form, such as protobasin, and two-ring basin, onset diameter. In a comprehensive review of crater and basin morphometry on the Earth, Mars, Moon, and Mercury, Pike (1988) found that neither average nor onset sizes of multiring basins on Mercury and the three other planets scale with gravity and concluded that surface gravity g, substrate rheology, and impactor velocity decrease in importance with increasing size of the impact, with g the last to disappear. Although much of the complexity of the interior of craters appears to be due to gravity-driven rim failure, inertially driven uplift of the crater bowl apparently played a major role in initiating the collapse. The apparent absence of clear influence of gravity on multiringed basin onset diameter led Pike (1988) to propose that multiringed basin formation is dominated by some combination of energy-scaled and hydrodynamic-periodic processes. Crater morphologic and morphometric characteristics were examined for craters on different substrates (e.g., smooth plains versus intercrater plains) in order to search for variations attributable to differences in the substrate physical properties. Although evidence for some variations was found, the effects were apparently minor.

MESSENGER data offer the opportunity to extend the study of crater morphometry globally, to increase the population and the statistical sample, to obtain more reliable quantitative measurements through altimeter observations and higher resolution images, and to search for substrate differences over larger areas.

The morphology of impact crater deposits added significant insight into the physics of the cratering process. For example, Gault et al. (1975) documented the role of gravity in the emplacement of ejecta relative to the Moon, illustrating the reduction in the range of the ballistic transport, the change in topography of the rim crest ejecta, and subsequent collapse and the formation of terraces. Furthermore, Schultz (1988) combined the Mercury observations with results from laboratory experiments and suggested that crater shapes intrinsically become flatter as the time for energy/momentum transfer increases, provided that a critical transfer time is exceeded. This resulted in the prediction that observed shallower craters on Mars relative to Mercury (at the same diameter) may be due to the low rms impact velocities at Mars relative to Mercury.

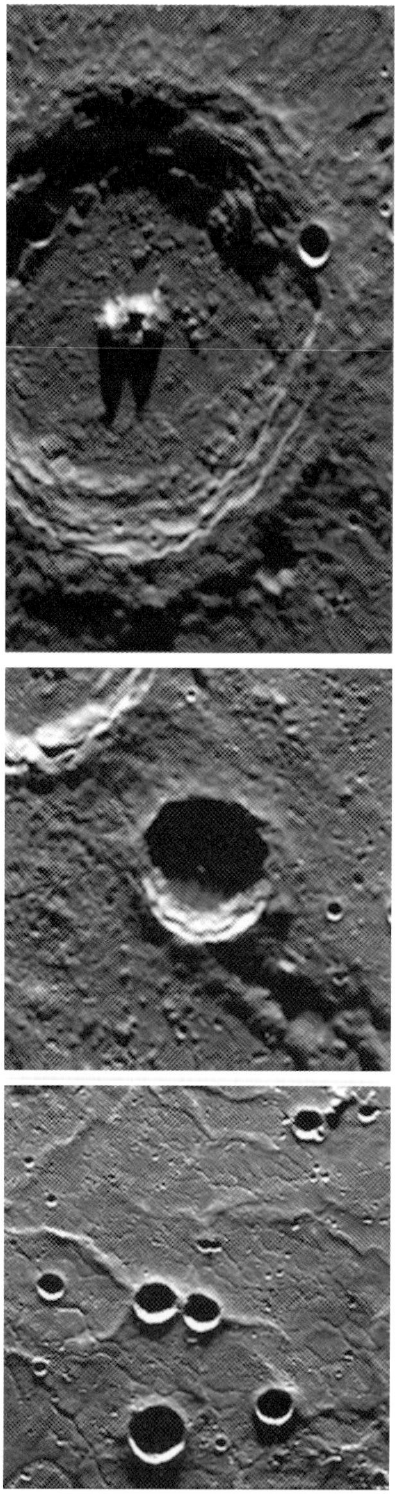

Fig. 11 Impact craters on Mercury revealed by Mariner 10. *Left*, a series of small circular, bowl-shaped simple craters less than ~15 km in diameter (width of field is ~175 km; Mariner 10 image 0000126). *Middle*, a crater in the diameter range 15–30 km, with a scalloped rim and the beginnings of wall slumping and flat floor, intermediate between simple and complex (width of field is 90 km; Mariner 10 image 0000098). *Right*, a large, complex crater greater than 30 km in diameter, with a polygonal outline, terraced walls, a flat floor, and a central peak (width of view is 110 km; Mariner 10 image 000080)

MESSENGER altimetry and imaging data, together with similar recently acquired data for Mars, will permit this prediction to be tested. Testing this hypothesis will permit the better understanding of potential differences in the impactor size-frequency distribution with time in different parts of the Solar System, a key parameter in assessing planetary chronology and interplanetary age comparisons (Schultz 1988). As pointed out by Schultz (1988), the cratering record on Mercury contains critical information for the true understanding of planetary bombardment history and distinguishing the effects of contrasting combinations of targets and impactors. Impact crater degradation processes on Mercury were also analyzed (e.g., Gault et al. 1975; Smith 1976) and shown to be very similar to the impact-caused degradation seen on the Moon (e.g., Head 1975) with important variations related to the more limited lateral ejecta dispersal on Mercury. MESSENGER image and altimetry data will provide the basis to quantify these degradation relationships and to assess the relative roles of impact degradation and viscous relaxation.

One of the most exciting discoveries of the Mariner 10 mission was the 1,300-km-diameter Caloris impact basin (Murray et al. 1974). This feature (Fig. 12), similar in morphology to lunar impact basins such as Orientale (e.g., McCauley 1977; McCauley et al. 1981), provided important insight into the nature of the surface of Mercury, the origin of circum-Caloris smooth plains (were they emplaced as impact ejecta, e.g., Wilhelms 1976a; or as volcanic plains, e.g., Trask and Strom 1976). The Caloris basin belongs to a class of features known as central peak and multiringed basins on the Moon (e.g., Wilhelms 1987) of which there are many more representatives on Mercury (e.g., Murray et al. 1974; Wood and Head 1976; Head 1978; McKinnon 1981; Pike and Spudis 1987; Pike 1988). Twenty central peak basins (protobasins) were identified from the Mariner 10 data (Pike 1988) with diameters between 72 and 165 km. Thirty-one two- ringed basins, between 132 and 310 km in diameter, and possibly as many as 23 multiringed basins, between 285 and 1,530 km, were also detected. Furthermore, many other, more degraded features may be basins poorly detected in the area imaged by Mariner 10. The degree of degradation and stratigraphic relationships of these large basins are a fundamental factor in the development of global stratigraphic relations on Mercury and other planets (e.g., Wilhelms 1987; Spudis and Guest 1988). Indeed, using Mariner 10 stereo image data, Watters et al. (2001) discovered a previously unknown impact basin. On the basis of the importance of high-resolution altimetry data in the detection of degraded craters and basins on Mars (e.g., Smith et al. 2001; Frey et al. 1999), it is obvious that the new MESSENGER image and stereo data, together with the altimeter data, will reveal many previously undetected basins in both the previously seen and unimaged areas of Mercury.

Related questions raised by the discovery of Caloris focus on how the interior of the planet might respond to such a huge event, both in the basin interior and its far exterior. For the far exterior, Mariner 10 discovered an unusual hilly and lineated terrain at the antipodal point of the Caloris basin. The hilly and lineated morphology disrupts crater rims and other pre-existing landforms, and stratigraphic relationships suggest that the texture formed at the same time as Caloris (Trask and Guest 1975; Spudis and Guest 1988). Similar terrains are seen on the Moon antipodal to the Imbrium and Orientale basins (e.g., Wilhelms 1987), and it is thought that intense seismic waves might have been focused on the far side during the basin-forming event, causing complex patterns of disruption (e.g., Schultz and Gault 1975; Hughes et al. 1977). Unknown is the relative role of surface and interior waves, and how different interior structure might influence the patterns and degree of development of the terrain, which differs on the Moon and Mercury. An alternative hypothesis is that the terrain formed by impact basin ejecta converging at the antipodal point (Moore et al. 1974; Wilhelms and El-Baz 1977; Stuart-Alexander 1978; Wieczorek and Zuber 2001). Furthermore, clusters of crustal magnetic anomalies have been mapped at the antipodes of some

Fig. 12 The Caloris basin on Mercury, compared with the Orientale and Imbrium lunar impact basins. *Left*, the Caloris basin, 1,340 km in diameter. *Center*, the lunar Orientale basin, about 900 km in diameter and almost completely unfilled by subsequent lavas (Lunar Orbiter photograph). *Right*, in contrast to Orientale and more similar in appearance to Caloris, the Imbrium basin on the Moon, about 1,200 km diameter, is filled with several kilometers of mare lavas (Lick Observatory photograph)

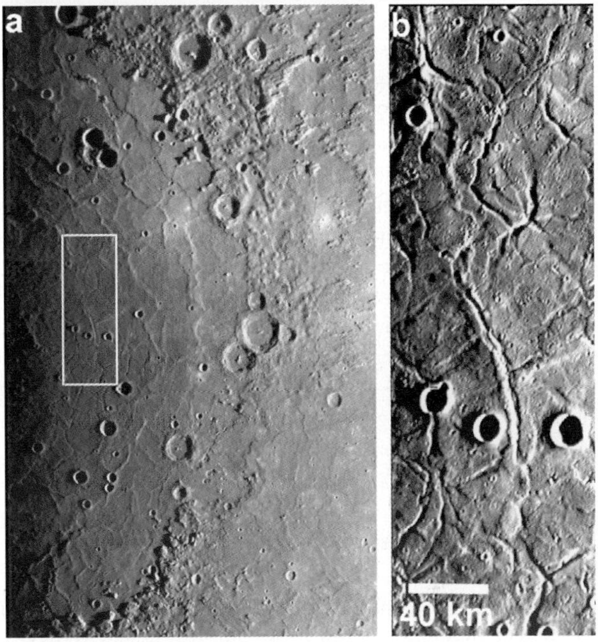

Fig. 13 The Caloris basin interior. (**a**) The interior plains of the Caloris basin have contractional (wrinkle ridges) and extensional (troughs) tectonic landforms. (**b**) High-resolution image of the interior plains showing extensional troughs that form giant polygons. Area is shown by *white box* in (**a**) (Mariner 10 image 0529055)

lunar impact basins, suggesting a relationship between crustal magnetization and antipodal basin effects (Hood 1987; Lin et al. 1988; Richmond et al. 2005). Thus, MESSENGER data on detailed unit characteristics, crustal magnetic anomalies, and the gravity and internal structure of Mercury will help to shed light on this significant but poorly known phenomenon.

The smooth plains that fill the interior of the imaged side of the Caloris basin have been heavily deformed. Basin-concentric and basin-radial wrinkle ridges are crosscut by a complex pattern of narrow extensional troughs (Fig. 13) (Strom et al. 1975; Dzurisin 1978; Melosh and McKinnon 1988; Watters et al. 2005). In plan view, the troughs are highly variable; some are linear while others are very sinuous, forming a polygonal pattern that strongly resembles giant polygons subsequently found in polygonal terrain on Mars and Venus (Carr et al. 1976; Pechmann 1980; McGill 1986; Hiesinger and Head 2000; Johnson and Sandwell 1992; Smrekar et al. 2002). The Caloris polygonal troughs are distributed in an arc ~180 km from the basin rim, extending inwards ~450 km towards the basin center (Fig. 13). How far the troughs extend into the unimaged hemisphere is currently unknown; however, the location of the most prominent polygonal troughs suggests that the peak extensional strain is ~180 to 450 km from the basin rim (Watters et al. 2005).

Giant polygons in the interior of the Caloris basin are in sharp contrast to lunar maria where trough-forming graben are found near the margins or outside the basins (McGill 1971; Strom 1972; Maxwell et al. 1975; Golombek 1979). This lunar pattern is thought to be due to loading from relatively dense, uncompensated volcanic-fill-inducing flexure of the lithosphere and resulting in interior compression and extension on the margins (Phillips et al. 1972; Melosh 1978; Solomon and Head 1979, 1980; Freed et al. 2001). Further, the crosscutting relationships between wrinkle ridges and polygonal troughs indicate that extension in Caloris postdates contraction (Strom et al. 1975; Dzurisin 1978; Melosh and McKinnon 1988). The wrinkle ridges in the interior fill material of Caloris and in the smooth plains exterior to the basin are likely to have formed in response to subsidence of the interior fill material (see Melosh and McKinnon 1988), possibly aided by a compressional stress bias in the lithosphere due to global contraction (see Watters et al. 2004, 2005). Basin-interior extension, however, is not consistent with mascon tectonic models (see Freed et al. 2001). Interior extensional stresses may have resulted from exterior annular loading due to the emplacement of the expansive smooth plains adjacent to Caloris (Melosh and McKinnon 1988). This annular load could cause basin-interior extension and concentric normal faulting. Alternatively, the Caloris troughs may have formed from lateral flow of a relatively thick crust toward the basin center (Watters et al. 2005). Lateral crustal flow causes late-stage basin uplift and extension consistent with the location and magnitude of the stresses inferred from the polygonal troughs. The MESSENGER mission will obtain imaging, mineralogy, and altimetry data (Table 1) to document the temporal and spatial relationships of these units and structures in order to assess their origin and evolution. Furthermore, the new MESSENGER data will provide extensive detection of other basins and craters, and their geological and geophysical characteristics, in the unimaged portion of Mercury (e.g., Harmon et al. 2007).

5 Geological Processes on Mercury: Tectonism

The style and evolution of tectonism on a planetary body provide important information on the lateral continuity, thickness, and lateral and vertical movement of the lithosphere in space and time (Head and Solomon 1981). The geological record of tectonism on planetary surfaces contains information on the style, timing, and magnitude of deformation, the candidate causative processes and the relation to global thermal evolution. Indeed, a well-constrained global history of tectonism may permit a much more refined understanding of the formation and evolution of Mercury's core, its spin-orbit history, and the origin of its magnetic field (e.g., Zuber et al. 2007).

Tectonic features are a manifestation of the stress history of crustal and lithospheric materials on solid planetary bodies. Compressional and extensional stresses result in a variety of tectonic landforms. Crustal extension results in normal faults, graben, and rifts [e.g., graben on the Moon (McGill 1974); rift zones on Mars (Lucchitta et al. 1992) and Venus (Solomon et al. 1992; Basilevsky and Head 2002)], while compression results in folds, thrust faults, and high-angle reverse faults [e.g., wrinkle ridges and lobate scarps on Mars (Watters 1988, 1991, 1993, 2003; Golombek et al. 2001), fold belts on Venus (Basilevsky and Head 2000)]. Furthermore, compressional and extensional features are often found in and around areas of inferred mantle upwelling and downwelling [e.g., circum-corona structures on Venus (Stofan et al. 1997)], or lithospheric loading [e.g., deformation surrounding mare loads on the Moon (Solomon and Head 1980) and the Tharsis Rise on Mars (Banerdt et al. 1992)]. The combination of knowledge of the style, timing, and magnitude of deformation has permitted the distinction between histories dominated by segmented and laterally interacting lithospheres, such as the plate tectonic system on Earth, and one-plate planetary bodies (Solomon 1978), such as the Moon, Mars, and Mercury, dominated by an unsegmented continuous global lithosphere. One-plate planets are characterized by evolutionary thickening of the lithosphere and predominantly vertical deformation (upwelling, loading) (Head and Solomon 1981).

One of the major surprises of the Mariner 10 mission was the presence of widespread evidence of hemisphere-scale crustal deformation (Strom et al. 1975). Tectonic landforms are distributed throughout highland and lowland plains and the floor of the Caloris basin, in the ancient intercrater plains and in the youngest smooth plains. The dominant form of

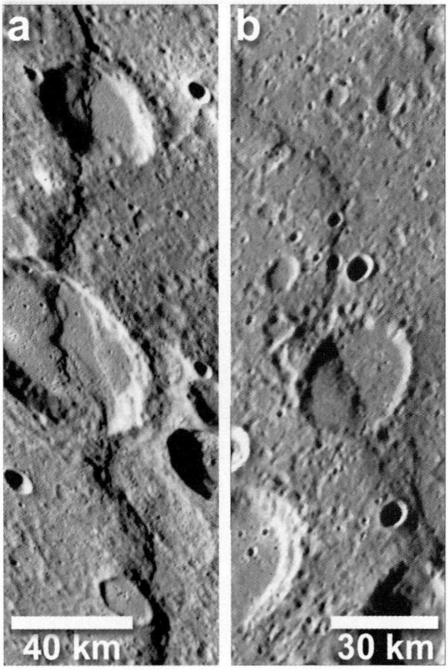

Fig. 14 Lobate scarps in the hemisphere of Mercury imaged by Mariner 10. (**a**) Discovery Rupes [~55° S, 323° E (37° W)] and (**b**) Santa Maria Rupes [~4° N, 340.5° E (19.5° W)] are two of the most prominent lobate scarps, landforms interpreted to be the surface expressions of thrust faults (Mariner 10 images 0528884 and 0027448)

deformation in the imaged hemisphere of Mercury is crustal shortening, expressed by three landforms: lobate scarps, wrinkle ridges, and high-relief ridges. Lobate scarps are linear or arcuate in plan view and asymmetric in cross-section, with a steeply sloping scarp face and a gently sloping back scarp (Strom et al. 1975; Cordell and Strom 1977; Dzurisin 1978; Melosh and McKinnon 1988; Watters et al. 1998, 2001, 2002). The asymmetric morphology and evidence of offset crater floors and walls indicate that lobate scarps are the expression of surface-breaking thrust faults (Fig. 14) (Strom et al. 1975; Cordell and Strom 1977; Melosh and McKinnon 1988; Watters et al. 1998, 2001, 2002, 2004). Topographic data derived from Mariner 10 stereo pairs indicate that the longest known lobate scarp, Discovery Rupes (~500 km), also has the greatest relief (~1.5 km) (Fig. 14) (Watters et al. 1998, 2001).

Wrinkle ridges are generally more complex morphologic landforms than lobate scarps (Fig. 15), often consisting of a broad, low-relief arch with a narrow superimposed ridge (Strom 1972; Bryan 1973; Maxwell et al. 1975; Plescia and Golombek 1986; Watters 1988). These two morphologic elements can occur independently of one another, and for wrinkle ridges in the imaged hemisphere of Mercury, this is the rule rather than the exception (see Strom et al. 1975). Although the consensus is that wrinkle ridges are the result of a combination of folding and thrust faulting, the number and the geometry of the faults involved are not obvious (see Schultz 2000; Gold et al. 2001; Watters 2004). Mercury's known wrinkle ridges are predominantly found in the floor material of the Caloris basin and in the smooth plains surrounding the basin.

High-relief ridges are the rarest of the contractional features (Watters et al. 2001). Commonly symmetric in cross-section, high-relief ridges have greater relief than wrinkle ridges (Fig. 16). Topographic data show that the high-relief ridge informally named Rabelais Dorsum (Fig. 16) has a maximum relief of ~1.3 km. Some high-relief ridges, like Rabelais Dorsum, transition into lobate scarps (Fig. 16), suggesting that they are also fault-controlled structures, possibly the surface expression of high-angle reverse faults (Watters et al. 2001).

Fig. 15 Wrinkle ridges on the plains of Mercury and similar features on the Moon and Mars. *Left*, wrinkle ridges in the plains of Mercury. View is ~385 km in width (Mariner 10 image 0000167). *Middle*, the southern part of lunar Mare Serenitatis showing the development of wrinkle ridges in the mare basalts. View is ~70 km in width (Apollo image). *Right*, wrinkle ridges in Lunae Planum on the eastern part of the Tharsis rise (MOLA digital topographic image). Note the similarities in the ridges in terms of general trends, separation, convergence, cross-cutting, and circularity around apparently buried craters

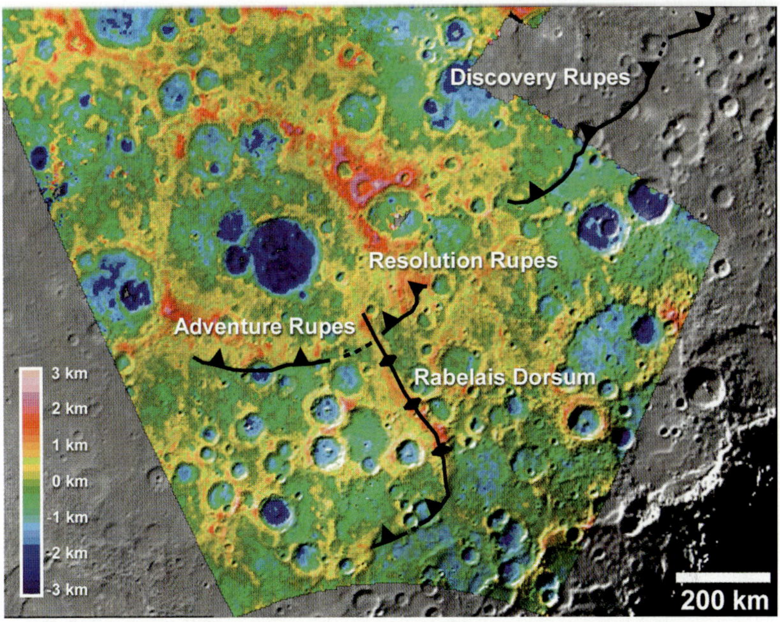

Fig. 16 Topographic expression of some prominent tectonic features in the southern hemisphere of Mercury. The digital elevation model was generated from Mariner 10 stereo pairs (see Watters et al. 2001) and is overlaid on an image mosaic. The locations of Discovery Rupes, Resolution Rupes, Adventure Rupes (all lobate scarps), and Rabelais Dorsum (a high-relief ridge) are shown. Thrust fault dip directions are indicated by *black triangles*. The mosaic covers 50°–75°S and 335°–280°E (25–80°W). Elevations are relative to a 2,439.0-km-radius sphere

One of the remarkable aspects of tectonics on Mercury is the absence of clear evidence of extension in the hemisphere imaged by Mariner 10 outside the Caloris basin. More subtle evidence of extension may occur in the form of a fabric of fractures that make up what has been described as a tectonic grid (Dzurisin 1978; Melosh and McKinnon 1988). This tectonic grid is expressed by lineaments that may reflect ancient lines of weakness in the lithosphere (Melosh and McKinnon 1988).

Of the tectonic features on Mercury, lobate scarps are the most widely distributed spatially (Fig. 17). An accurate assessment of the spatial distribution of the lobate scarps is difficult because the distribution may be strongly influenced by observational bias introduced by variations in the lighting geometry across the imaged hemisphere (see Cordell and Strom 1977; Melosh and McKinnon 1988; Thomas et al. 1988). The incidence angle of Mariner 10 images changes from 90° at the terminator to 0° at the subsolar point. Thus, only a small percentage of the imaged hemisphere has an optimum lighting geometry for the identification of lobate scarps or other tectonic features. However, recent mapping suggests that the distribution of lobate scarps is not uniform, even in areas where the incidence angle is optimum (>50°) (Watters et al. 2004). More than 50% of the area-normalized cumulative length of lobate scarps occurs south of 30°S, with the greatest cumulative length between 50°S and 90°S (Watters et al. 2004) (Fig. 17). The dip directions of the thrust faults inferred from the hanging wall-foot wall relationship suggests that there is no preferred thrust slip direction north of 50°S (Fig. 17). South of 50°S, however, the lobate scarp faults all dip to the north, NW, or NE; none dip southward (Fig. 17) (Watters et al. 2004). This indicates that

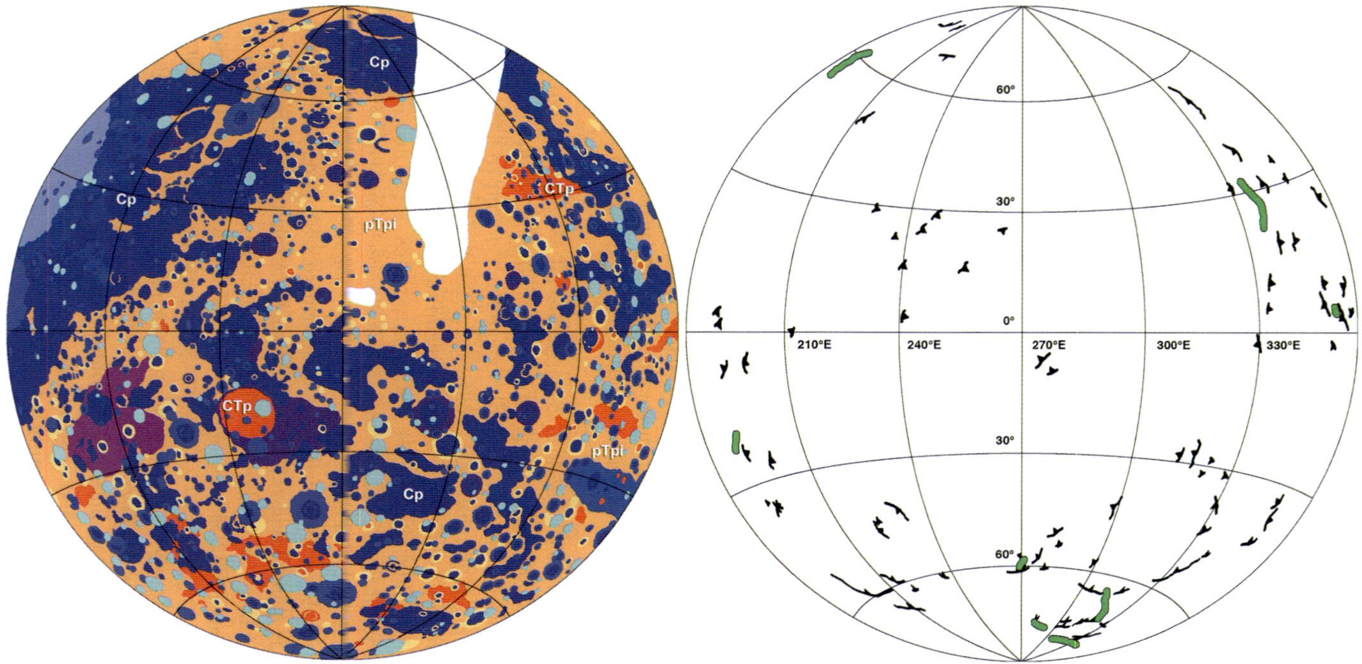

Fig. 17 Location of lobate scarps and high-relief ridges in the hemisphere of Mercury imaged by Mariner 10. The geologic map of Mercury (Spudis and Guest 1988) (*left*) provides context for the location of the lobate scarps and high-relief ridges (*right*). Major geologic units are intercrater plains material pTpi (*tan*), Calorian-Tolstojan plains material CTp (*red*), and Calorian plains material Cp (*blue*). Lobate scarps are *black* (thrust fault dip directions are indicated by *black triangles*), and high-relief ridges are in *green*

across a broad zone in the southern hemisphere, there is a preferred southward thrust dip direction.

The age of the lobate scarps is not well constrained. Lobate scarp thrust faults clearly deform the oldest plains material, pre-Tolstojan intercrater plains emplaced near the end of the period of heavy bombardment. Thus, the earliest preserved record of lobate scarp formation occurred near or after the end of heavy bombardment (Strom et al. 1975). Lobate scarps are also found in younger Tolstojan and Calorian-aged smooth plains units, suggesting that thrust faulting continued after the formation of the Caloris basin and the emplacement of the youngest smooth plains (Fig. 17) (Strom et al. 1975; Melosh and McKinnon 1988). If lobate scarps were uniformly distributed in the imaged hemisphere, their absence in hilly and lineated terrain antipodal to the Caloris basin would suggest that most of the scarps were pre-Caloris in age (Cordell and Strom 1977). Several lines of evidence, however, suggest a Calorian age of formation for the lobate scarps. First, lobate scarps are not uniformly distributed outside the hilly and lineated terrain (Fig. 17). Second, there is no evidence of embayment of scarps by ancient intercrater plains (Strom et al. 1975; Melosh and McKinnon 1988) or by younger Tolstojan and Calorian smooth plains materials (Watters et al. 2004). Third, while lobate scarp thrust faults often cut across and offset the floors and rim walls of large impact craters (Fig. 14), there are no incidences of large craters superimposed on scarps. Fourth, there is no apparent degradation or partial burial of lobate scarps by Caloris ejecta in the northern hemisphere (Watters et al. 2004). Thus, all the lobate scarps in the imaged hemisphere may have formed after the emplacement of the Calorian smooth plains (Watters et al. 2004).

Mechanisms for the formation of the lobate scarps include global contraction due to interior cooling, tidal despinning, a combination of thermal contraction and despinning, and the interaction of thermal stresses and stresses related to the Caloris basin (Strom et al. 1975; Cordell and Strom 1977; Melosh and Dzurisin 1978a, 1978b; Pechmann and Melosh 1979; Melosh and McKinnon 1988; Thomas et al. 1988). Slow thermal contraction of the planet from secular cooling of the interior is predicted to result in global, horizontally isotropic compression (Solomon 1976, 1977, 1978, 1979; Schubert et al. 1988; Phillips and Solomon 1997; Hauck et al. 2004). Thermal models predict the onset of lithospheric contraction before the end of the period of heavy bombardment (Solomon 1977; Schubert et al. 1988). Slowing of Mercury's rotation by despinning due to solar tides and the relaxation of an early equatorial bulge is predicted to induce stresses in the lithosphere (Melosh 1977; Melosh and Dzurisin 1978a; Pechmann and Melosh 1979; Melosh and McKinnon 1988). Stresses from tidal despinning predict E–W compression in the equatorial zone and N–S extension in the polar regions (Melosh 1977). The rapid spindown model suggests that despinning and thermal contraction thus may have been coincident and the stresses coupled (Pechmann and Melosh 1979; Melosh and McKinnon 1988). The formation of the Caloris basin may have influenced the pattern of tectonic features by introducing stresses that interacted with existing lithospheric stresses from thermal contraction (Thomas et al. 1988). This interaction might have temporarily reoriented stresses and resulted in the formation of Caloris-radial thrust faults.

All the models described here have limitations in explaining the spatial and temporal distribution of the lobate scarps. Although the orientation of wrinkle ridges in the smooth plains exterior to the Caloris basin may have been influenced by basin-related stresses, few lobate scarps in the imaged hemisphere are radial to Caloris (Fig. 17). Tidal despinning predicts a system of normal faults in Mercury's polar regions that have not been observed (Solomon 1978; Schubert et al. 1988; Melosh and McKinnon 1988; Watters et al. 2004). In the absence of other influences, thermal contraction would be expected to generate a uniform distribution of thrust faults with no preferred orientation and no preferred thrust slip direction (Watters et al. 2004). The amount of crustal shortening expressed by lobate scarps is another important constraint. Strom et al. (1975) estimated a reduction in planetary radius of $\sim 1-2$ km assuming an average displacement of 1 km for the total length of the lobate scarps mapped over an area covering $\sim 24\%$ of the surface. From displacement–length (*D–L*) relationships of the thrust faults, the strain expressed by the lobate scarps in an area covering $\sim 19\%$ of the surface has been estimated to be $\sim 0.05\%$, corresponding to a radius decrease of <1 km (Watters et al. 1998). This is consistent with estimates obtained using all the known lobate scarps in the imaged hemisphere (Fig. 17). Such low values of strain and radial contraction are difficult to reconcile with existing thermal contraction models and may indicate that only a fraction of the total strain due to interior cooling is expressed by the observed thrust faults, or that the earliest activity is obscured by the cratering flux or intercrater plains. Other tectonic features such as long-wavelength lithospheric folds (Dombard et al. 2001; Hauck et al. 2004) or small-scale faults that are difficult to detect with existing image data may account for the strain deficit. Similar, broad contractional belts are seen on Venus (e.g., Frank and Head 1990) but differ in morphology in that evidence for extensive thrusting and surface shortening is not as apparent.

Conspicuously absent from the portion of Mercury seen thus far are features that might be attributed to large mantle swells or voluminous mantle-derived magmatism (such as the Tharsis region of Mars) and intermediate-scale mantle activity (such as coronae on Venus). The relatively thin (~ 500 km) mantle of Mercury may have limited the length-scale of mantle-driven tectonism. With the exception of the interior of the Caloris basin, also absent is evidence for extensive lithospheric loading and flexural deformation (as seen on the Moon and Mars) and large-scale features indicative of extensional deformation (such as graben and rifts seen on the Moon, Venus, and Mars). The common occurrence of crustal heterogeneities and asymmetries on Mars, the Moon, and Venus suggest that it is unwise to conclude at present that the other half of Mercury will be the same as the hemisphere seen by Mariner 10.

The MESSENGER mission (Table 1) will provide regional- to global-scale altimetry and imaging data that will permit the quantitative characterization of the tectonic features on the part of Mercury unimaged by Mariner 10 and allow a quantitative global assessment of tectonic features in order to derive more rigorous estimates of the style, timing, and magnitude of deformation, the candidate causative processes, and the relation to global thermal and interior evolution (e.g., Nimmo 2002; Nimmo and Watters 2004).

6 Geological Processes on Mercury: Volcanism and Plains Formation

Volcanism, the eruption of internally derived magma, and its surface deposits, provide one of the most important clues to the location of interior thermal anomalies in space and time and to the general thermal evolution of the planet. Volcanism is among the dominant endogenic geologic process on other terrestrial planetary bodies and can produce significant resurfacing during the evolution of the body. Volcanism is a key element in the formation and evolution of secondary crusts (those derived from partial melting of the mantle) and tertiary crusts (those derived from remelting of primary and secondary crusts) (Taylor 1989).

Little is known concerning the history of volcanism on Mercury. In contrast to the Moon, where there are distinctive composition-related albedo variations between the cratered uplands (relatively high) and the smooth volcanic mare lowlands (relatively low), the albedo of Mercury is relatively uniform across the surface. Prior to the Apollo 16 mission to the Moon in 1972, a widely distributed smooth plains unit (the Cayley Formation) was mapped in the lunar uplands, lying stratigraphically between the younger, low-albedo maria, and the older, high-albedo impact basins and cratered terrain (Wilhelms and McCauley 1971). One of the purposes of the Apollo 16 mission (Hinners 1972) was to determine the petrology and absolute age of this unit, thought prior to the mission to represent a distinctive pre-mare, highland phase of volcanism (e.g., Trask and McCauley 1972). During Apollo 16 surface operations it became rapidly clear that the Cayley Formation consisted of impact breccias (Young et al. 1972), and later assessments suggested that the deposits were a combination of local (e.g., Head 1974) and regional, basin-related impact ejecta (Oberbeck et al. 1974, 1977). On the basis of the Apollo 16 results, lunar light plains were subsequently considered by most workers to have been emplaced by impact crater and basin ejecta processes (Oberbeck 1975), rather than by extrusive volcanism (Trask and McCauley 1972). The subsequent documentation of mare volcanic deposits buried by layers of highland crater ejecta (cryptomaria; Head and Wilson 1992; Antonenko et al. 1995) as well as some local moderate-albedo units thought to be of extrusive volcanic origin (e.g., the Apennine Bench Formation; Spudis and Hawke 1986) suggested that the interpretations of light plains might be more complex than simple ejecta emplacement.

Arriving at Mercury shortly after Apollo 16, Mariner 10 revealed the presence of two smooth Cayley-plains-like units alternately interpreted to represent effusive volcanic deposits or basin ejecta. These widespread plains deposits, occurring as relatively smooth surfaces between craters (intercrater plains), and as apparently ponded material (smooth plains; Fig. 3), were proposed by some to be volcanic in origin (Murray 1975; Murray et al. 1975; Trask and Guest 1975; Strom et al. 1975; Strom 1977; Dzurisin 1978). Others argued that the plains deposits might represent basin ejecta, similar to those found at the lunar Apollo 16 landing site (Wilhelms 1976a; Oberbeck et al. 1977). Part of the problem concerning interpretation of smooth plains on Mercury as volcanic or impact in origin is the relatively low resolution of the Mariner 10 data. Early on it was pointed out that the Mariner 10 image data do not have the resolution required to resolve lunar-like volcanic features such as flow fronts, vents, and small domes (Schultz 1977; Malin 1978), a problem further explored by Milkovich et al. (2002) (Fig. 18). Detailed examination of lunar images at resolutions and viewing geometries comparable to those of Mariner 10 readily showed that small shields and cones, elongate craters, sinuous rilles and flow fronts, all hallmarks of the identification of volcanism on the Moon (e.g., Head 1976; Head and Gifford 1980), would not be resolvable in most of the Mariner 10 images (Milkovich et al. 2002). Furthermore, larger features typical of volcanism on Mars [such as huge volcanic edifices and calderas (Carr 1973; Crumpler et al. 1996)], and not seen on the Moon (Head and Wilson 1991), were not observed by Mariner 10 on Mercury. Also not observed in the Mariner 10 data were examples of the large (10–30 km diameter) steep-sided domes suggestive of crustal magmatic differentiation processes seen on the Moon (Head and McCord 1978; Chevrel et al. 1999) and Venus (Pavri et al. 1992; Ivanov and Head 1999). Lobate fronts exposed at the edge of smooth plains occurrences on Mercury (Fig. 19; arrows) suggested that these might have been volcanic flow margins, but comparisons to marginal basin ejecta deposits on the Moon indicated that such features could also be a product of impact ejecta emplacement (e.g., Milkovich et al. 2002). Thus, although the surface features observed by Mariner 10 were most similar to lunar plains, there were also fundamental differences between volcanism occurring on the two bodies (Head et al. 2000).

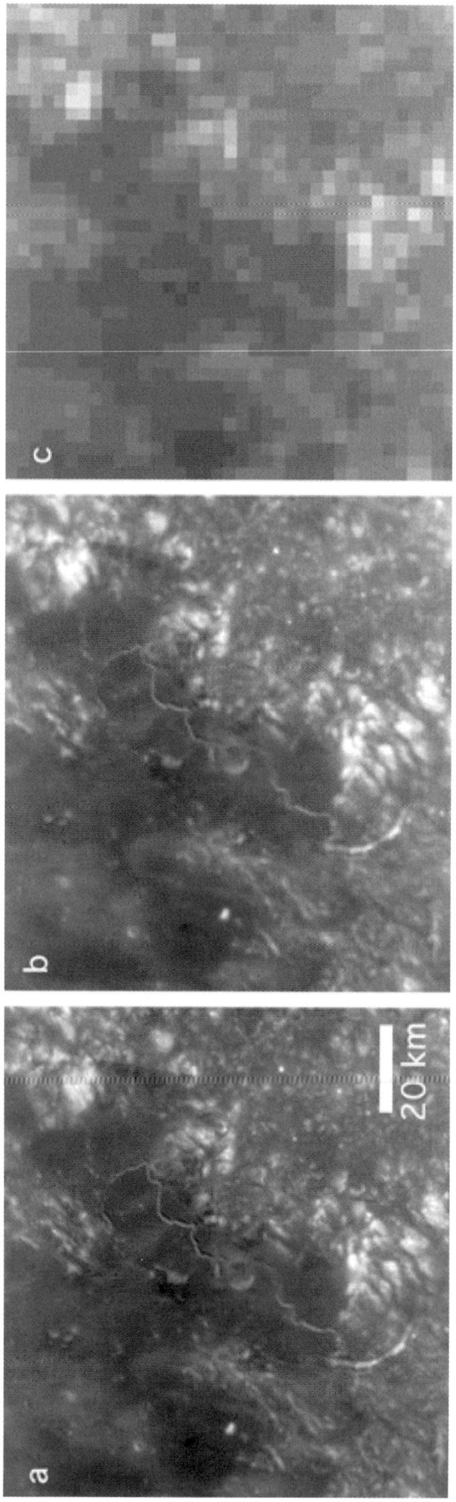

Fig. 18 Lunar volcanic features shown as a function of resolution. Hadley Rille (~25°N, 3°E) at multiple resolutions: (**a**) 100 m/pixel; (**b**) 500 m/pixel; (**c**) 2.5 km/pixel. After Milkovich et al. (2002)

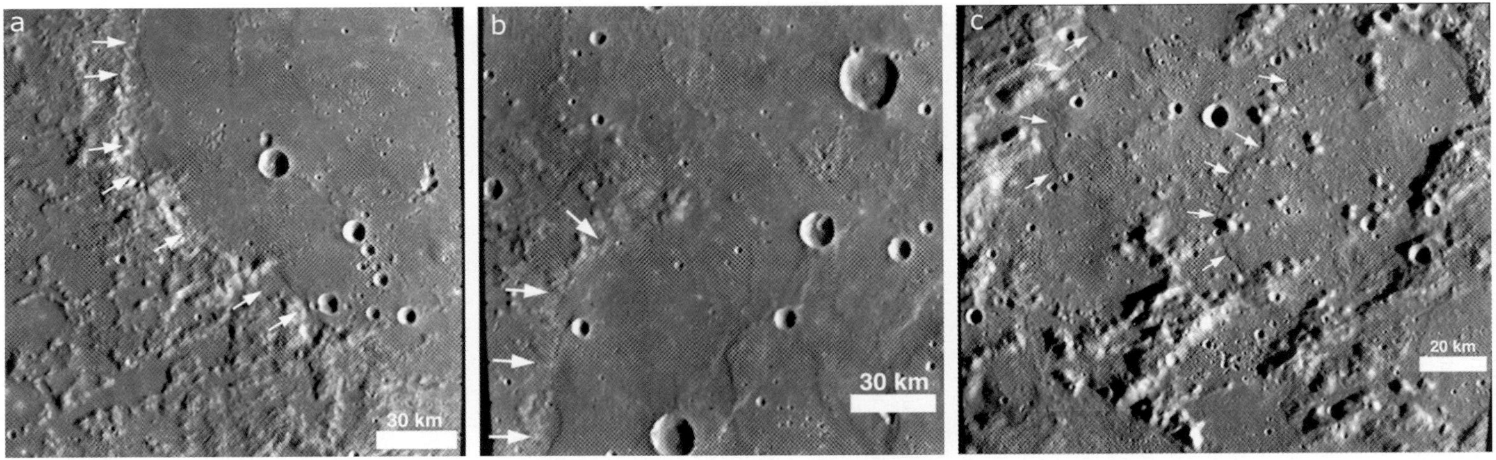

Fig. 19 Smooth plains deposits on Mercury. (**a**) Smooth plains (*middle* and *upper right* of image) embaying the inner rim of the crater Van Eyck; arrows indicate flow-like scarps. Mariner 10 image 0000077; 180 m/pixel. (**b**) Smooth plains (*lower middle* and *right* of image) in the Odin Formation. *Arrows* indicate the margin of plains that appear to embay the Odin Formation. Mariner 10 image 0000104; 280 m/pixel. (**c**) Lobate scarps in the Nervo Formation. The *arrows* indicate two lobate scarps that appear to have been emplaced by flow. Mariner 10 image 0000108; 300 m/pixel

These various observations raised the interesting possibility that there may be no identifiable volcanic units on Mercury. Crater counts of Caloris basin ejecta facies and smooth plains deposits, however, indicated that the smooth plains were emplaced after the Caloris basin (Spudis and Guest 1988) (Figs. 9 and 10), and on this basis they were interpreted to be the product of volcanic eruptions, not contemporaneous ejecta emplacement. On the other hand, the lunar Cayley plains also often showed younger ages than the adjacent textured ejecta deposits from major basins [for smaller crater diameters; see Wilhelms (1987)]. The counts of Spudis and Guest (1998) were carried out using carefully defined contiguous map units and only larger craters unaffected by secondaries. Reprocessed Mariner 10 color data (e.g., Robinson and Lucey 1997) (Figs. 4–7), as discussed earlier, provided additional evidence for the possible volcanic origin of the smooth plains.

Could extrusive volcanism not have occurred on Mercury, and if such advective cooling processes did not occur, how did the planet dissipate its accretional heat and that from subsequent decay of radioactive nuclides? Perhaps Mercury experienced styles of volcanism unknown on the Earth and Moon, and current data do not allow us to recognize such unusual deposits. Another possibility is that partial melting of the mantle may have occurred, but that extrusive volcanism did not. Investigating these possibilities, Head and Wilson (2001) assessed the ascent and eruption of magma under Mercury conditions for a range of scenarios and found that a thick low-density crust could, as with the Moon (e.g., Head and Wilson 1992), inhibit and potentially preclude dikes from rising to the surface and forming effusive eruptions. This, combined with an apparent global compressional net state of stress in the lithosphere (e.g., Strom et al. 1975), could produce a scenario in which rising magma intruded the crust but did not reach the surface to produce the level of resurfacing or the array of landforms seen on the Moon, Mars, and Venus. Indeed, Head and Wilson (2001) showed how easy it was, given the range of conditions known to occur in the history of terrestrial planetary bodies, to create a planet with little to no extrusive volcanic activity.

The fact that such fundamental questions remain concerning Mercury's thermal evolution underscores the importance of the MESSENGER mission to elucidating the early evolution of the terrestrial planets. The dominant endogenic geologic process on the Earth, Moon, Mars, and Venus is volcanism, characterized by massive extrusions of basaltic lavas, significant resurfacing of their surfaces, and emplacement of large volumes of intrusive magmas (e.g., Basaltic Volcanism Study Project 1981).

From these and related observations described earlier, it is possible to make some general inferences concerning the source regions of volcanic extrusions—the upper mantle. From terrestrial analyses it is known that FeO abundance of mantle source regions corresponds, to a first order, to the FeO content of the erupted magma (e.g., Longhi et al. 1992). The observation that candidate volcanic deposits identified on Mercury do not have FeO abundances differing from the hemispheric average indicates that the mantle source of such material is not enriched in FeO relative to the crust, or conversely that the ancient crust is not depleted in FeO relative to the upper mantle (Robinson et al. 1997, 1998). If the plains deposits had a significant increase (or decrease) in FeO relative to the basement rock that they overlie (ancient crust), then they would appear as a mappable unit in the iron-plus-maturity image and albedo (Figs. 4, 5, 6 and 7). In contrast, mare deposits found on the Moon (mare lavas versus anorthositic crust) have a significant contrast in FeO content relative to the ancient anorthositic crust they overlie [typically $>$15 wt% versus $<$6 wt%, respectively; see Lucey et al. (1998)]. The global crustal abundance of FeO on Mercury has been estimated to be less than 6 wt% from remote sensing data (McCord and Adams 1972; Vilas and McCord 1976; Vilas et al. 1984; Vilas 1985, 1988; Veverka et al. 1988; Sprague et al. 1994; Blewett et al. 1997). The lack of structures corresponding to the candidate volcanic plains units in the iron-plus-maturity image is consistent with mantle magma source regions approximately sharing the crustal FeO composition, and so supports the idea that Mercury is highly reduced and most of its iron is sequestered in a metallic core.

Thus, if there are no significant variations in iron abundance in the areas seen by Mariner 10 one must address the question "What could be the composition of candidate volcanics on Mercury?" Komatiitic volcanics are found on the Earth with FeO abundances of under 5 wt%, as are relatively low-iron mafic lavas on the Moon in the Apennine Bench Formation (Spudis and Hawke 1986). The most likely candidate is a high-magnesium, low-iron magma. The MESSENGER mission (Table 1) will therefore not only provide very important information on the possible volcanic origin of surface plains deposits from imaging and altimetry, but it will also permit assessment of mantle characteristics and mineralogy and core evolution processes from surface mineralogy and chemistry. Superficially, Mercury looks like the Moon, but Mariner 10 and terrestrial remote sensing data tell us that it must be very different in many fundamental respects. Could Mercury be an Earth's Moon that did not undergo surface evolution by endogenic processes (e.g., mare volcanism) subsequent to the period of large basin formation?

7 Geological Processes on Mercury: Polar Deposits

One of the most impressive discoveries from Earth-based observations is the detection of high radar backscatter, strongly depolarizing deposits in the near-polar regions of Mercury (e.g., Harmon and Slade 1992; Butler et al. 1993; Harmon et al. 1994; Harmon 1997). The obliquity of Mercury is near 0° so there are extensive areas of permanently shadowed regions within fresh craters that can act as cold traps for volatile compounds. On the basis of orbital geometries, from Earth we are able to view these areas on Mercury slightly better than the polar regions on the Moon. Earth-based radar observations (Harmon et al. 2001) detected highly radar reflective deposits in these areas at both the north and south poles. The deposits occur in fresh craters as low as 72°N latitude (Fig. 20). Degraded craters do not show the high radar backscatter deposits because their interiors are exposed to the Sun. The neutron spectrometer on the Lunar Prospector spacecraft discovered enhanced hydrogen signals in permanently shadowed craters in the polar regions of the Moon (Feldman et al. 1998). This has been interpreted as water ice with a concentration of $1.5 \pm 0.8\%$ weight fraction.

The radar depolarized, highly backscattered signal is essentially identical to the intensity and characteristics of the radar backscatter signals from the martian south polar water ice cap and the icy Galilean satellites (Harmon et al. 2001). This has been used as evidence that the deposits are water ice. The permanently shadowed cold traps are essentially full, and the strong radar signals indicate that if the material is water ice then it is quite pure. The estimated thickness of the deposits is believed to be at least 2 m, but radar observations cannot set an upper limit on the thickness. The area covered by the polar deposits (both north and south) is estimated to be $\sim(3 \pm 1) \times 10^{14}$ cm^2. This is equivalent to 4×10^{16} to 8×10^{17} g of ice, or 40–800 km^3 for a deposit 2–20 m thick (Vasavada et al. 1999).

Other material has been suggested for the polar deposits including sulfur, which has radar backscatter characteristics similar to water ice, but a higher stability limit (Sprague et al. 1995). A 1-m-thick layer of water ice is stable for 10^9 years at a temperature of -161°C, while sulfur is stable at a considerably higher temperature of -55°C. Much of the region surrounding permanently shadowed craters is less than -55°C, but there are no radar reflective deposits there (Vasavada et al. 1999). Very cold silicates have also been suggested as a possibility, with the high radar response of the polar regions attributed to the decrease of dielectric loss of silicate materials with lower temperature (Starukhina 2001).

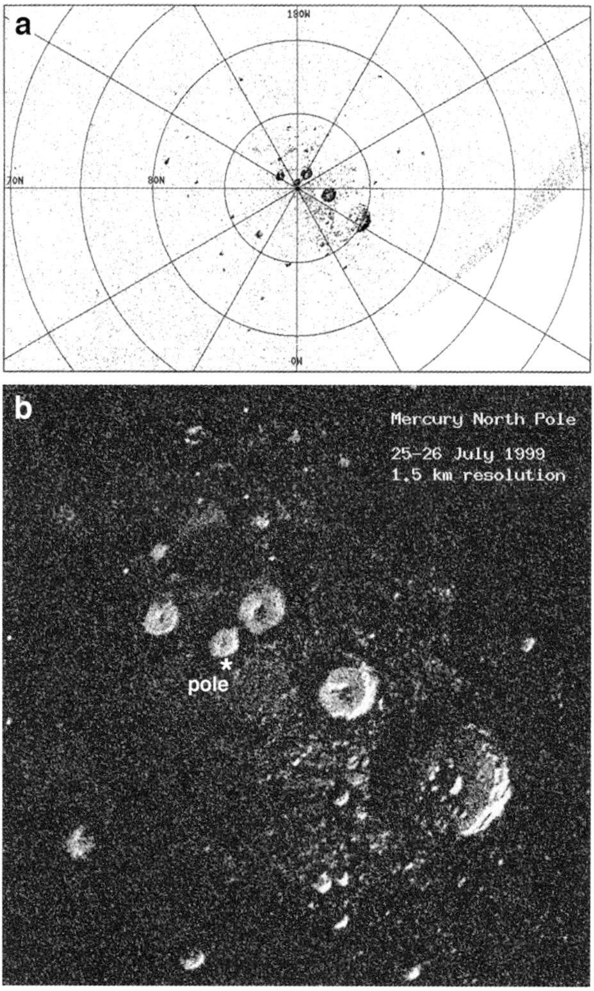

Fig. 20 Bright radar signals from localized regions in permanently shadowed craters at high northern latitudes on Mercury's surface. (**a**) Ten-microsecond north polar SC image from Arecibo radar observations on July 25–26, 1999, with a superimposed location grid. Radar illumination is from the upper left, and the region beyond the radar horizon is at lower right. (**b**) Details of the central portion of the radar image from July 25–26, 1999, in the vicinity of the north pole (see star). Resolution is 1.5 km. From Harmon et al. (2001)

If the deposits are water ice, the most likely sources of the water are micrometeoroid, comet, and water-rich asteroid impacts. If the current terrestrial influx of interplanetary dust particles is extrapolated to Mercury, over the last 3.5 billion years it could have delivered (3–60) $\times$ 10^{16} g of water ice to the permanently shadowed polar regions (an average thickness of 0.8–20 m). Impacts from Jupiter-family comets over the last 3.5 billion years could supply 0.1–200 $\times$ 10^{16} g of water to Mercury's polar regions (corresponding to an ice layer 0.05–60 m thick). Halley-type comets can supply 0.2–20 $\times$ 10^{16} g of water to the poles (0.1–8 m ice thickness). These sources alone provide more than enough water to account for the estimated volume of ice at the poles (Moses et al. 1999). The ice deposits could, at least in part, be relatively recent deposits, if the two radar features A and B were the result of recent comet or water-rich asteroid impacts. Crider and Killen (2005) estimated that if the polar ice deposits are both clean and buried by $\sim$20 cm of regolith, then they must have been emplaced less than $\sim$50 My ago.

Barlow et al. (1999) tested for the presence of subsurface ice by comparing the differences in depth/diameter relationships between high-latitude and low-latitude craters and found no evidence for variations or the presence of "terrain softening" ground ice. Vilas et al. (2005), using shadow measurements, examined the depth-diameter relationships of 12 near-north polar craters containing polar deposits; they found that these craters are shallower by about one third than craters of comparable diameter in the general population. For a single 30-km-diameter crater, the shallowing amounts to ~900 m, representing more than 600 km^3 of material. This volume of infilling material is significantly greater than that predicted by proposed mechanisms for the emplacement by either sulfur or water ice. If these measurements are correct, then these craters, dating from the Mansurian Period (perhaps ~3–3.5 Ga), have been preferentially shallowed and have accumulated radar-anomalous material on the permanently shadowed parts of their floors. A satisfactory mechanism for such a process is unknown. The MESSENGER mission will use laser altimetry, neutron spectrometry, high-resolution imaging, and elemental and mineralogical remote sensing to verify these measurements and assess processes of crater floor fill.

8 The Geological History of Mercury: Impact Cratering Rates and the Absolute Time Scale

The heavily cratered surfaces of the Moon, Mars, and Mercury all have similar crater size/frequency distributions that probably represent the period of late heavy bombardment (LHB) early in Solar System history (Fig. 21). The population of impactors during the LHB is widely thought to have been similar throughout the inner Solar System (to within differences in encounter probabilities and energies), and the same is thought to be true for the later and different population of impactors subsequent to the LHB. The LHB period ended at ~3.8 Ga on the Moon and may have ended about the same time on Mercury. A notable difference between the lunar curve and those for Mercury and Mars is that at diameters less than about 50 km there is a paucity of craters on Mercury and Mars compared with the

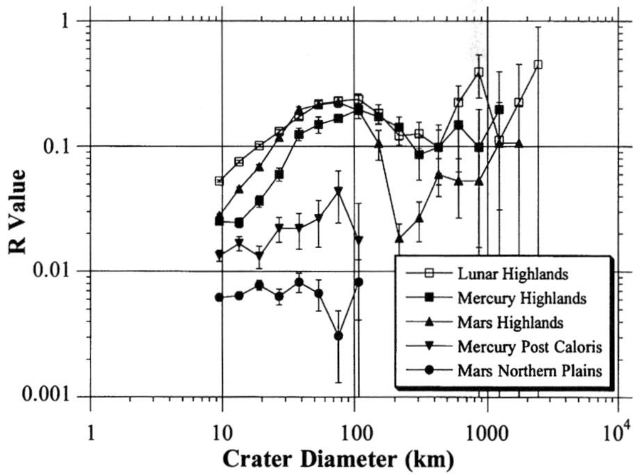

Fig. 21 Comparative impact crater size-frequency distributions. This "*R* plot," obtained by removing an inverse cubed power-law from the observed frequency versus diameter, is a comparison of the crater size/frequency distribution of the heavily cratered highlands on the Moon, Mercury, and Mars. All three have a similar shape, perhaps indicating a common origin. The steeper slopes for Mercury and Mars at smaller diameters reflect loss processes discussed in the text. Also shown is the size distribution of the post-Caloris crater population. The *bottom curve* represents the comparatively less cratered, relatively younger surfaces on Mars that have a distinctly different crater size/frequency distribution. After Strom et al. (2005)

Moon. This is usually interpreted, for Mars, as reflecting the loss of smaller craters due to erosion and infilling processes early in the post-LHB history of Mars. In the case of Mercury, it has been suggested that formation of thin but extensive so-called intercrater plains may be responsible for the loss of smaller craters. It is not yet clear why such plains should be so pervasive on Mercury, yet quite rare on the Moon. Intercrater plains are the most common terrain type on at least the part of Mercury viewed by Mariner 10 (Fig. 9). On Mercury only about 25% of the surface was viewed by Mariner 10 at Sun angles low enough to perform reliable crater counts. Unfortunately, Mariner 10 resolution is only about 1 km/pixel, so large counting areas are needed to obtain reliable statistics. These areas include the Mercury highlands and the plains within and surrounding the Caloris basin.

The smooth plains surfaces that surround and fill the Caloris basin may also show a crater size/frequency distribution similar to that of the lunar highlands, but at a lower density (Fig. 21). This post-Caloris curve may be less steep in part because it has not been affected by plains emplacement as has the highlands. Strom et al. (2005) interpreted the lower crater density to indicate that the post-Caloris surface is younger than the highlands and suggest that it formed near the end of the late heavy bombardment.

Surfaces younger than the LHB have a different crater size/frequency distribution on the Moon and Mars (Strom et al. 1992). The distribution is characterized by a single-slope −3 differential power-law distribution (Fig. 21). No region on Mercury imaged by Mariner 10 clearly shows the younger population, although error bars on the post-Caloris counts are nearly compatible with it and would certainly be compatible with a mixture of end-LHB craters and still more recent cratering such as that observed on the lunar maria. This could mean (e.g., Strom et al. 2005) that the surfaces available for crater counts in Mariner 10 images are all older than about 3.8 Ga. Perhaps there are more pristine, younger surfaces on the 55% of the planet that will be available for study from future MESSENGER images.

One can measure the relative ages of geological units on planetary surfaces from the spatial densities of superposed impact craters. "*R* values" represent a type of spatial density measurement (e.g., Strom et al. 2005). For example, an *R* value of 1 basically means that 100% of an area is covered by craters of that general diameter; an *R* value of 0.01 means that craters cover about 1% of the area. *R* values of craters on Mercury 40–100 km in diameter (Fig. 21) are roughly five times lower for post-Caloris surfaces than for the highlands. If one knows the rate at which craters are formed, and how that rate has changed with time, then the absolute age of a surface can be determined. The rate of crater formation depends on the relative proportions of different classes of impactor (e.g., comet or asteroid) that have impacted the planet. Estimates of these factors contain large uncertainties and, therefore, the estimated absolute ages are very uncertain. If the LHB was a cataclysmic event lasting only 50 or 100 My around 3.9 Ga that saturated the surfaces of the terrestrial planets, then the cratering record prior to that event has been lost (Tera et al. 1974; Ryder 1990; Kring and Cohen 2002).

More recently, the terrestrial planets have been impacted by a population of collisional fragments derived from the main asteroid belt by a variety of dynamical processes (including resonances and the Yarkovsky effect), plus some contribution of smaller comets, that presumably have impacted not only the Moon but all the terrestrial planets, although conceivably in somewhat different proportions. Another possibility for Mercury (Leake et al. 1987) is that a population of so-called vulcanoids, orbiting the Sun in the vicinity of Mercury's orbit, has preferentially cratered Mercury, but with little influence on the other terrestrial bodies. Searches for vulcanoids have not yet been successful. A vulcanoid population may be modest today, or even totally depleted, but vulcanoids could have cratered Mercury during post-LHB times, shifting much of Mercury's inferred chronology from epochs around the LHB to more recent times. Indeed, Vokrouhlicky et al. (2000) calculated that the depletion of a vulcanoid population mediated by the Yarkovsky effect might occur on the time scale of about a billion years. On the other hand, vulcanoids may never have existed at all. If vulcanoids never existed and the LHB was an inner-solar-system-wide phenomenon, then the crater age of Mercury's highlands is about 3.9 Ga (the absolute ages of the rocks may be older), while the more lightly cratered Caloris plains are probably closer to about 3.8 Ga, to the degree that they still show the signature of the LHB.

Another source of uncertainty concerning the origin and ages of Mercury's craters is the role of large secondary craters from Mercury's numerous basins. It is possible that many or most of Mercury's craters up to 25 km in size or larger are secondaries formed by massive ejecta from the dozens of large impact basins on Mercury, as has been advocated for the Moon by Wilhelms (1976b). Although this perspective is not widely accepted, it has not been disproven and should also be assessed in MESSENGER imaging.

Thus MESSENGER is absolutely crucial to determining the impactor flux in the inner solar system and both absolute and relative ages, and therefore the chronology of the geological evolution of the surface of Mercury. Some geological features (e.g., the lobate scarps) are believed to reflect the geophysical evolution of the planet (e.g., global contraction during cooling of the interior), while the wide range of plains deposits may hold the key to whether volcanism has played a significant role in the evolution of Mercury, and if so, on what time scales. It is vital that MESSENGER data be used to study the wide range of cratering-related issues that were raised and certainly not firmly settled by Mariner 10 and subsequent Earth-based studies of Mercury.

9 Geological Processes and Evolution and Outstanding Questions

This review of our current knowledge of the nature of the surface of Mercury, the geological processes operating there, and the geological history implied by their sequence of events and relative importance with time sets the stage for a series of outstanding questions that can be addressed by the MESSENGER mission and its suite of instruments (Table 1). Among these critical questions are: What is the distribution of geological features and units in the 55% of the planet not imaged by Mariner 10? Will new discoveries made in this current terra incognita change our view of the dominant processes on Mercury and the resulting geological history? Will the considerably higher-resolution MESSENGER images reveal the presence of extensive volcanic source regions that could confirm a magmatic source for the smooth plains of Mercury? Will there be distinctive mineralogical differences between different occurrences of smooth plains or between smooth plains and intercrater plains? Will studies of crater populations sort out contributions due to asteroids, comets, possible vulcanoids, and secondary craters? Will crater counts in the unimaged terrain show evidence for the presence of post-$\sim$3.7-Ga planetary resurfacing? Will multispectral images and spectrometer data of various deposits and landforms show evidence that can be interpreted in terms of crustal thickness, structure, and vertical and lateral heterogeneity? Will tectonic features discovered on the other half of the planet support the current view of a global net compressional state of stress in the lithosphere over most of geological time? Will any evidence for extensional deformation unrelated to basin deformation be found? Could the unimaged portion of Mercury be characterized solely by extensional features, balancing the contractional features on the imaged portion? Has the sign and magnitude of global stress changed with time? What clues to the nature and thickness of the lithosphere, and how it has changed with time, can be deduced from altimetric, gravity, and imaging data on crater, basin, and tectonic landforms? Will the presence of ice be confirmed in permanently shadowed zones in crater interiors of the polar regions? Will the unusual depth relationships of impact craters with near-polar radar anomalies determined from Mariner 10 shadow measurements be confirmed by MESSENGER altimetry? Will these same craters show any unusual mineralogical characteristics relative to those without radar anomalies? Can MESSENGER impact crater studies reveal details of impact melt generation and its fate in this high-velocity impact environment? Can the full complement of MESSENGER data extend morphologic and morphometric studies sufficiently to establish differences in substrate characteristics and to help further in distinguishing the relative roles of impact velocity, gravity, and substrate characteristics in the impact cratering process? Do any detected crustal magnetic anomalies relate to geologic features and structures? Can evidence distinguishing crustal and dynamo origins for the magnetic field be obtained (e.g., Stanley et al. 2005), and if so, what are the implications for crustal formation processes and the history of mantle convection? The complement of instruments on the MESSENGER spacecraft (Table 1), described elsewhere in this volume, will provide the data necessary to address and, in many cases answer, these fundamental questions. The MESSENGER mission, together with BepiColombo (Grard et al. 2000) and future missions (e.g., Schulze-Makuch et al. 2005), will bring new insights into key processes in planetary formation and evolution.

Acknowledgements We gratefully acknowledge the team of NASA and MESSENGER managers, engineers, and scientists who have worked together to make this mission a reality. The MESSENGER mission is supported by NASA's Discovery Program through a contract with the Carnegie Institution of Washington. This work has been supported in part by Carnegie Institution of Washington Contract DTM-3250-05. We also thank James Dickson, Anne Côté, and Peter Neivert for help in manuscript preparation. We thank Sean Solomon, Paul Spudis, Dave Blewett, Jeff Gillis-Davis, and an anonymous reviewer for excellent reviews that helped to improve the paper.

References

B.J. Anderson et al., Space Sci. Rev. (2007, this issue). doi:10.1007/s11214-007-9246-7
J.D. Anderson, R.F. Jurgens, E.L. Lau, M.A. Slade III, G. Schubert, Icarus **124**, 690–697 (1996)
G.B. Andrews et al., Space Sci. Rev. (2007, this issue). doi:10.1007/s11214-007-9272-5
I. Antonenko, J.W. Head, J.F. Mustard, B.R. Hawke, Earth, Moon Planets **69**, 141–172 (1995)
W.B. Banerdt, M.P. Golombek, K.L. Tanaka, in *Mars*, ed. by H.H. Kiefer, B.M. Jakosky, C.W. Snyder, M.S. Matthews (University of Arizona Press, Tucson, 1992), pp. 249–297
N.G. Barlow, R.A. Allen, F. Vilas, Icarus **141**, 194–204 (1999)
Basaltic Volcanism Study Project, *Basaltic Volcanism on the Terrestrial Planets* (Pergamon, New York, 1981), 1286 pp
A.T. Basilevsky, J.W. Head, J. Geophys. Res. **105**, 24583–24611 (2000)
A.T. Basilevsky, J.W. Head, J. Geophys. Res. **107**, 5041 (2002). doi:10.1029/2000JE001471
S.K. Bhattacharya, J.N. Goswami, D. Lal, P.P. Patel, M.N. Rao, *Proc. Lunar Science Conf. 6th*, 1975, pp. 3509–3526
D.T. Blewett, P.G. Lucey, B.R. Hawke, G.G. Ling, M.S. Robinson, Icarus **129**, 217–231 (1997)
D.T. Blewett, B.R. Hawke, P.G. Lucey, Meteorit. Planet. Sci **37**, 1245–1254 (2002)
E. Bowell, B. Hapke, D. Domingue, K. Lumme, J. Peltoniemi, A.W. Harris, in *Asteroids II*, ed. by R.P. Binzel, T. Gehrels, M.S. Matthews (University of Arizona Press, Tucson, 1989), pp. 524–556
W.V. Boynton et al., Space Sci. Rev. (2007, this issue). doi:10.1007/s11214-007-9258-3
W.B. Bryan, *Proc. Lunar Science Conf. 4th*, 1973, pp. 93–106
B.J. Butler, D.O. Muhleman, M.A. Slade, J. Geophys. Res. **98**, 15003–15023 (1993)
M.H. Carr, J. Geophys. Res. **78**, 4049–4062 (1973)
M.H. Carr et al., Science **193**, 766–776 (1976)
J.F. Cavanaugh et al., Space Sci. Rev. (2007, this issue). doi:10.1007/s11214-007-9273-4
S.D. Chevrel, P.C. Pinet, J.W. Head, J. Geophys. Res. **104**, 16515–16529 (1999)
M.J. Cintala, J. Geophys. Res. **97**, 947–973 (1992)

P.E. Clark, R.F. Jurgens, M.A. Leake, in *Mercury*, ed. by F. Vilas, C.R. Chapman, M.S. Matthews (University of Arizona Press, Tucson, 1988), pp. 77–100
B.M. Cordell, R.G. Strom, Phys. Earth Planet. Interiors **15**, 146–155 (1977)
D. Crider, R.M. Killen, Geophys. Res. Lett. **32**, L12201 (2005). doi:10.1029/2005GL022689
L.S. Crumpler, J.W. Head, J.C. Aubele, in *Volcano Instability on the Earth and Other Planets*, ed. by W.J. McGuire, A.P. Jones, J. Neuberg. Spec. Pub. 110 (Geological Society, London, 1996), pp. 307–348
A. Danjon, Bull. Astron. J. **14**, 315–345 (1949)
R.A. De Hon, D.H. Scott, J.R. Underwood, U.S. Geol. Surv. Misc. Inv. Ser., Map I-1233 (1981)
G. de Vaucouleurs, Icarus **3**, 187–235 (1964)
A. Dollfus, M. Auriere, Icarus **23**, 465–482 (1974)
A.J. Dombard, S.A. Hauck, II, S.C. Solomon, R.J. Phillips, Lunar Planet. Sci. **32** (2001), abstract 2035
D.L. Domingue, A.L. Sprague, D.M. Hunten, Icarus **128**, 75–82 (1997)
D. Dzurisin, J. Geophys. Res. **83**, 4883–4906 (1978)
W.C. Feldman, S. Maurice, A.B. Binder, B.L. Barraclough, R.C. Elphic, D.J. Lawrence, Science **281**, 1496–1500 (1998)
S.L. Frank, J.W. Head, Earth, Moon Planets **50/51**, 421–470 (1990)
A.M. Freed, H.J. Melosh, S.C. Solomon, J. Geophys. Res. **106**, 20603–20620 (2001)
H. Frey, S.E.H. Sakimoto, J.H. Roark, Geophys. Res. Lett. **26**, 1657–1660 (1999)
D.E. Gault, J.E. Guest, J.B. Murray, D. Dzurisin, M.C. Malin, J. Geophys. Res. **80**, 2444–2460 (1975)
R.E. Gold et al., Planet. Space Sci. **49**, 1467–1479 (2001)
J.O. Goldsten et al., Space Sci. Rev. (2007, this issue). doi:10.1007/s11214-007-9262-7
M.P. Golombek, J. Geophys. Res. **84**, 4657–4666 (1979)
M.P. Golombek, F.S. Anderson, M.T. Zuber, J. Geophys. Res. **106**, 23811–23822 (2001)
R. Grard, M. Novara, G. Scoon, ESA Bull. **103**, 11–19 (2000)
M.J. Grolier, J.M. Boyce, U.S. Geol. Surv. Misc. Inv. Ser., Map I-1660 (1984)
J.E. Guest, R. Greeley, U.S. Geol. Surv. Misc. Inv. Ser., Map I-1408 (1983)
B. Hapke, J. Geophys. Res. **86**, 3039–3054 (1981)
B. Hapke, Icarus **59**, 41–59 (1984)
B. Hapke, Icarus **67**, 264–280 (1986)
B. Hapke, *Theory of Reflectance and Emittance Spectroscopy* (Cambridge University Press, Cambridge, 1993), 469 pp
B. Hapke, Icarus **157**, 523–534 (2002)
B. Hapke, C. Christman, B. Rava, J. Mosher, *Proc. Lunar Planet. Science Conf. 11th*, 1980, pp. 817–821
J.K. Harmon, Adv. Space Res. **19**, 1487–1496 (1997)
J.K. Harmon, D.B. Campbell, in *Mercury*, ed. by F. Vilas, C.R. Chapman, M.S. Matthews (University of Arizona Press, Tucson, 1988), pp. 101–117
J.K. Harmon, M.A. Slade, Science **258**, 640–642 (1992)
J.K. Harmon, D.B. Campbell, D.L. Bindschadler, J.W. Head, I.I. Shapiro, J. Geophys. Res. **91**, 385–401 (1986)
J.K. Harmon, P.J. Perillat, M.A. Slade, Icarus **149**, 1–15 (2001)
J.K. Harmon, M.A. Slade, R.A. Vélez, A. Crespo, M.J. Dryer, J.M. Johnson, Nature **369**, 213–215 (1994)
J.K. Harmon, M.A. Slade, B.J. Butler, J.W. Head, M.S. Rice, D.B. Campbell, Icarus **187**, 374–405 (2007)
S.A. Hauck, II, A.J. Dombard, R.J. Phillips, S.C. Solomon, Earth Planet. Sci. Lett. **222**, 713–728 (2004)
S.E. Hawkins, III, et al., Space Sci. Rev. (2007, this issue). doi:10.1007/s11214-007-9266-3
J.W. Head, Moon **11**, 77–99 (1974)
J.W. Head, Moon **12**, 299–329 (1975)
J.W. Head, Rev. Geophys. Space Phys. **14**, 265–300 (1976)
J.W. Head, Lunar Planet. Sci. **9**, 485–487 (1978)
J.W. Head, Earth, Moon Planets **85–86**, 153–177 (2001a)
J.W. Head, in *The Century of Space Science*, ed. by J. Bleeker, J.H. Geiss, M.C.E. Huber, A. Russo (Kluwer, The Netherlands, 2001b), pp. 1295–1323
J.W. Head, Brown University-Vernadsky Institute Microsymposium 44, abstract m44_24 (2006)
J.W. Head, A. Gifford, Moon Planets **22**, 235–258 (1980)
J.W. Head, T.B. McCord, Science **199**, 1433–1436 (1978)
J.W. Head, S.C. Solomon, Science **213**, 62–76 (1981)
J.W. Head, L. Wilson, Geophys. Res. Lett. **18**, 2121–2124 (1991)
J.W. Head, L. Wilson, Geochim. Cosmochim. Acta **55**, 2155–2175 (1992)
J.W. Head, L. Wilson, in *Workshop on Mercury: Space Environment, Surface, and Interior* (Lunar and Planetary Institute, Houston, 2001), pp. 44–45
J.W. Head, L. Wilson, M. Robinson, H. Hiesinger, C. Weitz, A. Yingst, in *Environmental Effects on Volcanic Eruptions: From Deep Oceans to Deep Space*, ed. by T. Gregg, J. Zimbelman (Plenum, New York, 2000), pp. 143–178

H. Hiesinger, J.W. Head, J. Geophys. Res. **105**, 11999–12022 (2000)
H. Hiesinger, J.W. Head, U. Wolf, R. Jaumann, G. Neukum, J. Geophys. Res. **108**, 5065 (2003). doi:10.1029/2002JE001985
N.W. Hinners, in *Apollo 16 Preliminary Science Report*. SP-315 (NASA, Washington, DC, 1972), pp. 1-1–1-3
L.L. Hood, Geophys. Res. Lett. **14**, 844–847 (1987)
G.H. Hughes, F.N. App, T.R. McGetchin, Phys. Earth Planet. Interiors **15**, 251–263 (1977)
M.A. Ivanov, J.W. Head, J. Geophys. Res. **104**, 18907–18924 (1999)
R. Jeanloz, D.L. Mitchell, A.L. Sprague, I. de Pater, Science **268**, 1455–1457 (1995)
C.L. Johnson, D.T. Sandwell, J. Geophys. Res. **97**, 13601–13610 (1992)
W.S. Kiefer, B.C. Murray, Icarus **72**, 477–491 (1987)
D.A. Kring, B.A. Cohen, J. Geophys. Res. **107**, 5009 (2002). doi:10.1029/2001JE001529
L.V. Ksanfomality, Sol. Syst. Res. **38**, 21–27 (2004)
L.V. Ksanfomality, G. Papamastorakis, N. Thomas, Planet. Space Sci. **53**, 849–859 (2005)
M.A. Leake, C.R. Chapman, S.J. Weidenschilling, D.R. Davis, R. Greenberg, Icarus **71**, 350–375 (1987)
R.P. Lin, K.A. Anderson, L.L. Hood, Icarus **74**, 529–541 (1988)
J.E. Longhi, J.E. Knittle, J.R. Holloway, H. Wanke, in *Mars*, ed. by H.H. Kiefer, B.M. Jakosky, C.W. Snyder, M.S. Matthews (University of Arizona Press, Tucson, 1992), pp. 184–208
B.K. Lucchitta et al., in *Mars*, ed. by H.H. Kiefer, B.M. Jakosky, C.W. Snyder, M.S. Matthews (University of Arizona Press, Tucson, 1992), pp. 453–492
P.G. Lucey, G.J. Taylor, E. Malaret, Science **268**, 1150–1153 (1995)
P.G. Lucey, D.T. Blewett, B.R. Hawke, J. Geophys. Res. **103**, 3679–3700 (1998)
M.C. Malin, *Proc. Lunar Planet. Science Conf. 9th*, 1978, pp. 3395–3409
A. Mallama, D. Wang, R.A. Howard, Icarus **155**, 253–264 (2002)
J.L. Margot, S.J. Peale, R.F. Jurgens, M.A. Slade, I.V. Holin, Science **316**, 710–714 (2007)
T.A. Maxwell, F. El-Baz, S.H. Ward, Geol. Soc. Am. Bull. **86**, 1273–1278 (1975)
J.F. McCauley, Phys. Earth Planet. Interiors **15**, 220–250 (1977)
J.F. McCauley, J.E. Guest, G.G. Schaber, N.J. Trask, R. Greeley, Icarus **47**, 184–202 (1981)
W.E. McClintock, M.R. Lankton, Space Sci. Rev. (2007, this issue). doi:10.1007/s11214-007-9264-5
T.B. McCord, J.B. Adams, Icarus **17**, 585–588 (1972)
T.B. McCord, R.N. Clark, J. Geophys. Res. **84**, 7664–7668 (1979)
G.E. McGill, Icarus **14**, 53–58 (1971)
G.E. McGill, Icarus **21**, 437–447 (1974)
G.E. McGill, Geophys. Res. Lett. **13**, 705–708 (1986)
G.E. McGill, E.A. King, U.S. Geol. Surv. Misc. Invest. Ser., Map I-1409 (1983)
A.F. McGuire, B.W. Hapke, Icarus **113**, 134–155 (1995)
W.B. McKinnon, in *Multi-Ring Basins*, ed. by R.O. Merrill, P.H. Schultz (Geochimica et Cosmochimica Acta, Suppl. **15**, Pergamon Press, 1981), pp. 259–273
H.J. Melosh, Icarus **31**, 221–243 (1977)
H.J. Melosh, *Proc. Lunar Planet. Science Conf. 9th*, 1978, pp. 3513–3525
H.J. Melosh, D. Dzurisin, Icarus **35**, 227–236 (1978a)
H.J. Melosh, D. Dzurisin, Icarus **33**, 141–144 (1978b)
H.J. Melosh, W.B. McKinnon, in *Mercury*, ed. by F. Vilas, C.R. Chapman, M.S. Matthews (University of Arizona Press, Tucson, 1988), pp. 374–400
M. Mendillo, J. Warell, S.S. Limaye, J. Baumgardner, A. Sprague, J.K. Wilson, Planet. Space Sci. **49**, 1501–1505 (2001)
S.M. Milkovich, J.W. Head, L. Wilson, Meteorit. Planet. Sci. **37**, 1209–1222 (2002)
H.J. Moore, C.A. Hodges, D.H. Scott, *Proc. Lunar Science Conf. 5th*, 1974, pp. 71–100
J.I. Moses, K. Rawlins, K. Zahnle, L. Dones, Icarus **137**, 197–221 (1999)
B.C. Murray, J. Geophys. Res. **80**, 2342–2344 (1975)
B.C. Murray et al., Science **185**, 169–179 (1974)
B.C. Murray, R.G. Strom, N.J. Trask, D.E. Gault, J. Geophys. Res. **80**, 2508–2514 (1975)
G. Neukum, J. Oberst, H. Hoffmann, R. Wagner, B.A. Ivanov, Planet. Space Sci. **49**, 1507–1521 (2001)
F. Nimmo, Geophys. Res. Lett. **29**, 1063 (2002). doi:10.1029/2001GL013883
F. Nimmo, T.R. Watters, Geophys. Res. Lett. **31**, L02701 (2004). doi:10.1029/2003GL018847
S.K. Noble, C.M. Pieters, Sol. Syst. Res. **37**, 31–35 (2003)
V.R. Oberbeck, Rev. Geophys. Space Phys. **13**, 337–362 (1975)
V.R. Oberbeck, R.H. Morrison, F. Hörz, W.L. Quaide, D.E. Gault, *Proc. Lunar Planet. Science Conf. 5th*, 1974, pp. 111–136
V.R. Oberbeck, W.L. Quaide, R.E. Arvidson, H.R. Aggarwal, J. Geophys. Res. **82**, 1681–1698 (1977)
B. Pavri, J.W. Head, K.B. Klose, L. Wilson, J. Geophys. Res. **97**, 13445–13478 (1992)

J.C. Pechmann, Icarus **42**, 185–210 (1980)
J.B. Pechmann, H.J. Melosh, Icarus **38**, 243–250 (1979)
R.J. Phillips, S.C. Solomon, Lunar Planet. Sci. **28**, 1107–1108 (1997)
R.J. Phillips, J.E. Conel, E.A. Abbott, W.L. Sjogren, J.B. Morton, J. Geophys. Res. **77**, 7106 (1972)
R.J. Pike, in *Mercury*, ed. by F. Vilas, C.R. Chapman, M.S. Matthews (University of Arizona Press, Tucson, 1988), pp. 165–273
R.J. Pike, P.D. Spudis, Earth, Moon Planets **39**, 129–194 (1987)
J.B. Plescia, M.P. Golombek, Geol. Soc. Am. Bull. **97**, 1289–1299 (1986)
B. Rava, B. Hapke, Icarus **71**, 397–429 (1987)
N.C. Richmond, L.L. Hood, D.L. Mitchell, R.P. Lin, M.H. Acuña, A.B. Binder, J. Geophys. Res. **110**, E05011 (2005). doi:10.1029/2005JE002405
M.S. Robinson, P.G. Lucey, Science **275**, 197–200 (1997)
M.S. Robinson, G.J. Taylor, Meteorit. Planet. Sci. **36**, 841–847 (2001)
M. Robinson, B.R. Hawke, P.G. Lucey, G.J. Taylor, P.D. Spudis, Lunar Planet. Sci. **28** (1997), abstract 1189
M. Robinson, B.R. Hawke, P.G. Lucey, G.J. Taylor, P.D. Spudis, Lunar Planet. Sci. **29** (1998), abstract 1860
G. Ryder, Eos Trans. Am. Geophys. Union **71**, 313, 322–323 (1990)
A.G. Santo et al., Planet. Space Sci. **49**, 1481–1500 (2001)
G.G. Schaber, J.F. McCauley, U.S. Geol. Surv., Map I-1199 (1980)
C.E. Schlemm, II et al., Space Sci. Rev. (2007, this issue). doi:10.1007/s11214-007-9248-5
G. Schubert, M.N. Ross, D.J. Stevenson, T. Spohn, in *Mercury*, ed. by F. Vilas, C.R. Chapman, M.S. Matthews (University of Arizona Press, Tucson, 1988), pp. 429–460
P.H. Schultz, Phys. Earth Planet. Interiors **15**, 202–219 (1977)
P.H. Schultz, in *Mercury*, ed. by F. Vilas, C.R. Chapman, M.S. Matthews (University of Arizona Press, Tucson, 1988), pp. 274–335
P.H. Schultz, D.E. Gault, Moon **12**, 159–177 (1975)
R.A. Schultz, J. Geophys. Res. **105**, 12035–12052 (2000)
D. Schulze-Makuch, J.M. Dohm, A.G. Fairén, V.R. Baker, W. Fink, R.G. Strom, Astrobiology **5**, 778–795 (2005)
V.V. Shuvalov, N.A. Artemieva, Lunar Planet. Sci. **37** (2006), abstract 1168
D.E. Smith et al., J. Geophys. Res. **106**, 23689–23722 (2001)
E.I. Smith, Icarus **28**, 543–550 (1976)
S.E. Smrekar, P. Moreels, B.J. Franklin, J. Geophys. Res. **107**, 5098 (2002). doi:10.1029/2001JE001808
V.S. Solomatov, C.C. Reese, in *Workshop on Mercury: Space Environment, Surface, and Interior* (Lunar and Planetary Institute, Houston, 2001), pp. 92–93
S.C. Solomon, Icarus **28**, 509–521 (1976)
S.C. Solomon, Phys. Earth Planet. Interiors **15**, 135–145 (1977)
S.C. Solomon, Geophys. Res. Lett. **5**, 461–464 (1978)
S.C. Solomon, Phys. Earth Planet. Interiors **19**, 168–182 (1979)
S.C. Solomon, Earth Planet. Sci. Lett. **216**, 441–455 (2003)
S.C. Solomon, J.W. Head, J. Geophys. Res. **84**, 1667–1682 (1979)
S.C. Solomon, J.W. Head, Rev. Geophys. Space Phys. **18**, 107–141 (1980)
S.C. Solomon et al., J. Geophys. Res. **97**, 13199–13256 (1992)
S.C. Solomon et al., Science **307**, 1214–1220 (2005)
A.L. Sprague, R.W.H. Kozlowski, F.C. Witteborn, D.P. Cruikshank, D.H. Wooden, Icarus **109**, 156–167 (1994)
A.L. Sprague, D.M. Hunten, K. Lodders, Icarus **118**, 211–215 (1995)
A.L. Sprague, W.J. Schmitt, R.E. Hill, Icarus **135**, 60–68 (1998)
A.L. Sprague, J.P. Emery, K.L. Donaldson, R.W. Russell, D.K. Lynch, A.L. Mazuk, Meteorit. Planet. Sci. **37**, 1255–1268 (2002)
P.D. Spudis, in Repts. Planet. Geol. Prog. (NASA TM-87563, 1985), pp. 595–597
P.D. Spudis, J.E. Guest, in *Mercury*, ed. by F. Vilas, C.R. Chapman, M.S. Matthews (University of Arizona Press, Tucson, 1988), pp. 118–164
P.D. Spudis, B.R. Hawke, in *Workshop on the Geology and Petrology of the Apollo 15 Landing Site*. LPI Tech. Rept. 86-03 (Lunar and Planetary Institute, Houston, TX, 1986), pp. 105–107
P.D. Spudis, J.G. Prosser, U.S. Geol. Surv., Map I-1659 (1984)
D.K. Srinivasan, M.E. Perry, K.B. Fielhauer, D.E. Smith, M.T. Zuber, Space Sci. Rev. (2007, this issue). doi:10.1007/s11214-007-9270-7
S. Stanley, J. Bloxham, W.E. Hutchison, M.T. Zuber, Earth Planet. Sci. Lett. **234**, 27–38 (2005)
L. Starukhina, J. Geophys. Res. **106**, 14701–14710 (2001)
D.J. Stevenson, Science **287**, 997–1005 (2000)

E.R. Stofan, V.E. Hamilton, D.M. Janes, S.E. Smrekar, in *Venus II*, ed. by S.W. Brougher, D.M. Hunten, R.J. Phillips (University of Arizona Press, Tucson, 1997), pp. 931–965
R.G. Strom, Mod. Geol. **2**, 133–157 (1972)
R.G. Strom, Phys. Earth Planet. Interiors **15**, 156–172 (1977)
R.G. Strom, *Mercury: The Elusive Planet* (Smithsonian Inst. Press, Washington, 1987), 197 pp
R.G. Strom, G. Neukum, in *Mercury*, ed. by F. Vilas, C.R. Chapman, M.S. Matthews (University of Arizona Press, Tucson, 1988), pp. 336–373
R.G. Strom, A.L. Sprague, *Exploring Mercury: The Iron Planet* (Springer, New York, 2003), 216 pp
R.G. Strom, N.J. Trask, J.E. Guest, J. Geophys. Res. **80**, 2478–2507 (1975)
R.G. Strom, S.K. Croft, N.G. Barlow, in *Mars*, ed. by H.H. Kiefer, B.M. Jakosky, C.W. Snyder, M.S. Matthews (University of Arizona Press, Tucson, 1992), pp. 383–423
R.G. Strom, R. Malhotra, T. Ito, F. Yoshida, D.A. Kring, Science **309**, 1847–1850 (2005)
D.E. Stuart-Alexander, U.S. Geol. Surv., Misc. Invest. Ser., Map I-1047 (1978)
F. Tera, D.A. Papanastassiou, G.J. Wasserburg, Earth Planet. Sci. Lett. **22**, 1–21 (1974)
P.G. Thomas, P. Masson, L. Fleitout, in *Mercury*, ed. by F. Vilas, C.R. Chapman, M.S. Matthews (University of Arizona Press, Tucson, 1988), pp. 401–428
S. Tompkins, C.M. Pieters, Meteorit. Planet. Sci. **34**, 25–41 (1999)
N.J. Trask, D. Dzurisin, U.S. Geol. Surv. Misc. Invest. Ser., Map I-1658 (1984)
N.J. Trask, J.E. Guest, J. Geophys. Res. **80**, 2461–2477 (1975)
N.J. Trask, J.F. McCauley, Earth Planet. Sci. Lett. **14**, 201–206 (1972)
N.J. Trask, R.G. Strom, Icarus **28**, 559–563 (1976)
A.L. Tyler, R.W.H. Kozlowski, L.A. Lebofsky, Geophys. Res. Lett. **15**, 808–811 (1988)
A.R. Vasavada, D.A. Paige, S.E. Wood, Icarus **141**, 179–193 (1999)
J. Veverka, P. Helfenstein, B. Hapke, J.D. Goguen, in *Mercury*, ed. by F. Vilas, C.R. Chapman, M.S. Matthews (University of Arizona Press, Tucson, 1988), pp. 37–58
F. Vilas, Icarus **64**, 133–138 (1985)
F. Vilas, in *Mercury*, ed. by F. Vilas, C.R. Chapman, M.S. Matthews (University of Arizona Press, Tucson, 1988), pp. 59–76
F. Vilas, T.B. McCord, Icarus **28**, 593–599 (1976)
F. Vilas, M.A. Leake, W.W. Mendell, Icarus **59**, 60–68 (1984)
F. Vilas, P.S. Cobian, N.G. Barlow, S.M. Lederer, Planet. Space Sci. **53**, 1496–1500 (2005)
D. Vokrouhlicky, P. Farinella, W.F. Bottke Jr., Icarus **148**, 147 (2000)
J. Warell, Icarus **156**, 303–317 (2002)
J. Warell, Icarus **167**, 271–286 (2004)
J. Warell, D.T. Blewett, Icarus **168**, 257–276 (2004)
J. Warell, S.S. Limaye, Planet. Space Sci. **49**, 1531–1552 (2001)
J. Warell, A.L. Sprague, J.P. Emery, R.W.H. Kozlowski, A. Long, Icarus **180**, 281–291 (2006)
T.R. Watters, J. Geophys. Res. **93**, 10236–10254 (1988)
T.R. Watters, J. Geophys. Res. **96**, 15599–15616 (1991)
T.R. Watters, J. Geophys. Res. **98**, 17049–17060 (1993)
T.R. Watters, J. Geophys. Res. **108**, 5054 (2003). doi:10.1029/2002JE0001934
T.R. Watters, Icarus **171**, 284–294 (2004)
T.R. Watters, M.S. Robinson, A.C. Cook, Geology **26**, 991–994 (1998)
T.R. Watters, A.C. Cook, M.S. Robinson, Planet. Space Sci. **49**, 1523–1530 (2001)
T.R. Watters, R.A. Schultz, M.S. Robinson, A.C. Cook, Geophys. Res. Lett. **29**, 1542 (2002). doi:10.1029/2001GL014308
T.R. Watters, M.S. Robinson, C.R. Bina, P.D. Spudis, Geophys. Res. Lett. **31**, L04701 (2004). doi:10.1029/2003GL019171
T.R. Watters, F. Nimmo, M.S. Robinson, Geology **33**, 669–672 (2005)
M.A. Wieczorek, M.T. Zuber, J. Geophys. Res. **106**, 27853–27864 (2001)
D.E. Wilhelms, Icarus **28**, 551–558 (1976a)
D.E. Wilhelms, *Proc. Lunar Planet. Science Conf. 7th*, 1976b, pp. 2883–2901
D.E. Wilhelms, *The Geologic History of the Moon*. Prof. Paper 1348 (U.S. Geological Survey Washington, DC, 1987), 302 pp
D.E. Wilhelms, F. El-Baz, U.S. Geol. Surv. Misc. Geol. Invest. Ser., Map I-948 (1977)
D.E. Wilhelms, J. McCauley, U.S. Geol. Surv. Misc. Inv. Ser., Map I-703 (1971)
C.A. Wood, J.W. Head, *Proc. Lunar Planet. Science Conf. 7th*, 1976, pp. 3629–3651
J.W. Young, T.K. Mattingly, C.M. Duke, in Apollo 16 Preliminary Science Report, SP-315 (NASA, Washington, DC, 1972) pp. 5-1–5-6
M.T. Zuber et al., Space Sci. Rev. (2007, this issue). doi:10.1007/s11214-007-9265-4

Space Sci Rev (2007) 131: 85–104
DOI 10.1007/s11214-007-9258-3

MESSENGER and the Chemistry of Mercury's Surface

William V. Boynton · Ann L. Sprague · Sean C. Solomon · Richard D. Starr · Larry G. Evans · William C. Feldman · Jacob I. Trombka · Edgar A. Rhodes

Received: 23 October 2006 / Accepted: 7 August 2007 / Published online: 13 October 2007

Abstract The instrument suite on the MErcury Surface, Space ENvironment, GEochemistry, and Ranging (MESSENGER) spacecraft is well suited to address several of Mercury's outstanding geochemical problems. A combination of data from the Gamma-Ray and Neutron Spectrometer (GRNS) and X-Ray Spectrometer (XRS) instruments will yield the surface abundances of both volatile (K) and refractory (Al, Ca, and Th) elements, which will test the three competing hypotheses for the origin of Mercury's high bulk metal fraction: aerodynamic drag in the early solar nebula, preferential vaporization of silicates, or giant impact. These same elements, with the addition of Mg, Si, and Fe, will put significant constraints on geochemical processes that have formed the crust and produced any later volcanism. The Neutron Spectrometer sensor on the GRNS instrument will yield estimates of the amount of H in surface materials and may ascertain if the permanently shadowed polar craters have a significant excess of H due to water ice. A comparison of the FeO content of olivine and pyroxene determined by the Mercury Atmospheric and Surface Composition

W.V. Boynton · A.L. Sprague (✉)
Lunar and Planetary Laboratory, University of Arizona, Tucson, AZ 85721, USA
e-mail: sprague@lpl.arizona.edu

S.C. Solomon
Carnegie Institution of Washington, Washington, DC 20015, USA

R.D. Starr
The Catholic University of America, Washington, DC 20064, USA

L.G. Evans
Computer Sciences Corporation, Lanham-Seabrook, MD 20706, USA

W.C. Feldman
Planetary Science Institute, Tucson, AZ 85719, USA

J.I. Trombka
NASA Goddard Space Flight Center, Greenbelt, MD 20770, USA

E.A. Rhodes
The Johns Hopkins University Applied Physics Laboratory, Laurel, MD 20723, USA

Spectrometer (MASCS) instrument with the total Fe determined through both GRNS and XRS will permit an estimate of the amount of Fe present in other forms, including metal and sulfides.

Keywords Mercury · MESSENGER · Mercury's surface chemistry · Gamma-ray spectrometry · X-ray spectrometry · Space missions · Planetary surfaces

1 Introduction

Mariner 10 made no detailed measurements of surface elemental abundances or specific minerals or rock types on Mercury. All that is known about Mercury's surface composition comes from ground-based observations and inferences from color reconstructions of Mariner 10 images. Recalibrations of the Mariner 10 images and the ratioing of images obtained at different colors suggest that compositional boundaries and soil differences may be discerned (Robinson and Lucey 1997) but give no chemical information. As ground-based telescopic instrumentation has improved in sensitivity and efficiency, more data have been collected from visible to radio wavelengths. These data have resulted in some knowledge of the chemistry of Mercury's surface, and almost all of the observations have raised new questions. A major advance in our understanding of the composition of Mercury's surface will come from the suite of instruments onboard the MErcury Surface, Space ENvironment, GEochemistry, and Ranging (MESSENGER) spacecraft (Solomon et al. 2001; Gold et al. 2001; Santo et al. 2001), scheduled to begin mapping Mercury's surface from orbit in March 2011.

These instruments will contribute significantly to four of the six prime objectives of the MESSENGER mission (Solomon et al. 2001): (1) What planetary formational processes led to the high metal/silicate ratio in Mercury? (2) What is the geological history of Mercury? (3) What are the radar-reflective materials at Mercury's poles? (4) What are the important volatile species and their sources and sinks on and near Mercury? Mercury's high uncompressed density implies that a metal, iron-rich core occupies a much larger fraction of the total mass than for any of the other terrestrial planets. Proposed explanations for the high metal fraction include mechanical sorting of silicate and metal grains by aerodynamic drag in the early solar nebula (Weidenschilling 1978), vaporization of much of the outer silicate shell of a differentiated planet by a hot solar nebula (Cameron 1985; Fegley and Cameron 1987), and selective removal of silicate by a giant impact onto a differentiated planet (Wetherill 1988; Benz et al. 1988). These hypotheses make distinct predictions for the chemical make-up of Mercury's present crust (Lewis 1988), so geochemical remote sensing of Mercury's surface can distinguish among the explanations for the planet's unusually high metal fraction. Geochemical remote sensing will also elucidate compositional differences among geological units and, along with imaging and spectral measurements, will assist in the definition of the global geological evolution of the planet. Candidate materials for Mercury's polar deposits—including water ice, elemental sulfur, and cold silicates (Slade et al. 1992; Harmon and Slade 1992; Sprague et al. 1995; Starukhina 2001)—can potentially be distinguished through orbital observations. Because Mercury's exosphere contains a number of species that were derived from the planetary surface (Hunten et al. 1988), geochemical remote sensing can elucidate sources and sinks for exospheric species and their temporal and spatial variations.

In this paper we begin with a summary of current knowledge of the surface chemistry of Mercury derived from Earth-based telescopic observations at visible to radio wavelengths as well as from Mariner 10 color images. We briefly discuss suggestions that rocks from Mercury may be found among the world's meteorite collections. We then summarize the MESSENGER instruments that are most pertinent to observations of surface chemistry and mineralogy. We close with a discussion of how data from these instruments will address four of the scientific objectives of the MESSENGER mission.

2 Current State of Knowledge of Mercury's Surface Chemistry

A variety of Earth-based astronomical observations of Mercury's surface have yielded limited information on surface chemistry and mineralogy, including spectral reflectance measurements at visible and near-infrared wavelengths, mid-infrared spectroscopy, microwave emission observations, and radar imaging. Also relevant are observations of several surface-derived species in Mercury's tenuous exosphere and information on the color of Mercury's surface obtained from Mariner 10 images. Each type of observation is reviewed in turn.

2.1 Reflectance Spectra

One of the best-understood features in reflectance spectra of rocky bodies is an absorption caused by an electronic transition in an iron cation (Fe^{2+}). The electronic transition occurs when Fe is bound to O in a silicate lattice. This absorption band is often seen in spectra from asteroids and the Moon and indicates an FeO content of at least several weight percent in minerals comprising the rocks on those surfaces. On Mercury, however, despite many searches, the first unambiguous identification of the FeO band was obtained only recently in spectra obtained at the Infrared Telescope Facility (IRTF) on Mauna Kea, Hawaii, using the high-resolving-power échelle spectrograph, SpeX (Warell et al. 2006). Previous efforts to identify this feature in spectra of Mercury's surface (McCord and Clark 1979; Vilas et al. 1984; Vilas 1985, 1988; Warell and Blewett 2004) have often been suggestive but not convincing. The most recent spectra (Warell et al. 2006), covering the widest spectral range, clearly exhibit a shallow absorption feature from 0.8 to 1.3 μm with the center of the symmetric shallow band near 1.1 μm. Several locations have been measured, but only two spectra unambiguously exhibit an absorption characteristic of iron-bearing silicates (Fig. 1).

From this evidence we deduce that Mercury's surface materials contain at most a few percent FeO in surface materials at some locations; at other locations the FeO abundances are even lower. A very slight depression in the Mercury 2003 S spectrum centered near 1.85 μm (Fig. 1) may be indicative of a small amount of orthopyroxene, which can exhibit absorptions centered at 1 and 1.85 μm that are associated with the M2 lattice sites of Fe^{2+}. However, the absence of the 1.85 μm feature in the Mercury 2003 N spectrum indicates that if the 1 μm absorption is from FeO in pyroxene, it must be clinopyroxene, because the Fe^{2+} cation in the clinopyroxene occupies the M1 lattice site but not the M2 site. The shallow band centered at 1 μm and the absence of the band centered on 1.85 μm is also consistent with olivine, and the shape of the band is broadly similar to that indicative of forsterite, the Mg-rich end member.

For the spectra showing the FeO absorption, the spectroscopic slits traverse several different types of geological units, including intercrater plains, smooth plains, and heavily cratered terrain (Spudis and Guest 1988). Because of atmospheric blurring of the image at the telescope and the size of the slit sector footprint on the planet (about 800 km $\times$ 800 km), it is not possible to identify the absorption feature with a particular geological unit.

These measurements support the inference, derived from a comparison of visible and near-infrared reflectance measurements (0.4–0.8 μm) from Mercury with those from known

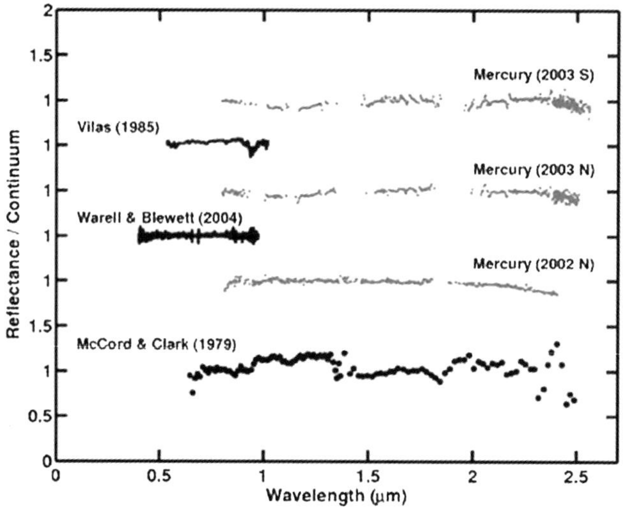

Fig. 1 Spectra obtained at the IRTF using SpeX (*gray data*) from three different locations on Mercury's surface after removal of solar reflectance and thermal emission components and division by a linear fit to the continuum at 0.77 and 1.6 μm. All spectra are normalized at 1 μm and then offset for easy comparison to one another and to spectra published by others. The 2002 Mercury spectrum from ~110°E longitude and high northern latitudes shows no evidence for the FeO absorption band. Shallow bands centered at 1.1 μm are present in the 2003 spectra from longitude ~200°E and mid-northern and southern latitudes. These two spectra are the first reflectance data to show an unambiguous FeO absorption band from 0.8 to 1.1 μm. From Warell et al. (2006)

anorthositic regions on the Moon (Blewett et al. 1997), that Mercury's surface has a low abundance of ferrous iron (<2–3 wt% FeO). Blewett et al. (1997) also demonstrated, with ratios of spectral reflectance at 410 nm to that at 750 nm, that near-infrared spectra from Mercury's surface have slopes that are steeper (redder) than lunar spectra and indicative of mature anorthosite and no more than ~1% TiO_2 as ilmenite. Warell and Blewett (2004) were able to constrain more tightly the FeO and TiO_2 contents on Mercury's surface from radiative-transfer models of laboratory spectra of mineral powders fit to visible (0.4–0.65 μm) spectra obtained at the Nordic Optical Telescope (NOT). The composition of the mineral mixture with the best fit had 1.2 wt% FeO and ~0 wt% TiO_2.

2.2 Microwave Emissivity

A detailed study of microwave emissivity (0.3–20.5 cm) from Mercury's regolith (Mitchell and de Pater 1994) demonstrated that Mercury's surface is more transparent to electromagnetic radiation at these wavelengths than the lunar surface and a suite of terrestrial basalts. Mercury's regolith materials must therefore be lower in Fe and Ti than those of the Moon. This result is illustrated in Fig. 2, adapted from Mitchell and de Pater (1994), where the specific loss tangent as a function of frequency for materials in Mercury's regolith is compared with those from samples of the lunar maria, lunar highlands, and terrestrial basalts.

That Mercury's regolith is more transparent to microwave radiation and at the same time exhibits a weak or absent FeO absorption band in reflectance spectra indicates that surface materials are very low in Fe, either in the form of FeO or of Fe metal. Because Fe metal would be expected from extensive space weathering of FeO-bearing silicates (Noble and Pieters 2003; Sasaki and Kurahashi 2004), the shallow 1-μm absorption band is

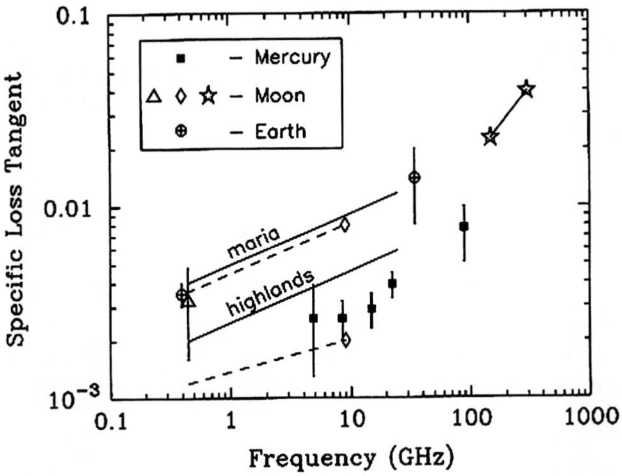

Fig. 2 Ground-based microwave imaging and modeling have demonstrated that the specific loss tangent of materials in Mercury's regolith is systematically lower than that of the lunar regolith and a suite of terrestrial basalts. The *solid* and *dashed lines* connect measurements made of the same sample at both 45 MHz and 9 GHz. From these results it has been inferred that Ti and Fe are not as abundant on Mercury's surface as on the Moon. From Mitchell and de Pater (1994)

indicative of an inherently low FeO mineralogy and is not entirely the result of extensive space weathering (Robinson and Taylor 2001). In addition, we know that Mercury's surface is more mature than that of the Moon because its spectral slope is reddened at visible and near-infrared wavelengths (McCord and Clark 1979; Vilas 1988; Robinson and Lucey 1997; Blewett et al. 1997; Warell 2002, 2003; Warell and Blewett 2004). From these observations, we may conclude that any volcanic units on Mercury are generally lower in FeO and TiO_2 than the lunar maria (Jeanloz et al. 1995).

2.3 Mercury's Exosphere as a Possible Indicator of Surface Composition

Mercury's surface-bounded exosphere contains six known species (H, He, O, Na, K, and Ca). The partitioning between exogenic and truly indigenous sources, however, is not known (e.g., Potter and Morgan 1997; Domingue et al. 2007; Koehn and Sprague 2007). The association of enhanced emissions from Na and K with specific regions on Mercury's surface (Sprague et al. 1990, 1997, 1998) may be indicative of enriched Na and K abundances in surface materials in these regions. Sprague et al. (1990), for instance, reported enhanced exospheric potassium emission near the longitude of the Caloris basin.

Enhanced Na emissions are associated in particular with areas of high radar backscatter sometimes termed Goldstein features (Goldstein 1970). Follow-on radar imaging (Slade et al. 1992; Harmon and Slade 1992; Butler et al. 1993; Harmon 1997; Harmon et al. 2007) has led to naming two of these regions "features A and B" with centers near 35°S, 10°E, and 55°N, 15°E, respectively. Both features have been interpreted as comparatively young impact craters with extensive systems of radar-bright rays (Harmon et al. 2007). The freshly excavated ejecta material may be the source of the Na seen in the enhanced emission observations. Figure 3 shows one example of ground-based images of Mercury's Na exospheric bright spots (Potter and Morgan 1990) and their association with radar bright spots on Mercury's surface (Sprague et al. 1998). The same phenomenon has been observed in slit spectroscopy over these regions (Sprague et al. 1997, 1998).

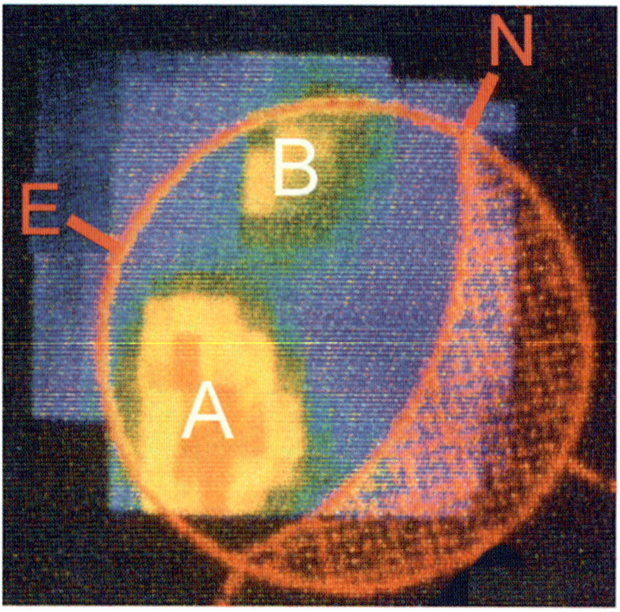

Fig. 3 Locations of enhanced Na emission in Mercury's surface-bounded exosphere are associated with several features on Mercury's surface. In this image, *bright orange regions* (enhanced atmospheric Na emission) fall near the radar-bright regions *A* and *B* (Slade et al. 1992; Harmon and Slade 1992; Butler et al. 1993; Harmon 1997; Harmon et al. 2007). Atmospheric smearing of the telescopic image and image rectification can account for the offsets of the Na emission highs from the mapped features. From Sprague et al. (1998)

2.4 Mid-Infrared Spectroscopy

Planetary surface spectra in the mid-infrared can contain important information on major minerals present in surface materials. Inferences on mineral composition can come from comparisons with spectra from surfaces of known composition or with spectra from powders of minerals or their mixtures. Major minerals have their fundamental molecular vibration (Reststrahlen) bands in the region from 7.5 to 11 μm. A transparency minimum (TM) between 11 and 13 μm is associated with the change from surface to volume scattering and is also an indicator of SiO_2 content. Another feature is the emissivity maximum (EM) of a silicate spectrum, which usually occurs between 7 and 9 μm and is a good diagnostic of bulk SiO_2 content in powdery mixtures of rocks, minerals, and glasses common in regoliths.

Mid-infrared spectra from Mercury exhibit variations with latitude and longitude that indicate considerable heterogeneity of surface chemistry (Sprague et al. 1994, 2002; Sprague and Roush 1998; Emery et al. 1998; Cooper et al. 2001). Comparisons of individual spectra with those of lunar surface material or mixtures of rock and mineral powders indicate that the dominant minerals at Mercury's surface are plagioclase feldspar and pyroxene with some spectral features suggestive of feldspathoids (Sprague et al. 1994; Jeanloz et al. 1995). Mid-infrared spectra of Mercury's surface from 240° to 348°E longitude exhibit an EM associated with the principal Christiansen frequency (Sprague et al. 1994; Emery et al. 1998). The areal extent of the spatial footprint in such measurements is about 200 km by 200 km for the best spatially resolved observations to date and as much as 1,000 km by 1,000 km for the least spatially resolved regions. The EM wavelengths in these spectra are generally indicative of intermediate silica content (∼50–57% SiO_2 by weight). An example spectrum is compared

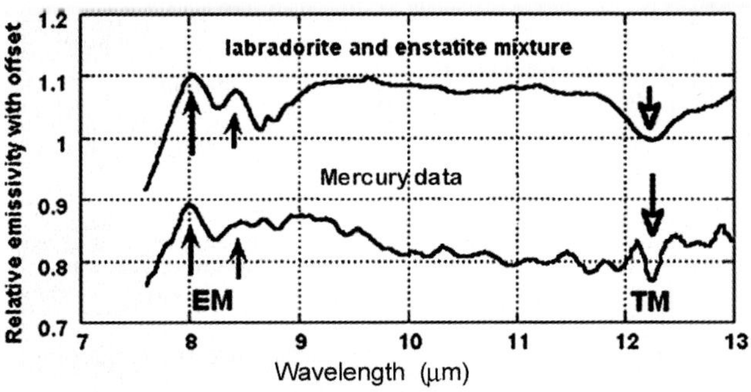

Fig. 4 (*Top*) Model spectrum created from two laboratory spectra—Na-bearing plagioclase feldspar (labradorite) and low-iron, Mg-rich orthopyroxene (enstatite). (*Bottom*) Mid-infrared spectroscopic measurement of Mercury's near-equatorial surface from longitudes ~240° to 250°E (~110° to 120°W). The Mercury spectrum (Sprague and Roush 1998) and the model both exhibit an emissivity maximum (EM) for both minerals and a transparency minimum (TM) for feldspar

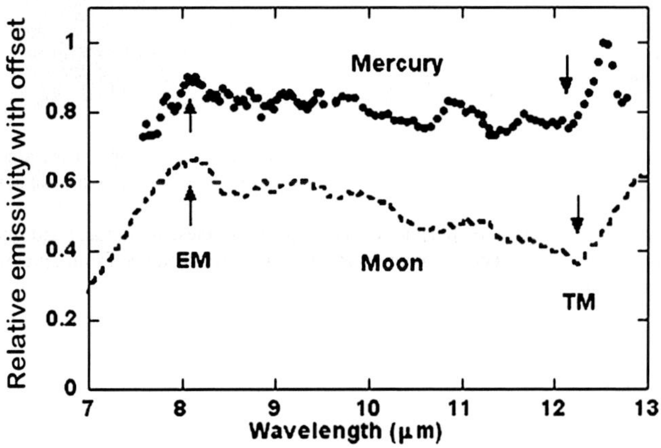

Fig. 5 A mid-infrared Mercury spectrum from equatorial regions at longitudes 335° to 340°E (20° to 25°W) is shown along with a laboratory lunar spectrum from Apollo 16 particulate breccia 67031 (90% anorthite, 10% pyroxene). Figure courtesy of A. Sprague

in Fig. 4 with a model spectrum created from laboratory spectra of plagioclase feldspar and Mg-rich pyroxene.

A spectrum from an equatorial region near 335° to 340°E longitude is compared with a laboratory spectrum from an Apollo 16 lunar breccia in Fig. 5. The lunar sample is approximately 90% anorthite (Ca-plagioclase) and ~10% pyroxene. The EM for both spectra is centered close to 8 μm and marked with an arrow. Other features in the Mercury spectrum are similar to the lunar sample spectrum and support inferences that Mercury's surface may be compositionally similar to low-FeO areas of the lunar highlands (e.g., Blewett et al. 2002).

A strong 5-μm emission feature in a spectrum from 275° to 315°E longitude (Fig. 6) closely resembles laboratory spectra of some pyroxene powders. The best fit is to diopside ($CaMgSi_2O_6$). The low FeO abundance indicated by near-infrared reflectance spectroscopy supports a low-iron pyroxene. The EM is also evident in the spectrum at a wavelength similar to the EM in Fig. 4.

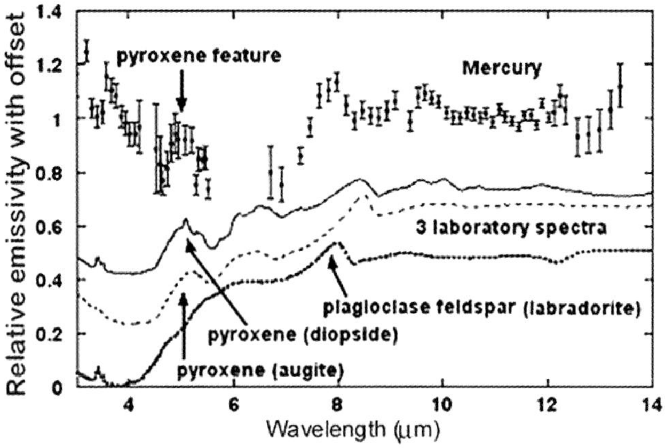

Fig. 6 A spectral feature at 5 μm in Mercury's spectrum (Sprague et al. 2002) from longitudes 275° to 315°E (45° to 85°W) resembles that exhibited in laboratory spectra (Salisbury et al. 1987, 1991) from two samples of clinopyroxene. Also exhibited is an EM at 7.9 μm indicative of intermediate SiO_2 content. Plagioclase feldspar intermediate between anorthite and albite also has an EM at this wavelength. Spectra such as these indicate that Mercury's surface at these locations has spectral characteristics similar to those of low-iron material rich in feldspar and calcic pyroxene

Many smaller fluctuations in the Mercury spectrum in Fig. 6 are not present in the spectrum from the laboratory sample. These features may be in the Mercury spectrum because Mercury's surface is much hotter than the environment of the chamber holding the laboratory sample and the regolith is interfacing to a vacuum (Hunt and Vincent 1968; Logan and Hunt 1970). Alternatively, the features may be noise in the spectrum or features contributed by minerals that were not in the model mixture. Spectra obtained with the Kuiper Airborne Observatory and taken above much of the Earth's attenuating atmosphere from longitudes 200° to 260°E have multiple EMs indicating a more complicated bulk composition or a more diverse mineralogy (Emery et al. 1998).

The transparency minimum (TM) between 11 and 13 μm in Figs. 4, 5, and 6 is associated with a change from surface to volume scattering. The wavelength where the minimum occurs is generally a good indicator of the SiO_2 weight fraction in a powder or soil sample. The spectrum in Fig. 4 has a clear TM at 12.3 μm that is at the same location as the TM in a laboratory spectrum of a mixture of feldspar and pyroxene powders. This agreement is consistent with the location of the EM in the same spectrum as described earlier.

Spectra of Mercury (Cooper et al. 2001) showing the transparency minimum measured using the McMath Pierce Solar telescope on Kitt Peak with a circular aperture over Mercury on six different days are displayed in Fig. 7. Wavelengths of the TM fall between 12 and 12.7 μm. The dominant spectral signature comes from the hottest regions in the field of view. A simple thermal model gives an estimate of the longitude responsible for the greatest flux at the detector. Middle spectra from ~180°, 170°, and 350°E longitude have probable transparency minima at 12 μm. On the basis of empirical relations between TM wavelength and silica content (Strom and Sprague 2003), the bulk composition associated with a transparency feature at this wavelength is intermediate to mafic (45–57 weight% SiO_2). The spectrum from ~275°E (second from the top in Fig. 7) has a TM at 12.5 μm, indicative of about 44 weight% SiO_2 or an ultramafic composition. The spectrum from ~215°E longitude (bottom spectrum in Fig. 7) has a doublet TM indicating two different dominant components in the regolith, one mafic and one ultramafic.

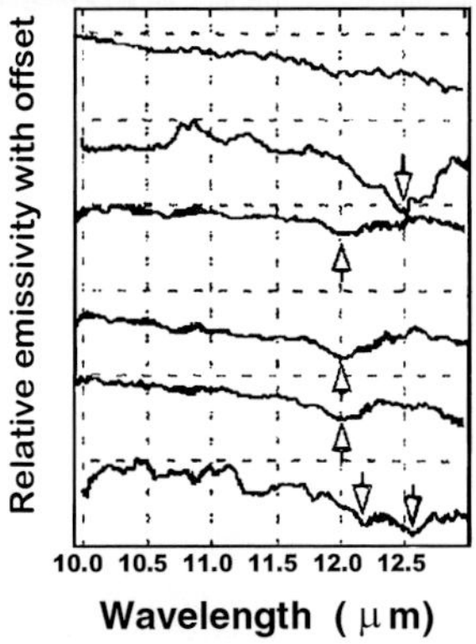

Fig. 7 Spectra from Mercury's surface for gibbous-phase, whole-disk measurements at six different locations show transparency minima indicative of SiO_2 content. Sub-solar longitudes, top to bottom, are 268°, 265°, 180°, 170°, 350°, and 205°E. Sub-Earth longitudes are, top to bottom, 350°, 345°, 104°, 94°, 280°, and 131°E. Wavelengths of TMs are between 12 and 12.7 μm. Locations contributing the greatest flux to the observations estimated from simple thermal models are, from top to bottom, ~280°, 275°, 180°, 170°, 350°, and 215°E. Figure is adapted from Cooper et al. (2001)

Spectral features indicative of mafic or ultramafic rocks point to the presence of olivine or pyroxene (or phases that are undersaturated in SiO_2). Mg-rich olivine is a good candidate because spectra from rocks with this mineral would not exhibit a strong FeO absorption band (see Fig. 1) in their near-infrared spectra. In addition, the band in Mg-olivine is centered near 1 μm rather than at 1.2 μm as in Fe-olivine because the Fe^{2+} cation in the M1 and M2 lattice sites absorbs at longer wavelength.

2.5 Color Variations

Recalibrated Mariner 10 images taken in the ultraviolet (375 nm) and orange (575 nm) indicate compositional variations across Mercury's surface consistent with those deduced by Earth-based spectroscopic observations. The color images show color boundaries between weathered surfaces and material excavated by fresh impacts. They also show color differences between smooth plains and the surrounding terrain indicating surface heterogeneity owing to space weathering, grain size, or compositional differences (Robinson and Lucey 1997). The region of the Rudaki plains (3°S, 304°E) and Tolstoj (16°S, 196°E) smooth plains display embayment relations indicative of lava flow boundaries (Fig. 8) and scattering properties similar to those of pyroclastic deposits and glasses on the Moon (Robinson and Taylor 2001).

In contrast to the Fe- and Ti-bearing basalts of the lunar maria, on Mercury there is no evidence for substantial FeO and the microwave observations and near-infrared spectral modeling and comparisons with the Moon appear to rule out a significant abundance of TiO_2. In at least two cases the smooth plains overlie material that is bluer (higher ultraviolet/orange ratio) and enriched in opaque minerals relative to the average for the hemisphere imaged by Mariner 10 (Robinson and Lucey 1997). Because the smooth plains are likely lava flows, and the FeO solid/liquid distribution coefficient is near unity during partial melting, it is thought that Mercury's mantle has an FeO content similar to plains materials and is $<3\%$ (Robinson and Taylor 2001).

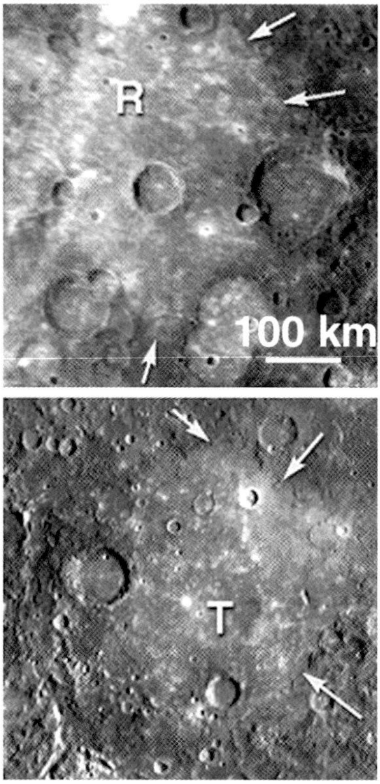

Fig. 8 The mid-infrared spectrum of Mercury shown in Fig. 6 comes from a portion of the region on Mercury's surface shown in the *top image*. In the Rudaki (*R*) plains, analysis of color images has revealed unit boundaries consistent with volcanic emplacement. The *lower image* is of the Tolstoj (*T*) smooth plains. Sites of apparent embayment by one unit over another are indicated by *arrows*. From Robinson and Taylor (2001)

2.6 Radar Observations of Polar Deposits

One of the most puzzling and exciting results from ground-based observations of Mercury was the discovery of regions of highly coherent backscattered radar signals from deposits in the floors of impact craters near Mercury's north and south poles (Slade et al. 1992; Harmon and Slade 1992; Butler et al. 1993; Harmon et al. 1994). The coherent backscatter property of the material is similar to the radar backscatter signature from water ice at the Martian south pole and on the Galilean satellites of Jupiter. Because of this agreement, as well as the fact that the floors of craters at the highest latitudes on Mercury are in permanent shadow and therefore very cold, water ice is a good candidate for polar deposit material. The water ice could be juvenile, released during volcanic emissions and quickly cold-trapped and covered by regolith gardening at polar regions. Alternatively, it could have been delivered by the impacts of comets or volatile-rich asteroids (Butler 1997; Moses et al. 1999; Crider and Killen 2005) or formed by surface chemical interactions driven by solar wind impingement on the surface and followed by poleward migration (Potter 1995).

Subsequent very high-resolution radar observations (Harmon et al. 2001) showed that the high-backscatter material is found at latitudes as low as 72°N, making the water ice interpretation more difficult to explain. Thermal models (Paige et al. 1992; Butler et al. 1993; Vasavada et al. 1999) and Monte Carlo models of volatile distribution and storage in permanently shadowed regions (Butler 1997; Moses et al. 1999; Crider and Killen 2005) have shown that water ice could be stable for several million years following deposition. A recent comet impact delivering water vapor to Mercury could satisfy the lifetime requirement.

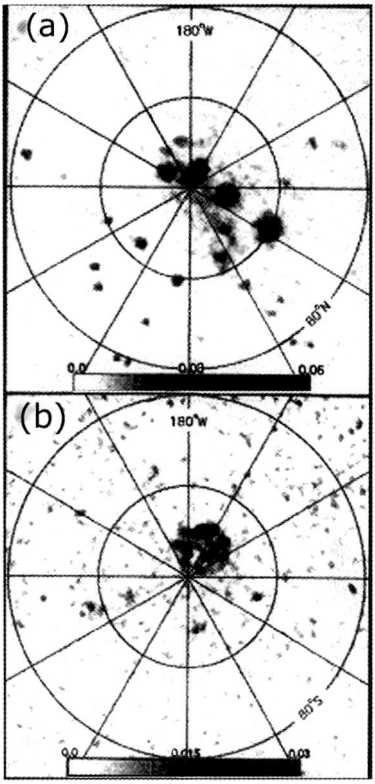

Fig. 9 Regions of high coherent radar backscatter and strong depolarization (dark areas in images) from the (**a**) north and (**b**) south polar regions of Mercury are attributed to deposits of water ice or some other radar-transparent material (e.g., sulfur, cold silicates) on the floors of permanently shadowed craters. Figure adapted from Harmon et al. (1994)

Sprague et al. (1995) argued that sulfur may be the backscattering material on grounds that it is not as volatile as H_2O ice and that it should be abundant in infalling and vaporizing micrometeoroids and may be sputtered from minerals such as sphalerite in the surface regolith. Migration and cold trapping would naturally follow. In addition, the latitude range of stability for S is greater than that of H_2O ice, and deposits would not require special conditions of permanently shadowed regions at polar latitudes. However, uncertainty exists with respect to the stability of S and its potential to form more deposits than those observed, for example extended polar caps (Butler 1997). The discovery of a substantial neutral sulfur atmosphere (resonant emission triplet centered at 181.4 nm) at Mercury would support the suggestion that the material is sulfur, either pristine or brought in by meteoroid infall. If the putative volatile is water ice, then an OH exosphere may be present in polar regions (Killen et al. 1997).

The physical property required of any surface material responsible for the high backscatter and depolarizing behavior is low dielectric loss at low temperatures. Starukhina (2001) argued that silicates at very low temperatures could also produce the observed radar characteristics. Silicates, of course, are ubiquitous on Mercury's surface, and extrapolations to low temperatures and radar wavelengths of loss tangent measurements made on silicates at other temperatures and frequency bands are consistent with the required low dielectric loss (Starukhina 2001). On these grounds, cold, low-iron and low-Ti silicates provide another candidate for polar deposit material.

3 Meteorites from Mercury?

Delivery of meteorites to Earth from Mercury is far less probable than from the Moon or Mars (Melosh and Tonks 1993; Gladman et al. 1996; Dones et al. 1999). The delivery must be made rapidly, within a few tens of millions of years, or the gravitational well of the Sun will circularize the orbit and eventually the ejecta will reimpact Mercury. Cratering models over a range of ejection speeds from 0.25 to 5 km/s predict that less than 0.5% of the ejecta from a Mercury impact should reach Earth (Melosh and Tonks 1993). Given this efficiency, Gladman et al. (1996) inferred from the number of known Martian meteorites that a few meteorites from Mercury may be expected on Earth.

Despite the predicted scarcity of such material, an awareness of the characteristics that might be expected for a sample from Mercury's surface or crust is important. On the basis of the apparent lack of FeO at Mercury's surface, differentiated Mg-rich silicates are suggested. Love and Keil (1995) gave a thorough discussion of the Mg-rich silicates found in the meteorite collection at that time and concluded that the best candidate was the anomalous aubrite, Shallowater. However, the I-Xe closure age of Shallowater precludes the scenario of delivery from Mercury to Earth in the short time required by dynamical calculations.

Another class of meteorites, the angrites, also originated in one or more differentiated parent bodies (Mittlefehldt et al. 2002). Irving et al. (2005) argued that angrites may have come from Mercury on the basis of petrographic evidence for decompression from pressures appropriate to a planetary mantle, low Na, distinctive oxygen isotopes, and a range of cosmic exposure ages suggestive of a large target body. The igneous formation ages (>4.55 Ga) are likely too ancient for these rocks to be from a terrestrial planet, however, and angrites are generally FeO-rich, contrary to the composition of most of Mercury's surface. Thus, to date, the weight of evidence suggests that known meteorites are not from Mercury.

4 MESSENGER's Instrumentation for Surface Chemistry Measurements

Although many of the observations described earlier are suggestive of Mercury's surface chemistry, they are limited in scope and accompanied by the uncertainties inherent in ground-based remote sensing. Interpretation of data from instrumentation designed for geochemical measurements on MESSENGER will face challenges as well. The most obvious of these is the effect of "space weathering," the cumulative changes to a pristine regolith surface caused by radiation effects, melting, and volatilization by meteoroid impact, solar wind implantation, and conversion of oxidized iron in rocks and minerals to nanophase iron ($npFe^0$) deposits (cf. Hapke 2001; Noble et al. 2001; Noble and Pieters 2003). In addition, the recycling of exospheric ions in Mercury's magnetosphere back to the surface may locally concentrate exospheric atoms (cf. Kallio and Janhunen 2003; Zurbuchen et al. 2004; Koehn and Sprague 2007). Together, space weathering and local concentration of exospheric constituents may affect the data and interpretation from surface science instruments.

The Gamma-Ray and Neutron Spectrometer (GRNS) (Goldsten et al. 2007), X-Ray Spectrometer (Schlemm et al. 2007), Mercury Atmospheric and Surface Composition Spectrometer (MASCS) (McClintock and Lankton 2007), and Mercury Dual Imaging System (MDIS) (Hawkins et al. 2007) are the principal MESSENGER instruments that will contribute to the study of Mercury's surface chemistry and should enable the mission to achieve its principal scientific objectives (Solomon et al. 2001). These instruments and their strengths and weaknesses are briefly discussed in the following sections.

4.1 GRNS

The GRNS instrument is composed of two subsystems (Goldsten et al. 2007)—the Gamma-Ray Spectrometer (GRS) and the Neutron Spectrometer (NS). Versions of both instruments have been highly successful in mapping the surface composition of the Moon (Feldman et al. 1998, 2002, 2004a) and Mars (Boynton et al. 2002, 2004, 2007; Feldman et al. 2004b, 2005; Mitrofanov et al. 2002; Prettyman et al. 2004; Tokar et al. 2002). Because of MESSENGER's highly eccentric orbit (Santo et al. 2001), the highest-resolution information will be obtained near periapsis (60–72°N) and in general the northern hemisphere will be much better mapped than the southern hemisphere. Both the GRS and NS subsystems require repeated orbits to build up a significant signal-to-noise ratio. In general the GRS signal is weaker than that of the NS, so it requires a greater number of passes over the same region to obtain a significant signal. Although the NS has a better signal-to-noise ratio than the GRS, interpretation of its data is more model-dependent (e.g., Feldman et al. 2000a). As discussed in the following, in some cases we can take advantage of the GRS and XRS data to constrain the interpretation of the NS data, giving us the advantage of the higher signal-to-noise ratio of the NS but without as much model dependence for its results.

4.1.1 GRS

Because it is based on a nuclear technique, the GRS maps elemental abundances without regard to the chemical compounds or minerals in which the elements may be located by measuring gamma rays whose energies can be identified with specific nuclear transitions. The GRS probes beneath the surface to $\sim$10–15 cm. Therefore, the signal is not altered by most space weathering mechanisms or by exospheric ion concentration. This feature has the advantage that the results are quantitative and relatively unambiguous in terms of the amount of an element present beneath the surface, with relatively small corrections needed for matrix effects due to the abundances of other elements. Obviously it would be advantageous to know not only the total amount of an element present but also its concentration in minerals. To some extent we shall be able to do this by combining data from the GRS with data from the UltraViolet and Visible Spectrometer (UVVS) and Visible and InfraRed Spectrograph (VIRS) sensors on the MASCS instrument (McClintock and Lankton 2007). The GRS measures gamma rays during both day and night and relies on long integration intervals to build adequate statistics. The nature of the GRS data is such that it is possible to make predictions of expected statistical uncertainties in the data with an assumed model composition. Table 1 summarizes the results of one such calculation (Solomon et al. 2001). The accuracies shown are sufficient to address the MESSENGER science objectives and are discussed in more detail later. Each element emits one or more gamma rays with a characteristic energy that is detected and identified by the GRS. The slight model-dependence of the result is sensitive mostly to uncertainties in the flux of neutrons, which are the excitation source for most of the gamma rays. The gamma rays from K and Th are due to radioactive decay, and they are not subject to the uncertainties in the neutron flux.

On Mars Odyssey it was possible to detect H via both gamma rays and neutrons from the Martian surface (Boynton et al. 2002; Mitrofanov et al. 2002; Feldman et al. 2004b). In the case of MESSENGER, however, the much lower expected H content—hundreds of ppm expected from the solar wind versus thousands of ppm at Mars from water as ice or water of hydration in minerals (Boynton et al. 2002, 2007; Feldman et al. 2004b)—and the shorter integration times close to the planet are expected to put H below the detection limit of the GRS. The prospects of GRS-detectable quantities of water-ice-equivalent H in the top 30 cm of Mercury's regolith, even in the permanently shadowed regions where water ice has been proposed to reside, are not strong.

Table 1 Expected statistical uncertainties of MESSENGER GRS data for selected elements

Element	Model composition	Statistical uncertainty[a]
Mg	22.6%	0.8%
Si	21.5%	0.8%
K	0.12%	0.005%
Ca	4.0%	1.8%
Ti	1.0%	0.2%
Fe	2.3%	0.4%
Th	2.0 ppm	0.2 ppm

[a]Absolute uncertainty (1-σ) for an eight-hour integration time. A measurement time of eight hours is considered the minimum necessary to carry out coarse mapping of surface elemental composition at spacecraft altitudes less than 1,000 km over Mercury's northern hemisphere (with perhaps tens of to more than a hundred pixels, depending on GRS element sensitivity). The total measurement time at such altitudes is only about 137 hours for the entire nominal mission. Table modified from Solomon et al. (2001)

4.1.2 NS

The NS, like the GRS, uses a nuclear-based technique, and it is sensitive to elemental abundances independent of their chemical or mineral associations. As mentioned, interpretation of the NS data in terms of elemental abundances is somewhat more model-dependent than that of the GRS data, because many elements can have a significant effect on the energy spectrum of the neutrons coming from Mercury's surface (Feldman et al. 2000a), and there are no spectral lines unambiguously identifying specific elements. The energy spectrum is generally divided into just three energy bands, thermal, epithermal, and fast, and from the flux in these bands the abundances of some elements can be estimated (Feldman et al. 1998, 2000a).

Unlike the GRS and the XRS (discussed later), it is more difficult to calculate quantitative detection limits a priori for the NS. The hydrogen abundance, however, can be determined with relatively little ambiguity from the neutron flux, because the equal masses of the hydrogen nucleus and the neutron allow H to "downscatter" neutrons to lower energies much more effectively than any other element. Polar hydrogen deposits were mapped at the Moon (Feldman et al. 2000b), and potentially the most significant contribution of the NS experiment at Mercury will be the establishment of the presence of H in the polar regions (Feldman et al. 1997) and the mapping of its abundance over most of the northern hemisphere. High abundances of H near the permanently shadowed craters at high northern latitudes will be strongly suggestive of near-surface water ice in these craters (and by inference in craters at high southern latitudes).

Though the elliptical orbit of MESSENGER is biased such that the periapsis is over northern latitudes, the distance from the north pole is such that the permanently shadowed craters are a small fraction of the instrument footprint (Solomon et al. 2001). This orbit will make it difficult to detect the signal from any polar ice deposits unless they are present in very high amounts or cover a significant fraction of the field of view of the NS. At this point it is unclear whether the MESSENGER NS will see a sufficiently strong signal to give confidence that hydrogen-rich areas have been found.

It is possible that the NS may also be able to map abundances of Gd and Sm. This technique depends on the very high thermal-neutron cross-sections of these elements and the fact that their presence will lower the thermal neutron flux measured in orbit compared

Table 2 Expected counting times and spatial resolutions of MESSENGER XRS data for selected elements

Element	Assumed abundances	Counting time		Spatial resolution (km)	
		Quiet Sun	Active Sun	Quiet Sun	Active Sun
Mg	22.5%	11 min	3 min	400	200
Al	3.0%	8 hr	50 min	2,800	900
Si	21.5%	18 min	3 min	500	200
Ca	4.0%	–	10 s		60
Ti	1.0%	–	3 min		200
Fe	2.3%	–	80 s		150
S	4.0%	–	50 s		120

Note: Quiet Sun and Active Sun refer to periods with and without solar flares. Counting times and the derived spatial resolution are those required to achieve a statistical uncertainty of 10% of the amount of element present

with that expected in their absence. For this approach to work, good knowledge of the other major neutron absorbers, e.g., Fe and Ti, is required, as are high abundances of Sm and Gd. An example where such a measurement was successful is the Moon. There maps of Gd+Sm were made from data obtained by the Lunar Prospector Neutron Spectrometer (Elphic et al. 2000; Maurice et al. 2004).

4.2 XRS

As with the data from the GRS sensor, the elemental abundances determined from the XRS instrument (Al, Mg, Si, S, Ca, Ti, and Fe) represent bulk values for surface material without regard to the distribution among different minerals. However, the relative elemental abundances will permit inference of plausible mineral type when combined with information from the other instruments. The XRS has much better spatial resolution than the GRS, but it determines abundances in the upper hundreds of microns versus tens of centimeters for the GRS, and thus XRS data must be interpreted in the context of possible local concentration of exospheric species. However, XRS data probe beneath many of the typical space weathering coatings that are known to affect the interpretations of near-infrared spectral data (Noble et al. 2001; Noble and Pieters 2003). The nature of the XRS data, as for GRS data, is such that it is possible to make model-dependent predictions of expected statistical uncertainties in the data. These calculations, given in detail by Schlemm et al. (2007), are summarized in Table 2.

The sensitivity of the XRS depends strongly on the activity of the Sun. MESSENGER will be in orbit about Mercury near the time of solar maximum, so we may expect a large amount of solar flare activity, which significantly shortens the time of data collection for a given statistical precision. Table 2 shows the extent of this reduction in data collection times. The calculation for active Sun is made for an M-class flare. The derived spatial resolution in Table 2 is calculated assuming an average orbital altitude of 560 km in the northern polar zone (Solomon et al. 2001).

4.3 MASCS and MDIS

The UVVS and VIRS sensors on MESSENGER's MASCS instrument are well suited to make spectral maps of Mercury's surface from 115 to 1,450 nm at a spatial resolution of less than 10 km near periapsis. The UVVS will map spectral features at ultraviolet (UV) and visible wavelengths that are diagnostic of mineralogy, especially discerning between types of feldspars, pyroxenes, olivines, and sulfides using distinct differences in spectral slope and features between 115 and 600 nm (Wagner et al. 1987). The VIRS sensor will map 0.8–1.2 μm Fe^{2+} charge transfer absorptions associated with the M2 lattice sites in common silicates such as pyroxene and olivine. Such measurements will determine the FeO abundance in the regolith and permit mapping of its variation with geological unit. Other features between 200 and 1,450 nm associated with sulfides and possible additional reduced regolith materials (Burbine et al. 2002) will be sought and mapped. Further, the UVVS sensor will be used to search for S and OH components in the exosphere. A discovery of either or both of these components at high latitudes could indicate the composition of the polar deposits even if the sensitivity for detection of H via the NS sensor is not sufficient. If neither of these components is detected, the results could provide support for the suggestion that the depolarized signals are a signature of cold silicates with low loss (high transmission) at radar wavelengths (Starukhina 2001). MDIS will give detailed color imaging and ratios that will be used to understand the degree of space weathering of the surface, which is critical to our ability to interpret spectral features found in the MASCS data.

4.4 Synergy between MESSENGER Instruments

One of the important characteristics of the MESSENGER payload with respect to the geochemical objectives is the ability to combine the data from several instruments. As noted earlier, both the GRNS and XRS determine elemental abundances independent of the chemical or mineralogical host for the element. MDIS and MASCS will both measure spectral slope and color ratios, map the major FeO absorption bands if present, and make distinctions among feldspar, pyroxene, and olivine compositions if space weathering has not obscured the spectral features that laboratory measurements have shown to be important. For example, as noted, both olivine and pyroxene have FeO bands that can be used to estimate oxidized Fe content. These species are likely to be the principal silicate minerals to contain Fe, so the GRS and XRS data on Fe can be compared with the mineralogical data to constrain the interpretation. Because of extensive space weathering of Mercury's surface, there may be Fe in metallic form ($npFe^0$), although ground-based microwave measurements indicate that such material must be much less abundant than on the surface of the Moon. Comparison of FeO determined via MASCS and total Fe determined via GRS and XRS could yield the amount of $npFe^0$ in the regolith. Likewise, determining the variation of Si and the Th/K ratio across the surface from the GRNS instrument (Boynton et al. 2007; Taylor et al. 2006, respectively, for Mars) will help to infer the nature of magmas and volcanic deposits on Mercury's surface.

5 Addressing MESSENGER Science Objectives

5.1 The Origin of Mercury's High Metal/Silicate Ratio

Data from the MESSENGER instruments will distinguish among the hypotheses for the high metal fraction of Mercury: aerodynamic drag by solar nebular gas (Weidenschilling 1978), preferential vaporization of silicates in a hot solar nebula followed by removal by a strong solar wind (Cameron 1985; Fegley and Cameron 1987), or selective removal of silicate as a result of a giant impact (Wetherill 1988; Benz et al. 1988). The different instruments on MESSENGER are well suited to address this question.

Under the aerodynamic drag hypothesis, the bulk composition of Mercury should be similar to that predicted on the basis of thermodynamic equilibrium in the solar nebula (Lewis 1988) except for a higher metal content. Subsequent differentiation of the planet would then leave the surface with a crust enriched in Ca, Al, and incompatible elements, e.g., K and Th, overlying the mafic mantle and metallic core. Under the preferential vaporization hypothesis, the surface of Mercury should be greatly reduced in volatile elements (e.g., K) and enriched in refractory elements (e.g., Th). Under the giant impact hypothesis, the removal of an early crust would leave a surface composition appropriate to the formation of a post-impact crust from partial melting of a generally depleted mantle. We would expect depletion of Ca, Al, and the incompatible elements that would have been enriched under the preferential vaporization hypothesis.

Clearly the elements determined by the GRNS and XRS instruments are well suited to testing these hypotheses. The NS can map regions of relatively high Fe, Ti, or Gd and Sm abundances. Depending on the activity of the Sun, coverage of Ca and Al may be limited with the XRS, but we can combine the XRS data with MASCS observations to advantage. MASCS will be able to map regions of the planet with different mineralogical composition, so even with only limited XRS coverage of Ca and Al we can tie those data to other regions through mineralogy and color imaging to estimate global abundances.

5.2 The Geological History of Mercury

It is expected that MIDIS imaging and spectral reflectance measurements will probably make the biggest contribution toward understanding the geological history of Mercury. However, the elemental abundances may provide significant insight into processes that may have been important in the formation of Mercury's surface as well as providing information useful for understanding the relative chronology of surface units. For example, it would be of interest to know if different volcanic units have different contents of incompatible elements and Si. If so, units having lower K and Th might indicate later evolution of basalts from a mantle already depleted by earlier melt extraction.

5.3 Mercury's Polar Deposits

Mercury's polar deposits appear, on the basis of Earth-based radar, to be present only within permanently shadowed crater floors, and if so their characterization by MESSENGER may be challenging. Differences between the results from radar imaging of Mercury and the Moon (Harmon 1997; Campbell et al. 2003) highlight some of the uncertainties in the interpretation of the origin of these deposits. Such observations argue for more prominent deposits of ice or other radar-anomalous species in the polar regions of Mercury, despite the facts that the area of the polar zone that can retain water ice is larger on the Moon than on Mercury (Ingersoll et al. 1992; Salvail and Fanale 1994) and the retention of water vapor released by comets and meteorites after impact should be greater at the Moon than on Mercury because of lesser impact speeds at the Moon (Moses et al. 1999). By way of counterbalance, observations by *Solar and Heliospheric Observatory* (SOHO) coronagraphs have revealed approximately 700 Sun-grazing comets since 1979 (Marsden 2005). Such bodies cross Mercury's orbital plane, so it is likely that some of this population of objects impact Mercury. Any water so delivered would quickly migrate to the polar regions.

If large areas of the north polar region of Mercury have high contents of ice or sulfur, the NS and the GRS should be able to detect this material and determine its composition.

5.4 Mercury's Volatile Species and Their Sources and Sinks

Mercury's tenuous exosphere contains very small amounts of H, He, Na, O, K, and Ca (Hunten et al. 1988; Bida et al. 2000). Although the exosphere will be studied primarily via the MASCS and Energetic Particle and Plasma Spectrometer (EPPS) (Andrews et al. 2007) instruments, the GRNS and XRS instruments can help to understand the nature of the sources and sinks of these and any newly discovered species. One of the exospheric constituents, K, is the element for which the GRS has the highest sensitivity. A GRS high-resolution map of the abundance of K in crustal material is expected. The abundances and variability of K found in the exosphere can then be related to different regions on the surface to see if there is any correlation between exospheric activity and surface composition. From Earth-based observations, Sprague et al. (1990) found enhanced column abundances of K over the Caloris basin and the antipodal hilly and lineated terrain. It may be that extensional troughs on the floor of Caloris (Watters et al. 2005) and the chaotic nature of hilly and lineated terrain (Strom and Sprague 2003) combined with the high diurnal temperature variations near Mercury's hot longitudes provide the conditions required to release K from crustal materials to the exosphere.

6 Conclusions

The instruments on the MESSENGER spacecraft are well suited to address the scientific objectives of the mission and many of the questions that have resulted from ground-based observations made during the past several years. A combination of data from several instruments will be needed to address fully several of these questions. MESSENGER is exemplary of the strong benefits of missions in NASA's Discovery Program, under which comprehensive investigations are formulated with substantial synergy among the different elements of the mission built from the outset into every element of mission and spacecraft design.

Acknowledgements This work has been supported by the NASA Discovery Program through contracts NASW-00002 to the Carnegie Institution of Washington and NAS5-97271 to The Johns Hopkins University Applied Physics Laboratory. Sprague gratefully acknowledges support from the National Science Foundation through grant AST-0406796000 for studies of Mercury's surface and atmosphere and support from the Infrared Telescope Facility operated by the University of Hawaii.

References

G.B. Andrews et al., Space Sci. Rev. (2007, this issue). doi:10.1007/s11214-007-9272-5
W. Benz, W.L. Slattery, A.G.W. Cameron, Icarus **74**, 516–528 (1988)
T.A. Bida, R.M. Killen, T.H. Morgan, Nature **404**, 159–161 (2000)
D.T. Blewett, P.G. Lucey, B.R. Hawke, G.G. Ling, M.S. Robinson, Icarus **129**, 217–231 (1997)
D.T. Blewett, B.R. Hawke, P.G. Lucey, Meteorit. Planet. Sci. **37**, 1245–1254 (2002)
W.V. Boynton et al., Science **297**, 81–85 (2002)
W.V. Boynton et al., Space Sci. Rev. **110**, 37–83 (2004)
W.V. Boynton et al., J. Geophys. Res. **112** (2007, in press)
T.H. Burbine, L.R. Nittler, G.K. Benedix, E.A. Cloutis, T.L. Dickinson, Meteorit. Planet. Sci. **37**, 1233–1244 (2002)
B.J. Butler, J. Geophys. Res. **102**, 19283–19291 (1997)
B.J. Butler, D.O. Muhleman, M.A. Slade, J. Geophys. Res. **98**, 15003–15023 (1993)
A.G.W. Cameron, Icarus **64**, 285–294 (1985)
B.A. Campbell, D.B. Campbell, J.F. Chandler, A.A. Hine, M.C. Nolan, P.J. Perillat, Nature **426**, 137–138 (2003)

B. Cooper, A. Potter, R. Killen, T. Morgan, J. Geophys. Res. **106**, 32803–32814 (2001)
D.H. Crider, R.M. Killen, Geophys. Res. Lett. **32**, L12201 (2005). doi:10.1029/2005GL022689
D.L. Domingue et al., Space Sci. Rev. (2007, this issue). 10.1007/s11214-007-9260-9
L. Dones, B. Gladman, H.J. Melosh, W.B. Tonks, H.F. Levison, M. Duncan, Icarus **142**, 509–524 (1999)
R.C. Elphic et al., J. Geophys. Res. **105**, 20333–20345 (2000)
J.P. Emery, A.L. Sprague, F.C. Witteborn, J.E. Colwell, R.W.H. Kozlowski, D.H. Wooden, Icarus **136**, 104–123 (1998)
B. Fegley Jr., A.G.W. Cameron, Earth Planet. Sci. Lett. **82**, 207–222 (1987)
W.C. Feldman, B.L. Barraclough, C.J. Hansen, A.L. Sprague, J. Geophys. Res. **102**, 25,565–25,574 (1997)
W.C. Feldman et al., Science **281**, 1489–1493 (1998)
W.C. Feldman, D.J. Lawrence, R.C. Elphic, D.T. Vaniman, D.R. Thomsen, B.L. Barraclough, J. Geophys. Res. **105**, 20347–20363 (2000a)
W.C. Feldman et al., J. Geophys. Res. **105**, 4175–4195 (2000b)
W.C. Feldman et al., J. Geophys. Res. **107**, 5016 (2002). doi:10.1029/2001JE001506
W.C. Feldman et al., J. Geophys. Res. **109**, E07S06 (2004a). doi:10.1029/2003JE002207
W.C. Feldman et al., J. Geophys. Res. **109**, E09006 (2004b). doi:10.1029/2003JE002160
W.C. Feldman et al., J. Geophys. Res. **110**, E11009 (2005). doi:10.1029/2005JE002452
B.J. Gladman, J.A. Burns, M. Duncan, P. Lee, H.S. Levison, Science **271**, 1387–1392 (1996)
R.E. Gold et al., Planet. Space Sci. **49**, 1467–1479 (2001)
R.M. Goldstein, Science **168**, 467–468 (1970)
J.O. Goldsten et al., Space Sci. Rev. (2007, this issue). doi:10.1007/s11214-007-9262-7
B. Hapke, J. Geophys. Res. **106**, 10039–10073 (2001)
J.K. Harmon, Adv. Space Res. **19**, 1487–1496 (1997)
J.K. Harmon, M.A. Slade, Science **258**, 640–642 (1992)
J.K. Harmon, M.A. Slade, R.A. Velez, A. Crespo, M.J. Dryer, J.M. Johnson, Nature **369**, 213–215 (1994)
J.K. Harmon, P.J. Perillat, M.A. Slade, Icarus **149**, 1–15 (2001)
J.K. Harmon, M.A. Slade, B.J. Butler, J.W. Head III, M.S. Rice, D.B. Campbell, Icarus **187**, 374–405 (2007)
S.E. Hawkins III, et al., Space Sci. Rev. (2007, this issue). doi:10.1007/s11214-007-9266-3
G.R. Hunt, R.K. Vincent, J. Geophys. Res. **73**, 6039–6046 (1968)
D.M. Hunten, T.H. Morgan, D. Shemansky, in *Mercury*, ed. by F. Vilas, C.R. Chapman, M.S. Matthews (University of Arizona Press, Tucson, 1988), pp. 562–612
A.P. Ingersoll, T. Svitek, B.C. Murray, Icarus **100**, 40–47 (1992)
A.J. Irving, et al., Eos Trans. Am. Geophys. Union **86** (Fall Meeting suppl.), F1198–F1199 (2005)
R. Jeanloz, D.L. Mitchell, A.L. Sprague, I. de Pater, Science **268**, 1455–1457 (1995)
E. Kallio, P. Janhunen, Geophys. Res. Lett. **30**, 1877 (2003). doi:10.1029/2003GL017842
R.M. Killen, J. Benkhoff, T.H. Morgan, Icarus **125**, 195–211 (1997)
P.L. Koehn, A.L. Sprague, Planet. Space Sci. **55**, 1530–1540 (2007)
J.S. Lewis, in *Mercury*, ed. by F. Vilas, C.R. Chapman, M.S. Matthews (University of Arizona Press, Tucson, 1988), pp. 651–667
L.M. Logan, G.R. Hunt, J. Geophys. Res. **75**, 6539–6548 (1970)
S.G. Love, K. Keil, Meteorit. Planet. Sci. **30**, 269–278 (1995)
B.G. Marsden, Annu. Rev. Astron. Astrophys. **43**, 75–102 (2005)
S. Maurice, D.J. Lawrence, W.C. Feldman, R.C. Elphic, O. Gasnault, J. Geophys. Res. **109**, E07S04 (2004). doi:10.1029/2003JE002208
W. McClintock, M.R. Lankton, Space Sci. Rev. (2007, this issue). doi:10.1007/s11214-007-9264-5
T.B. McCord, R.N. Clark, J. Geophys. Res. **84**, 7664–7668 (1979)
H.J. Melosh, W.B. Tonks, Meteoritis **28**, 398 (1993)
D.L. Mitchell, I. de Pater, Icarus **110**, 2–32 (1994)
D.W. Mittlefehldt, M. Killgore, M.T. Lee, Meteorit. Planet. Sci. **37**, 345–369 (2002)
D. Mitrofanov et al., Science **297**, 78–81 (2002)
J.I. Moses, K. Rawlins, K. Zahnle, L. Dones, Icarus **137**, 197–221 (1999)
S.K. Noble, C.M. Pieters, Sol. Syst. Res. **37**, 34–39 (2003)
S.K. Noble et al., Meteorit. Planet. Sci. **36**, 31–42 (2001)
D.A. Paige, S.E. Wood, A.R. Vasavada, Science **258**, 643–646 (1992)
A.E. Potter, Geophys. Res. Lett. **22**, 3289–3292 (1995)
A.E. Potter, T.H. Morgan, Science **248**, 835–838 (1990)
A.E. Potter, T.H. Morgan, Planet. Space Sci. **45**, 95–100 (1997)
T.H. Prettyman et al., J. Geophys. Res. **109**, E05001 (2004). doi:10.1029/2003JE002139
M.S. Robinson, P.G. Lucey, Science **275**, 197–200 (1997)
M.S. Robinson, G.J. Taylor, Meteorit. Planet. Sci. **36**, 841–847 (2001)
J.W. Salisbury, B. Hapke, J.W. Eastes, J. Geophys. Res. **92**, 702–710 (1987)

J.W. Salisbury, L.S. Walter, N. Vergo, D.M. D'Aria, *Infrared (2.1–25 μm) Spectra of Minerals* (Johns Hopkins University Press, Baltimore, 1991), 267 pp
J.R. Salvail, F.P. Fanale, Icarus **111**, 441–455 (1994)
A.G. Santo et al., Planet. Space Sci. **49**, 1481–1500 (2001)
S. Sasaki, E. Kurahashi, Adv. Space Res. **33**, 2152–2155 (2004)
C.E. Schlemm et al., Space Sci. Rev. (2007, this issue). doi:10.1007/s11214-007-9248-5
M.A. Slade, B.J. Butler, D.O. Muhleman, Science **258**, 635–640 (1992)
S.C. Solomon et al., Planet. Space Sci. **49**, 1445–1465 (2001)
A.L. Sprague, T.L. Roush, Icarus **133**, 174–183 (1998)
A.L. Sprague, R.W.H. Kozlowski, D.M. Hunten, Science **249**, 1140–1143 (1990)
A.L. Sprague, R.W.H. Kozlowski, F.C. Witteborn, D.P. Cruikshank, D.H. Wooden, Icarus **109**, 156–167 (1994)
A.L. Sprague, D.M. Hunten, K. Lodders, Icarus **118**, 211–215 (1995)
A.L. Sprague et al., Icarus **129**, 506–527 (1997)
A.L. Sprague, W.J. Schmitt, R.E. Hill, Icarus **135**, 60–68 (1998)
A.L. Sprague, J.P. Emery, K.L. Donaldson, R.W. Russell, D.K. Lynch, A.L. Mazuk, Meteorit. Planet. Sci. **37**, 1255–1268 (2002)
P.D. Spudis, J.E. Guest, in *Mercury*, ed. by F. Vilas, C.R. Chapman, M.S. Matthews (University of Arizona Press, Tucson, 1988), pp. 118–164
L.V. Starukhina, J. Geophys. Res. **106**, 14701–14710 (2001)
R.G. Strom, A.L. Sprague, *Exploring Mercury the Iron Planet* (Springer-Praxis, Chichester, 2003), 216 pp
G.J. Taylor et al., J. Geophys. Res. **111**, E03S06 (2006). doi:10.1029/2006JE002676
R.L. Tokar et al., Geophys. Res. Lett. **29**, 1904 (2002). doi:10.1029/2002GL015691
A.R. Vasavada, D.A. Paige, S.E. Wood, Icarus **141**, 179–193 (1999)
F. Vilas, Icarus **64**, 133–138 (1985)
F. Vilas, in *Mercury*, ed. by F. Vilas, C.R. Chapman, M.S. Matthews (University of Arizona Press, Tucson, 1988), pp. 59–76
F. Vilas, M.A. Leake, W.W. Mendell, Icarus **59**, 60–68 (1984)
J. Wagner, B. Hapke, E. Wells, Icarus **69**, 14–28 (1987)
J. Warell, Icarus **156**, 303–317 (2002)
J. Warell, Icarus **161**, 199–222 (2003)
J. Warell, D.T. Blewett, Icarus **168**, 257–276 (2004)
J. Warell, A.L. Sprague, J.P. Emery, R.W.H. Kozlowski, A. Long, Icarus **180**, 281–291 (2006)
T.R. Watters, F. Nimmo, M.S. Robinson, Geology **33**, 669–672 (2005)
S.J. Weidenschilling, Icarus **35**, 99–111 (1978)
G.W. Wetherill, in *Mercury*, ed. by F. Vilas, C.R. Chapman, M.S. Matthews (University of Arizona Press, Tucson, 1988), pp. 670–691
T.H. Zurbuchen, P. Koehn, L.A. Fisk, T. Gombosi, G. Gloeckler, K. Kabin, Adv. Space Res. **33**, 1884–1889 (2004)

Space Sci Rev (2007) 131: 105–132
DOI 10.1007/s11214-007-9265-4

The Geophysics of Mercury: Current Status and Anticipated Insights from the MESSENGER Mission

Maria T. Zuber · Oded Aharonson · Jonathan M. Aurnou · Andrew F. Cheng · Steven A. Hauck II · Moritz H. Heimpel · Gregory A. Neumann · Stanton J. Peale · Roger J. Phillips · David E. Smith · Sean C. Solomon · Sabine Stanley

Received: 24 July 2006 / Accepted: 10 August 2007 / Published online: 18 October 2007

M.T. Zuber (✉) · G.A. Neumann · S. Stanley
Department of Earth, Atmospheric and Planetary Sciences, Massachusetts Institute of Technology, Cambridge, MA 02139-4307, USA
e-mail: Zuber@mit.edu

O. Aharonson
Division of Geological and Planetary Sciences, California Institute of Technology, Pasadena, CA 91125, USA

J.M. Aurnou
Department of Earth and Space Sciences, University of California, Los Angeles, CA 90095, USA

A.F. Cheng
The Johns Hopkins University Applied Physics Laboratory, Laurel, MD 20723-6099, USA

S.A. Hauck II
Department of Geological Sciences, Case Western Reserve University, Cleveland, OH 44106, USA

M.H. Heimpel
Department of Physics, University of Alberta, Edmonton, AB, T6G 2J1, Canada

G.A. Neumann · D.E. Smith
Solar System Exploration Division, NASA Goddard Space Flight Center, Greenbelt, MD 20771, USA

S.J. Peale
Department of Physics, University of California, Santa Barbara, CA 93106, USA

R.J. Phillips
Department of Earth and Planetary Sciences, Washington University, St. Louis, MO 63130, USA

S.C. Solomon
Department of Terrestrial Magnetism, Carnegie Institution of Washington, Washington, DC 20015, USA

S. Stanley
Department of Physics, University of Toronto, Toronto, ON, M5S 1A7, Canada

Abstract Current geophysical knowledge of the planet Mercury is based upon observations from ground-based astronomy and flybys of the Mariner 10 spacecraft, along with theoretical and computational studies. Mercury has the highest uncompressed density of the terrestrial planets and by implication has a metallic core with a radius approximately 75% of the planetary radius. Mercury's spin rate is stably locked at 1.5 times the orbital mean motion. Capture into this state is the natural result of tidal evolution if this is the only dissipative process affecting the spin, but the capture probability is enhanced if Mercury's core were molten at the time of capture. The discovery of Mercury's magnetic field by Mariner 10 suggests the possibility that the core is partially molten to the present, a result that is surprising given the planet's size and a surface crater density indicative of early cessation of significant volcanic activity. A present-day liquid outer core within Mercury would require either a core sulfur content of at least several weight percent or an unusual history of heat loss from the planet's core and silicate fraction. A crustal remanent contribution to Mercury's observed magnetic field cannot be ruled out on the basis of current knowledge. Measurements from the MESSENGER orbiter, in combination with continued ground-based observations, hold the promise of setting on a firmer basis our understanding of the structure and evolution of Mercury's interior and the relationship of that evolution to the planet's geological history.

Keywords Mercury · MESSENGER · Core · Rotational state · Magnetic dynamos · Thermal history

1 Introduction

Mercury's internal structure and evolution collectively constitute one of the solar system's most intriguing geophysical enigmas. In terms of its size and surface geology, Mercury is often compared with Earth's Moon. But in striking contrast to the Moon, which is depleted in iron and has a small (if any) metallic core, Mercury's size and mass (Anderson et al. 1987, 1996) indicate a high metal/silica ratio and a metallic mass fraction more than twice that of Earth, Venus, and Mars (Wood et al. 1981). In addition, while the Moon is believed to have cooled rapidly subsequent to accretion (Zuber et al. 1994; Neumann et al. 1996), Mercury appears to possess a liquid outer core (Margot et al. 2007). Such an internal structure is puzzling, as simple thermal evolution models (Cassen et al. 1976; Solomon et al. 1981; Schubert et al. 1988) predict that a pure iron or iron-nickel core should have cooled and solidified by now. A liquid core would survive if there is a sufficient amount of a light alloying element such as sulfur to lower the melting point (Schubert et al. 1988).

Mercury's internal structure and its thermal evolution ultimately must be reconciled with the planet's surface geology. Mercury has a heavily cratered surface (Murray et al. 1974; Murray 1975; Trask and Guest 1975) with ancient compressional tectonic structures that have been taken to imply global contraction (Strom et al. 1975; Watters et al. 1998) associated with secular cooling (Siegfried and Solomon 1974). Ancient intercrater plains and somewhat younger smooth plains of possible volcanic origin (Strom et al. 1975; Trask and Strom 1976; Robinson and Lucey 1997) constrain the early history of the mantle and crustal magmatism.

The evolution of Mercury's core state with time has implications for the planet's spin evolution. Mercury currently displays a 3 : 2 spin–orbit resonance, and the presence of a fluid core would have enhanced considerably its probability of capture into this state (Counselman 1969).

It could be argued that the formation and dynamics of Mercury's core has had a greater influence on the geophysical evolution of the planet than for any other terrestrial planetary body. Consequently in this paper we treat the core as a point of focus as we review current understanding of Mercury's geophysics. In the context of this review we emphasize recent advances in measuring and interpreting Mercury's rotational state, in interpreting existing magnetic observations, and in convective modeling of the planet's mantle and core. We describe how anticipated future observations from NASA's MErcury Surface, Space ENvironment, Geochemistry, and Ranging (MESSENGER) mission will provide a means of unraveling the unusual characteristics of Mercury's evolution.

2 Physical and Chemical Characteristics

2.1 Geophysical Parameters

The size, shape, and mass of Mercury have been measured from radio tracking of the Mariner 10 spacecraft and Earth-based radar ranging. Current knowledge of these parameters, summarized in Table 1, is based on historical observations as well as more recent reanalysis of combined data sets (Anderson et al. 1987, 1996). Mercury has the largest uncompressed density of the planets (Ringwood 1979) and thus the largest metal/silicate ratio, exhibiting a fractional core mass $M_c/M = 0.65$ (Urey 1951; Siegfried and Solomon 1974), where M is Mercury's total mass. Although the interior has not been sampled, no heavy element other than iron has a cosmic abundance that can account for the observed density. It is believed that the planet has differentiated into an iron-nickel core of radius $R_c/R \sim 0.75$ (Siegfried and Solomon 1974), where R is Mercury's radius. The gravitational flattening (J_2) of Mercury measured by Mariner 10 exceeds the value consistent with hydrostatic equilibrium at Mercury's slow rotation rate. Thus the present J_2 must be either "frozen in" from a more rapid rotation rate in the past (Lambeck and Pullan 1980), or it must be dominated by nonhydrostatic contributions such as the response of a finite-strength lithosphere to the formation and modification of impact basins. Unfortunately the value of J_2 does not in and of itself yield a useful constraint on plausible geochemical models of the radial density distribution of the interior (Solomon 1976).

2.2 Bulk Composition

Beyond inference on fractional iron abundance from the mean density there is no direct information on Mercury's bulk composition (Wood et al. 1981). Solar system condensation models (Lewis 1972) suggest that Mercury is enriched in refractory relative to volatile

Table 1 Shape and bulk properties of Mercury

Parameter	Value
Mass, 10^{23} kg	3.302
Mean radius, km	$2{,}440 \pm 1$
Displacement of center of figure from center of mass in equatorial plane[1], m	640 ± 78
Mean density, kg m^{-3}	5,427
Uncompressed density[2], kg m^{-3}	5,300
Surface gravity, m s^{-2}	3.70
J_2	$(6 \pm 2) \times 10^{-5}$
C_{22}	$(1 \pm 0.5) \times 10^{-5}$

From Yoder (1995) except where noted

[1]Anderson et al. (1996); [2]Wood et al. (1981)

The C_{22} coefficient is unnormalized

elements compared with other terrestrial planets, and an equilibrium condensation scenario (Lewis 1972, 1973) suggests that the silicate fraction of Mercury is dominated by magnesium-rich pyroxene (Wood et al. 1981). Disk-averaged Earth-based visible and infrared spectral observations are consistent with the average composition of anorthositic materials in the lunar highlands (Sprague et al. 1994). The lack of persistent absorption bands identified with mafic minerals limits the average FeO content of surficial materials to less than a few percent (McCord and Clark 1979; Wood et al. 1981; Jeanloz et al. 1995). Explanations for Mercury's high metal content include the differential response of iron and silicate grains to gas drag in the early solar nebula (Weidenschilling 1978), preferential vaporization of silicate relative to metal in the hot solar nebula (Cameron 1985; Fegley and Cameron 1987), and preferential removal of silicate by a giant impact that occurred after Mercury had differentiated (Wetherill 1988). Implications of each of these hypotheses for the chemistry of Mercury's silicate fraction are described at greater length in a companion paper (Boynton et al. 2007).

2.3 Crustal and Mantle Structure

The crustal thickness on Mercury is not presently known. There are no observed surface structural features with regular length scales associated with a subsurface chemical or rheological discontinuity (Zuber 1987) that hint at the existence of a low-density crust (Solomon et al. 1981). Anderson et al. (1996) interpreted the offset between the centers of mass and figure offset for Mercury (Table 1) as indicative of a hemispheric asymmetry in crustal thickness. If interpreted in this context, along with assumptions about internal densities and compositions, the maximum crustal thickness difference between the imaged and unimaged hemispheres of Mercury would be about 13 km. Of course the interpretation of internal structure using the information currently available is highly nonunique, so the use of geophysical and geochemical data sets from MESSENGER will advance considerably the constraints on internal structure.

Barring the existence of diagnostic tectonic structures on the unimaged parts of Mercury's surface, the best chance of estimating crustal and mantle structure may come from future combined analysis of topography (Cavanaugh et al. 2007) and gravity (Srinivasan et al. 2007). Inversions in the spatial and frequency domain will be used to infer crustal thickness and effective elastic lithosphere thickness (cf. Zuber et al. 1994; Simons et al. 1997) on spatial scales comparable with data resolution.

2.4 Orbital and Rotational Parameters

Mercury's modern orbital ephemeris has been developed from radar time-delay and Doppler observations from the Goldstone, Arecibo, and Haystack radio observatories (Harmon et al. 1986; Harmon and Campbell 1988) as well as the three flybys of the Mariner 10 spacecraft (Standish 1990; Standish et al. 1992). These data, combined with radar tracking of the surface, revealed that Mercury exhibits a 3 : 2 spin–orbit resonance (Pettengill and Dyce 1965) that is a consequence of tidal dissipation (Colombo 1965; Colombo and Shapiro 1966; Goldreich and Peale 1966). Orbital information has also been used to refine the perihelion advance of Mercury (Shapiro et al. 1972; Roseveare 1982) predicted by general relativity (Einstein 1916). A summary of Mercury's orbital and rotational parameters is given in Table 2.

Table 2 Orbital and rotational parameters of Mercury

Parameter	Value
Semi-major axis, 10^6 km	57.91
Orbital eccentricity	0.2056
Perihelion, 10^6 km	46.00
Aphelion, 10^6 km	69.82
Sidereal orbital period, Earth days	87.97
Sidereal rotation period, Earth days	58.65
Synodic period with respect to Earth, Earth days	115.88
Mean solar day, Earth days	175.94
Rotation rate $\omega \times 10^5$, s	0.124
Obliquity[1], arc minutes	2.1 ± 0.1

From Yoder (1995) except where noted

[1]Margot et al. (2007)

3 Surface Constraints on Thermal Evolution

The record of impact, volcanism, and tectonism preserved on Mercury's surface places important constraints on models of thermal evolution, and to some extent, on the rotational and orbital evolution. Here we review salient aspects of Mercury's surface geology relevant to the planet's geophysical evolution. Additional details on Mercury's geology can be found in a companion paper (Head et al. 2007).

3.1 Major Impact Basins

The largest impacts experienced by a planet during the late stages of accretion represent a significant source of accretional energy (Safronov 1978; Kaula 1979), establish that crustal formation predated the end of heavy impact bombardment of the solar system, and provide information on the thermomechanical structure of the lithosphere at the time of and subsequent to formation of preserved impact structures. At 1,300 km in diameter, Caloris is the largest well-preserved impact basin on Mercury; other significant basins include Beethoven (625 km in diameter) and Tolstoj (400 km in diameter) (McKinnon 1981). Caloris and Beethoven were only partially imaged by Mariner 10.

Application of models of basin formation and response to loading (Melosh 1978; Melosh and McKinnon 1978; McKinnon and Melosh 1980) constrained by observed deformation has been of limited value in reconstructing Mercury's early thermal state because of the absence of gravity and altimetry, limited imaging coverage of the largest basins, and the paucity of large basins so far identified. The rarity of multiring basins on Mercury in comparison with the Moon may indicate that Mercury's lithosphere was thicker and the planet substantially cooler in comparison with the Moon during late heavy bombardment (Melosh and McKinnon 1988). A rough calculation indicates a lithosphere thickness >100 km for Mercury (McKinnon 1981) in comparison with 25 to >75 km for the Moon (Solomon and Head 1980) for the time of plains loading shortly after the formation of the youngest basins. But such rapid cooling for Mercury is not easily reconciled with preliminary recent evidence for a present-day liquid outer core (Margot et al. 2007). It has also been suggested that early multiring basins were obliterated by viscous relaxation, intercrater plains formation, and subsidence of lithospheric ring blocks (Leake 1982; Melosh and McKinnon 1988). If the paucity of major basins relative to the Moon is ultimately confirmed, other possible explanations, such as the presence of a shallow core–mantle boundary that would limit basin depth and possibly reduce the topographic relaxation time of such structures, should also be investigated.

Shock waves associated with the formation of Caloris may be responsible for disruption of the surface that produced the hilly and lineated terrain at the antipode of the impact (Schultz and Gault 1975b; Hughes et al. 1977; Strom 1984). Early simulations have shown that for a planet-scale event such as Caloris, shock waves can combine constructively at the planet's surface antipodal to the impact (Schultz and Gault 1975a, 1975b; Boslough et al. 1996). The Caloris event could have produced vertical ground movement of about 1 km at the antipode (Hughes et al. 1977), which would require significant acceleration of the surface. Improved modeling may ultimately provide information on the state of Mercury's interior at the time of the largest impacts.

3.2 Volcanism

Mariner 10 images of the surface of Mercury do not show obvious evidence of primary volcanic landforms (Strom et al. 1975; Spudis and Prosser 1984). Some small volcanic structures may have been identified, including domes, rimless pits, crater floor mounds, lineaments, and contrasting crater floor/rim morphology (Dzurisin 1978; Malin 1978; Head et al. 1981) in generally coarse-resolution Mariner 10 images. Ambiguous interpretation of surface structures has been attributed to resolution effects (Head et al. 1981).

Mercury contains two major plains units that have been interpreted by some workers to be a consequence of surface volcanism: older intercrater plains and younger smooth plains (Trask and Strom 1976; Cintala et al. 1977; Strom 1977; Adams et al. 1981; Spudis and Guest 1988). The volcanic origin of the intercrater plains units has been debated, and an alternative interpretation is that these units consist of impact ejecta (Wilhelms 1976; Oberbeck et al. 1977). The intercrater plains correspond to gently rolling terrain between and surrounding areas of heavily cratered terrain and contain craters <10 km in diameter (Trask and Guest 1975). The intercrater plains and heavily cratered terrain have a complex stratigraphic relationship and are not clearly distinguishable in relative age (Trask and Guest 1975; Spudis and Guest 1988). It has been hypothesized that the plains preserve the record of an early resurfacing event (Murray 1975). The intercrater plains are volumetrically significant, obliterating craters smaller than 300–500 km in diameter (Spudis and Guest 1988).

Smooth plains (Strom et al. 1975; Trask and Guest 1975) account for about 15% of Mercury's imaged surface. These units have been viewed as analogous to lunar maria, with the most obvious difference from lunar deposits being the lack of a strong albedo contrast with surrounding cratered terrain (Adams et al. 1981). The smooth plains have albedos comparable to the brightest lunar maria, consistent with that of low-iron, low-titanium basalts (Hapke et al. 1975; Sprague et al. 1994). Combined analysis of opaque mineral abundance, iron content and soil maturity in recalibrated Mariner 10 color mosaics (Robinson and Lucey 1997) has provided the most compelling evidence thus far for a volcanic origin of certain plains units, on the grounds that color boundaries between plains units display lobate geometries consistent with emplacement of surface flows.

3.3 Tectonics

The absence of structural features indicative of plate tectonics argues strongly that Mercury, like the Moon, exhibited a single, continuous mechanical lithosphere from the time of heavy bombardment (Solomon 1978). Mercury's tectonic features are collectively a consequence of secular cooling of the planet, tidal forces, lithospheric loading, and limited local stresses. In further analogy to the Moon, the large Caloris basin contains what appear to be contractional (thrust) and extensional (graben) structures that are likely a consequence of basin loading and viscoelastic relaxation (Maxwell and Gifford 1980; Fleitout and Thomas 1982; Melosh and McKinnon 1988; Thomas et al. 1988). If smooth plains are volcanic in origin, fractures in the basin floor may represent the surface expression of conduits that enabled the upward transport of magma (Burke et al. 1981).

Unlike the Moon, the part of Mercury's surface imaged by Mariner 10 shows evidence for a regional to global distribution of tectonic landforms that include ridges, troughs, and lineaments of ambiguous origin (Dzurisin 1978; Burke et al. 1981; Melosh and McKinnon 1988). The most prominently expressed features are lobate scarps (Strom et al. 1975), which have been interpreted as the surface expressions of large-offset thrust faults (Strom et al. 1975; Dzurisin 1978; Strom 1979). The scarps are arcuate to quasi-linear, 20–500 km-long structures that crosscut various terrains. Early work suggested a more or less uniform distribution over the imaged part of Mercury's surface (Strom et al. 1975), which favored an origin due to global contraction. More recent mapping (Watters et al. 2004) shows that there is a higher density of scarps in the imaged hemisphere of Mercury at latitudes poleward of 50°S, which suggests that regional stresses also played a role. The observation that some scarps are disrupted by large craters has been interpreted to indicate that scarp formation was ongoing during the later stages of heavy bombardment (Burke et al. 1981), but the presence of scarps on the smooth plains indicates that thrust faulting continued after smooth plains emplacement (Strom 1979). The shortening associated with lobate scarps in the imaged hemisphere of Mercury implies horizontal surface strains of $\sim 0.05-0.1\%$, corresponding to a $\sim 1-2$ km decrease in the planet's radius (Strom et al. 1975; Watters et al. 1998). Global contraction of this magnitude is also predicted by some thermal history models (Solomon 1976, 1977).

Mercury's lobate scarps have alternatively been interpreted as a consequence of despinning by solar tides (Burns 1976; Melosh and Dzurisin 1978) that also led to Mercury's current spin–orbit resonance (Goldreich and Peale 1966; Colombo and Shapiro 1966). Although predicted low-latitude scarps are observed, the expected high-latitude extensional features (Melosh 1977) have not yet been observed. The known distribution of tectonic features dictates that global contraction is required to explain the deformation whether or not despinning contributed to the lithospheric stress field.

4 Information from the Rotational State

4.1 Spin–Orbit Resonance

Mercury is unique among all solar system bodies in that its rotational angular velocity is stabilized at 1.5 times its orbital mean motion. Figure 1 shows schematically this commensurate spin. Here the stability against the secular tidal slowing of the spin is effected by the average torque on the permanent asymmetry, which tends to keep the axis of minimum moment of inertia aligned toward the Sun as Mercury passes through perihelion (Colombo 1965; Goldreich and Peale 1966). Tidal friction naturally slows an initially higher spin, and Mercury would have had opportunity to be captured into any of several stable resonant spin states with angular velocities that are half-integer multiples of the mean motion, n. The probability of capture into one of the spin–orbit resonances, as the spin rate is slowed by tides, increases as the order of the resonance decreases. The 3 : 2 resonance has the highest probability of capture of any of those previously encountered if tides were the only dissipative force acting and if the current orbital eccentricity prevailed (Goldreich and Peale 1966). The chaotic nature of the orbital motion in the solar system leads to much wider excursions in Mercury's orbital eccentricity ($\sim 0-0.325$) than were obtained in the previously

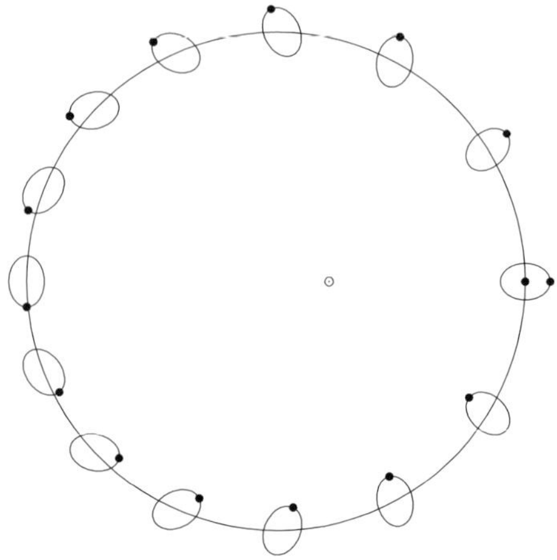

Fig. 1 Rotation of Mercury in the 3 : 2 spin–orbit resonance. The *dot on the orbital curve* marks the perihelion of the orbit, whereas the *dots on the small ellipses* representing Mercury mark one end of the axis of minimum moment of inertia. The Mercury ellipses are separated by equal time intervals

assumed quasi-periodic motion (0.11–0.24). Repeated passes of Mercury's spin through the spin–orbit resonances in both directions thereby yield capture into the 3 : 2 resonance after 4.5×10^9 yr in 55% of trial evolutionary simulations (Correia and Laskar 2004).

The probability for capture into the spin–orbit resonances increases substantially if Mercury's mantle were decoupled from the core by a liquid layer during the time of resonance passage (Counselman and Shapiro 1970). There is thereby a relatively large probability of capture into the 2 : 1 spin resonance that is a function of the core viscosity, the tidal dissipation function, the orbital eccentricity at the time of passage, and the value of $(B - A)/C_m$, where $A < B < C$ are the principal moments of inertia of Mercury and C_m is the moment of inertia of the mantle and crust about the spin axis. For a kinematic core viscosity of 10^{-6} m^2 s^{-1} and plausible choices for the other parameters, the probability of capture can exceed 0.5 (Peale and Boss 1977). However, even if capture into the 2 : 1 resonance were to have occurred, the chaotic evolution of Mercury's orbit can lead to escape from this resonance as the eccentricity falls below a critical value near 0.005, where the resonance becomes unstable (Correia and Laskar 2004).

4.2 Determination of Core State

A procedure for ascertaining the nature of Mercury's core from precise determination of the rotation state (Peale 1976; Peale et al. 2002) depends on

$$\left(\frac{C_m}{B-A}\right)\left(\frac{B-A}{MR^2}\right)\left(\frac{MR^2}{C}\right) = \frac{C_m}{C} \leq 1. \quad (1)$$

The following equations determine the factors in (1):

$$\begin{aligned}
\phi_0 &= \frac{3}{2}\frac{B-A}{C_m}\left(1 - 11e^2 + \frac{959}{48}e^4 + \cdots\right),\\
\frac{C}{MR^2} &= \frac{\left[\frac{J_2}{(1-e^2)^{3/2}} + 2C_{22}\left(\frac{7}{2}e - \frac{123}{16}e^3\right)\right]\frac{n}{\mu}}{(\sin I)/i_c - \cos I}, \\
\frac{B-A}{MR^2} &= 4C_{22}.
\end{aligned} \quad (2)$$

In (2), ϕ_o is the amplitude of the physical libration, which is the maximum deviation of the axis of minimum moment of inertia from the position it would have had if the rotation were uniform at $1.5n$, i_c is the obliquity of the Cassini state, the state that Mercury's spin axis is expected to occupy, J_2 and C_{22} are the second degree zonal and tesseral spherical harmonic coefficients, respectively, in the harmonic expansion of Mercury's gravitational field, e is the orbital eccentricity, and I is the inclination of the orbit plane to the Laplacian plane on which the orbit precesses at the uniform rate $-\mu$. [See Peale (1969, 1974) for a discussion of generalized Cassini's laws.] From virtually any initial obliquity, tidal and core–mantle dissipation drive Mercury's spin to the Cassini state (Peale 1974; Ward 1975), in which state the spin axis and orbit normal remain coplanar with the normal to the Laplace plane as both of the former precess around the latter with the ~300,000-year period of the orbital precession. The spin axis is fixed in and precesses with the orbit frame of reference if Mercury is in the Cassini state. Otherwise, it precesses around this state with a 500- or 1,000-year period, depending on whether the core follows the mantle precession (Peale 2005).

The forced physical libration, with a period of 88 days, is due to the reversing torque on Mercury as it rotates relative to the Mercury–Sun line, as shown in Fig. 1. The amplitude of this libration is inversely proportional to the moment of inertia C_m of the mantle and crust in the first part of (2) because the liquid core does not follow the short-period librations of the mantle. In addition, the full polar moment of inertia C appears in the second part of (2) because the core is likely to follow the mantle during its 300,000-year precession with the orbit (see Sect. 4.3). These two conditions are necessary for the success of the experiment in determining C_m/C, which will be near a value of 0.5 in most models of Mercury's interior (Siegfried and Solomon 1974). If C_m/C in (1) were equal to 1, it would mean that C_m would be replaced by C in the first factor in (1), and the core would be firmly coupled for the long and short timescales. The conditions for determining core state are satisfied for a wide range of core viscosities that include all current estimates of the viscosity of the Earth's core (Peale 1976, 1981, 1988; Peale et al. 2002).

The only unknowns in (2) are ϕ_o, i_c, J_2, and C_{22}. In the MESSENGER mission (Solomon et al. 2001), all of the required parameters will be estimated or improved via the altimetry (Cavanaugh et al. 2007) and radio science (Srinivasan et al. 2007) experiments, and the first two are also measurable by Earth-based radar (Holin 1988, 1992, 2003; Margot et al. 2007). In addition, measurement of the forced libration will be accomplished as part of the MESSENGER mission by estimating independently the libration of the surface from altimetry and the interior from gravity using long-wavelength shapes of the global fields (Zuber and Smith 1997). Table 3 summarizes the expected recovery of geophysical parameters using a simulation of the approximate MESSENGER mission scenario (Zuber and Smith 1997). Because of MESSENGER's elliptical orbit, coverage of the surface from the Mercury Laser Altimeter (MLA) (Cavanaugh et al. 2007) will be limited to the northern hemisphere; radio occultations of the spacecraft will be used to constrain planetary shape in the southern hemisphere (Srinivasan et al. 2007). The latitude-dependent resolution of the gravity field will permit study of crustal and lithospheric structure for much of the northern hemisphere where information at spherical harmonic degrees as high as 76 (spatial resolution 100 km) will be resolvable, while in the southern hemisphere only the lowest degree harmonics (no higher than 10; relevant for deep internal structure and libration recovery) will be obtained. Because of the nonuniform resolution expected, spherical harmonics are not optimum and other representations of the gravity field will be investigated.

Table 3 Simulated geophysical parameter recovery from MESSENGER

Parameter	A priori value	Recovery from gravity	Recovery from altimetry
Libration amplitude in longitude, radians	1.435×10^{-4}	1.542×10^{-4}	1.568×10^{-4}
Libration amplitude in longitude, m	350	376, 7% error	386, 9% error
Obliquity, arcsec	0.0	2.0	0.2
Gravity			
C_{20}	-2.7×10^{-5}	-2.688×10^{-5} (0.5% error)	–
C_{22}	1.6×10^{-5}	1.598×10^{-5} (0.2% error)	–
S_{22}	0.0	1.4×10^{-5}	–
Topography			
C_{20}, m	−733	–	−754 (3% error)
C_{22}, m	72	–	180 m (150% error)
S_{22}, m	395	–	447 m (13% error)

C_{lm} and S_{lm} are spherical harmonic coefficients, where l and m represent degree and order of the spherical harmonic expansion. All coefficients are normalized according to Kaula (1966). Note that $C_{20} = -J_2/(5)^{1/2}$. Simulation described by Zuber and Smith (1997)

4.3 Core–Mantle Coupling

Previous work (Peale 1976, 1981) has indicated that viscous coupling of a liquid core and mantle satisfies the criterion that the core not follow the mantle on the 88-day timescale of the physical libration but does follow the mantle on the 300,000-year timescale of the precession for a wide range of plausible viscosities. More recent work (Peale et al. 2002) shows that core–mantle coupling due to magnetic fields (Buffett 1992; Buffett et al. 2000), to topography on the core–mantle boundary (CMB) (Hide 1989), and to gravitational interaction between the mantle and an asymmetric inner core (Szeto and Xu 1997), are not likely to frustrate the condition that the liquid core not follow the mantle on the 88-day libration timescale.

Inertial (pressure) coupling of the core to the mantle could compromise the experiment by affecting either the libration in longitude or the spin precession. It is likely that any deviations of the circularity of the CMB equator will be no larger than possible topography induced by convection cells in the mantle whose horizontal extent is not significantly larger than the mantle thickness (Hide 1989). It is therefore necessary to consider only a possible effect of the pressure coupling on the precession of the spin axis. The criterion that pressure at an elliptical CMB couples the core to the mantle is (Stewartson and Roberts 1963; Toomre 1966)

$$\dot{\Omega}/\dot{\psi} < \varepsilon, \tag{3}$$

where $\dot{\psi}$ is the rotational angular velocity, $\dot{\Omega}$ is the precessional angular velocity $\varepsilon = (a-c)/a$ is the core ellipticity, and a and c are the equatorial and polar radii of the CMB, respectively. If we assume for the moment that the entire planet is in hydrostatic equilibrium, then

$$J_2 = \frac{k_f R^3 \dot{\psi}^2}{3GM}, \tag{4}$$

where k_f is the fluid Love number (Munk and MacDonald 1960), and G is the gravitational constant. We also note that, to lowest order (Kaula 1968),

$$J_2 \approx \frac{2}{3}\varepsilon - \frac{\dot{\psi} R^3}{3GM}. \tag{5}$$

The ε in (5) refers to the entire planet. The ellipticity of the CMB will depend on the radial density distribution, but since the core radius is $\sim 0.75R$, the ellipticity of the CMB will not be much smaller than this ε if hydrostatic equilibrium prevails as assumed here. The two expressions for J_2 may be combined to yield

$$\varepsilon = 4.5 \times 10^{-7} \frac{1 + k_f}{2} \frac{\dot{\psi}^2}{n^2}, \tag{6}$$

so pressure coupling would occur if

$$\frac{\dot{\Omega}}{\dot{\psi}} = 3.2 \times 10^{-4} \leq 4.5 \times 10^{-7} \frac{\dot{\psi}^2}{n^2}, \tag{7}$$

which is true if $\dot{\psi}/n > 26.7$. This ratio is so large that it allows some relaxation from the assumption of strict isostasy. Since $\dot{\psi}/n = 1.5$, pressure coupling of the core to the mantle is completely negligible.

4.4 Free Motions?

There are possible situations where Mercury would not occupy the precise equilibrium state as described that constrains the core properties. A large impact or other unspecified excitation mechanism could excite the three free rotational motions. These include a free libration in longitude, a free precession of the spin vector about the Cassini state, and a free wobble, where "free" means that the amplitudes and phases of these motions are arbitrary. The free libration in longitude results if Mercury's axis of minimum moment of inertia is displaced from the Mercury–Sun line when Mercury is at perihelion (Fig. 1). The gravitational torque on the asymmetric planet, averaged around the orbit, acts to restore this alignment so that the long axis will tend to librate around the solar direction at perihelion. The period of this libration is close to 12 y. The short-period torque reversals causing the 88-day physical libration average to zero in this application. A free precession is characterized by a displacement of the spin axis from the Cassini state position; this displacement leads to a precession of the spin axis about the Cassini state with an approximately constant angular separation. The period of this precession is about 500 y if the liquid core is not dragged along with the mantle, and double that time if it is. A free wobble is the precession of the spin vector around the axis of maximum moment of inertia in the body frame of reference. It is also called nonprincipal axis rotation. The period of spin vector excursion in the body frame of reference is about 300 years if only the mantle and crust participate. Because the duration of measurements by MESSENGER is much shorter than this period, and the proposed measurements will be unable to detect a free wobble, we consider it no further here.

Free motions are subject to dissipative damping, both from tides and from the relative motion of the liquid core and solid mantle. Whether we should expect to find free motions depends on the timescales for damping of such motions, which are given by (Peale 2005)

$$
\begin{aligned}
T_{\text{lib}} &= \frac{2.92 \text{ years}}{1.41 \times 10^{-4}\nu^{1/2} + 3.93 \times 10^{-4}\frac{k_2}{Q_0}} = 1.8 \times 10^5 \text{ years},\\
T_{\text{prec}} &= \frac{89 \text{ years}}{8.59 \times 10^{-3}\nu^{1/2} + 8.08 \times 10^{-3}\frac{k_2}{Q_0}} = 1.0 \times 10^5 \text{ years},
\end{aligned}
\tag{8}
$$

where T_{lib} and T_{prec} are the damping timescales for the free libration and the amplitude of the free precession, respectively. In (8), ν is the kinematic viscosity in cm^2/s, k_2 is the second-degree tidal Love number (Munk and MacDonald 1960), and Q_o is the tidal dissipation function at a tidal period corresponding to the orbit period. The tidal model is equivalent to the assumption that Q is inversely proportional to frequency. The torque between the liquid core and solid mantle is assumed to be proportional to the difference between the vector angular velocities of the core and mantle. The proportionality constant is related to the kinematic viscosity of the core fluid by equating the timescale for the damping of a differential velocity between core and mantle to the timescale for the relaxation of the differential motion of a fluid inside a rotating, closed container of radius $R_c[T = R_c/(\dot{\psi}\nu)^{1/2}]$ (Greenspan and Howard 1963). [Further details in calculating the timescales in (8) were given by Peale (2005).] The analytical expressions for the timescales shown in (8) are verified by numerical integration of the complete equations of motion for $\nu = 0.01\ \text{cm}^2\,\text{s}^{-1}$ and $k_2/Q_o = 0.004$, values that yield the numerical estimates on the right-hand sides of these equations. The value of ν is in the middle of a rather small range estimated for the Earth's core (de Wijs et al. 1998), and the value of k_2/Q_o is comparable to that appropriate to Mars (Smith and Born 1976; Yoder et al. 2003; Bills et al. 2005).

The timescales for damping the free libration in longitude and the free precession are both short compared with the age of the solar system, so ordinarily we would expect both to be damped to undetectable magnitudes. However, the small variations in the orbital elements due to the planetary perturbations induce long periods of forced librations dominated by a 5.93-year variation, which is half of Jupiter's orbital period (Peale et al. 2007). There will also be a small amplitude variation near the free libration period of about 12 years due to a near resonance of the orbital variations at Jupiter's orbital period. This latter variation may have been seen in the recent radar data (Margot et al. 2007).

Because the forced physical libration period is much shorter than that of the free libration, any measurable amplitude of the latter will not compromise the determination of the former's amplitude. The physical libration will simply be superposed on the longer-period free libration as shown in Fig. 2. However, a significant amplitude of the free precession would mean that the spin axis would not be coincident with the Cassini state, and the thereby uncertain position for the latter would make the determination of C/MR^2 via the second equation in (2) more uncertain. If Mercury's spin is in the Cassini state, it is coplanar with the orbit normal and the Laplace plane normal. The straight line in Fig. 3 represents the intersection of the plane defined by the orbit normal and the Laplace plane normal with the unit sphere centered on the coordinate system origin. Then the spin axis would intersect the unit sphere at a point on this line. If there is a finite-amplitude free precession, the spin axis would be offset from this position on a circular precession trajectory at some arbitrary phase. Figure 3 shows an example position determined observationally that would be indicative of a free precession along with a segment of its precession trajectory.

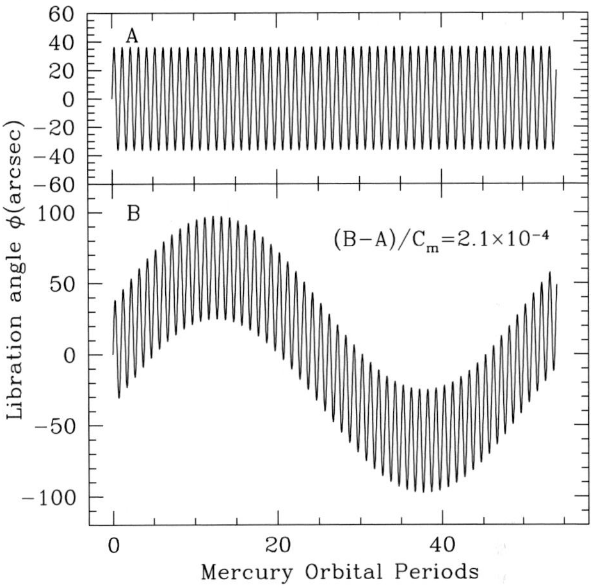

Fig. 2 **(A)** Physical libration of Mercury with no free libration. **(B)** Physical libration of Mercury superposed on a small free libration. The free libration does not hinder measurement of the physical libration amplitude

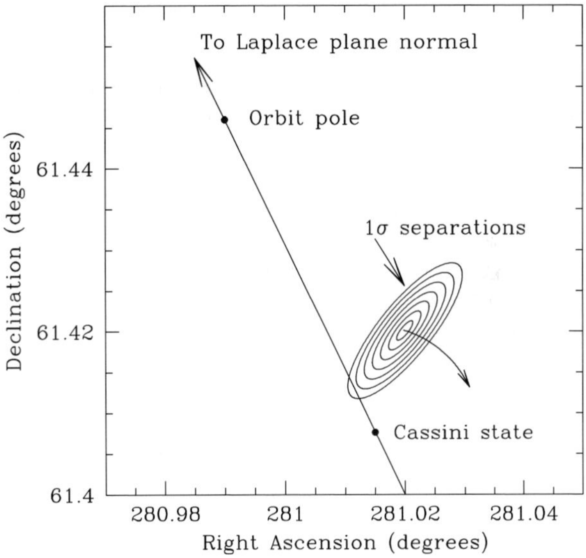

Fig. 3 Signature of a free precession showing the offset of the spin position from the Cassini state. The *straight line* is the intersection of the plane defined by the orbit normal and the Laplace plane normal with the unit sphere. The Cassini state is coplanar with the orbit normal and the Laplace plane normal, and a unit vector in the Cassini state direction intersects the unit sphere on this line as shown. The uncertainty contours surround the least squares fit to the hypothetical observed position of the spin axis. The *curved arrow* is a portion of the spin precession trajectory on the unit sphere

4.5 The Changing Cassini State Position

There is another possible deviation of the spin axis from the Cassini state in addition to a free precession. The second equation in (2) shows the dependence of the Cassini state on the orbital parameters, which includes the orientation of the Laplace plane. The variation in the orbital parameters due to planetary perturbations, and the similarly slow change in the geometry of the planets, means that the position of the Laplace plane and of the Cassini state will change on the same timescales. However, an action integral, which is proportional to the solid angle swept out by the spin vector as it precesses around the Cassini state, is an adiabatic invariant if the precession is fast relative to the slow changes in the parameters that define the Cassini state (Goldreich and Toomre 1969; Peale 1974). As the timescales for the slow variations usually exceed 10^5 y and the precession period is 500 or 1,000 y, one expects the adiabatic invariant to keep the spin close to the instantaneous value of the Cassini state as the latter's position slowly changes. The adiabatic invariant is not conserved for the short-period fluctuations in the orbital elements. However, these short-period fluctuations are of small amplitude, and one expects the deviations of the spin vector from the Cassini state to be commensurate with these amplitudes. The increasing precision of the radar determinations of Mercury's spin properties, and that anticipated for the MESSENGER mission, warrant a check on just how closely the spin axis follows the changing position of the Cassini state.

By following simultaneously the spin position and the Cassini state position during long-timescale orbital variations over the past 3 My (Quinn et al. 1991) and short-timescale variations for 20,000 y (Jet Propulsion Laboratory Ephemeris DE 408, E.M. Standish, private communication 2005), Peale (2006) showed that the spin axis remains within one arcsec of the Cassini state after it is brought there by dissipative torques. In Fig. 4 the variations of Mercury's eccentricity and inclination to the ecliptic of year 2000 are shown for the last 3×10^6 y from data obtained from a simulation by T. Quinn (ftp://ftp.astro.washington.edu/pub/hpcc/QTD). On this same timescale, the ascending node of the orbit plane on the ecliptic generally regresses, with fluctuations in the rate. The Quinn data have been filtered to eliminate periods less than 2,000 y. It is therefore necessary to check also the effect of short-period variations. Figure 5 shows the variation of e, I, and Ω over the 20,000-y time span of the JPL DE 408 ephemeris, where Ω is the longitude of the ascending node of the orbit plane on the ecliptic. Short-period fluctuations are superposed on the almost linear trend of these variables, where the amplitudes are comparable to the line widths. Generally, the parameters $e, I, \Omega, dI/dt$, and $d\Omega/dt$ affect the Cassini state position. The angular velocity of the orbit plane can be represented by the vector sum of dI/dt and $d\Omega/dt$ averaged over a suitable interval, say 2,000 y. This angular velocity, while suitable for use in the equations of motion, is not the instantaneous value of μ in the second equation of (2), since another con-

Fig. 4 Variation of Mercury's eccentricity, *e* (*solid line*), and orbital inclination, *I* (*dotted line*), to the ecliptic of year 2000, from T. Quinn (ftp://ftp.astro.washington.edu/pub/hpcc/QTD)

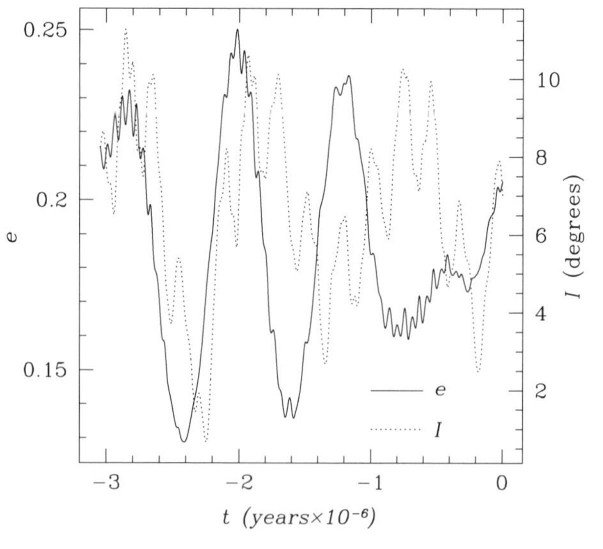

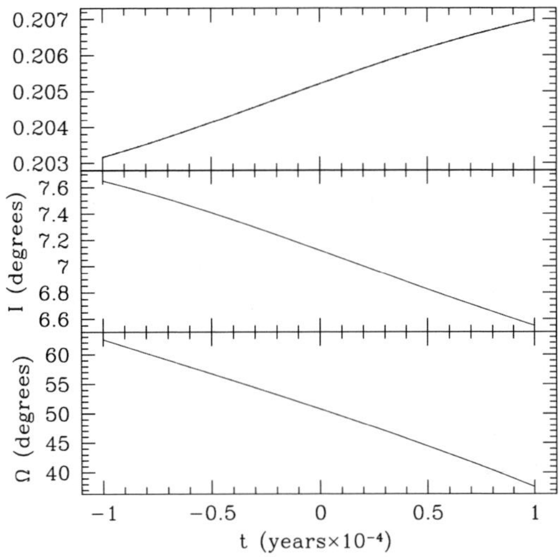

Fig. 5 Variation of e, I, and Ω relative to the ecliptic of J2000 for JPL Ephemeris DE 408. From Peale (2005)

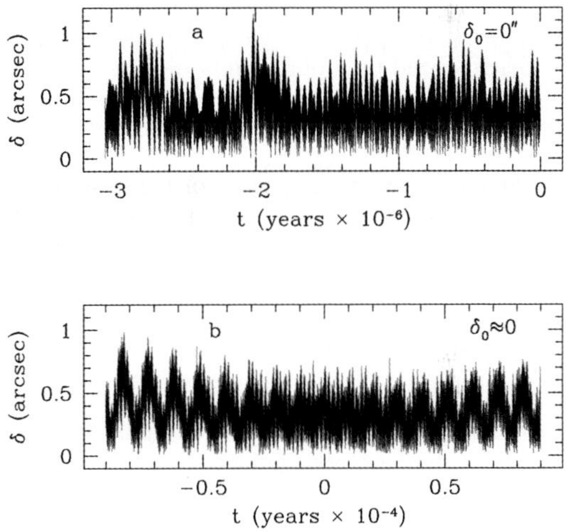

Fig. 6 Angular separation of Mercury's spin axis from Cassini state 1: **(a)** For long-timescale variations over 3×10^6 y from simulations by T. Quinn. **(b)** For short-time-scale variations over 20,000 y from JPL Ephemeris DE 408 provided by E.M. Standish. The periodic variation in δ results from the spin precession around the Cassini state

straint is necessary to determine the Laplace plane orientation (Yseboodt and Margot 2006; Peale 2006).

The equations of variation of the spin position in the orbit frame of reference are numerically integrated with the five parameters e, I, Ω, dI/dt, and $d\Omega/dt$ determined at arbitrary times from spline fits to the data. The position of the Cassini state is followed simultaneously, and the proximity of the spin to the Cassini state is determined as a function of time. Since dissipation will drive the spin to the Cassini state, we start the spin in this state initially and allow the evolution to proceed. The angle δ between the spin and Cassini state positions is shown in Fig. 6 for both the long- and short-timescale variations in the orbital parameters. The fluctuations in δ are generally $<1''$ as asserted earlier. If the spin is started somewhat displaced from the Cassini state, the initial displacement is maintained to within $1''$ as the spin precesses around the Cassini state, which is consistent with the adiabatic invariance of the solid angle described by the spin. Aside from superposed, small-amplitude fluctuations, an initial displacement for the short-period orbital variations is also maintained to the same precision.

It is apparent that the real orbital element variations will not cause the spin to deviate from the Cassini state by more than 1″, even with no damping. This conclusion leaves an unlikely free precession as the only reason the spin vector might be displaced from the Cassini state by a measurable amount unless tidal and core–mantle dissipation displaces the spin significantly such as that which occurs for the Moon (Williams et al. 2001). The experiment to determine the nature and extent of Mercury's core from the details of the rotation is thus likely to be successful.

5 Magnetic Field: Observations and Possible Explanations

5.1 Observations from Mariner 10

Two of the three flybys of Mercury by the Mariner 10 spacecraft resulted in the surprising measurement of a planetary magnetic field and magnetosphere (Ness et al. 1974, 1975, 1976). Mercury's magnetosphere, reviewed in a companion paper (Slavin et al. 2007), exhibits a bow shock and a magnetopause, which are manifest as well-resolved discontinuities in the Mariner 10 magnetometer and plasma observations (Russell et al. 1988). The magnetic field is characterized by a dominantly dipolar structure with the same polarity as Earth's present field. Mercury's field is aligned approximately with the ecliptic normal and has a moment of 300 nT-R_M^3 (Connerney and Ness 1988). Because of its intensity, the observed field cannot be a consequence of solar wind induction (Herbert et al. 1976) and is likely of internal origin (Ness et al. 1975, 1976). Two possible interpretations of an internal field include remanent magnetization of Mercury's crust and a present-day dynamo generated in a liquid iron core.

5.2 Remanent Magnetization

Thermoremanent magnetization acquired by igneous rocks that cooled below the Curie temperature of mineral magnetic carriers during the presence of an ambient field has been documented on Earth, the Moon, and Mars and can be expected for at least some areas of the crust of Mercury. Remanence has the appealing feature that it could explain the Mariner 10 observations in the event that Mercury's core has either solidified or is partially liquid but is not sufficiently well-stirred by convection to generate dynamo action (Stevenson 1983; Stevenson et al. 1983). A crustal source for Mercury's field was originally dismissed (Ness 1978, 1979; Connerney and Ness 1988), but new planetary magnetic observations combined with experiment and theory have collectively caused the possibility to be revisited (Aharonson et al. 2004).

First, large coherent structures with sufficiently high remanent magnetization to explain the Mariner 10 observations were not believed to exist in the terrestrial planets. However, the Mars Global Surveyor mission has since provided magnetic observations of Mars pointing to a crustal field of high specific magnetization ($\sim$20 A m^{-1}) (Acuña et al. 1998, 1999, 2001) with horizontal scales for coherence of magnetization direction of hundreds of kilometers. In addition, laboratory experiments have shown that single- and multidomain thermoremanent magnetization in some iron oxide-rich minerals can be significant (Kletetschka et al. 2000a, 2000b). Although highly oxidized minerals (hematite, magnetite) may not be present in the FeO-poor, water-poor crust of Mercury, other magnetic minerals such as pyrrhotite are more likely to be present and can potentially carry magnetic remanence (Rochette et al. 2001).

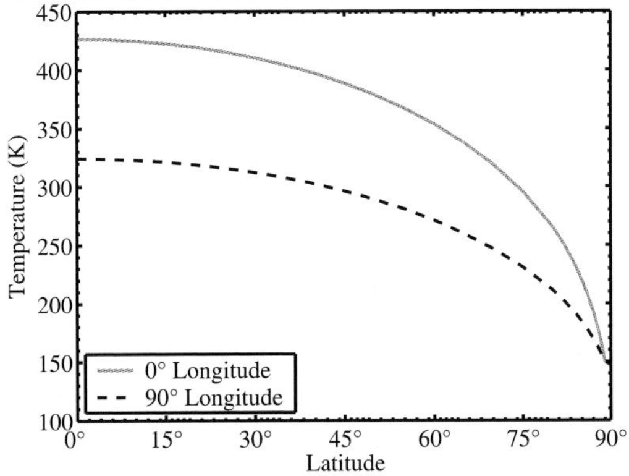

Fig. 7 Spherical harmonic fits to Mercury's near-surface temperature (diurnal variations averaged out) from the model of Aharonson et al. (2004). Strong latitudinal and longitudinal dependence is predicted and seen. A spherical planet has been assumed

Another reason for excluding remanence derived from a simple theorem in magnetostatics (Runcorn 1975a, 1975b): a uniform shell magnetized by an internal source subsequently removed has no external field. This theorem, known as Runcorn's theorem, dictates that even if magnetic minerals were present in Mercury's crust, they could not produce an external remanent field if they were distributed uniformly (Srnka 1976; Stephenson 1976). However, it has recently been shown (Aharonson et al. 2004) that latitudinally and longitudinally varying solar insolation could lead to long-wavelength variations in the depth to the Curie temperature, and thus the spatial distribution and intensity of remanent magnetization. As is apparent from Fig. 1, Mercury's 3 : 2 spin orbit resonance causes the planet to have two equatorial "hot poles," at 0° and 180° longitude. In addition, Mercury's surface temperature also varies latitudinally, with the greatest insolation at low latitudes. Figure 7 shows the expected latitudinal pattern of surface temperature on Mercury at two longitudes. Because of solar insolation and Mercury's rotational and orbital orientations, therefore, surface temperature is a maximum at the equator and decreases toward the north and south poles.

Such a crustal thermal structure breaks the previously assumed symmetry of Mercury's magnetized shell and allows for the possibility that the planet could have "frozen in" a magnetic field. Aharonson et al. (2004) recognized that the expected remanent signature would have a dipolar component, as observed for Mercury. While the analysis of Aharonson et al. does not prove that Mercury's magnetic field is a consequence of remanence, it removes previous objections to the idea and broadens the currently allowable thermophysical range of the current core state. Stanley et al. (2005) discussed how MESSENGER could resolve the internal vs. crustal origin of the field. The power spectrum of the field measured by MESSENGER will provide insight into its origin, as would the detection of any temporal variability, which would point to an internally generated dynamo. If the field structure correlates with gravity indicating topography at the CMB, then a thermoelectric dynamo (Stevenson 1987) would be favored (Giampieri and Balogh 2002). Finally, small-scale magnetic structure with a shallow source depth would point to a crustal remanent field.

The question of the timescales of magnetic reversal versus crustal cooling remains to be addressed in future testing of this hypothesis. Reversal timescales can in principle be estimated from numerical dynamo models (e.g., Glatzmaier and Roberts 1995), and these can be compared to models of crustal cooling (Turcotte and Schubert 1982).

5.3 Core Dynamo Models

An active dynamo source for Mercury's field has been viewed as problematic because of discrepancies between the observed field's magnitude and theoretical estimates of the magnetic field strength produced by an Earth-like dynamo. Specifically, an Earth-like dynamo, driven by thermo-compositional core convection, would be expected to produce a much stronger field than observed at Mercury (Stevenson 1987; Schubert et al. 1988).

There are two independent methods for estimating the magnetic field strength generated by a dynamo. The first method, energy balance, involves balancing the gravitational energy release driving the dynamo and the ohmic energy dissipated through electrical currents. Using thermal evolution models for Mercury to estimate the gravitational energy, magnetic field strengths of the order of 10^5-10^7 nT are obtained (Stevenson et al. 1983; Stevenson 1987).

The second method, magnetostrophic balance, relies on assuming that Mercury's dynamo operates in the strong-field regime where the Lorentz force balances the Coriolis force. This balance results in an estimate for the magnetic field of $B = (2\dot{\psi}\rho \mathrm{Re}_M/\sigma)^{1/2}$, where B is the magnetic field, $\dot{\psi}$ is the planetary rotation rate, ρ is density, σ is electrical conductivity, $\mathrm{Re}_M = UL/\eta$ is the magnetic Reynolds number, η is the magnetic diffusivity, U is a typical velocity scale, and L is a typical length scale. For a magnetic Reynolds number Re_M on the order of 10–1,000 (10 is the minimum Re_M value for dynamo action), the estimated field strength is in the same range as obtained using energetic arguments.

The magnetic field estimates provided by these two methods do not immediately conflict with the observed field value since the estimate is of the field strength in the fluid core, rather than at some distance outside the core where the observations are made. Because the magnetic field is solenoidal, the core field can be represented in spherical harmonics using the toroidal-poloidal decomposition

$$B = B_T + B_P = \nabla \times T\hat{r} + \nabla \times \left(\nabla \times P\hat{r}\right), \tag{9}$$

where B_T and B_P are the toroidal and poloidal field components, respectively, T and P are the toroidal and poloidal scalars, respectively, and $\hat{r}$ is a unit vector in the radial direction. Because the toroidal field has no radial component, it is not observable outside the conducting core and only the poloidal field is measured outside the dynamo generation region. The magnetic field estimate given by the energy and magnetostrophic balance arguments pertains to the dominant magnetic field component in the core. Since the poloidal field at the core mantle boundary inferred from the measured magnetic dipole intensity cannot match the magnetic field estimates, it is most likely that the toroidal field is dominant in the core (unless the poloidal field inside the core is significantly larger than both the toroidal field in the core and the poloidal field at the core–mantle boundary).

A problem arises when one compares the dipole field strength at Mercury's core–mantle boundary, which is assumed to be representative of the poloidal field strength at the CMB, to the toroidal field strength in Mercury's core, estimated using the energy or magnetostrophic balance arguments. The ratio of these fields is $B_{\mathrm{dip}}/B_T \sim 10^{-2}-10^{-4}$. From a similar analysis of Earth's dynamo, $B_{\mathrm{dip}}/B_T \sim 10^{-1}$. These dynamo solutions for Mercury are problematic, because it appears that Mercury's dynamo produces a much smaller B_{dip}/B_T ratio than Earth's dynamo and it is unclear why this should be the case.

Additional testing of the viability of Mercury's magnetic field as a consequence of a core dynamo requires consideration of observed and expected field properties for the end-member thin-shell and thick-shell core geometries (see Fig. 8), where shell size reflects the thickness of the liquid iron region relative to the total core radius.

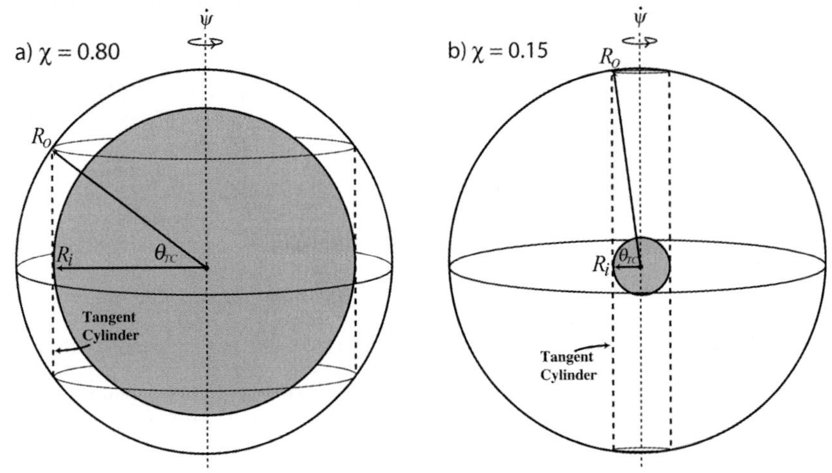

Fig. 8 Schematics of **(a)** thin-shell and **(b)** thick-shell dynamo models. Here R_i is the radius of the solid inner core, R_O is the radius of the liquid outer core, $\chi = R_i/R_O$, and θ_{TC} defines the angular size of the tangent cylinder

5.3.1 Thin-Shell Dynamo Models

One possible explanation for the difference in field partitioning is that Mercury's dynamo operates in a different core geometry than does Earth's. Thermal evolution models for Mercury (Stevenson et al. 1983; Stevenson 1987; Schubert et al. 1988) suggest that its core might consist of a thin fluid shell surrounding a large solid inner core. This thin-shell dynamo geometry is different from the thick-shell geometry of Earth's core ($R_i/R_o \approx 0.35$), a distinction that may lead to differences in the magnetic fields they produce.

A recent study (Stanley et al. 2005) of magnetic field partitioning and characteristics for various outer core shell thicknesses and Rayleigh numbers using a numerical dynamo model (Kuang and Bloxham 1997, 1999) demonstrated that some thin-shell dynamos can produce average B_{dip}/B_T ratios of $\sim 10^{-2}$. These ratios are variable in time and can be as low as $\sim 10^{-3}$, consistent with the theoretical values for Mercury. In these dynamos, both the ratio of poloidal field to toroidal field in the core (B_P/B_T) and the ratio of dipole field at the CMB to poloidal field in the core (B_{dip}/B_P) are smaller than the Earth-like case.

In these thin-shell dynamo models, illustrated in Fig. 9, a strong, stable toroidal field is maintained inside the tangent cylinder through differential rotation. Little convection occurs in this region, and hence little poloidal field is produced from the strong toroidal field inside the tangent cylinder. Outside the tangent cylinder, only a few convection columns are present that can produce poloidal magnetic field from toroidal field. However, the toroidal field is much weaker in this region and very time dependent. The combination of a few convection cells and a weak toroidal field results in a much weaker and smaller-scale poloidal field produced in this region.

These Mercury-like dynamos occur over limited ranges of shell thickness and Rayleigh number. Such dynamos require both a thin-shell geometry in order to restrict the convection columns to the small region outside the tangent cylinder, so that they are not effective at interacting with toroidal field, as well as a relatively low Rayleigh number (close to the critical value for the onset of strong field dynamo action), so that convection is not yet generated inside the tangent cylinder. In intermediate shell geometries, the region inside the

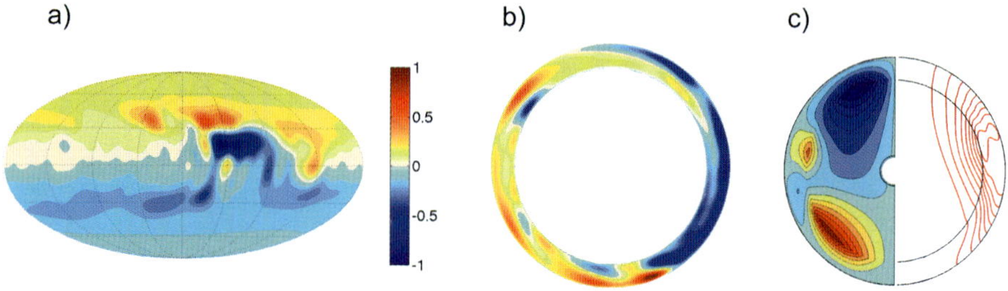

Fig. 9 Snapshot from the $R_i/R_o = 0.8$ case of Stanley et al. (2005). **(a)** Contours of the magnetic field on the CMB. **(b)** Contours of axial vorticity in the equatorial plane viewed from above. **(c)** Meridional slices showing (*left*) contours of the axisymmetric toroidal magnetic field and (*right*) the axisymmetric poloidal magnetic field lines. Because the poloidal field is far weaker than the toroidal field, the two vector fields are contoured independently in this figure. The *colorbar* in **(a)** applies to **(b)** and the left half of **(c)** as well

tangent cylinder is smaller and convection columns outside the tangent cylinder can interact with the majority of the toroidal field. At larger Rayleigh numbers, convection inside the tangent cylinder is effective at producing a strong poloidal field there, and hence Earth-like partitioning results.

5.3.2 Thick-Shell Dynamo Models

Recent thermal evolution modeling for Mercury has shown that the present ratio of the solid inner core radius to its outer core radius (R_i/R_o) is a sensitive function of core sulfur content and initial core temperature (Harder and Schubert 2001; Hauck et al. 2004). These results suggest that thick-shell dynamo models may be relevant to Mercury. Numerical simulations have been carried out to determine how dynamo action varies for a wide range of core geometries (Heimpel et al. 2005), covering $R_i/R_o = 0.15$ to 0.65. In these simulations, illustrated in Fig. 10, weak external magnetic fields, comparable to that observed for Mercury, are produced in strong-field, thick-shell cases with $R_i/R_o = 0.10$–0.20. The ratio of B_P/B_T is $\sim 10^{-2}$ in the case with $R_i/R_o = 0.15$. In contrast, cases with intermediate core geometries between $R_i/R_o = 0.25$ and 0.65 generate strong external dipole fields that are more Earth-like than the thick-shell cases.

For thick-shell dynamos (with $R_i/R_o < 0.25$), a single turbulent convection column is the preferred mode of convection. This single columnar structure dominates the poloidal flow in this simulation. Helical flow within the column acts locally to generate a poloidal magnetic field. A strong retrograde zonal flow also exists throughout the region outside the tangent cylinder. This strong zonal flow shears out the poloidal field into a toroidal magnetic field. Because poloidal field generation occurs only in a single, localized region, the total poloidal field energy is small. In contrast, the large-scale zonal shearing efficiently induces a strong large-scale toroidal magnetic field.

It should be noted that a similar mechanism operates in both the thin-shell $R_i/R_o = 0.80$ model of Fig. 9 (Stanley et al. 2005) and the thick-shell $R_i/R_o = 0.15$ model of Fig. 10 (Heimpel et al. 2005). Columnar convection generates the poloidal magnetic field in a relatively small portion of the core fluid, whereas toroidal magnetic field is induced by large-scale zonal flows that occur over a much larger fractional volume of the core. Thus, in both cases, the poloidal magnetic field is much weaker than the toroidal field because poloidal magnetic field generation is generated efficiently only in a small, localized region.

Magnetic field observations by MESSENGER should be able to distinguish between thick- and thin-shell dynamo scenarios. In both cases, localized strong radial magnetic

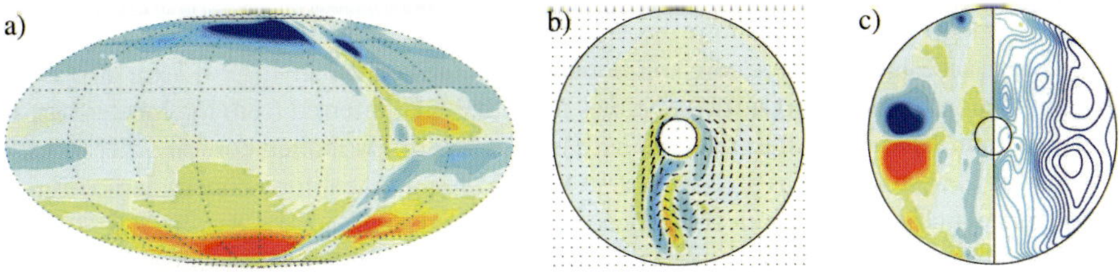

Fig. 10 Snapshot from the $R_i/R_o = 0.15$ case of Heimpel et al. (2005). **(a)** Contours of the magnetic field on the CMB. **(b)** Contours of axial vorticity in the equatorial plane, viewed from above. *Arrows* denote equatorial velocity vectors. The region of intense vorticity marks out the single turbulent convection column that develops in this model. The large-scale spiraling structure away from the convective region marks out the region of zonal shear. **(c)** Meridional slices showing (*left*) contours of the axisymmetric toroidal magnetic field and (*right*) the poloidal magnetic field lines. Because the poloidal field is far weaker than the toroidal field, the two vector fields are contoured independently in this figure

flux patches are produced outside the tangent cylinder where the convection cells generate poloidal field from toroidal field. Since the size of the tangent cylinder depends on the inner core size, the thin- and thick-shell core geometries have different tangent cylinders. The intersection of the tangent cylinder with the core–mantle boundary in a thick-shell geometry occurs at much higher latitudes (~81° in the Heimpel et al. models) than in the thin-shell case (~37° in the Stanley et al. models). If MESSENGER can determine the approximate latitude of these magnetic flux patches, it will be possible to determine the size of the tangent cylinder and hence whether the dynamo operates in a thick- or thin-shell geometry. Another difference between the thick- and thin-shell numerical models that may be of help in determining Mercury's core geometry is the tilt of the dipole component. The Stanley et al. thin-shell models have dipole tilts that vary significantly in time and can be quite large. In contrast, the Heimpel et al. models have a small dipole tilt that varies little in time. Finally, the tangent cylinder is the site of a shear boundary layer in many models of core flow. Such a large-scale shear can produce latitudinal gradients in secular variation that may be inverted to estimate the geometry of Mercury's core (e.g., Olson and Aurnou 1999; Pais and Hulot 2000).

6 Thermal Evolution Models

6.1 Mantle Structure and Dynamics

The possibility that Mercury's magnetic field is a consequence of a core dynamo would require that the planet's metallic core is at present at least partially molten. For thermal evolution models in which core–mantle differentiation occurred early and the core is either pure iron or an iron-nickel solid solution, an initially molten core should have frozen out by now (Siegfried and Solomon 1974; Cassen et al. 1976; Fricker et al. 1976).

Current understanding of Mercury's internal evolution has been developed in large part through simulation of long-term heat transfer through the planet's mantle. These models incorporate either, or both, conductive and convective mechanisms for heat loss during Mercury's history. Since the Mariner 10 flybys, modeling has provided insights into: (1) the planet's interior evolution within the constraint of (limited) global contraction as inferred from lobate scarps observed in images and interpreted as contractional in origin (Strom et al. 1975), and (2) the existence of an intrinsic magnetic field (Ness et al. 1975). Early models were based on the assumption that Mercury's mantle lost heat exclusively by conduction and were important in establishing the significance of an alloying element in Mercury's large core to forestall complete core solidification and the accompanying ~17 km of planetary contraction that would result [~15 times more than that inferred from the lobate scarps (e.g., Solomon 1976)]. Furthermore, the necessity of a partially molten core in order to satisfy contraction estimates was also consistent with the requirement for a core dynamo origin for the magnetic field (i.e., an electrically conductive fluid shell).

Widespread recognition that terrestrial planetary mantles could be unstable to convection (Schubert et al. 1979) led to further work on Mercury's thermal evolution (Stevenson et al. 1983; Schubert et al. 1988). As would be expected, models that included the more efficient planetary cooling provided by mantle convection confirmed that a modest amount, at least 2–3 weight%, of an alloying element such as sulfur is necessary to limit contraction. However, it has also been demonstrated that even the enhanced cooling of mantle convection might not be sufficient to drive the core convection necessary for a modern dynamo (Schubert et al. 1988).

6.2 Parameterized Mantle Convection

A recent reanalysis of the thermal evolution of Mercury (Hauck et al. 2004) extended the parameterized mantle convection technique used in earlier studies to include the energetics of mantle partial melting and temperature- and pressure-dependent flow laws for mantle materials. This class of model is also able to account for the conductive transport of heat in the mantle when the thermal Rayleigh number is subcritical for convection. Such a transition from early convective to later conductive heat transport is directly applicable to Mercury's thin mantle, a transition that tends to stabilize it against convection relative to the other terrestrial planets. The time of such a convection-to-conduction transition depends on the details of mantle material properties such as rheology and the concentration of heat-producing elements, though a principal result is that the relatively low efficiency of cooling via conduction is an important element of models capable of satisfying the small amount of contractional strain recorded in lobate scarps (Hauck et al. 2004). As illustrated in Fig. 11, that study also demonstrated that models that strictly satisfied the dual requirements of ~1−2 km of radial contraction since 4 Ga and a modern fluid outer core had >6.5 weight% sulfur content in the core, a creep-resistant (i.e., anhydrous) flow law for mantle material, and heat production provided primarily by the very long-lived isotope ^{232}Th. Figure 12 shows the timing of the transition from mantle convection to conduction as a function of bulk core sulfur content. At low sulfur contents most of the models have an inner core when the model starts; the size of the initial inner core decreases as sulfur content increases. Increasing sulfur content toward the peak at ~7 weight % sulfur results in a greater volume of inner core crystallizing over the age of the solar system. Past the peak, the effect of the adopted initial temperature profiles and core states is not apparent. The behavioral transition at ~7 weight % sulfur is probably controlled by the volume of core crystallized and underscores the need to couple core and mantle evolution in modeling.

The dominance of ^{232}Th in heat production in these models is predicted for the silicate vaporization scenario for Mercury's anomalously high iron-silicate ratio (Fegley and Cameron 1987). However, if the radial contraction is underestimated by even a factor of two, including unrecognized mechanisms such as low-amplitude, long-wavelength folding that might accommodate strain in addition to that localized and recorded in the lobate scarps, then the concentration of heat-producing elements is unconstrained, though the bulk sulfur content of the core would likely need to be greater than about 6 weight% to prevent greater contraction (Hauck et al. 2004). Thermal convection in the outer core, a prerequisite for some dynamo

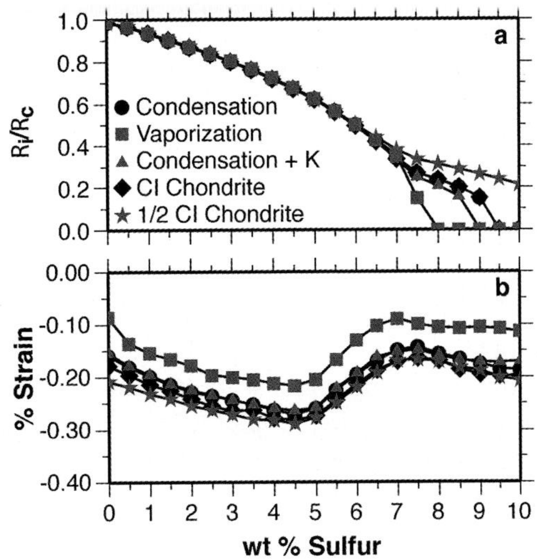

Fig. 11 Comparison of the effects of heat-producing element concentrations on **(a)** normalized inner core radius, and **(b)** surface strain as functions of bulk core sulfur content for a dry-olivine mantle rheology, mantle melt extraction and crustal growth, and an initial upper mantle temperature of 1,800 K. From Hauck et al. (2004)

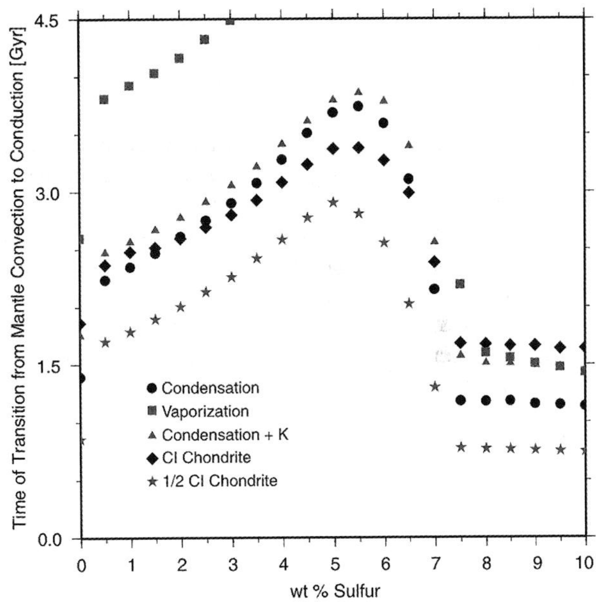

Fig. 12 Time of transition from mantle convection to conduction in Gy, where 4.5 Gy corresponds to the present, for several models of interior heat production

models, is possible only under limited conditions: a weak (e.g., wet) mantle flow law, large sulfur content of the core, and global contraction a factor of two or more greater than current estimates. The likelihood and importance of heat production from ^{40}K in the core (Murthy et al. 2003) have not been thoroughly assessed for Mercury. Compositionally induced convection, fueled by the sinking of solid iron-nickel in a cooling outer core in which light-element (e.g., sulfur) enrichment progressively increases, eases these restrictions considerably and may be favored if Mercury's magnetic field has a dynamo origin.

7 Looking Ahead

Constraints on obliquity and libration provided by Earth-based radar (Margot et al. 2007) prior to MESSENGER will allow improved a priori estimates of these parameters, permitting more accurate measurements to be recovered by MESSENGER. These, combined with improved estimates of the degree-two gravitational harmonic coefficients from radio tracking of the MESSENGER spacecraft (Srinivasan et al. 2007), along with refined geochemical models enabled by MESSENGER observations of surface chemistry (Boynton et al. 2007), will permit a direct estimate of Mercury's core size. Any difference (or not) in the forced libration of Mercury's surface and deep interior as determined by tracking the libration in the long-wavelength topography and gravity fields, respectively, will allow the viscous coupling of the lithosphere and deep interior to be estimated (Zuber and Smith 1997). An improved knowledge of the present core state will allow reconstructions of the planet's thermal evolution over time to be sharpened.

Recent observational evidence that Mercury has a liquid outer core (Margot et al. 2007) strengthens the hypothesis that the planet's magnetic field is generated by a dynamo, though it does not invalidate the possibility that a remanent crustal field could potentially contribute to the field observed by Mariner 10 (Stephenson 1976; Aharonson et al. 2004). The current state of dynamo modeling demonstrates that a dynamo solution for Mercury's magnetic field is possible on the grounds that both thin-shell and thick-shell dynamos can produce fields with Mercury-like partitioning of toroidal and poloidal fields. However, the success of these models does not rule out crustal magnetization or a thermoelectric dynamo (see Stevenson 1987) as the source of Mercury's field. Future measurements of Mercury's magnetic field by the MESSENGER mission should resolve the issue. If any field variability in time is observed, then an active dynamo source will be indicated. If the field structure is correlated with gravity signatures at wavelengths appropriate to topography at the core–mantle boundary, then a thermoelectric dynamo will be the most likely answer (Giampieri and Balogh 2002). Any small-scale structure with shallow source depths is crustal in origin. If no time variation is detected, such a result would not rule out a dynamo source; rather it would mean that the timescale of secular variation is longer than the length of time the observations have been carried out. Determining whether the field is crustal or dynamo generated in this case may be possible if evidence of an effect due to the tangent cylinder is seen. If the character of the magnetic field is different inside and outside the tangent cylinder due to different convection patterns in these regions, then a dynamo source for the field may be indicated.

The Mercury Planetary Orbiter spacecraft in the BepiColombo mission, expected to explore Mercury subsequent to MESSENGER, is planned to be in a comparatively low-eccentricity orbit with an equatorial periapsis (Grard and Balogh 2001). If implemented as currently planned, BepiColombo will provide much-improved altimetric and gravitational coverage of the southern hemisphere of Mercury, allowing global models of crustal and lithosphere structure and refinement of rotational state. Measurements of Mercury's magnetic field, coupled with those made by MESSENGER, will extend the temporal baseline over which temporal variations in the internal field may be discernable. These observations will collectively allow a fuller assessment of the relationship between Mercury's core state and the planet's thermal and geologic evolution.

Acknowledgements The MESSENGER mission is supported by NASA's Discovery Program through contract NASW-00002 with the Carnegie Institution of Washington.

References

M.H. Acuña et al., Science **279**, 1676–1680 (1998)
M.H. Acuña et al., Science **284**, 790–793 (1999)
M.H. Acuña et al., J. Geophys. Res. **106**, 23,403–23,417 (2001)
J.B. Adams et al., in *Basaltic Volcanism on the Terrestrial Planets*, ed. by T.R. McGetchin, R.O. Pepin, R.J. Phillips (Pergamon, New York, 1981), pp. 439–490
O. Aharonson, M.T. Zuber, S.C. Solomon, Earth Planet. Sci. Lett. **218**, 261–268 (2004)
J.D. Anderson, G. Colombo, P.B. Esposito, E.L. Lau, G.B. Trager, Icarus **71**, 337–349 (1987)
J.D. Anderson, R.F. Jurgens, E.L. Lau, M.A. Slade, III, G. Schubert, Icarus **124**, 690–697 (1996)
B.G. Bills, G.A. Neumann, D.E. Smith, M.T. Zuber, J. Geophys. Res. **110**, E07004 (2005). doi:10.1029/2004JE002376
M.B. Boslough, E.P. Chael, T.G. Trucano, D.A. Crawford, D.L. Campbell, in *The Cretaceous-Tertiary Event and Other Catastrophes in Earth History, Spec. Paper 307*, ed. by G. Ryder, D. Fastovsky, S. Gartner (Geol. Soc. Am., Boulder, 1996), pp. 541–550
W.V. Boynton et al., Space Sci. Rev. (2007, this issue). doi:10.1007/s11214-007-9258-3
B.A. Buffett, J. Geophys. Res. **97**, 19,581–19,597 (1992)
B.A. Buffett, E.J. Garnero, R. Jeanloz, Science **290**, 1338–1342 (2000)
K.C. Burke et al., in *Basaltic Volcanism on the Terrestrial Planets*, ed. by T.R. McGetchin, R.O. Pepin, R.J. Phillips (Pergamon, New York, 1981), pp. 803–898
J.B. Burns, Icarus **28**, 453–458 (1976)
A.G.W. Cameron, Icarus **64**, 285–294 (1985)
P. Cassen, R.E. Young, G. Schubert, R.T. Reynolds, Icarus **28**, 501–508 (1976)
J.F. Cavanaugh et al., Space Sci. Rev. (2007, this issue). doi:10.1007/s11214-007-9273-4
M.J. Cintala, C.A. Wood, J.W. Head, *Proc. Lunar Planet. Sci. Conf. 8th*, 1977, pp. 3409–3425
G. Colombo, Nature **208**, 575 (1965)
G. Colombo, I.I. Shapiro, Astrophys. J. **145**, 296–307 (1966)
J.E.P. Connerney, N.F. Ness, in *Mercury*, ed. by F. Vilas, C.R. Chapman, M.S. Matthews (Univ. Ariz. Press, Tucson, 1988), pp. 494–513
A.C.M. Correia, J. Laskar, Nature **429**, 848–850 (2004)
C.C. Counselman, Ph.D. thesis, Mass. Inst. Tech., Cambridge, 1969
C.C. Counselman, I.I. Shapiro, *Symposia Mathematica, Instituto Nazionale di Alta Matematica*, vol. III (Academic, London, 1970) pp. 121–169
G.A. de Wijs et al., Nature **392**, 805–807 (1998)
D. Dzurisin, J. Geophys. Res. **83**, 383–386 (1978)
A. Einstein, Annalen der Physik, 1916, 46 pp
B.J. Fegley, A.G.W. Cameron, Earth Planet. Sci. Lett. **82**, 207–222 (1987)
L. Fleitout, P.G. Thomas, Earth Planet. Sci. Lett. **58**, 104–115 (1982)
P.E. Fricker, R.T. Reynolds, A.L. Summers, P.M. Cassen, Nature **259**, 293–294 (1976)
G. Giampieri, A. Balogh, Planet. Space Sci. **50**, 757–762 (2002)
G.A. Glatzmaier, P.H. Roberts, Nature **377**, 203–209 (1995)
P. Goldreich, S.J. Peale, Astron. J. **71**, 425–438 (1966)
P. Goldreich, A. Toomre, J. Geophys. Res. **74**, 2555–2567 (1969)
R. Grard, A. Balogh, Planet. Space Sci. **49**, 1395–1407 (2001)
H.P. Greenspan, G.N. Howard, J. Fluid Mech. **17**, 17,385–17,404 (1963)
B. Hapke, G.E.J. Danielson, K. Klaasen, L. Wilson, J. Geophys. Res. **80**, 2431–2443 (1975)
H. Harder, G. Schubert, Icarus **151**, 118–122 (2001)
J.K. Harmon, D.B. Campbell, in *Mercury*, ed. by F. Vilas, C.R. Chapman, M.S. Matthews (Univ. Ariz. Press, Tucson, 1988), pp. 101–117
J.K. Harmon, D.B. Campbell, D.L. Bindschadler, J.W. Head, I.I. Shapiro, J. Geophys. Res. **91**, 385–401 (1986)
S.A. Hauck II, A.J. Dombard, R.J. Phillips, S.C. Solomon, Earth Planet. Sci. Lett. **22**, 713–728 (2004)
J.W. Head et al., in *Basaltic Volcanism on the Terrestrial Planets*, ed. by T.R. McGetchin, R.O. Pepin, R.J. Phillips (Pergamon, New York, 1981), pp. 701–800
J.W. Head et al., Space Sci. Rev. (2007, this issue). doi:10.1007/s11214-007-9263-6
M.H. Heimpel, J.M. Aurnou, F.M. Al-Shamali, N. Gomez Perez, Earth Planet. Sci. Lett. **236**, 542–557 (2005)
F. Herbert, M. Wiskerchen, C.P. Sonett, J.K. Chao, Icarus **28**, 489–500 (1976)
R. Hide, Phil. Trans. R. Soc. Lond. **A328**, 351–363 (1989)
L.V. Holin, Radiophys. Quantum Electron. **31**, 371–374 (1988)
L.V. Holin, Radiophys. Quantum Electron. **35**, 284–287 (1992)
L.V. Holin, Meteorit. Planet. Sci. **38**(Suppl.), A9 (2003)

H.G. Hughes, F.N. App, T.R. McGetchin, Phys. Earth Planet. Interiors **15**, 251–263 (1977)
R. Jeanloz, D.L. Mitchell, A.L. Sprague, I. de Pater, Science **268**, 1455–1457 (1995)
W.M. Kaula, *Theory of Satellite Geodesy: Applications of Satellites to Geodesy* (Blaisdell, Waltham, 1966), 124 pp
W.M. Kaula, *An Introduction to Planetary Physics: The Terrestrial Planets* (Wiley, New York, 1968), 490 pp
W.M. Kaula, J. Geophys. Res. **84**, 999–1008 (1979)
G. Kletetschka, P.J. Wasilewski, P.T. Taylor, Earth Planet. Sci. Lett. **176**, 469–479 (2000a)
G. Kletetschka, P.J. Wasilewski, P.T. Taylor, Meteorit. Planet. Sci. **35**, 895–899 (2000b)
W. Kuang, J. Bloxham, Nature **389**, 371–374 (1997)
W. Kuang, J. Bloxham, J. Comp. Phys. **153**, 51–81 (1999)
K. Lambeck, S. Pullan, Phys. Earth Planet. Interiors **22**, 29–35 (1980)
M.A. Leake, in *Advances in Planetary Geology*. NASA TM-84894, Washington, DC, 1982, pp. 3–535
J.S. Lewis, Earth Planet. Sci. Lett. **15**, 286–290 (1972)
J.S. Lewis, Ann. Rev. Phys. Chem. **24**, 339–351 (1973)
M.C. Malin, *Proc. Lunar Planet. Sci. Conf. 9th*, 1978, pp. 3395–3409
J.-L. Margot, S.J. Peale, R.F. Jurgens, M.A. Slade, I.V. Holin, Science **316**, 710–714 (2007)
T.A. Maxwell, A.W. Gifford, *Proc. Lunar Planet. Sci. Conf. 11th*, 1980, pp. 2447–2462
T.B. McCord, R.N. Clark, J. Geophys. Res. **84**, 7664–7668 (1979)
W.B. McKinnon, ed. by in R.B. Merrill, P.H. Schultz, Multi-ring basins, proc. Lunar Planet. Sci. **12**, Part A, 1981, Pergamon Press, New York, pp. 259–273
W.B. McKinnon, H.J. Melosh, Icarus **44**, 454–471 (1980)
H.J. Melosh, Icarus **31**, 221–243 (1977)
H.J. Melosh, *Proc. Lunar Planet. Sci. Conf. 9th*, 1978, pp. 3513–3525
H.J. Melosh, D. Dzurisin, Icarus **35**, 227–236 (1978)
H.J. Melosh, W.B. McKinnon, Geophys. Res. Lett. **5**, 985–988 (1978)
H.J. Melosh, W.B. McKinnon, in *Mercury*, ed. by F. Vilas, C.R. Chapman, M.S. Matthews (Univ. Ariz. Press, Tucson, 1988), pp. 374–400
W.H. Munk, G.J.F. MacDonald, *The Rotation of the Earth* (Cambridge Univ. Press, New York, 1960), 323 pp
B.C. Murray, J. Geophys. Res. **80**, 2342–2344 (1975)
B.C. Murray et al., Science **185**, 169–179 (1974)
V.R. Murthy, W. van Westrenen, Y. Fei, Nature **423**, 163–165 (2003)
N.F. Ness, Space Sci. Rev. **21**, 527–553 (1978)
N.F. Ness, Ann. Rev. Earth Planet. Sci. **7**, 249–288 (1979)
N.F. Ness, K.W. Behannon, R.P. Lepping, Y.C. Whang, K.H. Schatten, Science **185**, 151–160 (1974)
N.F. Ness, K.W. Behannon, R.P. Lepping, Y.C. Whang, J. Geophys. Res. **80**, 2708–2716 (1975)
N.F. Ness, K.W. Behannon, R.P. Lepping, Y.C. Whang, Icarus **28**, 479–488 (1976)
G.A. Neumann, M.T. Zuber, D.E. Smith, F.G. Lemoine, J. Geophys. Res. **101**, 16,841–16,863 (1996)
V.R. Oberbeck, W.L. Quaide, R.E. Arvidson, H.R. Aggarwal, J. Geophys. Res. **82**, 1681–1698 (1977)
P. Olson, J. Aurnou, Nature **402**, 170–173 (1999)
A. Pais, G. Hulot, Phys. Earth Planet. Interiors **118**, 291–316 (2000)
S.J. Peale, Astron. J. **74**, 483–489 (1969)
S.J. Peale, Astron. J. **79**, 722–744 (1974)
S.J. Peale, Nature **262**, 765–766 (1976)
S.J. Peale, Icarus **48**, 143–145 (1981)
S.J. Peale, in *Mercury*, ed. by F. Vilas, C.R. Chapman, M.S. Matthews (Univ. Ariz. Press, Tucson, 1988), pp. 461–493
S.J. Peale, Icarus **178**, 4–18 (2005)
S.J. Peale, Icarus **181**, 338–347 (2006)
S.J. Peale, A.P. Boss, J. Geophys. Res. **82**, 743–749 (1977)
S.J. Peale, R.J. Phillips, S.C. Solomon, D.E. Smith, M.T. Zuber, Meteorit. Planet. Sci. **37**, 1269–1283 (2002)
S.J. Peale, M. Yseboot, J.-L. Margot, Icarus **87**, 365–373 (2007)
G.H. Pettengill, R.B. Dyce, Nature **206**, 1240 (1965)
T.R. Quinn, S. Tremaine, M. Duncan, Astron. J. **101**, 2287–2305 (1991)
A.E. Ringwood, *The Origin of the Earth and Moon* (Springer, New York, 1979), 295 pp
M.S. Robinson, P.G. Lucey, Science **275**, 197–200 (1997)
P. Rochette, J.P. Lorand, G. Fillion, V. Sautter, Earth Planet. Sci. Lett. **190**, 1–12 (2001)
N.T. Roseveare, *Mercury's Perihelion from Le Verrier to Einstein* (Oxford Sci. Publications, Clarendon Press, 1982), 208 pp
S.K. Runcorn, Nature **253**, 701–703 (1975a)
S.K. Runcorn, Phys. Earth Planet. Interiors **10**, 327–335 (1975b)

C.T. Russell, D.N. Baker, J.A. Slavin, in *Mercury*, ed. by F. Vilas, C.R. Chapman, M.S. Matthews (Univ. Ariz. Press, Tucson, 1988), pp. 514–561
V.S. Safronov, Icarus **33**, 1–12 (1978)
G. Schubert, P.M. Cassen, R.E. Young, Icarus **38**, 192–211 (1979)
G. Schubert, M.N. Ross, D.J. Stevenson, T. Spohn, in *Mercury*, ed. by F. Vilas, C.R. Chapman, M.S. Matthews (Univ. Ariz. Press, Tucson, 1988), pp. 429–460.
P.H. Schultz, D.E. Gault, *Proc. Lunar Sci. Conf. 6th*, 1975a, pp. 2845–2862
P.H. Schultz, D.E. Gault, Moon **12**, 159–177 (1975b)
I.I. Shapiro, G.H. Pettengill, M.E. Ash, R.P. Ingalls, D.B. Campbell, R.B. Dyce, Phys. Rev. Lett. **28**, 1594–1597 (1972)
R.W. Siegfried, II, S.C. Solomon, Icarus **23**, 192–205 (1974)
M. Simons, S.C. Solomon, B.H. Hager, Geophys. J. Int. **131**, 24–44 (1997)
J.A. Slavin et al., Space Sci. Rev. (2007, this issue). doi:10.1007/s11214-007-9154-x
J.C. Smith, G.H. Born, Icarus **27**, 51–53 (1976)
S.C. Solomon, Icarus **28**, 509–521 (1976)
S.C. Solomon, Phys. Earth Planet. Interiors **15**, 135–145 (1977)
S.C. Solomon, Geophys. Res. Lett. **5**, 461–464 (1978)
S.C. Solomon, J.W. Head, Rev. Geophys. **18**, 107–141 (1980)
S.C. Solomon et al., in *Basaltic Volcanism on the Terrestrial Planets*, ed. by T.R. McGetchin, R.O. Pepin, R.J. Phillips (Pergamon, New York, 1981), pp. 1129–1233
S.C. Solomon et al., Planet. Space Sci. **49**, 1445–1465 (2001)
A.L. Sprague, R.W.H. Kozlowski, F.C. Witteborn, D.P. Cruikshank, D.H. Wooden, Icarus **109**, 156–167 (1994)
P.D. Spudis, J.E. Guest, in *Mercury*, ed. by F. Vilas, C.R. Chapman, M.S. Matthews (Univ. Ariz. Press, Tucson, 1988), pp. 118–164
P.D. Spudis, J.G. Prosser, Map I-1689, U.S. Geol. Survey, Denver, CO, 1984
D.K. Srinivasan, M.E. Perry, K.B. Fielhauer, D.E. Smith, M.T. Zuber, Space Sci. Rev. (2007, this issue). doi:10.1007/s11214-007-9270-7
L.J. Srnka, Phys. Earth Planet. Interiors **11**, 184–190 (1976)
E.M. Standish, Astron. Astrophys. **233**, 463–466 (1990)
E.M. Standish, X.X. Newhall, J.G. Williams, D.K. Yeomans, in *Explanatory Supplement to the Astronomical Almanac*, ed. by P.K. Seidelmann (Univ. Sci. Books, Mill Valley, 1992), pp. 279–323
S. Stanley, J. Bloxham, W.E. Hutchison, M.T. Zuber, Earth Planet. Sci. Lett. **234**, 27–38 (2005)
A. Stephenson, Earth Planet. Sci. Lett. **28**, 454–458 (1976)
D.J. Stevenson, Rept. Prog. Phys. **46**, 555–620 (1983)
D.J. Stevenson, Earth Planet. Sci. Lett. **82**, 114–120 (1987)
D.J. Stevenson, T. Spohn, G. Schubert, Icarus **54**, 466–489 (1983)
K. Stewartson, P.H. Roberts, J. Fluid Mech. **17**, 1–20 (1963)
R.G. Strom, Phys. Earth Planet. Interiors **15**, 156–172 (1977)
R.G. Strom, Space Sci. Rev. **24**, 3–70 (1979)
R.G. Strom, in *The Geology of the Terrestrial Planets*, ed. by M.H. Carr (NASA, Washington, 1984), pp. 13–55
R.G. Strom, N.J. Trask, J.E. Guest, J. Geophys. Res. **80**, 2478–2507 (1975)
A.M.K. Szeto, S. Xu, J. Geophys. Res. **102**, 27,651–27,657 (1997)
P.G. Thomas, P. Masson, L. Fleitout, in *Mercury*, ed. by F. Vilas, C.R. Chapman, M.S. Matthews (Univ. Ariz. Press, Tucson, 1988), pp. 401–428
A. Toomre, in *The Earth–Moon System*, ed. by B.G. Marsden, A.G.W. Cameron (Plenum, New York, 1966), pp. 33–45
N.J. Trask, J.E. Guest, J. Geophys. Res. **80**, 2462–2477 (1975)
N.J. Trask, R.G. Strom, Icarus **28**, 559–563 (1976)
D.L. Turcotte, G. Schubert, *Geodynamics: Applications of Continuum Physics to Geologic Problems* (Wiley, New York, 1982), 450 pp
H.C. Urey, Geochim. Cosmochim. Acta **1**, 209–277 (1951)
W.R. Ward, Astron. J. **80**, 64–69 (1975)
T.R. Watters, M.S. Robinson, A.C. Cook, Geology **26**, 991–994 (1998)
T.R. Watters, M.S. Robinson, C.R. Bina, P.D. Spudis, Geophys. Res. Lett. **31**, L04701 (2004). doi:10.1029/2003GL019171
S.J. Weidenschilling, Icarus **35**, 99–111 (1978)
G.W. Wetherill, in *Mercury*, ed. by F. Vilas, C.R. Chapman, M.S. Matthews (Univ. Ariz. Press, Tucson, 1988), pp. 670–691
D.E. Wilhelms, Icarus **28**, 551–558 (1976)

J.G. Williams, D.H. Boggs, C.F. Yoder, J.T. Ratcliff, J.O. Dickey, J. Geophys. Res. **10**, 27,933–27,968 (2001)
J.A. Wood et al., in *Basaltic Volcanism on the Terrestial Planets*, ed. by T.R. McGetchin, R.O. Pepin, R.J. Phillips (Pergamon, New York, 1981), pp. 634–699
C.F. Yoder, in *Global Earth Physics, A Handbook of Physical Constants*, ed. by T.J. Ahrens (American Geophys. Un., Washington, 1995), pp. 1–31
C.F. Yoder, A.S. Konopliv, D.-N. Yuan, E.M. Standish, W.M. Folkner, Science **300**, 299–303 (2003)
M. Yseboodt, J.-L. Margot, Icarus **181**, 327–337 (2006)
M.T. Zuber, J. Geophys. Res. **92**, E541–E551 (1987)
M.T. Zuber, D.E. Smith, Lunar Planet Sci. **27**, 1637–1638 (1997)
M.T. Zuber, D.E. Smith, F.G. Lemoine, G.A. Neumann, Science **266**, 1839–1843 (1994)

Space Sci Rev (2007) 131: 133–160
DOI 10.1007/s11214-007-9154-x

MESSENGER: Exploring Mercury's Magnetosphere

James A. Slavin · Stamatios M. Krimigis · Mario H. Acuña · Brian J. Anderson · Daniel N. Baker · Patrick L. Koehn · Haje Korth · Stefano Livi · Barry H. Mauk · Sean C. Solomon · Thomas H. Zurbuchen

Received: 22 May 2006 / Accepted: 1 February 2007 /
Published online: 21 June 2007

Abstract The MErcury Surface, Space ENvironment, GEochemistry, and Ranging (MESSENGER) mission to Mercury offers our first opportunity to explore this planet's miniature magnetosphere since the brief flybys of Mariner 10. Mercury's magnetosphere is unique in many respects. The magnetosphere of Mercury is among the smallest in the solar system; its magnetic field typically stands off the solar wind only ~1000 to 2000 km above the surface. For this reason there are no closed drift paths for energetic particles and, hence, no radiation belts. Magnetic reconnection at the dayside magnetopause may erode the subsolar magnetosphere, allowing solar wind ions to impact directly the regolith. Inductive currents in Mercury's interior may act to modify the solar wind interaction by resisting changes

J.A. Slavin (✉)
Heliophysics Science Division, Goddard Space Flight Center, Code 670, Greenbelt, MD 20771, USA
e-mail: james.a.slavin@nasa.gov

S.M. Krimigis · B.J. Anderson · H. Korth · S. Livi · B.H. Mauk
The Johns Hopkins University Applied Physics Laboratory, Laurel, MD 20723, USA

M.H. Acuña
Solar System Exploration Division, Goddard Space Flight Center, Code 690, Greenbelt, MD 20771, USA

D.N. Baker
Laboratory for Atmospheric and Space Physics, University of Colorado, Boulder, CO 80303, USA

P.L. Koehn
Physics and Astronomy Department, Eastern Michigan University, Ypsilanti, MI 48197, USA

S.C. Solomon
Department of Terrestrial Magnetism, Carnegie Institution of Washington, Washington, DC 20015, USA

T.H. Zurbuchen
Department of Atmospheric, Oceanic and Space Sciences, University of Michigan, Ann Arbor, MI 48109, USA

due to solar wind pressure variations. Indeed, observations of these induction effects may be an important source of information on the state of Mercury's interior. In addition, Mercury's magnetosphere is the only one with its defining magnetic flux tubes rooted beneath the solid surface as opposed to an atmosphere with a conductive ionospheric layer. This lack of an ionosphere is probably the underlying reason for the brevity of the very intense, but short-lived, ~1–2 min, substorm-like energetic particle events observed by Mariner 10 during its first traversal of Mercury's magnetic tail. Because of Mercury's proximity to the sun, 0.3–0.5 AU, this magnetosphere experiences the most extreme driving forces in the solar system. All of these factors are expected to produce complicated interactions involving the exchange and recycling of neutrals and ions among the solar wind, magnetosphere, and regolith. The electrodynamics of Mercury's magnetosphere are expected to be equally complex, with strong forcing by the solar wind, magnetic reconnection, and pick-up of planetary ions all playing roles in the generation of field-aligned electric currents. However, these field-aligned currents do not close in an ionosphere, but in some other manner. In addition to the insights into magnetospheric physics offered by study of the solar wind–Mercury system, quantitative specification of the "external" magnetic field generated by magnetospheric currents is necessary for accurate determination of the strength and multipolar decomposition of Mercury's intrinsic magnetic field. MESSENGER's highly capable instrumentation and broad orbital coverage will greatly advance our understanding of both the origin of Mercury's magnetic field and the acceleration of charged particles in small magnetospheres. In this article, we review what is known about Mercury's magnetosphere and describe the MESSENGER science team's strategy for obtaining answers to the outstanding science questions surrounding the interaction of the solar wind with Mercury and its small, but dynamic, magnetosphere.

Keywords Planetary magnetospheres · Reconnection · Particle acceleration · Substorms · Mercury · MESSENGER

1 Introduction: What Do We Presently Know and How Do We Know It?

Launched on November 2, 1973, Mariner 10 (M10) executed the first reconnaissance of Mercury during its three encounters on March 29, 1974, September 21, 1974, and March 16, 1975 (see reviews by Ness 1979; Russell et al. 1988; Slavin 2004; Milillo et al. 2005). All flybys occurred at a heliocentric distance of 0.46 AU, but only the first (Mercury I) and third (Mercury III) encounters passed close enough to Mercury to return observations of the solar wind interaction and the planetary magnetic field. The first encounter targeted the planetary "wake" and returned surprising observations that indicate a significant intrinsic magnetic field. The closest approach to the surface during this passage was 723 km where a peak magnetic field intensity of 98 nT was observed (Ness et al. 1974). During Mercury I the magnetic field investigation observed clear bow shock and magnetopause boundaries along with the lobes of the tail and the cross-tail current layer (Ness et al. 1974, 1975, 1976). The Mercury III observations were of great importance because they confirmed that the magnetosphere was indeed produced by the interaction of the solar wind with an intrinsic planetary magnetic field. Once corrected for the differing closest approach distances, the polar magnetic fields measured during Mercury III are about twice as large as those along the low-latitude Mercury I trajectory, consistent with a primarily dipolar planetary field. Magnetic field models derived using different subsets of the Mariner 10 data and various assumptions concerning the external magnetospheric magnetic field indicate that the tilt of the dipole relative to the planetary rotation axis is about 10°, but the longitude angle of the dipole is very poorly constrained (Ness et al. 1976).

The plasma investigation was hampered by a deployment failure that kept it from returning ion measurements. Fortunately, the electron portion of the plasma instrument did operate as planned (Ogilvie et al. 1974). Good correspondence was found between the magnetic field and plasma measurements as to the locations of the Mercury I and III bow shock and magnetopause boundaries. Plasma speed and density parameters derived from the electron data produced consistent results regarding bow shock jump conditions and pressure balance across the magnetopause (Ogilvie et al. 1977; Slavin and Holzer 1979a). Within Mercury's magnetosphere, plasma density was found to be higher than that observed at Earth by a factor comparable to the ratio of the solar wind density at the orbits of the two planets (Ogilvie et al. 1977). Similar correlations are observed between solar wind and plasma sheet density at Earth (Terasawa et al. 1997). Throughout the Mercury I pass plasma sheet-type electron distributions were observed with an increase in temperature beginning near closest approach coincident with a series of intense energetic particle events (Ogilvie et al. 1977; Christon 1987).

Several groups have estimated the magnetic moment of Mercury from the observations made during Mercury I and III. Conducting a least-squared fit of the Mercury I data to an offset tilted dipole, Ness et al. (1974) obtained a dipole moment of 227 $\mathrm{nT}R_M^3$, where R_M is Mercury's radius (1 $R_M = 2439$ km), and a dipole tilt angle of 10° relative to the planetary rotation axis. Ness et al. (1975) considered a centered dipole and an external contribution to the measured magnetic field and found the strength of the dipole to be 349 $\mathrm{nT}R_M^3$ from the same data set. From the Mercury III encounter observations these authors determined a dipole moment of 342 $\mathrm{nT}R_M^3$ (Ness et al. 1976). Higher-order contributions to the internal magnetic field were examined by Jackson and Beard (1977) (quadrupole) and Whang (1977) (quadrupole, octupole). Both sets of authors reported 170 $\mathrm{nT}R_M^3$ as the dipole contribution to Mercury's intrinsic magnetic field. The cause for the large spread in the reported estimates of the dipole term is the limited spatial coverage of the observations, which is insufficient for separating the higher-order multipoles (Connerney and Ness 1988), and variable magnetic field contributions from the magnetospheric current systems (Slavin and Holzer 1979b; Korth et al. 2004; Grosser et al. 2004).

Mercury's magnetosphere is one of the most dynamic in the solar system. A glimpse of this variability was captured during the Mercury I encounter. Less than a minute after M10 entered the plasma sheet during this first flyby there was a sharp increase in the B_z field component (Ness et al. 1974). The initial sudden B_z increase and subsequent quasi-periodic increases are nearly coincident with strong enhancements in the flux of >35 keV electrons observed by the cosmic ray telescopes (Simpson et al. 1974; Eraker and Simpson 1986; Christon 1987). Taken together, these M10 measurements are very similar to the "dipolarizations" of the near-tail magnetic field frequently observed in association with energetic particle "injections" at Earth (Christon et al. 1987). This energetic particle signature, and several weaker events observed later in the outbound Mercury I pass, was interpreted as strong evidence for substorm activity and, by inference, magnetic reconnection in the tail (Siscoe et al. 1975; Eraker and Simpson 1986; Baker et al. 1986; Christon 1987).

The stresses exerted on planetary magnetic fields by magnetospheric convection are transmitted down to the planet and its environs by Alfven waves carrying field-aligned current (FAC). At planets with electrically conductive ionospheres, such as the Earth, Jupiter, and Saturn, these current systems are well observed and transfer energy to their ionospheres, an important energy sink as well as serving as a "brake" that limits the speed and rate of change of the plasma convection (Coroniti and Kennel 1973). Mercury's atmosphere, however, is a tenuous exosphere, and no ionosphere possessing significant electrical conductance is present (Lammer and Bauer 1997). For these reasons, the strong variations in the east-west component of the magnetic field measured by M10 during the Mercury I pass several minutes following the substorm-like signatures may be quite significant. These perturbations were first examined by Slavin et al. (1997), who concluded that the spacecraft crossed three FAC sheets similar to those often observed at the Earth (Iijima and Potemra 1978). The path by which these currents close is not known, but their existence may indicate that the conductivity of the regolith is greater than is usually assumed (e.g., Hill et al. 1976) or other closure paths exist (Glassmeier 2000).

Another unique aspect of Mercury concerns the origin of its very tenuous, collisionless, neutral atmosphere (e.g., Goldstein et al. 1981). Three exospheric neutral species, Na, K, and Ca, have been measured spectroscopically from the Earth (Potter and Morgan 1985, 1986; Bida et al. 2000), and three other species, H, He, and O, were observed by Mariner 10 (Broadfoot et al. 1976). The large day-to-day variability in the sodium and potassium exosphere at Mercury, including changes in both total density and global distribution, are quite striking and may be linked to dynamic events in the solar wind and their effect on the magnetosphere. For example, Potter et al. (1999) suggested that the underlying cause of the large day-to-day changes in the neutral exosphere might be the modulation of the surface sputtering rates by variations in the spatial distribution and intensity of solar wind proton impingement on the surface. Our present understanding of Mercury's neutral atmosphere and the contributions that the MErcury Surface, Space ENvironment, GEochemistry, and Ranging (MESSENGER) mission will make to this discipline are the subject of a companion paper (Domingue et al. 2007).

2 MESSENGER Science Instruments

The MESSENGER spacecraft, instrument payload, and mission plan have been described elsewhere (Gold et al. 2001). Here we provide a brief overview to emphasize the nature of the measurements to be returned and how they will be used to achieve the mission's scientific objectives (Solomon et al. 2001, 2007). The MESSENGER spacecraft is shown in Fig. 1. A key aspect of its design is the presence of a large sunshade that faces sunward when the spacecraft is closer than 0.9 AU to the Sun. The spacecraft is three-axis stabilized, but rotations about the Sun-spacecraft axis will be carried out while in Mercury orbit to orient some of the instruments toward the surface.

The MESSENGER Magnetometer (MAG) is described in detail by Anderson et al. (2007). The triaxial sensor is mounted at the end of a 3.6-m boom to minimize the magnitude of stray spacecraft-generated magnetic fields at the MAG sensor location. Ground testing and in-flight calibration have shown that the intensity of uncorrectable (i.e., variable) stray fields will be less than 0.1 nT (Anderson et al. 2007). While the magnetometer is capable of measuring the full strength of the Earth's field for integration and check-out, it is designed to operate in its most sensitive field range of ± 1500 nT per axis when the spacecraft is orbit about Mercury. The 16-bit telemetered resolution yields a digital resolution of 0.05 nT. In the baseline mission plan, the sampling rate of the instrument will be varied according to a pre-planned schedule from 2 to 20 vectors s^{-1} once in orbit about Mercury. Additionally, 8-minute intervals of 20-vectors s^{-1} burst data will be acquired during periods of lower-rate continuous sampling. The accuracy of the MESSENGER magnetic field measurements is 0.1%.

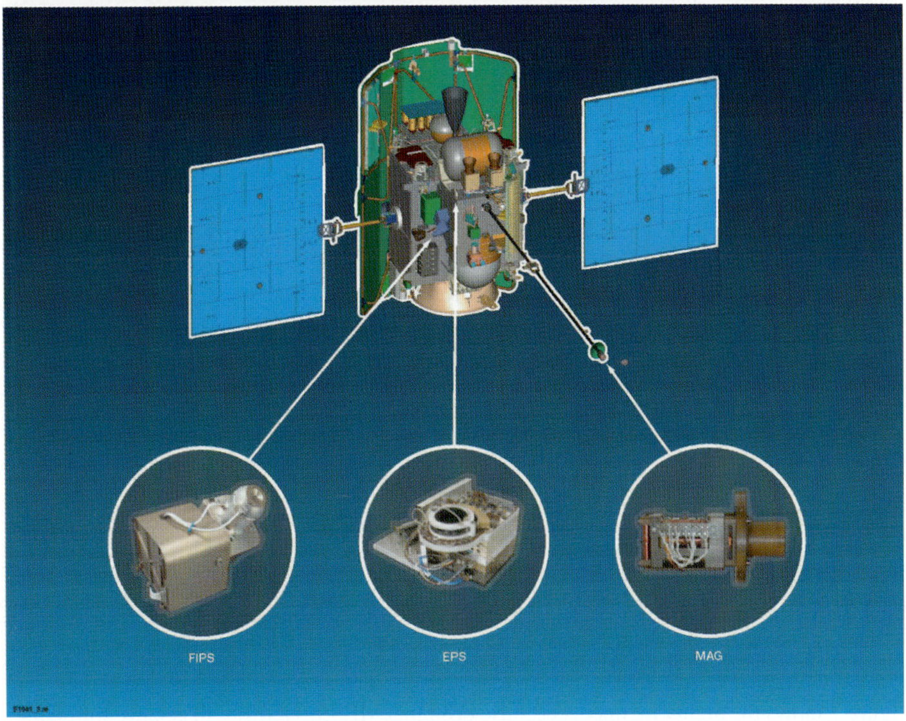

Fig. 1 The MESSENGER spacecraft behind its sunshade. Note the adapter ring at the bottom of the vehicle which encloses the planet-nadir-pointing instruments. The Magnetometer (MAG) is located at the end of a 3.6-m double-hinged boom. The FIPS and EPS sensors are shown along with arrows indicating their locations on the spacecraft

The MESSENGER Energetic Particle and Plasma Spectrometer (EPPS) is described in detail by Andrews et al. (2007). EPPS is composed of two charged-particle detector systems, the Fast Imaging Plasma Spectrometer (FIPS) and the Energetic Particle Spectrometer (EPS). FIPS has a near-hemispherical field of view and accepts ions with an energy-to-charge ratio from 0.05 to 20 keV/q. EPS has a $12^\circ \times 160^\circ$ field of view and accepts ions and electrons with energies of 10 keV to 5 MeV and 10 keV to 400 keV, respectively. The EPPS sensors are mounted between the attach points for the MAG boom and one of the solar arrays. The fields of view of both the FIPS and EPS are such that they will measure charged particles coming from the anti-sunward direction, as well as from above and below the spacecraft. The FIPS field of view also encompasses the sunward direction, but this portion of its field-of-view is blocked by the spacecraft and the sunshade. The EPPS measurements are central to resolving the issues that arose from the incomplete and ambiguous energetic particle measurements of M10. The energetic particle measurements of Simpson et al. (1974) were compromised by electron pileup in the proton channel (Armstrong et al. 1975). They were later reinterpreted as having been responding to intense fluxes of >35 keV electrons (Christon 1987).

3 MESSENGER Mission Plan

When evaluating the potential scientific impact of in-situ magnetospheric measurements, the spatial coverage of the critical boundaries and regions is one of the most important factors.

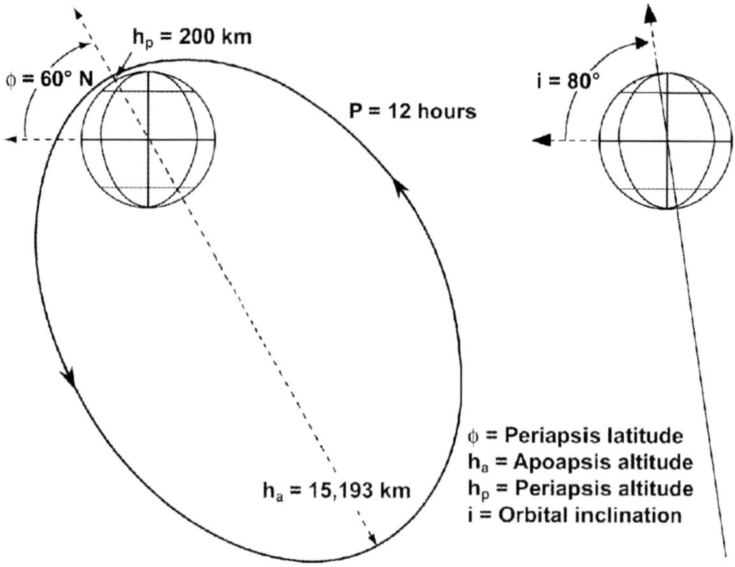

Fig. 2 Orthogonal views of the 12-hr-long MESSENGER orbit. The *left-hand image* shows the orbital plane; periapsis and apoapsis altitudes are 200 km and 15 193 km, respectively. The 80° inclination of the orbit is apparent in the orthogonal view in the *right-hand image*

Figure 2 shows the highly inclined, eccentric orbit that MESSENGER will achieve following insertion and trim maneuvers. This orbit represents a carefully considered trade between the sometimes competing requirements of the planetary interior, surface geology, atmospheric and magnetospheric science investigations and engineering constraints, particularly those related to the thermal environment (Solomon et al. 2001; Santo et al. 2001). This orbit satisfies the primary requirements of all of these planetary science disciplines and will enable an outstanding set of measurements to be gathered.

From the standpoint of the magnetosphere, the most informative view of the MESSENGER orbital coverage is to examine it relative to the bow shock and magnetopause surfaces. These boundaries are, to first order, axially symmetric with respect to the X axis in Mercury-Solar-Orbital (MSO) coordinates. This coordinate system is the Mercury equivalent of the familiar Geocentric-Solar-Ecliptic (GSE) system used at the Earth. In this system X_{MSO} is directed from the planet's center to the Sun, Y_{MSO} is in the plane of Mercury's orbit and positive opposite to the planetary velocity vector, and Z_{MSO} completes the right handed system. Mercury's rotation axis is normal to its orbital plane and, therefore, parallel to the Z_{MSO} axis. As discussed more fully in Anderson et al. (2007), Mercury's magnetic field is best described by a planet-centered dipole whose tilt relative to the Z_{MSO} axis is about 10° (Connerney and Ness 1988). However, the longitude of Mercury's magnetic poles are not well constrained by the Mariner 10 data (Ness et al. 1976). (Note: for many science applications the MSO coordinates will be "aberrated" using the relative speeds of the planet and the solar wind so that the X'_{MSO} axis is opposite to the mean solar wind velocity direction in the rest frame of Mercury. Due to the high orbital speed of Mercury, especially at perihelion, the aberration angle can approach, and for slow solar wind speeds exceed 10°.)

The efficacy of the MESSENGER orbit for magnetospheric investigations may be judged by plotting these boundaries and the trace of the orbit over a Mercury year in the $(Y_{\mathrm{MSO}}^2 + Z_{\mathrm{MSO}}^2)^{1/2}$ versus X_{MSO} and Z_{MSO} versus Y_{MSO} planes. In Figs. 3a and 3b, bow

Fig. 3a Projection of the first Mercury year of predicted orbits for MESSENGER onto the X_{MSO}–$(Y^2_{\mathrm{MSO}} + Z^2_{\mathrm{MSO}})^{1/2}$ plane. The conic traces in red are the expected mean locations of the magnetopause and bow shock boundaries on the basis of the two M10 encounters

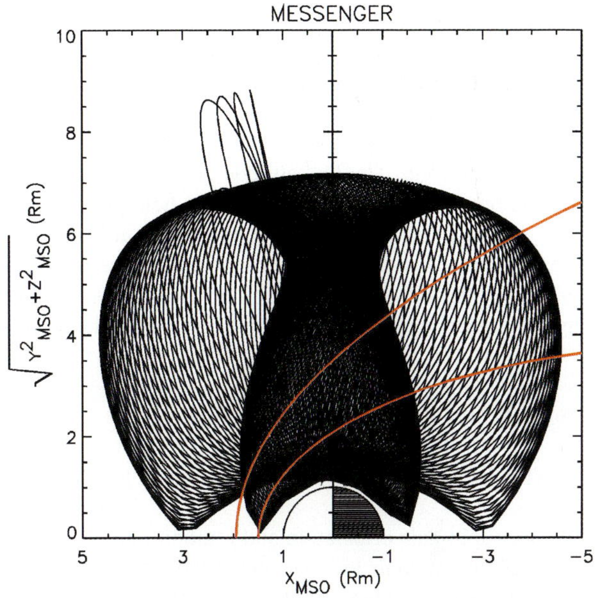

Fig. 3b Projection of the first Mercury year of predicted orbits for MESSENGER onto the Y_{MSO}–Z_{MSO} plane. The circular traces are the expected mean locations of the magnetopause and bow shock boundaries in the $X_{\mathrm{MSO}} = 0$ plane on the basis of two M10 encounters

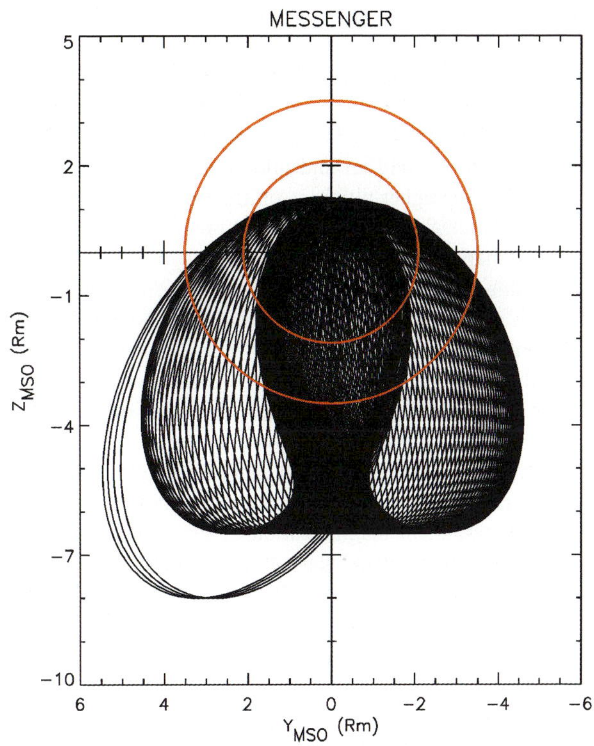

shock and magnetopause surfaces are displayed based upon the subsolar magnetopause altitudes for northward B_z for a 6×10^{-8} dyne/cm^2 solar wind pressure determined by Slavin and Holzer (1979a). Mariner 10 crossed these boundaries too few times, over too restricted of a range of solar zenith angles, to allow their shape to be accurately mapped. Hence, terrestrial bow shock and magnetopause shapes from Slavin and Holzer (1981) and Holzer and Slavin (1978), respectively, have been assumed. As shown, the MESSENGER orbit will provide dense sampling of all of the primary regions of the magnetosphere and its interaction with the solar wind. The low-altitude polar passes in the northern hemisphere provide an excellent opportunity to observe and map field aligned currents. The region south of Mercury's orbital plane is better sampled at high altitudes than those to the north. However, we expect that Mercury's magnetosphere possesses considerable north-south symmetry. In this manner, the MESSENGER orbit provides nearly comprehensive coverage of Mercury's magnetosphere and solar wind boundaries sunward of $X_{\mathrm{MSO}} \sim -3\ R_M$.

4 Solar Wind–Magnetosphere Interaction

4.1 What Is the Origin of Mercury's Magnetic Field?

A planet's spin axis orientation and rotation rate, its atmosphere, the existence and nature of any satellites, and its location within the heliosphere are all important factors influencing magnetospheric structure and dynamics. However, the single most important factor is the nature of its magnetic field, consisting of the magnetic field intrinsic to the planet and an external contribution due to magnetospheric currents. As shown in Fig. 4, the sum of the primarily dipolar planetary magnetic field and the fields due to the external currents produce a magnetospheric magnetic field that is very different from a vacuum dipole even quite close to the planet's surface. On the dayside, the magnetic field is greatly compressed; the intensity near the subsolar point is about twice that due to the planetary field alone. Conversely, the surface magnetic field around midnight is somewhat reduced from that due to the planet alone, while at higher altitudes on the nightside the local magnetic field is much stronger than the planetary dipole field would predict due to the current systems that form the long extended magnetotail.

Possible sources for Mercury's magnetic field are an active dynamo, thermoremanent magnetization of the crust, or a combination thereof. On the basis of analogy with the Earth, it is often assumed that the source of Mercury's magnetic field is an active dynamo. Although thermal evolution models predict the solidification of a pure iron core early in Mercury's history (Solomon 1976), even small quantities of light alloying elements, such as sulfur or oxygen, could have prevented the core from freezing (Stevenson et al. 1983). An active hydrodynamic (Stevenson 1983) or thermoelectric (Stevenson 1987; Giampieri and Balogh 2002) dynamo operating at Mercury is, therefore, a strong possibility.

Thermoremanent magnetization of the crust may have been induced either by a large external (i.e., solar or nebular) magnetic field or by an internal dynamo that existed earlier in the planet's history. The former possibility is implausible because any early solar or nebular field would presumably have decayed much faster than the timescale for thickening of Mercury's lithosphere (Stevenson 1987). The latter hypothesis of an early dynamo as the source for thermoremanent magnetization at Mercury faces additional requirements set forth by the magnetostatic theorem of Runcorn (1975a, 1975b). Runcorn showed that the symmetry of the magnetic field due to thermoremanent magnetization of a uniform, thin shell by a formerly active internal dynamo at the planet's center does not produce a magnetic field external to the planet. However, Runcorn's theorem is valid only under several ideal conditions, including that (1) the permeability of the magnetized shell was uniform

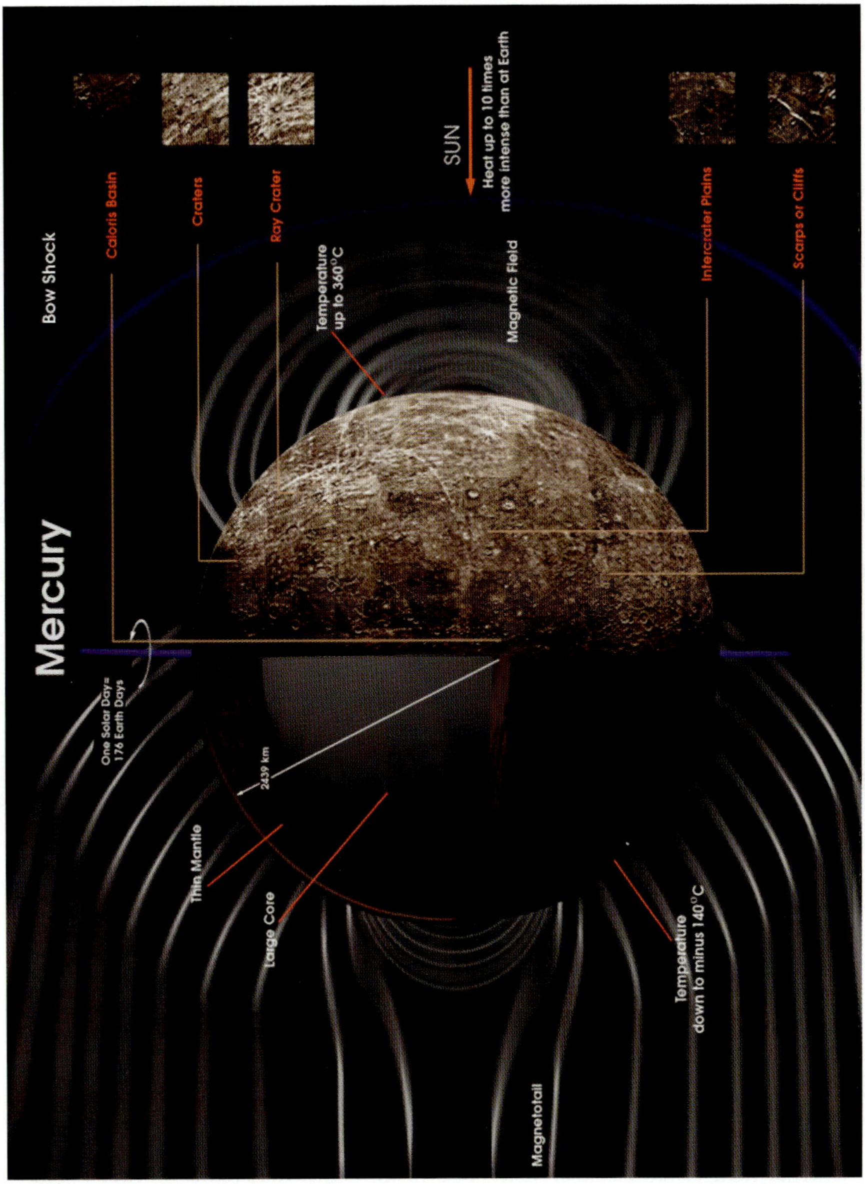

Fig. 4 Mercury's magnetic field and the strong asymmetries introduced by its interaction with the solar wind (Copyright European Space Agency)

(Stephenson 1976), (2) the cooling of the planetary interior occurred from the outermost layer progressively inward (Srnka 1976), and (3) the thermal structure of the lithosphere exhibited no asymmetries during the cooling process (Aharonson et al. 2004). Breaking any of the above stringent conditions could result in a net planetary magnetic moment. Hence, crustal magnetization cannot be excluded as a source for some or all of Mercury's planetary magnetic field. More comprehensive discussions of the important issues surrounding the origins of Mercury's magnetic field and the contributions to their solution to be made by MESSENGER can be found in companion papers by Anderson et al. (2007) and Zuber et al. (2007).

4.2 How Will MESSENGER Measurements Be Used to Determine the Origin of Mercury's Intrinsic Magnetic Field?

Determining the origin of Mercury's magnetic field is one of MESSENGER's prime objectives. The approach to addressing this objective will be to produce an accurate representation, or "map," of Mercury's intrinsic magnetic field and use it to distinguish among the several hypotheses for the field's origin. This process, combined with the MESSENGER gravity and altimetry investigations, should ultimately yield considerable insight into the interior structure and evolution of this small planet. Clues to the origin of the planetary magnetic field are also expected to be found in the multipole decomposition of the planetary field, which will be retrieved from an inversion of the magnetic field measurements.

The principal external current systems are the magnetopause current that confines much of the magnetic flux originating in the planet to the magnetospheric cavity and the cross-tail current layer that separates the two lobes of the tail. A "ring current" due to the drift motion of trapped energetic ions and electrons, observed at Earth during geomagnetic "storms," is not expected because of the absence of closed drift paths in Mercury's magnetosphere. However, a "partial" ring current may exist at times (see Glassmeier 2000). Finally, Slavin et al. (1997) have reported evidence of high-latitude field-aligned currents at Mercury, but owing to the absence of a conducting ionosphere, their global distribution may differ significantly from those at Earth.

Two methods of accounting for the external field contribution are typically used when inverting the measured magnetic field to create model descriptions of the intrinsic magnetic field. In the first, a spherical harmonic expansion series is derived for the planetary field and the external field is treated by adding a scalar potential function. Whether a scalar representation best captures the external contribution is not clear. The second approach applies our present understanding of magnetospheric current systems to model the individual magnetospheric current systems and subtract their contribution prior to evaluating the structure of the intrinsic field. Several workers have adapted geometric descriptions of the magnetic fields from magnetopause currents and tail currents in the Earth's magnetosphere to Mercury's magnetosphere (Whang 1977; Korth et al. 2004). Our ability to characterize reliably the structure of Mercury's intrinsic magnetic field is, therefore, determined by the extent to which the external field can be understood and accurately modeled.

The extensive spatial and temporal coverage of the MESSENGER observations will yield a number of important benefits. First, the residuals remaining after fitting for different external field conditions will vary more distinctively, thus allowing better determination of the quality of the inversion solutions. Second, cross-correlation among the spherical harmonic coefficients will be significantly reduced, allowing for the derivation of improved quasi-linearly independent higher-order moments of the field representation (see Connerney and Ness 1988). Simulations of the magnetic field environment at Mercury have shown that the dipole moment should be recoverable to within

10% without applying any corrections for the external field (Giampieri and Balogh 2001; Korth et al. 2004). Further, the magnetic field data will provide significant clues about the occurrence of dynamic magnetospheric processes, so it will be possible to pre-select the data to be included in the inversion and reduce dynamic effects to a minimum. It is expected that the most reliable solutions will be afforded by the most carefully chosen "northward IMF—non-substorm" observations when the magnetospheric currents are weakest. We expect that the ultimate accuracy will be determined by a trade-off between statistical uncertainty, which grows as the number of observations is reduced, and systematic error, which decreases as the data are more carefully selected. In any case, the ultimately achievable accuracy for the dipole term will be fairly high, on the order of a few percent, and many higher-order terms should also be reliably recovered.

Additional analyses will examine the fine structure of Mercury's crustal magnetic field. The altitudes of the MESSENGER orbit in the northern hemisphere are sufficiently low (200-km minimum altitude) that field structures due to crustal anomalies, if present, can be directly mapped. The closest approach points of the three flybys are also at 200 km altitude but at low latitudes. Large crustal remanent fields were found at Mars (Acuña et al. 1998, 1999) and may also be present at Mercury, although the carriers of the remanence and the internal field history are probably very different for the two bodies. If only those magnetic features having a lateral extent larger than the spacecraft altitude can be resolved, then the effective longitudinal and latitudinal resolution is determined by the spacecraft orbit. Accordingly, we expect to be able to resolve magnetic features with horizontal dimensions of 5° (about 200 km) near the orbit-phase periapsis at ~60–70° N latitude and near the closest approach points of the flybys.

In summary, the MESSENGER data can be used to discriminate between the various hypotheses for Mercury's magnetic field only to the extent that the competing theories make differing predictions involving quantities that can be measured directly or inferred from the data. Unfortunately, the knowledge regarding the interior of Mercury is so limited that it is difficult to forecast now how specific hypotheses will be validated or ruled out simply through the generation of a more complete and accurate mapping of the planetary magnetic field. The more likely scenario is that all of MESSENGER's measurements taken together will reveal unexpected features of the planet, its interior, and magnetic field that cannot be accommodated by the present hypotheses for the origin of its intrinsic magnetic field – thus, allowing some or most to be discarded and replaced by new theories and models.

4.3 How, When, and Where Does the Solar Wind Impact the Planet?

The manner, flux, energy spectrum, and location of solar wind and solar energetic particle (SEP) impact upon the surface is important because of the role that these processes play in sputtering neutrals out of the regolith into the exosphere and their contribution to changing the appearance and physical properties of the surface (Killen et al. 1999, 2001; Lammer et al. 2003; Sasaki and Kurahashi 2004). Solar wind and SEP charged particles may intercept the surface by two mechanisms. First, finite gyroradius effects can result in ions being lost to collisions with regolith material wherever the strength of the magnetospheric magnetic field and the height of the magnetopause is such that their centers of gyration are within one Larmor radius of the surface (Siscoe and Christopher 1975; Slavin and Holzer 1979a). Second, "open" magnetospheric flux tubes with one end rooted in the planet and the other connected to the upstream interplanetary magnetic field will act as a "channel" that guides charged particles down to the surface, except for those that "mirror" prior to impact (Kabin et al. 2000; Sarantos et al. 2001; Massetti et al. 2003).

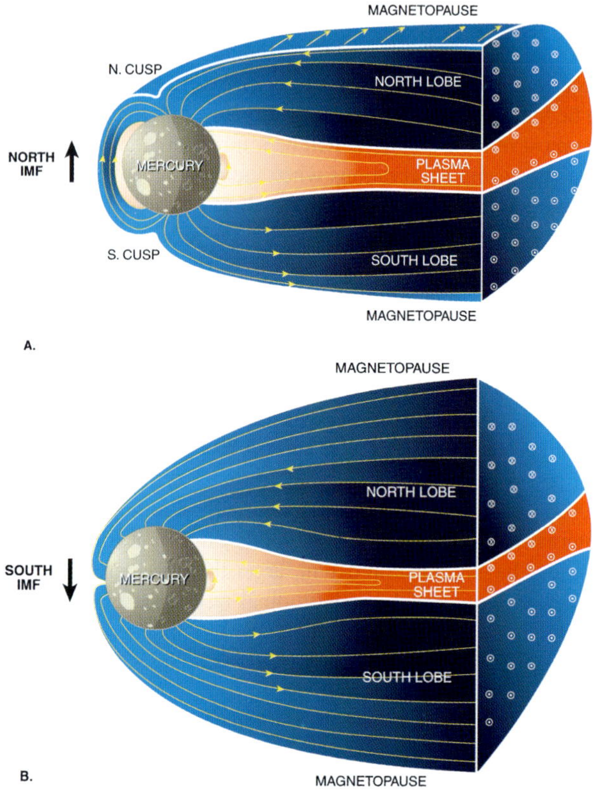

Fig. 5 Schematic view of the magnetosphere of Mercury. Regions with low plasma temperature (solar wind and tail lobes) are colored *blue* while the hotter regions (inner magnetosphere and plasma sheet) are shown in *redder hues*. The two images illustrate the extreme cases of minimal (**A**) and maximal (**B**) tail flux expected for strongly northward and southward interplanetary magnetic field, respectively

An idealized view of Mercury's magnetosphere under a northward interplanetary magnetic field (IMF), based on the Mariner 10 measurements, is presented in Fig. 5(a). It has been drawn using an image of the Earth's magnetosphere and increasing the size of the planet by a factor of ~7–8 to compensate for the relative weakness of the dipole field and the high solar wind pressure at Mercury (Ogilvie et al. 1977). The mean ~1.5 R_M distance from the center of Mercury to the nose of the magnetosphere inferred from the Mariner 10 measurements (Siscoe and Christopher 1975; Ness et al. 1976; Russell 1977; Slavin and Holzer 1979a) corresponds to 10–11 R_E, where R_E is Earth's radius and 1 $R_E = 6378$ km.

Whether or not the solar wind is ever able to compress the dayside magnetosphere to the point where solar wind ions directly impact the surface at low latitudes remains a topic of considerable interest and controversy. Siscoe and Christopher (1975) were the first to take a long time series of solar wind ram pressure data taken at 1 AU, scale it by $1/r^2$ inward to Mercury's perihelion, and then compute the solar wind stand-off distance using a range of assumed planetary dipole magnetic moments. They found that only for a few percent of the time would the magnetopause will be expected to fall below an altitude of ~0.2 R_M, the point where solar wind protons begin to strike the surface due to finite gyro-radius effects.

Rapid large-amplitude changes in solar wind ram pressure associated with high-speed streams and interplanetary shocks might be expected easily to depress the magnetopause close to the surface of planet. However, induction currents will be generated in the planetary interior (Hood and Schubert 1979; Suess and Goldstein 1979; Goldstein et al. 1981;

Glassmeier 2000; Grosser et al. 2004), and these currents will act to resist rapid magnetospheric compressions. Hence, the sudden solar wind pressure increases associated with interplanetary shocks and coronal mass ejections may not be as effective depressing the dayside magnetopause as a very slow, steady pressure increase of comparable magnitude. Mercury's interaction with the solar wind may, therefore, also provide a unique opportunity to study this planet's large electrically conductive core via its inductive reactance to externally imposed solar wind pressure variations.

The "erosion", or transfer, of magnetic flux into the tail is well studied at Earth, where the distance to the subsolar magnetopause is reduced by ~10–20% during a typical interval of southward IMF (Sibeck et al. 1991). Analysis of the Mariner 10 boundary crossings, after scaling for upstream ram pressure effects, by Slavin and Holzer (1979a) indicated that the subsolar magnetopause extrapolated from the individual boundary encounters varied from 1.3 to 2.1 R_M, with the larger values corresponding to IMF $B_z > 0$ and the smaller to $B_z < 0$. Similar variations in dayside magnetopause height have been found in MHD simulations of Mercury's magnetosphere under southward IMF conditions by Kabin et al. (2000) and Ip and Kopp (2002). Further evidence that reconnection operates at Mercury's magnetopause comes in the form of the "flux transfer events" identified in the Mariner 10 data by Russell and Walker (1985). These flux rope-like structures have been studied extensively at the terrestrial magnetopause where they play a major role in the transfer of magnetic flux from the dayside to the nightside magnetosphere.

In the limit that all of the magnetic flux in the dayside magnetosphere of Mercury were to reconnect quickly, the north and south cusps are expected to move equatorward and merge to form a single cusp as displayed in Fig. 5(b). All of the flux north (south) of this single cusp will map back into the northern (southern) lobe of the tail. Direct solar wind impact on the surface will take place in the vicinity of the single, merged low-altitude cusp. However, such extreme events are not necessary. As shown by Kabin et al. (2000) and Sarantos et al. (2001), the strong radial IMF near Mercury's orbit should always be conducive to solar wind and SEP particles being channeled to the surface along reconnected flux tubes that connect to the upstream solar wind. For the completely eroded dayside magnetosphere shown in Fig. 5(b), the solar wind and SEP charged particles would impact a large fraction of the northern (southern) hemisphere of Mercury for IMF $B_x > 0$ ($B_x < 0$). Whether or not the fully reconnected dayside magnetosphere shown in Fig. 5(b) is ever realized will be determined by the rate of reconnection at the magnetopause and how long it takes for Mercury's magnetosphere to respond by reconnecting magnetic flux tubes in the tail and convecting magnetic flux back to the dayside. However, it is notable that Slavin and Holzer (1979a) have argued that the high Alfven speeds in the solar wind at 0.3 to 0.5 AU may produce very high magnetopause reconnection rates and lead to strong erosion of the dayside magnetosphere even if the timescale for the magnetospheric convection cycle is only ~1–2 min.

4.4 How Will MESSENGER Determine the Extent of Solar Wind Impact?

The two critical factors controlling the impact of the solar wind and SEP flux to the surface are the height of the magnetopause and the distribution of "open" magnetic flux tubes that are topologically connected to the upstream region. MESSENGER will encounter and map the principal magnetospheric boundaries and current sheets, i.e., the bow shock, magnetopause, magnetic cusps, field-aligned currents, and the cross-tail current layer, throughout the mission. Typically, these surfaces are modeled by identifying "boundary crossings" and then employing curve fitting techniques to produce 2- or 3-dimensional surfaces. If such encounters can be collected for a variety of solar wind and magnetospheric conditions, then parameterized models may be produced. The essential requirement for this technique to be successful is the availability of crossings over a wide range of local times and latitudes along trajectories that provide good spatial coverage above and below the mean altitude of the surfaces (e.g., see Slavin and Holzer 1981). Inspection of the first Mercury-year of MESSENGER orbits, displayed in Figs. 3a and 3b, indicates that the modeling of bow shock, magnetosphere, and cross-tail current layer using boundary crossings should work very well sunward of $X \sim -3.5\ R_M$. The lack of coverage of the northern halves of the bow shock and magnetopause surfaces should not be a significant problem because of the expected symmetry between the two hemispheres. The models of the magnetopause and magnetic cusps will be used to infer the extent and frequency with which the magnetopause altitude becomes so low that a given population of interplanetary charged particles may find itself within one Larmor radius of the surface.

However, the measurements of the charged particle distribution functions and pitch angle distributions by the FIPS and EPS sensors when MESSENGER is within the magnetosphere will provide the most direct information regarding the ion and electron fluxes reaching the surface of the planet. Charged particles on magnetic flux tubes that connect to the planet will be lost if their magnetic mirror points are below the surface of the planet. This effect produces a "loss cone" signature in the particle pitch angle distributions, which is a definitive indication of particles impacting the surface. The EPPS instrument will return charged particle distribution functions according to particle composition, charge state, and energy that will be inverted to infer the flux of particles impacting the surface of Mercury. The results are expected to vary greatly depending upon where the spacecraft is located, the topology of the local magnetic field, and the state of the magnetosphere (i.e., IMF direction and substorm versus non-substorm conditions).

5 Magnetospheric Dynamics

5.1 What Are the Principal Mechanisms for Charged Particle Acceleration at Mercury?

Charged particle acceleration is one of the most fundamental processes occurring in space plasmas. Planetary magnetospheres are known to accelerate particles from thermal to high energies very rapidly via a range of processes. The plasma in Mercury's magnetosphere is expected to come from two sources, the solar wind and the ionization of the neutral exosphere. Solar wind plasma enters the magnetosphere by flowing along "open" flux tubes that connect to the interplanetary medium as shown in Fig. 6. After reconnection splices together an interplanetary and a planetary flux tube, the solar wind particles are channeled down into the cusp region where either they mirror and reverse their direction of motion or they impact the regolith and are absorbed. The solar wind particles that mirror and then flow tailward find themselves in the "plasma mantle" region of the tail lobe. Due to the dawn-to-dusk electric field that the solar wind interaction impresses across the magnetosphere, the plasma in the mantle will "$E \times B$" drift toward the equatorial regions of the tail where it will be assimilated into the plasma sheet. Delcourt et al. (2003) showed that the large Larmor radii of the newly created sodium ions will result in significant "centrifugal" acceleration as the ions $E \times B$ drift at lower altitudes over the polar regions of Mercury. Similarly, at higher altitudes Delcourt et al. found that these large Larmor radii will result in ion motion that is generally non-adiabatic and follows "Speiser-type" trajectories near the cross-tail current layer with the ions rapidly attaining energies of several keV.

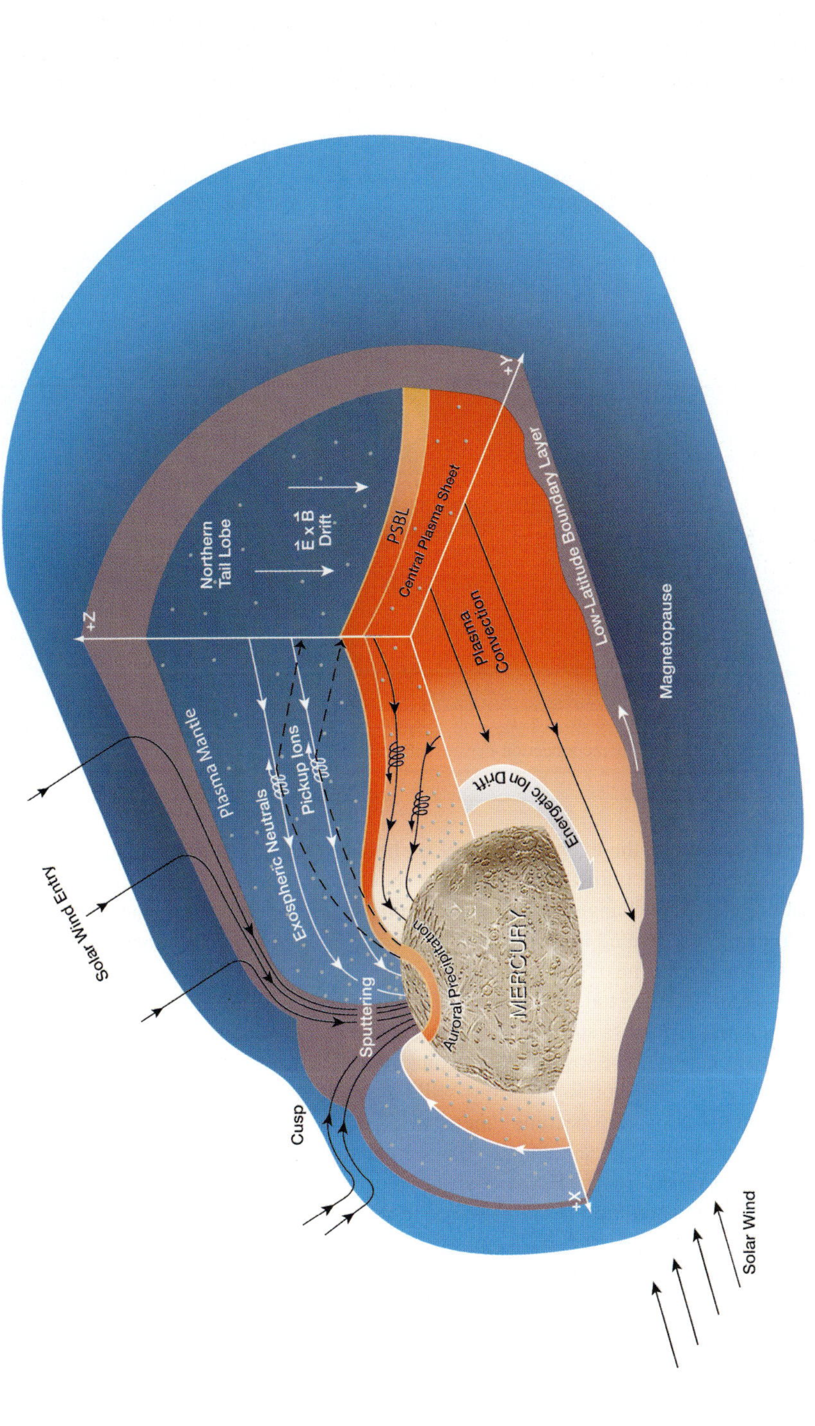

Fig. 6 Magnetospheric convection and the pick-up of newly ionized exospheric particles. Note the relatively straight equatorial convection paths expected at Mercury due to the planet's extremely slow rotation rate

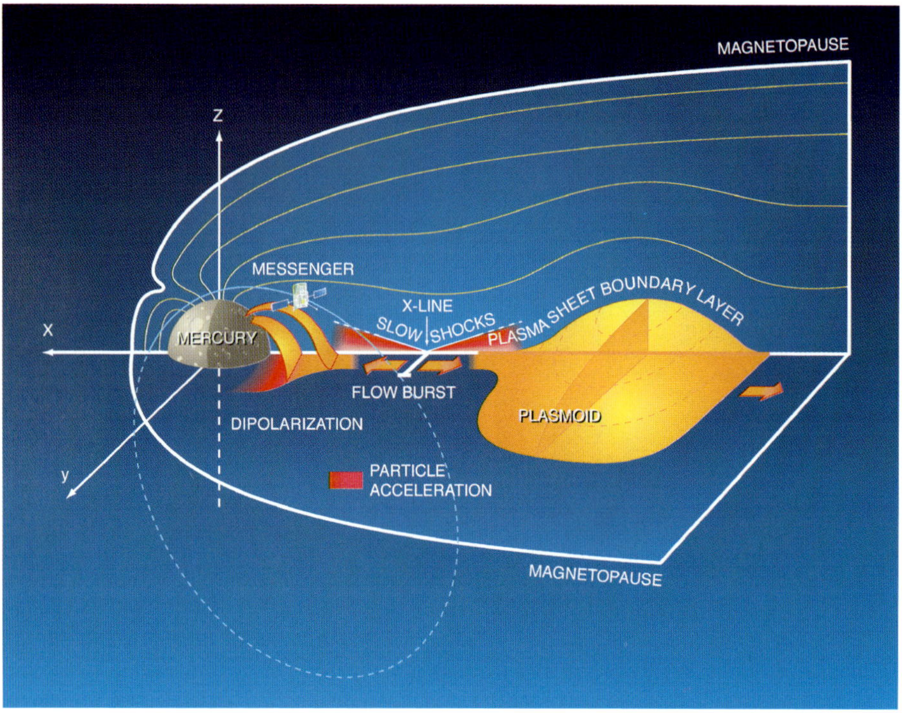

Fig. 7 Schematic depiction of a reconnection-driven substorm within Mercury's magnetosphere (Slavin 2004)

The neutral species in the exosphere travel on ballistic trajectories determined only by gravity and light pressure until the point where they become ionized by solar ultraviolet (UV) radiation, charge exchange with a magnetospheric ion, or electron impact ionization. At that point the newly created ion will begin to execute single particle motion according to its velocity vector at the time of creation and the ambient electric and magnetic fields within the magnetosphere (Cheng et al. 1987; Ip 1987; Delcourt et al. 2002, 2003). Alternatively, some of the ions may possess sufficiently large Larmor radii to intersect quickly the magnetopause or the planet and be lost. For those pickup ions remaining in the magnetosphere, their non-Maxwellian distribution functions will cause plasma waves to be excited, grow, and scatter the ions until they become "thermalized." The determination of the extent to which planetary pick-up ions can actually be thermalized within Mercury's small magnetosphere is a major objective of MESSENGER. Since Mercury takes 59 days to spin once about its axis, planetary rotation is not expected to play any role in particle acceleration or transport. Hence, the $E \times B$ drift or "convection" path for magnetospheric plasma is expected to follow relatively straight lines from the plasma sheet sunward toward the nightside of the planet and the forward magnetopause, as shown in Fig. 6.

Some of the most energetic charged particles in the tail are thought to be accelerated by the intense electric fields driven by the reconnection of magnetic flux tubes from the lobes of the tail (Hill 1975). At Earth, recently reconnected flux tubes are observed to be bounded by "magnetic separatrices" populated with newly accelerated ions and electrons (Cowley 1980; Scholer et al. 1984). The particles possessing the highest V_{parallel} are found farthest from the current sheet and closest to the separatrix boundary. These regions of sunward and tailward streaming energetic charged particles are colored red in Fig. 7. Indeed, short-lived "spikes" in energetic ions and electron flux extending up to at least several MeV have been seen in Earth's distant magnetotail (Krimigis and Sarris 1979) and have been associated with episodes of X-line formation and reconnection (Sarris and Axford 1979; Richardson et al. 1996). Non-adiabatic processes are necessary to explain these acceleration events, usually attributed to the effect of extreme thinning of the cross-tail current sheet relative to the Larmor radii of the ions and electrons (Büchner and Zelenyi 1989; Delcourt et al. 2003; Hoshino 2005).

Many of these accelerated charged particles are immediately lost as they flow down the tail to the interplanetary medium. Others, however, are carried sunward and undergo further acceleration due to first invariant conservation. At Earth, ions convected from the inner edge of the tail may have their energy increased by a factor of 100 by the time they reach the "ring current" region at a radial distance of $\sim 3\ R_E$ from the center of the planet. By contrast, the weak planetary magnetic field at Mercury greatly limits this type of adiabatic acceleration. Indeed, Mercury may be ideal for the direct observation of acceleration associated with X-line formation. As these charged particles approach Mercury and experience stronger magnetic fields, the ions and electrons will begin to experience gradient and curvature drift that causes the ions to drift about the planet toward dusk while electrons are diverted toward dawn, as indicated in Fig. 6. The loss of these energetic particles via intersection with the surface of Mercury or the magnetopause is expected to limit severely the flux of quasi-trapped particles that complete a circuit about the planet (Lukyanov et al. 2001; Delcourt et al. 2003), but their loss constitutes an additional source of surface sputtering.

Charged particles also experience acceleration during interactions with ultra-low-frequency (ULF) waves (e.g., Blomberg 1997). Ion pickup due to photo-ionization of neutrals sputtered from the surface is expected to be a persistent feature of Mercury's magnetosphere (Ip 1987). These newly created ions will then experience the convection electric field and be picked up in the plasma flow much as newly ionized atoms are swept up in the solar wind flow near comets (e.g., Coates et al. 1996). The resulting pickup ion distributions contain significant free energy and can be unstable to various cyclotron wave modes, many of which have magnetic signatures in the vicinity of the ion gyro-frequencies (Gomberoff and Astudillo 1998). Cyclotron waves may also be generated by ions accelerated in the magnetotail as they convect sunward and are commonplace at Earth (Anderson et al. 1992). At Earth ion populations can also drive long-wavelength, low-frequency waves which, in turn, couple to field-line resonances (e.g., Southwood and Kivelson 1981). While the ion gyro-frequencies and field-line resonance frequencies at Earth are separated by a factor of 10 to 100, at Mercury the gyro-frequencies are fairly low because of the low magnetic field strength at Mercury, while the field-line resonance periods should be relatively high owing to the small size of the magnetosphere (Russell 1989). The wave-particle physics at Mercury may, therefore, be particularly interesting, because the kinetic and longer wavelength waves should be coupled (Othmer et al. 1999; Glassmeier et al. 2003).

5.2 How Will Energetic Particle Acceleration Processes Be Measured at Mercury?

The EPPS and MAG instruments will be used in concert to explore Mercury's magnetosphere, map out its different regions, and determine the spatial and temporal variations in the charged particle populations peculiar to the different parts of the magnetosphere (see Mukai et al. 2004). For example, the magnetic field and plasma measurements will be used to calculate the ratio of thermal particle pressure to magnetic pressure, termed the "β" value of the plasma. The inner regions close to the planet and the lobes of the tail are typically dominated by the magnetic field pressure and have very low β values, i.e., <0.1. The plasma sheet region (see Fig. 6), in contrast, is dominated by thermal pressure. At Earth the plasma sheet has β values that vary from a few times 0.1 in the outer layers to $\gg 10$ in the central portion where the cross-tail electric current density peaks. The most dynamic events observed in the Earth's magnetosphere, such as "bursty bulk flows" and "dipolarizations," are generally found in the plasma sheet region (Angelopoulos et al. 1992). Streaming energetic particles accelerated in the separatrices emanating from X-lines are most frequently observed in the outer layers of the plasma sheet where $\beta \sim 0.1$.

The MESSENGER EPPS instrument will provide comprehensive observations of ions and electrons from low altitudes over the north polar regions (see Fig. 3b) out through the lobes and into the cross-tail current layer. The flux of ions moving up and down these magnetic flux tubes will be measured directly and used to infer the rate at which mass is exchanged between the surface of the planet and the magnetosphere. Furthermore, any attendant acceleration of the charged particles will also be observed. The natural tendency of energetic particles to disperse, with faster particles reaching an observer before the slower particles, is a strong modeling constraint for determining the source populations, drift paths, and magnetic conjugacies. Modeling and analysis of dispersed ion-injection events at various distances down the tail at the Earth have shown that it is often possible to specify the time and location where the initial acceleration event took place (Mauk 1986; Sauvaud et al. 1999; Kazama and Mukai 2005).

The MESSENGER Magnetometer is also designed to characterize waves and wave-particle interactions at Mercury. The MAG instrument provides coverage up to 10 Hz, a band that spans all of the relevant ion gyro-frequencies including protons throughout the planetary magnetosphere. Even during periods of low telemetry allocations the magnetospheric sampling will be no coarser than 2 vectors s^{-1} providing coverage over all ULF and heavy-ion gyro-frequencies. Moreover, the MAG provides a burst detection capability that will allow capture of large-amplitude wave events. In concert with FIPS and EPS observations of ion distributions and composition such measurements will provide a comprehensive survey of wave activity and determine their correspondence with the local ion populations.

5.3 Do Terrestrial-Style Substorms Occur at Mercury?

Mercury is expected to be one of the best places to test and extend our understanding of magnetospheric substorms. Because of its closeness to the Sun, this magnetosphere is subject to the most intense solar wind pressure and IMF intensity in the solar system (Burlaga 2001). The MESSENGER measurements will give detailed observations of substorms in a magnetosphere where planetary rotation is negligible and no ionosphere is present. The slow rotation of Mercury will result in sunward convection in the equatorial region being dominant throughout the forward magnetosphere. This is in stark contrast with Saturn, the other planet where terrestrial-type substorms are thought to occur. Saturn has a rapid rotation that dominates magnetospheric convection to the point where even the tail magnetic field may be twisted into a helical configuration (Mitchell et al. 2005; Cowley et al. 2005).

The absence of a collisional ionosphere at Mercury also has important consequences for global electric currents and plasma convection. At Earth and Saturn it is believed that the timescale for the substorm growth phase is determined by ionospheric line-tying that in turn limits the rate of magnetic flux circulation from the dayside magnetosphere to the nightside and back again (Coroniti and Kennel 1973). Furthermore, some theories of the substorm expansion phase at Earth require active feedback between the magnetosphere and an ionosphere whose conductivity varies at least somewhat with the rate of charged particle precipitation (Baker et al. 1996). Such feedback is presumably absent at Mercury, although we shall evaluate the situation using the MESSENGER observations. In this manner, it will be determined whether or not active feedback between an ionosphere and the equatorial magnetosphere is a necessary condition for magnetospheric substorms. Finally, the flow of field-aligned currents to low altitudes produces auroras in the Earth's upper atmosphere when the charge carriers impact neutral atoms. It is unlikely that classical auroras would occur at Mercury. Nonetheless, Joule heating due to magnetospheric field-aligned currents closing at very shallow depths beneath Mercury's surface may create a "warm" auroral oval that might be visible at infrared wavelengths (Baker et al. 1987b).

Siscoe et al. (1975) and Ogilvie et al. (1977) showed that the energetic particle bursts detected by Mariner 10 tended to occur at times when the magnetic field exhibited the disturbed behavior characteristic of substorms at Earth. Mariner 10 entered the near-tail plasma sheet on the dusk side of the tail during its first Mercury encounter. The magnetic field observed inside the magnetopause was very tail-like and relatively quiet during the inbound half of the encounter. Shortly after closest approach, $|\mathbf{B}|$ decreased rapidly, and the field inclination increased markedly, becoming less tail-like and more dipole-like. Such magnetic field variations are a classic signature of substorm expansive phase onset at the Earth where they are termed "dipolarization events" (Baker et al. 1996). Christon et al. (1987) conducted comparative studies of the magnetic field changes observed in association with the Mercury and Earth magnetosphere energetic particle events in the near-tail and found them to be extremely similar.

Siscoe et al. (1975) also called attention to the fact that the IMF switched from northward to southward while Mariner 10 was in the magnetosphere. These authors suggested, by analogy to Earth, that this change in IMF direction initiated reconnection at the dayside magnetopause, magnetic flux transfer to the tail, and, finally, tail reconnection. As shown schematically in Fig. 7, tail reconnection is believed to drive fast plasma flows and energetic particle acceleration. Indeed, Slavin and Holzer (1979a) found that the altitude of the dayside magnetopause for both M10 encounters was reduced whenever the IMF B_z component was southward, consistent with the reconnection model. Siscoe et al. further used scaling arguments to suggest that if substorms occurred at Mercury, then their timescales should be of order 1–2 min, similar to the M10 energetic particle events, in contrast with the $\sim$1 hr typical of the Earth's magnetosphere.

Eraker and Simpson (1986) and Baker et al. (1986) developed this scenario further and suggested that the substorms in Mercury's magnetotail resulted from magnetic reconnection in the range of 3–6 R_M on the nightside, as shown in Fig. 7. During substorms in Earth's magnetosphere, the plasma sheet has been observed to be severed by magnetic reconnection quite close to Earth, i.e., $\sim$20–30 R_E or $\sim$2–3 times the solar wind stand-off distance. The reconnection process produces a magnetically confined structure (i.e., loop-like or helical magnetic topology) termed a "plasmoid" (Hones et al. 1984; Slavin et al. 1984) that is ejected down the tail at high speed, as schematically depicted in Fig. 7. As at the Earth, the observation of plasmoids in Mercury's magnetotail would provide direct information regarding the time of onset and intensity of magnetic reconnection (e.g., Baker et al. 1987a; Slavin et al. 2002).

5.4 How Will Substorm Activity Be Identified in the MESSENGER Measurements?

Given our present understanding of the Mariner 10 observations, we expect that substorms in the MESSENGER data will appear whenever the IMF upstream of Mercury becomes persistently southward. When the spacecraft is located in the tail lobes on the night side of Mercury, MAG should measure increases in the magnetic field strength and a "tail-like" stretching of the field lasting for some tens of seconds. We expect that these energy storage or "growth" phases (McPherron et al. 1973) would be soon followed by one or more rapid dipolarizations of the magnetic field. If MESSENGER finds itself on the sunward side of the site for magnetic reconnection in Mercury's magnetotail, there should be strong sunward plasma flow and intense energetic particle bursts. On the other hand, if MESSENGER were located tailward of the magnetic reconnection site, then strong anti-sunward plasma flow and magnetic signatures of plasmoids or flux ropes are anticipated (Fig. 7). However, it would not be surprising if the small dimensions of this magnetosphere produce some unique and unexpected substorm features.

Beyond these basic expectations, there are many important "system response" characteristics that will be determined. For example, substorms at Earth are known to exhibit both "driven" responses that can be predicted using linear "filters" and knowledge of the upstream solar wind and IMF (Bargatze et al. 1985) and an unpredictable, "spontaneous" component related to the magnetic energy stored in the lobes. Lacking a conductive ionosphere, it may be that Mercury's magnetosphere cannot store significant amounts of energy and the spontaneous component of substorm energy dissipation will be small or absent (Luhmann et al. 1998). In this case MESSENGER's magnetic field and charged particle instruments may frequently observe "continuous dissipation events" in the tail that are directly driven by the solar wind. At Earth, in contrast, this type of intense, relatively featureless magnetospheric convection is seen only in response to many-hour-long intervals of intense, strongly southward IMF (Tanskanen et al. 2005).

We also expect to use the Mercury Atmospheric and Surface Composition Spectrometer (MASCS) instrument to study key aspects of the solar wind-magnetosphere-exosphere interaction. This spectrometer is described in a companion article (McClintock and Lankton 2007). As portrayed in Fig. 6, the strong dissipation of energy in substorm-like events at Mercury is expected to produce powerful bursts of plasma and energetic particles directed along magnetic field lines down onto the cold nightside surface of Mercury (Baker et al. 1987b). It is part of our observation strategy with MESSENGER to use the infrared detection capabilities of MASCS to look for evidence of heating of the nightside regolith of Mercury due to substorm energy precipitation.

6 Magnetosphere–Planetary Coupling

6.1 How Much Mass Is "Recycled" Between the Regolith, Exosphere, and Magnetosphere?

Extensive analysis and modeling have been devoted to the investigation of how ion impact sputtering, aided by solar photon desorption and micrometeoroid vaporization, may result in the injection and acceleration of newly created ions followed by further re-circulation, as diagrammed in Fig. 8 (Killen et al. 2001). Trajectory analyses conducted by Killen et al. (2001) indicate that perhaps 60% of these photo-ions may subsequently impact the surface, where they are adsorbed and become available for release via sputtering and impact vaporization. Furthermore, if reconnection at the dayside magnetopause frequently exposes significant fractions of the surface directly to impact by charged particles from the interplanetary medium, then the contribution of neutrals sputtered by solar wind ions (Sarantos et al. 2001) and solar energetic particles (Leblanc et al. 2003) into the exosphere may be a major

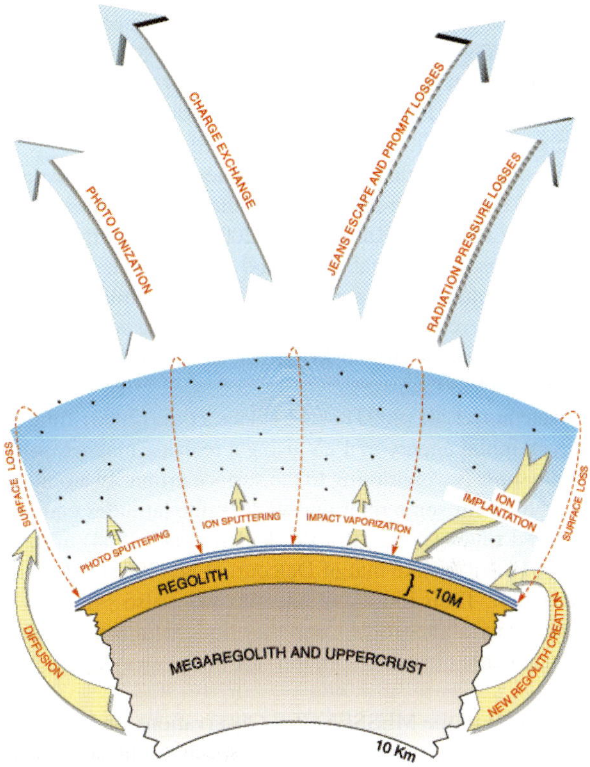

Fig. 8 Mass exchange between the solar wind, magnetosphere, exosphere, and regolith at Mercury (Killen et al. 2001)

driver for this system. The relatively short times, i.e., hours, required for photo-ionization and charge exchange will lead to sputtered neutrals being quickly ionized and picked-up by the convective flow within the magnetosphere to produce a coupled system (see also Fig. 6).

The nature of this complex chain of coupled processes that link the planet to the atmosphere and the magnetosphere has become a major focus for the Mercury research community. Killen et al. (2001) found that Mercury's atmosphere is sufficiently tenuous that it would soon be depleted by losses, if it were not being continuously replenished from below as depicted in Fig. 8. The creation of exospheric neutrals involves several competing processes including photon desorption, ion sputtering, and meteoritic impact (e.g., Killen and Ip 1999; Milillo et al. 2005). All of these processes and how the MESSENGER measurements will contribute to our understanding of them are discussed in detail in a companion article (Domingue et al. 2007).

The dynamic nature of Mercury's magnetosphere is expected to complicate the measurement of the rate of recycling of magnetospheric ions and neutrals. After a newly liberated neutral leaves the surface, it follows a ballistic trajectory until it impacts the surface or becomes ionized (see review by Hunten et al. 1988). If a given particle becomes ionized then it will be accelerated by the magnetospheric electric fields until it either collides with the planetary surface or is thermalized and swept along by the convective flow in the equatorial magnetosphere. The exospheric neutrals available for ionization depend heavily on the composition of the planetary surface. Charged particle and photon sputtering work on the first few monolayers of the surface grains, so pre-sputtered atoms must first make their way to the monolayers by diffusion. Gardening rates for the crust are much faster than the time required to deplete a grain of a given species, providing a constant supply of neutrals to the exosphere (see Killen et al. 2004). On the basis of the Mariner 10 observations, the magnetospheric reconfiguration time is only a few minutes (Siscoe et al. 1975; Slavin and Holzer 1979a; Luhmann et al. 1998). Hence, the trajectories of magnetospheric ions through the magnetosphere may be quite complex (Delcourt et al. 2002).

Several studies have been performed that examine the importance of ion recycling in Mercury's exosphere and magnetosphere (e.g., Zurbuchen et al. 2004). Koehn (2002) used the MHD model of Kabin et al. (2000) to study the surface-to-surface transport of OH^+ and S^+ ions. For normal solar wind conditions, he found that ions created at mid-latitudes tended to return to equatorial dusk regions, while ions formed elsewhere were lost to the magnetosphere and solar wind. For very strong solar wind conditions, returning ions tended to move poleward and duskward, enhancing mid-latitude regions. Recycling rates for this study were less than 10%.

Delcourt et al. (2003) and Leblanc et al. (2003) followed the trajectories of Na^+ ions with an initial energy of 1 eV using a realistic magnetospheric magnetic field model. Their results show Na^+ returning to the surface primarily along two mid-latitude regions centered on $\pm 30°$, with some returning to duskside latitudes equatorward of $\pm 20°$. Recycling rates for these studies were 10–15%. Killen et al. (2004) utilized a new model (see Sarantos et al. 2001) that, unlike that of Delcourt et al. (2003), takes into account radial magnetic field orientation. Their ion initial energies were also ~ 1 eV. The measured recycling rates are significantly higher (60%), and they find that dawnside-generated ions tend to return to the surface, while duskside-born ions are swept away by the solar wind.

6.2 How Will the MESSENGER Observations Be Used to Discover the Extent of Mass Exchange Between Mercury's Regolith, Exosphere, and Magnetosphere?

Perhaps no science objective will so fully utilize the MESSENGER instruments as the study of the mass exchange between the planetary surface, exosphere, and magnetosphere. The MAG instrument will map the magnetic field, providing insight into magnetospheric dynamics and supporting improved field models. The Gamma-Ray and Neutron Spectrometer (GRNS) and X-Ray Spectrometer (XRS) instruments (Goldsten et al. 2007; Schlemm et al. 2007) will provide elemental composition maps of the surface, from which exospheric neutrals arise, forming the seed population for magnetospheric ions. MASCS will measure the composition of the neutral atmosphere, recently liberated from the regolith. EPPS will detect pickup ions in the magnetosphere and then map detected ions back to surface regions from which they escaped. The Mercury Dual Imaging System (MDIS) instrument (Hawkins et al. 2007) will then image the surface from which these neutrals and ions are sputtered, thereby tying atmospheric and magnetospheric measurements back to surface features.

In addition, EPPS and MAG will allow us to understand the complex interplay between magnetospheric plasmas and the magnetic field. In the event that the magnetopause is compressed sufficiently such that the solar wind can come into direct contact with much of the surface, EPPS will monitor the likely large increase in exospheric neutrals and newly created magnetospheric ions. Large increases in the rate of ion sputtering from the surface has been offered as a likely explanation for the high degree of temporal and spatial variation in Mercury's atmosphere as observed from the Earth (Potter et al. 1999). Such increases would soon lead to a large number of new photo-ions that can modify the magnetospheric configuration, which MAG can also detect. In summary, most aspects of the recycling of magnetospheric ions between the exosphere and surface are still very much open issues.

The MESSENGER instrument payload will make the critical measurements that will resolve the most important questions regarding the mass exchange within this closely coupled system.

6.3 Do Field-Aligned Currents Couple Mercury to Its Magnetosphere, and How Do They Close?

Among the fundamental aspects of all planetary magnetospheres visited thus far are field-aligned electric currents (Kivelson 2005). When magnetospheric magnetic fields reconnect with the IMF and are pulled back into the tail at Earth, sheets of field-aligned current, termed "Region 1" currents, flow down into the high-latitude ionosphere on the dawn side of the polar cap and outward on the dusk side. These Region 1 currents are also expected to be present at Mercury, as schematically depicted in Fig. 9, though their intensity and temporal evolution may be greatly modified depending upon the nature of current closure path and the electrical conductivity of the regolith (Slavin et al. 1997). When the magnetic flux tubes in the tail reconnect again and high-speed plasma jets are generated toward and away from the planet (see Fig. 7), another set of field-aligned currents are generated, termed the "substorm current wedge (SCW)" (McPherron et al. 1973; Hesse and Birn 1991; Shiokawa et al. 1998). These currents are also shown in Fig. 9. The SCW currents connect the midnight region of the polar cap to the plasma sheet and transfer to the planet a significant fraction of the total energy being released in the tail (e.g., Fedder and Lyon 1987). These SCW currents are the most likely source of the field-aligned currents in the Mariner 10 measurements reported by Slavin et al. (1997). Numerical simulations by Janhunen and Kallio (2004) and Ip and Kopp (2004) suggest that Region 1 and SCW field-aligned currents will have important consequences for the structure of Mercury's magnetosphere as they do for that of Earth (cf. Fedder and Lyon 1987). However, in order for these currents to reach a steady-state, they must have a conductive path for closure. As Mercury possesses no such conductive ionosphere, other mechanisms or paths must be found.

A moderately conductive regolith is a likely candidate for FAC closure at Mercury. Hill et al. (1976) suggested a conductance value of 0.1 mho, which is not unreasonable based upon the lunar measurements. If this value were indeed appropriate, however, the high rate of joule heating in the regolith would severely limit the duration of the current flow as the available magnetospheric energy would be quickly dissipated. Cheng et al. (1987) showed that sputtering is a possible means to generate the neutral sodium atmosphere of Mercury and a source population for magnetospheric ions. They further pointed out that the new ions created by photo-ionization, electron impact ionization, and charge exchange are available to be "picked-up" by the convective motion of the magnetospheric flux tubes (Fig. 6). In doing so, they would give rise to a "pickup" or "mass loading" conductance (see Kivelson 2005) that might contribute to the generation and/or closure of FACs at Mercury. Photoelectrons have also been suggested as a source of current carriers (Grard et al. 1999; Grard and Balogh 2001). However, the pick-up of planetary ions and the photoelectron sheath over Mercury's sunlit surface provide conductances that are only slightly greater than the lunar values. Measurements of surface characteristics, as well as neutral and ion populations near the surface of Mercury, are necessary for a better understanding of magnetosphere-surface coupling. Glassmeier (2000) argued that field-aligned currents at Mercury may close as diamagnetic currents in regions of enhanced plasma density near the planet; for example, at low altitudes over the nightside of the Mercury where sunward-directed, reconnection-driven fast flows encounter strong planetary magnetic field. Janhunen and Kallio (2004) considered possible surface materials and mineralogy and concluded that

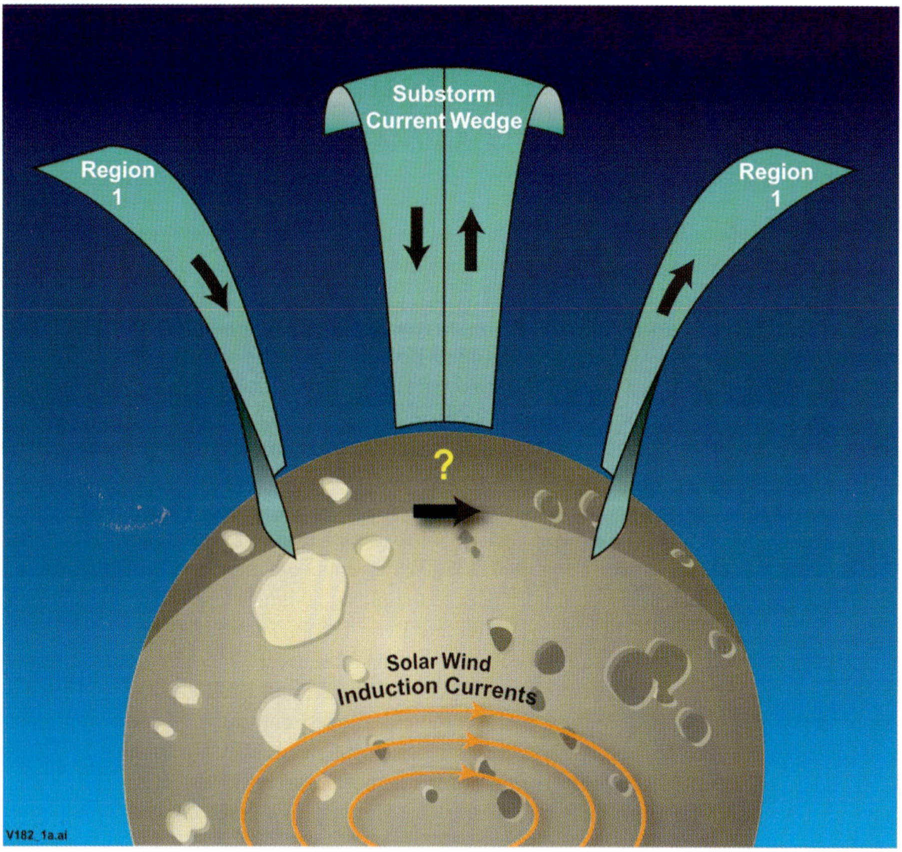

Fig. 9 Depiction of possible Region 1 and substorm current wedge field-aligned currents at high latitudes. Subsurface solar wind induction currents, flowing in the planetary interior, are shown at lower latitudes

the effective height-integrated conductance could fall within a wide range. In summary, the new, more detailed measurements to be returned by MESSENGER are necessary before the nature of the electrodynamic coupling between Mercury, its atmosphere, and its magnetosphere can be determined.

6.4 How Will MESSENGER Detect and Map Field-Aligned Currents and Determine Their Closure Path?

The large-scale field-aligned currents at Mercury should be readily detected by MESSENGER's instrumentation. The signatures of currents at low altitudes in the Earth system are well known (Iijima and Potemra 1978; Zanetti and Potemra 1982). The low-altitude northern hemisphere portion of the MESSENGER orbit, as depicted in Figs. 3a and 3b, is ideal for detecting field-aligned currents since the magnetic perturbation scales as $r^{-3/2}$ (Rich et al. 1990). Experience at the Earth shows that single spacecraft measurements provide accurate field-aligned current determinations using the infinite-current-sheet assumption (Rich et al. 1990; Anderson et al. 1998, 2000). Although this approach breaks down more than a few hours from dawn and dusk and under other circumstances (e.g., northward

IMF) when fringing effects dominate the signatures, the magnetic signatures unambiguously indicate the presence of the currents even if one cannot confidently infer the current density distribution from the data (cf. Fung and Hoffman 1992; Waters et al. 2001; Korth et al. 2005). In addition, particle data provide useful correlative observations of the current carriers, and the FIPS and EPS sensors which measure composition, velocity, and density for both ions and electrons will, in principle, give a direct measurement of current. As most current carriers are of relatively low energy, the broad energy range of EPPS will be of particular importance. Finally, GRNS, XRS, and other instruments will provide data about the makeup of the regolith, allowing better estimates of the surface conductivity.

7 Summary

A common paradigm describing the accumulation of knowledge about a planetary body suggests that advances come in four mission phases: "reconnaissance," "exploration," "intensive study," and "understanding." Applied to Mercury, the Mariner 10 mission can be said to have contributed to our progress by providing a reconnaissance-level characterization. In particular, those measurements showed that Mercury possesses an intrinsic magnetic field that interacts strongly with the solar wind, especially when the IMF is southward, and produces short-duration, intense variations in the tail magnetic field that are well correlated with energetic particle acceleration events.

As described here, the MESSENGER mission will explore Mercury's magnetic field and its magnetosphere for the first time. The mission will determine whether the planet's magnetic field is the result of an ongoing convective dynamo, some other type of dynamo, or crustal magnetization. MESSENGER will also characterize the structure and dynamics of Mercury's magnetosphere and its response to average and extreme interplanetary conditions. In particular, it will determine whether Earth-like "substorms" occur, how the lack of an ionosphere affects magnetospheric dynamics, and the processes responsible for Mercury's intense energetic particle acceleration events. Moreover, the mission will determine the nature and importance of the coupling between this magnetosphere, the exosphere, and the regolith and, via induction effects, the planetary interior. The importance of magnetospheric charged particle precipitation for the maintenance and variability of the exosphere will be determined. Conversely, the impact of newly formed heavy ions due to the ionizing effects of solar extreme-UV radiation on the exosphere will also be measured. Finally, the existence of field-aligned currents and the nature of their low-altitude closure will be revealed. The success of MESSENGER in achieving these exploration-level, and some intensive-study, objectives will, in turn, produce a foundation for the comprehensive investigations to be carried out by future missions, such as BepiColombo (Grard and Balogh 2001), that will yield a detailed understanding of this most intriguing magnetosphere.

Acknowledgements The authors express their great appreciation to all of those who have contributed to the MESSENGER mission. Useful comments and discussion with J. Eastwood, M. Sarantos, and S. Boardsen, are also gratefully acknowledged.

References

M.H. Acuña et al., Science **279**, 1676–1680 (1998)
M.H. Acuña et al., Science **284**, 790–793 (1999)
O. Aharonson, M.T. Zuber, S.C. Solomon, Earth Planet. Sci. Lett. **218**, 261–268 (2004)

B.J. Anderson et al., Space Sci. Rev. (2007, this issue)
B.J. Anderson, R.E. Erlandson, L.J. Zanetti, J. Geophys. Res. **97**, 3075–3088 (1992)
B.J. Anderson, J.B. Gary, T.A. Potemra, R.A. Frahm, J.R. Sharber, J.D. Winningham, J. Geophys. Res. **103**, 26323–26335 (1998)
B.J. Anderson, K. Takahashi, B.A. Toth, Geophys. Res. Lett. **27**, 4045–4048 (2000)
G.B. Andrews et al., Space Sci. Rev. (2007, this issue)
V. Angelopoulos et al., J. Geophys. Res. **97**, 4027–4039 (1992)
T.P. Armstrong, S.M. Krimigis, L.J. Lanzerotti, J. Geophys. Res. **80**, 4015–4017 (1975)
D.N. Baker, J.A. Simpson, J.H. Eraker, J. Geophys. Res. **91**, 8742–8748 (1986)
D.N. Baker, R.C. Anderson, R.D. Zwickl, J.A. Slavin, J. Geophys. Res. **92**, 71–81 (1987a)
D.N. Baker et al., J. Geophys. Res. **92**, 4707–4712 (1987b)
D.N. Baker, T.I. Pulkkinen, V. Angelopoulos, W. Baumjohann, R.L. McPherron, J. Geophys. Res. **101**, 12975–13010 (1996)
L.F. Bargatze, D.N. Baker, R.L. McPherron, E.W. Hones Jr., J. Geophys. Res. **90**, 6387–6394 (1985)
T.A. Bida, R.M. Killen, T.H. Morgan, Nature **404**, 159–161 (2000)
L.G. Blomberg, Planet. Space Sci. **45**, 143–148 (1997)
A.L. Broadfoot, D.E. Shemansky, S. Kumar, Geophys. Res. Lett. **3**, 577–580 (1976)
J. Büchner, L.M. Zelenyi, J. Geophys. Res. **94**, 11821–11842 (1989)
L.F. Burlaga, Planet. Space Sci. **49**, 1619–1627 (2001)
A.F. Cheng, R.E. Johnson, S.M. Krimigis, L.J. Lanzerotti, Icarus **71**, 430–440 (1987)
S.P. Christon, Icarus **71**, 448–471 (1987)
S.P. Christon, J. Feynman, J.A. Slavin, in *Magnetotail Physics*, ed. by A.T.Y. Lui (Johns Hopkins University Press, Baltimore, 1987), pp. 393–402
A.J. Coates, A.D. Johnstone, F.M. Neubauer, J. Geophys. Res. **101**, 27573–27584 (1996)
J.E.P. Connerney, N.F. Ness, in *Mercury*, ed. by F. Vilas, C.R. Chapman, M.S. Matthews (University of Arizona Press, Tucson, 1988), pp. 494–513
F.V. Coroniti, C.F. Kennel, J. Geophys. Res. **78**, 2837–2851 (1973)
S.W.H. Cowley, Space Sci. Rev. **25**, 217–275 (1980)
S.W.H. Cowley et al., J. Geophys. Res. **110**, A02201 (2005). doi: 10.1029/2004JA010796
D.C. Delcourt, T.E. Moore, S. Orsini, A. Millilo, J.-A. Sauvaud, Geophys. Res. Lett. **29**, 1591 (2002). doi: 10.1029/2001GL013829
D.C. Delcourt et al., Ann. Geophys. **21**, 1723–1736 (2003)
D.L. Domingue et al., Space Sci. Rev. (2007, this issue)
J.H. Eraker, J.A. Simpson, J. Geophys. Res. **91**, 9973–9993 (1986)
J.A. Fedder, J.G. Lyon, Geophys. Res. Lett. **14**, 880–883 (1987)
S.F. Fung, R.A. Hoffman, J. Geophys. Res. **97**, 8569–8579 (1992)
G. Giampieri, A. Balogh, Planet. Space Sci. **49**, 1637–1642 (2001)
G. Giampieri, A. Balogh, Planet. Space Sci. **50**, 757–762 (2002)
K.-H. Glassmeier, in *Magnetospheric Current Systems*, ed. by S.-I. Ohtani, R. Fujii, M. Hesse, R.L. Lysak. Geophysical Mon., vol. 118 (American Geophysical Union, Washington, 2000), pp. 371–380
K.-H. Glassmeier, N.P. Mager, D.Y. Klimushkin, Geophys. Res. Lett. **30**, 1928 (2003). doi: 10.1029/2003GL017175
R.E. Gold et al., Planet. Space Sci. **49**, 1467–1479 (2001)
J.O. Goldsten et al., Space Sci. Rev. (2007, this issue)
B.E. Goldstein, S.T. Suess, R.J. Walker, J. Geophys. Res. **86**, 5485–5499 (1981)
L. Gomberoff, H.F. Astudillo, Planet. Space Sci. **46**, 1683–1687 (1998)
R. Grard, A. Balogh, Planet. Space Sci. **49**, 1395–1407 (2001)
R. Grard, H. Laakso, T.I. Pulkkinen, Planet. Space Sci. **47**, 1459–1463 (1999)
J. Grosser, K.-H. Glassmeier, S. Stadelmann, Planet. Space Sci. **52**, 1251–1260 (2004)
S.E. Hawkins III, et al., Space Sci. Rev. (2007, this issue)
M. Hesse, J. Birn, J. Geophys. Res. **96**, 19417–19426 (1991)
T.W. Hill, J. Geophys. Res. **80**, 4689–4699 (1975)
T.W. Hill, A.J. Dessler, R.A. Wolf, Geophys. Res. Lett. **3**, 429–432 (1976)
R.E. Holzer, J.A. Slavin, J. Geophys. Res. **83**, 3831–3839 (1978)
E.W. Hones Jr. et al., Geophys. Res. Lett. **11**, 5–7 (1984)
H. Hoshino, J. Geophys. Res. **110**, A102154 (2005). doi: 10.1029/2005JA011229
L.L. Hood, G. Schubert, J. Geophys. Res. **84**, 2641–2647 (1979)
D.M. Hunten, T.H. Morgan, D.E. Schemansky, in *Mercury*, ed. by F. Vilas, C.R. Chapman, M.S. Matthews (University of Arizona Press, Tucscon, 1988), pp. 562–612
T. Iijima, T.A. Potemra, J. Geophys. Res. **83**, 599–615 (1978)
W.-H. Ip, Icarus **71**, 441–447 (1987)

W.-H. Ip, A. Kopp, J. Geophys. Res. **107**, 1348 (2002). doi: 10.1029/2001JA009171
W.-H. Ip, A. Kopp, Adv. Space. Res. **33**, 2172–2175 (2004)
D.J. Jackson, D.B. Beard, J. Geophys. Res. **82**, 2828–2836 (1977)
P. Janhunen, E. Kallio, Ann. Geophys. **22**, 1829–1830 (2004)
K. Kabin, T.I. Gombosi, D.L. DeZeeuw, K.G. Powell, Icarus **143**, 397–406 (2000)
Y. Kazama, T. Mukai, J. Geophys Res. **110**, A07213 (2005). doi: 10.1029/2004JA010820
R.M. Killen, W.-H. Ip, Rev. Geophys. Space Phys. **37**, 361–406 (1999)
R.M. Killen et al., J. Geophys. Res. **106**, 20509–20525 (2001)
R.M. Killen, M. Sarantos, A.E. Potter, P.H. Reiff, Icarus **171**, 1–19 (2004)
M.G. Kivelson, Space Sci. Rev. **116**, 299–318 (2005)
P.L. Koehn, Ph.D. Thesis, University of Michigan, Ann Arbor, 2002
H. Korth et al., Planet. Space Sci. **54**, 733–746 (2004)
J. Korth, B.J. Anderson, Frey, C.L. Waters, Ann. Geophys. **23**, 1295–1310 (2005)
S.M. Krimigis, E.T. Sarris, in *Dynamics of the Magnetosphere*, ed. by S.-I. Akasofu (Reidel, Dordrecht, 1979), pp. 599–630
H. Lammer, S.J. Bauer, Planet. Space Sci. **45**, 73–79 (1997)
H. Lammer et al., Icarus **166**, 238–247 (2003)
F. Leblanc, J.G. Luhmann, R.E. Johnson, M. Lui, Planet. Space Sci. **51**, 339–352 (2003)
J.G. Luhmann, C.T. Russell, N.A. Tsyganenko, J. Geophys. Res. **103**, 9113–9119 (1998)
A.V. Lukyanov, S. Barabash, R. Lundin, P.C. Brandt, Planet. Space Sci. **49**, 1677–1684 (2001)
S. Massetti et al., Icarus **166**, 229–237 (2003)
B.H. Mauk, J. Geophys. Res. **91**, 13423–13431 (1986)
W. McClintock, M.R. Lankton, Space Sci. Rev. (2007, this issue)
R.L. McPherron, C.T. Russell, M.P. Aubry, J. Geophys. Res. **78**, 3131–3149 (1973)
A. Milillo et al., Space Sci. Rev. **117**, 397–443 (2005)
D.G. Mitchell et al., Geophys. Res. Lett. **32**, L20S01 (2005). doi: 10.1029/2005GL022647
T. Mukai, K. Ogasawara, Y. Saito, Adv. Space Res. **33**, 2166–2171 (2004)
N.F. Ness, in *Solar System Plasma Physics, vol. II*, ed. by C.F. Kennel, L.J. Lanzerotti, E.N. Parker (North-Holland, New York, 1979), pp. 185–206
N.F. Ness, K.W. Behannon, R.P. Lepping, Y.C. Whang, K.H. Schatten, Science **185**, 151–160 (1974)
N.F. Ness, K.W. Behannon, R.P. Lepping, J. Geophys. Res. **80**, 2708–2716 (1975)
N.F. Ness, K.W. Behannon, R.P. Lepping, Y.C. Whang, Icarus **28**, 479–488 (1976)
K.W. Ogilvie et al., Science **185**, 145–150 (1974)
K.W. Ogilvie, J.D. Scudder, V.M. Vasyliunas, R.E. Hartle, G.L. Siscoe, J. Geophys. Res. **82**, 1807–1824 (1997)
C. Othmer, K.-H. Glassmeier, R. Cramm, J. Geophys. Res. **104**, 10369–10378 (1999)
A.E. Potter, T.H. Morgan, Science **229**, 651–653 (1985)
A.E. Potter, T.H. Morgan, Icarus **67**, 336–340 (1986)
A.E. Potter, R.M. Killen, T.H. Morgan, Planet. Space Sci. **47**, 1441–1448 (1999)
F.J. Rich, D.A. Hardy, R.H. Redus, M.S. Gussenhoven, J. Geophys. Res. **95**, 7893–7913 (1990)
I.G. Richardson, C.J. Owen, J.A. Slavin, J. Geophys. Res. **101**, 2723–2740 (1996)
S.K. Runcorn, Nature **253**, 701–703 (1975a)
S.K. Runcorn, Phys. Earth Planet. Inter. **10**, 327–335 (1975b)
C.T. Russell, Geophys. Res. Lett. **4**, 387–390 (1977)
C.T. Russell, R.J. Walker, J. Geophys. Res. **90**, 11067–11074 (1985)
C.T. Russell, D.N. Baker, J.A. Slavin, in *Mercury*, ed. by F. Vilas, C.R. Chapman, M.S. Matthews (University of Arizona Press, Tucscon, 1988), pp. 514– 561
C.T. Russell, Geophys. Res. Lett. **16**, 1253–1256 (1989)
A.G. Santo et al., Planet. Space Sci. **49**, 1481–1500 (2001)
M. Sarantos, P.H. Reiff, T.W. Hill, R.M. Killen, A.L. Urquhart, Planet. Space Sci. **49**, 1629–1635 (2001)
E.T. Sarris, W.I. Axford, Nature **77**, 460–462 (1979)
S. Sasaki, E. Kurahashi, Adv. Space. Res. **33**, 2152–2155 (2004)
J.-A. Sauvaud et al., J. Geophys. Res. **104**, 28565–28586 (1999)
C.E. Schlemm II et al., Space Sci. Rev. (2007, this issue)
M. Scholer, G. Gloecker, B. Klecker, F.M. Ipavich, D. Hovestadt, E.J. Smith, J. Geophys. Res. **89**, 6717–6727 (1984)
K. Shiokawa et al., J. Geophys. Res. **103**, 4491–4507 (1998)
D.G. Sibeck, R.E. Lopez, E.C. Roelof, J. Geophys. Res. **96**, 5489–5495 (1991)
J.A. Simpson, J.H. Eraker, J.E. Lamport, P.H. Walpole, Science **185**, 160–166 (1974)
G.L. Siscoe, L. Christopher, Geophys. Res. Lett. **2**, 158–160 (1975)
G.L. Siscoe, N.F. Ness, C.M. Yeates, J. Geophys. Res. **80**, 4359–4363 (1975)

J.A. Slavin, Adv. Space Res. **33**, 1587–1872 (2004)
J.A. Slavin, R.E. Holzer, J. Geophys. Res. **84**, 2076–2082 (1979a)
J.A. Slavin, R.E. Holzer, Phys. Earth Planet. Inter. **20**, 231–236 (1979b)
J.A. Slavin, R.E. Holzer, J. Geophys. Res. **86**, 11401–11418 (1981)
J.A. Slavin, C.J. Owen, J.E.P. Connerney, S.P. Christon, Planet. Space Sci. **45**, 133–141 (1997)
J.A. Slavin et al., Geophys. Res. Lett. **11**, 657–660 (1984)
J.A. Slavin et al., J. Geophys. Res. **107**, 1106 (2002). doi: 10.1029/2000JA003501
S.C. Solomon, Icarus **28**, 509–521 (1976)
S.C. Solomon et al., Planet. Space Sci. **49**, 1445–1465 (2001)
S.C. Solomon, R.L. McNutt Jr., R.E. Gold, D.L. Domingue, Space Sci. Res. (2007, this issue)
D.J. Southwood, M.G. Kivelson, J. Geophys. Res. **86**, 5643–5655 (1981)
L.J. Srnka, Phys. Earth Planet. Inter. **11**, 184–190 (1976)
A. Stephenson, Earth Planet. Sci. Lett. **28**, 454–458 (1976)
D.J. Stevenson, Rep. Prog. Phys. **46**, 555–620 (1983)
D.J. Stevenson, Earth Planet. Sci. Lett. **82**, 114–120 (1987)
D.J. Stevenson, T. Spohn, G. Schubert, Icarus **54**, 466–489 (1983)
S.T. Suess, B.E. Goldstein, J. Geophys. Res. **84**, 3306–3312 (1979)
T. Terasawa et al., Geophys. Res. Lett. **24**, 935–938 (1997)
E.I. Tanskanen et al., J. Geophys. Res. **110**, A03216 (2005). doi: 10.1029/2004JA010561
C.L. Waters, B.J. Anderson, K. Liou, Geophys. Res. Lett. **28**, 2165–2168 (2001)
Y.C. Whang, J. Geophys. Res. **82**, 1024–1030 (1977)
L.J. Zanetti, T.A. Potemra, Geophys. Res. Lett. **9**, 349–352 (1982)
M.T. Zuber et al., Space Sci. Rev. (2007, this issue)
T.H. Zurbuchen, P. Koehn, Fisk, T. Gombosi, G. Gloeckler, K. Kabin, Adv. Space Res. **33**, 1884–1889 (2004)

Space Sci Rev (2007) 131: 161–186
DOI 10.1007/s11214-007-9260-9

Mercury's Atmosphere: A Surface-Bounded Exosphere

Deborah L. Domingue · Patrick L. Koehn · Rosemary M. Killen · Ann L. Sprague · Menelaos Sarantos · Andrew F. Cheng · Eric T. Bradley · William E. McClintock

Received: 28 August 2006 / Accepted: 7 August 2007 / Published online: 24 October 2007

Abstract The existence of a surface-bounded exosphere about Mercury was discovered through the Mariner 10 airglow and occultation experiments. Most of what is currently known or understood about this very tenuous atmosphere, however, comes from ground-based telescopic observations. It is likely that only a subset of the exospheric constituents have been identified, but their variable abundance with location, time, and space weather events demonstrate that Mercury's exosphere is part of a complex system involving the planet's surface, magnetosphere, and the surrounding space environment (the solar wind and interplanetary magnetic field). This paper reviews the current hypotheses and supporting observations concerning the processes that form and support the exosphere. The outstanding questions and issues regarding Mercury's exosphere stem from our current lack of knowledge concerning the surface composition, the magnetic field behavior within the local space environment, and the character of the local space environment.

Keywords Atmospheres · Exosphere · Mercury · Space physics · Space weathering · MESSENGER

D.L. Domingue (✉) · A.F. Cheng
The Johns Hopkins University Applied Physics Laboratory, Laurel, MD 20723, USA
e-mail: deborah.domingue@jhuapl.edu

P.L. Koehn
Department of Physics and Astronomy, Eastern Michigan University, Ypsilanti, MI 48197, USA

R.M. Killen · M. Sarantos
University of Maryland, College Park, MD 20742, USA

A.L. Sprague
Lunar and Planetary Laboratory, University of Arizona, Tucson, AZ 86721, USA

E.T. Bradley · W.E. McClintock
Laboratory for Atmospheric and Space Physics, University of Colorado, Boulder, CO 80303, USA

1 Introduction

The discovery of an atmosphere, or more accurately an exosphere, around Mercury was made through the ultraviolet airglow and occultation experiments on the Mariner 10 spacecraft during its three flybys of the planet in 1974 and 1975. The Mariner 10 occultation experiment set an upper limit on Mercury's atmospheric density of approximately 10^5 atoms/cm^3, corresponding to a pressure of about 10^{-12} bar (Broadfoot et al. 1976; Hunten et al. 1988), thus defining it as a collisionless exosphere with its exobase coincident with Mercury's surface: a surface-bounded exosphere. Ultraviolet (UV) emissions of the three atomic elements, hydrogen (H), helium (He), and oxygen (O), were detected with the UV airglow spectrometer (Broadfoot et al. 1976; Kumar 1976). Since the Mariner 10 flybys, exploration of Mercury's exosphere has been conducted by means of ground-based telescopic observations. Three additional elements, sodium (Na), potassium (K), and calcium (Ca), have been detected through their resonance scattering emission lines (Potter and Morgan 1985, 1986; Bida et al. 2000). Because the combined pressures of the known species are much less than the total exospheric pressure measured by the Mariner 10 occultation experiment, other species are expected to exist in this tenuous atmosphere. Additional constituents, such as carbon (C), carbon monoxide (CO), carbon dioxide (CO_2), lithium (Li), argon (Ar), neon (Ne), and xenon (Xe) have been sought but not detected (Broadfoot et al. 1976; Fink et al. 1974; Hunten et al. 1988; Sprague et al. 1996). Other species, such as hydroxyl (OH) and sulfur (S), have been suggested (Slade et al. 1992; Butler et al. 1993; Sprague et al. 1995) and modeled (Killen et al. 1997; Koehn 2002; Koehn et al. 2002) as related to the radar-bright deposits near Mercury's poles (Harmon and Slade 1992; Slade et al. 1992).

Telescopic observations from the mid-1980s to today have shown that there is temporal and spatial variability in Mercury's exosphere. The elements have both high- and low-velocity components and are influenced by the thermal and radiative environments in addition to the interstellar medium. For example, the variability in exospheric Na has been mapped to variability in the solar wind (Killen et al. 1999, 2004a, 2004b) and its effects on Mercury's magnetosphere.

Mariner 10 also made the first in situ measurements of the planet's magnetic field (Ness et al. 1974; Simpson et al. 1974) and the space environment around Mercury (Ogilvie et al. 1977). During the first flyby the spacecraft passed through the magnetotail of the planet and provided the first hint that Mercury may have a magnetic field similar to, though of lower amplitude than, the Earth's. The second flyby passed across the dayside of the planet, and the third again crossed the tail, this time closer to the planet's surface. Analysis of these data showed that the planetary magnetic field was probably a dipole with a moment of 350 to 400 nT-R_M^3, oriented within 10° of the rotational axis (Connerney and Ness 1988). Additional details concerning Mercury's magnetosphere can be found in a companion paper (Slavin et al. 2007).

The fundamental observation, however, was that the magnetic field of Mercury is able to stand off the solar wind, at least under nominal solar wind conditions. This implies a dynamical coupling to the planet that is mediated by magnetospheric current systems that must close near or within the planet. At Earth, the corresponding current systems close in Earth's ionosphere, but Mercury has no ionosphere. How do the required current systems close at Mercury? One hypothesis is that Mercury's exosphere provides a so-called "pick-up conductance," derived from the ionization and electric field acceleration of atmospheric species, that enables the formation of an Earth-like magnetosphere despite the absence of an ionosphere (Cheng et al. 1987; Ip 1993). Another proposed mechanism for closing current systems is based on the assumption that the surface of Mercury is itself conducting

(Janhunen and Kallio 2004). As will be discussed, the exosphere, magnetosphere, and surface of Mercury form a complex, interacting system whose properties and dynamics are still incompletely understood.

Exosphere–surface interactions for many of the exospheric constituents are not well understood, since many properties of Mercury's regolith, such as porosity and composition, are still poorly known (Head et al. 2007; Boynton et al. 2007). Given the variability of solar wind conditions at Mercury's orbit, and the relative weakness of Mercury's magnetic field, the solar wind at times can drive the magnetopause down to the surface of the planet. Under these unusual conditions the surface is exposed directly to solar wind plasma and particles. Even under normal solar wind conditions, solar wind plasma and particles can access the magnetosphere and surface via a variety of processes, such as dayside reconnection creating open field lines or boundary layer processes. The composition of Mercury's exosphere, with its abundant H and He, clearly indicates a strong solar wind source. Once solar wind plasma and particles gain access to the magnetosphere, they predominantly precipitate to the surface, where solar wind species are neutralized, thermalized, and released again into the exosphere. Moreover, bombardment of the surface by solar wind particles, especially energetic ions, contributes to ejection of neutral species from the surface into the exosphere (via "sputtering") as well as other chemical and physical surface modification processes. Details concerning the resulting "space weathering" of the regolith from scouring by solar wind particles are given by Head et al. (2007).

This paper summarizes our current state of knowledge concerning the exosphere composition, especially in terms of sources, sinks, and processes. It discusses the observed structure and density distributions within the exosphere and their association with the local environment. Exosphere–surface interactions and modeling efforts are compared with the current set of observations. The complex interaction and interconnections between the space environment (solar wind and interplanetary magnetic field, or IMF) and Mercury's magnetic field, exosphere, and surface are examined. Last, predictions are summarized for what the MErcury Surface, Space ENvironment, GEochemistry, and Ranging (MESSENGER) mission may observe and discover.

2 Composition: Sources, Sinks, and Processes

Table 1, adapted from Strom and Sprague (2003), summarizes the currently known constituents in Mercury's atmosphere and their approximate abundances. The abundance of H in Mercury's exosphere is at least 10 times the abundance seen in the Moon's exosphere

Table 1 Mercury's exospheric species

Constituent	Discovery reference	Column abundance (atoms per cm^2)
Hydrogen (H)	Broadfoot et al. (1976)	$\sim 5 \times 10^{10}$
Helium (He)	Broadfoot et al. (1976)	$\sim 2 \times 10^{13}$
Oxygen (O)	Broadfoot et al. (1976)	$\sim 7 \times 10^{12}$
Sodium (Na)	Potter and Morgan (1985)	$\sim 2 \times 10^{11}$
Potassium (K)	Potter and Morgan (1986)	$\sim 1 \times 10^{9}$
Calcium (Ca)	Bida et al. (2000)	$\sim 1 \times 10^{7}$

(Hunten and Sprague 1997). This large difference may be connected to the presence of a magnetic field on Mercury (Goldstein et al. 1981; Hunten and Sprague 1997). Goldstein et al. (1981) demonstrated that it is possible for Mercury's magnetosphere to stand off the normal solar wind (except at the high-latitude cusps). More recent hybrid models show that Mercury's magnetosphere is open a large part of the time (Kallio and Janhunen 2003a, 2003b, 2004a, 2004b). The distribution of Na atoms (discussed in the next section) is commensurate with these more open models. The abundances of Na and K have been observed to vary diurnally (Sprague 1992; Sprague et al. 1997; Hunten and Sprague 2002; Schleicher et al. 2004), with latitude (Potter and Morgan 1990, 1997; Sprague et al. 1997), and in association with surface features such as the Caloris basin and radar-bright spots (Sprague et al. 1990, 1998). Localized enhancements of Na have been observed with both imaging and spectroscopic techniques (Potter and Morgan 1990, 1997; Sprague et al. 1990, 1997; Hunten and Sprague 1997). More details on these variations are addressed in the next section, which describes the structure and density distribution of materials within the exosphere.

Measurements of many of the exospheric constituents show evidence for two-component velocity (temperature) distributions (e.g., H), or for Weibull (a continuous probability distribution function) distributions (Na and K). The vertical distribution of H is best modeled by two components: a dominant cold component with a temperature characteristic of the cold nightside surface and a smaller component with a temperature distribution more commensurate with the hot dayside surface temperature (Shemansky and Broadfoot 1977; Hunten and Sprague 1997). Observations of Na indicate that this species is emitted into the exosphere at higher velocities (commensurate with sputtering from the surface by charged particles or energetic solar photons) and that thermalization to the surface temperature is inefficient (Hunten et al. 1988; Hunten and Sprague 1997; Killen et al. 1999). Models of the exosphere sources and sinks are constrained by the measurements of these multiple velocity or temperature components.

Table 2 outlines the possible sources, sinks, and processes producing Mercury's exosphere. The sources and sinks are related to a complex interplay between exosphere–surface interactions (discussed in more detail in Sect. 4) and solar wind, IMF, and planetary magnetospheric interactions (discussed in more detail in Sect. 5) with the surface.

The observed abundances of H and He can be explained by solar wind capture and radiogenic decay (^{4}He). These species are lost from Mercury's exosphere through thermal escape. Heavier atoms, such as Na and K, are lost by photoionization. Once ionized, roughly half are carried away by the solar wind (Killen et al. 2004a; Killen and Crider 2004). Most of those recycled return on the dayside depending on the scale height (Sarantos 2005). However, there is a small population of these ions that could be trapped in Mercury's magnetotail, accelerated back to the nightside of the planet, and recycled through interactions with the surface (Ip 1993; Hunten and Sprague 1997; Sarantos 2005).

The most extensively studied species in Mercury's exosphere is Na. Sodium has been observed to be highly variable, with column densities varying on timescales smaller than 24 hours. Distributions and enhancements of Na in the exosphere have been correlated with solar wind magnetospheric interactions and variations in the regolith composition (e.g., Potter and Morgan 1990, 1997; Killen et al. 1990, 1999, 2001; Sprague et al. 1997, 1998; Potter et al. 1999; Lammer et al. 2003). Killen et al. (2001) examined how much of the Na population could be attributed to photon-stimulated desorption (PSD), ion sputtering, and impact vaporization of micrometeoroids. They concluded that the impact vaporization process could provide up to 25% of the Na seen. The relative importance of each process is variable, both

Table 2 Mercury exosphere sources and sinks

Source processes	Relevant species	Recycling process	References
Direct to exosphere			
Solar wind capture	H, He		a, f
Radiogenic decay and outgassing	He		a, f
Meteoroid volatilization	Na, K, Ca		b, d, f, g
Delivery to surface			
Diffusion	H, He, O, Na, K		d, e, h, j, k, l, m
Regolith turnover	H, He, O, Na, K, Ca		d, e, f
Magnetotail or ion recycling	H, He, O, Na, K, Ca		a, c, e, k
Release from surface			
Sputtering: physical	Na, K, Ca	Yes	c, d, e, f, j
Sputtering: chemical	Na, K, Ca, OH	Yes	l, o
Thermal desorption (evaporation)	H, He, O, Na, K	Yes	d, f, k, l
Photon stimulated desorption (PSD)	Na, K	Yes	d, l
Impact vaporization	All		n
Sink processes			
Photoionization	H, He, O, Na, K, Ca		e, f, d, k
Thermal escape	H, He		f
Surface implantation: adsorption	H, He, O, Na, K, Ca	Yes	e, f, k
Surface implantation: chemical bonding	H, O, Na, K, Ca	Yes	f, m

a. Goldstein et al. (1981). b. Potter and Morgan (1985). c. Ip (1986). d. McGrath et al. (1986). e. Cheng et al. (1987). f. Hunten et al. (1988). g. Morgan et al. (1988). h. Tyler et al. (1988). i. Killen (1989). j. Sprague (1990). k. Sprague (1992). l. Hunten and Sprague (1997). m. Potter (1995). n. Cintala (1992). o. Potter (1995)

with true anomaly angle and with solar activity, and is highly uncertain. The remainder of the Na in Mercury's exosphere is considered to come from a combination of processes that deliver Na to the surface followed by a set of surface release processes.

The three processes listed in Table 2 that are believed to deliver exospheric material to the surface are diffusion, regolith turnover, and ion recycling. Diffusion processes can be subdivided into three types: regolith diffusion, volume diffusion, and grain-boundary diffusion. Regolith diffusion, as examined by Sprague (1990), is the diffusion of sodium and/or potassium along cracks and voids between grains and rock fragments. Volume diffusion, as studied by Killen (1989), is the solid-state diffusion through crystalline lattices. Grain-boundary diffusion is the diffusion of material across grain surfaces. Killen (1989) found that volume diffusion is too slow to produce the Na abundance observed in Mercury's exosphere. Sprague (1990) suggested that the combination of regolith and grain-boundary diffusion can provide the requisite amounts of Na and K over the solar system lifetime without relying on efficient recycling of material. These diffusion processes can also explain the relative observed abundances of Na and K at the Moon (Sprague 1990). Mercury's highly variable Na/K abundance ratio, which is also much higher than the Na/K ratio observed for the Moon, remains unexplained (Potter et al. 2002a). However, potassium column densities have been observed to decrease with increasing levels of solar activity (Potter et al. 2002a), which led Ip and Kopp (2004) to postulate that the Na/K ratio variability, and high value compared with lunar values, may be caused by the favored removal of K through its preferential acceleration (resulting from its lower gyrofrequency compared with Na) by ion cyclotron waves generated in the polar caps due to solar wind interactions. However, Sarantos (2005) tested the hypothesis that there exist more fractional losses for potassium ions and found no statistical difference for the fractional recycling of Na^+ and K^+ due to acceleration by the large-scale electric field. A more likely explanation for the variable Na/K ratio is thermal diffusion acting faster on bound sodium than on potassium (Sprague 1992; Killen et al. 2004a). It is also quite possible that the observed variation in the Na/K ratio results from the fact that the two species were not observed concurrently and that spatial and temporal variations are the cause of the differences.

Regolith turnover is another mechanism for bringing exospheric material to the surface, namely through impact gardening. Ion recycling has been studied by Killen et al. (2004a), Killen and Crider (2004), and Sarantos (2005). Work by Ip (1993) showed that exospheric material, after ionization by solar UV photons, could be trapped in Mercury's magnetotail and transported to the nightside. Sprague (1992) suggested that these ions could be implanted into the nightside surface, where they can be neutralized and adsorbed into the surface. However, Sarantos' (2005) modeling shows that approximately twice as many ions recycle to the dayside, but this result is highly variable with IMF conditions. Dayside recycled photoions can be reemitted at short timescales due to PSD, but nightside recycled ions can be reemitted only by meteoritic vaporization, ion sputtering, and electron-impact sputtering. The emission rate on the nightside is probably one-third to one-half the dayside rate, depending on the impact vaporization rate and nightside sputtering rate.

Although meteoroid volatilization is listed in Table 2 as a "direct to exosphere" source process, volatiles released during the impact process may also be trapped within the surface for later release to the exosphere. Conversely, the impact process could also release exospheric species already present in the surface.

Once these materials are brought to the surface, a process is needed to release them to the exosphere. The release processes listed in Table 2 include sputtering (both physical and chemical), evaporation, meteoritic vaporization, and PSD. Physical sputtering is the release of material through impact by energetic particles. Chemical sputtering involves a chemical reaction between the surface material and the energetic particle, where the reaction product is desorbed. Chemical sputtering has been suggested as a source of Na and OH (Potter 1995). The high-velocity component observed in the Na emission lines indicates a high-energy mechanism for release of materials from the surface. This mechanism is commensurate with sputtering by either charged particles from the solar wind or solar photons. Thermal desorption, or evaporation, as proposed by McGrath et al. (1986), is too rapid to characterize the atmosphere alone, but it is part of the exosphere recycling process. PSD is the desorption

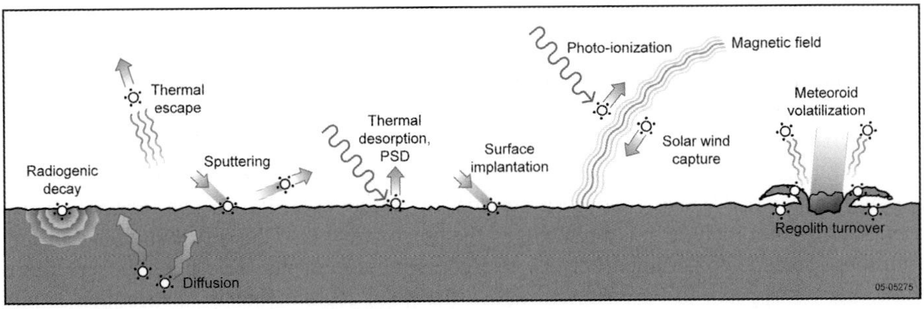

Fig. 1 The sources, sinks, and processes within the exosphere, surface, and magnetopause of Mercury

of particles due to UV photon bombardment. Laboratory studies by Yakshinskiy and Madey (1999) demonstrated that Na can be released via PSD from lunar soil simulate. Killen et al. (2001) modeled particle release from Mercury's surface by PSD and found that PSD could be an efficient particle release process on the dayside for volatile species such as Na and K. The strong observed dependence of Na abundance with solar zenith angle supports PSD as a major source process for Na (Killen et al. 2001). Refractory elements such as Ca would not be thermally desorbed or sputtered by UV photons. Refractory species are most efficiently released by impact vaporization (e.g., Mangano et al. 2007). Very hot calcium has been observed in the exosphere, a large fraction of which directly escapes (Killen et al. 2005).

The possible sinks for depleting Mercury's exosphere include thermal escape (H and He only), photoionization and entrainment in the solar wind, and surface implantation. Neutral atoms in the exosphere can be ionized and removed to the interplanetary medium via the solar wind and magnetosphere (Killen et al. 2004a; Killen and Crider 2004; Sarantos 2005). Thermal escape is an efficient process for the lighter elements H and He. Thermal escape assisted by solar radiation pressure occurs for sodium and presumably also for potassium. Up to 10% of the total sodium production rate is lost by escape into the Mercury "tail" during periods of maximum solar radiation pressure (Potter et al. 2002b). Removal of material from the exosphere via surface implantation can occur either through adsorption onto surface grains or through chemical interactions, such as those that produce space weathering effects in the regolith. Adsorption is more productive on the nightside, where evaporation is less effective at releasing the material. Adsorption also includes the process of cold-trapping material, such as in shadowed areas at high latitudes, or under outcroppings (Yan et al. 2006). The accumulation of material on the nightside for later release during Mercury's day is supported by the diurnal variations seen in the Na abundances (Sprague 1992; Sprague et al. 1997; Hunten and Sprague 2002; Schleicher et al. 2004).

The life cycle (Fig. 1) of an exospheric species can be followed by starting its journey to the surface from either the subsurface (by diffusion or regolith turnover) or from the exosphere (by collision with the surface) (Killen and Crider 2004). If a neutral atom collides with the surface it can be either scattered back into the exosphere or implanted into the surface, where it is either adsorbed or chemically bonded in a surface space-weathering process. A photoionized species can also collide with the surface (Killen et al. 2004a; Sarantos 2005), where it will be either neutralized and adsorbed (such as with magnetotail or ion recycling) or chemically bonded into the surface, also as part of the space-weathering process. Species from the surface can be introduced, or reintroduced, into the exosphere by sputtering, evaporation, PSD, meteoroid volatilization, impact vaporization, or scattering from collision at the surface.

3 Structure and Density Distributions

Several sets of observations constrain the six known constituents (H, He, O, Na, K, and Ca). Atomic hydrogen, observed in two height distributions, is characterized by the day and night surface temperatures, respectively, with scale heights above the surface of about 1,330 and 230 km. Helium emission, greatest over the dayside and above the sunward limb, was measured as far as 3,000 km above the surface (Broadfoot et al. 1976). The Mariner 10 observations did not include measurements in the polar regions, so the distribution associated with those areas is constrained only by telescopic measurements. Telescopic observations show that Ca appears to be enhanced above the polar regions, but its distribution is not fully known (Killen et al. 2005). Calcium is observed at extreme temperatures, and a large fraction is above escape velocity (Killen et al. 2005). The mechanism for imparting these energies is unknown. For O, observations by Mariner 10 provided only an upper limit on the abundance estimate.

Ground-based observations of Na and K show variable abundances and distributions of these elements. Possible associations between bright-ray craters and regions of freshly overturned regolith with enhanced Na abundance, as indicated by emission at the Na D2 line at 589 nm (Sprague et al. 1998), are shown in Fig. 2. The brightest Na emission falls over the Kuiper-Muraski crater complex (designated K) and the location of features with notably bright albedo at both visible and radar wavelengths (A). Both of these geologic features are associated with regions of freshly excavated material (Robinson and Lucey 1997; Harmon 1997). The brightest region in the figure appears offset from the associated geologic features. Atmosphere turbulence effects, or atmospheric seeing, would move the bright region toward the center of the image; thus it is more likely associated with the features closer

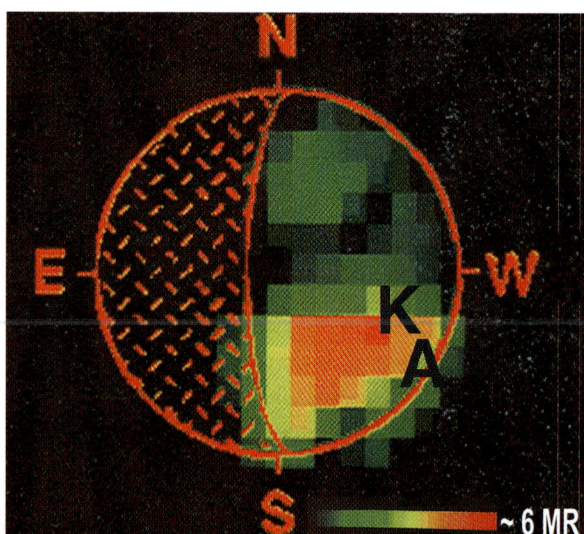

Fig. 2 This Na emission image, adapted from Potter and Morgan (1997), was taken with an image slicer on the McMath-Pierce Solar Telescope at Kitt Peak. It shows an example of enhanced Na emission observed over a freshly cratered region (*K*, Kuiper-Muraski crater complex). A radar-bright spot (*A*) is at the longitude of the limb. Both areas *K* and *A* are associated with regions of freshly overturned regolith. Radar-bright spot B is at the same longitude in the northern hemisphere but shows no enhancement of Na. Localized Na sources near the planet limb observed from ground-based telescopes appear offset in the direction of the planet center due to atmospheric seeing effects. Thus the bright red region is likely associated with geologic features near the limb. The *color bar* codes the intensity of emission, with *red* the brightest (~6 MRayleighs for the date of observation; December 7, 1990) and *blue-green* the minimum

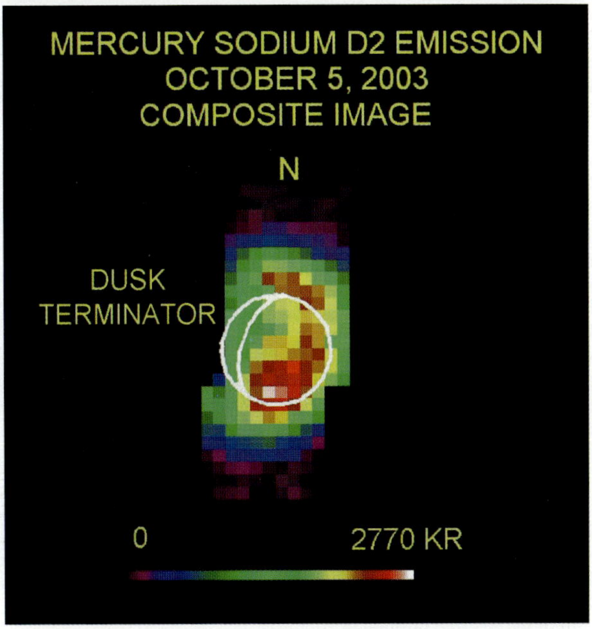

Fig. 3 An example of the asymmetric distribution of Na in Mercury's atmosphere (Potter et al. 2006). This observation, made near Mercury's perihelion, is a composite of three $10'' \times 10''$ images taken with the image slicer on the echelle spectrograph at Kitt Peak's McMath-Pierce Solar Telescope. Each pixel is $1''$ square and is produced by extracting the D2 emission line from a high-resolution spectrum. The figure shows integrated intensity along the line of sight in kRayleighs (kR) and is not in a 1 : 1 relationship to zenith column abundance. The tail is not imaged here since only three $10'' \times 10''$ fields are shown. Limb brightening is apparent on the dayside due to line of sight effects

to the limb. In addition, the processing of image data obtained from an image slicer to a representative two-dimensional map also shifts the location of the bright region. With this said, it should be noted that the distribution of these elements changes on a daily basis and this distribution could be a function of magnetosphere–solar wind interactions (Killen et al. 2001). Another example of the uneven distribution of atmospheric sodium (Fig. 3) is the observation of brightening along the sunlit limb and excess sodium at high southern latitudes (Potter et al. 2006). In this case peak emission is observed at ~2.7 MRayleighs (2,700 kR).

Sodium and potassium atoms are massive enough to be mostly bound to the planet, and thermal components with scale heights roughly approximated by the surface temperature (30–60 km) are present (Sprague et al. 1997). However, several ground-based observations have found an extended Na component that varies in distance according to Mercury's true anomaly, solar activity, and the orientation of magnetic fields in the solar wind (Potter et al. 2006). Scale heights of 150 km have been used to model the observed Na distributions seen by Schleicher et al. (2004). Figure 4, adapted from Potter et al. (2002) and obtained from Kitt Peak's McMath-Pierce Solar Telescope, shows three regimes of the Na exosphere. Extended coronae sometimes exist above both the north and south polar regions; a lower-scale height region can be observed at the subsolar point; and there is a "tail" of material streaming in the anti-sunward side. All of these three regimes are expected as a result of solar radiation pressure. The abundance of the distant tail population is controlled by pressure from solar photons and various source mechanisms. Thus the tail's extent is expected to vary with the position of Mercury relative to the Sun and has been modeled by Smyth and Marconi (1995) and Leblanc and Johnson (2003).

Figures 5 and 6 show two extreme examples taken from the models of Leblanc and Johnson (2003). The exosphere model simulation by Leblanc and Johnson (2003) includes a surface source of atoms from an initially adsorbed layer of 4×10^{12} atoms cm^{-2} as well as what is provided by infalling meteoritic material. As discussed in Sect. 4, these researchers do not consider such other sources as diffusion and meteoritic vaporization. In their model

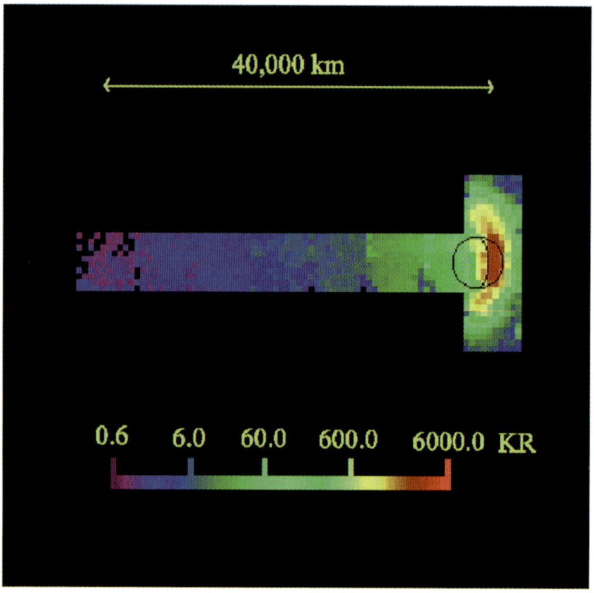

Fig. 4 Emission from Na atoms in Mercury's exosphere and anti-sunward tail are shown along with a *color bar* indicating emission intensity in kRayleighs (Potter et al. 2002b). This image was taken on May 26, 2001, at a Mercury true anomaly of 130°. This is a composite of ten $10'' \times 10''$ images taken with an image slicer on the echelle spectrograph at the McMath Pierce Solar Telescope. Resonance scattering reveals sodium only in sunlight; therefore, the relatively low emission to the left, on the nightside, does not mean that there is a lack of sodium there. The Na distribution and emission intensity are variable and depend on true anomaly as well as variable source processes. Line-of-sight velocity was determined by measuring the shift of the centroid position of the emission line from the rest position on Mercury. These velocities were converted to heliocentric velocities by dividing by the sine of the phase angle. This image, made close to aphelion, shows the Na extended high above the polar regions similar to the model shown in Fig. 5

atoms accumulate on the nightside, and as a consequence the morning terminator is the main source of sodium atoms to the atmosphere. Leblanc and Johnson (2003) concluded that the north–south asymmetries can be attributed to maxima in surface Na concentration at high latitudes just before the dawn terminator due to cold trapping.

Alternative models (as discussed in Sect. 4), such as that proposed by Killen et al. (2001, 2004a), incorporate additional processes. In these models cold-trapping is less efficient on the nightside because impact vaporization and ion sputtering deplete the surface of adsorbed atoms, and because many of the atoms that reach the nightside are already above escape velocity. As a result, the exosphere is most dense on the dayside, either at the subsolar point or with local enhancements where the magnetosphere is open to the solar wind.

Measurements of the distribution and abundance of the Na atmosphere above the limb of Mercury were made during Mercury's transit of the Sun in 2003 (Schleicher et al. 2004). Using a triple etalon system with adaptive optics, the absorption profile of the Mercury Na D2 line was measured. Na emission was observed to be greater above both polar regions than above the equatorial limb of the planet. In addition, measurements above the morning side of the planet exhibited greater Na emission than on the evening side. The predictive model of Leblanc and Johnson (2003) shown in Fig. 5 is consistent with the observations, which also occurred near aphelion. Schleicher et al. (2004) give Na scale heights for four geometries

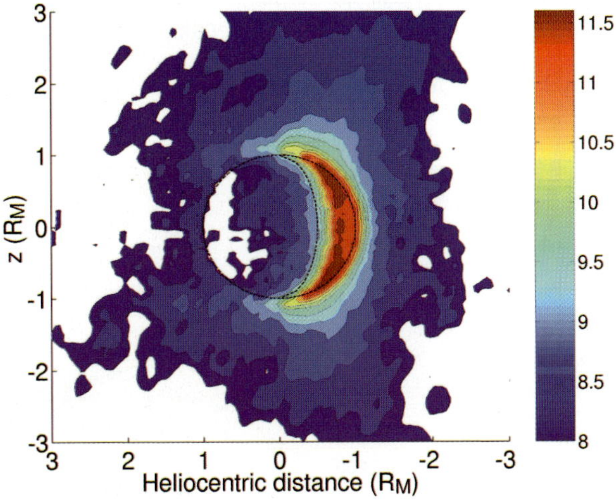

Fig. 5 At aphelion, at a true anomaly angle of 181°, the Na exosphere is extended above the polar regions on the Sun side in this model (Leblanc and Johnson 2003). The abundance units on the *color bar* are logarithmic in Na/cm^2 and distances are in units of planet radius

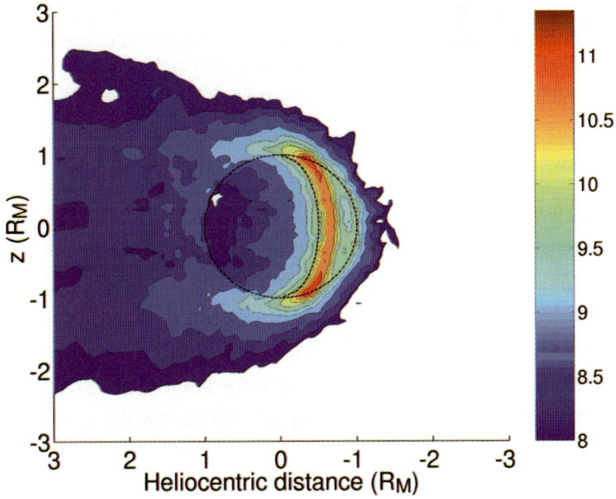

Fig. 6 At a true anomaly of 252° the maximum Na emission appears near the morning terminator in this model from Leblanc and Johnson (2003). As described in the text, the Na streams behind the planet under the increased influence of solar radiation pressure. This geometric configuration corresponds closely to that shown in Fig. 4. Column abundance units on the color bar are logarithmic in Na/cm^2

Table 3 Estimate of sodium scale heights

Location	Scale height (km)
Northern maximum	135 ± 30
Southern maximum	130 ± 30
Western planetary limb	(150)
Eastern planetary limb	(150)

at Mercury's limb, corresponding to the equivalent widths shown in Fig. 7. Good model fits to the northern and southern maximum measurements give scale heights with error bars as shown in Table 3. For the eastern and western planetary limbs, a scale height of 150 km was assumed for the model (shown in parentheses). Note that equivalent atmospheric temperatures corresponding to these scale heights are between 1,350 and 1,520 K, consistent with the results of Na D2 line width modeling by Killen et al. (1999).

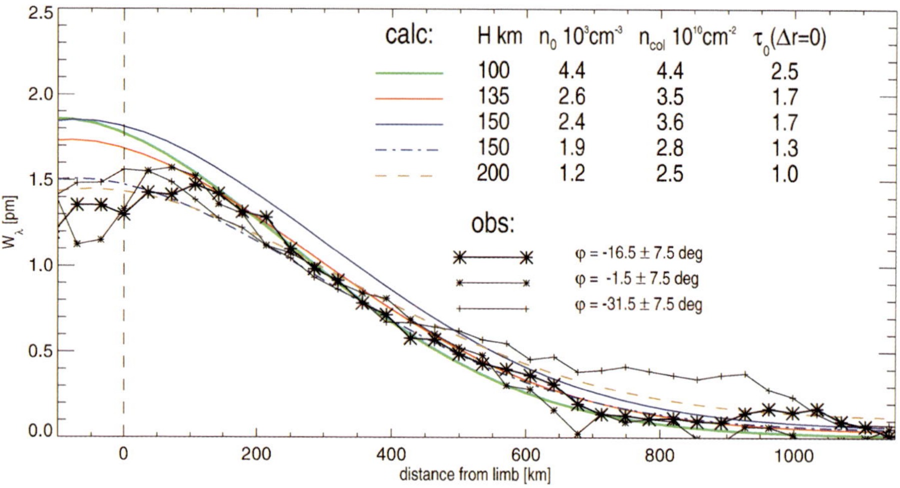

Fig. 7 Equivalent widths of the Mercury Na D2 line are plotted as functions of the distance from Mercury's limb. Models estimating fits to the equivalent width distributions yield estimates of the scale height (H) of the Na distribution (Schleicher et al. 2004). φ is the azimuthal angle from the north pole of Mercury measured in a clockwise direction around the observed limb of the Earth-facing disk, τ_0 is the optical depth at the line's center along the gazing line of sight, n_0 is the Na particle density, and n_{col} is the Na column density

An estimate of about 1.6 $\mathrm{km\,s^{-1}}$ is made for the velocity along the line-of-sight of the measurement. This estimate is based on the Doppler width of the Na D2 line convolved with the spectral resolution of the instrument.

Some information regarding the height distribution of Ca above Mercury's surface has been obtained by several spectroscopic measurements (Killen et al. 2005). Figure 8a shows intensity of Ca 422.6 nm emission at the slit locations during one such observing period, with the Ca intensity color coded. Figure 8b shows the corresponding line of sight Ca velocity corresponding to the slit locations shown in Fig. 8a. Clearly a high-energy process is placing Ca up to 3,000 km above Mercury's surface. No near-surface Ca emission has yet been found on the sunlit limb due to the difficulty of observing near the bright limb. The data shown in Fig. 8 indicates that Ca is moving at extreme velocity in the direction of orbital motion of Mercury, with a large fraction probably moving above escape velocity. These velocities cannot be due to impact vaporization or ion sputtering but must be caused by a secondary process such as dissociation or charge exchange (Killen et al. 2005).

No actual measured information regarding the height distribution of O is available (the detection and estimated abundance of O made by Mariner 10 had no associated height distribution information, and ground-based observations cannot detect the emission line at 130.4 nm because of telluric atmospheric opacity in the UV). It is possible that Ca may be vaporized in molecular form, and that CaO is dissociated after ejection from the surface, leaving both Ca and O at high energy (Killen et al. 2005). In such a case it could be expected that O will be escaping. Koehn and Sprague (2007) suggested that O^{6+} and Ca^{11+} delivered by the solar wind to Mercury's surface and subsequently ejected as neutral atoms into an energetic distribution in the Mercury space environment or exosphere are adequate to explain the observations.

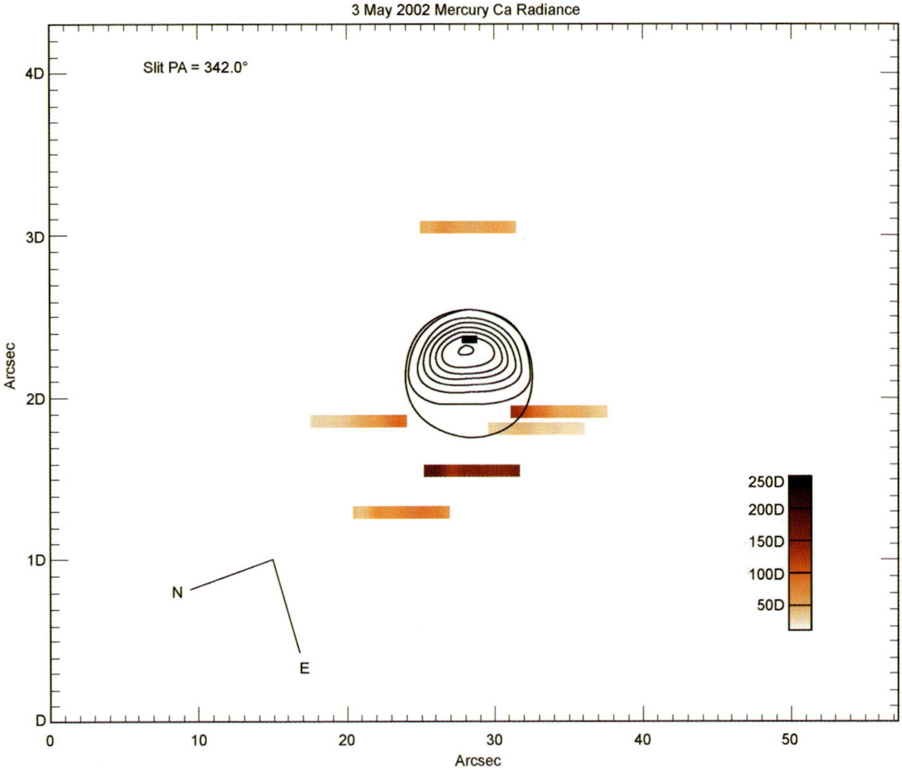

(a)

Fig. 8 (**a**) Intensity of Ca emission at various slit locations of the High Resolution Echelle Spectrometer (HIRES) placed from the limb to about one Mercury diameter away from the planet during observations at the Keck I telescope on May 3, 2002. (**b**) Line-of-sight velocity components of calcium of 1 to 3 km/s are color coded, corresponding to the slit locations in (**a**). They indicate extreme velocity in the direction of orbital motion of the planet, probably on the order of 1.5 to 4. km/s. A large fraction of the calcium is probably above escape velocity

4 Exosphere–Surface Boundary Interactions

Early theoretical treatments of Mercury's (and the Moon's) tenuous atmosphere assumed the surface was saturated with an adsorbate (e.g., Hodges 1973; Hartle and Thomas 1974; Hartle et al. 1973). This assumption defined a process in which every impact of an atom is followed by the release of another similar atom at the impact site. Hartle and Thomas (1974) assumed that the source of the lunar atmosphere is the solar wind, and that the flux of solar wind particles hitting the surface is balanced by the rising flux. Therefore an equilibrium is assumed between incoming solar wind ions and outgoing neutrals. A similar model was derived for Mercury (Hartle et al. 1973), but it was invalidated by Mariner 10's discovery of an intrinsic magnetic field.

Classical exospheric models, derived for an exosphere whose exobase is in contact with an atmosphere, are based on the assumption that the exospheric constituents are derived from a reservoir of atoms whose source is freely evaporating with a Maxwellian velocity distribution equivalent to the local surface temperature. The resulting altitude distributions follow a barometric law with certain deviations due to loss by escape (Chamberlain 1963). How-

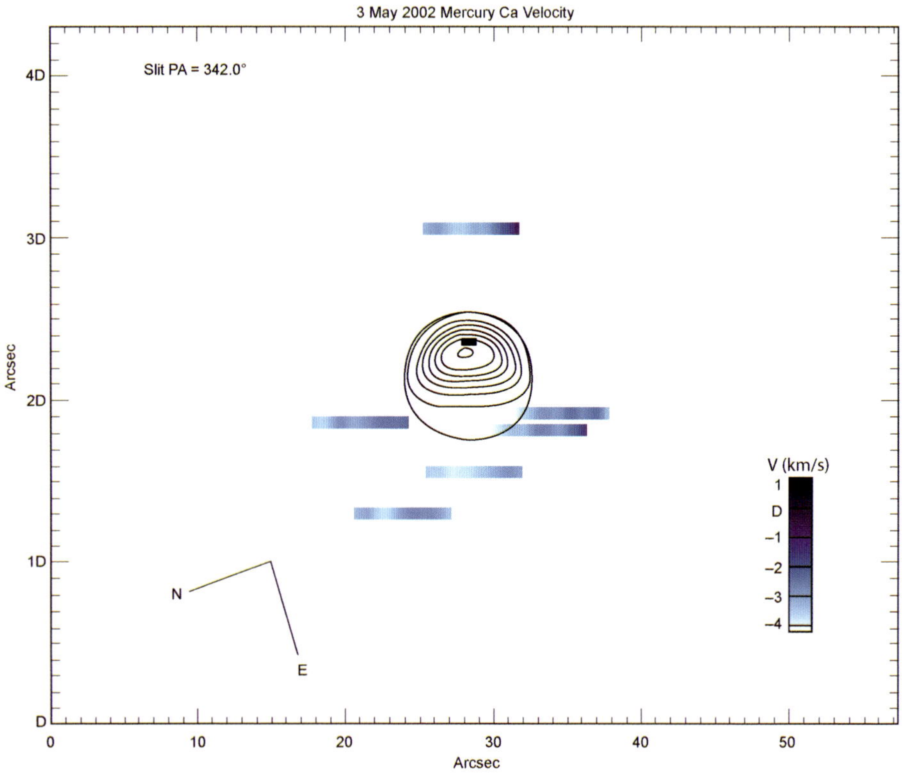

(b)

Fig. 8 (*Continued*)

ever, it is evident that many details of these models are not consistent with surface-bounded exospheres. For a surface-bounded exosphere, the expected source processes include PSD, ion sputtering, and meteoritic vaporization (see Table 2). The first two processes produce a source whose velocity distribution is described by a statistical Weibull distribution at an equivalent temperature of about 1,200 K (Madey et al. 1998). Unlike a Maxwellian velocity distribution, this distribution is much hotter than the surface and includes a tail distribution of very hot atoms. Vapor derived from meteoritic vaporization produces a gas that is thermal, but very hot, about 3,000 to 5,000 K (Kadono and Fujiwara 1996; Kadono et al. 2002; Sugita et al. 2003). Therefore classical exospheric models are not directly applicable to Mercury's exosphere.

The degree to which atoms thermally accommodate to the surface on contact is not well known and is probably species dependent. The global distribution of H and He in Mercury's exosphere as measured by Mariner 10 is incompatible with thermal accommodation. Models based on the assumption of thermal accommodation predict an anti-solar/subsolar He number density ratio of 200 at the surface. However, the measured ratio is smaller, by a factor of about 10. Altitude distributions of the known gases in Mercury's exosphere, when fit to a scale height, give temperatures that are different for each constituent. The H seen above the subsolar point has both a very cold component and a warm component, which at 420 K is colder than the surface at that location. Helium was fit to a 575 K barometric atmosphere above the subsolar point, which is closer to the subsolar temperature. However, the anti-solar density observed for He is much less than that predicted on the basis of thermal accommodation. Sodium line profiles measured near the surface (not in the tail, which is accelerated by radiation pressure) are consistent with a much hotter gas than the surface. The Na exosphere is measured to be 1,200 K at the equator and 750 K at the poles, or at least 550 K hotter than the surface (Killen et al. 1999). The original Na D2 line width measured by Potter and Morgan (1985) was shown to be consistent with $\sim$90% of the Na atoms at 500 K and the remaining $\sim$10% at even higher temperatures (Hunten et al. 1988), suggesting that different locations and times of measurement have different equivalent temperature distributions. The only other known constituent of Mercury's atmosphere, Ca, has been observed and modeled with temperatures consistent with a very hot gas (6,000 to 12,000 K) (Bida et al. 2000). These temperatures indicate that nonthermal processes are important in producing and maintaining Mercury's surface-bounded exosphere, and that none of the known constituents is in thermal equilibrium with the surface.

Another variable in the exosphere–surface boundary models is the sticking time, or the time between impact of the downwelling atom and its rerelease into the atmosphere. Long sticking times imply chemisorption (absorption resulting in a chemical bond) or physisorption (absorption into an atomic potential well resulting in a physical bond). It has been suggested that atoms that stick to the surface become bound and do not have lower binding energies than atoms intrinsic to the rock (Madey et al. 1998). The binding energy of sodium is somewhat controversial, but there may be a variety of binding sites on a silicate surface, particularly for a radiation-damaged surface (Yakshinskiy et al. 2000). In addition, the sticking efficiency on a porous surface such as a regolith is greater than the sticking efficiency on a flat surface due to multiple interactions with the surface on each impact. On silicate surfaces, the sticking efficiency is about 0.5 at $T_s = 250$ K and decreases to a value of 0.2 at $T_s = 500$ K (Johnson 2002).

While Johnson (2002) argued that most collisions with the surface are free–bound (the atom is adsorbed or goes through a bound state prior to scattering), Shemansky and Broadfoot (1977) argued that the surface of Mercury is not saturated with gas and that most collisions of light atoms with the planetary surface are "free–free" (the atom is unbound before and after the collision and does not go through a bound state), not free–bound. Measured energy accommodation coefficients for He and Ne are quite small (0.009–0.07 and 0.06–0.3, respectively), while accommodation coefficients for Ar can be large (0.29–0.67). Shemansky and Broadfoot (1977) argued that collisions that do not involve adsorption are limited in the amount of energy that can be gained or lost to the substrate at each collision. This situation is consistent with a gas that is derived from an energetic process such as ion sputtering or meteoritic vaporization and remains hotter than the surface. It also explains the uniformity of the H atmosphere relative to that predicted on the basis of thermal accommodation.

Thermal vaporization rates are critical to the resulting source rates and velocity distributions of the constituents in Mercury's exosphere. The role of thermal vaporization rates is especially controversial in how they are treated in various exosphere–surface interaction models. Thermal vaporization rates are governed by the binding energy of the atom balanced against the temperature. The more tightly bound the atom, the lower the thermal vaporization rate, but the higher the temperature, the higher the vaporization rate. The controversy over the effect of thermal vaporization on the surface of Mercury arises from the recognition that vaporization at equatorial latitudes near the subsolar region will deplete a monolayer of sodium atoms quite quickly. Some models, such as that of Leblanc and Johnson (2003), do not include any additional source of atoms at the exobase, e.g., through meteoritic vaporization or diffusion (including regolith, volume, and grain-boundary diffusion). Given these assumptions, a rapid depletion of sodium at equatorial latitudes on the sunlit side of Mercury will occur. Leblanc and Johnson (2003) concluded that the hot equatorial surface becomes depleted in adsorbed sodium, thus shutting down the supply of atoms to the exosphere.

Sprague (1990) and Killen et al. (2004a) showed through modeling that the source of Na atoms through grain-boundary diffusion and meteoritic vaporization taps a deeper source and is therefore not depleted through thermal vaporization. Killen et al. (2004a) concluded that a source of atoms to the extreme surface from the grains is robust enough to maintain PSD, ion sputter, and meteoritic vapor rates, and therefore these source rates for these processes are independent of the abundance of adsorbed atoms. In other words, thermal vaporization taps a different reservoir of atoms than the other three source processes.

Killen et al. (2004a) also showed that regolith gardening is fast enough on Mercury to replenish the supply of sodium to the surface for the age of Mercury. Regolith gardening rates are known for the Moon (Heiken et al. 1991). Killen et al. (2004a) assumed that the regolith gardening rate can be scaled linearly to the meteoroid influx, which they assume is roughly an order of magnitude larger at Mercury than that at the Moon. The scaled micrometeoroid flux is based on measurements at Earth orbit (Cintala 1992), and the flux of larger meteoroids estimated from Spacewatch measurements (Marchi et al. 2005). These results, however, are quite uncertain and will not be fully characterized without in situ measurements. Meteoroid impacts supply the surface with fresh rock from below and bury the top layers of regolith.

In addition to regolith gardening, meteoroid and micrometeoroid impacts affect the composition and storage of volatiles through vapor deposition. Although some of the vapor derived from meteoroid impact is lost to the system, much of it is retained (Butler et al. 1993; Crider and Killen 2005). The regolith grains at Mercury are most likely coated with glass derived from impact vapor. Since glass is amorphous, atoms more readily diffuse through it and evaporate from it (Shih et al. 1987). Thus source rates for the atmosphere from the surface-boundary exobase are intimately dependent on meteoroid impact rates and on the distribution of the impact vapor.

In modeling exosphere–surface boundary interactions, another effect that should be considered is the impact of surface charging on sputtering. Surface charging has been shown to be an important process at both Mercury (Grard 1997) and the Moon (Vondrak et al. 2004) due to efficient photoionization and interaction of the surface with solar wind ions and possibly with ions of planetary origin. Surface charging affects the trajectories of photoions (Grard 1997) and hence recycling rates for photoions (Killen et al. 2004b). There is a possibility that charged dust levitates on Mercury as well as on the Moon (Stubbs et al. 2005). Surface charging also will affect the rates and energies of ion sputtering since about 10% of the sputter products will be ions (Elphic et al. 1993). For example, low-energy electrons can cause negative charging and enhance outward diffusion of sodium (Madey et al. 2002).

The rates and effectiveness of ion sputtering as a process for producing and maintaining Mercury's exosphere are strongly affected by the local space environment. Mercury's surface is shielded from the solar wind by its magnetosphere to an extent dependent on the solar wind dynamic pressure and on the magnitude and direction of the IMF. This magnetosphere is a dynamic system that can vary dramatically between being closed or open (Kabin et al. 2000; Sarantos et al. 2001; Kallio and Janhunen 2003a, 2004a, 2004b). The solar wind can impinge on a large fraction of the dayside surface when the magnetosphere is in its open configuration. The effect of the solar wind impinging on the surface is controversial, and much is based on studies of the lunar exosphere, some of which is contradictory. While early models were based on the assumption that the lunar surface simply acts as a sponge for solar wind particles (Hartle and Thomas 1974), the rapid variation in the observed sodium atmosphere of Mercury prompted some observers to postulate that the heavy ions in the solar wind or interplanetary medium are quite effective in sputtering atoms, particularly sodium atoms, from the surface (Potter and Morgan 1990; Potter et al. 1999; Killen et al. 2001). Recent work has shown that highly charged heavy ions can be orders of magnitude more effective at sputtering insolator targets than singly charged ions of the same species (Shemansky 2003; Aumayr and Winter 2004). Thus heavy ions in the solar wind can be as effective or even more effective at sputtering than the more abundant H^+ and He^{++} in the solar wind. The resulting atmosphere would be asymmetric, as is observed, since the magnetosphere is expected to be asymmetric north/south due to the sunward component of the IMF (Sarantos et al. 2001; Leblanc and Johnson 2003). In addition to directly sputtering atoms from the surface, the solar wind ions hitting the surface can produce radiation damage, which enhances the efficiency of PSD (Potter et al. 2000). The role of magnetospheric shielding and its effects on the ion-sputtered production of Na can be tested by observations of the lunar atmosphere as the Moon moves inside and outside of the Earth's magnetosphere. Potter et al. (2000) showed evidence in lunar atmospheric Na observations that even though the Moon is fully in sunlight, its atmospheric density begins to decline as soon as it enters the Earth's magnetosphere, where it is shielded from the solar wind and continues to decline until the Moon reemerges from the magnetosphere. However, lunar corona observations by Mendillo and Baumgardner (1995) and Mendillo et al. (1999) of the Moon during eclipse show consistent atmospheric Na abundances regardless of the Moon's position within the Earth's magnetosphere. From an analysis of these observations Mendillo et al. (1999) suggested that the Na is from a blend of sources, 15% uniform micrometeoroid impact over the surface and the remainder (85%) from photon-induced desorption. More recently, Wilson et al. (2006) suggested that Na in the Moon's exosphere from more ubiquitous sources, such as solar and micrometeoroid bombardment, is augmented by plasma impact from the solar wind and Earth's magnetotail. These processes may also dominate at Mercury.

5 Solar Wind and Magnetospheric Interactions

Regardless of the relative roles of the processes listed in Table 2 the interactions among the solar wind, Mercury's magnetosphere, and Mercury's surface all have a strong influence on the production, maintenance, and character of Mercury's exosphere. In order to understand the generation and processes that maintain Mercury's exosphere, an understanding of the solar wind properties at Mercury and how the solar wind interacts with the planet's magnetosphere and surface is required.

As the solar wind approaches the planet, it first encounters the bow shock: the transition between the supersonic flow of interplanetary space to the subsonic, slower flows of the magnetosphere. The flow is held off by the magnetic field of the planet, but this field is not an impassable barrier. The dynamic pressure of the solar wind plasma compresses the planetary field and is balanced by the resulting magnetic pressure on the other side of the magnetopause. If the upstream dynamic pressure is sufficiently high, the magnetopause can come within a proton gyroradius of the surface, allowing direct precipitation to the surface (Goldstein et al. 1981; Kabin et al. 2000). As the IMF comes into contact with the planetary field, reconnection can occur—solar field lines join with the planetary lines (assuming an anti-parallel field configuration), and the resulting new line is dragged tailward by the solar wind. Once in the tail, another reconnection event recloses the field line with its conjugate in the opposing hemisphere. These reconnection events have several repercussions. First, they result in the peeling away of magnetic flux from the dayside of the planet, bringing the magnetopause closer to the subsolar point on the planet's surface (Slavin and Holzer 1979). Second, these events can open the magnetic cusp regions to the inflow of solar wind plasma, allowing its interaction with the regolith along open field lines. Reconnection can also accelerate impinging charged particles, leading to enhanced sputtering or scouring of the regolith. Reconnection during the passage of a strong coronal mass ejection (CME) or magnetic cloud can literally rip away the planetary magnetic field, giving the solar wind full access to the surface (Koehn 2002). If a solar energetic particle event accompanies the CME, these energetic particles increase the sputtering effect (Leblanc et al. 2003). The only information available on Mercury's magnetosphere comes from the Mariner 10 measurements of the magnetic field (Ness et al. 1974; Simpson et al. 1974) and of the thermal plasma environment (Ogilvie et al. 1977). The thermal plasma observations provide estimates of the standoff distances of the bow shock and magnetopause consistent with the estimates based on the magnetometer measurements (Ness et al. 1975; Ogilvie et al. 1977). Any refinements in our understanding of Mercury's magnetic field come from various modeling and simulation efforts. These models and simulations are constrained by the Mariner 10 magnetic field and plasma observations, by their predictions for the generation and maintenance of the exosphere, and how these predictions for the exosphere compare with observations (specifically the neutral Na measurements). An in-depth and focused discussion of Mercury's space environment and magnetosphere is provided in a companion paper (Slavin et al. 2007), but this section provides a generalized overview as it pertains to the generation and maintenance of Mercury's exosphere.

Models of Mercury's magnetic field, its interactions with the solar wind and IMF, and the resulting possible interactions with the surface can be grouped into four basic categories: analytic models (Luhmann et al. 1998; Sarantos et al. 2001; Delcourt et al. 2002, 2003), semi-empirical models (Luhmann et al. 1998; Massetti et al. 2003), a quasi-neutral hybrid (QNH) model (Kallio and Janhunen 2003a, 2004a, 2004b), and magnetohydrodynamic (MHD) models (Kabin et al. 2000; Ip and Kopp 2002). Each model is an attempt to describe the behavior of the magnetosphere as it interacts with the passing solar wind and IMF. Of interest are the relative openness of the magnetosphere, how the solar wind and IMF orientation drive this effect, and how the solar wind gains access to the planetary surface. Calculations of the standoff distance of the magnetopause are important, as the direct collision of the solar wind with the planetary regolith affects the composition of the exosphere. Qualitatively, the predictions of these models all agree. Since Mercury's magnetic field is small and the exosphere is tenuous, solar wind ions can collide with the planet's surface along open field lines. The region of open field lines, or "cusp," varies in response to changes in, and the orientation of, the IMF. The models diverge in their predictions of the extent of the cusp region and the amount of plasma interacting with the surface on open and along closed field lines. And, of course, it is the size of the cusp region and the plasma flux that are correlated to the amount of material observed within the exosphere.

An early attempt to model the magnetosphere of Mercury was made by Siscoe and Christopher (1975). They modeled the standoff distance of the magnetopause as a function of solar wind dynamic pressure, including the effects of the distance of the planet from the Sun. They estimated that the magnetopause would be compressed to the surface of the planet less than 1% of the time. Slavin and Holzer (1979) improved this model by including the effects of magnetic flux erosion in their calculations, allowing dayside reconnection to enhance the planetward motion of the magnetopause subsolar point. They demonstrated that the standoff distance is highly variable and predicted an upper bound on the subsolar standoff distance consistent with the Mariner 10 findings. Sarantos et al. (2001) used a modified Toffoletto-Hill model (Toffoletto and Hill 1993) to characterize the behavior of the magnetosphere and open field lines as functions of the IMF. They found that a strong B_X (radial component of the IMF), much more important at Mercury than at the Earth, controls the north–south asymmetry of the magnetosphere: for a southward IMF, a strong "positive" B_X (sunward) results in precipitation primarily in the southern hemisphere, and vice versa for a "negative" (antisunward) B_X. For a negative value of B_X, a turning B_Z regulates the size and latitude of the cusps. A strong negative B_Z drives the cusps equatorward and increases the open areas mapped by the cusps. B_Y controls the dusk–dawn asymmetry of the cusps, with a positive B_Y driving the open regions duskward, and negative B_Y producing a dawnward open region. Killen et al. (2001) used the same model to explore the effects of space weather on Mercury and to demonstrate the contribution of magnetospheric effects to the sodium variability seen by Potter et al. (1999).

Luhmann et al. (1998) used a scaled version of the Tsyganenko model (Tsyganenko and Stern 1996) of the Earth's magnetosphere to simulate that of Mercury. The Tsyganenko model allows only the Y- and Z-components of the IMF to be variables and has limited applicability to Mercury since B_X dominates there. This model predicted a very "open" magnetosphere during periods of southward IMF, which would allow the solar wind easy access to the surface. They further predicted that the relative weakness of the magnetic field of Mercury would render the magnetosphere highly sensitive to variations in the solar wind. Massetti et al. (2003) also used a modified Tsyganenko model to simulate the magnetosphere of Mercury. This model was explicitly used to map plasma precipitation onto the surface of the planet along magnetically open regions of the magnetosphere. They found that the cusps tend to map to a region ranging from 45° to 65° in latitude, with a longitudinal extent and position based on B_Y, as expected and shown in Sarantos et al. (2001). They showed a weak dependence of the open field lines on upstream dynamic pressure. They also found that the polar regions of the planet were relatively closed, as the solar wind tended to drag field lines with polar footprints into the tail.

Kallio and Janhunen (2003a, 2003b) have modeled Mercury's magnetosphere using a Quasi Neutral Hybrid (QNH) code, and they also examined the possible variation in interactions between the solar wind and planet surface based on different configurations of the IMF and dynamic pressure. Qualitatively their results are similar to MHD models, but there are distinct differences in the magnitude of the interactions predicted. Figure 9, taken from Kallio and Janhunen (2003b), maps the particle flux of impacting protons and the open/closed magnetic field line region for the following IMF configurations: (a) northward IMF, (b) southward IMF, (c) Parker spiral IMF, and (d) a high dynamical pressure case. In each case there are notable dawn–dusk particle flux asymmetries, but a north–south asymmetry develops only in the Parker-like scenario when the IMF B_X is dominant as it is at Mercury (Kallio and Janhunen 2003b). Three separate high-impact regions are apparent: (1) an "auroral" impact region equatorward of the open/closed field-line boundary, (2) a "cusp" impact region associated with the dayside noon–midnight meridian plane, and (3) a subsolar impact region apparent with high solar wind dynamical pressure. Kallio and Janhunen's (2003b) QNH model results predict that a southward IMF orientation produces a larger open field-line region than a northward IMF orientation. In addition, they predict that any north–south asymmetry in Mercury's magnetic field caused by the radial component of the IMF will also produce an asymmetry in the solar wind ion impact region on the surface, such that there will be a higher particle flux on the hemisphere that is magnetically connected to the solar wind.

MHD-based models include the acceleration effects of magnetic reconnection at Mercury, which is well-suited for studying the impact of solar-wind scouring of the regolith. Initial MHD-modeling by Kabin et al. (2000) looked at solar wind conditions with a Parker spiral IMF and showed that the magnetosphere was highly susceptible to the driving force of the solar wind. They found that the magnetic field lines for Mercury were closed at latitudes equatorward of 50° and that contact between the surface and magnetopause occurred

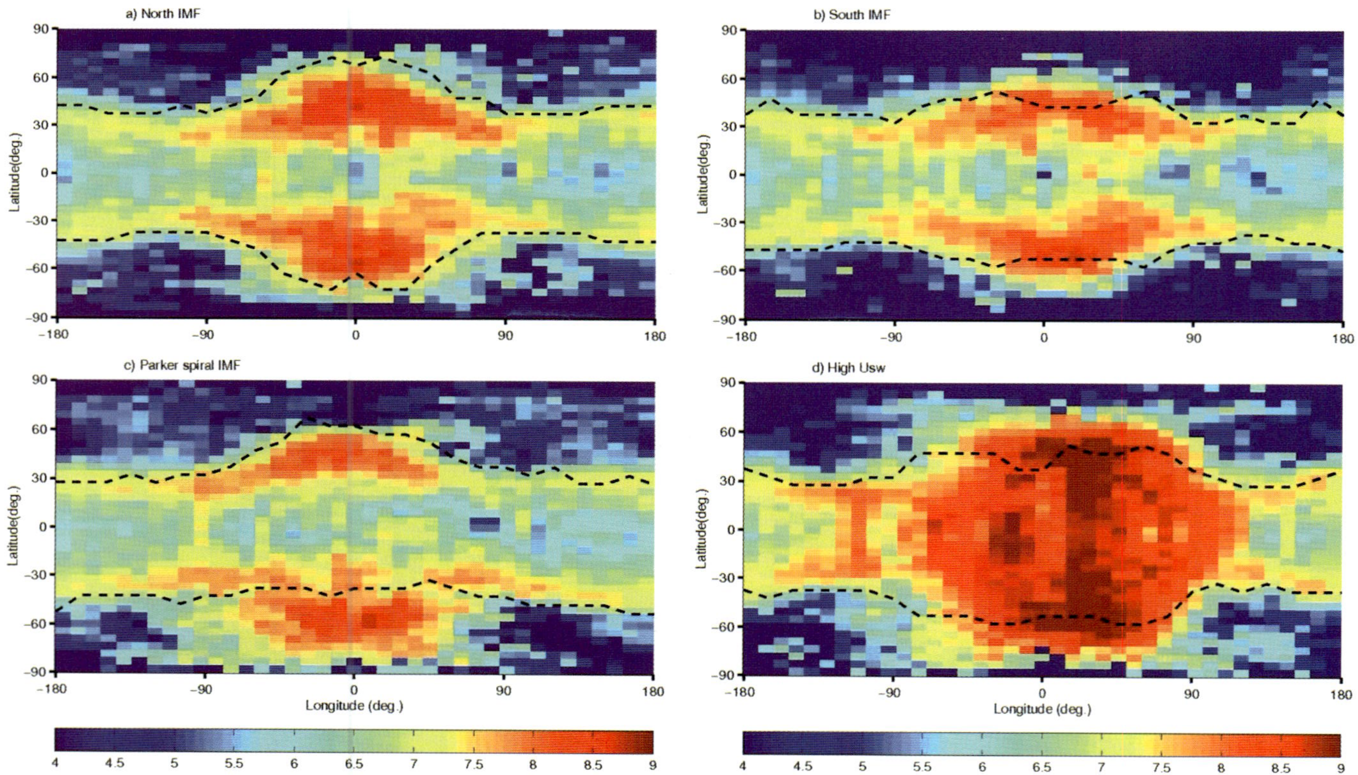

Fig. 9 Maps of the particle flux ($cm^{-2}\,s^{-1}$) of impacting H^+ ions and the open/closed magnetic field line region for (**a**) a pure northward IMF, (**b**) a pure southward IMF, (**c**) a Parker spiral IMF, and (**d**) a high solar wind event (Usw) (Kallio and Janhunen 2003b). The subsolar point is at the center of the image, and the *dashed lines* show the open/closed field line boundaries. The longitudes 0° and 180° correspond to local times of 0:00 and 12:00, respectively

at the subsolar point only when dynamic pressures increased by a factor of nine over the nominal conditions. They concluded that direct interaction between the solar wind and Mercury's surface is a rare phenomenon. Koehn (2002) extended the use of Kabin's model by incorporating data from the Helios spacecraft to generate upstream solar wind conditions as input for the model. This set of solar wind data includes a passing CME or magnetic cloud at 0.3 AU. This event compressed the simulated magnetopause to the planetary surface for several hours and opened the magnetopause to latitudes equatorward of 15°. During this time period, the solar wind had relatively unimpeded access to Mercury's surface. Studies by Ip and Kopp (2002) using a similar MHD model showed that there are differences in Mercury's magnetosphere, where interactions between the solar wind and surface vary as a function of the orientation of the IMF. They demonstrated that in a northward IMF configuration, Mercury's magnetosphere adopts a closed configuration and the size of the polar cap is at a minimum. In contrast, a southward IMF configuration opens up the polar cap region to its maximum size (from the pole equatorward to 20° latitude), even though the bow shock distance from the surface remains nearly the same for both configurations (Ip and Kopp 2002). Increases in dynamic pressure, such as those associated with CMEs, also push the polar cap boundary down to very low latitudes, supporting the hypothesis that solar wind variations could induce rapid temporal changes in the exospheric Na abundances (Killen et al. 2001; Ip and Kopp 2002).

There are limitations within each of these models, including the applicability of any model of the Earth's magnetosphere to Mercury. One of the major differences between the two magnetospheres is the mechanism for closing current systems. Within the Earth this is accomplished in the ionosphere, which does not exist for Mercury. Pick-up conductance within the exosphere (Cheng et al. 1987) and a conducting surface layer (Janhunen and Kallio 2004) have been proposed as possible mechanisms for Mercury. Many analytic and data-based models (e.g., Luhmann et al. 1998; Delcourt et al. 2002, 2003; Massetti et al. 2003) do not include the radial component (B_X), which is dominant at Mercury. MHD models (Kabin et al. 2000; Ip and Kopp 2002) also include assumptions that are invalid at Mercury, such as the existence of a thin shock boundary and the assumption of thermal equilibrium. The gyroradii of heavy ions can be as large as one planetary radius at Mercury, which contradicts the assumptions of a thin shock boundary and collective ion behavior. Mariner 10 measured a nonthermal electron distribution (Criston 1987), which further invalidates the assumption of thermal equilibrium. These models do not include the effects of induced surface currents, which will generate magnetic flux that opposes or counters efforts to change Mercury's magnetic field (Hood and Schubert 1979). Sudden jumps in solar wind pressure will therefore not be as effective as one would otherwise expect in pushing the magnetopause to the surface. The induction currents add magnetic flux to the dayside magnetosphere to oppose the compression.

These models all predict that the extent of the cusps is determined by the orientation and direction of the IMF and to a lesser extent by the dynamical pressure of the solar wind. Many of these models show that dynamical pressures associated with energetic solar events (such as CMEs) can compress the magnetopause to within a gyroradius of the surface, thus allowing solar ion interaction with the regolith. However, in order to correlate the optical emission of the exosphere with these solar wind–magnetosphere interactions, an estimate of the solar wind particle flux reaching the surface is required. Flux estimates (Massetti et al. 2003; Kallio and Janhunen 2003a; Sarantos et al. 2007) are comparable to the rates needed for PSD, implying that ion sputtering is an important but variable source for Mercury's exosphere. Killen et al. (2001) demonstrated that up to 32% of the exospheric Na content can be generated from ion sputtering along open field lines when the IMF is orientated southward, but ion-sputter yields are uncertain.

6 What's Next?

To resolve many of the outstanding issues regarding the generation, maintenance, and character of Mercury's exosphere in situ measurements are required, including: (1) mapping of the exospheric constituents and the variation of column density with location and time; (2) mapping of the magnetic field correlated to the solar plasma environment; (3) mapping of the plasma environment with time; and (4) mapping the elemental and mineralogical properties of the surface. Temporal and spatial correlations of these types of measurements and observations will provide a better understanding of the system that supports the existence of an exosphere.

The MESSENGER spacecraft science payload has the capability to provide these in situ measurements. The Mercury Atmospheric and Surface Composition Spectrometer (MASCS) (McClintock and Lankton 2007) will measure and map constituents within the atmosphere with its ultraviolet-visible spectrometer while also mapping the mineral spectral properties of the surface with its visible-infrared spectrograph. The MESSENGER Magnetometer (Anderson et al. 2007) will map the magnetic field, while the Energetic Particle and Plasma Spectrometer (EPPS) (Andrews et al. 2007) will observe and map the particle and plasma environment about Mercury. The X-Ray Spectrometer (XRS) (Schlemm et al. 2007) and the Gamma-Ray and Neutron Spectrometer (GRNS) (Goldsten et al. 2007) will map the surface elemental abundances. Collectively the measurements from this suite of instruments will go a long way toward resolving many of the unknown attributes of Mercury's exosphere and related processes.

MASCS, the key instrument targeted to study the exosphere, will provide pivotal data for investigating the neutral exosphere. It will use the standard limb-scanning technique employed by Mariner 10 (Hunten et al. 1988) to measure altitude profiles of resonantly scattered sunlight by exospheric species. The resulting column emission rates will be inverted to yield density as a function of altitude that will have 25-km vertical resolution and ~300- to 500-km horizontal resolution. MASCS will study the spatial distribution and temporal behavior of the known atmospheric species (Na, Ca, K, O, and H) and will search for additional exospheric constituents (e.g., S, Al, Fe, Mg, and Si). Measurements of seasonal and geographic changes in composition and structure will provide important input for models to constrain the source and sink processes for the exosphere.

For example, simulations of MASCS operations show that the data collected will be able to test such hypotheses as correlations between local surface features and exospheric sources. These simulations use a dayside model that predicts integrated zenith column abundances for the ambient exosphere. The model, including an additional localized source, was used to simulate the exosphere for two cases: one for which thermal vaporization (TV) was assumed as the dominant source for the ambient exosphere and another for which photon-stimulated desorption was the dominant ambient source. In both cases, output column abundances from the model were distributed with altitude in accordance with density distributions associated with a particular source process. MASCS observational geometries were then calculated using the MESSENGER trajectory database and used to generate observed tangential column emission rates for each spacecraft orbit. These were combined with measured instrument performance to produce an "instrument observation."

Figure 10 summarizes the results of a simulated observation for which the ambient sodium densities are enhanced by a localized thermal vaporization source from the Caloris basin. Two cases are shown; one for which the primary source of the ambient atmosphere is thermal vaporization and the other for which the primary source is photon-stimulated desorption. Due to the $3/2$ spin–orbit resonance of Mercury, the Caloris basin is observable by

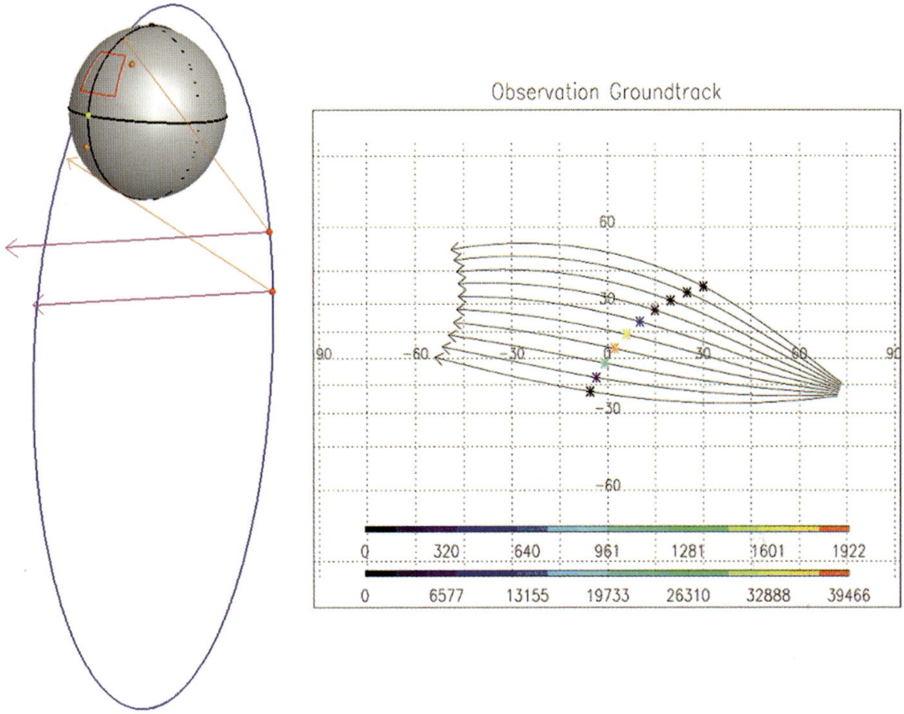

Fig. 10 (*Left*) Orbit of the spacecraft during the observation of the Caloris basin (*red box*). The *orange arrows* denote the line of sight at the beginning and end of the observation, while the *magenta arrows* denote the direction of the Sun. (*Right*) Ground track with the *arrows* denoting the line of sight. The stars represent the tangent points of the lines of sight and are color-coded to represent the instrument signal enhancement observed by MASCS above the ambient atmosphere for TV and PSD (*upper and lower color bars*, respectively)

MASCS only over a limited range of planet true anomaly values and also only for every other Mercury year. Observational opportunities for pointing MASCS above the region of the Caloris basin have been determined by using the MESSENGER trajectory database. The results presented in Figure 10 are for a 21-minute observation when the planet true anomaly is near 23°. During the observation the spacecraft is rotated 18.5° about the spacecraft–Sun vector, resulting in a scan that encompasses the Caloris basin and regions extending into the southern hemisphere. Note that surface temperature effects and observing geometry skew the peak emission away from the center of the enhanced region toward the subsolar point.

For these two cases, typical ambient zenith column abundances near Mercury true anomaly (TA) ~23° in the model are 2.8×10^9 Na cm^{-2} with a 720 K characteristic temperature and 4.35×10^{11} Na cm^{-2} with a 1,200 K characteristic temperature for thermal vaporization (TV) and PSD, respectively. Thus, the first case may be considered a low-energy, low-density ambient exosphere and the second considered a higher-energy, higher-density ambient exosphere. This simulation assumes that the minimum detectable column for a Caloris basin enhancement produces an instrument signal 10% above that from the ambient atmosphere signal, requiring ~5×10^{10} Na cm^{-2} and 7.6×10^{11} Na cm^{-2} for TV- and PSD-generated exospheres, respectively. These values can be compared with a column of 8.0×10^{11} Na cm^{-2} above a southern latitude radar-bright spot reported by Sprague et al. (1998). Repeat observations on successive orbits will reduce the minimum detectable column by a factor of 2 to 5.

7 Summary

Observations to date have demonstrated that Mercury's exosphere is highly variable and that this variability is a complicated function of location relative to the surface, Mercury true anomaly, time of Mercury day, and solar activity. Current modeling shows that the interactions among the exosphere, surface, magnetosphere, and space environment are complex. Our knowledge of these interrelated components is limited to snapshots in time, mostly constrained by the in situ observations of Mariner 10 or Earth-based telescopic observations. These data are insufficient to define fully the nature of Mercury's exosphere and provide only moderate constraints on models proposed for exosphere formation and maintenance. Neither the relative source strengths for exospheric components nor the partitioning among various release mechanisms are definitively established by the current data sets and models. One of the scientific goals for MESSENGER is to acquire an exospheric data set as a function of location relative to the surface, Mercury true anomaly, time of Mercury day, and solar activity of sufficient duration and resolution to distinguish among exospheric models and bound the relative source strengths and release mechanisms.

Many of the models and interpretations of the exospheric source and sink processes are based on assumptions regarding surface composition. What is known about the mineralogical and elemental composition of Mercury's surface is limited, however, to color data obtained by Mariner 10 (for less than half the planet) and ground-based telescopic observations (with low spatial resolution and filtered by the Earth's atmosphere). With the mineralogical and elemental compositional information to be returned by MESSENGER (Boynton et al. 2007; Head et al. 2007), additional constraints on exospheric processes and models will be possible. Equally limited is our knowledge of the magnetosphere and how it interacts with the solar environment. Measurements of Mercury's magnetic field to date are restricted to two comparatively high-altitude flybys of the planet by Mariner 10 (Slavin et al. 2007). MESSENGER will fly by the planet at low equatorial latitudes and through the magnetotail three times prior to orbit insertion. MESSENGER's orbit about Mercury will be highly elliptical, providing low-altitude information at high latitudes and multiple magnetospheric boundary crossings. All of this information will provide much-needed new constraints on the formation and maintenance mechanisms for Mercury's exosphere.

Acknowledgements The MESSENGER mission is supported by the NASA Discovery Program under contract NAS5-97271 to The Johns Hopkins University Applied Physics Laboratory. We thank Esa Kallio, Francois Leblanc, Andrew Potter, and Helmold Schleicher for the use of their figures. We thank Andrew Potter, Sean Solomon, and an anonymous reviewer for their insight and comments.

References

G.B. Andrews et al., Space Sci. Rev. (2007, this issue). doi:10.1007/s11214-007-9272-5
B.J. Anderson betal , Space Sci. Rev. (2007, this issue). doi:10.1007/s11214-007-9246-7
F. Aumayr, H. Winter, Phil. Trans. R. Soc. Lond. A **362**, 77–102 (2004)
T.A. Bida, R.M. Killen, T.H. Morgan, Nature **404**, 159–161 (2000)
W.V. Boynton et al., Space Sci. Rev. (2007, this issue). doi:10.1007/s11214-007-9258-3
A.L. Broadfoot, D.E. Shemansky, S. Kumar, Geophys. Res. Lett. **3**, 577–580 (1976)
B.J. Butler, D.O. Muhleman, M.A. Slade, J. Geophys. Res. **98**, 15003–15023 (1993)

J.W. Chamberlain, Planet. Space Sci **11**, 901–960 (1963)
A.F. Cheng, R.E. Johnson, S.M. Krimigis, L.J. Lanzerotti, Icarus **7**, 430–440 (1987)
M.J. Cintala, J. Geophys. Res. **97**, 947–973 (1992)
J.E.P. Connerney, N.F. Ness, in *Mercury*, ed. by F. Vilas, C.R. Chapman, M.S. Matthews (University of Arizona Press, Tucson, 1988), pp. 494–513
D. Crider, R.M. Killen, Geophys. Res. Lett. **32**, L12201 (2005). doi:10.1029/2005GL022689
S.P. Criston, Icarus **71**, 448–471 (1987)
D.C. Delcourt, T.E. Moore, S. Orsini, A. Millilo, J.-A. Sauvaud, Geophys. Res. Lett. **29**, 1591 (2002). doi:10.1029/2001GL073829
D.C. Delcourt et al., Ann. Geophysicae **21**, 1723–1736 (2003)
R.C. Elphic, H.O. Funsten III, R.L. Hervig, Lunar Planet. Sci. **24**, 439 (1993)
U. Fink, H.P. Larson, R.F. Poppen, Astrophys. J. **287**, 407–415 (1974)
B.E. Goldstein, S.T. Suess, R.J. Walker, J. Geophys. Res. **86**, 5485–5489 (1981)
J.O. Goldsten et al., Space Sci. Rev. (2007, this issue). doi:10.1007/s11214-007-9262-7
R. Grard, Planet. Space Sci. **45**, 67–72 (1997)
J.K. Harmon, Adv. Space Res. **19**, 1487–1496 (1997)
J.K. Harmon, M.A. Slade, Science **258**, 640–642 (1992)
R.E. Hartle, G.E. Thomas, J. Geophys. Res. **79**, 1519–1525 (1974)
R.E. Hartle, K.W. Ogilvie, C.S. Wu, Planet. Space Sci. **21**, 2181–2191 (1973)
J.W. Head et al., Space Sci. Rev. (2007, this issue). doi:10.1007/s11214-007-9263-6
G. Heiken, D. Vaniman, B.M. French, *Lunar Sourcebook: A User's Guide to the Moon* (Cambridge University Press, New York, 1991), 753 pp
R.R. Hodges Jr., J. Geophys. Res. **78**, 8055–8064 (1973)
L. Hood, G. Schubert, J. Geophys. Res. **84**, 2641–2647 (1979)
D.M. Hunten, A.L. Sprague, Adv. Space Res. **19**, 1551–1560 (1997)
D.M. Hunten, A.L. Sprague, Meteorit. Planet. Sci. **37**, 1191–1195 (2002)
D. Hunten, T.H. Morgan, D.E. Shemansky, in *Mercury*, ed. by F. Vilas, C.R. Chapman, M.S. Matthews (University of Arizona Press, Tucson, 1988), pp. 562–612
W.-H. Ip, Geophys. Res. Lett. **13**, 423–426 (1986)
W.-H. Ip, Astrophys. J. **418**, 451–456 (1993)
W.-H. Ip, A. Kopp, J. Geophys. Res. **107**, 1348 (2002). doi:10.1029/2001JA009171
W.-H. Ip, A. Kopp, Adv. Space Res. **33**, 2172–2175 (2004)
P. Janhunen, E. Kallio, Ann. Geophysicae **22**, 1829–1837 (2004)
R.E. Johnson, in *Atmospheres in the Solar System: Comparative Aeronomy*, ed. by M. Mendillo, A. Nagy, J.H. Waite. Geophysical Monograph, vol. 130 (American Geophysical Union, Washington, 2002), pp. 203–219
K. Kabin, T.I. Gombosi, D.L. DeZeeuw, K.G. Powell, Icarus **143**, 397–406 (2000)
T. Kadono, A. Fujiwara, J. Geophys. Res. **101**, 26097–26110 (1996)
T. Kadono et al., Geophys. Res. Lett. **29**, 1979 (2002). doi:10.1029/2002GL015694
E. Kallio, P. Janhunen, Ann. Geophysicae **21**, 2133–2145 (2003a)
E. Kallio, P. Janhunen, Geophys. Res. Lett. **30**, 1877 (2003b). doi:10.1029/2003GL017842
E. Kallio, P. Janhunen, Eos Trans. Am. Geophys. Union **85** (Fall meeting suppl.) (2004a), abstract P23A0241K
E. Kallio, P. Janhunen, Adv. Space Res. **33**, 2176–2181 (2004b)
R.M. Killen, Geophys. Res. Lett. **16**, 171–174 (1989)
R.M. Killen, D. Crider, Eos Trans. Am. Geophys. Union **85** (Fall meeting suppl.) (2004), abstract P23A-0248
R.M. Killen, A.E. Potter, T.H. Morgan, Icarus **85**, 145–167 (1990)
R.M. Killen, J. Benkhoff, T.H. Morgan, Icarus **125**, 195–211 (1997)
R.M. Killen, A.E. Potter, A. Fitzsimmons, T.H. Morgan, Planet. Space Sci. **47**, 1449–1458 (1999)
R.M. Killen et al., J. Geophys. Res. **106**, 20509–20525 (2001)
R.M. Killen, M. Sarantos, A.E. Potter, P. Reiff, Icarus **171**, 1–19 (2004a)
R.M. Killen, M. Sarantos, P. Reiff, Adv. Space Res. **33**, 1899–1904 (2004b)
R.M. Killen, T.A. Bida, T.H. Morgan, Icarus **173**, 300–311 (2005)
P.L. Koehn, Ph.D. thesis, University of Michigan, Ann Arbor, 2002, 183 pp
P.L. Koehn, A.L. Sprague, Planet. Space Sci. **55**, 1530–1540 (2007)
P.L. Koehn, T.H. Zurbuchen, G. Gloeckler, R.A. Lundgren, L.A. Fisk, Meteorit. Planet. Sci. **37**, 1173–1190 (2002)
S. Kumar, Icarus **28**, 579–591 (1976)
H. Lammer et al., Icarus **166**, 238–247 (2003)
F. Leblanc, R.E. Johnson, Icarus **164**, 261–281 (2003)
F. Leblanc, J.G. Luhmann, R.E. Johnson, M. Liu, Planet. Space Sci. **51**, 339–352 (2003)

J.G. Luhmann, C.T. Russell, N.A. Tsyganenko, J. Geophys. Res. **103**, 9113–9119 (1998)
T.E. Madey, B.V. Yakshinskiy, V.N. Ageev, R.E. Johnson, J. Geophys. Res. **103**, 5873–5887 (1998)
V. Mangano et al., Planet. Space Sci. **55**, 1541–1556 (2007)
T.E. Madey, R.E. Johnson, T.M. Orlando, Surf. Sci. **500**, 838–858 (2002)
S. Marchi, A. Morbidelli, G. Cremonese, Astron. Astrophys. **431**, 1123–1127 (2005)
S. Massetti et al., Icarus **166**, 229–237 (2003)
W.E. McClintock, M.R. Lankton, Space Sci. Rev. (2007, this issue). doi:10.1007/s11214-007-9264-5
M.A. McGrath, R.E. Johnson, L.J. Lanzerotti, Nature **323**, 694–696 (1986)
M. Mendillo, J. Baumgardner, Nature **377**, 404–406 (1995)
M. Mendillo, J. Baumgardner, J. Wilson, Icarus **137**, 13–23 (1999)
T.H. Morgan, H.A. Zook, A.E. Potter, Icarus **75**, 156–170 (1988)
N.F. Ness, K.W. Behannon, R.P. Lepping, Y.C. Whang, K.H. Schatten, Science **185**, 151–160 (1974)
N.F. Ness, K.W. Behannon, R.P. Lepping, Y.C. Whang, K.H. Schatten, J. Geophys. Res. **80**, 2708–2716 (1975)
K.W. Ogilvie, J.D. Scudder, V.M. Vasyliunas, R.E. Hartle, G.L. Siscoe, J. Geophys. Res. **82**, 1807–1824 (1977)
A.E. Potter, Geophys. Res. Lett. **22**, 3289–3292 (1995)
A.E. Potter, T.H. Morgan, Science **229**, 651–653 (1985)
A.E. Potter, T.H. Morgan, Icarus **67**, 336–340 (1986)
A.E. Potter, T.H. Morgan, Science **248**, 835–838 (1990)
A.E. Potter, T.H. Morgan, Planet. Space Sci. **45**, 95–100 (1997)
A.E. Potter, R.M. Killen, T.H. Morgan, Planet. Space Sci. **47**, 1141–1148 (1999)
A.E. Potter, R.M. Killen, T.H. Morgan, J. Geophys. Res. **105**, 15073–15084 (2000)
A.E. Potter, C.M. Anderson, R.M. Killen, T.H. Morgan, J. Geophys. Res. **107**, 5040 (2002a). doi:10.1029/2000JE001493
A.E. Potter, R.M. Killen, T.H. Morgan, Meteorit. Planet. Sci. **37**, 1165–1172 (2002b)
A.E. Potter, R.M. Killen, M. Sarantos, Icarus **181**, 1–12 (2006)
M.S. Robinson, P.G. Lucey, Science **275**, 197–198 (1997)
M. Sarantos, Ph.D. dissertation, Rice University, Houston, TX, 2005, 81 pp
M. Sarantos, P.H. Reiff, T.W. Hill, R.M. Killen, A.L. Urquhart, Planet. Space Sci. **49**, 1629–1635 (2001)
M. Sarantos, R.M. Killen, D. Kim, Planet. Space Sci. (2007). doi:10.1016/j.pss.2006.10.011
H. Schleicher, G. Wiedemann, H. Wohl, T. Berkefeld, D. Soltau, Astron. Astrophys. **425**, 1119–1124 (2004)
C.E. Schlemm II et al., Space Sci. Rev. (2007, this issue). doi:10.1007/s11214-007-9248-5
D.E. Shemansky, in *AIP Conf. Proc. 63, Rarefied Gas Dynamics 23rd Int. Symp.*, ed. by A.D. Ketsdever, E.P. Muntz, Chap. 9, Topics in Astrophysics (2003), pp. 687–695
D.E. Shemansky, A.L. Broadfoot, Rev. Geophys. Space Phys. **15**, 491–499 (1977)
C.-Y. Shih, L.E. Nyquist, D.D. Bogard, E.J. Dasch, B.M. Bansal, H. Wiesman, Geochim. Cosmochim. Acta **51**, 3255–3271 (1987)
J.A. Simpson, J.H. Eraker, J.E. Lamport, P.H. Walpole, Science **185**, 160–166 (1974)
G. Siscoe, L. Christopher, Geophys. Res. Lett. **2**, 158–160 (1975)
M.A. Slade, B.J. Butler, D.O. Muhleman, Science **258**, 635–640 (1992)
J.A. Slavin, R.E. Holzer, J. Geophys. Res. **84**, 2076–2082 (1979)
J.A. Slavin et al., Space Sci. Rev. (2007, this issue). doi:10.1007/s11214-007-9154-x
W.H. Smyth, M.L. Marconi, Astrophys. J. **441**, 839–864 (1995)
A.L. Sprague, Icarus **84**, 93–105 (1990)
A.L. Sprague, J. Geophys. Res. **97**, 18257–18264 (1992)
A.L. Sprague, R.W. Kozlowski, D.M. Hunten, Science **129**, 506–527 (1990)
A.L. Sprague, D.M. Hunten, K. Lodders, Icarus **118**, 211–215 (1995)
A.L. Sprague, D.M. Hunten, F.A. Grosse, Icarus **123**, 345–349 (1996)
A.L. Sprague et al., Icarus **129**, 506–527 (1997)
A.L. Sprague, W.J. Schmitt, R.E. Hill, Icarus **136**, 60–68 (1998)
R.G. Strom, A.L. Sprague, *Exploring Mercury: The Iron Planet* (Springer, New York, 2003), 216 pp
T.J. Stubbs, R.R. Vondrak, W.M. Farrell, Lunar Planet. Sci. **36** (2005), abstract 1899
S. Sugita, K. Hamano, T. Kadono, P.H. Schultz, T. Matsui, in *Impact Cratering; Bridging the Gap between Modeling and Observations* (Lunar and Planetary Institute, Houston, 2003), p. 68
F.R. Toffoletto, T.W. Hill, J. Geophys. Res. **98**, 1339–1344 (1993)
N.A. Tsyganenko, D.P. Stern, J. Geophys. Res. **101**, 27187–27198 (1996)
A.L. Tyler, R.W. Kozlowski, L.A. Lebofsky, Geophys. Res. Lett. **15**, 808–811 (1988)
R.R. Vondrak, T.J. Stubbs, W.M. Farrell, Eos Trans. Am. Geophys. Union **85** (Fall meeting suppl.) (2004), abstract SM53A0401
J. Wilson, M. Mendillo, H.E. Spence, J. Geophys. Res. **111**, A07207 (2006). doi:10.1029/2005JA011364
B.V. Yakshinskiy, T.E. Madey, Nature **400**, 642–644 (1999)
B.V. Yakshinskiy, T.E. Madey, V.N. Ageev, Surf. Rev. Lett. **7**, 75–87 (2000)
N. Yan, F. Leblanc, E. Chassefiere, Icarus **181**, 348–362 (2006)

Space Sci Rev (2007) 131: 187–217
DOI 10.1007/s11214-007-9269-0

The MESSENGER Spacecraft

James C. Leary · Richard F. Conde · George Dakermanji · Carl S. Engelbrecht · Carl J. Ercol · Karl B. Fielhauer · David G. Grant · Theodore J. Hartka · Tracy A. Hill · Stephen E. Jaskulek · Mary A. Mirantes · Larry E. Mosher · Michael V. Paul · David F. Persons · Elliot H. Rodberg · Dipak K. Srinivasan · Robin M. Vaughan · Samuel R. Wiley

Received: 24 July 2006 / Accepted: 15 August 2007 / Published online: 16 November 2007

Abstract The MErcury Surface, Space ENvironment, GEochemistry, and Ranging (**MESSENGER**) spacecraft was designed and constructed to withstand the harsh environments associated with achieving and operating in Mercury orbit. The system can be divided into eight subsystems: structures and mechanisms (e.g., the composite core structure, aluminum launch vehicle adapter, and deployables), propulsion (e.g., the state-of-the-art titanium fuel tanks, thruster modules, and associated plumbing), thermal (e.g., the ceramic-cloth sunshade, heaters, and radiators), power (e.g., solar arrays, battery, and controlling electronics), avionics (e.g., the processors, solid-state recorder, and data handling electronics), software (e.g., processor-supported code that performs commanding, data handling, and spacecraft control), guidance and control (e.g., attitude sensors including star cameras and Sun sensors integrated with controllers including reaction wheels), radio frequency telecommunications (e.g., the spacecraft antenna suites and supporting electronics), and payload (e.g., the science instruments and supporting processors). This system architecture went through an extensive (nearly four-year) development and testing effort that provided the team with confidence that all mission goals will be achieved.

Keywords MESSENGER · Mercury · Spacecraft · Subsystem · Mass · Power

1 Introduction

The MErcury Surface, Space ENvironment, GEochemistry, and Ranging (MESSENGER) mission to Mercury requires a spacecraft capable of at least eight years of operation over a

Larry E. Mosher passed away during the preparation of this paper.

J.C. Leary (✉) · R.F. Conde · G. Dakermanji · C.S. Engelbrecht · C.J. Ercol · K.B. Fielhauer · D.G. Grant · T.J. Hartka · T.A. Hill · S.E. Jaskulek · M.A. Mirantes · M.V. Paul · D.F. Persons · E.H. Rodberg · D.K. Srinivasan · R.M. Vaughan
Space Department, The Johns Hopkins University Applied Physics Laboratory, 11100 Johns Hopkins Road, Laurel, MD 20723-6099, USA
e-mail: james.leary@jhuapl.edu

S.R. Wiley
Aerojet, P.O. Box 13222, Sacramento, CA 95813, USA

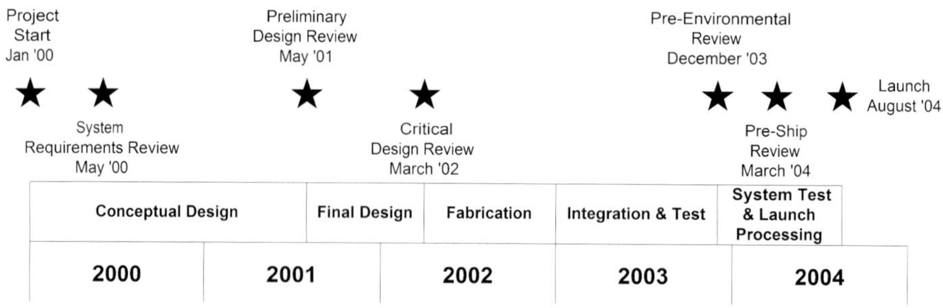

Fig. 1 MESSENGER spacecraft development timeline

Table 1 Spacecraft mass by subsystem

Subsystem	Mass (kg)
Payload	47.2
Avionics	11.6
Power	93.9
Communications	31.6
Guidance and control	34.1
Thermal	52.2
Propulsion	81.7
Structure	129.4
Harness	26.1
Dry mass total	*507.9*
Propellant	599.4
Total spacecraft mass	*1,107*

broad range of extreme environments. The spacecraft (S/C), designed and built by The Johns Hopkins University Applied Physics Laboratory (APL), combines innovative engineering techniques and technologies that yield an efficient design while satisfying all mission requirements.

Spacecraft development extended over 55 months, from project start (January 2000) to launch (August 2004). The design and fabrication phase formally started in July 2001 (at project confirmation) and culminated with the start of spacecraft integration and testing in January 2003. Spacecraft integration and testing (including sine vibration) was performed at APL, with acoustics, mass properties, and thermal-vacuum testing performed at the Goddard Space Flight Center (GSFC). System testing continued through shipment to the Kennedy Space Center launch site until integration with the Delta II 7925-H launch vehicle. The spacecraft was launched successfully on August 3, 2004. See Fig. 1 for a graphical representation of the spacecraft development timeline.

2 System Overview

Spacecraft requirements flow directly from the mission design (McAdams et al. 2007) and science requirements (Solomon et al. 2007). The Delta II 7925-H launch vehicle was the largest available to a Discovery-class mission. This vehicle provided 1,107 kg of lift mass to achieve the necessary heliocentric orbit. This fact, coupled with the complex trajectory

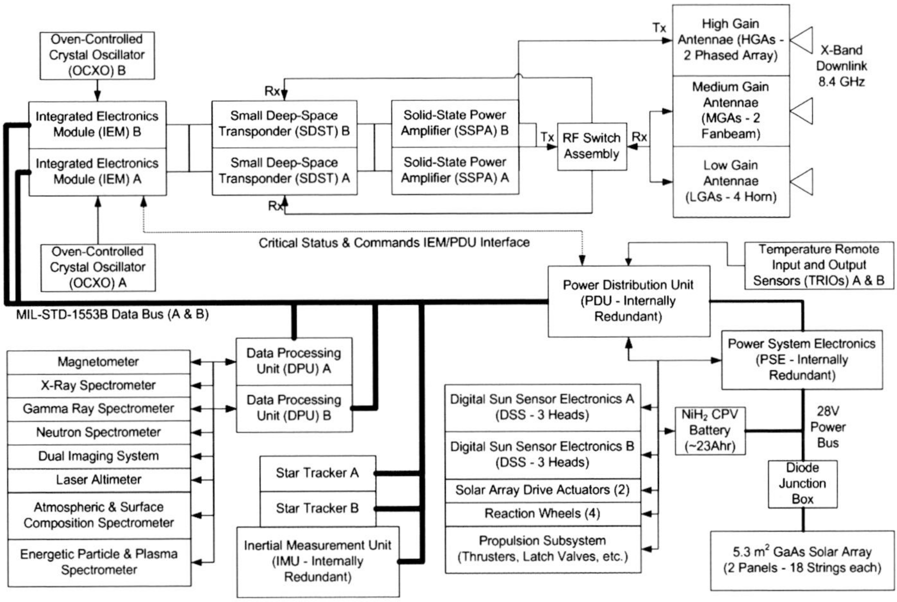

Fig. 2 System block diagram. Tx and Rx denote transmit and receive, respectively

requiring almost 600 kg (54%) of the spacecraft to be propellant, limited spacecraft dry mass—a very challenging constraint for a fully redundant spacecraft. Table 1 shows mass distribution by subsystem. Further mass details can be found in the following sections.

The MESSENGER design (Santo et al. 2001, 2002), while fully redundant, is simple. The structure is constructed primarily of lightweight composite material (Sect. 3) housing a complex dual-mode propulsion system (Sect. 4). Primarily passive thermal management techniques are used (Sect. 5) to minimize the amount of input required from the solar-array-based power subsystem (Sect. 6) while protecting the spacecraft from the harsh environment. A single processor performs all nominal spacecraft functions, while two other processors monitor spacecraft health and safety (Sects. 7 and 8). The spacecraft attitude control system (ACS) is three-axis stable and momentum biased (Sect. 9) to ensure Sun pointing while allowing instrument viewing by rotation around the Sun line. Spacecraft communications are performed through one of two antenna suites at X-band (7.2–8.4 GHz) via the Deep Space Network (DSN, Sect. 10). All of these subsystems support an extensive instrumentation package (Sect. 11) capable of achieving all science goals. Figure 2 shows the schematic block diagram for the spacecraft. Details can be found in the following sections.

Because the spacecraft is solar powered (except for a battery needed for eclipses) power generation increases with sunward motion, toward Mercury. Early in the mission ("outer cruise," see Table 2) spacecraft operation maintained the sunshade pointed away from the Sun, allowing a substantial reduction in heater power. Peak power demand occurs in orbit during science operations. Table 2 details the minimum power design point that occurs during the orbital phase when the spacecraft is experiencing eclipses. As the spacecraft orbits Mercury, there are seasons of eclipses of varying length. Operations are constrained (much of the instrumentation is off) during periods of eclipse greater than 35 minutes. Power subsystem details can be found in Sect. 6.

The engineering team used existing technologies wherever possible. The design, while building from many heritage components, included several key developmental technologies

Table 2 Spacecraft power budget by subsystem (worst-case values), in watts

	Outer cruise[a] (>0.85 AU)	Inner cruise (<0.85 AU)	Orbit[b]	Eclipse >35 min	Eclipse <35 min
Science instruments and heaters	*13.2*	*15.4*	*95.1*	*42.6*	*58.9*
IEM	*32.7*	*34.2*	*33.5*	*30.2*	*30.2*
Power system	*25.6*	*25.6*	*105.6*	*20.9*	*20.9*
Telecommunications	*65.5*	*65.5*	*110.5*	*22.8*	*25.4*
Guidance and control	*70.3*	*70.3*	*119.7*	*119.7*	*119.7*
Thermal	*10.2*	*62.2*	*58.4*	*10.0*	*23.8*
Propulsion	*21.9*	*95.6*	*74.6*	*34.0*	*47.8*
Harness	*3.7*	*7.6*	*12.0*	*4.4*	*5.5*
Total	**247.8**	**382.7**	**597.4**	**284.6**	**332.2**
S/C power input from solar array	490.0	528.2	720.0[c]	—	—
Maximum depth of battery discharge	—	—	—	55%	36%
Power margin	49%	28%	17%	—	—

[a]At 1.0765 AU

[b]Sun period during eclipsing orbits

[c]Limited by power system electronics output capability

that allowed the spacecraft to meet the mission goals. State-of-the-art lightweight fuel tanks (Sect. 4), ceramic-cloth sunshade (Sect. 5), and phased-array antennas (Sect. 11; Srinivasan et al. 2007) were necessary for the spacecraft to meet requirements.

Spacecraft integration and testing was accomplished during 2003. Preparations began in January 2003 for delivery of the integrated structure and propulsion system in February. Electrical components were integrated and verified throughout spring and summer. System testing commenced in the fall, and environmental testing (first vibration testing at APL, then thermal-vacuum testing at GSFC) continued through winter. See Fig. 3 for front (sunshade) and rear (panels stowed) views of the spacecraft during this timeframe.

The spacecraft was tested thoroughly following delivery to the Astrotech processing facility in Titusville, FL, for launch vehicle integration. Testing culminated with the final system functional test performed once the spacecraft was mated to the launch vehicle third stage and erected on the pad. The development period accumulated more than 3,000 hours of integrated spacecraft test time.

3 Structures and Mechanisms

Major mechanical assemblies are shown by finite element model (FEM) representations in Fig. 4. The spacecraft core is the integrated structure/propulsion system (Persons et al. 2000) supplied by Composite Optics, Inc. (COI), and GenCorp Aerojet (Sect. 4). MESSENGER optimizes structural load paths using state-of-the-art, lightweight titanium fuel tanks designed specifically for the structural configuration chosen. Figure 5 shows the spacecraft front end, graphite-cyanate ester (GrCE) panels, and part of the tank-support system. The

Fig. 3 MESSENGER during testing

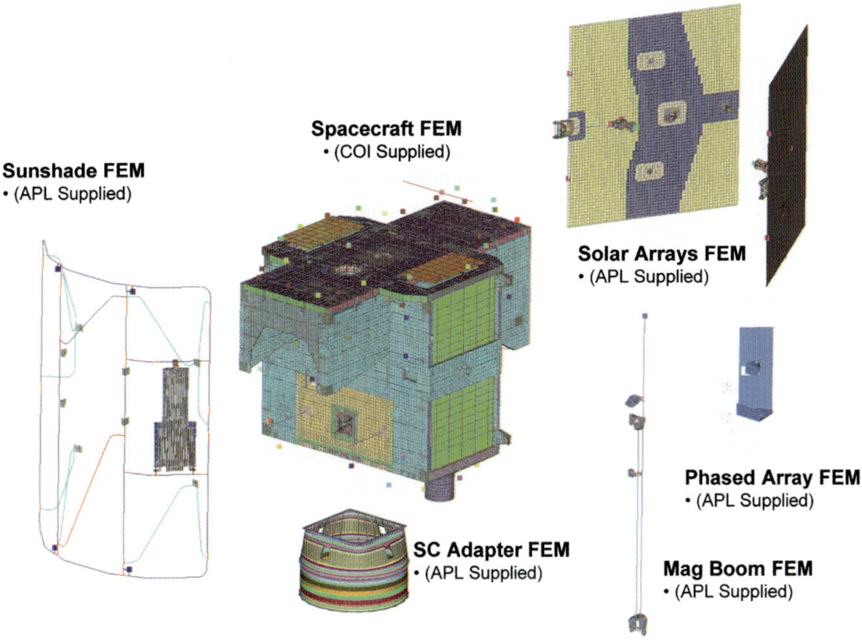

Fig. 4 Major mechanical finite element models

tank struts transfer lateral loads to the structure corners, allowing the composite panels to be thin relative to their size. The thin copper conductive ground plane added to the composite panels improves conductivity for grounding.

With the composite structure designed to channel all loads into the center column, a square-to-round adapter was necessary to match with the Delta II 3712C separation clamp band interface (see Fig. 4). The solution chosen was to machine an aluminum forging, carefully tailored to distribute evenly structural loads from the corners of the center column to the round vehicle interface. The forward adapter flange was slotted between each bolt to ac-

Fig. 5 Integrated structure and propulsion system, main tank bottom views

commodate thermal expansion mismatch between the aluminum adapter and the near-zero coefficient of thermal expansion composite structure.

Solar panel design and development proved to be challenging from a material engineering prospective (Wienhold and Persons 2003). Extensive, ply-by-ply structural analysis was required for the solar array substrate due to the large cantilever in the stowed configuration and the use of high-conductivity, but relatively low-strength, GrCE materials in sandwich face sheets. Overall solar array structural analysis was performed using FEMAP and NASTRAN commercial finite element analysis codes, while structural analysis of the sandwich laminate was completed using a Composite-Pro commercial laminate analysis tool. Note how stress patterns shown in Fig. 6 match the dark blue doubler areas of the solar array FEM shown in Fig. 4.

The sunshade support structure was chosen to be welded titanium tubing construction. Tubing supports five antennas, the X-Ray Spectrometer (XRS) solar monitoring sensor, four digital Sun sensors, and the sunshade. The sunshade final projected area was required to be tailored late in the program to bring the center of solar pressure as close to the measured center of spacecraft mass as possible. By using solar pressure, tilting the spacecraft relative to the Sun can unload the momentum wheels without expending fuel. Figure 7 shows the sunshade backside view during spacecraft mass properties testing at GSFC. Note the sunshade titanium tubing, ceramic-cloth sunshade material, antennas, and digital Sun sensors.

Three mechanical assemblies required deployment: two solar panels and the 3.6-m Magnetometer (MAG) boom. The hot solar arrays are thermally isolated from the spacecraft body by short arms made of titanium tubing. Solar array deployment was designed to be as mechanically simple as possible. The panels were released first and allowed to over-travel and the "saloon door" hinges settle; the arms were then released and settled into position. The MAG boom deployment was performed by the same sequence, with separately commanded release and settle events for each hinge line. After confirmation of full deployment, all six hinge lines are pinned in place to prevent hinge rotation during high-thrust maneuvers.

4 Propulsion

The MESSENGER propulsion system (MPS)—shown in layout form in Fig. 8 and schematically in Fig. 9—is a pressurized bipropellant, dual-mode system using hydrazine (N_2H_4) and nitrogen tetroxide (N_2O_4) in the bipropellant mode and N_2H_4 in the monopropellant

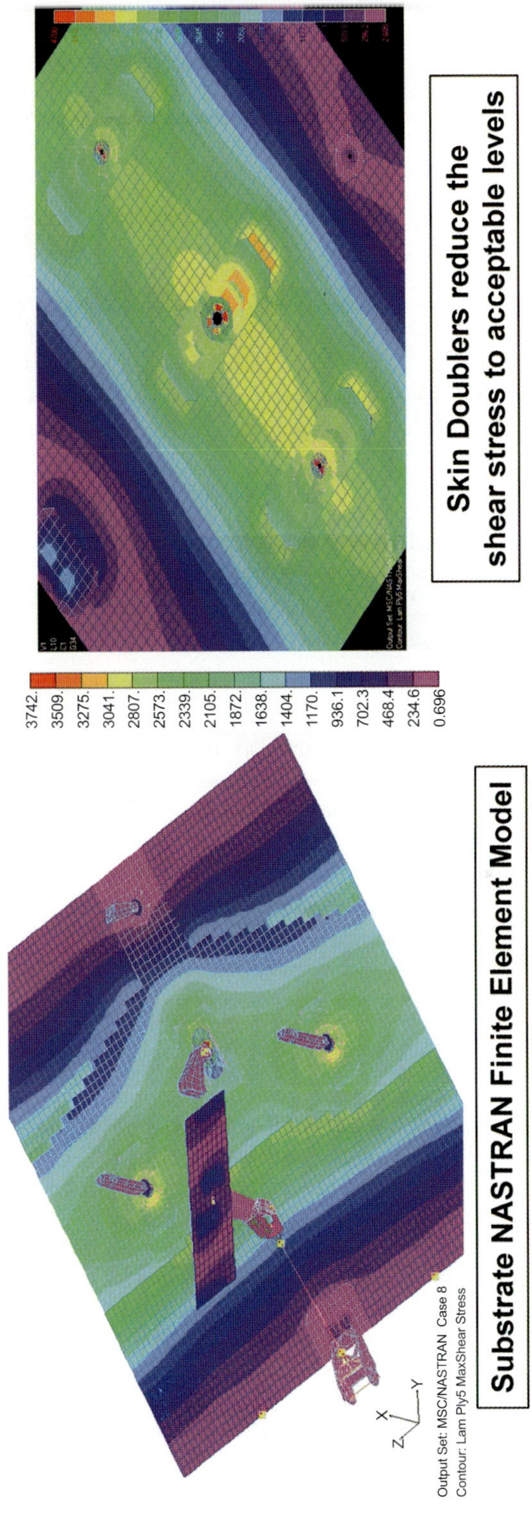

Fig. 6 Solar array finite element models of maximum shear stress

Fig. 7 MESSENGER in post-launch configuration during mass properties testing

Table 3 Propulsion subsystem mass

Subsystem	Mass (kg)
Propellant and helium tanks	39.17
Thrusters	10.07
Valves and plumbing	15.55
Electrical subsystem	6.30
Thermal subsystem	3.21
Secondary structures and misc. hardware	7.44
Total propulsion system mass	81.74

mode. Three main propellant tanks, a refillable auxiliary fuel tank, and a helium pressurant tank provide propellant and pressurant storage. These tanks provide propellant storage for approximately 368 kg of fuel and 231 kg of oxidizer.

The MPS hydraulic schematic consists of four main elements: the pressurization system, fuel feed system, oxidizer feed system, and thruster modules (Fig. 9, Wiley et al. 2003). Additional MPS constituents include secondary structures, the electrical subsystem, and the thermal management subsystem. Propulsion subsystem masses are summarized in Table 3.

The helium tank is a titanium-lined composite over-wrapped leak-before-burst pressure vessel (COPV) based on the flight-proven A2100 helium tank. A second outlet was added to the existing helium tank to provide a dual pressurization capability for the fuel and oxidizer systems.

A new lightweight main propellant tank was developed and qualified for MESSENGER (Tam et al. 2002). The tank configuration is an all-titanium, hazardous-leak-before-burst design with a measured mass less than 9.1 kg, including all attachment features. The tank

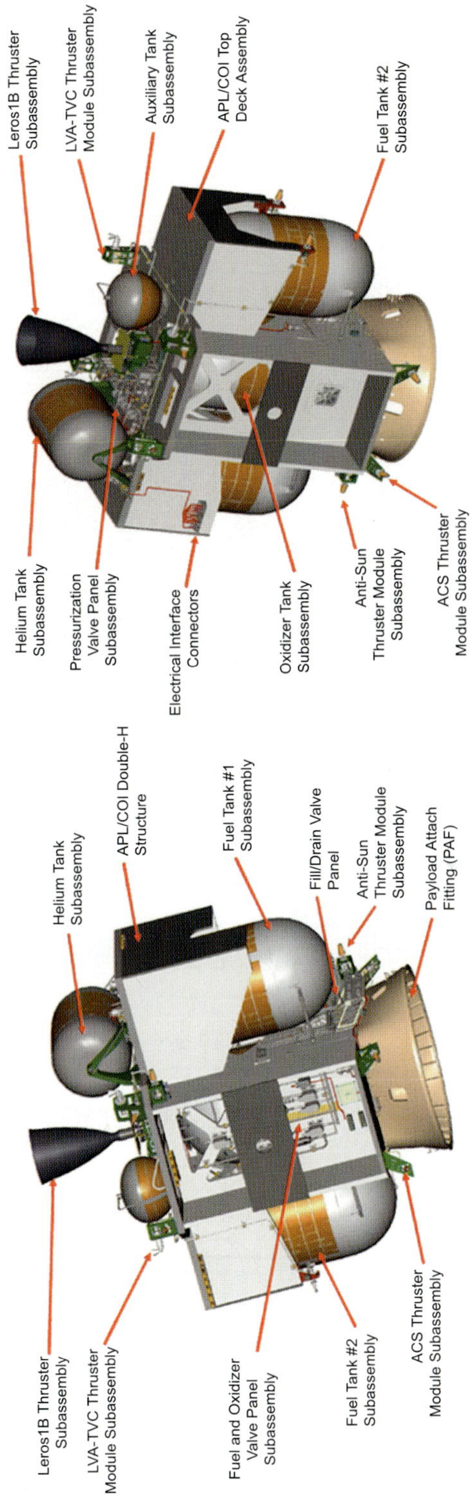

Fig. 8 Propulsion system layout

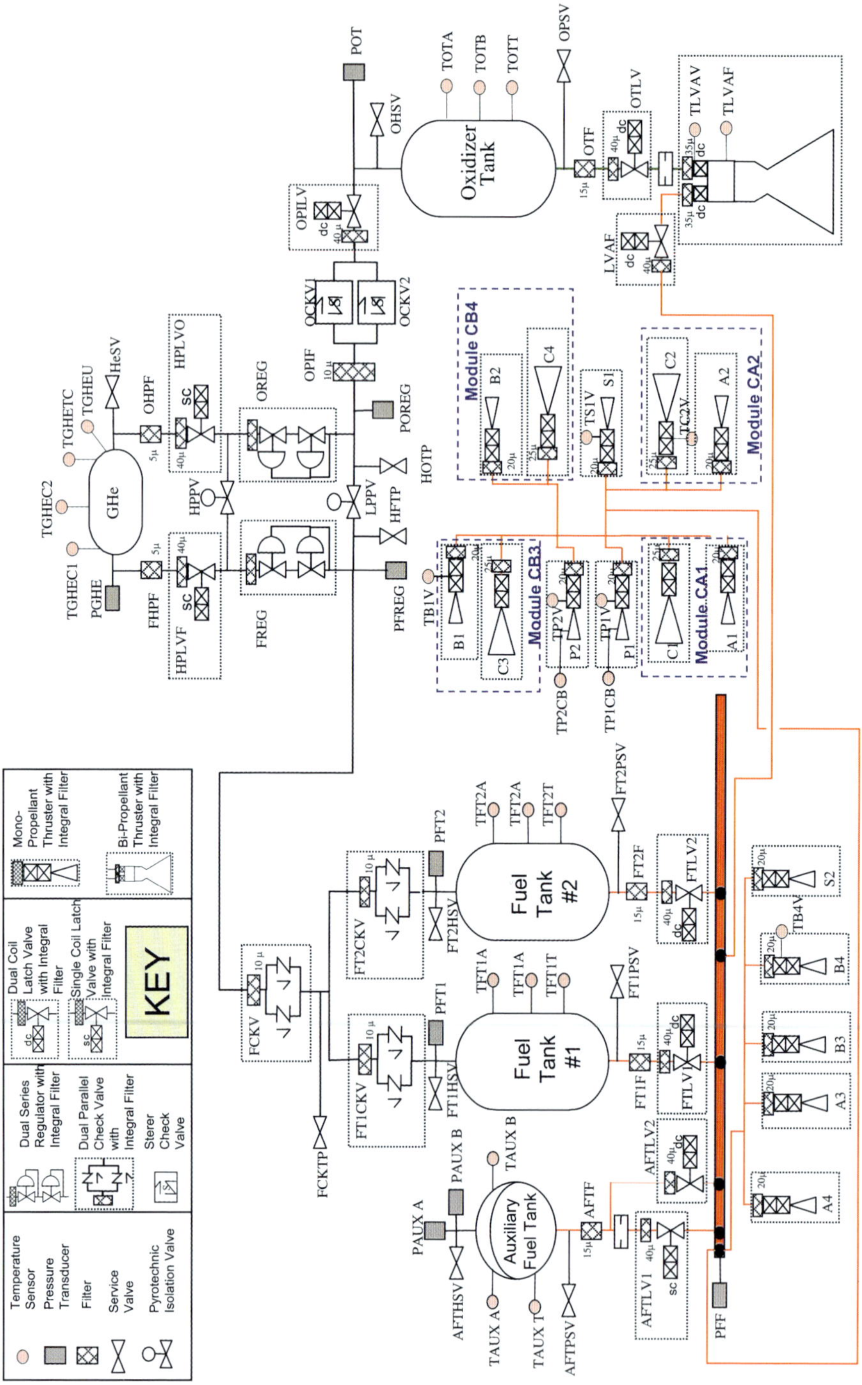

Fig. 9 Propulsion system hydraulic schematic

CONTROL AXIS	CONTROL AXIS	A THRUSTERS	TORQUE N-m	B THRUSTERS	TORQUE N-m	C THRUSTERS	TORQUE N-m
+ AROUND Y	+ PITCH	A3A4	4.084	B1B2	3.941	C1C2	26.624
- AROUND Y	- PITCH	A1A2	-3.941	B3B4	-4.084	C3C4	-26.624
+ AROUND Z	+ YAW	A1A3	4.008	B2B4	4.008	-	-
- AROUND Z	- YAW	A2A4	-4.008	B1B3	-4.008	-	-
+ AROUND X	+ ROLL	A2A3	2.187	B2B3	2.187	C2C4	21.963
- AROUND X	- ROLL	A1A4	-2.187	B1B4	-2.187	C1C3	-21.963

FORCE DIRECTION	THRUSTERS USED	FORCE N	EFFICIENCY %
+ X	A1A2A3A4	16.601	93.30
- X	B1B2B3B4	16.601	93.30
+ Y	A2A4B2B4	4.448	25.00
+Y	P1P2	8.896	100.00
- Y	A1A3B1B3	4.448	25.00
- Y	S1S2	8.896	100.00
+ Z	A1A2B1B2	4.605	25.88
+ Z	C1C2C3C4	97.861	100.00
- Z	A3A4B3B4	4.605	25.88

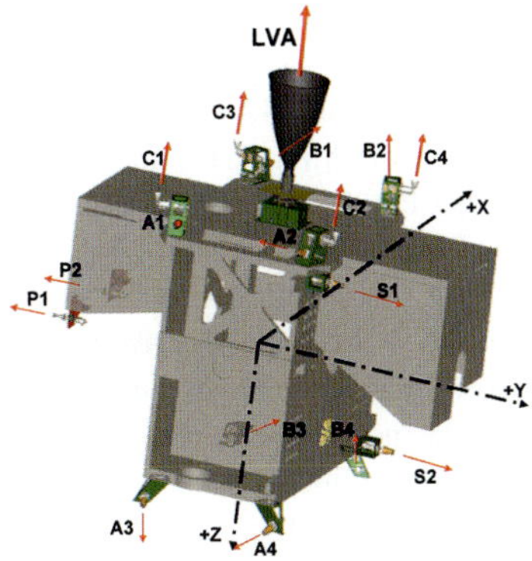

Fig. 10 MPS thruster arrangement

shell is fabricated from solution-treated and aged (STA) 6Al-4V titanium. The tank is approximately 559 mm in diameter by 1,041 mm in length. The hemispherical domes are approximately 0.5 mm thick. The cylindrical section is approximately 1.0 mm thick. Each tank design includes two 178-mm wide, 0.25-mm thick, 6Al-4V titanium baffles used for spacecraft nutation control. A 6Al-4V-titanium vortex suppressor is provided at each tank outlet to delay vortex formation. The main propellant tanks are symmetrically positioned about the spacecraft centerline to maintain center of mass control during propellant expulsion in flight. Two fuel tanks flank the center oxidizer tank. Each tank main load path is through a Custom 455 steel bearing pin that interfaces to a titanium receiver fitting in the composite center box structure. The four titanium struts, two boss mounted and two side mounted, provide for tank lateral support.

A small 6Al-4V titanium auxiliary tank is a hazardous-leak-before-burst design. It has an internal diaphragm to allow positive expulsion of propellant for use in small burns (e.g., attitude control, fine velocity adjustment, and settling burns). The tank operates in blowdown mode between 280 and 110 pounds-per-square-inch absolute pressure (psia) (between 1930 and 758 kPa) and is recharged in flight.

The MPS includes a total of 17 thrusters. Three thruster types, arranged in five different thruster module configurations, provide the required spacecraft forces. The MPS thruster arrangement is shown in Fig. 10 and detailed in the following paragraph.

The large velocity adjustment (LVA) thruster is a flight-proven Leros-1b provided by Atlanta Research Corporation—United Kingdom Division (ARC UK). The LVA operates at a nominal mixture ratio (MR) of 0.85, provides a minimum of 667 N of thrust, and operates at a specific impulse of 316 s. For the MESSENGER program, the Leros 1b was qualified to operate with fuel and oxidizer inlet pressures up to 290 psia, which would be required to accommodate the thermal conditions of the spacecraft as well as some possible feed system failures (e.g., the failure of one segment of a pressure regulator) and/or subtle changes in component flow characteristics over time. The results of this testing represented an increase in the engine's "operating box" above the previous limit of approximately 280 psia, established in qualification tests performed for previous uses of the engine. Tests of the MESSENGER engine showed that the engine operated stably and with a relatively low chamber temperature over the entire tested range of inlet pressures, demonstrating the engine's robustness to the MESSENGER flight conditions. Four 22-N, monopropellant LVA-thrust vector control (TVC) thrusters (also identified as C-thrusters) provide thrust vector steering forces during main thrust burns and primary propulsion for some of the smaller velocity adjustment (ΔV) maneuvers. The C-thrusters are flight-proven Aerojet P/N MR-106Es. They are fed with N_2H_4 in both the pressurized and blow-down modes. Twelve monopropellant thrusters provide 4.4 N of thrust at 220-s average specific impulse for fine attitude control burns, small ΔV burns, and momentum management. The 4.4-N thrusters are flight-proven Aerojet P/N MR-111Cs. These thrusters are also fed with N_2H_4 in both the pressurized and blow-down modes. Eight 4.4-N thrusters (A and B) are arranged in double canted sets of four for redundant three-axis attitude control. Two 4.4-N thrusters (S) are used to provide velocity changes in the sunward direction. The final two 4.4-N thrusters (P) are used to provide velocity changes in the anti-Sun direction. The P thrusters are located on the spacecraft −Y side and protrude through the sunshade.

The MPS includes secondary structures and brackets to support thrusters, fill and drain valves, and electrical interface connectors. The majority of these secondary structures were fabricated using magnesium to reduce mass. Propulsion system integrating components include all the required high- and low-pressure latch valves, regulators, pyro isolation valves, check valves, filters, fill and drain valves, pressure transducers, bimetallic joints, fittings, and tubing necessary to control propellant to the system thrusters. The MPS includes a dual-string electrical system with harnesses, diode terminal boards, and connectors that are used to control individual propulsion elements. The MPS interfaces to the spacecraft through 12 electrical connectors.

The MPS thermal system employs heaters to maintain acceptable system temperatures. Heaters are used during the cruise phase to maintain propellant temperatures and in the operational phases to preheat thrusters in preparation for operation. These heaters can total 143 W. Cruise-phase heaters are installed on the propellant and pressurant tanks, thruster valves, valve panel, fill and drain valve bracket, and various propellant manifolds. Cruise-phase heaters are controlled by spacecraft software (helium and main propellant tanks) and with mechanical thermostats (all remaining cruise-phase heaters). Operational-phase (when the MPS is in active use) heaters include monopropellant thruster catalyst bed heaters and the LVA flange heater. When enabled, all catalyst bed heaters are time controlled, while the LVA flange heater is controlled with mechanical thermostats. Operational-phase heaters can total 69 W.

5 Thermal Control

The thermal design of the MESSENGER spacecraft relies upon a ceramic-cloth sunshade to protect the vehicle from the intense solar environment encountered when inside of Venus orbit. Creating a benign thermal environment behind the sunshade allowed for the use of essentially standard electronics, components, and thermal blanketing materials (e.g., multi-layer insulation, MLI). Most components were designed to −34 °C to +65 °C limits. Non-standard thermal designs were required for the solar arrays, sunshade, digital Sun sensors, three of the seven instruments, and phased-array antennas (Fig. 11). These components have been designed to operate at Mercury perihelion (Mercury closest to the Sun) and also during orbits that cross over one of the Mercury hot poles. For example, the solar panels can experience temperatures as low as −135 °C during the one-hour eclipse near Venus during cruise and as high as +270 °C during orbit if left normal to the Sun.

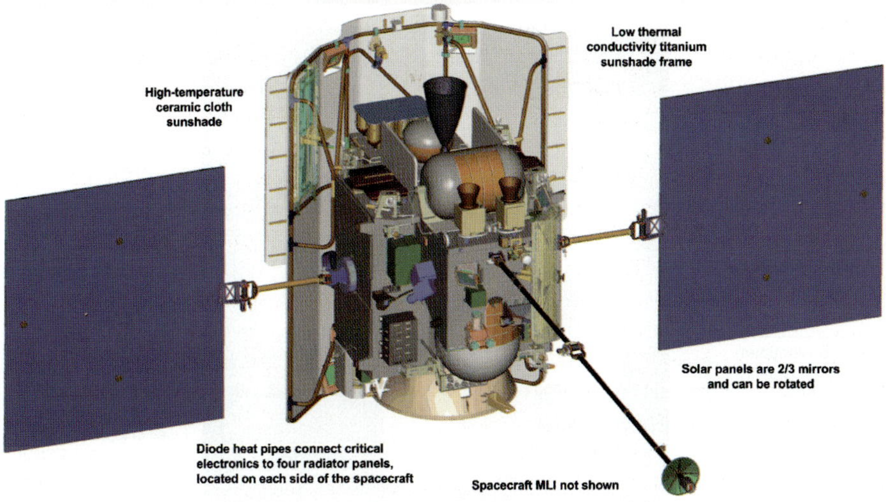

Fig. 11 MESSENGER orbital configuration with thermal control design highlights

When at perihelion, the sunshade, solar arrays, sunshade-mounted digital Sun sensors, and sunshade-mounted antenna suite will experience as much as 11 times the solar flux near Earth and the sunshade temperature will rise to over 300 °C. During certain orbits around Mercury, the spacecraft will be between the Sun and illuminated planet for approximately 30 minutes. During this period the sunshade will protect the spacecraft from direct solar illumination, but the spacecraft rear will be exposed to the hot Mercury surface. Components such as the battery and star trackers are positioned such that the spacecraft body blocks a substantial portion of the planet view, minimizing direct radiation from the planet surface. Planet-viewing instruments such as the Mercury Dual Imaging System (MDIS) required a specialized thermal design to allow full operation during this hot transient period. Diode heat pipes were employed in both the spacecraft and imager thermal designs to protect attached components when radiator surfaces are exposed to thermal radiation emitted by Mercury. Diode heat pipes effectively stop conducting when the radiator surface begins to get hot and return to conduction when the radiator surface cools, restoring normal thermal control. Analysis of the orbiting environment as a function of orbit geometry and planet position was integrated into the mission design and has helped to phase the orbit plane relative to solar distance, minimizing planet infrared heating of the spacecraft and thus minimizing required mass (Ercol and Santo 1999).

The MESSENGER spacecraft and associated Sun-illuminated component testing presented a challenging set of problems from both technical and cost/schedule aspects. It was decided early in development that the system-level thermal vacuum test would be done at GSFC in a nonsolar simulator environment. Specialized solar simulation testing of engineering model solar arrays, Digital Sun Sensor Heads (DSSHs), radio frequency (RF) antennas, solar monitor, and sunshade were performed at the Glenn Research Center (GRC) Tank 6 thermal-vacuum chamber (Ercol et al. 2000). Originally designed to simulate near-Earth solar conditions for the Solar Dynamic Power Experiment, the original test setup was modified to produce an 11-Sun equivalent solar environment over approximately 1.5 m^2. This illuminated area proved large enough to test the various solar array designs, DSSH configurations, sunshade design, and antenna components simultaneously (Figs. 12 and 13) while keeping the ratio of testing cost to components tested small.

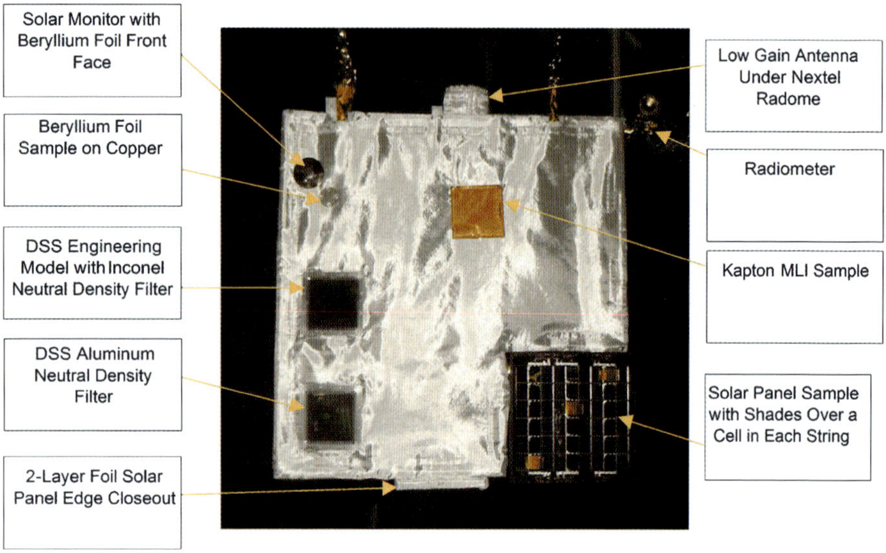

Fig. 12 Typical GRC solar simulation test setup

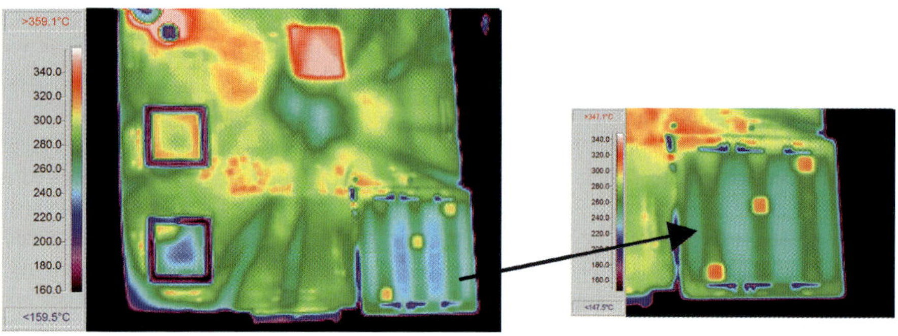

Fig. 13 Infrared mapping of steady-state temperatures for the hardware shown in Fig. 12 during 11-Sun solar simulation

Because all solar simulation testing was done in vacuum with the chamber liner near room temperature, solar simulation tests represented worst-case thermal conditions for various components. Individual thermal models were correlated to the test conditions for each component and then integrated into the system flight thermal model. Since all Sun-illuminated components are thermally isolated, it was not necessary to perform an expensive system-level solar simulation test. Instead, three banks of heating elements were designed to contour the flight sunshade and produce an equivalent temperature environment on the Sun side of the sunshade that represented the 11-Sun thermal condition, with 50 °C margin, measured during GRC solar simulation testing. During the spacecraft thermal-vacuum test, the flight sunshade was heated to 350 °C and held at this temperature for nearly 15 days. Also during the spacecraft thermal-vacuum testing, on/off operation of the diode heat pipes used to maintain spacecraft electronics temperature control was verified (typical operation for the

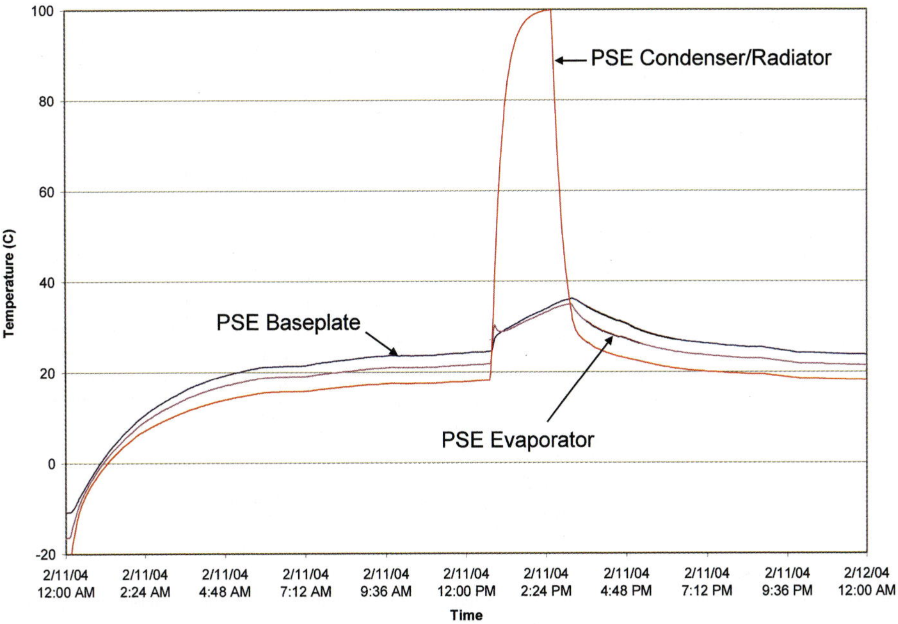

Fig. 14 Typical on/off operation of the PSE diode heat pipe system

Power System Electronics, PSE, is shown as an example in Fig. 14). All MESSENGER instruments were fully qualified during subsystem-level thermal-cycle and design verification testing. All thermal environments used by instruments were provided by a detailed model representing appropriate instrument configurations. Thermal interactions between each instrument and the spacecraft were quantified, and worst-case transient environments with 20% margin were created during each instrument thermal vacuum test.

Much emphasis has been placed on the extreme heat and high temperatures associated with a spacecraft orbiting Mercury. Near the mission beginning, with the sunshade pointing sunward, the spacecraft will use heater power to make up for the maximum solar and planet heating. In order to reduce heater power consumption and increase solar array power margin, the MESSENGER design allows the spacecraft to be flipped such that the anti-Sun side can be illuminated during outer cruise (configured to fly with the sunshade pointed away from the Sun). MESSENGER flew in this reversed orientation between launch and March 2005, and based upon measurements made during cold system-level thermal vacuum balance testing that simulated outer cruise with the shade pointing toward the Sun, heater power savings averaged 110 W. Since launch, the MESSENGER thermal control system has worked as expected.

6 Power

A peak power tracker topology with strong heritage from the APL-designed Thermosphere, Ionosphere, Mesosphere, Energetics and Dynamics (TIMED) spacecraft power system was selected for the MESSENGER power subsystem (Dakermanji et al. 2005). This architecture isolates the battery and power bus from variations of solar array voltage and current characteristics and optimizes solar array power output over the highly varied operating conditions

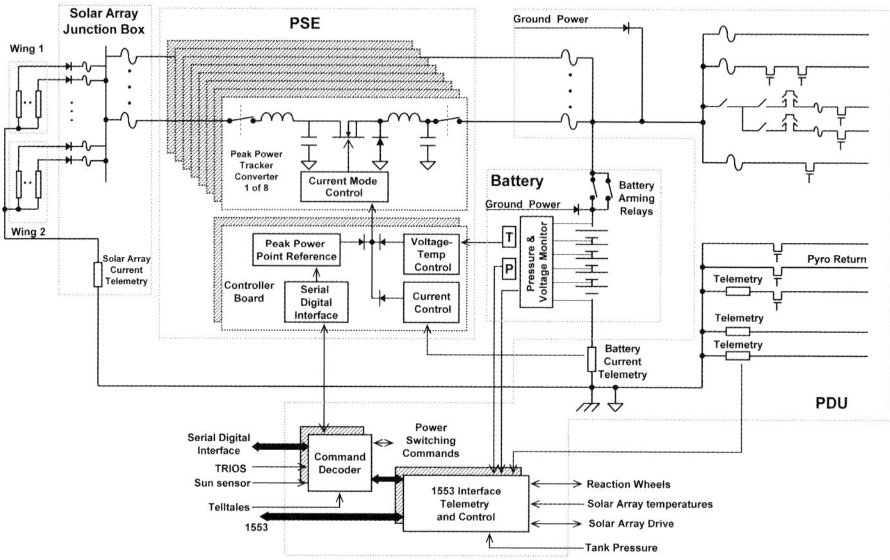

Fig. 15 Power subsystem block diagram. TRIOS denotes Temperature Remote Input Output Sensors, and T and P denote temperature and pressure, respectively

of the mission. The power system is designed to support about 390 W of load power near Earth and 640 W during Mercury orbit. The power system (Fig. 15) consists of the PSE, Power Distribution Unit (PDU; Le et al. 2002), Solar Array Junction Box (SAJB), battery, two solar array panels, and Solar Array Drive Assembly (SADA).

The large solar distance variations impose severe requirements on the solar array design. The maximum solar array voltage during normal operations is expected to vary between 45 and 95 V, but this range does not include the higher transient voltages expected on the cold solar arrays at exits from eclipses. Triple junction solar cells are used on the solar array. Solar cell strings are placed between Optical Solar Reflector (OSR) mirrors with a cell-to-OSR ratio of 1:2 to reduce panel absorbance (see Fig. 3). Thermal control is performed by tilting the panels away from normal incidence with increased solar intensity. In case of an attitude control anomaly near Mercury, the solar array temperature may reach 270 °C. All material and processes used in the solar panels are designed to survive worst-case predicted temperatures. Array strings are isolated with decoupling diodes placed inside the SAJB to protect them from the expected high temperatures. The two solar panels are maintained normal to the Sun until the panel temperature reaches a preset value (maximum 150 °C), at which point the panels are rotated by the SADA to limit temperature. In operation, panel rotation can be performed either by periodic ground commands or by onboard software algorithms. During eclipsing orbits, panel rotation will be performed by ground commands. Solar-flare protons can cause radiation damage on the solar cells. The estimated total dosage on the solar cells with 0.15-mm microsheet cover glass is 4×10^{14} equivalent 1-MeV/cm^2 electrons.

The loads are connected directly to the 22-cell, 23-Ahr NiH_2 battery used to support spacecraft loads during launch and eclipse periods. Each battery pressure vessel contains two battery cells and can be bypassed by a contactor that is automatically activated to short the vessel in case of an open circuit failure of that common pressure vessel (CPV) cell.

Nominal bus voltage is 28 V and can vary between 20 and 35 V depending on battery state. To minimize overall spacecraft weight, the battery will not support the full spacecraft load during predicted eclipses, and therefore load management is planned.

Primary battery charge (C) control is ampere-hour integration Charge-to-Discharge (C/D) ratio control performed by flight software within the Integrated Electronics Module (IEM). The battery is charged at high rate with available solar array power until battery State-of-Charge (SoC) reaches 90%. Commands from the IEM then lower battery charge current to C/10. When the selected C/D ratio is reached, the IEM commands the charge current to a ground selectable C/100 or C/150 trickle charge rate.

Battery voltage is controlled to preset safe levels with temperature-compensated voltage (V/T) limits that are implemented in hardware. Whenever battery voltage reaches the V/T limit, battery charge current will taper. This battery charge control technique reduces battery overcharge and associated heat dissipation and extends battery life.

The PSE is based on TIMED spacecraft power system electronics. It contains eight buck-type peak power tracking converter modules that are controlled by redundant controller boards. Only one controller is active at a time. Four analog control loops for peak power tracking, battery V/T limit, battery current limit, and maximum battery voltage run simultaneously. The dominant signal drives the peak power tracking converters. When the load and battery recharge power demand exceed solar array power capabilities, a software algorithm in the IEM recursively selects the solar array operating voltage that will generate maximum power.

The PDU contains circuitry for pyrotechnic firing control, power distribution switching, load current and voltage monitoring, fuses, external relay switching, reaction wheel relay selection, power system relays, Inertial Measurement Unit (IMU) reconfiguration, IEM selection relays, solar array drive control, propulsion thruster firing control, and propulsion latch valve control. There are two sides to the PDU; each side can command all of the internal circuitry.

The commands to control the PDU under normal circumstances are sent by the IEM over the MIL-STD-1553 bus to one of the PDU 1553 cards. The commands are then sent internally through the PDU Motherboard to either one or both PDU Command Decoder (CD) cards. There are also serial critical command interfaces between each IEM and each PDU CD card that allow reconfiguration of critical spacecraft components even if the Main Processor (MP) is inoperable. These commands can be initiated by either FPP or by an uplinked command.

Solid-state, radiation-hardened, power Metal Oxide Semiconductor Field Effect Transistors (MOSFETs) are used for power switching and distribution. Load switches consist of two P-channel MOSFETs in series that allow for high-side switching that ensures load turn-off, even with a failure in one MOSFET. Each load has redundant current and voltage monitoring, and the spacecraft is protected from load faults by redundant FM-12-type solid-body fuses from the main bus. Each PDU side can control the power MOSFETs through independent, isolated circuitry and share only the power MOSFET gates.

Telemetry that is collected by the PDU consists of external relay status information, load currents, load voltages, battery and solar array currents, solar array and battery temperatures, solar array voltages, reaction wheel speeds, propulsion tank and battery pressures, battery cell and total battery voltages, solar array position, temperatures, and digital Sun sensor information. All telemetry interfaces to both PDU sides. Analog telemetry is digitized in 12-bit A/D converters, and all telemetry can be sent to either IEM from either PDU side through MIL-STD-1553 busses.

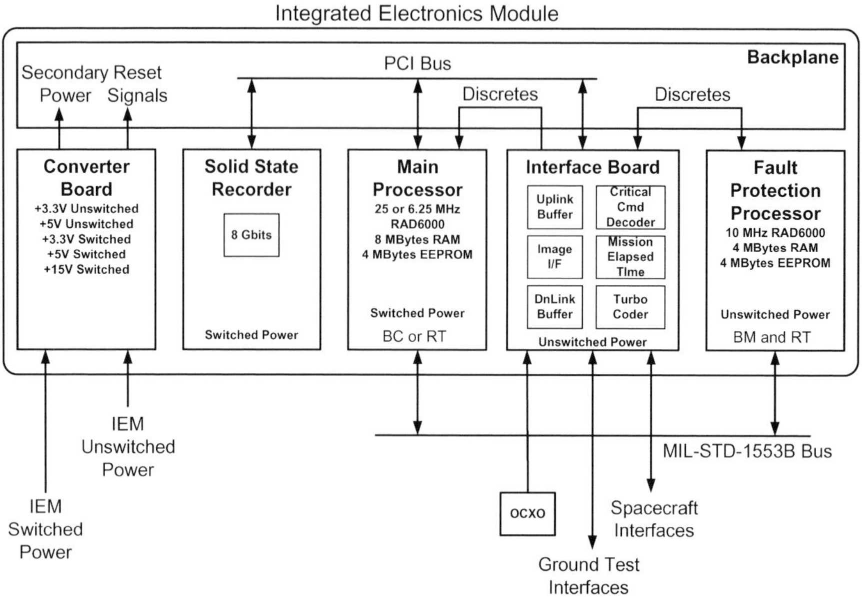

Fig. 16 IEM block diagram

7 Avionics

The IEM implements command and data handling (C&DH), guidance and control (G&C), and fault protection functions. These functions are implemented in five 6U compact peripheral component interconnect (cPCI)-compatible cards. The use of the cPCI standard (as opposed to a proprietary, custom standard) reduced development schedule and cost (Conde et al. 2002). It was recognized that the 6U cPCI board format was not optimal from a mechanical and thermal standpoints, but all thermal and mechanical requirements could still be satisfied. Conductive heat sinks and board stiffeners were used to satisfy mechanical and thermal requirements and retain 6U cPCI compatibility.

The IEM design is partitioned into five daughter cards, a backplane, and a chassis (Ling et al. 2002). A block diagram of the IEM is shown in Fig. 16. Three of the five daughter cards communicate over a 32-bit PCI bus operating at 25 MHz. Three cards, the MP, Fault Protection Processor (FPP), and Solid-State Recorder (SSR), were designed and manufactured by BAE Systems to APL specifications. These boards were specified to be as generic as possible, incorporating few features that would be mission-unique so that their specifications could be generated as early as possible and put out for bid. The MP and FPP boards are nearly identical. The Interface (I/F) Card, Converter Card, and backplane were designed and built by APL. These boards capture the MESSENGER-unique requirements. The chassis was designed by APL and fabricated by Nu-Cast. The chassis is cast aluminum with a 1.5-mm wall thickness. The chassis includes thermal plugs on the bottom that connect to a heat pipe running under the spacecraft deck. An Oven Controlled Crystal Oscillator (OCXO) supplied by Symmetricom is located outside the IEM chassis and provides precise timing.

The two IEMs use box-level redundancy. Only one IEM is in control of the spacecraft. The MP in the controlling IEM is configured via a discrete signal from the PDU to be the MIL-STD-1553 Bus Controller. The MP in the other IEM is configured to be a Remote

Terminal and is normally left powered off as a cold spare. The FPPs in both IEMs are powered continuously.

The MP and FPP boards use RAD6000 processors operating at 25 MHz and 10 MHz, respectively. The MP is populated with 8 Mbytes of Random Access Memory (RAM) and 4 Mbytes of Electrically Erasable Programmable Read-Only Memory (EEPROM); the FPP is populated with 4 Mbytes each of RAM and EEPROM. Single-bit errors in RAM and EEPROM are corrected. The MP includes a MIL-STD-1553 interface that is configured either as a Bus Controller (BC), if primary, or Remote Terminal (RT), if backup. The FPP includes a MIL-STD-1553 interface that is configured as a simultaneous RT and Bus Monitor (BM).

The SSR implements 8 Gbits of memory with upscreened commercial Synchronous Dynamic RAM (SDRAM). The memory is organized into 80-bit words (consisting of 64 bits of user data and 16 bits of error code correction data) organized as 20 nibbles. Any single nibble in error is corrected. Memory contents are retained through an IEM reset but are lost if IEM switched power is turned off.

The Interface Card implements the mission-unique functions required. A Critical Command Decoder implements a small set of uplink commands directly in hardware. A mission elapsed time is generated and time tags data with a resolution of 1 μs and an accuracy of 1 ms. Downlink framing hardware has a highly adjustable bit rate that can be tuned to the capabilities of the RF link. The downlink hardware implements an efficient turbo code to encode downlink data (Srinivasan et al. 2007). An uplink buffer accepts serial command data from both transponders and implements sync detection and a serial-to-parallel conversion function in order to provide the uplink data to the MP. An image interface circuit receives 4 Mbits/s image data that are transferred to the SSR.

A primary driver of the IEM architecture was to simplify spacecraft fault protection. Fault protection is centralized in the two FPPs. Each FPP independently collects spacecraft health information over the MIL-STD-1553 bus and over dedicated links to the Power Distribution Unit. The health data are continuously evaluated by a rule-based autonomy system. The FPP corrects faults by sending commands to the MP and by sending a small subset of reconfiguration commands directly to the PDU. The IEM Interface board includes hardware limits to prevent a failed FPP from continuously sending commands that would disrupt spacecraft operation.

8 Flight Software

The MESSENGER spacecraft flight software operates within each MP and FPP. These RAD6000 processors execute all spacecraft software applications. The flight software is implemented as C code (Heiligman et al. 2003) that operates under the VxWorks 5.3.1 real-time operating system.

8.1 Main Processor Software

MP software implements all C&DH and G&C (see Sect. 9) functionality in a single flight code application. Only one MP is designated "active" or "primary" and executes the full MP flight application. The "redundant" or "backup" MP will typically remain unpowered due to MESSENGER mission power constraints and does not serve as a "hot spare." The backup MP, if powered, remains in boot mode and supports rudimentary command processing and telemetry generation for the purpose of reporting the health status of that processor and to support uploads of code and parameters to EEPROM. It operates as an RT on the 1553 data bus. The primary MP serves as 1553 bus controller and manages all communication with devices on that bus.

C&DH functionality in the primary MP includes uplink and downlink management using the Consultative Committee for Space Data Systems (CCSDS) protocol, command processing and dispatch to other spacecraft processors and components, support for stored commands (command macros) and time-tagged commands, management of the 8-Gbit SSR and file system, science data collection, image compression, telemetry generation, memory load and dump functions, and support for transmission of files from the SSR on the downlink using CCSDS File Delivery Protocol (CFDP). The uplink and downlink functions include control of two transponders via the 1553 bus. C&DH software also collects analog temperature data from Temperature Remote Input Output (TRIO) sensors, via a 1553 interface to the PDU, and implements a peak power tracking algorithm to optimize charging of the spacecraft battery via the PDU interface. To support operational autonomy actions, the MP incorporates the same autonomy rule engine that is implemented in the FPP software. A number of C&DH functions interface to the spacecraft through an APL-built Interface Card that is in the IEM. For example, the uplink/downlink data buffers are on that card. The interface card also allows for critical hardware commands to be sent from ground or the FPP to force resets of spacecraft processors.

Much of the MP C&DH software has design and development history from previous missions; however, the SSR data management, storage, and playback design was new for MESSENGER. A new feature designed for MESSENGER was the storage of science and telemetry on the SSR using a file system (Krupiarz et al. 2002, 2003). The MP requirement was to locate files selectively on the SSR for downlink to optimize science return. Contingency files can be stored and downlinked only if needed. Raw images are stored on the SSR and can be compressed using the Integer Wavelet compression algorithm or reduced using subframing techniques. Downlinking of files to the ground is accomplished using the CFDP. This protocol provides guaranteed delivery of all file data. MESSENGER is the first NASA mission to fly CFDP and the first APL mission to use an SSR file system.

The primary MP interfaces to two Data Processing Units (DPUs) and the two FPPs via the 1553 data bus. The DPUs provide the interface to all other instrument processors. G&C software in the MP passes data to the primary DPU to route attitude data to the imager and laser altimeter instruments and to steer the imager pivot motor.

8.2 Fault Protection Processor Software and Autonomy

The two FPPs are on unswitched power, so both are always powered, although a critical hardware command allows ground to hold either in boot mode or reset if needed. Each FPP executes an identical flight code application that supports a command and telemetry interface to the MPs via the 1553 data bus. The main purposes of each FPP are to perform fault detection and to isolate fault correction responses within these processors (Moore 2002). Each FPP implements an autonomy rule engine, which accepts uploadable health and safety rules that can operate on data collected from the 1553 data bus or a state message transmitted by the primary MP. In addition to being an RT on the 1553 bus, each FPP serves as a 1553 bus monitor to collect spacecraft data that can be monitored by autonomy rules. The rules are expressed in Reverse Polish Notation (RPN), and the action of each rule can dispatch a command (or a series of commands from a stored FPP macro) to the primary MP for subsequent execution by the MP to correct faults. Fault correction can include actions such as switching to redundant components, demotion to lower spacecraft modes (Safe

Hold or Earth Acquisition), or shedding power loads. The spacecraft has two safing modes. During safing, all time-tagged command execution is halted and the spacecraft (including instruments) is taken to a predefined simple state. Safe Hold is the first level of safing and assumes knowledge of ephemeris time, orbit, and attitude. Earth Acquisition is the lowest level of safing responding to the most critical faults (e.g., battery at low state of charge), and no knowledge of ephemeris time, orbit, or attitude (with respect to the inertial reference frame) is assumed. The spacecraft is put into a slow rotation (one revolution every 3.5 hours) allowing the antenna suite to sweep past the Earth periodically regardless of location.

Additionally, the FPPs have a custom serial interface via the Interface Card to the PDUs, to receive PDU critical status or send special commands in the event of loss of 1553 bus communications or a failed MP. The PDU command interface allows the FPPs to swap the Bus Controller functionality between MPs, power on and switch to the redundant MP and declare it primary, or power up the redundant PDU remote terminal. Each FPP exercises some control over the MP in its own IEM; it can select which of two stored flight applications the MP loads and executes, and it can reset the MP.

9 Guidance and Control

The primary functions of the MESSENGER G&C subsystem are to maintain spacecraft attitude and to execute propulsive maneuvers for spacecraft trajectory control. Attitude knowledge is required to be maintained to better than 350 μrad and attitude control better than 0.1° at one standard deviation for each axis during normal operations. Software algorithms run in the MP to coordinate data processing and commanding of sensors and actuators to maintain a three-axis stabilized spacecraft and to implement desired velocity changes. Multiple options are available for pointing the spacecraft sunshade, antennas, thrusters, and science instruments at designated targets. The orientation of the two solar panels is also software controlled to maintain a Sun offset angle that provides sufficient power at moderate panel temperatures. The system enforces two attitude safety constraints. The most important of these is the Sun Keep-In (SKI) constraint that keeps the sunshade pointed sunward to protect the spacecraft bus from extreme heat and radiation. The hot-pole keep-out constraint protects spacecraft top deck components from additional thermal extremes due to reradiation of sunlight from the planet surface once in orbit (O'Shaughnessy and Vaughan 2003; Vaughan et al. 2005).

The sensor suite consists of star trackers, an IMU, and Sun sensors as shown in Fig. 17, along with the coordinate axes defining the spacecraft body frame. Inertial attitude reference is provided by two co-boresighted star trackers mounted on the top deck and looking out along the $-Z$ axis. The two star trackers from Galileo Avionica in Florence, Italy, have a total mass of 6.37 kg, including baffles. Each uses a maximum of 12.3 W in the normal tracking mode. The trackers do image processing to identify star patterns internally, and the attitude solution in the form of quaternion and rate is output to the flight software. Typically, only a single tracker is powered with the other acting as a cold spare. Spacecraft rotation rates and translational accelerations are provided by an inertial measurement unit with four hemispherical resonance gyroscopes (HRGs) and Honeywell QA3000 accelerometers. The Scalable Space Inertial Reference Unit (S-SIRU) from Northrop-Grumman in Woodland Hills, CA, has a mass of 6.85 kg and consumes a maximum of 32 W with all gyros and accelerometers active. The S-SIRU has two processor boards providing internal redundancy with the second board acting as a cold spare. Typically, one processor board and all four gyros are powered at all times, while the four accelerometers are powered only when performing a trajectory correction maneuver (TCM). Spacecraft attitude is estimated by the

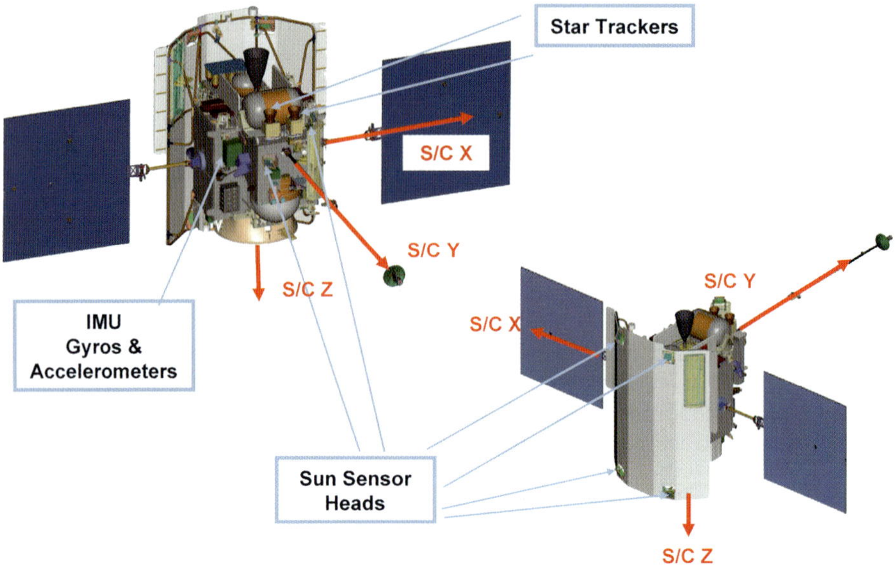

Fig. 17 MESSENGER coordinate frame and G&C sensor locations

MP software using a Kalman filter algorithm to combine star tracker and gyro measurements. A simpler filter is used to estimate accumulated velocity change from accelerometer measurements when executing TCMs.

MESSENGER also carries a set of digital Sun sensors (DSSs) to provide Sun-relative attitude knowledge if there is a failure in the primary attitude sensors. There are two separate Sun sensor systems consisting of a digital Sun sensor electronics (DSEE) box connected to three sensor heads (DSSHs), two of which are located on opposite corners of the sunshade and one on the back of the spacecraft (Fig. 17). These DSS systems, from Adcole Corporation in Marlborough, MA, have a total mass of 5.93 kg (two boxes and six heads) and consume 4.2 W. Modifications were made to the heads, and special filters were designed by APL to accommodate the high temperatures and intensity of sunlight near Mercury. The Sun sensors are always powered, providing two independent Sun direction readings at all times. Sun direction is independently computed from the sensor readings by the MP software and by the FPP software (Sect. 8.2). Fault protection logic initiates a reset of the MP if a SKI violation is detected, supplementing the SKI constraint checks in MP software.

The primary actuators for maintaining attitude control are four reaction wheels provided by Teldix in Heidelberg, Germany. These RSI 7-75/601 wheels were specially modified to have a total mass of 16.61 kg (<4.2 kg each) and can consume a maximum of 80 W each. All four wheels operate continuously, and typical total power consumption is in the range 20–30 W. Each wheel provides a maximum torque of 0.075 Nm and can store up to 7.5 Nms of momentum. The wheels are mounted on the spacecraft deck behind the sunshade as shown in Fig. 18. MESSENGER is the first APL mission to use a nonlinear control law to compute wheel torques (Wie et al. 2002; Shapiro 2003). A more traditional time-optimal slew-proportional integral derivative (PID) control law is also available as a backup. The nonlinear algorithm minimizes "chattering" of the commanded torques seen with the slew-PID formulation while following a path that deviates slightly from a pure eigen-axis turn. The flight software monitors system momentum and can autonomously perform thruster

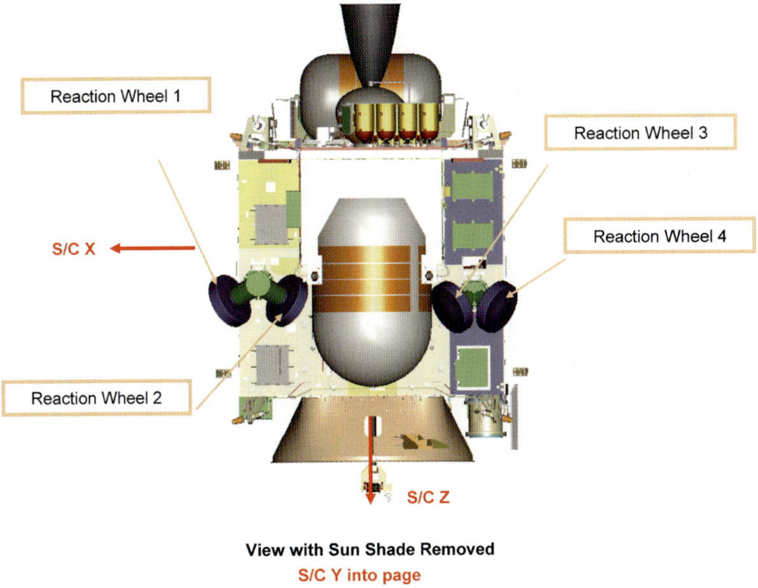

Fig. 18 MESSENGER reaction wheel locations

momentum dumps when needed to desaturate the wheels. Passive momentum dumping using solar torque is available using one of the pointing options to alter spacecraft attitude and solar panel orientation automatically to reach a specified momentum target (Vaughan et al. 2001). This capability has been used recently in flight and will be used in orbit to minimize the frequency of momentum dumps that generate small perturbations to the spacecraft trajectory from thruster firings.

Thrusters in the propulsion system (see Sect. 4) are used for attitude control during TCMs and momentum dumps and may also be used as a backup system for attitude control in the event of multiple wheel failures. The propulsion system has a large bi-propellant engine and two sets of mono-propellant thrusters, twelve 4-N thrusters, and four 22-N thrusters. Eight of the 4-N thrusters are used for attitude control. The remaining four 4-N thrusters are used for small velocity changes, while the four 22-N thrusters are used for larger velocity changes. The bi-propellant main engine is used for very large velocity changes such as the five deep-space maneuvers (DSMs) and Mercury orbit insertion (MOI). The flight software coordinates the operation of the propulsion system components for each of these three sizes of velocity changes. Heater and latch valve configuration is automatically controlled along with thruster firings during the burns. The software also implements a set of initiation and abort checks that ensure proper MPS and G&C configurations prior to and during thruster firing. The software can abort a burn if allowable operational ranges are violated.

The G&C system also interfaces with actuators for three other spacecraft components to position them properly based on knowledge of the Sun, Earth, and target planet directions relative to the spacecraft. Solar panel rotation is performed using two SADAs from MOOG in Chatsworth, CA. These assemblies have a total mass of 7.84 kg (single, internally redundant electronics box and two drives) and consume a maximum of 56 W when moving the panels. The drives can rotate in two directions about the X-axis through an angular arc of 228° in the Y–Z plane centered at the $-Z$ axis. Panels are rotated in steps of 0.02° at a constant rotation rate of 2°/s (100 steps per second). Panel positions are computed from the

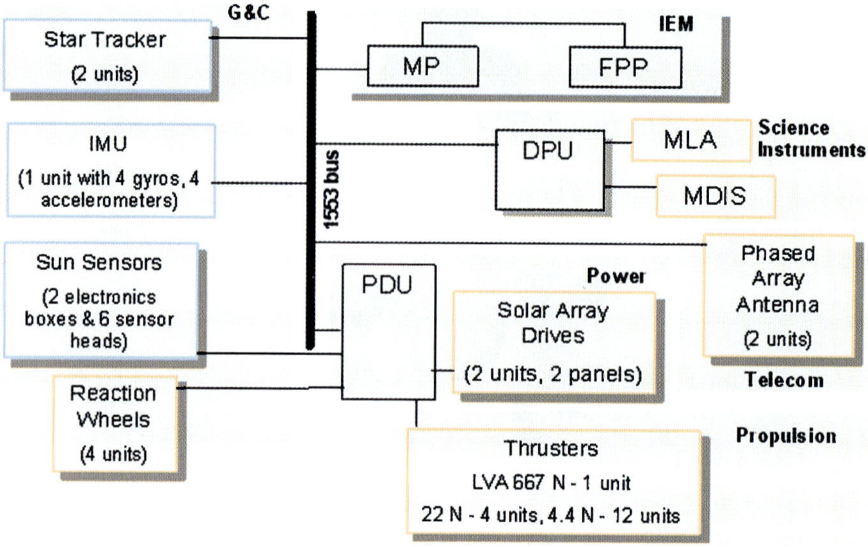

Fig. 19 MESSENGER G&C system block diagram

Sun direction to maintain the desired solar incidence angle on the cell side. The beam width (or field-of-view) of the two phased-array antennas is 12° in the X–Y plane and 3° normal to it. The boresight, centered in this beam, can be steered through an angular arc of 120°. The steering angle within this range is computed based on the projection of the Earth line in the X–Y plane. The antennas are mounted with boresights centered in the $+X$, $+Y$ and $-X$, $-Y$ quadrants. Full 360° coverage of Earth direction in the X–Y plane is obtained by rotating the spacecraft about the Y axis when necessary. Pointing options are available to place the Earth line in the proper quadrant of the X–Y plane when communicating with the phased-array or fanbeam antennas. MDIS is mounted on a pivot platform with a rotary drive that provides an operational range of travel of 90° in the Y–Z plane, 40° from the $+Z$ axis towards the sunshade ($-Y$), and 50° towards the back of the spacecraft ($+Y$). Both pivot position and spacecraft attitude can be varied to align the MDIS camera boresights with the desired target object. The interface with the Mercury Laser Altimeter (MLA) provides a range and "slant angle" by solving for the intersection of the MLA boresight direction with the target planet's surface; these quantities are used to set the instrument's internal configuration parameters for surface observations.

The block diagram in Fig. 19 shows the communication links between the flight computers, dedicated G&C hardware components, and other components with G&C interfaces. The large number of software functions and interfaces with other spacecraft components presented a significant challenge for prelaunch testing of the G&C system. Three main series of tests were performed to verify system functions: phasing or polarity tests, mission operations simulations, and autonomy tests.

The phasing tests were used to verify proper orientation, data interpretation, and commanding between the flight software and various hardware components. In addition to the dedicated G&C components, interfaces with the solar array drives, the phased-array antennas, the propulsion system, MDIS, and MLA were verified. Mission operations simulations covered nominal execution of all three types of TCMs, momentum dumps, and typical turns in cruise and in Mercury orbit for engineering and science activities. Autonomy tests exer-

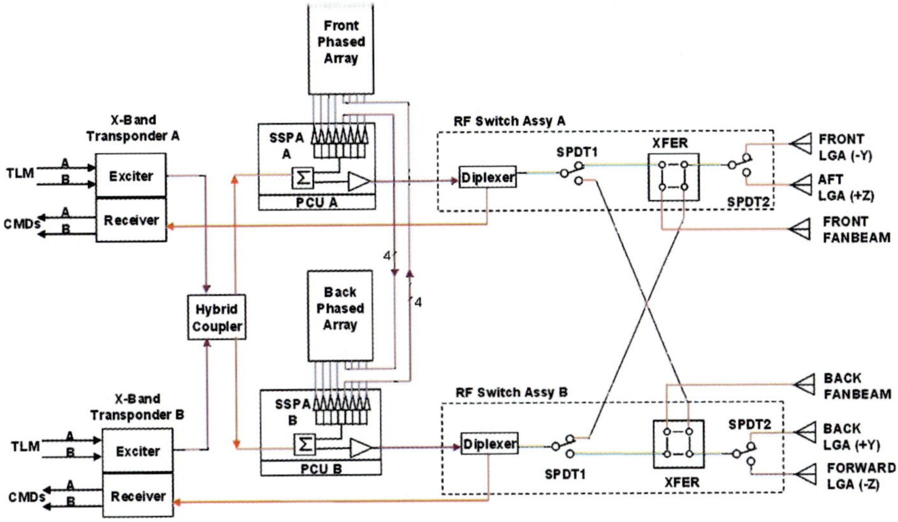

Fig. 20 MESSENGER RF subsystem block diagram. TLM denotes telemetry; CMD denotes command; PCU denotes power conversion unit; SPDT denotes single-pole double-throw switch; and XFER denotes transfer switch

cised the fault protection features included in the MP software such as momentum dumps, aborting TCMs, and safing turns in response to an attitude constraint violation. A small number of system-level tests were also conducted to cover special events such as launch and detumble.

10 RF Telecommunications

The RF telecommunications subsystem consists of redundant General Dynamics Small Deep Space Transponders (SDSTs), Solid-State Power Amplifiers (SSPAs), phased-array antennas, and medium- and low-gain antennas (Srinivasan et al. 2007), as shown in Fig. 20. The SSPAs and all antennas were manufactured, assembled, and tested by APL.

The goals of the RF telecommunications subsystem are: (1) to provide the highest quality and quantity of spacecraft housekeeping telemetry and scientific data return possible, (2) to provide highly accurate Doppler and range data for navigation and science, and (3) to provide spacecraft command capability. Furthermore, the RF telecommunications subsystem is integral to the geophysical objectives of the mission to be carried out by the MLA and radio science experiments. The MESSENGER spacecraft communicates at X-band and uses DSN antennas located in Goldstone, California; Madrid, Spain; and Canberra, Australia, for spacecraft commanding and telemetry reception. The spacecraft is controlled directly from the Mission Operations Center (MOC) at APL.

The phased-array antennas were developed specifically for the challenges of Mercury orbit and reliable scientific data return; these antennas were a mission-enabling technology. These electronically scanned antennas have no mechanical components that could fail in the challenging thermal environment at Mercury. The phased-array antennas are designed to work at the 350 °C extreme temperatures. MESSENGER is also the first spacecraft to use turbo encoding for downlink forward error correction. This coding results in an extra 0.9 dB margin in the RF downlink, which corresponds to nearly a 25% increase in data return

Table 4 Simplified MESSENGER traceability matrix

Map the elemental and mineralogical composition of Mercury's surface	MDIS, XRS, GRNS, MASCS
Image globally the surface at a resolution of hundreds of meters or better	MDIS
Determine the structure of the planet's magnetic field	MAG, EPPS
Measure the libration amplitude and gravitational field structure	MLA, RS
Determine the composition of the radar-reflective materials at Mercury's poles	GRNS, EPPS
Characterize exosphere neutrals and accelerated magnetosphere ions	MASCS/UVVS, EPPS

during the mission. In addition a unique data return optimization technique has been developed during flight using the MESSENGER RF communications system and onboard file storage system. During interplanetary and orbital operations, MESSENGER will downlink files stored on the SSR containing instrument data via CFDP. CFDP allows the automatic retransmission of mission Protocol Data Units and thus operation at lower link margins, enables a variable downlink symbol rate, and allows the targeting of individual DSN antenna performance. These three techniques together could double the science return rate over prelaunch estimates. This increase can be accomplished without penalty to the mission or science operations, relieving the need to identify and request retransmission of missing data units. MESSENGER will return >139 Gb of data over the life of the mission.

The RF telecommunications subsystem went through a very rigorous level of testing throughout the design, assembly, and test phases of development (Srinivasan et al. 2007). At the spacecraft-level, the RF system was fully tested during five comprehensive performance tests as well as DSN tests under thermal vacuum and ambient conditions.

11 Payload

The process of selecting the scientific instrumentation for a mission is typically a balance between answering as many science questions as possible and fitting within the available mission resources for mass, power, mechanical accommodation, schedule, and cost. In the case of MESSENGER (Table 4), the mass and mechanical accommodation issues were very significant constraints (see Table 5). Payload mass was limited to 50 kg because of the propellant mass needed for orbit insertion. The instrument mechanical accommodation was difficult because of the unique thermal constraints faced during the mission; instruments had to be mounted where Mercury would be in view but the Sun would not, and they had to be maintained within an acceptable temperature range in a very harsh environment. To address the many unanswered questions about Mercury, a wide variety of instrumentation was required to study the planet and its tenuous atmosphere in the infrared, visible, ultraviolet,

Table 5 Payload mass and power

	Instrument												
Power per Mission Phase	EPPS	GRS	GRS Anneal/Cooler	MAG	MASCS	DPU+MDIS	DPUs Only	MDIS Only	MLA	NS	XRS	Mag Shielding, Purge System, & Harness	Total
Power required for operation/heat (non survival heater)													
Nominal Operations (avg)	7.8	6.6	16.5	4.2	6.7	19.9	12.3	7.6	16.4	6.0	6.9	0	84.4
Nominal Operations (peak)	7.8	6.6	29.0	4.2	8.2	19.9	12.3	7.6	38.6	6.0	11.4	0	125.1
Average Instrument Power Required for Survival Heaters													
Outer Cruise	3.5	13.5	2.0	2.5	0	8.5	0	0	13.0	0	4.2	0	47.2
Non-Eclipse Orbit	2.0	11.0	2.0	2.5	0	2.5	0	0	0	0	0.3	0	20.3
Eclipse Orbit	3.1	0	0	2.5	0	6.0	0	0	0	0	0.6	0	12.2
Payload Mass Summary													
Mass Allocation (@CDR)	3.1	9.3	--	3.8	3.4	--	3.5	7.0	7.3	4.1	3.7	4.9	50.0
Final Mass	3.1	9.2	--	4.4	3.1	--	3.1	8.0	7.4	3.9	3.4	1.7	47.2

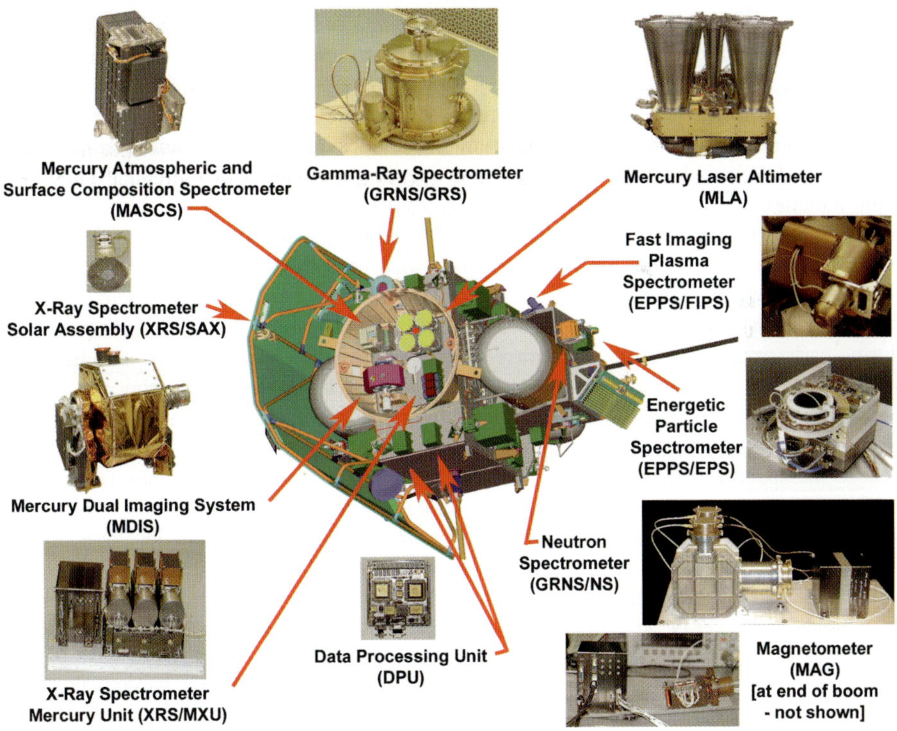

Fig. 21 MESSENGER payload accommodation

X-ray, and gamma-ray energy ranges, to measure the planet's shape, size, and topography, and to characterize its surrounding fields and particles environment (Fig. 21).

Table 6 Payload accommodation issues

	MDIS	MASCS	GRNS	MAG	EPPS	MLA	XRS
High data rates	X						
Magnetic cleanliness				X			
High power usage						X	
Complexity of thermal control	High	Med	Med	Med	Med	High	Med
FOV accommodation	X				X		
Contamination concerns	X	X			X	X	
Special spacecraft pointing	X	X				X	
On-board pointing knowledge	X					X	
On-board timing knowledge						X	
Alignment accuracy	X					X	

The science measurements were a direct result of the MESSENGER concept study science traceability matrix (Table 4), which map the prioritized science goals to the measurements needed to meet those goals.

The selected payload (Gold et al. 2001, 2002) was very ambitious (Fig. 21 and Table 5), because of both its breadth of capabilities and the limited amount of mass allocated to build it (50 kg for seven instruments). In addition to MDIS, MAG, MLA, and XRS, payload instruments include a Gamma-Ray and Neutron Spectrometer (GRNS), a Mercury Atmospheric and Surface Composition Spectrometer (MASCS), and an Energetic Particle and Plasma Spectrometer (EPPS), as well as a Radio Science (RS) experiment. The GRNS instrument includes both Gamma-Ray Spectrometer (GRS) and Neutron Spectrometer (NS) sensors, the MASCS instrument includes an Ultraviolet and Visible Spectrometer (UVVS) and a Visible and Infrared Spectrograph (VIRS), and the EPPS instrument includes both the Energetic Particle Spectrometer (EPS) and a Fast Imaging Plasma Spectrometer (FIPS). MDIS has separate wide-angle and narrow-angle cameras (WAC and NAC, respectively), and the XRS has a separate solar assembly for X-rays (SAX) mounted on the sunward side of the sunshade. Payload power requirements (Table 5) were also limited, but not in a typical manner for science missions. The most limited power period is during early cruise, when the solar arrays are generating their lowest power; this limit restricted the size of instrument heaters that could be used. In contrast, during the orbital phase of the mission, the solar arrays generate ample power, but during eclipse the battery power is still limited. The instrument designs were restricted by the ability to dissipate heat to the spacecraft deck or to the space environment. The demanding thermal requirements to stay warm enough during cruise and eclipse periods, but cold enough on orbit, were significant constraints on instrument designs throughout the payload development period.

The payload imposed a number of accommodation issues on the overall spacecraft design (e.g., field of view, FOV, accommodation); Table 6 provides an overview of these issues. As can be seen, the MLA and MDIS instruments were the biggest design drivers, in that they had significant thermal, pointing, and timing requirements. These issues were addressed early in the spacecraft design cycle by adding improved spacecraft timing knowledge and additional temperature sensors on the structure.

Also at the system level, there were several early decisions made concerning the overall interface architecture of the payload. The primary choices centered on using distributed versus centralized power distribution and data processing and involved tradeoffs among overall payload reliability, mass, power, and cost. In the end, it was decided to use both distributed power and data processing, meaning that each instrument had its own power supply and

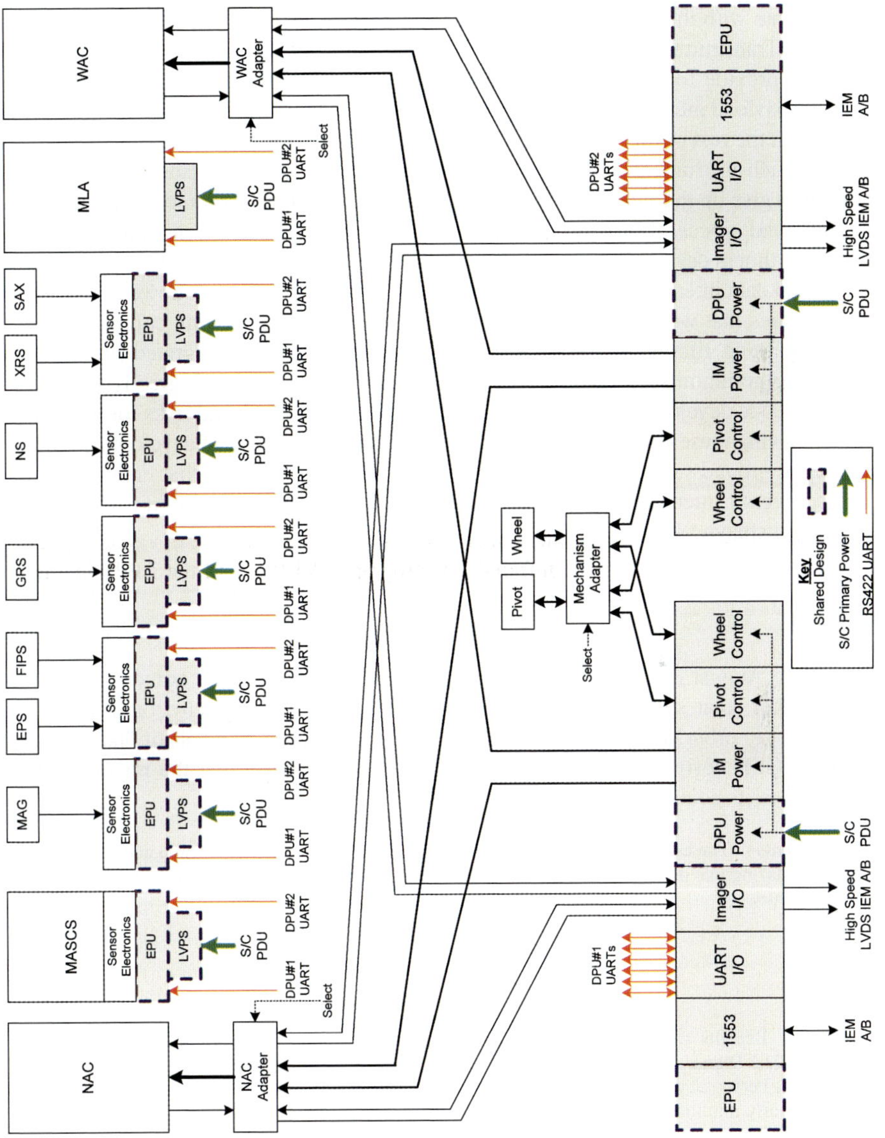

Fig. 22 Payload block diagram. I/O denotes input/output, LVPS low-voltage power supply, and LVDS low-voltage differential signaling

microprocessor. This decision greatly reduced the risk of discovering software and noise problems late in the integration schedule.

Redundant DPUs were created to buffer all data interfaces between the payload elements and spacecraft; one DPU is powered on whenever a payload element is active, while the other DPU is maintained off as a cold spare. The DPUs generally communicate with the spacecraft MPs via the spacecraft 1553 busses (plus a 4 Mbit/s serial link for image data) but communicate with the instruments via separate dedicated RS-422 Universal Asynchronous Receiver Transmitter (UART) interfaces (Fig. 22). The DPUs and the requirement that the spacecraft provide for simultaneous operation of all instruments greatly simplified the spacecraft-to-payload interface issues, allowing payload development and testing to proceed separately from the rest of the spacecraft.

Although each instrument has an individual power supply and processor, the program did not want to give up the savings associated with common designs. Common, flight-ready power supply and processor boards were developed and provided to each instrument team; all but one instrument design uses these boards. This commonality facilitated the use of identical power and data interfaces for every instrument in the payload. It also allowed common software modules that would handle a number of common tasks, including command and telemetry processing, timing, macro tools, data compression, and inter-integrated circuit bus mastering. This common software set, provided with the processor board, greatly reduced the amount of code development required for each instrument and reduced risks for payload integration. Having these common power and data interfaces also allowed development of a common set of ground support equipment (GSE) hardware and software that emulated the DPU and spacecraft interfaces and provided a graphical environment for command menus and telemetry displays.

12 Summary

With launch on August 3, 2004, the MESSENGER spacecraft started an arduous journey through the inner solar system. The spacecraft will encounter extreme environments and become the first to orbit Mercury and provide initial views of more than half of that planet. The MESSENGER design and testing program provides confidence that the mission will accomplish all of its goals.

Acknowledgements The authors thank the entire MESSENGER team for their dedication and hard work, resulting in an amazing feat of engineering that is currently well on its way to becoming the first spacecraft ever to orbit Mercury.

References

R.F. Conde et al., Benefits and lessons learned from the use of the compact PCI standard for spacecraft avionics, in *21st Digital Avionics Systems Conference*, Paper 9B5, 11 pp., Irvine, CA, 2002

G. Dakermanji, C. Person, J. Jenkins, L. Kennedy, D. Temkin, The MESSENGER spacecraft power system design and early mission performance, in *Proceedings of the Seventh European Space Power Conference, Special Publication SP-589* (European Space Agency, Noordwijk, 2005), CD-ROM

C.J. Ercol, A.G. Santo, Determination of optimum thermal phase angles at Mercury perihelion for an orbiting spacecraft, in *29th International Conference on Environmental Systems, Society of Automotive Engineers*, Tech. Paper Ser., 1999-01-21123,10 pp., Denver, CO, 1999

C.J. Ercol, J.E. Jenkins, G. Dakermanji, A.G. Santo, L.S. Mason, Prototype solar panel development and testing for a Mercury orbiter spacecraft, in *35th Intersociety Energy Conversion Engineering Conference*, American Institute of Aeronautics and Astronautics, Paper AIAA-2000-2881, 11 pp., Las Vegas, NV, 2000

R.E. Gold et al., The MESSENGER mission to Mercury: Scientific payload. Planet. Space Sci. **49**, 1467–1479 (2001)

R.E. Gold, S.C. Solomon, R.L. McNutt Jr., A.G. Santo, The MESSENGER spacecraft and payload, in *International Astronautical Congress, World Space Congress*, American Institute of Aeronautics and Astronautics, Paper IAC-02-Q. 4.1.02, 9 pp., Houston, TX, 2002

G.M. Heiligman, T.A. Hill, R.L. LeGrys, S.P. Williams, An incremental strategy for spacecraft flight software reuse, in *1st International Conference on Space Mission Challenges for Information Technology*, 8 pp., Pasadena, CA, 2003

C.J. Krupiarz et al., The use of the CCSDS file delivery protocol on MESSENGER, in *Space Operations 2002 Conference, World Space Congress*, American Institute of Aeronautics and Astronautics, Paper T5-35, 8 pp., Houston, TX, 2002

C.J. Krupiarz et al., File-based data processing on MESSENGER, in *Proceedings of the 5th International Academy of Astronautics International Conference on Low-Cost Planetary Missions, Special Publication SP-542*, ed. by R.A. Harris (European Space Agency, Noordwijk, 2003), pp. 435–442

B.Q. Le, S.X. Ling, L.R. Kennedy, G. Dakermanji, S.C. Laughery, The MESSENGER power distribution unit packaging design, in *21st Digital Avionics Systems Conference*, Paper 9B3, 8 pp., Irvine, CA, 2002

S.X. Ling, R.F. Conde, B.Q. Le, A light weight integrated electronics module (IEM) packaging design for the MESSENGER spacecraft, in *21st Digital Avionics Systems Conference*, Paper 9B4, 9 pp., Irvine, CA, 2002

J.V. McAdams, R.W. Farquhar, A.H. Taylor, B.G. Williams, MESSENGER mission design and navigation. Space Sci. Rev. (2007). doi: 10.1007/s11214-007-9162-x

R.C. Moore, Safing and fault protection for a mission to Mercury, in *21st Digital Avionics Systems Conference*, Paper 9A4, 8 pp., Irvine, CA, 2002

D.J. O'Shaughnessy, R.M. Vaughan, MESSENGER spacecraft pointing options, in *13th American Astronautical Society/American Institute of Aeronautics and Astronautics Space Flight Mechanics Conference*, Paper AAS-03-149, 20 pp., Ponce, Puerto Rico, 2003

D.F. Persons, L.E. Mosher, T.J. Hartka, The NEAR Shoemaker and MESSENGER spacecraft: Two approaches to structure and propulsion design, in *41st Structures, Structural Dynamics and Materials Conference*, American Institute of Aeronautics and Astronautics, Paper AIAA-00-1406, 10 pp., Atlanta, GA, 2000

A.G. Santo et al., The MESSENGER mission to Mercury: Spacecraft and mission design. Planet. Space Sci. **49**, 1481–1500 (2001)

A.G. Santo et al., MESSENGER: The Discovery-class mission to orbit Mercury, in *International Astronautical Congress, World Space Congress*, American Institute of Aeronautics and Astronautics, Paper IAC-02-U. 4.1.04, 11 pp., Houston, TX, 2002

H.S. Shapiro, Implementation of a non-linear feedback control algorithm for large angle slew, and simulations from MESSENGER applications, Internal Memo SRM-02-032, The Johns Hopkins University Applied Physics Laboratory, Laurel, MD, 2003

S.C. Solomon, R.L. McNutt Jr., R.E. Gold, D.L. Domingue, MESSENGER mission overview. Space Sci. Rev. (2007). doi: 10.1007/s11214-007-9247-6

D.K. Srinivasan, M.E. Perry, K.B. Fielhauer, D.E. Smith, M.T. Zuber, The radio frequency subsystem and radio science on the MESSENGER mission. Space Sci. Rev. (2007). doi: 10.1007/s11214-007-9270-7

W. Tam, S. Wiley, K. Dommer, L. Mosher, D. Persons, Design and manufacture of the MESSENGER propellant tank assembly, in *38th American Institute of Aeronautics and Astronautics/American Society of Mechanical Engineers/Society of Automotive Engineers/American Society for Engineering Education Joint Propulsion Conference and Exhibit*, Paper AIAA 2002-4139, 17 pp., Indianapolis, IN, 2002

R.M. Vaughan, D.R. Haley, D.J. O'Shaughnessy, H.S. Shapiro, Momentum management for the MESSENGER mission, in *American Astronautical Society/American Institute of Aeronautics and Astronautics Astrodynamics Specialist Conference*, Paper AAS 01-380, 22 pp., Quebec City, Quebec, Canada, 2001

R.M. Vaughan et al., MESSENGER guidance and control system performance during initial operations, in *28th Annual American Astronautical Society Guidance and Control Conference*, Paper AAS 05-083, 20 pp., Breckenridge, CO, 2005

B. Wie, D. Bailey, C. Heiberg, Rapid multitarget acquisition and pointing control of agile spacecraft. J. Guid. Control Dyn. **25**, 96–104 (2002)

P.D. Wienhold, D.F. Persons, The development of high-temperature composite solar array substrate panels for the MESSENGER spacecraft. SAMPE J. **39**(6), 6–17 (2003)

S. Wiley, K. Dommer, L. Mosher, Design and development of the MESSENGER propulsion system, in *American Institute of Aeronautics and Astronautics/Society of Automotive Engineers/American Society of Mechanical Engineers Joint Propulsion Conference*, Paper AIAA-2003-5078, 20 pp., Huntsville, AL, 2003

Space Sci Rev (2007) 131: 219–246
DOI 10.1007/s11214-007-9162-x

MESSENGER Mission Design and Navigation

James V. McAdams · Robert W. Farquhar · Anthony H. Taylor · Bobby G. Williams

Received: 22 May 2006 / Accepted: 15 February 2007 /
Published online: 16 June 2007

Abstract Nearly three decades after the Mariner 10 spacecraft's third and final targeted Mercury flyby, the 3 August 2004 launch of the MESSENGER (MErcury Surface, Space ENvironment, GEochemistry, and Ranging) spacecraft began a new phase of exploration of the closest planet to our Sun. In order to ensure that the spacecraft had sufficient time for pre-launch testing, the NASA Discovery Program mission to orbit Mercury experienced launch delays that required utilization of the most complex of three possible mission profiles in 2004. During the 7.6-year mission, the spacecraft's trajectory will include six planetary flybys (including three of Mercury between January 2008 and September 2009), dozens of trajectory-correction maneuvers (TCMs), and a year in orbit around Mercury. Members of the mission design and navigation teams optimize the spacecraft's trajectory, specify TCM requirements, and predict and reconstruct the spacecraft's orbit. These primary mission design and navigation responsibilities are closely coordinated with spacecraft design limitations, operational constraints, availability of ground-based tracking stations, and science objectives. A few days after the spacecraft enters Mercury orbit in mid-March 2011, the orbit will have an 80° inclination relative to Mercury's equator, a 200-km minimum altitude over 60°N latitude, and a 12-hour period. In order to accommodate science goals that require long durations during Mercury orbit without trajectory adjustments, pairs of orbit-correction maneuvers are scheduled every 88 days (once per Mercury year).

Keywords Gravity assist · MESSENGER · Mercury · Mission design · Navigation

J.V. McAdams (✉) · R.W. Farquhar
The Johns Hopkins University Applied Physics Laboratory, Space Department, Laurel, MD 20723, USA
e-mail: Jim.McAdams@jhuapl.edu

A.H. Taylor · B.G. Williams
Space Navigation and Flight Dynamics Practice, KinetX, Inc., Simi Valley, CA 93065, USA

1 Introduction

The launch of the MESSENGER (MErcury Surface, Space ENvironment, GEochemistry, and Ranging) spacecraft on 3 August 2004 began a new phase of exploration for the closest planet to our Sun. The 7.6-year trajectory the MESSENGER spacecraft will follow includes one Earth flyby, two Venus flybys, three Mercury flybys, and a one-year Mercury orbit phase. The three Mercury flybys will offer the opportunity to obtain images of most of the 55% of Mercury's surface not imaged by Mariner 10 in 1974 and 1975. As the exploration phase transitions from "flyby" to "orbit", MESSENGER will use a suite of seven science instruments to acquire data that will help answer six fundamental questions regarding the formation, composition, and field structure of Mercury and the region of space influenced by Mercury (Solomon et al. 2001).

Engineers at The Johns Hopkins University Applied Physics Laboratory (JHU/APL), numerous subcontractors, and universities worked closely with the mission's Principal Investigator, located at the Carnegie Institution of Washington, to design and assemble a spacecraft and science payload that would meet the mission's science objectives. The spacecraft's design includes features that ensure redundancy of most mission-critical functions, provide autonomy during time-critical activities, and offer reliable operation during a mission that places the probe between 30% and 108% of Earth's average distance from the Sun. Some of these features include a fixed sunshade to protect delicate spacecraft parts from direct sunlight, a dual-mode (fuel only or a fuel/oxidizer mix) propulsion system for course-correction maneuvers, two high-temperature-tolerant articulating solar arrays and a battery for power, state-of-the-art attitude control and telecommunications subsystems, and seven science instruments for data collection. While characteristics of the spacecraft's trajectory and propulsive maneuvers affected the design of many of these spacecraft components, spacecraft and ground-station functional limitations impose many operational constraints and guidelines on the spacecraft.

In order to fulfill all mission objectives, the spacecraft must follow a complex trajectory with many planned trajectory-correction maneuvers (TCMs) and planetary flybys. The mission design team at JHU/APL and the navigation team at KinetX, Inc., work together to design the most fuel-efficient, lowest-risk trajectory and TCMs possible. The remaining trajectory is re-optimized after each propulsive maneuver and planetary flyby to accommodate execution errors and trajectory uncertainties. In addition, each portion of the completed trajectory is determined at the highest possible precision. Much effort is required to ensure success for events that have the greatest change of the trajectory (i.e., planetary flybys and Mercury orbit insertion). The navigation team uses observed changes in the spacecraft's orbit to improve the accuracy of models (e.g., solar radiation pressure and Mercury gravity) for orbit perturbations in order to improve trajectory predictions.

The early operations activity of the mission design and navigation teams has been a key part of the early mission success. The design and implementation of TCMs, along with trajectory optimization and orbit determination, have both corrected errors introduced at launch and targeted the first two planetary flybys (of Earth and Venus). Highlights of these activities and the resulting TCM performance are provided.

2 Historical Background

A direct trajectory to Mercury requires substantial launch energy (i.e., a minimum C_3 of about 50 km^2/s^2). However, by using a Venus gravity-assist maneuver, the launch energy required is about 70% less than that needed for a direct transfer. This important result was originally obtained by Michael Minovitch (Minovitch 1963; Dowling et al. 1997). The Venus gravity-assist concept was developed further by Sturms and Cutting (1966) and eventually led to a proposal for a Mariner spacecraft mission to Mercury in 1973 (Bourke and Beerer 1971). The original flight plan called for a launch in November 1973, a Venus flyby in February 1974, and one Mercury flyby in March 1974. However, as suggested by Guiseppe Colombo in 1970 (Murray 1989), the Mercury flyby aim point can be chosen so that the spacecraft's heliocentric orbital period is exactly twice Mercury's orbital period, allowing the spacecraft to return to Mercury 176 days after the first encounter. Used twice during the flight of Mariner 10, this technique led to three successful Mercury flybys (Dunne and Burgess 1978). These flybys, which provided images of 45% of Mercury's surface, occurred on 29 March 1974, 21 September 1974, and 16 March 1975 at altitudes of 703 km, 48,069 km, and 327 km, respectively.

The spectacular success of the Mariner 10 mission generated considerable interest in further exploration of Mercury. An orbiter-class mission was the next logical step. However, it was soon realized that the energy requirements to place a meaningful payload into orbit around Mercury were rather formidable. A variety of ballistic mission modes utilizing near-perihelion propulsive maneuvers, powered Venus flybys, and multiple Venus flybys were studied in some detail (Hollenbeck et al. 1973). Because these results were less than satisfactory, Friedlander and Feingold (1977) concluded that a Mercury orbiter mission would require the use of a low-thrust delivery system such as solar-electric propulsion. This situation prevailed until 1985 when Chen-wan Yen (1989) showed that a "reverse delta-VEGA (ΔV Earth Gravity Assist) process" using multiple Venus and Mercury flybys along with selected propulsive maneuvers could produce energy-efficient ballistic trajectories for a Mercury orbiter.

Since 1990 Mercury orbiter mission studies continued this trend of lower propulsive requirements. These studies include a Hermes Orbiter study by Jet Propulsion Laboratory (JPL) and TRW (Cruz and Bell 1995), one by the European Space Agency (Grard et al. 1994), one by McAdams et al. (1998), and another by Yamakawa et al. (2000) from Japan's Institute of Space and Astronautical Science (ISAS). The 1998 and 1999 Mercury orbiter studies described an August 2005 launch with 16.0 km^2/s^2 launch energy that required two Venus and two Mercury flybys as part of the 4.2-year heliocentric transfer to Mercury orbit insertion. Preceding this trajectory with a one-year Earth–Earth transfer, and following it with a major propulsive maneuver and third Mercury flyby, forms the basis for the trajectory followed by the MESSENGER mission. More recent studies of ballistic mission modes for a Mercury orbiter have shown that ΔV costs for this mission could be reduced even further (Langevin 2000; Yen 2001; McAdams et al. 2002). The BepiColombo mission, planned by the European Space Agency to launch as early as 2013, will rely on high-thrust propulsion and multiple flybys of Venus and Mercury during its multi-year heliocentric transfer to Mercury.

3 Science and Engineering Background

The science requirements, spacecraft configuration, and launch vehicle performance influence the selection of the Sun-centered (cruise phase) and Mercury-centered (orbit phase) portions of the MESSENGER mission's trajectory. Other factors such as funding limitations and the allowable range of launch dates, while also important, are beyond the scope of this article. A complex interdependency among launch vehicle performance (e.g., the maximum spacecraft weight that can be delivered to a particular Earth escape speed and direction), available technology for the spacecraft subsystems and science instruments, propulsive requirements for various trajectories, and space environment conditions must be addressed early in the spacecraft design process. Early in the mission's conceptual design it became clear that the most capable launch vehicle allowed by NASA's Discovery Program would be required for MESSENGER. This requirement led to the selection of the Boeing Corporation's Delta II 7925H-9.5 launch vehicle.

Characteristics of the trajectory, specifications for TCMs, and requirements for navigation all led to spacecraft design features with inherent operational constraints. Examples of these design features include the type, number, and placement of propulsive thrusters, as well as placement and field-of-view of the spacecraft's visible-light-wavelength cameras. Mission design and navigation team members work with mission operations and science team members during the mission's cruise and orbit phases to ensure that all spacecraft operational constraints are met with adequate margin for all planned and contingency events that alter the spacecraft's trajectory.

3.1 Science Requirements

Mercury is the only planet in the inner solar system that has yet to be orbited by a spacecraft. During its three Mercury flybys in 1974 and 1975, Mariner 10 imaged about 45% of the surface at an average resolution of about 1 km and <1% of the surface at better than 500-m resolution (Murray 1975a, 1975b). Mercury has the highest known uncompressed density of any planet or satellite in the solar system. This observation, coupled with the determination of surface composition from analysis of MESSENGER science data, could reveal clues that would help to understand the processes by which planetesimals in the primitive solar nebula accreted to form planets.

The spacecraft's seven science instruments will acquire data in order to address six important questions on the nature and evolution of Mercury (Solomon et al. 2001). Answers to these questions, which will offer insights well beyond increased knowledge of the planet Mercury, are the basis for the science objectives:

1. Map the elemental and mineralogical composition of Mercury's surface.
2. Image globally the surface at a resolution of hundreds of meters or better.
3. Determine the planet's magnetic field structure.
4. Measure the libration amplitude and gravitational field structure.
5. Determine the composition of radar-reflective materials at Mercury's poles.
6. Characterize exosphere neutrals and accelerated magnetosphere ions.

The science instruments include the wide-angle and narrow-angle field-of-view imagers of the Mercury Dual Imaging System (MDIS), Gamma-Ray and Neutron Spectrometer (GRNS), X-Ray Spectrometer (XRS), Magnetometer (MAG), Mercury Laser Altimeter (MLA), Mercury Atmospheric and Surface Composition Spectrometer (MASCS), Energetic Particle and Plasma Spectrometer (EPPS), and an X-band transponder for the Radio Science (RS) experiment. Table 1 shows the role each instrument has in linking science objectives to the orbit at Mercury. Other articles in this issue offer a more comprehensive examination of the structure and function of all science instruments (Anderson et al. 2007; Andrews et al. 2007; Cavanaugh et al. 2007; Goldsten et al. 2007; Hawkins 2007; McClintock and Lankton 2007; Schlemm et al. 2007).

Table 1 Mapping of science objectives into Mercury orbit design

Mission objectives	Mission design requirements	Mission design features
Globally image surface at 250-m resolution	Provide two Mercury solar days at two geometries for stereo imaging of entire surface; near-polar orbit for full coverage (MDIS)	Orbital phase of one Earth year (13 days longer than two Mercury solar days) with periapsis altitude controlled to 200–500 km; 80° inclination
Determine the structure of Mercury's magnetic field	Minimize periapsis altitude; maximize altitude-range coverage (MAG)	Mercury orbit periapsis altitude of 200–500 km; apoapsis altitude near 15 200 km for 12-hour orbital period
Simplify orbital mission operations to minimize cost and complexity	Choose orbit with period of 8, 12, or 24 hours	
Map the elemental and mineralogical composition of Mercury's surface	Maximize time at low altitudes (GRNS, XRS)	
Measure the libration amplitude and gravitational field structure	Minimize orbital-phase thrusting events (RS, MLA) Orbital inclination 80°; latitude of periapsis near 60°N (MLA, RS)	Orbital inclination drifts from 80° to 82°; periapsis latitude drifts[1] from 60°N to 72°N; primarily passive momentum management; two orbit-correction ΔVs (30 hours apart) every 88 days
Determine the composition of radar-reflective materials at Mercury's poles	Orbital inclination 80°; latitude of periapsis maintained near 60°N (GRNS, MLA, MASCS, EPPS)	
Characterize exosphere neutrals and accelerated magnetosphere ions	Wide altitude range coverage; visibility of atmosphere at all lighting conditions	Extensive coverage of magnetosphere; orbit cuts bow shock, magnetopause, and upstream solar wind

[1]Solar and Mercury gravity perturb the spacecraft orbit away from some science requirements. Orbit-correction ΔVs can correct periapsis latitude drift

3.2 Spacecraft Configuration

The spacecraft's physical characteristics (Fig. 1) and software configuration were designed to accommodate all aspects (both extremes and routine activity) of MESSENGER's heliocentric cruise and Mercury orbit phases. Only those features that equip the spacecraft to function during trajectory-altering events (e.g., planetary flybys or TCMs) or trajectory-related environmental extremes (e.g., solar eclipse) will be discussed. The MESSENGER spacecraft's design emphasizes simple, proven techniques and functional redundancy, along with a combination of carefully selected advanced technologies and minimal moving parts. Key design features include a fixed ceramic-cloth sunshade, a dual-mode (bipropellant-

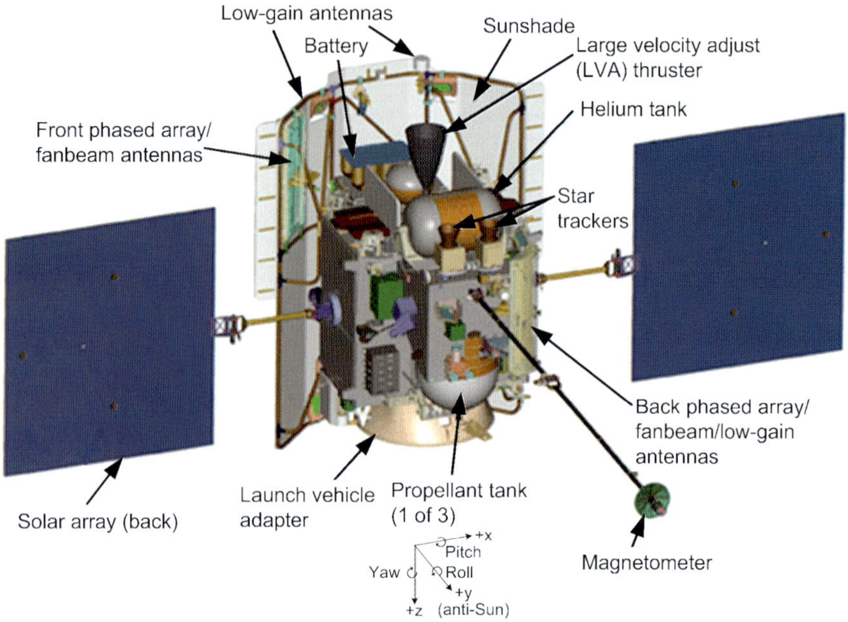

Fig. 1 Fully deployed, false-color view of the MESSENGER spacecraft. This view shows spacecraft details that are hidden by layers of thermal blankets or heat-resistant, Nextel ceramic cloth. This fully-deployed configuration and Sun-relative orientation is typical during much of the mission after late-March 2005

monopropellant) propulsion system, two rotating solar arrays, three-axis stabilization, and a state-of-the-art telecommunications system (Leary et al. 2007).

The propulsion system includes 17 thrusters (Fig. 2) arranged to allow small TCMs (<20 m/s ΔV) in any direction and small-to-large TCMs for ΔV in the $+z$ direction using either the four 26-N "C" thrusters or the 672-N large-velocity-adjust (LVA) thruster. The telecommunications subsystem includes four low-gain antennas that were mounted on the spacecraft such that planned and contingency course-correction maneuvers would be able to be monitored from Earth-based tracking stations with sufficient link margin. In addition, a 23-Ah NiH_2 battery will provide survival power for over an hour during eclipses. Software features include spacecraft autonomy that enables the spacecraft to operate independently (without commands from ground-based operators) for weeks at a time, a feature that will be used during multiple, long-duration solar conjunctions.

3.3 Operational Requirements and Constraints

After identifying those spacecraft subsystems with significant design criteria for ensuring safe spacecraft function throughout the mission, it is useful to describe the corresponding operational requirements and constraints. The trajectory and propulsive maneuvers designed by the mission design team and reconstructed by the navigation team must demonstrate compliance with every requirement and constraint defined in the JHU/APL-produced System Requirements Document. A brief description of these requirements and constraints will increase understanding of the complexity inherent in MESSENGER's mission design. It

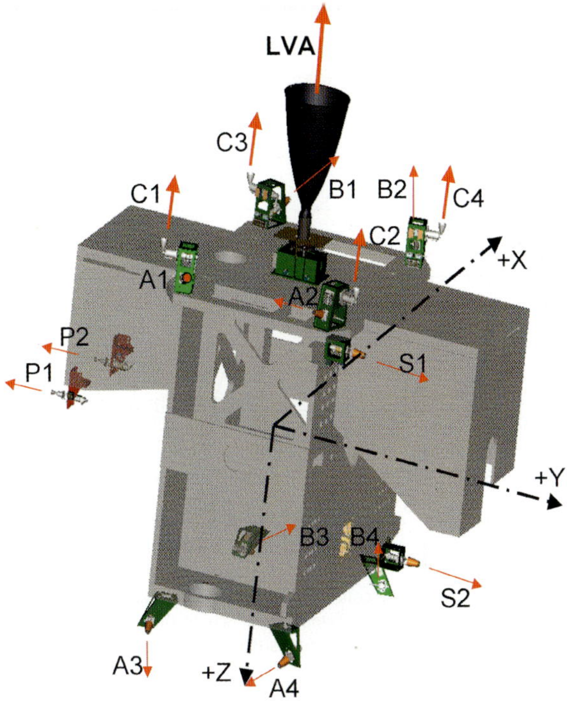

Fig. 2 View showing thruster orientation for the MESSENGER spacecraft. There are 12 4-N thrusters (P1 and P2 that poke through the sunshade, S1 and S2 on the opposite side of the spacecraft, and four thrusters A1 to A4 and B1 to B4 on each side of the spacecraft parallel to the Y–Z plane), four 26-N thrusters (C1 to C4), and one 672-N thruster (LVA)

may also become apparent that some orbit or maneuver options described in other Mercury orbiter studies would violate MESSENGER requirements or constraints.

With a severe thermal environment during Mercury orbit phase, the spacecraft's thermal subsystem had the most issues directly related to the spacecraft trajectory and propulsive maneuver design. Even though a circular orbit around Mercury would offer the opportunity for uniform, global images of Mercury's surface, Nelson et al. (1995) note that such an orbit would subject the spacecraft to an unmanageable thermal load. The 200-km-minimum-altitude by 12-hour orbit followed by MESSENGER at Mercury is stable (periapsis altitude and latitude increase, but orbit period is almost constant) and thermally manageable (Ercol and Santo 1999). Thermal analysis of this orbit at Mercury revealed acceptable worst-case temperatures for the sunshade, spacecraft bus, solar arrays, and science instruments. This thermal analysis related a spacecraft orbit orientation angle to the maximum acceptable spacecraft internal temperature. The 3 August 2004 launch brings the spacecraft's orbit to within 6° of the upper limit for this orbit orientation angle. This constraint effectively places the spacecraft orbit periapsis near the day/night terminator when Mercury is closest to the Sun.

Since the sunshade must protect the spacecraft bus from direct sunlight exposure during propulsive maneuvers that are performed <0.7 AU (Astronomical Unit, which equals the average Earth–Sun distance) from the Sun, a spacecraft maneuver attitude constraint is necessary. For TCMs that require either the bi-propellant LVA thruster and/or the C thrusters mounted on the same deck as the LVA thruster, the Sun–spacecraft–ΔV angle must be between 78° and 102°. This is equivalent to requiring the spacecraft–Sun direction to be $<12°$ from the $-y$-axis direction. This spacecraft attitude constraint during propulsive maneuvers provides two opportunities for performing orbit-correction maneuvers (OCMs) per 88-day

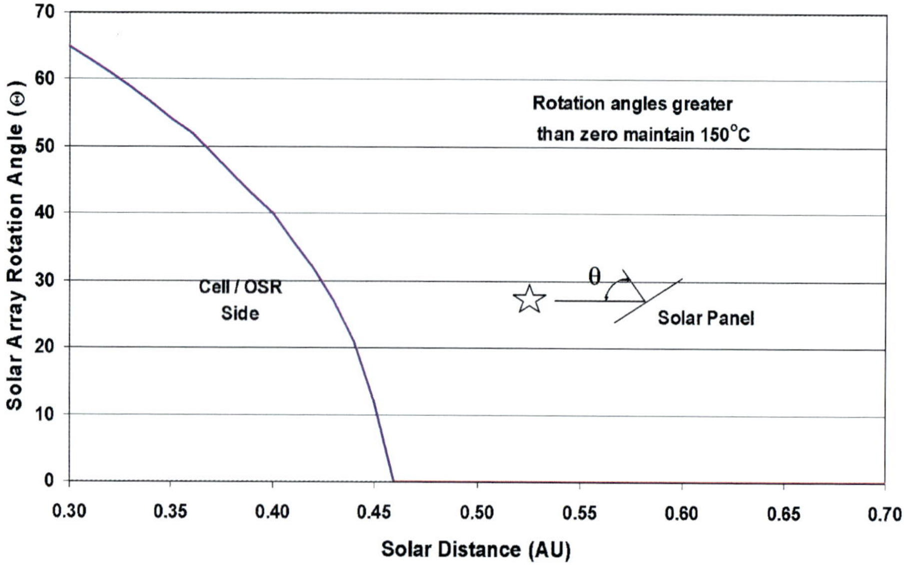

Fig. 3 Solar array angle constraint for preventing damage due to high temperature. Values of 0.30 AU and 0.46 AU, the range of solar distance over which the solar array must be tilted, correspond closely to MESSENGER's minimum and maximum solar distance in Mercury orbit

Mercury year. These two opportunities occur when the spacecraft orbit plane and Sun–Mercury line are nearly perpendicular. These OCM opportunities arise shortly after Mercury's minimum distance from the Sun (where Mercury orbit insertion occurs) and one-half Mercury orbit later. Because science objectives require long intervals between large spacecraft orbit adjustments, the time between OCMs will be maximized. Since the spacecraft's periapsis altitude nears the 500-km upper limit shown in Table 1 about one Mercury year (88 days) after Mercury orbit insertion (MOI), all OCM pairs will occur once every 88 days, soon after Mercury is closest to the Sun.

In order to maintain solar array temperature below an upper limit, the tilt angle of the solar array surface normal relative to the Sun direction must be carefully controlled. Mission design and navigation software requires knowledge of solar array orientation to predict solar pressure perturbations on the spacecraft's orbit. The thermal requirement for solar array rotation, shown in Fig. 3, is to keep the solar array surface, populated with 30% solar cells and 70% optical surface reflectors (OSRs), below 150°C. The size of the solar arrays was chosen to provide sufficient power to perform a TCM at a solar distance of 1.1 AU.

The spacecraft's battery must be able to supply the spacecraft's reduced power level requirement when the solar arrays do not receive sunlight during solar eclipse. Mass margin concerns early in the development phase limited the battery size to meet spacecraft power requirements for up to 65 minutes. With the planned initial orbit size and orientation, the spacecraft should experience a 61.5-minute maximum-duration eclipse in late-May 2011. Another requirement is that no ΔV may be performed within two hours of a solar eclipse.

Trajectory optimization and maneuver design must meet telecommunications subsystem requirements. These requirements include the goal (not a strict requirement) to monitor 100% of every planned propulsive maneuver, the responsibility to determine start and stop times for data transmission during Mercury orbit phase, and the need to define times when interference from the Sun prevents reliable spacecraft communication. The mission design team scheduled all OCMs to avoid Earth occultation, when Mercury blocks the spacecraft–Earth line of sight. As the Sun–Earth–spacecraft angle drops below 3°, the spacecraft enters solar conjunction—a region where solar interference degrades spacecraft communication with Earth ground stations. Time for data downlink is scheduled for eight hours on every other orbit and five additional 2- to 8-hour tracks per week. Knowledge of the spacecraft attitude, including solar array orientation, during these data downlink times is useful for precise modeling of the solar radiation pressure acting on the spacecraft and thus for reducing trajectory estimation and propagation error.

Navigation requirements and the maneuver design process time place more constraints on Mercury orbit insertion and on the spacing between OCMs. In order to lower risk by improving Mercury approach orbit determination, the navigation team must verify that a bright star is visible in the MDIS narrow-angle field-of-view for optical navigation images of Mercury. In addition, the quick-look maneuver performance assessment, subsequent maneuver update planning, and spacecraft upload time needed between OCMs 1 and 2, 3 and 4, and 5 and 6 must be accomplished within about 30 hours (2.5 orbits).

Analyses performed by the navigation team confirmed that the planetary flyby minimum-altitude constraints are conservative. Planetary protection requirements that establish a maximum allowable probability of impact (1×10^{-6}) for the Venus flybys are met with significant margin with MESSENGER's 300-km minimum-altitude constraint. Using planned Mercury approach navigation techniques, a 200-km minimum-altitude constraint for Mercury flybys provides a minute probability of impact.

4 Interplanetary Cruise Phase

The 6.6-year trip from launch to Mercury orbit insertion is one of the longest interplanetary cruise phase options considered for MESSENGER. The spacecraft system design lifetime accounted for a seven-year journey to Mercury followed by a one-year Mercury orbit phase. During interplanetary cruise phase the primary gravitational body changes from Earth (launch and flyby), to the Sun (between planetary flybys), to Venus (flybys), and to Mercury (flybys and arrival). As many as 40 TCMs, including five large deep-space maneuvers (DSMs), may be required during this mission phase.

Orbit determination (OD) during the cruise phase relies primarily on the Deep Space Network (DSN) Doppler and ranging tracking data. On approach to the Venus and Mercury flybys, these data will be augmented with both DSN Delta Differential One-Way Ranging (ΔDOR) and optical navigation measurements. After each OD solution, the navigation team will perform a mapping of the trajectory and its uncertainties to the aim point in the b-plane at the next planet flyby. The b-plane is the plane normal to the incoming asymptote of the hyperbolic flyby trajectory that passes through the center of the target body (Earth in the case of the Earth flyby). The 'S-axis' is in the direction of the incoming asymptote and hence is normal to the b-plane. The 'T-axis' is parallel to the line of intersection between the b-plane and the Earth Mean Ecliptic plane of J2000 (and is positive in the direction of decreasing right ascension). The 'R-axis' (positive toward the South Ecliptic Pole) completes the mutually orthogonal, right-handed Cartesian coordinate axes with origin at the center of the Earth such that $\hat{T} \times \hat{R} = \hat{S}$, where $\hat{T}$, $\hat{R}$, and $\hat{S}$ are unit vectors in each axis direction. The ideal aim point is determined by optimizing the trajectory over the remaining mission from the current epoch up to Mercury orbit insertion and on into the orbit about Mercury. The goal of maneuver design for TCMs is to stay on or near the optimum trajectory while minimizing the use of propulsive fuel.

A single ΔDOR measurement determines the angular position of the spacecraft relative to a baseline between two DSN sites, and hence it contains unique information compared with the line-of-sight measurements of Doppler and ranging from a single station. The baselines typically used are the Goldstone–Canberra and the Goldstone–Madrid station pairs. Optical navigation (opnav) provides a direct measure of the spacecraft position relative to the target body by using MDIS to take images of the target body against a field of background stars.

The DSN long-range planning schedule currently includes periods for obtaining MESSENGER ΔDOR tracking that includes measurements from two different baselines each week (for four weeks total) starting five weeks prior to each Venus and Mercury flyby. ΔDOR and opnav tracking taken before the first Venus flyby, which occurs when MESSENGER is at superior conjunction, will be used for after-the-fact data flow and processing tests since the aim point is over 3,000 km altitude. At this first Venus flyby, Doppler and ranging data types were sufficient for navigation and the enhanced accuracy available from ΔDOR and opnav was not required. This allowed the navigation team to validate the ΔDOR and opnav processing on the first Venus flyby in a parallel, non-critical process without affecting the regular navigation orbit determination and delivery process. Subsequent Venus and Mercury flybys will use ΔDOR during the approach phase to determine the targeting maneuvers and estimate the resulting flyby trajectory. For the Mercury flybys, orbit solutions including opnav images will be used as independent data types to resolve any apparent conflicts between the solutions based on radiometric tracking.

4.1 Launch

MESSENGER had three launch opportunities in 2004, but schedule delays and additions to spacecraft testing plans led to the choice of launching in the latest opportunity. Table 2 shows selected mission performance parameters for all three launch opportunities during 2004. Chen-wan Yen of JPL discovered all three heliocentric transfer trajectories used as launch options for MESSENGER in 2004, with engineers at KinetX and JHU/APL performing trajectory optimization, maneuver design, and navigation.

On 3 August 2004, at 06:15:56.537 UT, the MESSENGER spacecraft left Earth aboard a Delta II 7925H launch vehicle from Cape Canaveral Air Force Station in east central Florida. With launch services provided by Boeing and NASA Kennedy Space Center, the JHU/APL-built 1107.25-kg spacecraft departed Earth orbit with a 16.388 km^2/s^2 launch energy at a $-32.66°$ declination of launch asymptote (DLA) relative to the Earth mean equator at the standard J2000 (1 January 2000 at 12:00:00 ET) epoch. While there was little deviation from the planned trajectory during the first hour after launch (Fig. 4), the larger-than-expected 2.0-σ under burn required that the navigation team provide early orbit determination solutions to DSN tracking stations to improve their antenna pointing to the spacecraft. Prior to stage 3-spacecraft separation, Boeing's nearly perfect third-stage despin slowed the spacecraft from about 58 revolutions per minute (rpm) to 0.015 rpm. After separation the solar panels deployed and the spacecraft pointed its sunshade away from the Sun, thereby saving heater power by warming the spacecraft bus with solar energy when the spacecraft–Sun distance is near 1 AU.

4.2 Heliocentric Trajectory

Between launch and Mercury orbit insertion the spacecraft flies by Earth once, Venus twice, and Mercury three times. Five additional major course-correction maneuvers or DSMs are

Table 2 MESSENGER launch options for 2004

Month	March	May	August[1]	August (launch day)
Launch dates	10–29	11–22	30 Jul–13 Aug	3 Aug 2004
Launch period (days)	20	12	15	–
Launch energy (km^2/s^2)	≤15.700	≤17.472	≤16.887	16.388
Earth flybys	0	0	1	1
Venus flybys	2	3	2	2
Mercury flybys	2	2	3	3
Deterministic ΔV (m/s)	≤2026	≤2074	≤1991	1965[2]
Total ΔV (m/s)	2300	2276	2277	2250
Orbit insertion date	6 Apr 2009	2 Jul 2009	18 Mar 2011	18 Mar 2011

[1] The final launch period started on 2 August due to delays in the launch facility's availability

[2] Lower total ΔV on launch day reflects a reduced propellant load required to meet the spacecraft launch weight limit

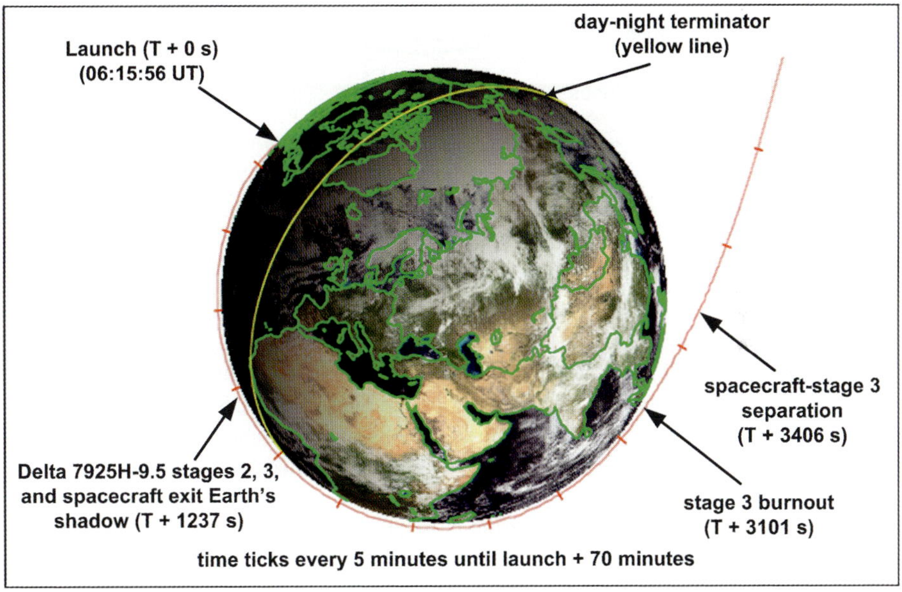

Fig. 4 Launch trajectory with final confirmed times for selected events. All confirmed launch event times were <10 s from the times predicted two weeks before launch

needed during the 6.6-year ballistic trajectory to Mercury. Figure 5 shows the Earth-to-Venus flyby 1 transfer orbit, which includes an Earth flyby one year after launch and the largest DSM, which is just before the spacecraft reaches the following perihelion. A month-long solar conjunction will begin shortly before Venus flyby 1.

The Venus flyby 1-to-Mercury flyby 1 transfer trajectory (Fig. 6) includes a second Venus flyby at almost exactly the same point in Venus' orbit and a DSM one orbit later. The second

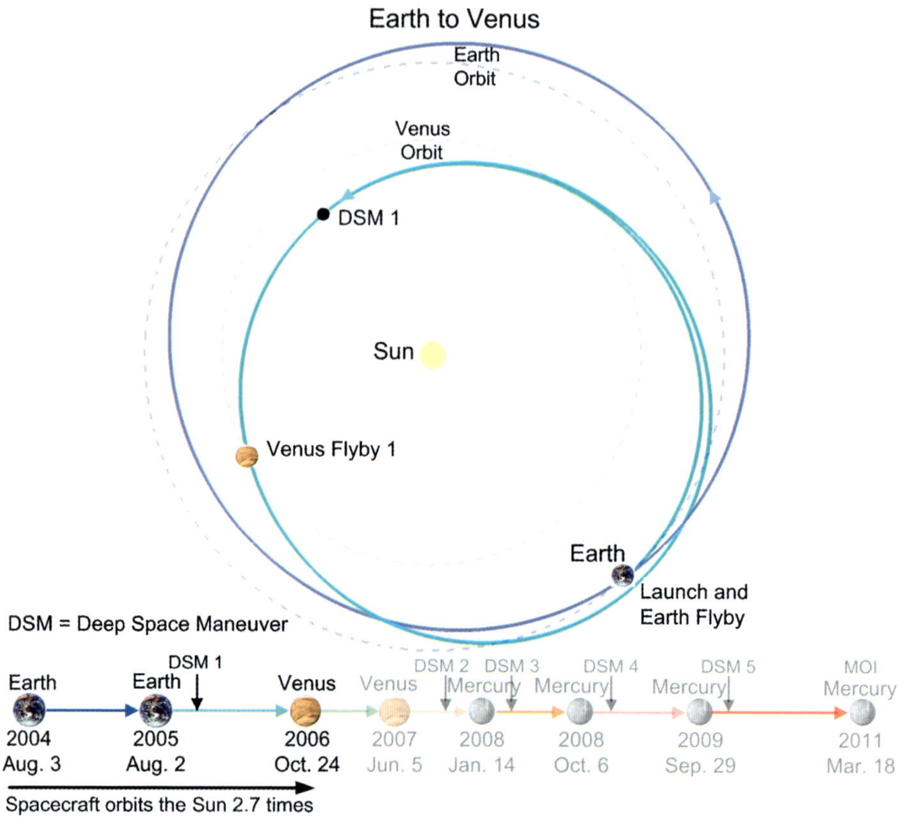

Fig. 5 North ecliptic pole view of the Earth-to-Venus flyby 1 trajectory. *Dashed lines* depict the orbits of Earth and Venus. *Timeline* fading helps emphasize primary events

Venus flyby and DSM-2 bring the spacecraft to Mercury at the right date and velocity to begin the "reverse delta-VEGA" process described in Sect. 2. The DSM-2 date was shifted earlier to place the maneuver before the mission's longest solar conjunction.

The Mercury flyby 1-to-Mercury orbit insertion transfer trajectory, shown in Fig. 7, includes three Mercury flyby-DSM segments that lower the spacecraft speed relative to Mercury while lowering spacecraft orbit period closer to Mercury's heliocentric orbit period. During this portion of the heliocentric transfer the longest solar conjunctions each last about two weeks. The spacecraft completes more than ten orbits of the Sun during this mission phase, or $2/3$ of the 15.3 orbits of the Sun between launch and MOI.

4.3 Planetary Flybys

If successful, MESSENGER will be the first spacecraft to use more than five planetary gravity-assist flybys. A detailed description of the trajectory-shaping contribution of each planetary flyby is given by McAdams et al. (2005). Without each planetary flyby, the spacecraft would require far too much propellant and a larger launch vehicle. Although impractical for a ballistic-trajectory Mercury orbiter, a direct Earth–Mercury transfer would take four to five months.

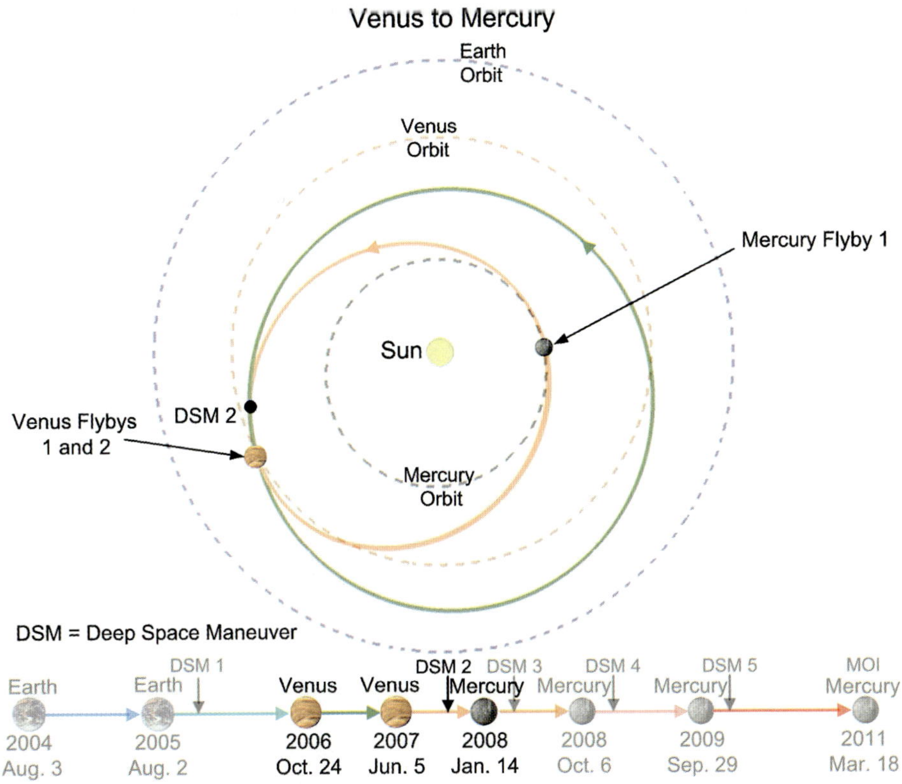

Fig. 6 North ecliptic pole view of the Venus flyby 1-to-Mercury flyby 1 trajectory. *Dashed lines* depict the orbits of Earth, Venus, and Mercury

One year after launch an Earth flyby (Fig. 8) lowered the spacecraft orbit's perihelion to 0.6 AU from the Sun and moved the perihelion direction more than 60° closer to Mercury's perihelion direction. Since the maximum DLA for the August 2005 launch period would be slightly higher than for August 2004, and since launch energy is nearly identical for the August 2004 and August 2005 launch periods, the Earth flyby enables the Delta II 7925H-9.5 to launch a slightly heavier spacecraft. The Earth flyby provides science instrument calibration opportunities using the Moon, thereby removing science observations from the early post-launch operations schedule. Close approach for Earth occurred about 2348 km over central Mongolia, high enough to avoid solar eclipse.

The first Venus flyby (Fig. 9) increased the spacecraft orbit's inclination and reduced the spacecraft's orbit period to almost exactly the heliocentric orbit period of Venus. The second Venus flyby will then occur at the same point in Venus' orbit 225 days later. As the spacecraft approached a brightly illuminated Venus, a 1.4° Sun–Earth–spacecraft angle limited the reliability of data transmission to or from the spacecraft near close approach. The spacecraft not only relied on battery power during a 56-minute solar eclipse just after the flyby but also received commands from flight controllers for only one day after the eclipse (until near the end of solar conjunction). The second Venus flyby (Fig. 9) is the first to lower perihelion enough to enable a Mercury flyby. However, use of a 313-km minimum altitude for the second Venus flyby leads to the need for a subsequent maneuver (DSM-2) to target the next planetary flyby (Mercury flyby 1). Each Venus flyby moves the spacecraft's

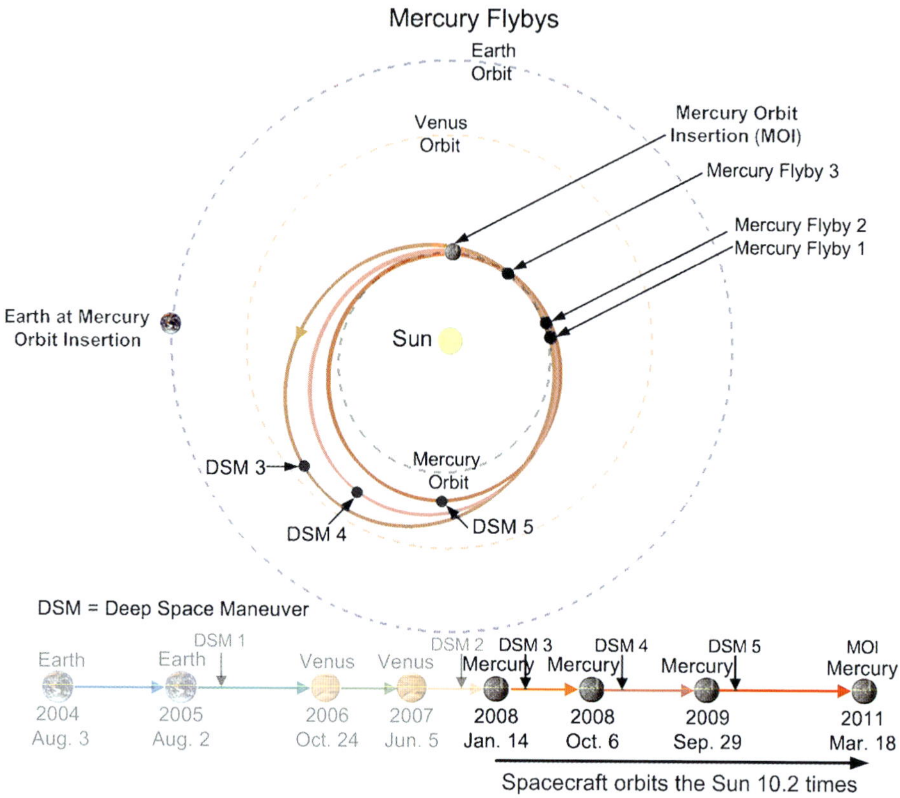

Fig. 7 North ecliptic pole view of the Mercury flyby 1-to-Mercury orbit insertion trajectory. *Dashed lines* depict the orbits of Earth, Venus, and Mercury

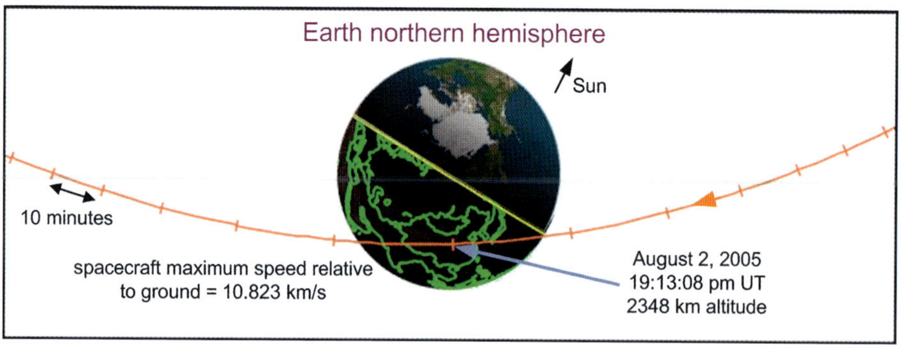

Fig. 8 View of the Earth flyby trajectory from above northern Asia. Major country borders are outlined with *green lines* on Earth's night side. The *yellow line* marks the position of the day/night or dawn/dusk terminator

aphelion and perihelion much closer to Mercury's perihelion and aphelion, thereby reducing the velocity change required for MOI.

By achieving progressively lower Mercury encounter velocities, the three Mercury flybys (Fig. 10) and subsequent DSMs are mission enabling. For direct transfers (no gravity-assist

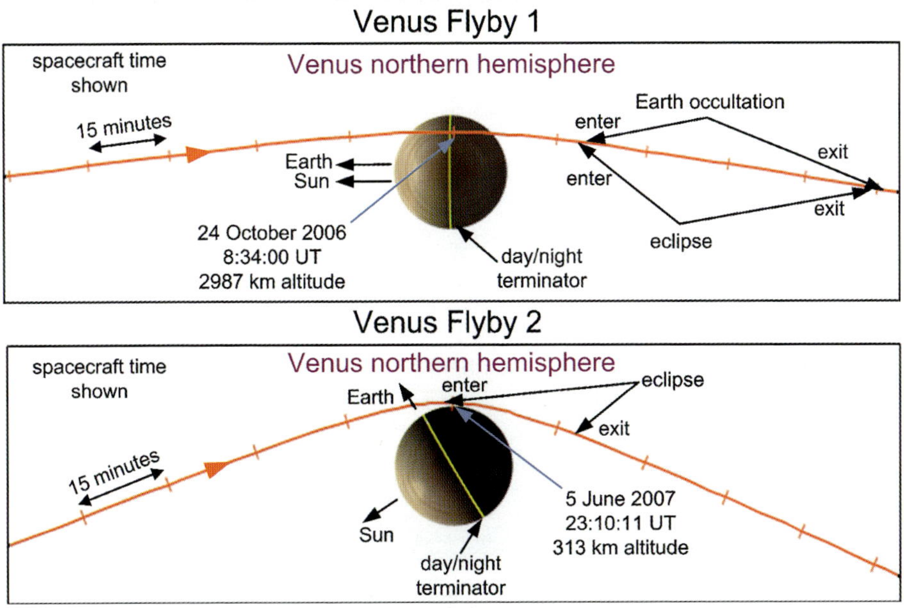

Fig. 9 View of both Venus flyby trajectories from above Venus' north pole

Table 3 Mercury encounter summary

Event	Phase, C/A[1] – 1 day (deg)	Phase, C/A + 1 day (deg)	V_∞ (km/s)	Sun–Earth–S/C (deg)	Earth range (AU)
Mercury 1	117	51	5.817	16.5	1.157
Mercury 2	127	36	5.177	2.3	0.659
Mercury 3	104	40	3.381	15.4	0.798
MOI	94	–	2.201	17.4	1.029

[1] Phase is the Sun–Mercury center-spacecraft angle. Close approach (C/A) denotes the minimum spacecraft–Mercury altitude

flybys) from Earth to Mercury at a minimum launch energy near 50 km^2/s^2, the MOI ΔV is about 10 km/s. Accounting for trajectory shaping by both Venus flybys, the MOI ΔV required for zero, one, two, and three Mercury flybys is at least 3.13 km/s, 2.40 km/s, 1.55 km/s, and 0.86 km/s, respectively. Trajectories utilizing fewer than two Mercury gravity assists would overheat the spacecraft during MOI and the following orbit phase.

Three 200-km minimum-altitude Mercury flybys, each followed by DSMs near the first aphelion after each flyby, are needed to reduce the Mercury arrival velocity enough to enable MOI in March 2011. The Mercury flybys and subsequent DSMs will yield successive orbits having spacecraft : Mercury orbital resonances of about 2 : 3, 3 : 4, and 5 : 6 (i.e., the spacecraft orbits the Sun five times while Mercury completes six orbits). Table 3 shows how this strategy reduces spacecraft hyperbolic excess velocity relative to Mercury, also known as V_∞.

Figure 10 shows how the spacecraft can observe opposite sunlit sides of Mercury's never-before-imaged hemisphere soon after close approach. With 1.5 Mercury solar days (176

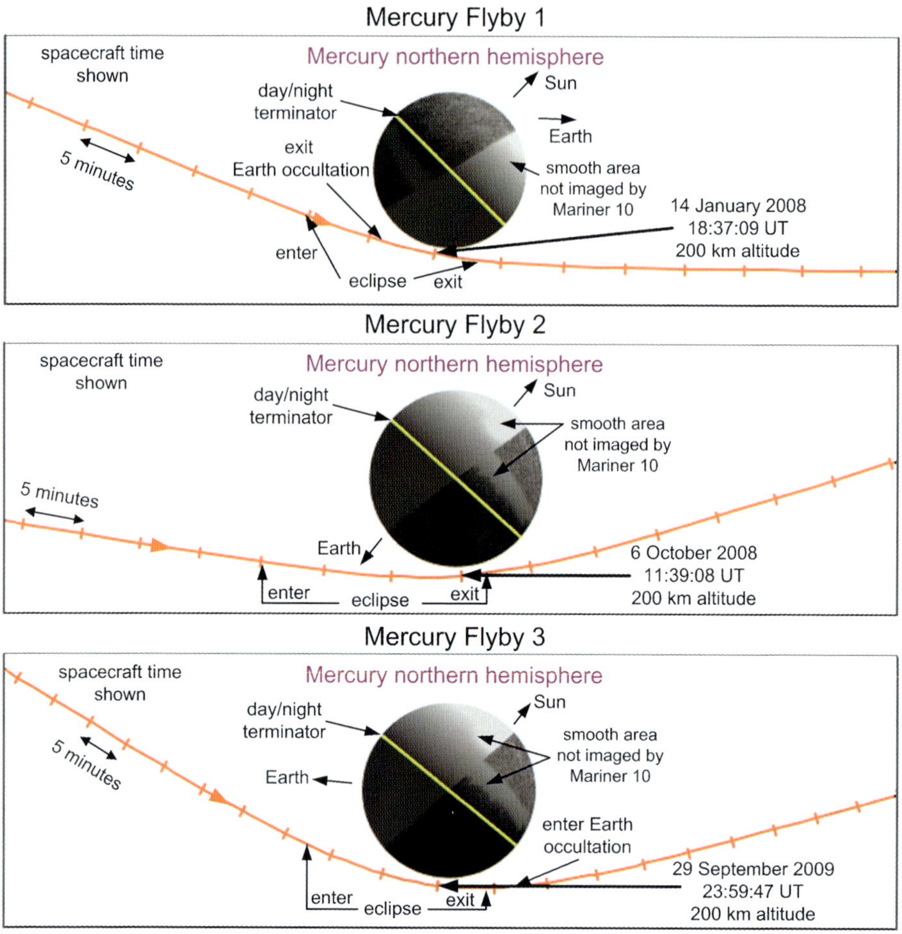

Fig. 10 View of all Mercury flyby trajectories from above Mercury's north pole

Earth days per Mercury solar day) between the first two Mercury flybys, the spacecraft will view opposite sunlit hemispheres. With two Mercury solar days between the second and third Mercury flybys, the same hemisphere is sunlit during these flybys. Figure 11 shows the timing and geometry for the Mercury flybys with Earth occultations.

4.4 Trajectory Correction Maneuvers

In order to remain on course for MOI, the MESSENGER spacecraft will need to perform five deterministic (i.e., characteristics are known with reasonable uncertainty prior to launch) maneuvers with ΔV magnitude >40 m/s (DSMs), two much smaller deterministic maneuvers, and numerous statistical TCMs. The statistical TCMs will correct errors associated with maneuver execution, planetary flyby aim point, and trajectory perturbation force models. The more efficient LVA bipropellant thruster will be the primary thruster for each DSM. Most maneuvers requiring ΔVs from 3 to 20 m/s with Sun–spacecraft–ΔV angle between 78° and 102° will use the four 26-N C thrusters as primary thrusters. These four thrusters also serve as attitude control thrusters during DSMs. A pair of 4-N thrusters mounted on the

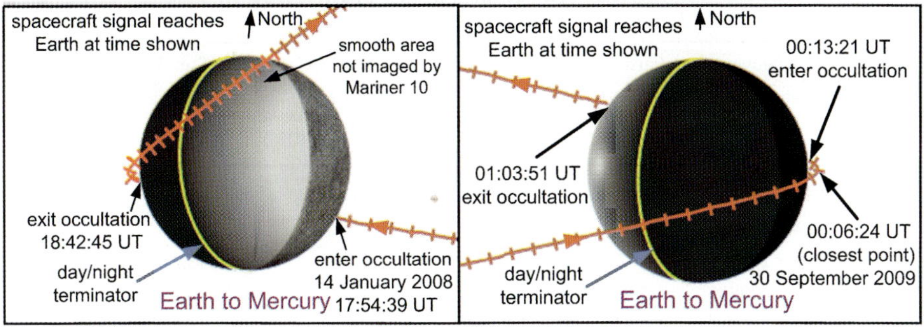

Fig. 11 Earth view of Mercury flyby trajectories occulted by Mercury

Table 4 ΔV budget after the first 35% of the heliocentric cruise phase

Maneuver category	ΔV (m/s)
Deep-space maneuvers	1032
Launch vehicle errors, navigation (99%)	115
Mercury orbit insertion	862
Mercury orbit-correction maneuvers	84
Contingency	136
Total	2229

spacecraft's sunward or anti-Sun side will provide the needed course correction when ΔV direction is within 12° of the Sun-to-spacecraft direction. Maneuvers with ΔV directions outside the above guidelines will require vector components from a combination of two of the above maneuver types.

The ΔV budget for MESSENGER is shown in Table 4. The 21 m/s reduction in ΔV capability (2,250 m/s in Table 2) since launch is attributable to increases in corrections after the first Venus flyby, propulsion system performance updates, and a six-day shift in DSM-2. While it is ideal to direct the spacecraft along a "minimum ΔV" trajectory, the MESSENGER mission will perform contingency plans and risk mitigation studies for each major propulsive maneuver. The ΔV budget for launch vehicle errors and navigation is a 99th percentile value derived from Monte Carlo analyses conducted by the navigation team. Since this ΔV budget category does not include the Mercury orbit phase, the "contingency" category includes variations in ΔV for MOI and OCMs.

MESSENGER's DSMs help target the spacecraft during the Earth–Venus 1, Venus 2–Mercury 1, Mercury 1–Mercury 2, Mercury 2–Mercury 3, and Mercury 3–MOI legs. The first two DSMs complete the partial targeting from the previous planetary flyby to deliver the spacecraft to the required aim point at the next planetary flyby. The last three DSMs move the next Mercury encounter closer to Mercury's MOI location. The pre-perihelion DSM-1 increases the spacecraft's Sun-relative speed, thereby raising aphelion and setting the Venus flyby 1 arrival conditions. The first DSM near aphelion, DSM-2, reduces the spacecraft's Sun-relative speed, lowering perihelion enough to target the first Mercury flyby. Placement of DSM-2 7.5 days before solar conjunction entry provides opportunity for recovery from either delays or errors, while limiting direct sunlight exposure on the LVA thruster. The ΔV magnitude for DSM-3 to DSM-5 (Table 5) is directly proportional to the change in

Table 5 Deep-space maneuvers during heliocentric transfer

Name	Maneuver date	Earth range (AU)	Sun range (AU)	Sun–S/C–ΔV (deg)	Sun–Earth–S/C (deg)	ΔV (m/s)
Requirement →				(78° to 102°)	(>3°)	
DSM-1	Dec 2005	0.688	0.604	92.4	37.5	315.6
DSM-2	Oct 2007	1.670	0.680	73.6	4.1	223.7
DSM-3	Mar 2008	0.677	0.685	87.4	43.3	73.4
DSM-4	Dec 2008	1.596	0.626	89.1	6.3	241.8
DSM-5	Nov 2009	1.529	0.565	89.5	7.4	177.5

Mercury's position (Fig. 7) between the previous and next Mercury encounters. These final three DSMs shift the upcoming Mercury encounter position counterclockwise with small increases in the spacecraft's Sun-relative speed.

5 Mercury Orbit Phase

By the time MESSENGER completes final preparations for both the mission-critical Mercury orbit insertion and the Mercury orbit phase, the mission design and navigation teams will be ready to conduct maneuver design, trajectory optimization, orbit determination, landmark tracking, and Mercury gravity field determination activities. A carefully devised Mercury approach navigation and TCM plan will precede a two-part conservative plan for Mercury orbit insertion. After Mercury orbit insertion, many factors such as early orbit determination and spacecraft health assessments will offer clues concerning what, if any, adjustment may be necessary to the orbit phase flight plan.

5.1 Mercury Orbit Insertion

The MESSENGER spacecraft's initial primary science orbit is required to have an 80° (±2°) orbit inclination relative to Mercury's equator, 200-km (±25 km) periapsis altitude, 12-hour (±1 minute) orbit period, 118.4° argument of periapsis (60°N periapsis latitude with 56°N to 62°N acceptable), and a 348° (169° to 354° acceptable) longitude of ascending node. These requirements are expressed in Mercury-centered inertial coordinates of epoch January 1.5, 2000.

The MOI strategy is a low-risk approach for placing the spacecraft into the primary science orbit as soon as possible within mission planning process constraints. This strategy uses two "turn while burning" variable thrust-direction maneuvers (MOI-1 and MOI-2) with the bipropellant LVA thruster providing the vast majority of each ΔV. To allow sufficient time for a rapid MOI-1 performance assessment and some post-MOI-1 orbit determination, the spacecraft will complete four full 16-hour orbits of Mercury (Fig. 12) between MOI-1 and MOI-2. For each MOI maneuver the thrust vector is nearly opposite to the spacecraft velocity vector. The time of MOI-2 is chosen to place daily data downlink periods (beginning about 40 minutes after periapsis on every other 12-hour orbit) between 8:10 am and 4:10 pm EDT (Eastern Daylight Time, 12:10 to 20:10 UT). During the second half of each Mercury year (e.g., 45 to 88 days after MOI-2) the daily downlink period shifts to 10:50 pm to 6:50 am EDT (02:50 to 10:50 UT). Even with time zone changes twice per year in

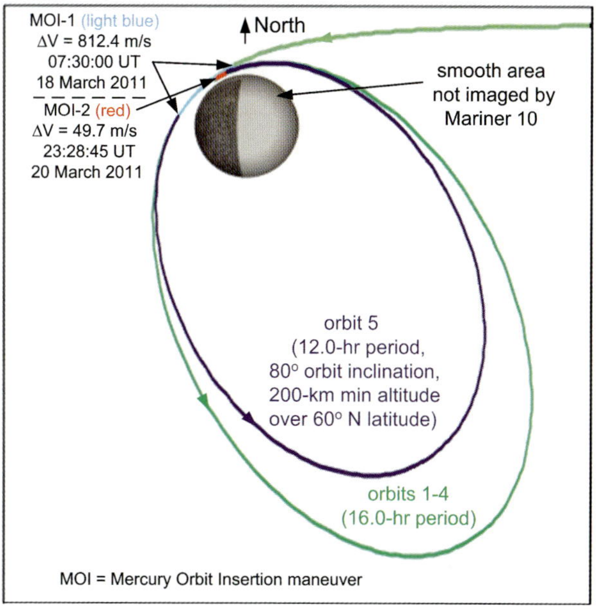

Fig. 12 View from the Sun of both Mercury orbit insertion maneuvers and the initial Mercury orbits. The longer-duration MOI-1 lasts nearly 14 minutes

the United States, and with time variations to the spacecraft's apoapsis introduced between OCM pairs, the daily downlink times remain close to standard daylight working shift hours for the JHU/APL-based (eastern United States) mission operations team about 50% of the time. A Sun–Earth–spacecraft angle $>17°$ ensures that solar interference will not disrupt spacecraft communications during the orbit insertion process. In addition, the spacecraft attitude enables a low-gain antenna to maintain contact between the spacecraft and Earth throughout both MOI maneuvers.

With MOI occurring when Mercury is near its perihelion, Mercury's rapid heliocentric orbital motion would, without a correction, quickly rotate the Sun-relative spacecraft orbit orientation (which moves far more slowly) until the sunshade is unable to shield the spacecraft bus at the required MOI-2 maneuver attitude. Although MOI-2 can occur less than three days after MOI-1, adding a small MOI-2 ΔV component normal to the orbit plane will meet the sunshade protection (Sun–spacecraft–ΔV angle range) constraint but will have little effect on orbit characteristics such as inclination.

5.2 Orbit Phase Navigation

A preliminary navigation accuracy analysis has shown that the planned DSN tracking coverage for MESSENGER's orbit provides sufficient navigation performance and margin throughout all mission phases, including the orbit phase. The preliminary navigation studies for the orbit phase used DSN Doppler and ranging that were conservatively weighted at 0.5 mm/s and 30 m, respectively, according to a DSN schedule of ten 8-hour tracks per week. Current DSN coverage includes 12 tracks per week. In addition, optical landmark tracking (planned during the orbit phase) was not included in the results below. Optical images of landmarks (e.g., craters) from MDIS enhance navigation performance by improving accuracy in the gravity field determination and by decreasing the time needed after OCMs or

Table 6 MESSENGER orbit uncertainties (1-σ) for a 10-day arc solution[1]

Orbit phase	Cross track (km)	Down track (km)	Out of plane (km)
Post Mercury orbit insertion	2	7	1
Mercury orbit (face-on viewing geometry)	17	100	5
Mercury orbit [2] (edge-on viewing geometry)	17	32	10

[1] Uncertainties include the effect of gravity uncertainties and gravity tuning, mapped to 10 days after the end of the arc

[2] Includes data outage due to spacecraft occultations by Mercury

momentum dumps for precise orbit determination. This navigation technique was first used successfully on the Near Earth Asteroid Rendezvous project (Miller et al. 2002), the first asteroid rendezvous mission.

Table 6 shows navigation orbit uncertainties in three orthogonal directions along the orbit: cross track (radial at periapsis and apoapsis), down track (in the direction of the orbit velocity vector), and out of plane (normal to the plane of the orbit). Table 6 includes uncertainties for three different orbit phases. The phases chosen for the study included the orbits immediately following MOI when the navigation team first tunes a model of the Mercury gravity field, the period where the orbits are nearly face-on when viewed from the Earth (near zero inclination to the plane-of-sky), and the period where the orbits are nearly edge-on when viewed from the Earth (near 90° inclination to the plane-of-sky).

The two viewing geometries represent near worst-case viewing angles for DSN Doppler tracking data, so these cases help bound the expected errors over the MESSENGER orbit. When viewing the orbit nearly face-on, the Doppler data are less sensitive to errors in the down-track directions because the line-of-sight to Earth is almost normal to the orbit velocity. This effect is seen in Table 6, where the down-track error for the face-on geometry is larger than that for the other two viewing angles. When viewing the orbit nearly edge-on, the Doppler data are less sensitive to errors in the out-of-plane direction, as the last column in Table 6 indicates. For other intermediate viewing angles to the plane-of-sky, experience has shown that the orbit errors are generally less than those at the bounding cases for face-on or edge-on geometry. Note that all three cases in Table 6 assume the same starting a priori gravity uncertainties. In reality the last two cases will occur some time after MOI; so some gravity tuning will have occurred and the errors should be less.

For the typical MESSENGER periapsis velocity of 3.8 km/s, the down-track errors in Table 6 imply a 1-σ (worst orbit viewing case) navigation timing uncertainty of about 26 s. A typical timing uncertainty will probably be less than 10 s (1-σ). Even when combined with maneuver execution error, the result is within the requirement for maintaining the MESSENGER orbit period to ± 1 minute (2-σ). The cross-track uncertainties at periapsis indicate the uncertainty in periapsis altitude. Note that the cross-track uncertainties from Table 6 meet the requirement to control periapsis altitude between 125 km and 225 km (3-σ) after orbit insertion and after each maneuver to lower periapsis to within 60 km (3-σ) of the nominal target altitude of 200 km.

5.3 Gravity Tuning for Navigation

The navigation strategy for the orbit about Mercury includes gravity estimation and localized tuning of a twentieth degree and order spherical harmonic field in addition to the usual spacecraft dynamic parameters and orbit state parameters. Such a representation of the gravity field is at higher degree and order than can be determined from the available tracking so that high-frequency signatures in the Doppler data will not alias the estimates of the low degree and order harmonic coefficients. The uncertainties shown in Table 6 include the a priori gravity uncertainties derived from a scaled lunar field. Since sub-spacecraft periapsis longitude on Mercury moves 360° in 59 days, it will take at least this long to obtain measurements for a global gravity model; hence, determining a series of locally tuned gravity models will be part of navigation operations during the initial orbit phase.

The navigation strategy for orbit determination when MESSENGER is first in Mercury orbit is to perform incremental gravity field tuning as periapsis moves through 360° in longitude. A complete circulation of Mercury by the periapsis point takes about 118 orbits. During this time, the navigation team will estimate spacecraft state and other parameters along with a harmonic gravity field over 10-day-long arcs of Doppler and range data. The harmonic coefficient estimates from these fits will be highly correlated, but experience with other planetary orbiters has shown that this technique fits the short-period gravity perturbations with the least amount of aliasing (Christensen et al. 1979; Williams et al. 1983). The gravity field harmonic coefficients from each 10-day arc represent a local fit to the particular orbit data (especially periapsis) and are not valid globally.

This short-arc technique of gravity field estimation uses the Square Root Information Factorization (SRIF) method to separate arc-dependent parameter information from the gravity field information. This technique was developed and used on early planetary orbiters when the planet gravity field was mostly unknown and the sampling was sparse due to high-eccentricity orbits. Examples are Vikings I and II at Mars (eccentricity ~0.8), Pioneer Venus (eccentricity ~0.8), and Magellan (eccentricity ~0.4). These early planetary orbiter missions had relatively large orbit eccentricities, in most cases even larger than that of MESSENGER (eccentricity ~0.7).

Once solutions are available from several 10-day arcs, that information will be combined with subsequent 10-day arcs to "bootstrap" a gravity solution that is valid for MESSENGER over an increasing range of longitudes. This technique extracts both short-period information from the velocity change at periapsis and longer-period orbit evolution information over the 10-day arc. As more data are accumulated, longer or shorter arc fits will be evaluated and included if appropriate to obtain the best navigation gravity model on an ongoing basis. After the periapsis has completed a circulation of Mercury, a tuned gravity solution will be obtained by combining tracking data from the first 118 orbits. This first circulation solution produces a locally tuned field for navigation that will be best determined for a band of latitudes around the mean periapsis latitude for that circulation. Gravity tuning will continue on subsequent data arcs and subsequent circulations of Mercury by the orbit periapsis. Each subsequent gravity tuning will use the a priori gravity information from the previous "global" field solution to constrain the final gravity solution.

A covariance analysis was also performed on a typical 10-day arc in the orbit phase by assuming a twentieth degree and order lunar field scaled to Mercury by the ratio of the radii. The resulting gravity field harmonic coefficients were assumed to have an a priori uncertainty of 50%. After processing 10 days of Doppler and range data, the low degree and order terms (up to degree and order 3) improved by factors of 2 to 3, while the higher degree and order terms did not improve much ($<1\%$) and were highly correlated. The results should improve as data arcs are combined and a more globally valid gravity field estimate emerges.

5.4 Orbit Correction Maneuvers

After entering the initial primary science orbit, the spacecraft coasts for more than 12 weeks without OCMs. This is sufficient time to refine Mercury's gravity model and update perturbing force models in order to reduce trajectory propagation errors. Occasional thruster firings for adjusting spacecraft angular momentum will perturb the trajectory by at most a few mm/s of unintentional ΔV.

Each pair of OCMs will return the spacecraft to the initial primary science orbit's size and shape. Solar gravity, solar radiation pressure, and spatial variations in Mercury's gravity field will move periapsis north, increase orbit inclination, and rotate the low-altitude descending node away from the Sun (relative to the orbit orientation at Mercury's perihelion). The first OCM of each pair will impart a ΔV parallel to the spacecraft velocity direction at periapsis, placing the spacecraft on a transfer orbit (Fig. 13) with apoapsis altitude matching the apoapsis altitude of the 200-km periapsis altitude by 12-hour orbit. Two-and-a-half orbits (30.6 hours) later, at apoapsis, the second OCM of each pair lowers periapsis altitude to 200 km and returns orbit period to 12 hours with a much larger (primary thruster is the LVA) ΔV opposite the spacecraft velocity direction. Selecting 2.5 orbits between the OCMs provides enough time to assess the first OCM's performance, determine the new orbit, design the next maneuver, and test, upload, verify, and enable the ΔV update. Keeping periapsis altitude below 500 km while meeting the sunshade orientation requirement and science constraints requires that OCM pairs occur once per 88-day Mercury year. Each OCM pair occurs about 0.31 AU from the Sun, when the spacecraft orbit's line of nodes (connecting the spacecraft orbit's southern and northern equatorial plane crossings) is nearly perpendicular to the spacecraft–Sun direction. Table 7 provides date, ΔV, and spacecraft orientation data for all six planned OCMs.

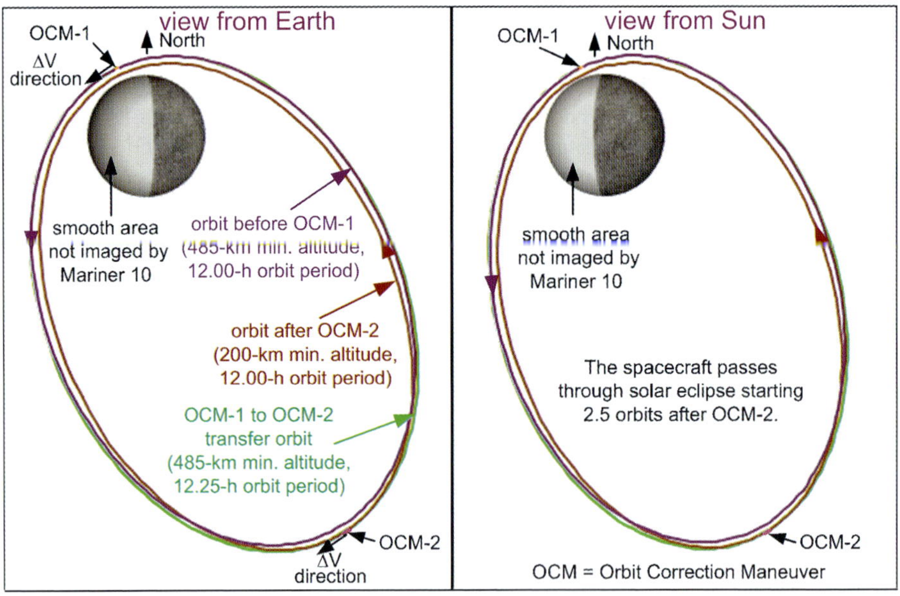

Fig. 13 View from the Earth and Sun of the first two orbit correction maneuvers, along with the preceding and subsequent spacecraft orbits

Table 7 Orbit correction maneuvers meet operational constraints

Orbit correction maneuver segment	Maneuver date (year month day)	ΔV (m/s)	Earth–S/C range (AU)	Sun–S/C–ΔV max. angle (deg)	Sun–Earth –S/C angle (deg)
Requirement →				(78° to 102°)	(>3°)
OCM-1 start	2011 Jun 15	4.22	1.319	96.1	3.3
OCM-2 start	2011 Jun 16	26.35	1.314	99.6	4.8
OCM-3 start	2011 Sep 09	3.92	1.097	86.0	16.1
OCM-4 start	2011 Sep 10	24.18	1.129	93.8	15.3
OCM-5 start	2011 Dec 05	3.59	0.683	82.1	3.2
OCM-6 start	2011 Dec 06	22.22	0.693	89.9	6.0

Deterministic ΔV for orbit phase = 84.48 m/s

5.5 Orbit Evolution

During the Mercury orbital phase the mission design and navigation teams will incorporate knowledge of past spacecraft attitude and predictions of spacecraft attitude to provide highly accurate orbit propagation and design of future OCMs. Trajectory perturbations caused by solar pressure, Mercury gravity field variation, solar gravity, end-of-life sunshade surface reflectance, and sunlight reflected off Mercury's surface (albedo effect) will be carefully coordinated with the spacecraft's planned attitude profile. All of these factors (except for Mercury albedo) are accounted for during Mercury orbit-phase propagations.

Solar gravity and Mercury oblateness, J_2, are major contributors to periapsis altitude increasing to 485, 467, and 444 km before OCM-1, -3, and, -5, respectively; periapsis latitude drifting north by $\sim$12°; orbit inclination increasing by about 2°; and longitude of ascending node (orbit plane-to-spacecraft–Sun line orientation) decreasing 6°. Figures 14 and 15 show the variation of solar eclipse duration, periapsis altitude, and periapsis latitude during the nominal one-year Mercury orbit phase.

Current orbit phase design uses a propagator that assumes a Mercury gravity model with normalized coefficients $C_{20} = -2.7 \times 10^{-5}$ and $C_{22} = 1.6 \times 10^{-5}$ (Anderson et al. 1987). Spacecraft attitude rules assume a daily 8-hour downlink period, up to 16 hours of science observation (sunshade toward the Sun with $+z$ aligned with Mercury nadir when possible) each day, and solar array tilt varying as a function of solar distance (Fig. 3).

6 Mission Status

For the first 26 months after launch, Mission Design and Navigation have not encountered any significant change from the initial schedule of post-launch activities. The excellent functional health of the spacecraft has enabled the accomplishment of nominal pre-launch planned activities such as maneuver design, trajectory optimization, orbit determination, and solar radiation pressure model adjustment. Careful planning and testing of the maneuver prior to implementation produced six highly successful TCMs with average ΔV errors of 1.2% in magnitude and 1.0° in direction. In addition to the navigation and mission design teams, the guidance and control and mission operations teams have significant responsibility in the maneuver design process.

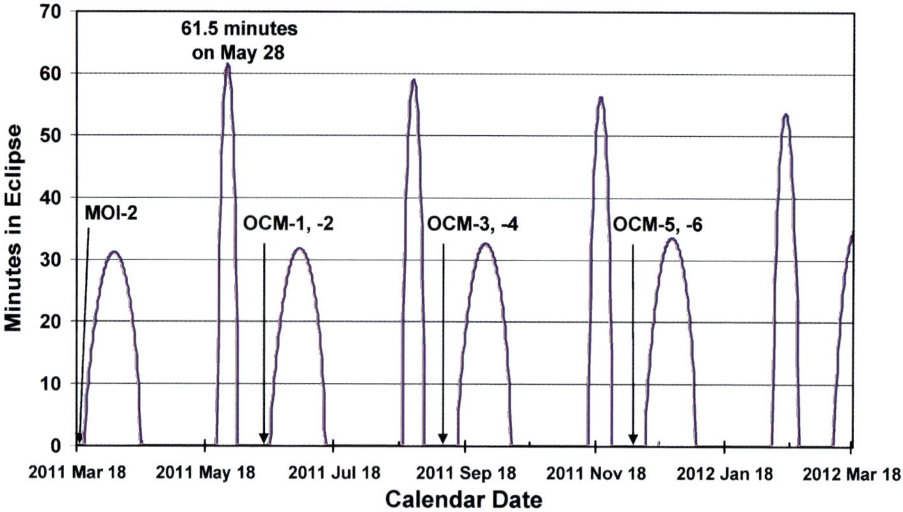

Fig. 14 Solar eclipse duration (short eclipses include periapsis) during Mercury orbit. Maximum eclipse duration decreases as periapsis latitude moves north

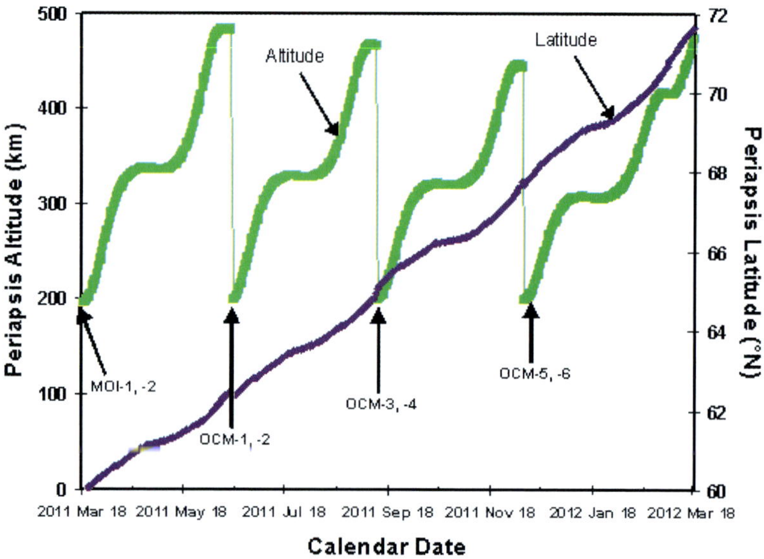

Fig. 15 Periapsis evolution during Mercury orbit phase

6.1 Launch and Early Operations

After launch, the goal of the navigation team has been to establish that the MESSENGER spacecraft is on the proper trajectory to return to the optimal Earth flyby condition. OD analysis is performed to calibrate the non-gravitational models for the spacecraft, and trajectory maneuvers are designed, executed and reconstructed to ensure that the predicted trajectory reaches the target. Prelaunch statistical analysis determined that the 99th-percentile launch error could be as large as 38 m/s, but after the first couple of days of DSN tracking were processed, the total correction required was only 21 m/s.

To account for the effects of solar radiation pressure (SRP), the navigation team uses a model made up of ten idealized flat plates that are oriented so as to approximate the SRP characteristics of the MESSENGER spacecraft bus (six plates) and the two articulating solar arrays (four plates for front and back of the independently pointed solar panels). During the OD process, scale factors for each of the plates in the SRP model are estimated along with the other spacecraft and observation model parameters. The SRP scale factor estimates depend on the spacecraft orientation model, which is driven by the attitude history and prediction file provided by the Mission Operations Center. The initial post-launch spacecraft attitude oriented the $+y$ spacecraft axis (opposite the sunshade) with the sunshade pointed away from the Sun. The spacecraft orientation has varied since launch, including an early slow spin about the Sun direction, and also subsequent pointing with small attitude rates about the Sun line to keep the appropriate antenna pointed toward Earth.

Solar radiation pressure scale factor estimates during the first few OD deliveries were about ninety percent of their a priori values. A scale factor of 1.0 indicates that the SRP model for that component is 100% effective. After a few weeks of data collection, however, the estimates began changing with data arc length and became sensitive to data weighting experiments. This result indicated a possible corruption of the solution due to the presence of un-modeled acceleration; so the navigation team began an analysis to check for early outgassing accelerations that are common immediately after launch. Figure 16 shows the effect of delaying the beginning of the data arc on the SRP scale factor estimates for the parts of the SRP model illuminated during cruise with the spacecraft $+y$ axis oriented toward the Sun. These are flat plate Area6, representing the $+y$ projection of the spacecraft main body, and Area7 and Area8 representing the front side of the two solar panels. The solutions tend

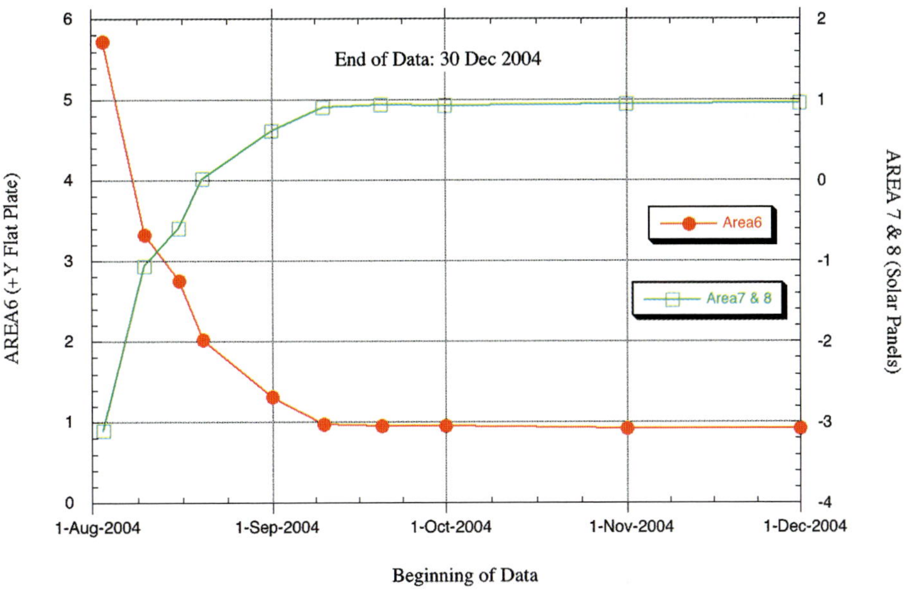

Fig. 16 Change in solar radiation pressure scale factor as a function of cutoff date for trajectory data. Effective area change prior to 10 September 2004 indicates that other non-gravitational accelerations (such as out-gassing) are present

to stabilize if the data arc excludes all data before about 10 September 2004, indicating that other un-modeled accelerations present before that time are aliasing into estimates of the scale factors. With such exclusion, the SRP acceleration estimates for the illuminated flat plates stabilized at between 90 and 96% of their pre-launch a priori values.

6.2 Maneuver Performance

Maneuver reconstruction results for the first seven TCMs appear in Table 8. The total ΔV expended thus far includes the 21 m/s required to correct launch injection errors and deterministic maneuvers in November 2004 and December 2005. By restricting the maximum size of the first burn to 18 m/s, more time became available for analysis and checkout of the

Table 8 Maneuver reconstruction for the first seven trajectory correction maneuvers

Parameter	Estimated	Planned	Reconstructed uncertainty (1-σ)	Difference from planned
TCM-1 at 21:00:07 UT, 24 August 2004, spacecraft event time				
ΔV (m/s)	17.9009	18.0000	0.0017	−0.099 (−0.55%)
RA (deg)	242.5414	242.8288	0.0032	−0.2874
Dec (deg)	−18.4960	−18.6426	0.0032	0.1466
TCM-2 at 18:00:00 UT, 24 September 2004, spacecraft event time				
ΔV (m/s)	4.5886	4.5899	0.0024	−0.001 (−0.03%)
RA (deg)	271.1789	271.3752	0.0096	0.1963
Dec (deg)	−24.1171	−24.3248	0.0171	0.2077
TCM-3 at 19:30:00 UT, 18 November 2004, spacecraft event time				
ΔV (m/s)	3.2473	3.2365	0.0006	0.011 (0.33%)
RA (deg)	315.7591	316.0507	0.0030	−0.2916
Dec (deg)	−15.2193	−15.4143	0.0152	0.1950
TCM-5 at 14:30:00 UT, 23 June 2005, spacecraft event time				
ΔV (m/s)	1.1033	1.1451	0.0011	−0.0418 (−3.65%)
RA (deg)	239.3472	239.6439	0.0438	−0.2967
Dec (deg)	3.7469	3.9750	0.0391	−0.2281
TCM-6 at 18:00:00 UT, 21 July 2005, spacecraft event time				
ΔV (m/s)	0.1505	0.0002	0.1468	0.0037 (2.51%)
RA (deg)	255.6900	252.1045	0.1548	3.5854
Dec (deg)	3.4969	6.3595	0.1503	−2.8626
TCM-9 at 11:30:00 UT, 12 December 2005, spacecraft event time				
ΔV (m/s)	315.6334	315.7200	0.0004	−0.0866 (−0.03%)
RA (deg)	217.2755	217.2570	0.0001	0.0185
Dec (deg)	−4.8828	−4.8652	0.0004	−0.0177
TCM-10 at 16:00:00 UT, 22 February 2006, spacecraft event time				
ΔV (m/s)	1.2807	1.4071	0.0005	−0.1264 (−8.98%)
RA (deg)	166.2108	168.8002	0.1273	−2.5894
Dec (deg)	9.2538	9.2222	0.3218	0.0316

[1] ΔV spherical angles, right ascension (RA), and declination (Dec) are in Earth mean equator of J2000 (1 Jan 2000 at 12:00:00 ET) coordinates

spacecraft systems prior to the completion of launch injection error correction. The design of the first maneuver, TCM-1, was split into two burns: the first part on 24 August 2004, and the remainder of about 4 m/s (renamed as TCM-2) on 24 September 2004. The first deterministic maneuver, re-optimized for execution on 18 November 2004, was renamed TCM-3. Contingency maneuvers before the Earth flyby (TCM-4 and TCM-7) and after the Earth flyby (TCM-8) were cancelled because it was more prudent to wait until the next planned TCM. TCM-5 and TCM-6 provided precise targeting for the 2 August 2005 Earth flyby. TCM-9 (also called DSM-1) included a 1-m/s over burn bias (compared with the optimal design) in order to maximize the potential for having a small clean-up maneuver (TCM-10) at a favorable spacecraft attitude. Additional maneuvers completed during the year 2006 include the 2.28-m/s TCM-11 (which used two components to maintain sunshade orientation) on September 12 and 0.50-m/s TCM-12 on October 5 for targeting the first Venus flyby. The final maneuver during 2006, the 36-m/s TCM-13 on December 2, corrected target offsets at the first Venus flyby and directed the spacecraft to a re-optimized aim point at the second Venus flyby. The co-location of solar conjunction ($<3^\circ$ Sun–Earth–spacecraft angle) from 18 October to 17 November 2006 and a 2987-km-altitude Venus flyby on 24 October 2006 contributed to the need for a large correction at TCM-13.

The time shown for each TCM is the actual maneuver execution time on the spacecraft. Due to initialization and fuel settling burns, the onset of acceleration observed in the received Doppler tracking data is typically delayed by several seconds. The reconstruction accounted for these delays when modeling the TCM acceleration in the OD filter. Table 8 compares the estimated ΔV magnitude and direction, and their 1-σ uncertainty, to the planned design values.

For most maneuvers performed thus far, the ΔV magnitude is within 0.1 m/s of the goal and the direction is within a few tenths of a degree. Not shown in Table 8 is the total directional error for each TCM, which for TCM-1 was 0.309°, for TCM-2 was 0.274°, for TCM-3 was 0.342°, for TCM-5 was 0.374°, for TCM-6 was 4.577°, for TCM-9 was 0.026°, and for TCM-10 was 2.556°. The largest TCM performance variations (difference between the achieved and goal ΔV vectors) are associated with use of 4-N thrusters as primary thrusters.

7 Summary

A little over four decades after the first Mercury mission trajectory design studies, and about three decades after Mariner 10's three Mercury flybys, MESSENGER, the first Mercury orbiter mission, launched safely on 3 August 2004. While there are many innovative trajectory options for a Mercury orbiter, only a small subset are compatible with the MESSENGER spacecraft configuration, science objectives, and operational constraints. With launch delayed to the third launch opportunity of 2004, the spacecraft will achieve most of the required trajectory alteration using flybys of Earth (one), Venus (two), and Mercury (three). Mercury orbit insertion on 18 March 2011 will mark the start of a one-year Mercury orbit phase.

The launch and early operations phase has been successful for the MESSENGER mission. The mission design and navigation teams have performed the planned post-launch trajectory re-optimization and designed the first ten trajectory correction maneuvers to deliver the MESSENGER spacecraft to the October 2006 and June 2007 Venus flybys and then on to Mercury. The cost of correcting the launch injection error (21 m/s) was less than the

99th percentile, worst-case correction computed pre-launch (~38 m/s). The early ΔV "savings" over what could have happened in the worst case, coupled with indications of precise maneuver capability, bode well for having adequate fuel for the remainder of the mission.

Acknowledgements The authors express gratitude to C.L. Yen of the Jet Propulsion Laboratory for designing the original heliocentric trajectory that became the basis for the current MESSENGER trajectory design. The MESSENGER mission is supported by the NASA Discovery Program under contracts to the Carnegie Institution of Washington (NASW-00002) and JHU/APL (NAS5-97271). MESSENGER navigation is carried out by the Space Navigation and Flight Dynamics Practice of KinetX, Inc., under subcontract with the Carnegie Institution of Washington.

References

B.J. Anderson et al., Space Sci. Rev. (2007, this issue)
J.D. Anderson, G. Colombo, P.B. Esposito, E.L. Lau, G.B. Trager, Icarus **71**, 337–349 (1987)
G.B. Andrews et al., Space Sci. Rev. (2007, this issue)
R.D. Bourke, J.G. Beerer, Astronaut. Aeronaut. **9**, 52–59 (1971)
J.F. Cavanaugh et al., Space Sci. Rev. (2007, this issue)
E.J. Christensen et al., J. Guid. Control **2**, 179–183 (1979)
M.I. Cruz, G.J. Bell, Acta Astronaut. **35**(Suppl.), 427–433 (1995)
R.L. Dowling, W.J. Kosmann, M.A. Minovitch, R.W. Ridenoure, Am. Astronaut. Soc. Hist. Ser., Hist. Rocket. Astronaut. **20**, 27–103 (1997)
J.A. Dunne, E. Burgess, *The Voyage of Mariner-10*, SP-424 (NASA, 1978), 222 pp.
C.J. Ercol, A.G. Santo, Dermination of optimum thermal phase angles at Mercury perihelion for an orbiting spacecraft. 29th International Conference on Environmental Systems, Society of Automotive Engineers, Tech. Paper Ser., 1999-01-21123, Denver, CO, 1999, 10 pp.
A.L. Friedlander, H. Feingold, Mercury orbiter transport study. Report No. SAI 1-120-580-T6, Science Applications, Inc., Schaumburg, IL, 1977, 85 pp.
J.O. Goldsten et al., Space Sci. Rev. (2007, this issue)
R. Grard, G. Scoon, M. Coradini, Eur. Space Agency J. **18**, 197–205 (1994)
S.E. Hawkins III et al., Space Sci. Rev. (2007, this issue)
G.R. Hollenbeck, D.G. Roos, P.S. Lewis, Study of ballistic mode Mercury orbiter missions. Report NASA CR-2298, Martin Marietta Corporation, Denver, CO, 1973, 104 pp.
Y. Langevin, Acta Astronaut. **47**, 443–452 (2000)
J.C. Leary et al., Space Sci. Rev. (2007, this issue)
J.V. McAdams, J.L. Horsewood, C.L. Yen, Discovery-class Mercury orbiter trajectory design for the 2005 launch opportunity. American Institute of Aeronautics and Astronautics/American Astronautical Society Astrodynamics Specialist Conference and Exhibit, paper AIAA-98-4283, Boston, MA, 1998, 7 pp.
J.V. McAdams, R.F. Farquhar, C.L. Yen, Adv. Astronaut. Sci. **109**, 2189–2204 (2002)
J.V. McAdams, D.W. Dunham, R.W. Farquhar, A.H. Taylor, B.G. Williams, Adv. Astronaut. Sci. **120**, 1185–1204 (2005)
W.E. McClintock, M.R. Lankton, Space Sci. Rev. (2007, this issue)
J.K. Miller et al., Icarus **155**, 3–17 (2002)
M. Minovitch, The determination and characteristics of ballistic interplanetary trajectories under the influence of multiple planetary attractions. JPL Technical Report 32-464, Jet Propulsion Laboratory, Pasadena, CA, 1963, 40 pp.
B.C. Murray, J. Geophys. Res. **80**, 2342–2344 (1975a)
B.C. Murray, Sci. Am. **233**(3), 58–68 (1975b)
B. Murray, *Journey into Space: The First Thirty Years of Space Exploration* (Norton, New York, 1989), 382 pp.
R.M. Nelson, L.J. Horn, J.R. Weiss, W.D. Smythe, Acta Astronaut. **35**(Suppl.), 387–395 (1995)
C.E. Schlemm II et al., Space Sci. Rev. (2007, this issue)
S.C. Solomon et al., Planet. Space Sci. **49**, 1445–1465 (2001)
F.M. Sturms, E. Cutting, J. Spacecr. Rockets **3**, 624–631 (1966)
B.G. Williams, N.A. Mottinger, N.B. Panagiotacopulos, Icarus **56**, 578–589 (1983)
H. Yamakawa, H. Saito, J. Kawaguchi, Y. Kobayashi, H. Hayakawa, T. Mukai, Acta Astronaut. **45**, 187–195 (2000)
C.L. Yen, J. Astronaut. Sci. **37**, 417–432 (1989)
C.L. Yen, Adv. Astronaut. Sci. **108**, 799–806 (2001)

Space Sci Rev (2007) 131: 247–338
DOI 10.1007/s11214-007-9266-3

The Mercury Dual Imaging System on the MESSENGER Spacecraft

S. Edward Hawkins, III · John D. Boldt · Edward H. Darlington · Raymond Espiritu · Robert E. Gold · Bruce Gotwols · Matthew P. Grey · Christopher D. Hash · John R. Hayes · Steven E. Jaskulek · Charles J. Kardian, Jr. · Mary R. Keller · Erick R. Malaret · Scott L. Murchie · Patricia K. Murphy · Keith Peacock · Louise M. Prockter · R. Alan Reiter · Mark S. Robinson · Edward D. Schaefer · Richard G. Shelton · Raymond E. Sterner, II · Howard W. Taylor · Thomas R. Watters · Bruce D. Williams

Received: 24 July 2006 / Accepted: 10 August 2007 / Published online: 23 October 2007

Abstract The Mercury Dual Imaging System (MDIS) on the MESSENGER spacecraft will provide critical measurements tracing Mercury's origin and evolution. MDIS consists of a monochrome narrow-angle camera (NAC) and a multispectral wide-angle camera (WAC). The NAC is a 1.5° field-of-view (FOV) off-axis reflector, coaligned with the WAC, a four-element refractor with a 10.5° FOV and 12-color filter wheel. The focal plane electronics of each camera are identical and use a $1{,}024 \times 1{,}024$ Atmel (Thomson) TH7888A charge-coupled device detector. Only one camera operates at a time, allowing them to share a common set of control electronics. The NAC and the WAC are mounted on a pivoting platform that provides a 90° field-of-regard, extending 40° sunward and 50° anti-sunward from the spacecraft $+Z$-axis—the boresight direction of most of MESSENGER's instruments. Onboard data compression provides capabilities for pixel binning, remapping of 12-bit data into 8 bits, and lossless or lossy compression. MDIS will acquire four main data sets at Mercury during three flybys and the two-Mercury-solar-day nominal mission: a monochrome

S.E. Hawkins, III (✉) · J.D. Boldt · E.H. Darlington · R.E. Gold · B. Gotwols · M.P. Grey · J.R. Hayes · S.E. Jaskulek · C.J. Kardian, Jr. · M.R. Keller · S.L. Murchie · P.K. Murphy · K. Peacock · L.M. Prockter · R.A. Reiter · E.D. Schaefer · R.G. Shelton · R.E. Sterner, II · H.W. Taylor · B.D. Williams
The Johns Hopkins University Applied Physics Laboratory, Laurel, MD 20723, USA
e-mail: ed.hawkins@jhuapl.edu

R. Espiritu · C.D. Hash · E.R. Malaret
Applied Coherent Technology, Herndon, VA 20170, USA

M.S. Robinson
School of Earth and Space Exploration, Arizona State University, Box 871404, Tempe, AZ 85287-1404, USA

T.R. Watters
Center for Earth and Planetary Studies, National Air and Space Museum, Smithsonian Institution, Washington, DC 20013, USA

global image mosaic at near-zero emission angles and moderate incidence angles, a stereo-complement map at off-nadir geometry and near-identical lighting, multicolor images at low incidence angles, and targeted high-resolution images of key surface features. These data will be used to construct a global image base map, a digital terrain model, global maps of color properties, and mosaics of high-resolution image strips. Analysis of these data will provide information on Mercury's impact history, tectonic processes, the composition and emplacement history of volcanic materials, and the thickness distribution and compositional variations of crustal materials. This paper summarizes MDIS's science objectives and technical design, including the common payload design of the MDIS data processing units, as well as detailed results from ground and early flight calibrations and plans for Mercury image products to be generated from MDIS data.

Keywords MESSENGER · Mercury · Imaging · Camera · Imager · CCD · Heat pipe · Wax pack · Photometry · Stereo

1 Introduction

Mariner 10, from its three flybys of Mercury in 1974–1975, provided a reconnaissance view of one hemisphere and measured the planet's magnetic field and interaction with the space environment. No spacecraft has returned in the intervening 30 years, however, and our knowledge of Mercury's composition, origin, and evolution is therefore limited. From its high bulk density, Mariner 10 observations (Murray 1975), and Earth-based remote sensing, Mercury is known to have a high metal-to-silicate ratio, a crust low in FeO (Rava and Hapke 1987; Vilas 1988; Blewett et al. 1997; Robinson and Taylor 2001), and an exosphere with such species as Na and K (Potter and Morgan, 1985, 1986). Even though our understanding of Mercury's bulk composition is limited, some constraints on models of planetary formation and evolution are possible. If Mercury condensed from the inner refractory portion of a hot early nebula, it should be strongly deficient in volatiles and FeO (e.g., Lewis 1972, 1974). The possibility that Mercury's semimajor axis experienced large excursion during growth of the inner planets (Wetherill 1994) is permissive of Mercury having greater fractions of volatiles and FeO.

Mariner 10 images showed a heavily cratered surface grossly similar to that of the Earth's Moon (Murray et al. 1975; Spudis and Guest 1988). One of the more distinctive morphologic features discovered by Mariner 10 is a class of tectonic features known as lobate scarps, interpreted to reflect large-scale contractional deformation of Mercury's crust. Lobate scarps are thought to be the surface expression of thrust faults formed as the planet's interior cooled and contracted, possibly during a period in which tidal despinning was also occurring (Strom et al. 1975; Cordell and Strom 1977; Melosh and Dzurisin 1978; Pechmann and Melosh 1979; Melosh and McKinnon 1988; Watters et al. 2004). Another distinguishing feature of Mercury is the smooth plains. Smooth plains are comparable in morphology to lunar mare deposits, but they lack the distinctive low albedo of their lunar counterparts, because of the very low FeO content (Trask and Guest 1975; Strom 1977; Kiefer and Murray 1987; Rava and Hapke 1987; Spudis and Guest 1988). Whether the smooth plains are volcanic or impact deposits is still debated (Wilhelms 1976; Kiefer and Murray 1987; Robinson and Lucey 1997).

Earth-based radar images of Mercury rival those obtained by Mariner 10 (Harmon et al. 2001). More importantly, radar led to the discovery of an anomalous class of materials inside permanently shadowed crater interiors in both polar regions. These materials exhibit high radar reflectivity and a circular polarity inversion consistent with a volume scatterer (Slade et al. 1992; Harmon and Slade 1992). Water ice remains the leading candidate material to explain the shadowed deposits, but many unanswered issues remain and final resolution must await orbital observations (Harmon et al. 2001).

The MErcury Surface, Space ENvironment, GEochmistry, and Ranging (MESSENGER) spacecraft was conceived, designed, and built to address six fundamental science questions regarding the formation and evolution of Mercury (Solomon et al. 2001). (1) What planetary formational processes led to the planet's high metal-to-silicate ratio? (2) What is Mercury's geological history? (3) What are the nature and origin of Mercury's magnetic field? (4) What are the structure and state of Mercury's core? (5) What are the radar-reflective materials at Mercury's poles? (6) What are the important volatile species and their sources and sinks on and near Mercury?

The process of selecting the scientific instrumentation to investigate these diverse questions balanced the available mission resources for mass, power, mechanical accommodation, schedule, and money. For MESSENGER, the mass and mechanical accommodation issues were very significant design constraints. The payload mass was limited because of the large amount of propellant needed for orbital insertion. The mechanical accommodation was difficult because of the unique thermal constraints faced in the mission. Taking into account all these factors, MESSENGER carries seven miniaturized instruments (Gold et al. 2003): the Mercury Dual Imaging System (MDIS), Gamma-Ray and Neutron Spectrometer (GRNS), X-Ray Spectrometer (XRS), Mercury Laser Altimeter (MLA), Magnetometer (MAG), Mercury Atmospheric and Surface Composition Spectrometer (MASCS), and Energetic Particle and Plasma Spectrometer (EPPS). Additionally, a radio science investigation will address key measurements such as Mercury's physical libration and gravity field.

The MESSENGER spacecraft was launched from Cape Canaveral on August 3, 2004, in a spectacular nighttime launch. On August 1, 2005, the spacecraft successfully completed an Earth gravity assist to slow the spacecraft and redirect it toward the inner solar system. En route to its primary mission at Mercury, MESSENGER experiences two Venus flybys and three Mercury flybys. The first Venus flyby occurred on October 24, 2006, and the second occurred on June 5, 2007. The three Mercury flybys will take place on January 14, 2008, October 6, 2008, and September 29, 2009, during which regions unexplored by Mariner 10 will be imaged by MDIS. Mercury orbit insertion will occur on March 18, 2011, and the spacecraft will begin the orbital phase of its mission, which is one Earth year in duration. The orbital mission is slightly longer than two Mercury solar days.

2 MDIS Measurement Objectives and Design Implementation

MDIS consists of two cameras, a monochrome narrow-angle camera (NAC) and a multispectral wide-angle camera (WAC), coaligned on a common pivot platform. The passively cooled detectors in each camera are thermally tied to its complex thermal system. This arrangement allows the detectors to be maintained within their operating temperature, even during the hottest portion of the orbit at Mercury. The pivot platform provides an added degree of freedom to point the dual cameras with minimal impact on the spacecraft. The full design details of the instrument are given in Sect. 3. Specifications of the two cameras, given in Table 1, are tailored to the orbit and imaging requirements of the MESSENGER mission.

MESSENGER will be placed in a highly eccentric orbit with a periapsis altitude of 200 km, a periapsis latitude of ~60°N, and an apoapsis altitude of 15,200 km. The orbit has a 12-hour period, is inclined 80° to the planet's equatorial plane, and is not Sun synchronous. During one Mercury solar day (noon to noon), the planet completes three full rotations

Table 1 MDIS camera specifications

	Narrow angle		Wide angle
Field of view	1.5° × 1.5°		10.5° × 10.5°
Pivot range		−40° to +50°	
(observational)		(Sunward) (Planetward)	
Exposure time	1 ms to ∼10 s		
Frame transfer time	3.84 ms		
Image readout time[a]	1 s		
Spectral filters	1		12 positions
Spectral range	700–800 nm		395–1,040 nm
Focal length	550 mm		78 mm
Collecting area	462 mm^2		48 mm^2
Detector-TH7888A	CCD 1024 × 1024, 14-μm pixels		
IFOV	25 μrad		179 μrad
Pixel FOV	5.1 m at 200-km altitude		35.8 m at 200-km altitude
Quantization	12 bits per pixel		
Hardware compression	Lossless, multi-resolution lossy, 12-to-*n* bits		
	MDIS Assembly		MDIS DPU-A or -B
Mass	7.8 kg		1.5 kg
Power[b]	7.6 W		12.3 W
Footprint	398 × 270 × 318 mm		157 × 117 × 104 mm
Data rate	16 Mbps (to DPU)		3 Mbps (to SSR)

[a]Transfer to DPU

[b]Nominal power configuration (DISE + NAC or WAC; DPU + MDIS motor + resolvers)

relative to the spacecraft orbital plane. At times the ground track is near the terminator (the "dawn–dusk orbit"); 22 days later it passes over the subsolar point (the "noon–midnight" orbit).

The two primary imaging objectives during the flybys are (1) acquisition of near-global coverage at ∼500 m/pixel, and (2) multispectral mapping at ∼2 km/pixel. During the flyby departures, large portions of the planet will be viewed at uniform low phase angles.

From orbit, gaps in color imaging acquired during the flybys will be filled with images taken at a wide variety of lighting geometries. Total flyby coverage will exclude only the polar regions and two narrow longitudinal bands. The flybys each have one of two basic geometries (Table 2), and similar observation strategies will be used for each (Table 3). During the flyby phase, 85% of the planet will be imaged in monochrome at a resolution averaging ∼500 m/pixel, and greater than 60% will be imaged in color at about 2 km/pixel. Half of the planet will be covered in color at ∼1 km/pixel. High-resolution NAC swaths will contain monochrome images at better than 125 m/pixel.

During the orbital phase of the mission the MDIS observation strategy will shift to acquisition of four key data sets: (1) a nadir-looking monochrome (750-nm) global photomosaic at moderate solar incidence angles (55°–75°) and 250 m/pixel or better sampling; (2) a 25°-off-nadir mosaic to complement the nadir-looking mosaic for stereo; (3) completion of the multispectral mapping begun during the flybys; and (4) high-resolution (20–50 m/pixel) image strips across features representative of major geologic units and structures.

Table 2 Key parameters describing the three MESSENGER Mercury flybys

Date	CA altitude	CA lon	Inbound lon	Outbound lon	Illuminated lon	Key features
1/14/08	200	40°	308°	132°	276° to 96°	Caloris, EUH
10/06/08	200	230°	136°	324°	94° to 274°	Kuiper, WUH, MG
09/29/09	200	212°	108°	315°	90° to 270°	Kuiper, WUH, MG

All closest approach (CA) latitudes are near-equatorial, and range is listed in kilometers; closest approach occurs on the nightside of the planet for all three flybys. The columns "Inbound lon" and "Outbound lon" indicate subspacecraft longitudes at 20,000 km range during the inbound and outbound legs of the respective flyby. Comments indicate the portion of Mercury imaged during each flyby (Caloris = Caloris basin, EUH = eastern half of hemisphere unseen by Mariner 10, Kuiper = Kuiper crater, WUH = western half of hemisphere unseen by Mariner 10, MG = Mariner 10 gore). All longitudes are positive east. During the three Mariner 10 flybys Mercury was illuminated from 350°E to 170°E

Both the nadir and off-nadir image mosaics will be acquired with the NAC for southern latitudes when altitude is high and with the WAC at lower altitudes over the northern hemisphere. This two-camera strategy results in near-uniform global coverage with an average spatial resolution of 140 m/pixel. The off-nadir mosaic will be acquired under nearly identical lighting geometries to the nadir map to facilitate automated stereo matching. The global digital elevation model derived from stereo imaging will have a spatial resolution of 1–4 km horizontally and 100–500 m vertically, depending on latitude. MDIS stereo imaging will be the main source of surface elevation mapping for the southern hemisphere, as MLA's 1,000-km slant range (Cavanaugh et al. 2007) largely limits laser altimetry to the northern hemisphere. Filling gaps in color coverage is a relatively simple matter of pointing at and imaging a particular location during times of favorable lighting, except at low altitudes over high northern latitudes. At northern mid-latitudes, low spacecraft altitudes will limit viewing opportunities and probably require gap-filling images to be taken in long strips. At the time of writing, the strategy for gap-filling of flyby color mapping is still being defined. High-resolution NAC imaging is effectively limited by ground motion smear to about 20 m/pixel in the along-track direction; accurate postprocessing correction for electronics artifacts (Sect. 4.3) requires exposure times of $\sim$7 ms or longer, equivalent to $<$20 m of along-track motion smear.

2.1 Science Traceability

The flowdown of requirements from science objectives to instrument design and data acquisition strategy are summarized in Table 4, which also compares the instrument requirements to as-built performance. Key constraints on the design of the MDIS investigation are the spacecraft's thermal environment, its highly eccentric orbit, the vertical accuracy required of stereo imaging, the low downlink rates, and the need to support optical navigation (opnav) prior to each Mercury flyby.

At Mercury, the intensity of the solar radiation varies from about 7 to 10 times the total irradiance falling on the Earth. A large sunshade protects MESSENGER from this intense solar illumination but constrains the spacecraft pointing ability (Leary et al. 2007). To support optical navigation during the flybys, Mercury will be imaged against the star background; MDIS thus had to be designed to image at phase angles as high as 140°. At the other extreme, flyby science observations require imaging at phase angles of 32°. To compensate for the restricted pointing capability of the spacecraft imposed by the sunshade,

Table 3 Mercury flyby imaging plan

Description	km/ pix	Filters	Pixel binning	Approx. hardware comp. ratio	Bits/ pixel	Wavelet comp. ratio	Cum. images	Cum. Gib, uncom- pressed	Cum. Gib, hardware com- pressed	Cum. Gib, com- pressed[b]
Approach opnavs		1	–	1.3	12	1	6	0.070	0.054	0.054
Approach movie		1	–	2.5	8	12	36	0.305	0.148	0.062
Approach color image	5	11	–	1.3	12	2	47	0.434	0.247	0.111
Approach mosaic	0.5	1	–	2.5	8	8	87	0.746	0.372	0.127
High-resolution mosaic 1	0.06	1	–	2.5	8	8	147	1.215	0.559	0.151
High-resolution color mosaic	1	11	–	1.3	12	2	246	2.375	1.452	0.597
Color photometry sequence[a]	–	11	2 × 2	1.3	12	2	345	2.665	1.675	0.708
High-resolution mosaic 2	0.2	1	–	2.5	8	8	499	3.868	2.156	0.768
Departure color mosaic	2	11	–	1.3	12	2	598	5.028	3.049	1.215
1st departure mosaic	0.4	1	–	2.5	8	6	697	5.802	3.358	1.266
2nd departure mosaic	0.5	1	–	2.5	8	6	796	6.575	3.667	1.318
Departure color image	5	11	–	1.3	12	2	807	6.704	3.767	1.367
3rd (stereo) mosaic	0.7	1	–	2.5	8	6	842	6.978	3.876	1.386
4th (stereo) mosaic	0.8	1	–	2.5	8	6	877	7.251	3.985	1.404
Departure opnavs		1	–	1.3	12	1	883	7.321	4.039	1.458
Departure movie		1	–	2.5	8	12	913	7.556	4.133	1.466

[a]Same spot at phase angles 50°–130° in 10° increments

[b]Total space available on solid-state recorder is 8 Gib

Table 4a Derived MDIS requirements and as-built performance for field-of-view, pointing, and spatial resolution

Measurement objective	Measurement requirement	Instrument / spacecraft requirement	As-built performance	Method of verification
Flyby near global monochrome map at 500 m/pixel	Image Mercury outbound on all three flybys; maximize area imaged in orbit at near-0° emission angle	Point from nadir to >40° antisunward, >50° sunward maintaining unobstructed FOV	Unobstructed FOV to 64° antisunward, 52° sunward using pivot and spacecraft slew	By design
Provide optical navigation support for Mercury flybys	Opnavs earlier than encounter (E) – 2. 5 days at flybys	Availability of imaging at 31°–142° phase angles at center of FOV, from combination of pivoting and spacecraft slewing, provides: • NAC imaging at E-6.0 days @ flyby 1, E-3.7 days @ flyby 2, E-7.2 days @ flyby 3 • WAC imaging at E-7.0 days @ flyby 1, E-4.7 days @ flyby 2, E-8.0 days @ flyby 3		Trajectory analysis
Near-nadir global monochrome map at 250 m/pixel	FOV wide enough for cross-track continuity, average dayside altitude	>9.3° WAC FOV for baseline orbit (>160 μrad pixel)	10.5° WAC FOV (179 μrad pixel)	Geometric calibration
	Cross-track continuity maintained at lowest altitudes using pivoting of FOV	1.1°/s pivot rate to stagger WAC FOVs to produce 15°wide composite image strips	1.1°/s	Pivot testing
	Acquire image strips without cross-track gores or excessive cross-track overlap	Acquire rectangular subframes at up to 1 Hz to manage downlink volume	Arbitrarily defined rectangular subframes available at 1 Hz	Spacecraft functional testing
	<250 m/pixel sampling, globally, with low emission angle	WAC and NAC IFOVs provide <250 m/pixel average sampling for nominal orbit	140 m/pixel average sampling of surface at optimal lighting	Coverage simulations
		WAC and NAC PSF support sampling at <2× average requirement	Projected FWHM of NAC PSF ~700 m in southern high latitudes; onground PSF measurement to be used for image restoration	Radiometric calibration and analytical modeling
		Main processor and downlink volume allow average spatial resolution of binned pixels <250 m/pixel with moderate compression	25.9 Gb MDIS downlink (d/l) allocation allows d/l of 140 m/pixel using 8:1 compression; MP loading requires on-chip 2 × 2 binning for portion of data, degrading resolution by up to 2×	Coverage and spacecraft performance simulations
	Point with sufficient accuracy to allow mosaicking with <10% image overlap	<0.15° pointing accuracy for spacecraft to maintain <10% NA image overlap	0.02° pointing accuracy for spacecraft; 0.03° including 1-σ uncertainty	Simulation based on uncertainties from component tests
		<0.15° step for MDIS pivot mechanism	0.01° step for MDIS pivot mechanism	By design

Underlined items incur additional data processing or revisions to the data acquisition strategy to meet requirements, as indicated

Table 4b Derived MDIS requirements and as-built performance for spacecraft slewing, stability, memory, and downlink

Measurement objective	Measurement requirement	Instrument / spacecraft requirement	As-built performance	Method of verification
Near-nadir global monochrome map at 250 m/pixel	Roll spacecraft sufficiently rapidly to track nadir at the minimum 200-km orbital altitude	1.35 mrad/s	1.7 mrad/s	Spacecraft functional testing
	Ability to track nadir from any orbit	Pivot and spacecraft roll coordinate to point MDIS boresight to local nadir	Spacecraft roll to keep nadir in pivot plane, spacecraft supplies MDIS with pivot attitude	Spacecraft functional testing
	Image smear by spacecraft jitter small compared with image smear due to downtrack motion	Jitter <25 μrad in 100 ms (<1 NAC pixel, typical exposure)	<1 μrad in 100 ms	Analytical modeling
	Hold orbital data in recorder	Hold volume of DPU-compressed images	Baseline mission requires storing <2.3 Gb; usable recorder vol. 6.2 Gb	Data coverage simulation
		Hold number of MP-compressed images	Baseline mission requires storing < 2500 compressed images; >8000 available	
Flyby near global monochrome map at 500 m/pixel	Hold flyby data in recorder	Hold volume of DPU-compressed images	Nominal flyby scenario 5.4 Gb; usable recorder volume of 6.2 Gb	
		Hold number of MP-compressed images	Nominal flyby scenario 913 images; uncompressed image directory holds 2046	
Multispectral map at 2 km/pixel	Downlink compressed flyby data to ground	>2.3 Gb downlink allocation to MDIS following each Mercury flyby (core data)	Current downlink >6 Gb after each flyby	
Targeted high-resolution imaging	Transfer data to SSR at rate greater than image acquisition	Highest image acquisition rate is 1 Hz (3.2 Mb/s for image binned 2 × 2 on-chip)	3.2 Mb/s link speed	Instrument and spacecraft functional testing

Table 4b (*Continued*)

Measurement objective	Measurement requirement	Instrument / spacecraft requirement	As-built performance	Method of verification
Near-nadir global monochrome map at 250 m/pixel	Downlink compressed orbital data to ground	5.6 Gb/yr minimum allocation to MDIS during orbital mission as per Concept Study	~25.9 Gb/yr allows improvement in average base map sampling from 250 to 140 m/pixel, color to 1 km/pixel	Data coverage simulation
Global multispectral map at 2 km/pixel		Pixels 2 × 2 or 4 × 4 without overflow at 7 ms exposure	2 × 2 binning on-chip sums signal, does not overflow at min. solar distance and phase angle; 2 × 2, 4 × 4 binning in MP averages signal	Radiometric calibration; instrument and spacecraft functional testing
		Lossy compression, >6 : 1 with artifacts near noise	Wavelet compression in MP; artifacts near noise at 8 : 1	Simulation using NEAR 16-bit images
Provide optical navigation support for Mercury flybys	Promptly downlink star-planet-star triplets	Return jailbars from Mercury images	Jailbars at regular period in image	Instrument and spacecraft functional testing
		Return ≥ 3 subframes of stars per image (to 140 × 140 pixels for NA, allowing for 0.1° pointing control)	Up to 5 subframes per image, sized up to full image	
	Locate Mercury against star background	Detect ≥3 stars in FOV at max exposures	Met for WAC at longest exposure; marginal in NAC	Radiometric calibration and analytical modeling
	Take unsmeared star frames	Jitter <25 μrad in 10 s (<1 NAC pixel)	<25 μrad in 10 s >99% of time	Analytical modeling

Table 4c MDIS requirements and as-built performance for multispectral imaging

Measurement objective	Measurement requirement	Instrument / spacecraft requirement	As-built performance	Method of verification
Multispectral map at 2 km/pixel	Move filter wheel to allow 11-color imaging strip from lowest altitude over day side	Worst-case FOV motion by 1 footprint in 15.5 s requires 14 s per cycle (allowing 10% overlap), 1 Hz imaging in 11 filters + 3 s for reposition	1 Hz imaging supported for exposures <500 ms; 2 s for reposition; 13 s total	Instrument functional testing
	<2 km/pixel spatial sampling	<2 km/pixel during flyby, from orbit to fill gaps in flyby coverage	2 km/pixel resolution at 4.5 planet radii, or within 10,000 km during orbit	Geometric calibration, trajectory analysis
	Spectral filters for mapping of olivine, pyroxene, glass, opaques	7 WAC spectral filters, violet to 1050 nm	11 WAC spectral filters to measure key features, plus clear filter	Radiometric calibration
	Map abundance variations in opaque and mafic minerals: 750/415 nm, 750/650 nm, and 750/950 nm ratios measured to 1% precision	MTF >0.62 at 1 cycle/8 pixels to preserve variations above noise level	MTF ~0.75 1 cycle/8 pixels for typical filter	Analysis and radiometric calibration
		Model dark current to ≪system noise (≪1 DN)	Dark current model residuals at read-noise level at <5°C, consistent with model accuracy ≪noise	Radiometric calibration
		Response linearity to 1%	Departures of ≤2% at low DN levels correctable during ground processing	Radiometric calibration
		Characterize responsivity to 5% absolute, 2% relative between filters	On-ground calibration at −34, −30, +24°C insufficient for interpolation to higher (−10°C) CCD operating temperatures; need flight measurement of Mercury-illuminated calibration target	Radiometric calibration
		Model flat field to 0.1% precision (0.4 of noise at full well)	Composite of 100 images provides 0.04% precision; flight measurements needed due to mobility of dust donuts	Radiometric calibration
	Protect optics from contamination that increases scattered light	Stow MDIS in position with optics protected from contaminants during burns, thermal spikes on sunshade	Front optic stowable in instrument base	Instrument functional test

Underlined items incur additional data processing or revisions to the data acquisition strategy to meet requirements, as indicated

Table 4d MDIS requirements and as-built performance for high-resolution and stereo imaging

Measurement objective	Measurement requirement	Instrument / spacecraft requirement	As-built performance	Method of verification
Targeted high-resolution imaging	Acquire continuous strip of images with NAC from minimum orbital altitude	At minimum altitude of 200 km, NA footprint moves 1 FOV in 1.4 s. Requires 1 Hz imaging.	1 Hz imaging supported using 2 × 2 pixel binning or quarter frames	Instrument functional testing
	Acquire <20 m/pixel NAC frames from lowest altitude expected over day side (280 km) with <20 m smear due to spacecraft motion	NAC IFOV <71 μrad	NAC IFOV 20 μrad	Geometric and radiometric calibration
		Minimize time for CCD frame transfer (minimum useful exposure time ~2× frame transfer time)	3.7 ms frame transfer yields ~7 ms exposure. At 3.3 km/s, 18 m linear smear. Pixel footprint 12 m/pixel after 2 × 2 binning required for 1 Hz imaging	
	No saturation at minimum exposure time	No saturation in 7 ms NAC image at 33° phase, minimum solar distance, 2 × 2 pixel binning	Worst-case saturation time 16 ms	Radiometric calibration and photometric model
Off-nadir imaging to complement nadir geometry for stereo	Image areas both at nadir and off-nadir with similar lighting	Consistent observation strategy meeting other spacecraft constraints	N hemisphere nadir first solar day, off-nadir second solar day; S hemisphere nadir and off-nadir on adjacent orbits first solar day. Meets spacecraft constraints as given in Table 4b.	Coverage simulation
	Ability to take off-nadir images an arbitrary orbit	Pivot and spacecraft roll usable together to point MDIS boresight offset along-track from local nadir	Roll spacecraft to keep nadir offset from pivot plane, supply MDIS with pivot attitude	Spacecraft functional testing
	Stereo vertical accuracy goal is ±2 km from 6000-km altitude (root sum squared pointing knowledge 240 μrad)	180 μrad boresight knowledge with in-flight calibrations (1.4× worse than NEAR with in-flight pointing cals. using star images; allows for scan plane vs. fixed pointing)	Ability to extract multiple subframes from starfield images for pointing calibrations within downlink; 200 μrad expected pointing knowledge	Instrument and spacecraft functional testing; flight calibration required for pointing knowledge

Table 4d *(Continued)*

Measurement objective	Measurement requirement	Instrument / spacecraft requirement	As-built performance	Method of verification
		150 μrad knowledge of pivot position	85 μrad	Pivot position calibration
		Knowledge of center of MDIS exposure to ±10 ms	<1 ms uncertainty	Analysis of image acquisition timing using external strobe and Universal Time clock
		Acquire images from orbits with minimal thermal disturbances to instrument deck	Nadir and off-nadir coverage planned from near-terminator orbit with lesser thermal variation	By design

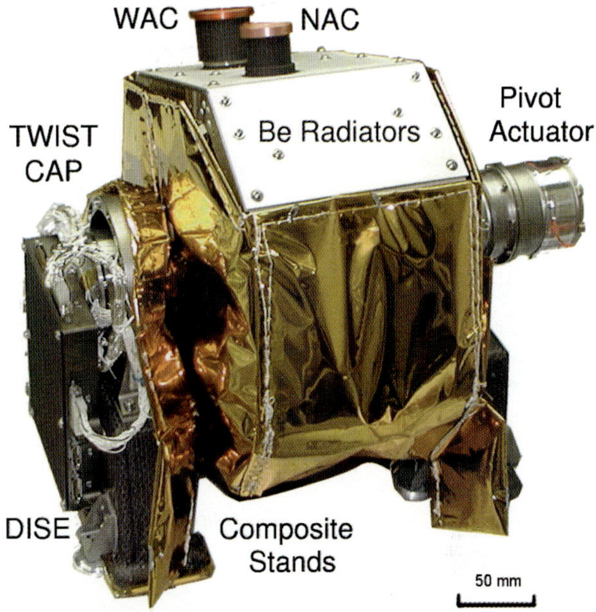

Fig. 1 Photograph of the Mercury Dual Imaging System (MDIS) instrument just prior to integration with the spacecraft (S/C). Redundant Data Processing Units (DPUs, not shown) connect to MDIS through the DPU Interface Switching Electronics (DISE). Red-tag covers were used to protect apertures during handling and were removed before flight. Some thermal blankets are not shown to reveal structure

the dual cameras of MDIS are able to pivot about a common axis (Fig. 1). The nominal operational scan range of the platform is 40° in the sunward direction and 50° anti-sunward. With spacecraft slewing, phase angle coverage of 26°–142° is possible in both cameras at the center of each field of view (FOV). Imaging is available in the WAC 5.3° farther in each direction because of the wide FOV; in the NAC imaging is possible 0.75° farther. During launch and key orbital maneuvers, the camera can be placed in a stowed position, providing contamination protection for the optics (Fig. 2).

The thermal environment poses challenges for MDIS performance, because the cameras must view the hot surface (>400°C) on some orbits for ~120 minutes. Although this thermal environment presents issues for all parts of the instrument, the most stressing case is maintaining nominal temperature of the charge-coupled device (CCD). Wide swings in detector temperature potentially degrade signal-to-noise ratio (SNR), calibration accuracy, and the value of the acquired images.

The thermal environment also poses a challenge for stereo imaging. Stereo provides measurement of both relief (the elevation difference between stereo resolution cells, about 5 × 5 to 7 × 7 pixels in size) and elevation relative to mean planetary radius. In the southern hemisphere, beyond the range of the MLA, the primary knowledge of elevation to mean planetary radius will be from photogrammetric analysis of MDIS images (plus occultations of radio signals from the spacecraft to Earth). Accuracy in an elevation determination from stereo is proportional to $h\sigma/\tan(e)$, where h is the orbital altitude, σ is the uncertainty in pointing knowledge between image pairs, and e is the emission angle of the off-nadir image. The goal for elevation accuracy is ±2 km. Assuming that accuracy can be improved by a factor of two using photogrammetric techniques at the corners of four stereo pairs, the required accuracy is ±4 km. For a 6,000-km orbital altitude, which is appropriate to southern high latitudes, and a 25° emission angle of the off-nadir images, the required pointing knowledge is ±240 μrad. This requirement is budgeted between uncertainty in the knowledge of image acquisition time (which translates into downtrack position error), uncertainty in pivot

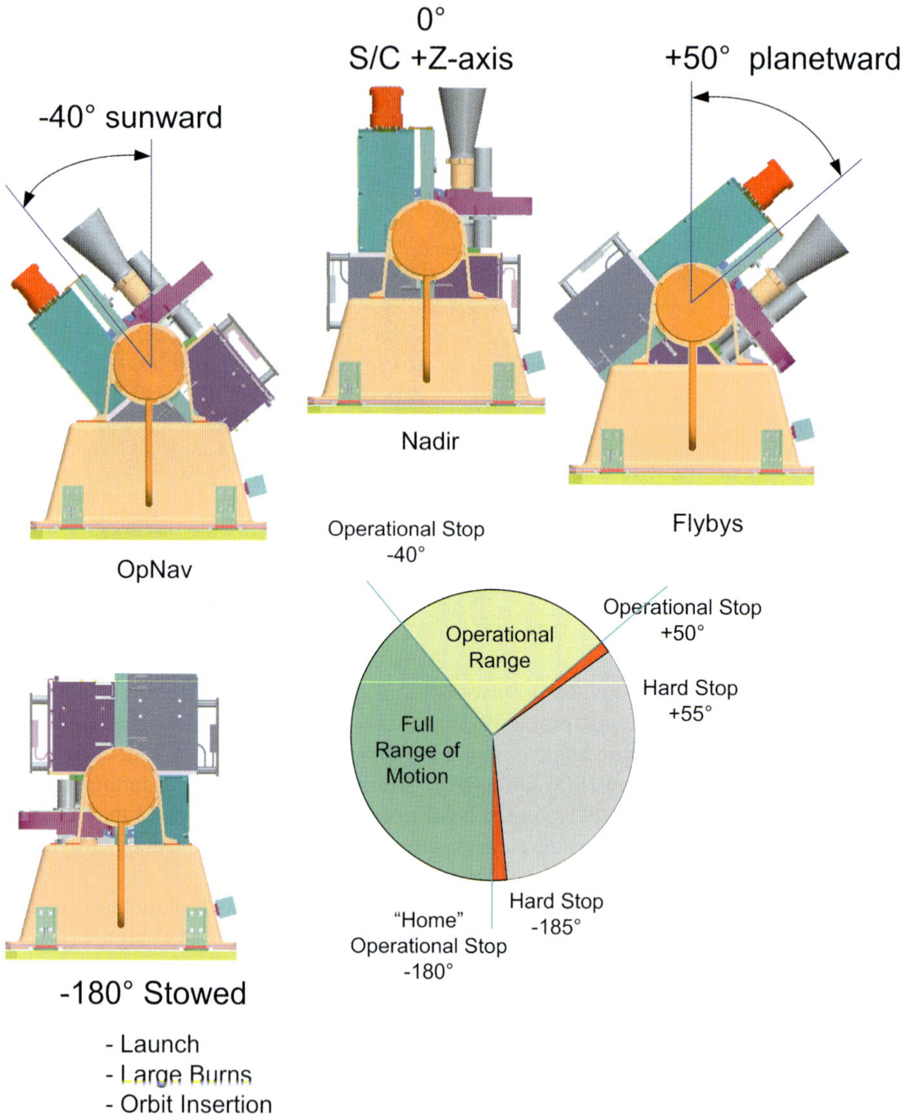

Fig. 2 Range of motion of the MDIS pivot platform. Operational range is −40° sunward to +50° antisunward (planetward). When stowed, the sensitive first optic of each telescope is protected

position within its plane of motion, and variation in orientation of the pivot plane relative to thermal distortion of the spacecraft. The largest term is due to variation in pivot plane orientation relative to the star camera (Table 4d). On the Near Earth Asteroid Rendezvous (NEAR) mission, orientation of its fixed-pointed camera was modeled to ±130 μrad as a function of temperature of the spacecraft structure, using star-field images to calibrate pointing. Allowing for the fact that MDIS moves in a plane, and that the plane may shift in orientation with temperature, the budget is increased by $\sqrt{2}$ to 180 μrad, leaving 140 μrad for uncertainty in position within the pivot plane and along-track errors. To facilitate pointing calibrations with MESSENGER's limited downlink, the spacecraft main processor provides a subframing capability that allows up to five subframes per image encompassing stars.

The eccentric orbit results in challenges associated with strongly variable spacecraft range and velocity. Spacecraft altitude above the surface ranges between 300 and 15,000 km on the dayside. This variation, combined with the requirement for imaging at near-nadir geometries, drove the selection of camera optics. The WAC's 10.5° FOV is sufficient that overlap occurs between nadir-pointed image strips taken on adjacent orbits, even at northern mid-latitudes where low altitudes occur. The NAC's 1.5° FOV is sufficiently narrow that 375 m/pixel sampling is attained at 15,000 km altitude. Low emission-angle geometries are available each solar day for all parts of the planet from altitudes of less than 10,000 km, providing 250 m/pixel sampling or better. The low altitude at periapsis also drives the speed of image acquisition. For multispectral imaging, acquisition of 11-color data from the minimum dayside altitude ($\sim$300 km) requires the WAC to take images at 1-Hz cadence. Along-track continuity of high-resolution imaging also requires 1-Hz imaging. In neither case is full resolution of either the WAC or NAC required; 2×2 pixel binning on-chip meets the spatial sampling requirements both for WAC color and for NAC high-resolution imaging from low altitude. Meeting low-altitude imaging requirements thus drives the speed of the WAC filter wheel (1-Hz imaging in adjacent filters) and link speed from either camera to the recorder (12-bit, 512×512 frames at 1 Hz) as described in Tables 4c and 4d.

The MESSENGER mission requires compression to meet its science objectives within the available downlink. Figure 3 summarizes the compression options available to MDIS at the instrument level using the spacecraft main processor (MP). At the focal plane, 2×2 binning is available on-chip to reduce the $1{,}024 \times 1{,}024$ images to 512×512 format, 12-bit data number (DN) levels can be converted to 8 bits, and data can be compressed losslessly. The strategy for image compression is to acquire all monochrome data in 8-bit mode, and color data in 12-bit mode, and to compress all data losslessly to conserve recorder space. After data are written to the recorder, they can be uncompressed and recompressed by the MP more aggressively using any of several options: additional pixel-binning, subframing, and lossy compression using an integer wavelet transform. The strategy for MP compression is that all data except flyby color imaging will be wavelet compressed, typically 8:1 for monochrome data and to a lower ratio ($\leq 4:1$) for orbital color data. Color imaging but not monochrome imaging may be further pixel-binned. For the special case of optical navigation images, there is a "jailbar" option that saves selected lines of an image at a fixed interval for optical navigation images of Mercury during flyby approaches.

Compression performance was extensively modeled prior to launch. The 12-to-8 bit look-up tables have been designed to retain preferentially information at low, medium, or high 12-bit DN values, for a nominal detector bias or for one that has decreased with time (Fig. 4). Compression ratios to be used for flight have been based on a study of the magnitude and spatial coherence of compression artifacts using NEAR images (Fig. 5). For expected loading of the main processor, simulations have shown that the MP can compress the equivalent of 82 full $1{,}024 \times 1{,}024$ images per day (or 330 512×512 images per day). The actual number of images has also been simulated, based on orbital trajectory simulations and the imaging plan described in the following. The MP image compression capabilities are consistent with the mission-average number of images per day. However, on days when lighting is favorable for global mapping, a peak of $\sim$260 images per day may be expected, requiring on-chip binning of most of the data on peak days. The full implications for average imaging resolution are still being assessed.

The final set of requirements involves the responsivity required for mapping Mercury and for optical navigation. For WAC spectral filters, passband widths were selected to provide

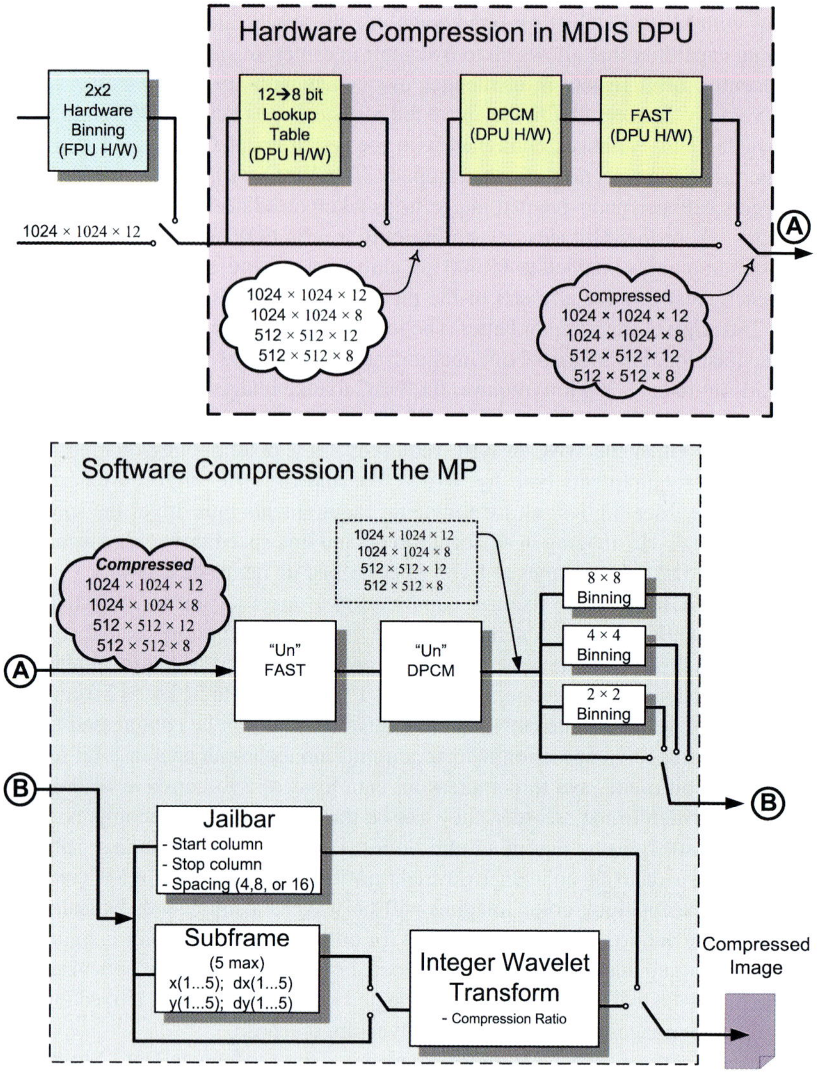

Fig. 3 Hardware (H/W) compression options available for MDIS images in the instrument DPU and software compression options available in the spacecraft main processor (MP). Note: (1) If 2×2 binned mode is selected as the camera mode, further binning options are not available in the MP. (2) A compression ratio of 0 results in a losslessly transformed image (the resulting image size actually grows due to transforming from 12- to 16-bit representations of each pixel)

required SNR in exposure times sufficiently short to prevent linear smear by along-track motion, yet sufficiently long (>7 ms) to avoid excessive artifacts from removal of frame transfer smear during ground processing. SNR is not an issue, as sufficient light is available for SNRs >200, but saturation is a concern at low phase angles. At the same time, both cameras must be sufficiently sensitive to provide star images for optical navigation and display adequate rejection of stray light. A sequence of long and short exposures will be used for optical navigation. A short exposure in which the planet is not saturated will permit determi-

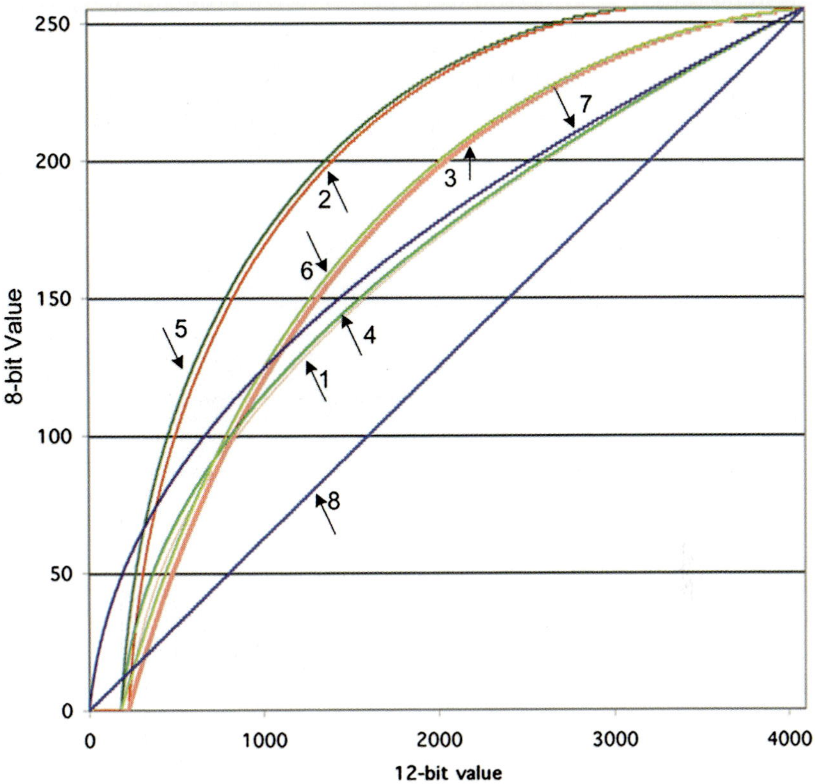

Fig. 4 Mapping of 12 bits to 8 bits will be accomplished using onboard look-up tables. The tables are designed to preserve preferentially information at different DN ranges, and they can accommodate a nominal detector dark level as well as one that has changed with time. (*1*) Low noise, high bias, SNR proportional. Usage: Typical imaging with varied brightness. (*2*) Low noise, high bias, DN-weighted, SNR proportional. Usage: Faint-object imaging. (*3*) High noise, high bias, DN-weighted, SNR proportional. Usage: Black and White (B/W), low brightnesses. (*4*) Low noise, medium bias, SNR proportional. (*5*) Low noise, medium bias, DN-weighted, SNR proportional. Usage: Faint objects. (*6*) High noise, medium bias, DN-weighted, SNR proportional. Usage: B/W, mostly low brightness. (*7*) Zero-bias, SNR proportional. Usage: Typical imaging, varied brightness. (*8*) Linear. Usage: High-brightness mapping, preserves high-DN information

nation of Mercury's location within the image using centroiding techniques. When imaging a saturated Mercury against a star background (as will be the case for long exposures), at least three stars must be visible per image at $\geq 7\times$ noise. For the WAC this requirement was easily met (Table 5a) using a clear filter. For the NAC, its single filter was designed with a first priority of not saturating on bright crater ejecta while imaging Mercury at low phase angles using pixel binning. As a consequence, sensitivity to stars is limited. Detection of three stars per frame for a typical patch of sky is only marginal (Table 5b).

2.2 Common Payload Design

In order to satisfy the science requirements and the design constraints of the MESSENGER mission, many aspects of the science payload were implemented in common. The need to share resources among the instruments played a significant part in the design implementation. The complexity of the MDIS instrument required a separate electronics box, shown in

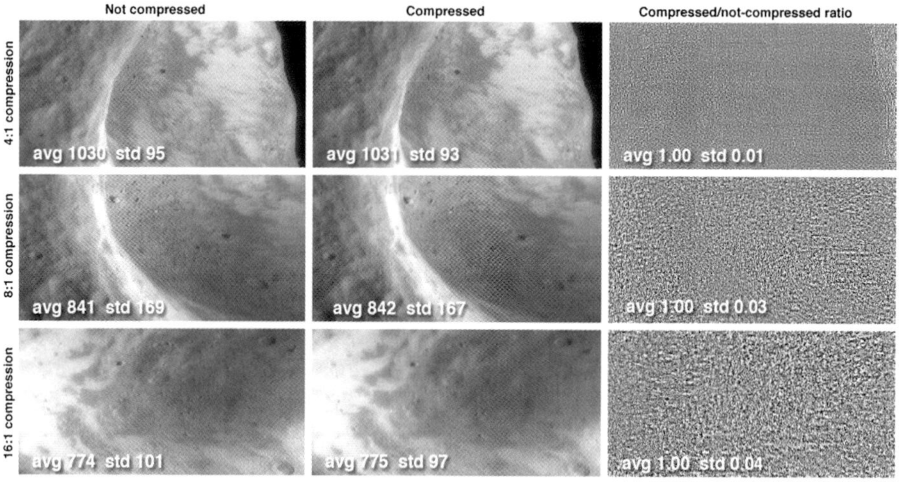

Fig. 5 Effects of compression to different ratios using the MESSENGER integer wavelet transform. Well-exposed 12-bit (DN peaks near 3,000 out of 4,095) NEAR MSI images simulate well the properties of MDIS raw data. The *left column* shows the image prior to compression, and the *middle column* after compression and decompression. The *right column* shows the ratio of the decompressed image to the original; the standard deviation of the ratio is a measure of the artifacts for typical illumination

Table 5a WAC sensitivity to stars

M_v	10 s exposure, no pixel sum, ensquared energy = 70% DN	10 s exposure, 2 × 2 pixel sum, ensquared energy = 90% DN	10 s exposure, no pixel sum, ensquared energy = 22% DN	# stars of ≥ mag in WA FOV
0	21,000	27,000	6,500	0.06
1	8,200	11,000	2,600	0.16
2	3,300	4,200	1,000	0.37
3	1,300	1,700	410	0.9
4	520	670	160	2.2
5	210	270	65	5.2
6	82	110	26	12
7	33	42	10	30
8	13	17	4.1	71
9	5.2	6.7	1.6	180
10	2.1	2.7	0.66	420
11	0.84	1.1	0.26	1,000

Light gray boxes represent visual magnitude (M_v) values with <3 stars in FOV; dark gray boxes represent DN levels below 7× read noise or 7 DN for the WAC. White boxes show the range of M_v values meeting optical navigation requirements (≥3 stars per FOV at 7× read noise)

Fig. 6, a Data Processing Unit (DPU) for image processing, motor control, and commanding. By adding some additional boards to the DPU to support the other payload science instruments, a significant savings was achieved in spacecraft resources, design effort, and testing.

Table 5b NAC sensitivity to stars

M_v	10 s exposure, no pixel sum, ensquared energy = 43% DN	10 s exposure, 2 × 2 pixel sum, ensquared energy = 70% DN	10 s exposure, no pixel sum, ensquared energy = 30% DN	# stars of ≥ mag in WA FOV
0	19,700	32,000	14,000	0
1	8,000	13,000	5,600	0
2	3,100	5,100	2,200	0.01
3	1,300	2,000	890	0.01
4	500	820	350	0.03
5	200	320	140	0.08
6	80	130	56	0.2
7	31.33	51	22	0.48
8	12.29	20	8.8	1.2
9	5.04	8.2	3.5	2.8
10	2.03	3.3	1.4	6.7

Light gray boxes represent visual magnitude (M_v) values with <3 stars in FOV; dark gray boxes represent DN levels below 7× read noise or 14 DN for the NAC. White boxes show the range of M_v values meeting optical navigation requirements (≥3 stars per FOV at 7× read noise)

Fig. 6 Redundant Data Processing Units (DPUs) support the entire science payload. Photograph shows DPU-A and DPU-B with their flight harnesses configured for thermal-vacuum qualification tests

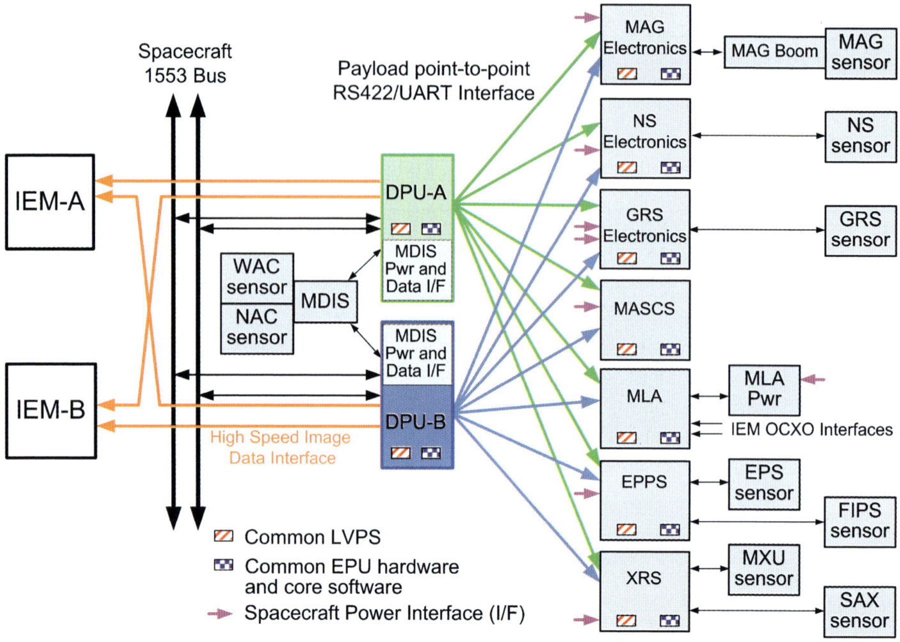

Fig. 7 High-level flow diagram showing relationship between the MDIS DPUs, spacecraft IEMs, and other science payload instruments (see Leary et al. 2007, and references therein). The redundant DPUs provide all processing and power (pwr) interfaces for MDIS, as well as the spacecraft interface for the other instruments. Note: SAX = Solar Assembly for X-rays; MXU = Mercury X-ray Unit; FIPS = Fast Imaging Plasma Spectrometer; EPS = Energetic Particle Spectrometer; GRS = Gamma-Ray Spectrometer; NS = Neutron Spectrometer

To increase the reliability of the payload, redundant DPUs were created to buffer all data interfaces between the payload elements and the spacecraft; one DPU is powered whenever a payload element is active, while the other DPU is maintained unpowered as a cold spare. The DPUs communicate with the spacecraft processors via the spacecraft MIL-STD-1553 busses (Leary et al. 2007), but they communicate with the instruments via separate dedicated RS-422 Universal Asynchronous Receiver Transmitter (UART) interfaces. The DPUs greatly simplified the spacecraft-to-payload interface issues, allowing payload development and testing separate from the rest of the spacecraft. Figure 7 provides a high-level flow diagram showing the relationship between the MDIS DPUs, the spacecraft Integrated Electronics Modules or IEMs (Leary et al. 2007), and the other science payload instruments.

Payload power and mass limitations impacted the design and operational constraints of the science instruments. The most limited power period for MESSENGER occurred during early cruise, when the solar arrays generated their lowest power, restricting the size of instrument heaters that could be used. In contrast, during the orbital phase of the mission, the solar arrays generate ample power, but during eclipse the battery power is still very limited. The instrument designs were limited by their ability to dissipate heat to the spacecraft deck or to the space environment.

The payload employs both distributed power and data processing for each of the instruments. Each instrument (other than MDIS) has its own power supply and microprocessor, thereby greatly reducing the risk of noise or software problems that could have impacted the spacecraft integration schedule. This distributed interface architecture provided a balanced tradeoff between payload reliability, power, mass, and cost constraints.

Although each instrument has individual power supplies and processors, all but one of the instruments use a common power supply board and processor board. This commonality facilitated the use of identical power and data interfaces for the science payload. It also allowed common software modules that would handle a number of common tasks, including command and telemetry processing, timing, macro tools, data compression, and inter-integrated circuit (I^2C) bus mastering. This common software set greatly reduced the amount of code development required for each instrument and reduced risks for payload integration. Having these common power and data interfaces also allowed development of a common set of ground support equipment (GSE) hardware and software that emulated the DPU and power bus interfaces and provided a graphical environment for command menus and telemetry displays. The detailed description of the common low-voltage power supply (LVPS) and event processing unit (EPU) electronics is given in Sect. 3.4.3; the common software description is provided in Sect. 3.5.1.

3 Instrument Design

The full MDIS instrument includes the pivoting dual camera system as well as the two redundant external DPUs. The dual camera assembly without the DPUs is usually simply referred to as "MDIS." The overall design and look of the MDIS, shown in Fig. 1, was driven by mass limitations, the severe thermal environment, and the requirement for a large field-of-regard needed for optical navigation and off-nadir pointing. A functional block diagram of MDIS is shown in Fig. 8.

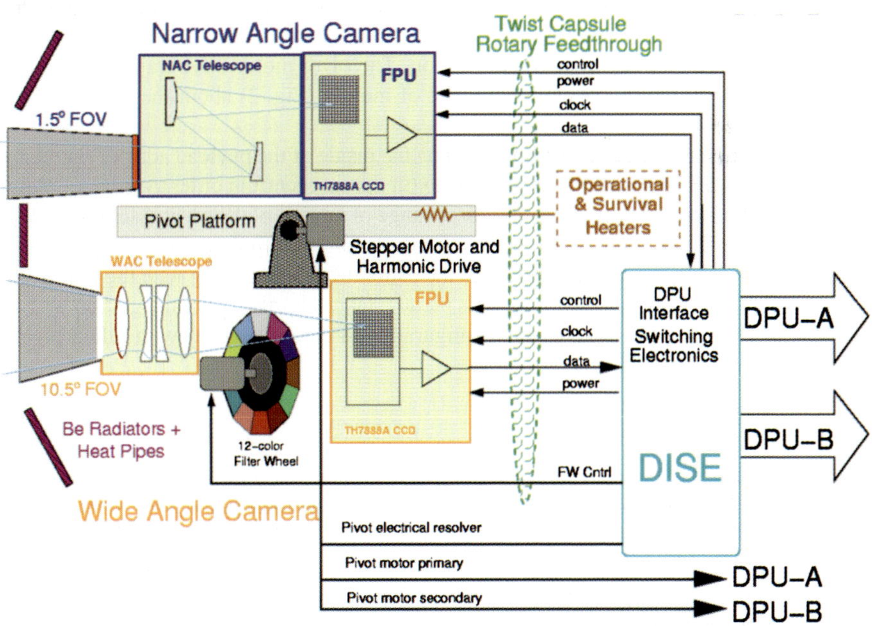

Fig. 8 MDIS does not communicate directly with the spacecraft, other than for spacecraft temperature sensors and heaters. All power, control signals, and data are cross-strapped through the redundant DPUs, as indicated in this functional block diagram for MDIS

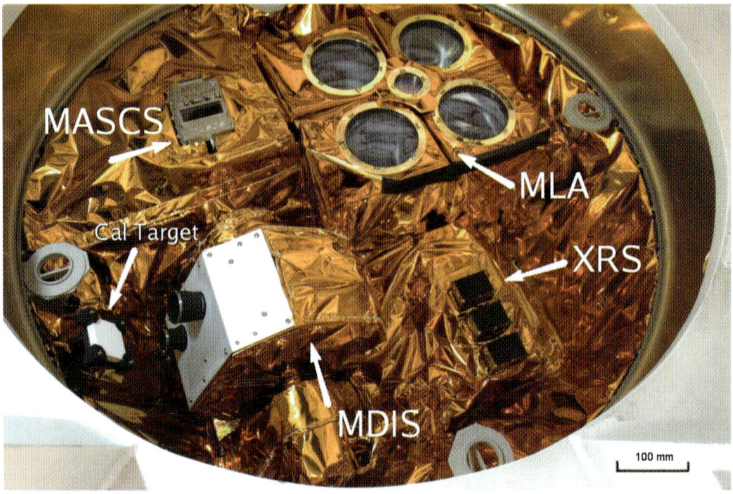

Fig. 9 Science instruments located within the payload adapter ring. The white calibration target was used during in-flight radiometric calibrations

On the pivot platform are the multispectral WAC and the monochrome NAC. A passive thermal design maintains the CCD detectors in the WAC and NAC within their operating temperature range of $-45°C$ to $-10°C$. Only one DPU may be active at a time, and due to thermal constraints only one camera will operate at a time; however, observations with the two cameras can be interleaved at 5-s intervals. A separate electronics assembly accommodates switching between the various modes of operating with the redundant DPUs. The pivot platform has a large range of motion ($\sim 240°$) to allow the cameras to be "tucked away" to protect the optics from contamination. The pivot motor drive-train provides precision rotation over the 90° operational range of motion (Fig. 2) about the spacecraft $+Z$ axis.

A spectral calibration target was mounted on the inside of the payload adapter ring. Early in the MESSENGER mission it was possible to tilt the spacecraft in order to provide solar illumination on the calibration target. The large range of motion of the pivot assembly enables either camera to point at the target, permitting an absolute in-flight radiometric calibration and flat-field measurement. Figure 9 shows the calibration target, along with the four instruments mounted inside the adapter ring. This picture, taken shortly before integration with the launch vehicle, shows the final blanket configuration of the instruments with MDIS pivoted for NAC observations of the target.

3.1 Optical Design

The WAC consists of a four-element refractive telescope having a focal length of 78 mm and a collecting area of 48 mm^2. The detector located at the focal plane is an Atmel (Thomson) TH7888A frame-transfer CCD with a $1{,}024 \times 1{,}024$ format and 14-μm pitch detector elements that provide a 179-μrad pixel (instantaneous) field-of-view (IFOV). A 12-position filter wheel (FW) provides color imaging over the spectral range of the CCD detector. Eleven spectral filters spanning the range 395–1,040 nm are defined to cover wavelengths diagnostic of different potential surface materials. The twelfth position is a broadband filter for optical navigation. The filters are arranged on the filter wheel in such a way as to provide complementary passbands (e.g., for three-color imaging, four-color imaging) in adjacent positions.

The NAC is an off-axis reflective telescope with a 550-mm focal length and a collecting area of 462 mm^2. The NAC focal plane is identical to the WAC's, providing a 25-μrad IFOV. The NAC has a single medium-band filter (100 nm wide), centered at 750 nm to match to the corresponding WAC filter for monochrome imaging.

One of several impacts of the thermal environment on calibration accuracy is the relative response of the CCDs at wavelengths longer than about 700 nm (Sect. 4.6.2). Response at longer wavelengths increases strongly with temperature (Janesick 2001). If data are acquired over a large temperature range, inaccuracies in correction for temperature-dependent response will introduce systematic errors in spectral properties at 850–1,000 nm that ultimately could lead to false mineralogic interpretations. To protect the MDIS CCDs from wide temperature swings, incoming thermal infrared (IR) radiation is rejected in the optics by heat-rejection filters on the first optic of each camera. In the WAC, this rejection is accomplished using a short-pass filter as the outer optic; for the NAC the bandpass filter has a specially designed heat-rejection coating on its first surface.

3.1.1 Wide-Angle Camera

Lens Design. The wide-angle camera (WAC) consists of a refractive telescope, dictated by the required wide FOV and short focal length. The design approach was to select the simplest lens design that gives acceptable image quality over the field; however, an important constraint on the design is the limited selection of glasses because of the radiation environment. The MESSENGER mission is expected to be subjected to a total dose of <15 krad (Si) (R. H. Maurer, personal communication, 2002). The design requirements were achieved by starting from a simple Cooke triplet and splitting the central negative element. The resulting design, a Dogmar (Fig. 10), gave good image quality although not over the full 395–1,040 nm spectral range. The uncorrected axial chromatic aberration was reduced by varying the thickness of each filter, a technique used in the NEAR Multi-Spectral Imager (MSI) (Hawkins et al. 1997).

The WAC is focused at infinity with a 78-mm focal length to spread the 10.5° FOV across the 14.3-mm detector. The 14-μm square pixels of the CCD provide an IFOV of 179 μrad. The selected aperture of $f/10$ means that the lenses, manufactured by Optimax, are quite small. Radiation-resistant glass was required, but only about 16 types were available. Fewer than half of these have adequate transmission below 400 nm. Of these, a combination of K5G20 and LF5G15 gave the best design solution. The optical transmissions of the coated samples of the two glasses used are given in Fig. 11.

Because of the mapping requirement for the WAC, the optical system must have very low distortion. The field curvature of the WAC produces a small focus variation across the field, but it has a negligible effect on the final modulation transfer function (MTF). The distortion has a maximum value of 0.06% at the corner of the field, which is less than a pixel. Figure 12 shows a representative MTF for the WAC. The two curves labeled "On-axis+diffraction" and "7°+diffraction" combine diffraction and the optical aberrations at these two field locations. Here, the wavelength used for diffraction is 700 nm. Because the design is diffraction limited

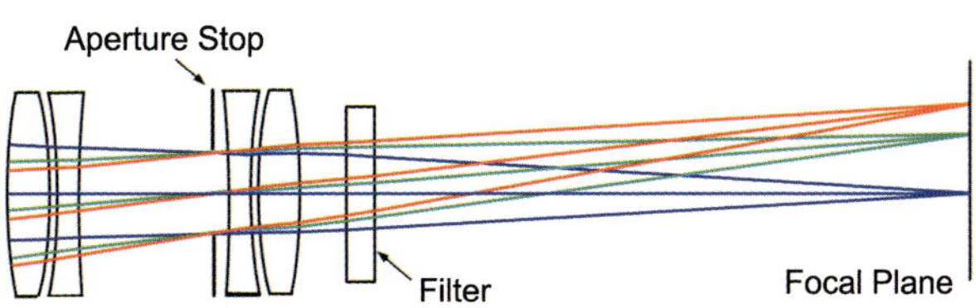

Fig. 10 Optical schematic of the WAC

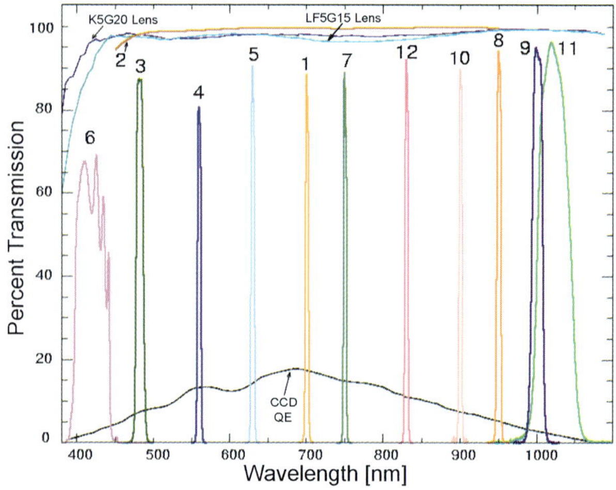

Fig. 11 Nominal passbands and lens transmissivities for the WAC along with the quantum efficiency (QE) of the CCD

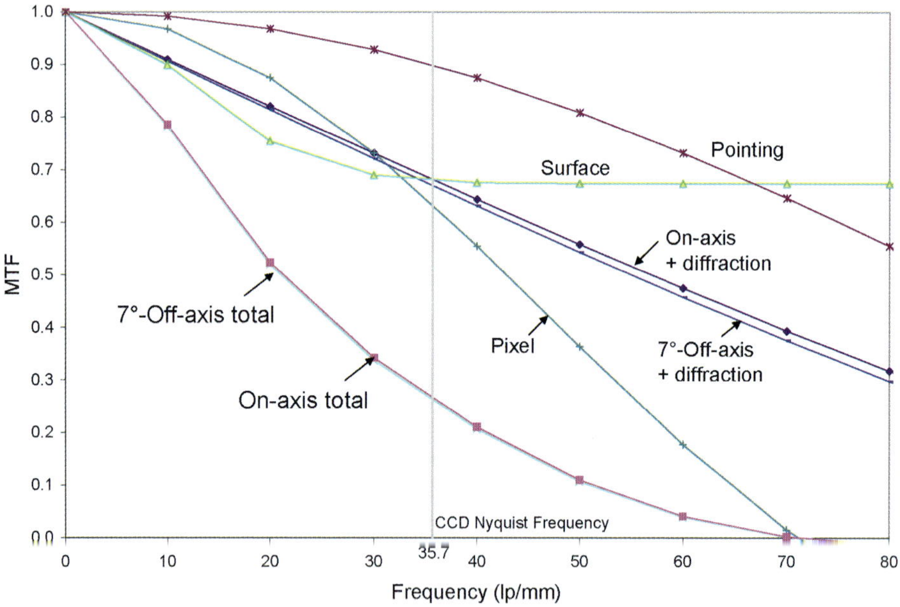

Fig. 12 Theoretical modulation transfer function (MTF) of the WAC. The various modeled contributors to the MTF are shown, with the product of these labeled "Total" for an on-axis ray and 7°-off-axis ray

over the entire field there is very little difference between the two lines. The figure shows that the MTF degradation is dominated by the pixel size at high frequencies (note the Nyquist sampling frequency of the detector). The surface degradation is dominant at low frequencies. The actual surface wavefront degradation here assumes a root mean square (rms) wavefront error of 0.1 waves through the system, including refractive index variations of the filter and lenses. Pointing has a small effect if the movement is limited to less than one-half pixel during an exposure.

The on-axis MTF value of 0.1 occurs at 50 line pairs per mm (lp/mm), or 3.9 lp/mrad, and is comparable to the Rayleigh criterion at which two features can just be distinguished. The resolution defined in this manner changes across the FOV from 0.26 mrad on-axis to 0.30 mrad in the corner.

Spectral Filter Design. A 12-position multispectral filter wheel provides color imaging over the spectral range of the CCD detector (395–1,040 nm). Eleven spectral filters are defined to cover wavelengths diagnostic of common crustal silicate materials and have full-width half maximum (FWHM) bandwidths of 5–40 nm (Table 6). A broadband clear filter was included for optical navigation imaging of stars. Because the optical signal at Mercury will be too high through the clear filter, the quality of the image through this filter was a secondary requirement.

Each filter consists of two or three pieces of glass using a radiation-resistant substrate of BK7G18 glass in combination with a long-pass filter glass. These colored glasses transmit efficiently over a specified wavelength and have a sharp cutoff at shorter wavelengths. The long-pass filter glasses are needed to block short wavelengths in the narrow bandpass filters of the WAC. Two filters required an additional layer of S8612 to achieve the desired passband. The designation G** added to the glass type identifies it as being radiation resistant. The number after the G gives the percent cerium times 10 used as the dopant. Standard glasses typically darken when exposed to radiation. Because of the uncertainty of the transmission degradation by radiation effects on the long-pass filter glasses used in the WAC filters, it was necessary to test the radiation effects on sample colored glasses provided by the filter manufacturer, Andover Corporation. The samples were made of high-quality Schott Glass that matched those used in the flight filters. The experimental setup and the results of the radiation tests are provided in Appendix 1.

The variation in filter thickness used to remove residual chromatic aberration results in a small variation in the focal length of the camera between filters. The extreme filters give a focal length of about 78 mm at 480 nm and about 78.5 mm at 1020 nm, respectively. Table 6

Table 6 Detailed specifications for the WAC filters and effective focal lengths

Filter number	System wavelength (measured at −26°C) (nm)	System bandwidth (measured at −26°C) (nm)	Peak transmission	Total thickness (mm)	Focal length (mm)	Scale change (%)
6	430	18	0.694	6.00	78.075	−0.216
3	480.4	8.9	0.875	6.30	77.987	−0.329
4	559.2	4.6	0.810	6.30	78.023	−0.283
5	628.7	4.4	0.898	6.20	78.109	0.173
1	698.8	4.4	0.892	6.00	78.218	−0.104
2	700	600	–	6.00	78.163	−0.104
7	749.0	4.5	0.896	5.90	78.218	−0.033
12	828.6	4.1	0.921	5.60	78.308	0.082
10	898.1	4.3	0.898	5.35	78.390	0.186
8	948.0	4.9	0.942	5.20	78.449	0.262
9	996.8	12.0	0.952	5.00	78.510	0.340
11	1010	20	0.964	4.93	78.535	0.372

lists the filters in order of increasing wavelength and identifies the number assigned to each filter, and the effective focal lengths for each one. This difference results in a variation in the image scale of 0.7%.

By positioning the filter in front of the detector (Fig. 10), the size of the filters is minimized. However, the incident angle θ of the beam on the filter varies with the FOV from 0° at the center to 14.9° at the corners of the 10.5°-square FOV. This angle will cause a shift in the spectral passband wavelength of the interference filters across the field according to the theoretical expression

$$\lambda = \lambda_o \sqrt{1 - \left(\frac{n_o}{n^*} \sin\theta\right)^2}, \tag{1}$$

where n_o is the refractive index of the external medium ($n_o = 1$ in vacuum), and n^* is the refractive index of the filter substrate. The effect of the incident angle is much more serious than the spatial variations across the surface of the filter. With a maximum angle of 7.6° at the corner of the field, spectral shifts of ~3 nm are expected. The problem with this shift is that it is a variation across the FOV and not a constant offset; these small variations in passband, however, are not expected to limit mineralogic identification or the mapping of surface abundance variations.

Thermal Effects on Optical Design. The change of refractive index with temperature of glass K5G20 varies between -0.9 and $+1.0 \times 10^{-6}$/K, depending on the temperature and the wavelength. For LF5G15 the range is -0.9 to 2.0×10^{-6}/K. Thus, for a temperature change of 40 K the refractive index will change by a maximum of 8×10^{-5}. This is a negligible change, as the index changes by this amount with a wavelength change of only a few nanometers.

The performance of the WAC lens is almost constant with field angle over the design range, so it is not necessary to show the variation with field angle. It does, however, vary with wavelength, so compensation is added by varying the thickness of the filter between 1.7 and 2.9 mm. The results shown in Fig. 13 are for the spectral range 0.45–0.6 μm, over which there is very little change in the image quality with a single filter thickness. The lowest curve is for a nominal temperature of 20°C and vacuum operation. At atmospheric pressure the refractive index of air has a small effect as shown by the highest curve. The effect of temperature is negligible as the curves show the change in the spot sizes at temperatures of −20°C and +20°C. The rms spot radii in this figure are only 0 to 4 μm, which is very small compared with the radius of the diffraction disk, 8.5 μm at a wavelength of 0.7 μm. These results indicate that the WAC optical performance is insensitive to the operational temperatures of the instrument.

Stray Light Analysis. The optical design of the WAC took into account various sources of stray light including scatter from the optical surfaces, intrascene scatter, spurious reflections between optical surfaces, and scatter from the CCD detector. Scatter from external surfaces such as the optical housings, WAC light-shade, and surface contamination were also considered. However, at Mercury, MDIS will be exposed to a radiance source of large angular extent. On orbit, the planet itself will be the dominant source of stray light.

The light-shade of the WAC was constrained in size and complexity because of mass limitations. In addition, because the pivot platform was required to rotate 180° to stow the cameras for launch and other possible contamination events (e.g., large thruster firings and orbital insertion), the overall length of the shade was constrained. A two-piece shade was constructed in order to minimize the size of the hole in the beryllium radiator needed for

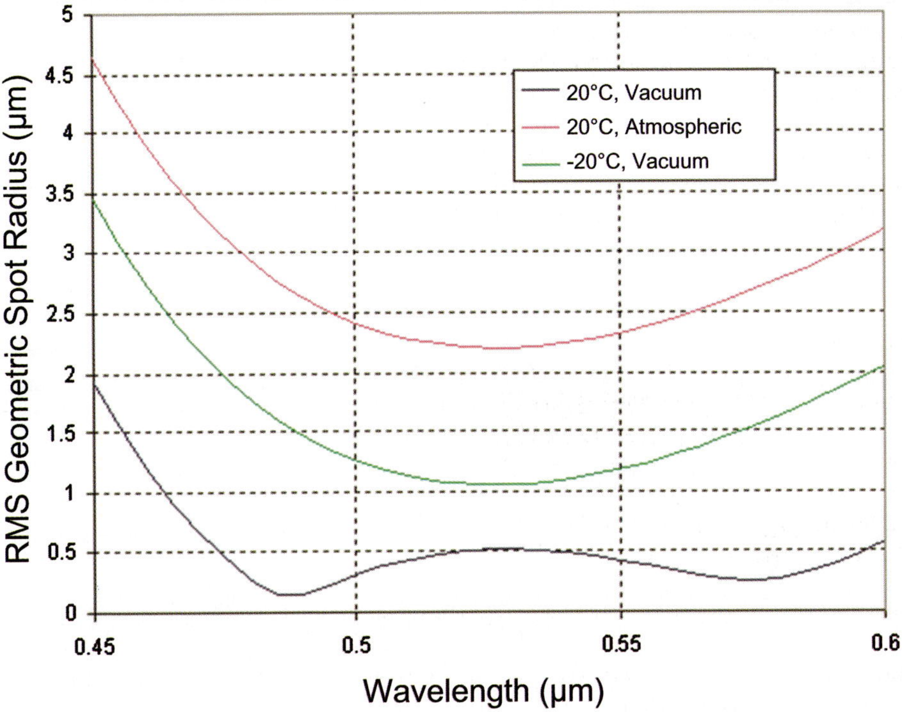

Fig. 13 Variation of WAC geometric spot radius with temperature as a function of wavelength. Residual chromatic aberration over wavelength has been compensated by varying the filter thickness from 1.7 to 2.9 mm. The variations shown are much less than a pixel (14 μm square)

the light-shade to pass through, yet be large enough not to constrain the FOV. Concentric grooves were machined into the titanium light-shade to minimize direct scatter into the system from the shade.

Extensive measurements were made during ground calibration of the WAC to characterize the effect of stray light in the camera. However, limitations of the experimental setup made separating chamber-induced stray light from internal scatter in the optical system of the WAC difficult. In either case, all of the ground-based observations to characterize stray light in the WAC showed the contribution to be small, on the order of 0.1% of the response of the small extended source (lamp filament ~7 pixels) viewed on axis. In-flight calibrations will be made to characterize the stray light performance of the WAC further.

3.1.2 Narrow Angle Camera

The primary purpose of the NAC is for high-resolution imaging of Mercury. Because of the very bright optical signal from the planet, a large collecting area is not required for sensitivity; the need to reduce blurring from diffraction, however, dictated a large (24 mm) aperture. The reflective design has a long focal length that required folding the optical path in order to fit in the available volume. The all-aluminum mirrors and telescope housing were assembled and tested by SSG Precision Optronics, Inc. The monochromatic design has a single medium-band filter centered at 750 nm with a FWHM of 100 nm. The center wavelength was chosen to match filter 1 of the WAC (cf. Table 6).

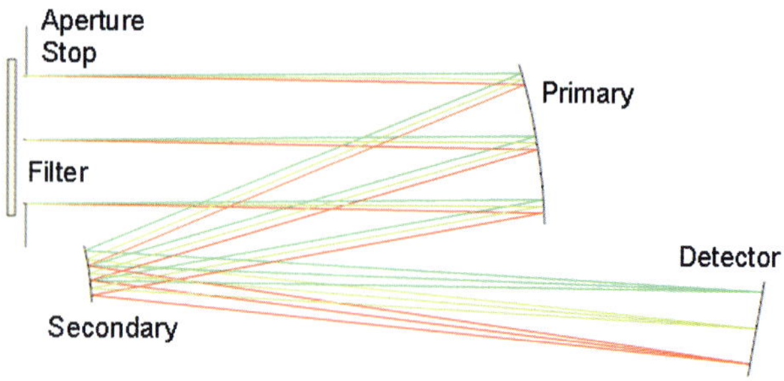

Fig. 14 Optical schematic of the NAC

Telescope Design. The NAC has a 1.5° FOV that is spread across the 14.3-mm detector, requiring a focal length of 550 mm. To limit the diffraction blur, the NAC has a 24-mm aperture, resulting in an $f/22$ system. An off-axis Ritchey-Chretien design was selected over a simple Cassegrain in order to avoid the central obscuration of the secondary mirror. The optical ray trace is shown in Fig. 14. The ellipsoidal primary mirror and hyperboloidal secondary mirrors are gold-coated aluminum with a surface roughness of 0.1 nm and 0.2 nm, respectively. In this design, the image plane is tilted at an angle of approximately 9° for optimal image quality.

A bandpass filter is the first optical component of the assembly and defines the spectral range of the instrument. A specially designed interference coating serves as a heat-rejection filter. Mercury absorbs solar radiation and reradiates this energy back into space. Because of its slow rotation, the dayside of the planet can be modeled as a 400°C blackbody (Hansen 1974). The radiation from this blackbody results in a significant amount of IR radiation that would pass through the NAC interference filter and heat the CCD. Figure 15 shows the transmission of the interference filter for the NAC, with the response of the CCD superposed on it. In addition, the reflectance spectrum of the heat-rejection coating is plotted with a normalized 400°C blackbody spectrum.

Figure 16 shows the spot diagrams corresponding to selected field angles for the NAC. The location of the focal plane was selected to balance the image quality across the field. The circle shows the Airy disk. Low distortion of the image was an important design specification for the NAC. At the higher orbital altitudes, the NAC will be used to complete the nadir and off-nadir mosaics at high resolution. The theoretical distortion of the NAC is 0.25% at the corner of the field. At the edge of the FOV this amounts to 1.28 pixels. The all-aluminum assembly of the NAC makes it insensitive to thermal distortions over the operational temperature range of the instrument. Figure 17 shows a representative MTF for the NAC in the same format described earlier for the WAC.

The aperture stop is located in front of the primary mirror. Figure 18 shows the geometrical spot radius out to an angle of 0.75°, the edge of the FOV. Note that the corners of the field are 1.06° off axis. The geometrical performance is excellent, and the upper line shows the limit imposed by diffraction.

Stray Light Analysis. The narrow FOV makes the NAC less sensitive than the WAC to off-axis sources, making the need for a long light-shade less critical. Mass and size limitations further constrained the light-shade design to be simple, without any internal baffling of the light-shade. The effects of this approach for the NAC light-shade were analyzed by considering a full hemispherical illumination of the NAC, representative of imaging at Mercury. This calculation shows that the scattered radiance as a fraction of the surface radiance of the out-of-field stray light is negligible for this system, even with no light-shade at all. The two

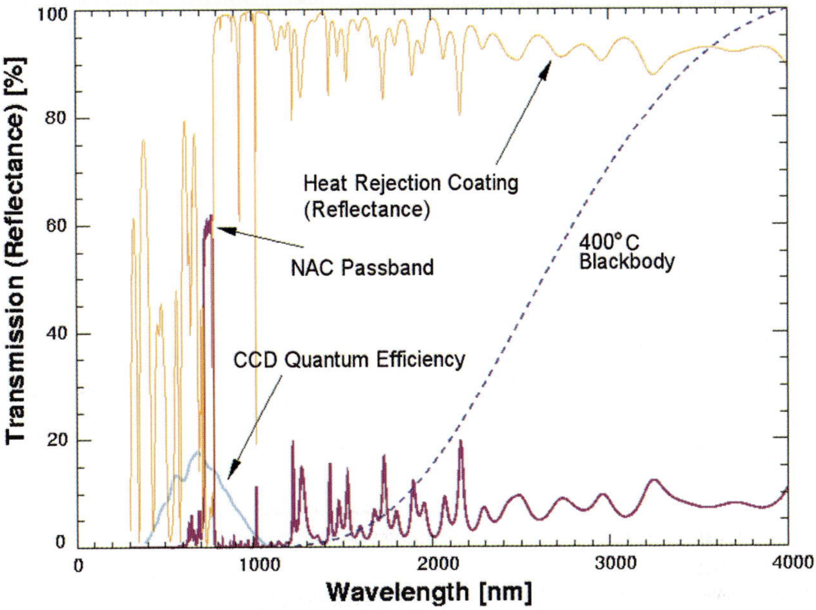

Fig. 15 The monochromatic NAC spectral passband (700–800 nm) is shown overlaid on the quantum efficiency response of the TH7888A CCD. Also shown is the heat-rejection coating specifically designed to reflect the longer IR wavelengths radiated from the hot planet, assumed to be a 400°C blackbody (shown in the figure in dimensionless units)

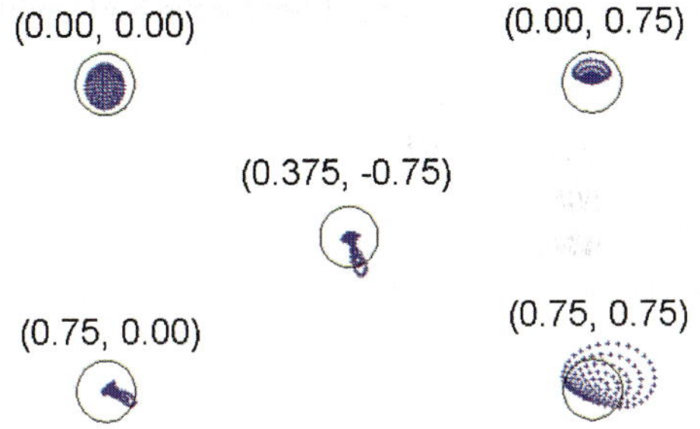

Fig. 16 Spot diagrams for selected field angles (azimuth, elevation, both in degrees) for the NAC. The diffraction-limited performance of the telescope is apparent when compared with an Airy diameter of about 38 μm or 2.7 pixels (*solid circle* in each diagram)

main reasons for including the light-shade were to provide a surface to project through the thermal system and to provide protection from stray light for optical navigation.

The NAC filter is mounted in the NAC light-shade and is tilted at 1° from boresight to prevent direct scatter back onto the detector. With the final light-shade design and filter in place, the calculated bidirectional transmittance distribution function (BTDF) has a value of 0.01 at 1° with a slope of −2; this value is typical for an optical surface. A combination of analytical and empirical methods was employed to identify and minimize scattered light in the NAC telescope. Off-axis light entering the imagers will hit the internal walls or baffles and be mostly absorbed by the black paint. As the absorption is not perfect some residual scattering will result. No direct path exists to the detector, and most scatter requires striking a minimum of two surfaces.

Off-axis rejection in the NAC was modeled extensively using the software package TracePro® and the detailed computer-aided design model of the optics. The calculated level

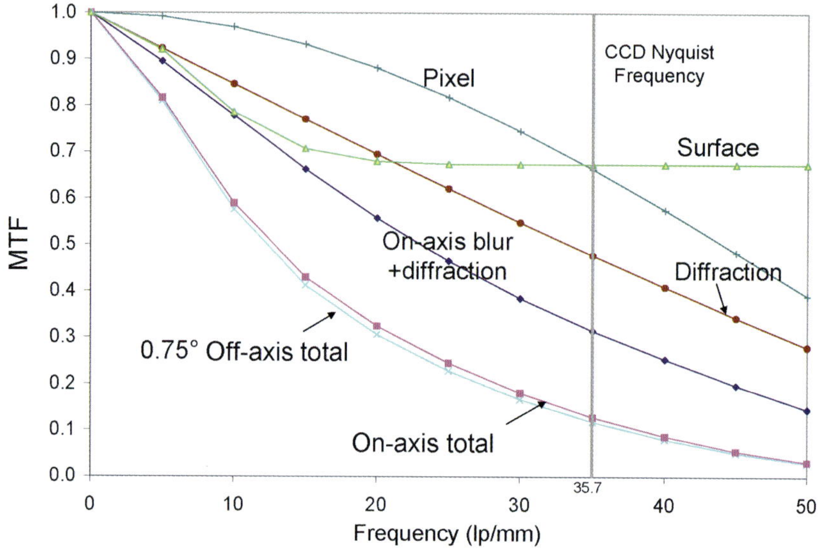

Fig. 17 Theoretical modulation transfer function (MTF) of the NAC. The various modeled contributors to the MTF are shown, with products of these labeled total for an on-axis ray and a 0.75°-off-axis ray

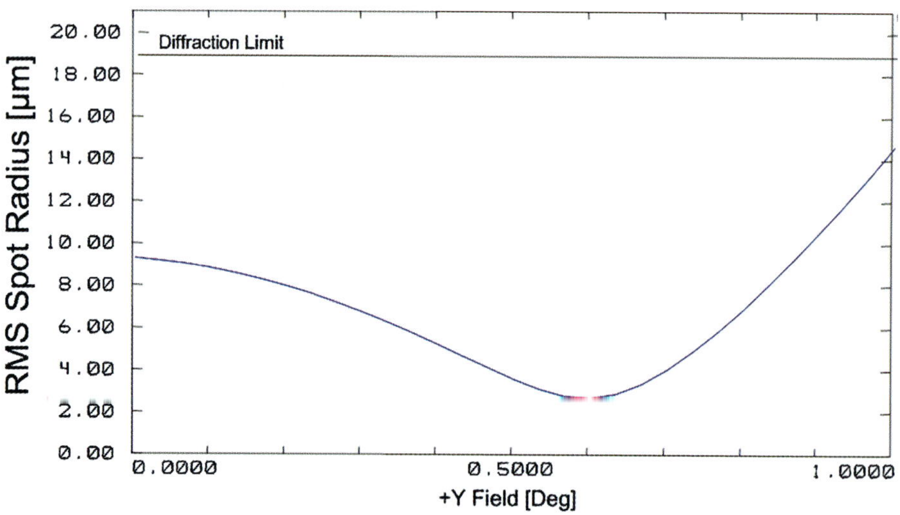

Fig. 18 Geometrical spot radius as a function of field angle for the NAC

of rejection in the NAC is below 0.01%. Measurements to characterize stray light in the NAC were acquired during ground calibration. However, as with the WAC, difficulties in separating out internal camera scatter from scatter induced by the test chamber made measuring the scattered light performance difficult. The measured results still show scattered light rejection below 0.1% of the response of the small extended source (lamp filament ~60 pixels) viewed on axis in the NAC.

Thermal Effects on Optics. The all-aluminum structure of the NAC telescope and gold-coated aluminum mirrors was selected to minimize any thermal distortions to the optical system. The telescope mechanical interface to the Focal Plane Unit (FPU) is a highly polished surface, and the input end of the telescope is supported by a titanium flexure mount. To minimize heat entering the system, a heat-rejection coating was applied to the NAC filter as discussed earlier. The filter housing is also the light-shade and is made of titanium to reduce heat flow into the telescope. The light-shade is painted black, as are the internal surfaces of the telescope and FPU. The optical performance was verified over temperature by measuring the wavefront distortion of the NAC telescope.

3.2 Electronics

The electronics systems of MDIS are fundamental to all aspects of the MESSENGER payload. Not only do the fully redundant Data Processing Units (DPUs) provide the interface between MDIS and the spacecraft, but each DPU also provides the interface for all the other payload instruments as well. Because of all the redundant systems built into the spacecraft, cross-strapping of these systems proved to be a significant task. The main electronics systems of MDIS include the DPUs, the DPU Switching Interface Electronics (DISE) box, and the FPU camera electronics.

3.2.1 Focal Plane Electronics

The detector electronics for both the WAC and NAC are identical. The top-level block diagram of the FPU is shown in Fig. 19. However, each CCD is bonded to a camera-specific mounting bracket (heat-sink) prior to assembly into the FPU electronics. The NAC heat-sink is tilted 9° to match the optimal orientation of the NAC focal plane, whereas the WAC heat-sink has no tilt. Because of this unique mounting configuration for each camera, once the CCD-heat-sink assembly is integrated into a set of FPU electronics, the electronics are no longer mechanically identical and are uniquely defined as a NAC or WAC FPU.

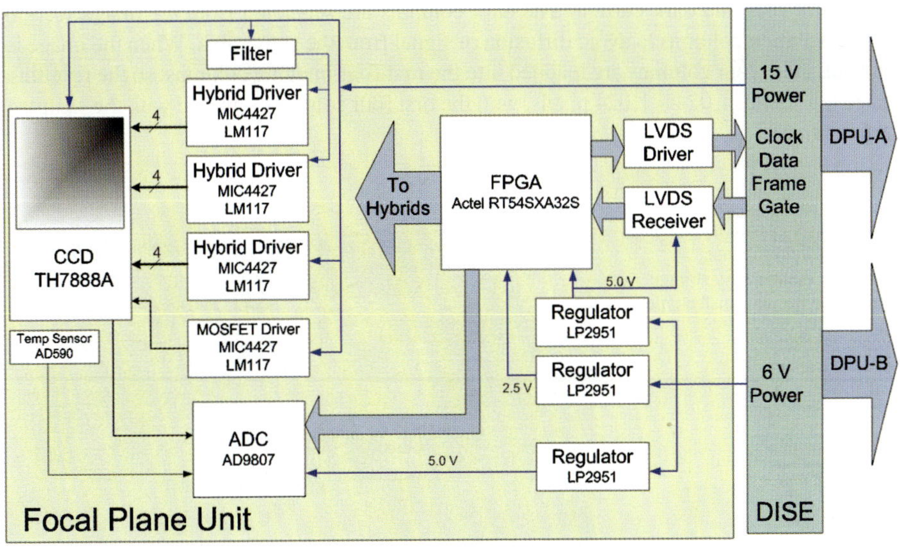

Fig. 19 Block diagram of the MDIS Focal Plane Units and the DPU Interface Switching Electronics (DISE) box. The NAC and WAC FPU electronics are identical in design

Performance Flowdown. Each modular FPU contains a 1-megapixel CCD that has 14-μm square pixels and antiblooming control. Due to thermal, power, and operational constraints, only one camera operates at a time. The frame rate of each FPU is fixed at 1 Hz; however, the frame rate to the spacecraft is not fixed but cannot exceed 1 Hz. Manual and autoexposure control from 1 ms to $\sim$10 s permits imaging over a broad range of intensities.

The requirement for a $1{,}024 \times 1{,}024$ format imager with electronic shutter and antiblooming dictates the general design of CCD that must be used. On-chip binning is required in order to achieve the 1-Hz data throughput to the spacecraft solid-state recorder (SSR) and for data compression. The sensitivity required is not very difficult to achieve, although the relatively short exposures, typically >7 ms, require a commercial rather than scientific type of CCD. The frame transfer time is 3.84 ms, limiting the minimum useful exposure to about 8 ms, and this limit requires post processing to recover a clean image (Murchie et al. 2002).

General Design Considerations. A CCD from Atmel (formerly Thomson-CSF), the TH7888A, was chosen because it was a good match for the requirements and also enabled design experience gained with the Comet Nucleus Tour (CONTOUR) Remote Imager/Spectrograph (CRISP) instruments (Darlington and Grey 2001) to be used. The nominally $1{,}024 \times 2{,}048$ pixel array of the TH7888A has two on-chip output amplifiers (only one amplifier is used in the MDIS design) and uses frame transfer to obtain electronic exposure control. The active image forms an array of $1{,}024 \times 1{,}024$ in which optical energy is accumulated. At the end of the exposure period, the accumulated charge in each pixel is quickly transferred to the masked-off $1{,}024 \times 1{,}024$ memory zone. A tantalum radiation shield not only protects the CCD from ionizing radiation but also blocks illumination to the memory zone as shown in Fig. 20.

The CCD is operated in two modes: binned and full-frame. In the full-frame mode for either the WAC or NAC, the first four columns of each image are taken from a region of the CCD that is never exposed to light and, thus, represents a dark level that is purely a function of bias and dark current. The dark columns are separated from the image section by five isolation columns to avoid diffusion of signal from the active area. When the image is read out, these four columns are mapped into the first four imaging columns, so the resulting image is a square $1{,}024 \times 1{,}024$ pixels, with the first four columns replaced with the sampled dark columns.

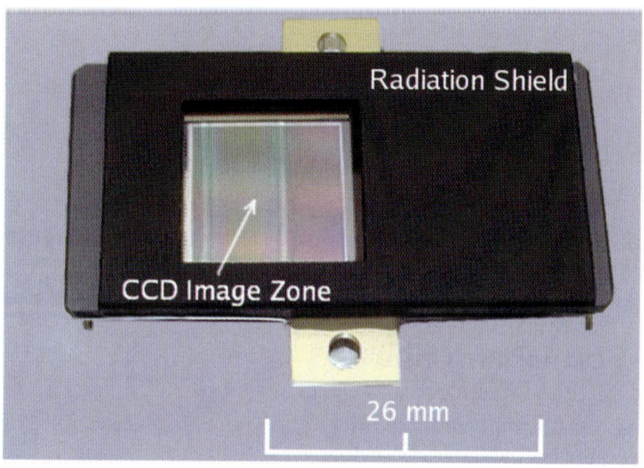

Fig. 20 The Atmel (Thomson) TH7888A CCD with its tantalum radiation shield in place protecting the memory storage zone of the frame-transfer-type device

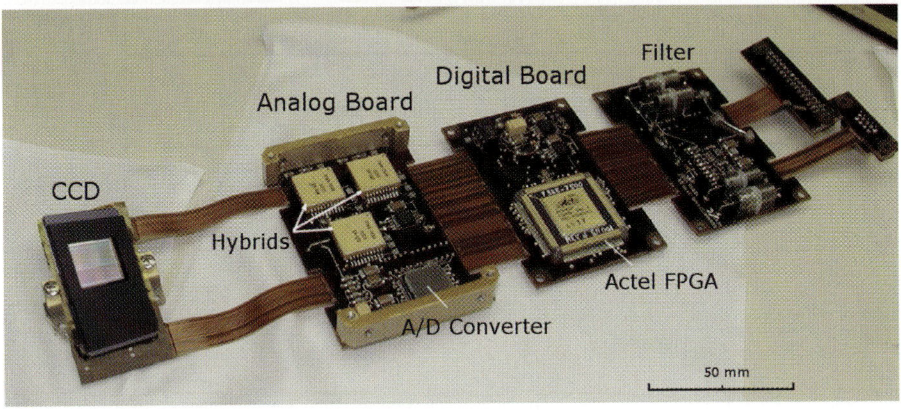

Fig. 21 Focal Plane Unit electronics, comprised of multiple rigid flex boards that fold into a compact design

Because of the severe mass limitations imposed on all aspects of the MESSENGER mission, the camera electronics were required to be smaller than the CONTOUR/CRISP FPU electronics on which the design was based (Darlington and Grey 2001). A hybrid incorporating considerable parts of the clock driver circuits was developed by The Johns Hopkins University Applied Physics Laboratory (APL). A type of construction of the printed circuit boards known as rigid-flex enabled the electronics to be folded into a small space (Fig. 21). The clock drivers use an integrated circuit (IC) designed to drive the gates of power metal oxide semiconductor field effect transistors (MOSFETs) and are almost ideal to drive the capacitive gates of a CCD. The Atmel CCD requires drive levels, speeds, and currents that are compatible with the drivers. The TH7888A has a relatively high internal gain of 6 μV/e. This value is large enough to drive the correlated double sampler analog-to-digital converter (ADC) integrated circuit, made by Analog Devices, directly while meeting the system noise requirements. The output is a 12-bit digital word for each CCD pixel value. A field-programmable gate array (FPGA) is used to provide clocks for the CCD and ADC derived from an input clock from the DPU. The DPU also provides a coded exposure time signal for the camera, as the autoexposure algorithm is in the DPU. Figure 22 shows the detailed timing for short exposures (<1 s) in the top panel and long exposures ($1 <$ exposure < 10 s) in the bottom panel. The FPGA also formats the data for transfer back to the DPU. The interface between cameras and DPU uses low-voltage differential signaling for low noise and to tolerate differences in grounding between the units.

During normal operation the CCD is read out once per second whether or not images are being saved. This action sweeps out accumulated dark current in the detector. The only time this sequence is interrupted is when exposures of longer than one second are being made. The CCD also has an antiblooming gate which removes accumulated dark current from the image area before the start of an integration by effectively setting the full well size to zero.

Hybrid and Power Conditioning. The hybrid incorporates two driver ICs and a linear regulator with various decoupling and bias components into a package a little larger than a single-driver IC. Figure 23 shows the MDIS hybrid with its lid removed and the two driver ICs and linear regulator magnified. One hybrid can provide four CCD clocks, and a total of three hybrids is needed by each camera. The frame transfer process, which takes place in 3.84 ms, repeatedly drives high capacitance lines to transfer an exposed image from the image area to the storage area of the CCD. The result is a large current requirement every

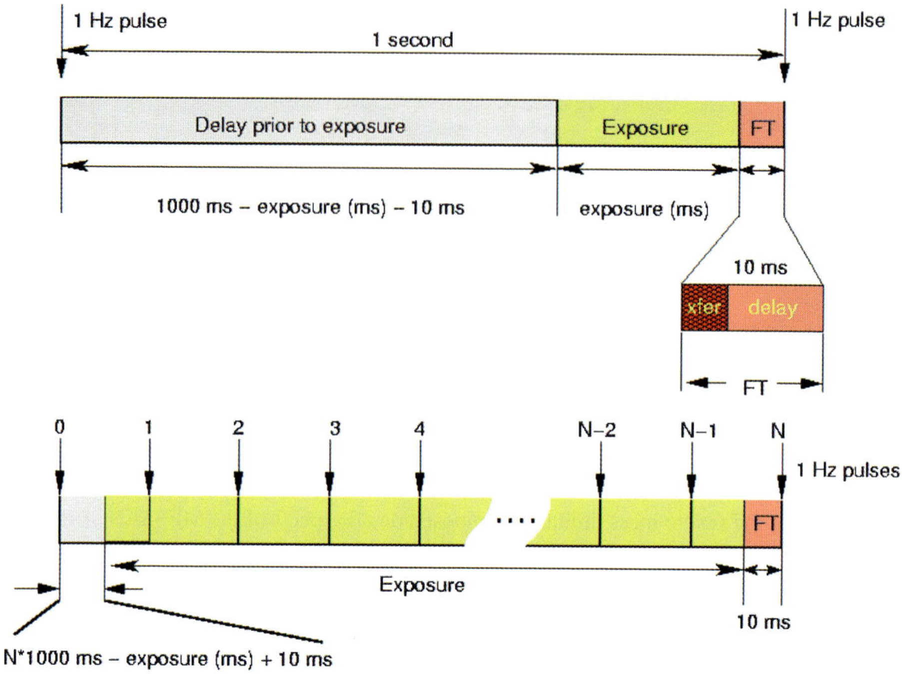

Fig. 22 Detailed frame timing for the FPUs. All clocking, control, and readout of the CCDs is generated within the FPGA in the FPU. The actual frame transfer (FT) time (xfer) is 3.84 ms

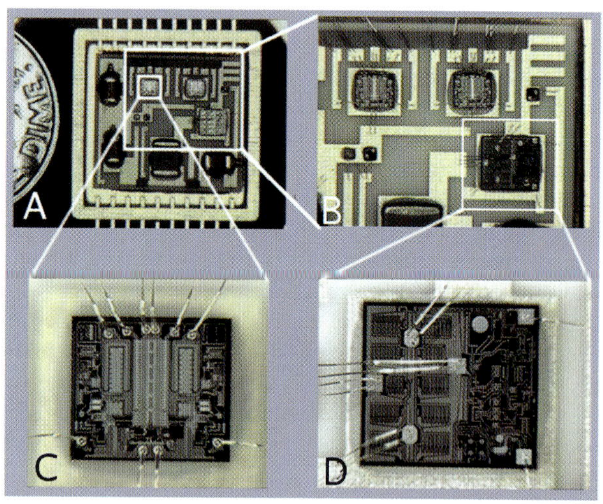

Fig. 23 The development of a hybrid driver component significantly reduced the overall size of the MDIS FPU electronics over earlier designs. (**A**) Custom hybrid component with lid removed showing external leads and relative size. (**B**) Enlargement of the two MOSFET-driver integrated circuits and linear regulator. (**C**) Detailed view of one MOSFET driver. (**D**) Enlargement of the LM117 linear regulator

frame, occurring for a few milliseconds each second. This current would be enough to overload the converter driving the camera if it appeared at the power input. To avoid overload, simple passive filtering is incorporated in the camera (see Fig. 19) to reduce the current surge to an acceptable level.

Table 7 Requirement flowdown for the MDIS camera detector electronics

Description	Requirement
Size of image plane	1,024 × 1,024 pixels, 14 × 14 μm
Other formats	2 × 2 binned (512 × 512)
Spectral coverage	400–1,100 nm
Electrical dynamic range	Up to 4,000 : 1
Noise level	50 electrons
Exposure range	1 ms to 10 s
Linearity	<1 %
Operating temperature, CCD	−40°C to −10°
Operating temperature, housing	−40°C to +30°C
Power consumption	<2 W

Test and Performance. The completed camera boards underwent a variety of electrical tests before the CCDs were installed to ensure that they had the expected electrical performance. After CCD installation, the camera design was tested with optical input to verify the performance requirements.

In general the cameras met the requirements shown in Table 7 without difficulty. The dark current for long exposures causes a small shift in dark level. The CCD provides specially shielded columns to provide an accurate measure of the dark level. Four of these dark pixels are sampled per line and substituted during readout, for four of the 1,024 active pixels in a full-resolution image (1,024 × 1,024). The four reference columns provide a measure of the dark background. In the case of 2 × 2 binning, no valid dark reference pixels are available.

The antiblooming feature of the CCD causes a scene-dependent nonlinearity. This nonlinearity is an artifact of the earlier design on which this one was based. The extent of the nonlinearity was measured to be below the required upper bound. When large areas of the focal plane array are illuminated, there is no measurable nonlinearity. However, there is some evidence that when small regions of the array are brightly illuminated, the linear response could be outside the specification.

3.2.2 DISE

The MDIS instrument contains six major electrical subsystems: two FPUs, a filter wheel motor, a filter wheel motor resolver, a platform pivot motor, and a platform pivot motor resolver. The pivot motor contains redundant windings with one set going to each DPU. The remaining systems must be controllable by either of the two redundant DPUs. Selection between these units is provided by a DISE box. The DISE box is identified in Fig. 1, and the overall signal flow is shown in Fig. 24.

The DISE box switches between the DPU and MDIS electronics while minimizing the connections through the rotary twist-capsule feedthrough. There are two switching functions: one selects which DPU is master and the other selects which camera will be active. For the DPU selection, latching relays are used. When a DPU powers up, it asserts itself as master by pulsing primary voltage on a control line to the DISE board. This pulse switches relays that control the power and signal interfaces to the cameras. This pulse also switches relays that pass DPU commands to the filter wheel, filter wheel resolver, and pivot motor resolver. There is one relay for each control line needed for these interfaces. After the DPU asserts itself as master, it then applies power to the camera of interest. Each DPU provides separate power lines for each camera. The DISE board uses these power connections to

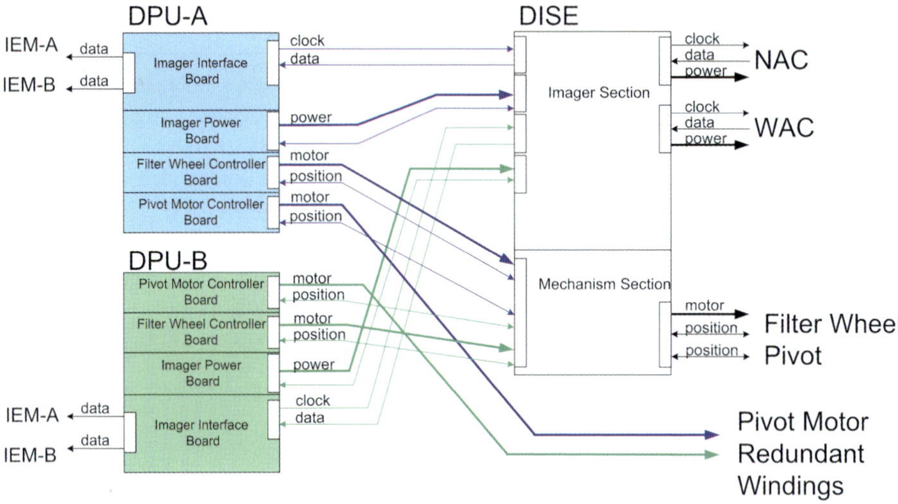

Fig. 24 Detailed interface showing power and signals from redundant DPUs through the DISE box. Mechanical relays in the DISE box are switched depending on which DPU provides power to the DISE box. The DISE box also selects which camera communicates with the DPU. The LVDS protocol used for all control signals and data largely eliminated any length constraints for the harness. However, cable runs between the DPUs and DISE were kept to a minimum for mass considerations

power its internal circuitry that selects the correct camera as well as powering the camera of interest. This selection is done in a master/slave configuration with the NAC having priority. If power is supplied to both cameras at the same time, signals from the NAC will be transmitted. This arrangement ensures that the master DPU communicates with the intended camera.

In addition to delivering each camera with the 15 V provided by the master DPU, the DISE box linearly regulates a 6-V supply. This regulation is done in the DISE box to limit power dissipation and reduce mass in the cameras. An FPU draws significant current at 6 V; the linear regulators therefore dissipate the majority of the 1.7 W of the DISE box. To ensure good thermal conduction of these components so that they can safely dissipate their heat, they are mounted directly to the lower corners of the DISE box frame. Since the DISE box physically mounts onto a thermally nonconducting composite stand of the MDIS bracket assembly, a copper thermal strap conductively ties the DISE box to the spacecraft deck. A Nusil thermal gasket was used to ensure excellent thermal contact between the copper heat-sink and the deck of the spacecraft.

The DISE box switches control lines using the low-voltage differential signaling (LVDS) protocol to and from each camera. Each camera has a single LVDS interface, and each DPU has a single LVDS interface. Switching between them is done with cold-sparing LVDS receivers and cold-sparing complementary metal oxide semiconductor (CMOS) multiplexers. The power and enable signals for the LVDS and multiplexer chips, shown schematically in Fig. 25, are derived from the active 15-V DPU camera supply power.

3.2.3 Data Processing Unit

The DPU electronics provide two distinct services: they act as a communications router and interface between all the instruments and the spacecraft, and they provide complete support

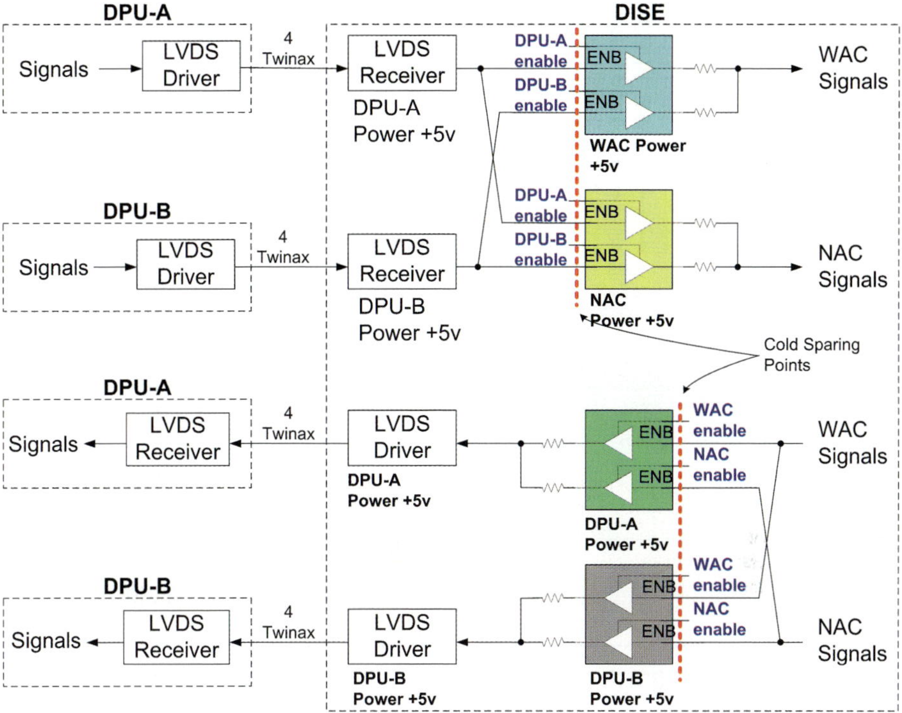

Fig. 25 Low-voltage differential signaling (LVDS) interface between the redundant DPUs and MDIS. Cross-strapping takes place in the DISE box, which selects either a WAC or a NAC through the enable (ENB) of a tri-state bus transceiver when commanded from the active DPU

for the MDIS camera electronics. Two separate sets of electronics boards are physically stacked and packaged together. Figure 26 shows a block diagram of the DPU. The first set of boards, comprised of the LVPS, EPU, and UART hub board, provides basic power conditioning, data processing, and communications functions. The second set supports only MDIS and is comprised of two motor controllers, WAC and NAC power supply boards, and a single imager interface board. The EPU and LVPS boards are common designs that are used in all instruments (except MLA).

Event Processing Unit (EPU) Board. The EPU board, one of the two boards of the common payload design, provides data processing capability and creates the telemetry data packets passed (through the DPU) to the spacecraft. The design was based heavily on versions that were flown on earlier instruments built at APL, including the Imager for Magnetopause-to-Aurora Global Exploration (IMAGE) High Energy Neutral Atom (HENA) and CONTOUR CRISP and Contour Forward Imager (CFI) instruments. All these designs are based on the Intersil (formerly Harris) RTX2010 16-bit processor. This processor, used in many APL instruments and spacecraft subsystems for the past 15 years, directly executes Forth language instructions. On MESSENGER the processor is operated at 6 MHz and can provide approximately 6 million instructions per second of processing power.

In addition to the RTX2010, the EPU board contains 64 KiB of fuse-link programmable read-only memory (PROM) for boot code, 256 KiB of electrically erasable PROM (EEPROM), and 256 KiB of SRAM. An Actel RT54SX32S FPGA provides all the logic

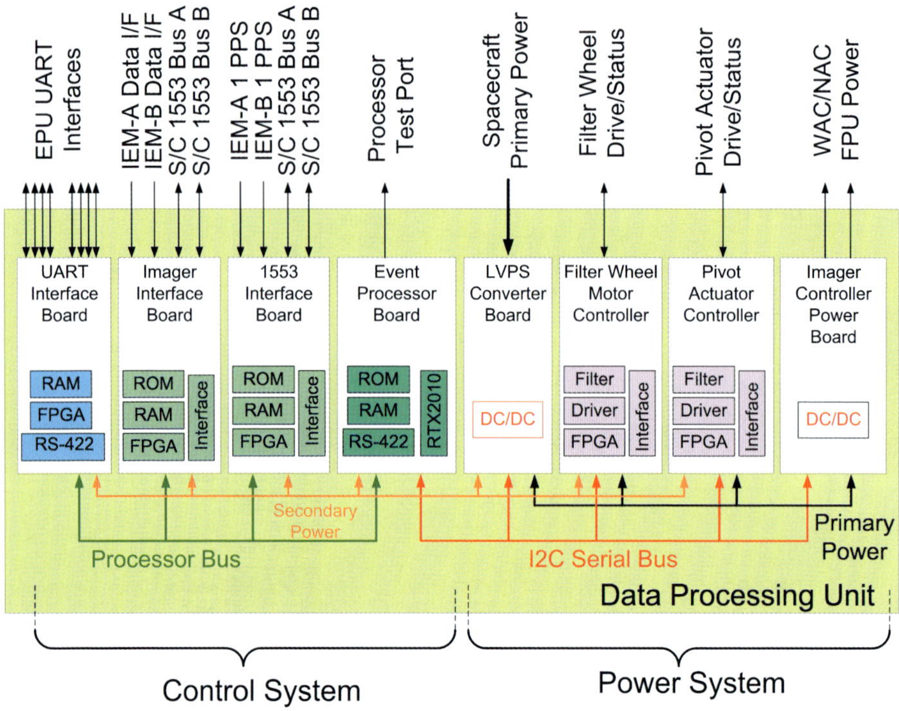

Fig. 26 Block diagram of the MESSENGER DPU

necessary to interconnect the processor with the stacking connector data bus, as well as implement the interface UART (running at 38.4 kbaud) to the DPU, the test port UART (running at 19.2 kbaud), and an I^2C bus to control off-board serial peripherals. The EPU also receives (via the DPU) a once per second timing pulse from the spacecraft. This timing "SYNC" signal originates at the spacecraft IEM oven-controlled crystal-oscillator (OCXO or coarse clock) and is used as an interrupt to the processor to maintain timing synchronization throughout the payload. This synchronization pulse is a key part of the overall timekeeping design to maintain <1 ms end-to-end timing error between the mission operations center and the payload.

Low-Voltage Power Supply. The LVPS board converts spacecraft primary power, ranging from 20 to 36 V, to isolated secondary power. The common payload LVPS design was based heavily on versions that were flown on the CONTOUR CRISP and CFI instruments.

Most of the instruments operate from only four output voltages: ± 5 V and ± 12 V. The EPPS instrument (Andrews et al. 2007) uses ± 15 V instead of ± 12 V. In addition to providing four secondary outputs, the converters also provide power switches for auxiliary primary power loads and some analog housekeeping capability.

A design tradeoff was made between the use of a customized converter design and one based on commercially available hybrid converters. Although a customized approach would have provided better overall power efficiency, the cost and schedule limitations had a significant impact on the decision to use the VPT, Inc., family of converters, based primarily on their small size and low noise output.

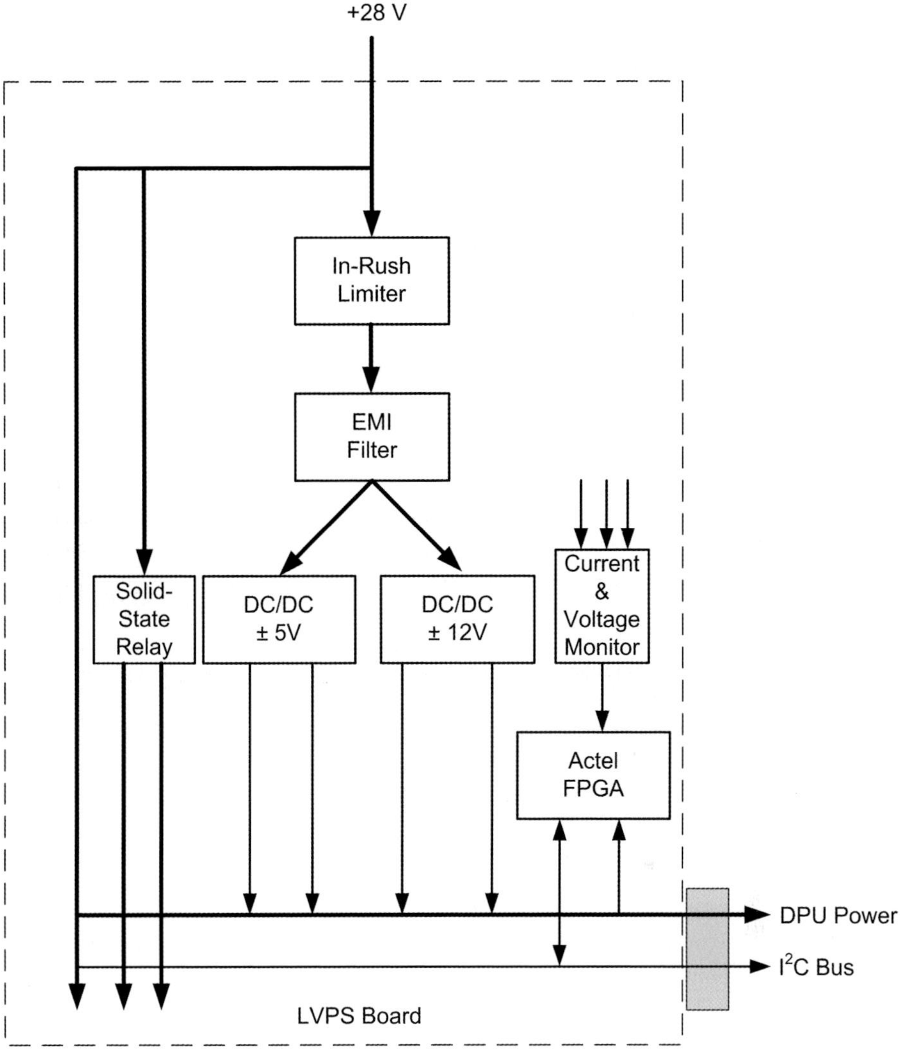

Fig. 27 Block diagram of the low-voltage power supply board, one of the two boards common to all the instruments (except MLA)

The functional block diagram for the LVPS board is shown in Fig. 27. The LVPS board uses an in-rush limiter, a VPT electromagnetic interference (EMI) MIL-STD-461 filter, and two VPT hybrid dual-output converters (one for ±5 V and the other for ±12 V). These converters provide better than 1% load and line regulation for nominal operation and good step response. Two radiation-hardened opto-field effect transistor (FET) relays were used to switch up to 1 A of primary power to off-board loads. An analog monitoring circuit, comprised of a 14-bit ADC, instrumentation amplifiers, a 16-1 multiplexer, and an Actel FPGA, provided 16 input analog channels, which were used to monitor four voltages and six currents on the board and five user-provided analog voltages from off the board; the remaining channel measured board temperature.

Table 8 MESSENGER payload requirements on the common low-voltage power supply board design

Instrument	Load on ±5 V	Load on ±12 V	Total power out (mW)	Estimated efficiency on ±5 V	Total efficiency
XRS	2,550	480	3,030	65%	55%
GRNS	3,250	1,440	4,690	67%	65%
EPPS	2,950	1,440	4,390	67%	65%
MAG	1,250	720	1,970	55%	48%
MDIS	1,000	480	1,480	50%	41%
DPU	3,250	0	3,250	67%	67%
MASCS	0	480	480	67%	30%

A major challenge in using a single-power-supply design for multiple instruments was the ability to handle varying load requirements. Table 8 shows the range of output loads between the payload elements that used the common LVPS board. A drawback of these dual-output hybrid converters is the need to provide a minimum load on each output to ensure that both outputs remained in regulation; however, this constraint did not result in any significant power losses.

UART Hub Board. Communication between each DPU and the instruments (other than MDIS) takes place through a Universal Asynchronous Receiver/Transmitter hub. This board uses an Actel RT54SX32S FPGA, 1 MiB of static read-only memory (SRAM), and a number of RS-422 receiver and transmitter circuits to implement eight 38.4-kbaud, RS-422, full duplex UART channels. Each channel has a separate first circuit interface for maximum fault tolerance. Independent SRAM-based receive buffers indexed by FPGA-based registers are used to offload byte-by-byte operations from the DPU software. Data may be received simultaneously from all eight channels. A single memory-mapped transmitter is steered to any of the eight output channels, such that only one channel is actively being commanded at a time. However, all output channels can be selected simultaneously for broadcast messages, such as time distribution.

The local EPU board for each instrument (other than MLA) formats the instrument telemetry packets. The telemetry is copied, byte by byte, into the SRAM buffers in the DPU as the data are received from the payload. Data may be transmitted continuously or in bursts and are instrument specific. The SRAM buffer is operated in a ping-pong fashion, where one set of buffers is receiving data, while the other is made available to the DPU expansion bus for processing by the DPU. Once per second, the DPU switches receive buffers, thus introducing a 1-s latency between when instruments send data and when the DPU forwards them to the spacecraft. The instruments are not allowed to send telemetry during a 300-μs period once per second to accommodate this buffer swap operation. The DPU also sends each instrument a differential one pulse per second (PPS) SYNC signal (also RS-422) to help define this telemetry "quiet time," as well as provide a very good time reference for the instruments to synchronize with the spacecraft 1-PPS clock.

1553 Interface Board. Communication between each DPU and the spacecraft is accomplished through MIL-STD-1553 busses. The 1553 Interface Board is an exact copy of a design used on the CONTOUR CRISP and CFI instruments. The design is based on the UTMC SUMMIT UT69151DXE protocol/transceiver chip and also includes an Actel RT14100A

FPGA and 64 KiB of SRAM. Each DPU is a remote terminal on the spacecraft 1553 bus. All payload commands and telemetry, except MDIS image data, are communicated via the 1553 bus.

The DPU receives a 1553 bus message from the spacecraft once per second containing the time corresponding to the next hardware 1-PPS signal used to synchronize systems on the spacecraft. This message is received at a fixed time before the 1-PPS signal. The DPU transmits this time message to all instruments prior to the arrival of the 1-PPS signal via the UART data links. This approach eliminates the software latency and jitter of processor interrupt response and avoids the latency of UART transmission.

Imager Interface Board. All image acquisition control and hardware image processing occur in the Imager Interface Board. The detailed signal interfaces for this board are shown in Fig. 28. This DPU slice is a modified version of a design used on the CONTOUR CRISP/CFI instruments. It uses two Actel RT54SX72S FPGAs, 2 MiB of SRAM, and a number of LVDS receiver and transmitter circuits to communicate both with the FPUs and the spacecraft high-speed IEM interface card. The Imager Interface Board can control only a single FPU at a time. Camera selection and signal multiplexing occur in the DISE box (see Sect. 3.4.2).

The Imager Interface Board contains two FPGA designs: an imager-interface FPGA and an RTX-bus FPGA. Image data are continuously received from the active FPU and buffered

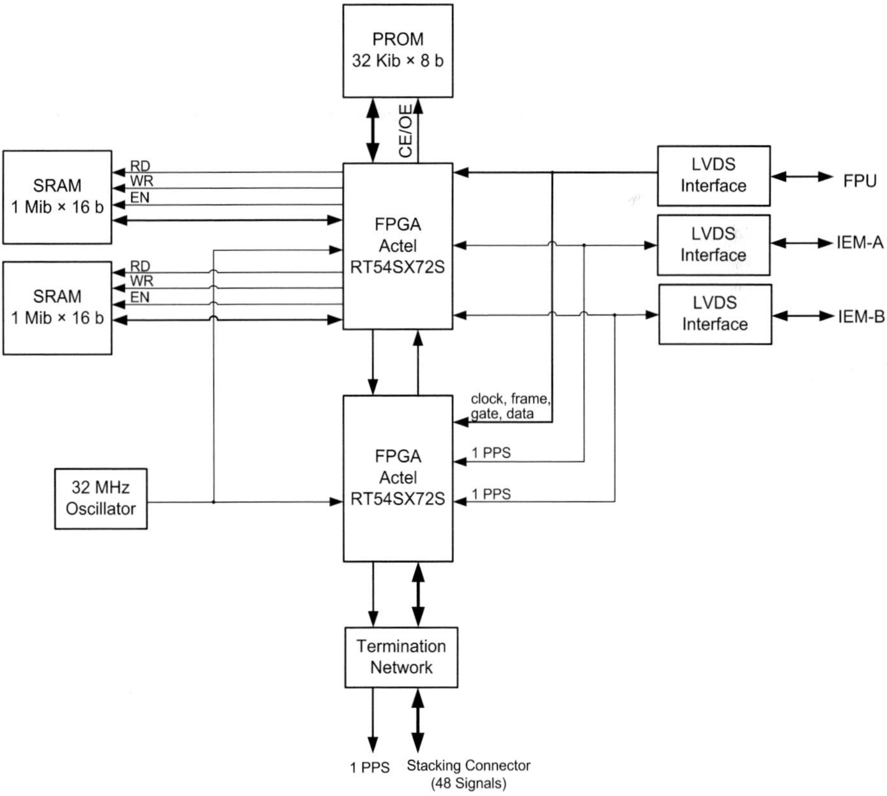

Fig. 28 Block diagram showing detailed signal interfaces for the imager interface board

in SRAM. If the DPU receives a command to record an image, the image data are transferred to either of the two redundant IEMs. The main function of the imager-interface FPGA is to read images from the FPU and send them to the IEM. There are two possible image sizes transmitted from the FPU: 512 × 512 (CCD on-chip) binned images and 1,024 × 1,024 full-frame images. Both formats use 12-bit pixels. A lossy compression method available through eight selectable look-up-tables (Fig. 4) reduces the 12-bit image data to 8 bits. A lossless algorithm, referred to as "Fast" compression, is a second hardware compression mode selectable by the imager-interface FPGA. With the two hardware compression options, a total of eight possible image formats may be sent to the IEM.

The image data read from the focal plane are received from a serial LVDS interface at 16 MHz. A full 1,024 × 1,024 image requires approximately 850 ms to be received in the DPU. The data are sent at 4 MHz to the IEM, limiting the image frame rate. Binned images may be sent to the IEM at 1 Hz; however, full-frame images can be sent only at 0.25 Hz. This interface limitation to the IEM requires the imager-interface FPGA to buffer the images in SRAM. The design consists of two main processes: one reads from the FPU and one writes to the IEM. The FPU read and IEM write processes are completely separate, and each has a handshaking signal to indicate that the respective process has completed. A third SRAM control process determines which SRAM is being accessed by which process by monitoring the 1-PPS and handshaking signals from the other processes.

Fast compression takes advantage of the fact that it is not usually necessary to represent a pixel value by the full 12 bits. This technique can be further optimized by recognizing that adjacent pixels often tend to be correlated. By differencing adjacent pixels, with a technique known as differential pulse code modulation (DPCM), fewer bits are required to represent the original pixel value. This technique, implemented in the imager-interface FPGA, permits the maximum number of images to be stored on the SSR without paying the time penalty for a more rigorous compression method. Each Fast-compressed image will be of variable length, and its overall size will be scene dependent. Typically, a higher scene entropy will result in a lower compression ratio.

The main functions of the RTX-bus FPGA are to store a 1,024-bit image header and the FPU command words generated by software running on the RTX processor in the DPU. This FPGA is also capable of generating a 32-bin histogram created from the pixel values produced by the FPU. This histogram is made available to the RTX processor and is integral to the MDIS autoexposure algorithm. The FPGA allows the RTX processor to control the flow of the 1-PPS signal, and it makes the FPU status data available to the RTX processor.

A dedicated memory address on the RTX bus backplane provides the RTX processor access to the RTX-bus FPGA. Through this interface, the RTX processor can write the 1,024-bit header and FPU control words, read the histogram data, and write the 1-PPS source and output registers. There are three valid sources of the 1-PPS signal. The first two are from each of the IEMs through the LVDS channels. The third is a register in the RTX-bus FPGA to which the RTX processor can write. Another register to which the RTX processor writes determines which 1-PPS source is sent to the FPU and down the DPU backplane.

Imager Power Board. The MDIS Imager Power Board provides switched +15 V to the NAC and WAC electronics. The board is a modified version of the common LVPS board already described and shares the same in-rush current limiter, EMI and secondary filter, FPGA, and telemetry monitoring design. The supply has only one VPT DVSA2815S DC/DC converter model, which produces a regulated 15-V output. A separate MOSFET-switched output is used for each set of NAC and WAC FPU electronics.

Table 9 Characteristics of the actuators used on MDIS

Characteristic	Pivot actuator	Filter wheel actuator
Manufacturer	Starsys	CDA InterCorp
Primary drive	24 to 35 V	24 to 35 V
Two-phase stepper	yes	yes
Reference frequency	400 Hz	2.5 kHz
Maximum step rate	1.1°/s	75°/s
Electrical resolver	8 speed	1 speed
Motor input step size	5°	30°
Gear system	Hybrid	In-line planetary
Gear reduction	580 : 1	50:1
Output step size	<0.01°	0.6°
Backlash	<0.01°	<0.6°
Mass (actuator + resolver)	916 g	139 g
Redundant windings	yes	no
Range of motion	240°	360°

Motor Controller Board. Each actuator in the MDIS design has a dedicated motor controller board. Although the characteristics of the pivot and filter wheel actuators are different (see Table 9), the board designs are effectively identical. Because most of the low-level motor stepping functions are directly controlled by the DPU processor, the boards are fairly simple. Each board uses an EMI filter, an Actel A1020 FPGA, and a bipolar UDS2998 motor driver to generate the appropriately phased motor drive waveforms. The angular position of each actuator is measured via an AD2S80A motor resolver readout hybrid. The DPU processor generates the motor phase drive timing signals and monitors the shaft position via the DPU I^2C bus. Opto-isolators are used to keep the primary power motor drive and control signal grounds separate.

3.3 Software

3.3.1 Common Software

The APL-developed MESSENGER instruments share common software, an instrument domain library. The library was created in 1994 to serve the NEAR instruments. It is currently used in MESSENGER as well as the Compact Reconnaissance Imaging Spectrometers for Mars (CRISM) instrument on the Mars Reconnaissance Orbiter and the LOng Range Reconnaissance Imager (LORRI) and Pluto Energetic Particle Spectrometer Science Investigation (PEPSSI) instruments on New Horizons.

The common software runs as a layer below the application-specific software. The first layer provides standard services, for example, command handling and telemetry packet queuing. Another layer below the service layer is a host abstraction layer. This layer abstracts the input/output (I/O) interfaces used by the instrument to communicate with the spacecraft. The host layer must be customized for each mission given that spacecraft interfaces are not standardized. For the MESSENGER instruments, two different host layers were needed. The MDIS instrument is connected directly to the spacecraft via a 1553 bus. The other instruments use RS422 serial lines to the UART hub board in the MDIS DPU. Software in the DPU acts as an intermediary between the spacecraft and the instrument.

The common software includes a boot program. Its most frequent use is to boot application code by copying it from EEPROM to RAM and then starting the application. However, the boot program also has commands for loading memory from the ground, dumping memory back to the ground, and copying memory to and from EEPROM. These features can be used to upload new instrument programs. The boot program is stored in nonvolatile ROM. In the event that an instrument's program in EEPROM gets corrupted, the boot program will allow the instrument software to be reloaded.

The common software provides command handling and telemetry queuing services. Application-specific software registers command handlers with the common software. Registration provides the command's operation code (opcode), the expected number of arguments, and some code to handle the command. When the common software receives a command from the spacecraft that is properly formatted and has the correct opcode and sufficient arguments, the provided handler code is executed. The common command software can store and run sequences of commands, called macros. Macros can loop and call other macros. Up to 64 macros can be executing at any time. The common software's telemetry service provides formatting and queuing of telemetry packets. The software gathers the packet's data and constructs a Consultative Committee for Space Data Systems (CCSDS) header with the appropriate flags, sequence number, and other information. The packet is queued for transmission to the spacecraft. A 64-KiB buffer allows instrument software to queue large bursts of telemetry that the common software then trickles out to the spacecraft. The common software also includes status/housekeeping packet generation, time management, voltage and current monitoring, and I^2C bus management services.

3.3.2 MDIS-Specific Software

The MDIS-specific software controls the imager hardware, exposure time, and mechanisms. It also serves as a communications hub for the other MESSENGER instruments. Time messages and timing pulses are broadcast to the other instruments. Commands are accepted from the spacecraft and forwarded to the appropriate instrument. Telemetry is collected from each instrument and forwarded to the spacecraft.

Imager Control. Only one of the MDIS cameras, either the NAC or WAC, can be used at any given time. The MDIS software selects the camera to use based on user command. Images are collected in three stages: expose the image in the FPU, read the image from the FPU to the DPU, and finally transmit it from the DPU to the spacecraft. Different processing steps are applied to the image during each of these stages. The steps include exposure, 2×2 binning, lossy 12 : 8-bit compression, and lossless Fast compression. All of these steps are done in hardware; in fact, the software cannot access the image at all. The software does control the processing, for example, by enabling or disabling the image compression at the correct time based on user commands. The software also manages the three stages of exposure, read, and transmit as a pipeline. In the simplest pipeline, each stage takes 1 s and therefore one image can be taken every second. More complex pipelines result when image binning is disabled (unbinned images take 4 s to be transferred to the spacecraft) or when multisecond exposures are commanded.

Exposure Control. The exposure time of images can be set manually by command or automatically by the software. In manual mode, the range of available exposure times is nearly 10 s. In automatic mode, the exposure time of the next image is computed by the software. This computation has two distinct steps. The first step computes a new exposure time based

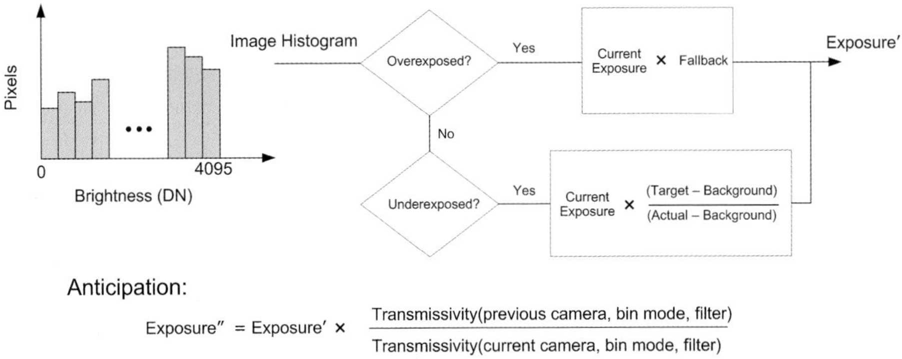

Fig. 29 Autoexposure algorithm decision tree. A 64-bin histogram is computed in hardware for each image. If an image is determined to be underexposed, the actual exposure is computed as Actual = minimum brightness such that the sum of the pixels above this brightness < saturation threshold

on the brightness of a previous image. The second step anticipates filter wheel motions and adjusts the computed exposure time accordingly.

During the read stage of the image pipeline, the hardware generates a histogram of the image. The histogram is analyzed by the software to determine if the image is overexposed or underexposed. First, the histogram is scaled by a factor of four if it comes from a 2×2 binned image. If the brightest histogram value exceeds a specified saturation threshold, the image is considered overexposed and the exposure time is scaled back by a specified fallback factor. Otherwise the image is considered underexposed. Histogram values are accumulated starting from the brightest bin down towards the dimmest bin, until the specified saturation threshold is exceeded. The brightness value that causes the sum to exceed the threshold is defined to be the actual image brightness. The exposure time is scaled by the ratio of the commanded target brightness, to the actual brightness, after a background brightness is removed. The algorithm is characterized by uploadable parameters for the commanded target brightness saturation threshold (the number of allowed saturated pixels), overexposure fallback, and background brightness. The algorithm is depicted in Fig. 29.

The algorithm described so far compensates for changes in scene brightness and filter wheel changes. The next step adjusts the exposure time further if the imager, binning mode, or filter selected for the next exposure does not match what was used in the test exposure. The exposure time is scaled by the ratio of the transmissivity of the old setup to the transmissivity of the new setup. An uploadable table of transmissivities for the WAC filters and for the NAC in either binning mode are used. Finally, the computed exposure time is forced to fall within an uploadable range but is always less than 1 s.

Mechanism Control. The MDIS pivot platform and filter wheel are controlled by stepper motors. The motor controller boards are on an I^2C bus, and each motor step must be commanded over this bus. There are also power supply boards on the I^2C bus; power switches must be commanded and analog housekeeping read from these boards. The common software provides a mechanism for building and executing an I^2C bus schedule. The MDIS software uses this code to construct a four-slot schedule that repeats every 8 ms, i.e., 2 ms per slot. One slot is used to step the pivot platform, another slot is used to step the filter wheel, and the other slots are used to control the power supply boards.

Given the pivot motor gear ratio and the step rate above, the platform can move about 1.1°/s. Similarly, the filter wheel can move 75°/s. Given 12 filters, an adjacent filter can be reached in 0.4 s and any filter within 2.4 s. Note that the I^2C scheduling method allows the pivot platform and filter wheel to move simultaneously.

3.3.3 Software Metrics

The MDIS software, excluding the boot program, has about 8,000 lines of source code (5,000 lines if only noncommand and nonblank lines are counted). Approximately 60% of this total is shared/reused code in the common software. The compiled application is about 63 KiB.

3.4 Electro-Mechanical Design

The mechanical assembly of MDIS consists of two main components, a bracket assembly that interfaces to the spacecraft and a pivoting platform on which both cameras and the thermal control system are located. One leg of the bracket assembly supports the DISE box and a rotary interface through which all electrical connections to the pivot platform pass. The opposite leg of the bracket supports the precision pivot motor.

3.4.1 Bracket Assembly

The bracket assembly, made of the same advanced composite material (graphite/cyanate ester fabric prepreg) as the MESSENGER spacecraft structure (Leary et al. 2007), was chosen to minimize mechanical stresses due to mismatches in coefficients of thermal expansion. The composite bracket assembly, being a very poor thermal conductor, isolates the pivot platform and thereby the cameras from the payload instrument deck. Eight titanium feet form the mechanical interface to the spacecraft. Four feet were bonded to each stand using a precision fixture, then bolted to the stand. A trim-cut machined from a top layer of epoxy on each composite stand ensured that the pivot axis of rotation was parallel to the spacecraft deck.

Pivot Actuator. The pivot motor assembly is located on one of the composite stands and is comprised of a stepper motor with a hybrid gear-train and electrical resolver. The stepper motor contains redundant windings so that the two fully redundant DPUs can independently control the motion of the pivot, enhancing the overall reliability of the pivot mechanism.

The total range of motion of MDIS is about 240°, limited by hard mechanical stops in the pivot motor. Pointing knowledge is determined by first "homing" the instrument, which is accomplished by driving the actuator into one of the mechanical hard stops for a period of time sufficient to ensure the orientation of the instrument if it had been previously stopped at the opposite extreme of travel. The rotational speed of the pivot platform is ~1.1°/s, so homing requires about four minutes. Once the location of the pivot actuator is known, the flight software retains this knowledge, and subsequent pointing commands are achieved by counting pulses (steps) to the motor.

The MDIS pivot actuator (Fig. 30) is capable of accurately stepping in intervals of 0.01° (~150 μrad) per step, achieved through a hybrid gear train within the actuator. The 5° stepper motor is coupled to a 5 : 1 in-line planetary gear train, used to drive the wave generator of a harmonic drive. Harmonic drives permit very precise pointing and high torque ratios and weigh less than comparable conventional gear systems. These devices use a radial motion to engage the teeth of a flex spine and a circular-toothed spline, rather than the rotating motion of other gear systems. The harmonic drive in the pivot actuator provides an additional factor

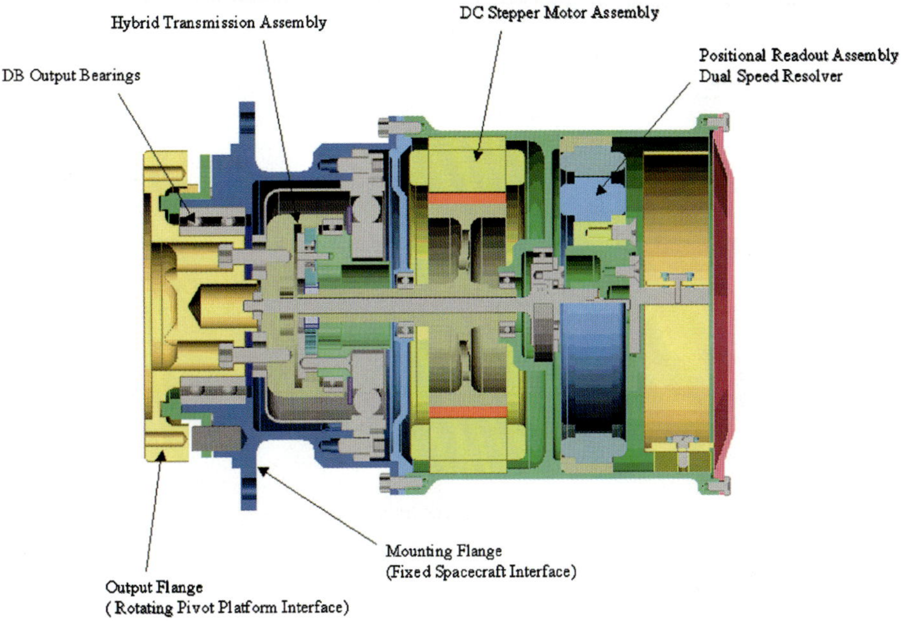

Fig. 30 Schematic view of the pivot actuator, with callouts identifying the hybrid transmission system and duplex pair radial bearings with outer races back-to-back (DB). An in-line planetary gear system drives a harmonic drive, achieving the accurate small-step pointing of MDIS. Individual motor windings are routed to each DPU, making the pivot actuator fully redundant

of 120 : 1 mechanical advantage, resulting in a total gear ratio of about 580 : 1. The hybrid stepper provides for very high gear ratios and near zero-backlash performance resulting in pointing knowledge <75 μrad. The pivot actuator assembly also includes an angular position sensor (electrical resolver). However, because of mechanical distortions in the drive shaft and electrical noise in the device itself, the output resolution of the electrical resolver is worse than a step size.

Protective covers for optical components are very desirable during ground-based testing, launch, and trajectory maneuvers requiring large thruster burns. However, due to mass limitations, conventional protective covers for MDIS were not practical. Instead, the critical first optic of each telescope is protected by rotating the platform 180° from the spacecraft's $+Z$-body axis such that both cameras look downward into the deck. This innovative approach acts like a reclosable cover and ensures a circuitous path for any particulate or molecular contamination.

Drive Train and Pointing. A challenging requirement on the mechanical design was to limit rotation-plane errors of the pivot platform to <100 μrad. The pivot actuator mounts on a titanium bracket, designed to isolate the platform assembly thermally from the composite bracket stands and not introduce any pointing errors due to thermal gradients from the stands to the platform. In order to minimize the error build up of the large number of mating surfaces making up the drive train of the pivot axis, a significant effort was made to hold very tight mechanical tolerances on all components affecting pointing of the platform. These components include a bearing assembly, a hollow shaft, the pivot platform, the motor shaft, and the pivot actuator. The tightest tolerances were placed on the mating surfaces of

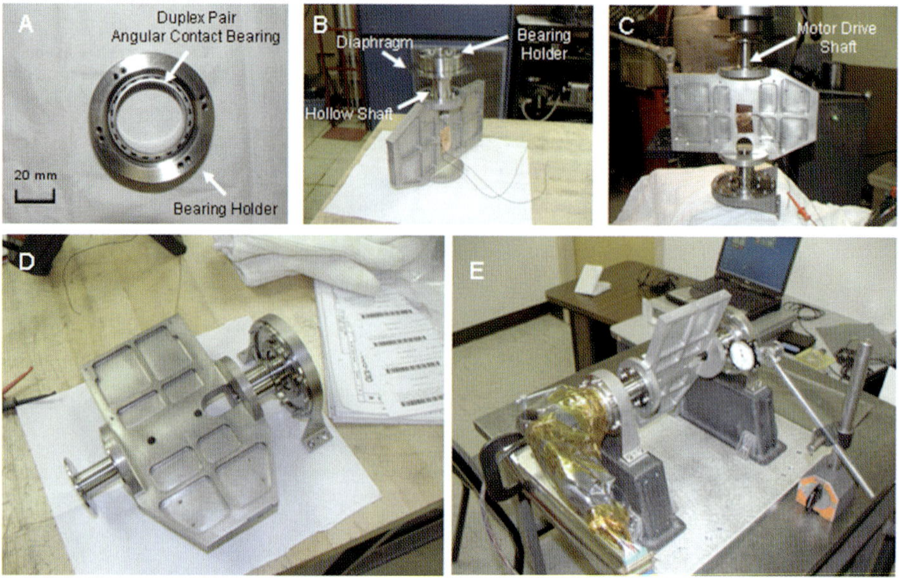

Fig. 31 Sequence of steps resulting in the precision drive train of the MDIS pivot assembly. (**A**) Duplex pair of angular contact bearings pressed into bearing holder. (**B**) Hollow shaft press fitted into the diaphragm and bearing assembly. Shaft then press fitted into platform assembly. (**C**) Motor shaft pressed into platform. (**D**) Completed drive train of platform, prior to integrating onto composite stands. (**E**) Verification of rotational trueness (runout)

these components, and interference fits were chosen to minimize error propagation in the drive train. Once assembled, all mechanical interfaces were drilled and pinned to minimize the chance of a change in the alignment due to thermal variations or vibration during launch.

The titanium bracket opposite the pivot actuator supports a bearing mount, flexible diaphragm, hollow shaft, and electrical feed-through. A duplex pair of angular contact bearings (Fig. 31A) was chosen to support the hollow shaft of the pivot platform. The stainless steel races and ceramic balls of these bearings ensure minimal molecular diffusion between the two surfaces and reduce the risk of "cold welds" that over time could decrease the torque margin of the motor. The angular contact bearings use a dry lubricant, MoS_2, and underwent a run-in period prior to installation. After preliminary cleaning, the bearings were vacuum soaked for 24 hours to eliminate moisture in the MoS_2. The bearings were exercised in a dry-N_2 environment for 10,000 cycles (revolutions) clockwise, followed by 10,000 cycles in the counterclockwise direction.

Installation of the angular contact bearings was accomplished by establishing a clearance of $\sim$12 μm between the outer race of the bearings and the inner diameter of the bearing holder. The bearings were pressed into the titanium bearing holder by heating the holder while keeping the bearings at room temperature (Fig. 31A). The hollow shaft was fabricated to have a line-to-line fit, i.e., zero clearance to within limits of mechanical measurement. Assembly of the hollow shaft required cooling the shaft, then pressing it into the room-temperature bearing and holder. A flexible titanium diaphragm is used to maintain the position of the hollow shaft in the radial direction; the diaphragm is secured to a bracket that mounts to the composite stand. The diaphragm was designed to be sufficiently stiff to support the pivot assembly while remaining flexible in the axial direction, permitting limited motion resulting from differences in thermal expansion coefficients between the composite spacecraft deck and the magnesium/titanium platform drive train.

Perpendicularity of the mating surfaces on the hollow shaft, motor shaft, and pivot platform were maintained to within machining limits (<25 μm). The interference fits of the shafts with the platform were achieved by heating the magnesium pivot platform before the hollow shaft was pressed into place (Fig. 31B). The rotational orientation of the motor shaft established the final alignment of the pivot platform relative to the spacecraft. Special keying tools ensured proper orientation of the motor shaft prior to it being pressed into the platform (Fig. 31C).

Once the rotating portion of the MDIS instrument was determined to satisfy the pointing requirement in terms of trueness of rotation to within the allowed tolerance, it was necessary to maintain the perpendicularity of the rotation axis to the mounting feet. A high-precision fixture was fabricated to emulate the spacecraft instrument deck. Each composite stand was mounted to the fixture, and the pivot platform assembly was loosely attached to the stands. Using a Coordinate Measurement Machine (CMM), two parallel planes were established relative to the mounting feet. The hardware attaching the motor drive train (bracket/pivot assembly) to the composite stands was slowly tightened. An iterative process of measuring the relative parallelism of each bracket to the mounting feet was performed. The diaphragm bracket measured parallel to the motor stand feet to within 13 μm. The assembly was bagged, and the motor/diaphragm brackets were drilled and pinned to the composite stands. Runout measured at the bearing end of the hollow shaft as shown in Fig. 31D was determined to be <20 μm (0.8 mils), corresponding to a rotational deviation of the pivot plane of motion from normal of <85 μrad.

Twist Capsule. A stainless steel rotary feed-through (twist capsule) mounts to the bearing holder and passes through the hollow shaft. Set screws position the rotating shaft of the twist capsule inside the hollow shaft. All electrical connections to the platform pass through the 110 conductors of the twist capsule. The angular extent of the twist capsule is +90° to −220° and exceeds that of the pivot motor. The combination of rigid-flex circuitry inside the twist capsule and stainless steel bearings requires extremely low torque to rotate. The connectors for each FPU were assembled after the twist capsule was installed. Most signal connections through the twist capsule are redundant and spliced just before each potted connector. The harness at the fixed end of the twist capsule largely connects to the DISE box with the few exceptions used for spacecraft thermal control and monitoring.

3.4.2 Platform Assembly

Because of mass constraints, mechanical housings were fabricated from magnesium instead of aluminum wherever possible. The pivot platform, the two FPUs, and the filter wheel housing were fabricated from magnesium stock. To minimize the stress on optical components due to thermal contraction, the WAC telescope, filter wheel, and the NAC and WAC light baffles are titanium. The NAC telescope is aluminum with aluminum mirrors. The supporting structure for the radiators and heat pipes is stainless steel, to match the coefficient of thermal expansion (CTE) of the stainless steel heat pipes and to minimize parasitic heat leaks from the radiators to the platform. The radiator panels are made of beryllium, selected for its light weight and high heat capacity. Each beryllium panel is passivated with a thin aluminum layer, which is then clear anodized to optimize its thermal properties.

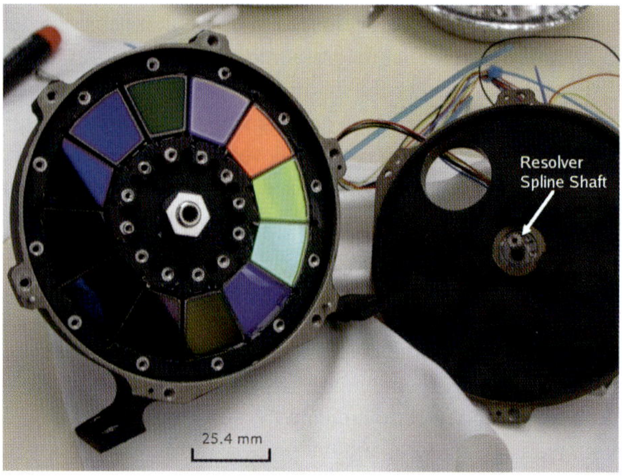

Fig. 32 Mechanical assembly of WAC filter wheel. Unique design of using an electrical resolver with a spline shaft eliminated the need for an additional bearing or bushing opposite the motor. The unusual shape of the filters reduced the overall size of the assembly

Filter Wheel Assembly. The 12 optical filters were bonded in place using a low-outgassing epoxy. By optimizing the physical shapes of the filters, the overall size and mass of the wheel assembly (Fig. 32) were reduced.

The filter wheel actuator for the WAC consists of a CDA InterCorp DS9-A stepper motor with output step angle of 30°, and an inline gear train with 50 : 1 reduction resulting in an output step size of 0.6°. The step rate is 75°/s, permitting 1-Hz imaging using adjacent filters, provided the exposure is less than 490 ms. The output shaft of the motor has a precision double-D interface to the wheel assembly and an internal spline to accept the shaft of the CDA InterCorp electrical resolver (DT1-A) angular position sensor. The motor mounts to the filter wheel housing, and the resolver mounts to the filter housing cover, eliminating the need for a bearing and providing a simple robust design.

Focal Plane Unit. In addition to mechanically supporting the detector electronics, the mechanical assembly of the FPU was designed to maintain the position of the CCD at the focal plane of the optics over the range of operating temperatures. The complex thermal design (Sect. 3.5) required that the CCDs be thermally isolated from the electronics located in the same housing.

The front panel of the FPU housing is a critical component in each camera, providing the precision mechanical support for the CCD detector and the mechanical interface for each camera's telescope. The NAC front panel was fabricated from aluminum to provide the best match to the physical characteristics of the telescope. The WAC front panel is magnesium. Best focus in each camera was achieved using precision spacers providing a three-point mount of the CCD to the front panel. Each focus spacer is pinned to the front panel to prevent shifting of the CCD during environmental qualification and launch. The focus spacers also thermally isolate the cold detector from the front panel of the FPU. Figures 33A and 33B show the inside surface of the front panel and the three focus spacers supporting the CCD. Epoxy-glass isolators increase the thermal resistance between the CCD heat sink and the front panel. The CCD itself is bonded to an aluminum heat sink with a thermally conductive epoxy. Panel C of Fig. 33 shows the assembled FPU with its electronics folded in place. An internal multilayer insulation (MLI) blanket surrounding the CCD further minimizes radiative coupling between the FPU electronics and the CCD.

Internal surfaces of the optical housing were painted black to minimize scatter. However, because of the small size of the lenses and lens spacers required by the optical prescription

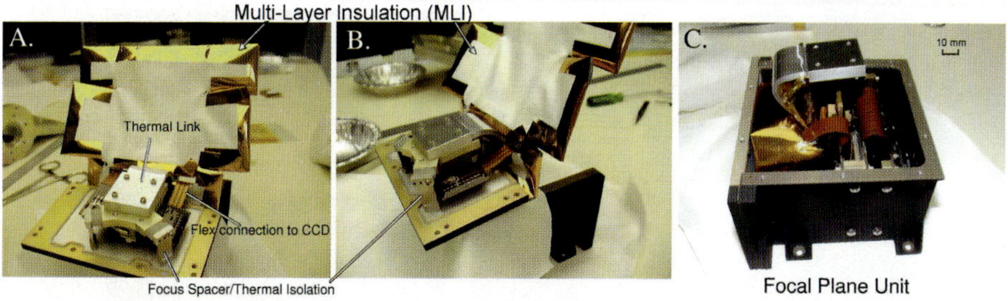

Fig. 33 Detailed integration of the NAC FPU housing. (**A**) Front panel, which supports the CCD on one side and telescope on opposite side. (**B**) Three-point mount provides correct focal position and isolates cold CCD from FPU front panel. Thermal blanket further isolates CCD from FPU electronics. (**C**) Assembled FPU without top panel

in the WAC, painting of these small components was not practical. Instead, the titanium lens spacers used in the WAC were anodized, resulting in a dark dull finish. Because these spacers are not directly in the optical path, the trade between the titanium anodization versus black paint was deemed acceptable.

3.4.3 DPU Assembly

The DPU design was optimized to keep overall mass low, accomplished primarily by using the 102 mm × 102 mm (4″ × 4″) slice architecture first developed on the CONTOUR CRISP and CFI instruments. The slice architecture consists of a simple mechanical frame that supports the printed circuit board. Both sides of the board are accessible for testing, even after assembly into the mechanical frame. The generic slice design provides for multiple configurations of the DPU, which greatly facilitated the common payload elements of MESSENGER.

3.5 Thermo-Mechanical Design

The MESSENGER spacecraft is protected from direct solar illumination by the spacecraft sunshade. Strict rules govern the allowable attitude of the spacecraft in order to ensure that Sun-keep-in limits are not violated. As a result, the thermal environment of the instruments is largely benign with the exception of short intervals during the 12-hour orbital period when Mercury's illuminated surface subtends a large solid angle. The passive thermal design of MDIS uses radiators to cool the instrument, but this radiative cooling system must be disconnected during the short heat impulse that occurs near periapses at Mercury to prevent the instrument from exceeding the CCD's operational high-temperature limit (−10°C). This thermal switching is accomplished by means of diode heat pipes. Survival heaters ensure that cold operational limits are not violated.

The DPU is directly coupled to the spacecraft deck and blanketed with MLI. The −40°C to +50°C operating temperature of the DPU follows the deck temperature. The temperatures of the imager interface board and the LVPS board are measured internal to the DPU and reported in the instrument housekeeping data. A number of external spacecraft temperature sensors report the deck temperature at the location of the DPUs and MDIS.

The MDIS temperature sensors are mounted at key locations on the instrument (Fig. 34) and are reported in image headers. The CCD temperature is the most important temperature for calibration (Sect. 4.2.1). The MDIS platform temperature and pivot actuator temperature

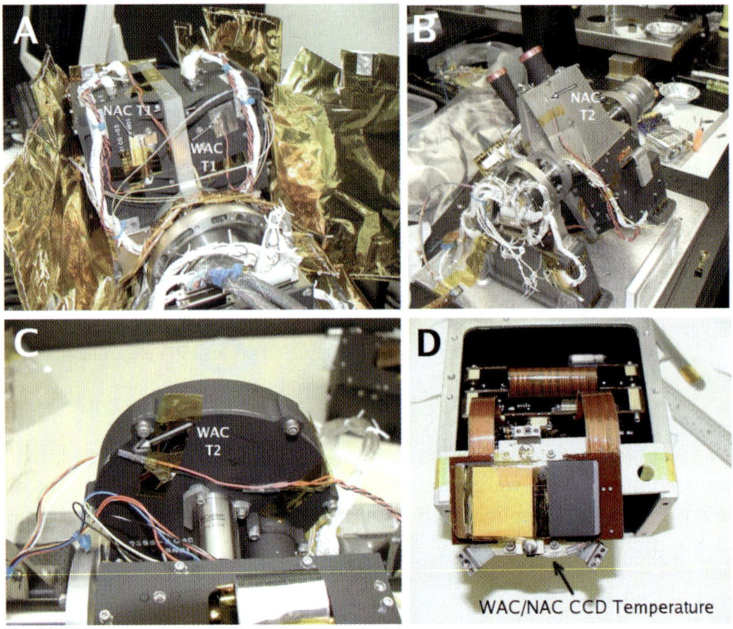

Fig. 34 Location of the MDIS instrument temperature sensors. (**A**) Location of FPU housing temperature, "CAM_T1," for each unit. (**B**) Telescope temperature, "CAM_T2," for the NAC. (**C**) WAC filter wheel housing temperature, "CAM_T2". (**D**) Location of the CCD temperature sensor for each camera

are reported in the spacecraft housekeeping. These measurements are a spacecraft function and are reported independently of the power state of the instrument.

3.5.1 Thermal Design of the MDIS Platform Assembly

Each camera has its own thermal control system, and each CCD is mounted to a heat sink that bolts onto a bracket that is thermally isolated from the FPU. Precision titanium spacers minimize heat conduction from the FPU to the CCD while maintaining focus over the operating temperatures of the instrument. A hole in the front panel of the housing allows light to pass through the assembly and onto the CCD detector, and the surface of the FPU closest to the CCD is unfinished to provide a lower emissivity surface to minimize radiative coupling between the CCD and the front panel (Fig. 34).

The thermal link passes through a hole in the top of the housing and ties the CCD to a phase-change material (PCM or wax pack) mounted to the top of the FPU housing. The wax pack is also thermally isolated from the FPU housing using epoxy-glass standoffs and MLI blankets. A thermal strap, secured to the top of the wax pack using a conductive epoxy, thermally ties the wax pack to the evaporator portion of a stainless steel diode heat pipe. Each camera has its own wax pack and heat pipe assembly. The two heat pipes mount inside the radiator structure, and both pipes connect to each of the three radiator plates (Fig. 35A). Each heat pipe is thermally connected to the inner surface of each of the three beryllium radiator plates with copper saddles. A photograph and schematic of the thermal system are shown in Figs. 35A and 35B, respectively.

Radiator Assembly. The three radiator panels are physically mounted to a stainless steel structure, supported by thin titanium rods to form a spoke-like assembly that minimizes heat

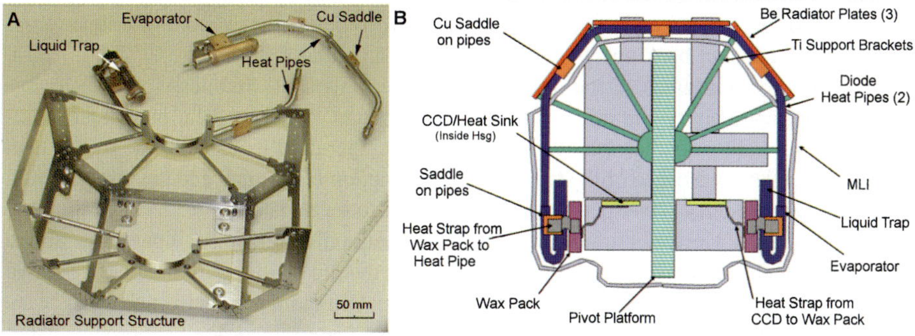

Fig. 35 (**A**) Photograph of the diode heat pipes and radiator support structure. (**B**) Schematic diagram showing key design elements that maintain the MDIS CCDs at nominal operating temperatures. Each CCD (NAC on *left*, WAC on *right*) is thermally coupled to phase-change wax pack that buffers the detector temperature. The wax packs are cooled by the radiator at times when the hot surface of Mercury does not occupy a large part of the radiator FOV. The diode nature of the heat pipes acts as a switch and provides a thermal pathway when the radiator FOV is cold and thermal isolation when it is hot. Hsg denotes housing

conduction from the radiators back to the platform (Fig. 35A). The structure also supports the two diode heat pipes. Copper pads, or saddles, are soldered to the pipe at the location of each radiator panel. The surface flatness of each saddle and radiator panel was controlled to ensure optimum contact area. MLI blankets enclose the entire CCD-to-heat-pipe thermal path, isolating the components from the rest of the instrument.

The three radiator panels are made of beryllium with a total area of about 340 cm^2. Beryllium was chosen for its large specific heat, and the panels were plated with aluminum in order to obtain a finish (clear-anodization) with a low solar absorptivity (α) and high emissivity. The low α was required to protect against overheating of the instrument by direct solar illumination as might occur in the event of a loss of attitude control of the spacecraft, resulting in a tumble.

Diode Heat Pipe Description. The heat pipes, fabricated by Swales Aerospace, are made from 9.5-mm-diameter stainless steel tubing filled with about 5 g of butane. During normal operation, heat flowing from the CCD, and through the wax pack, will be absorbed at the evaporator. Liquid butane is evaporated, and the gaseous butane flows down the center of the tube to be condensed on the cold walls of the condenser. An internal stainless steel wire mesh acts as a wick to transport the liquid from the condenser back to the evaporator, continuing the fluid loop.

A separate pipe, the liquid trap, is attached to the evaporator section that allows the heat pipe to act as a diode, i.e., allowing heat to flow only in one direction. A small tube connects the liquid trap to the evaporator. The diode action occurs when the condenser/radiators become warmer than the evaporator and the butane liquid in the condenser vaporizes. The now gaseous butane flows down the pipe and condenses onto the walls of the relatively cool ($-10°C$) evaporator and liquid trap. Since there is no wick in the small tube between the liquid trap and the evaporator, the liquid cannot flow out of the trap, back to the evaporator, and up to the condenser for reevaporation. As the radiator plates continue to heat up, more liquid is vaporized and flows down to the liquid trap area. This process, referred to as "burning out the pipe," continues until most of the liquid is located in the liquid trap and no more liquid can flow back up the pipe to the condenser. In this state, the heat pipe is effectively shut off and there is no longer any heat transfer between the evaporator and condenser. Once the radiators cool back down and the condenser again drops below the evaporator temperature, the heat pipe will start back up and conduct heat like a regular heat pipe.

The heat pipe was made of stainless steel for several reasons. Stainless steel has a very low thermal conductivity and a high melting temperature. When the heat pipes and radiators/condensers are near their predicted maximum temperature (~230°C) due to the heat impulse from the hot surface of the planet, the diode action of the pipe and the poor thermal conductivity of stainless steel permits very little heat to be transferred back down the pipe into the wax pack. Unlike an aluminum heat pipe that could soften at temperatures above 190°C, the mechanical properties of steel ensure that the pipe will retain its integrity above 300°C. Because the heat pipe is a pressure vessel, a leak in the pipe would cause it to cease to function. A final essential element in the selection of materials for the heat pipe was that the thermal interface of the mechanical connections between the beryllium radiator plates and heat pipe condenser had to be maintained up to 230°C. The integrity of this interface was maintained by soldering copper saddles to the stainless steel tube at both the condenser and evaporator ends. The copper saddles bonded in this way were empirically determined to tolerate the thermal stresses and maintain excellent thermal contact over the wide range of temperatures (−77°C to +230°C).

PCM Wax Pack Description. The wax pack is an aluminum enclosure filled with paraffin, which was selected for its melting point of about −10°C. The camera CCDs are required to be maintained between −45°C and −10°C throughout the mission. The bottom of the wax pack has threaded inserts that provide a mounting interface to the CCD thermal link. One side of a thermal strap connects to the top of the wax pack with a thermally conductive epoxy and is overlaid with aluminum tape.

Thermal Strap and Link Description. The thermal link between the CCD and wax pack was designed to carry 1 W of heat with a minimum drop of temperature. The thermal link was made of multiple sheets of thin aluminum foil, to provide the required flexibility and conductivity. Aluminum end blocks were swaged to each end of the foils to allow a mechanical attachment to the CCD interface plate and to the bottom of the wax pack. The mating surfaces on the thermal link end blocks, the CCD interface plate, and the wax pack were polished to provide a good thermal connection.

The thermal strap between the wax pack top and the heat pipe evaporator was created from two copper braids. One end of the braids was bonded into an aluminum end block, and that block was then bonded to the evaporator of the heat pipe. The other end of the braids was left free for permanent attachment to the top of the wax pack with thermally conductive epoxy.

Functional Operation of Thermal Control System. In orbit about Mercury, when the spacecraft is away from the hot planet, the MDIS radiator plates do not see the planet and are exposed only to deep space. In this condition, the radiator plates become very cold (−77°C) and the diode heat pipes operate as normal heat pipes. The gas internal to the pipe moves freely along the length of the pipe, transporting heat from one end to the other and radiating that heat out to space. The heat from the CCD will flow from the CCD through the thermal link, through the wax pack and thermal strap, to the evaporator of the heat pipe. As the liquid in the pipe evaporates, heat flows along the heat pipe to the condensers and finally to the radiator panels. When the paraffin in the wax pack is all frozen, the wax pack, thermal link, and CCD will continue to cool below the freeze point of the paraffin. As the spacecraft starts to approach the planet, the radiator plates begin to absorb heat from the hot planet. When the temperature of the condenser rises above the temperature of the evaporator, most of the heat-conducting gas in the pipe condenses in the liquid trap. In the burnt-out pipe condition, very few gas molecules are present and the pipe is shut down, effectively disconnecting the condensers and radiator plates from the rest of the thermal path.

With the heat pipe in its diode mode, namely, shut off, the heat that is still flowing from the CCD will now be absorbed into the wax. The paraffin will warm up to its melting point and remain at that temperature until all 240 g of paraffin have melted. The amount of wax was selected so that all the wax would not completely melt during the hours surrounding periapsis. The melting wax clamps the temperature of the CCD at its operating warm-temperature limit (−10°C). As the spacecraft moves farther away from the planet, the MDIS radiator and condenser temperatures will fall below the evaporator temperature and allow the heat pipe to operate normally again. With the condenser temperature below the evaporator temperature, heat will flow from wax pack to the radiators and begin refreezing the paraffin. The partially melted wax pack will remain at its melting (freezing) point until all the paraffin solidifies. At that point, the wax pack temperature will continue to drop down to −45°C where heaters and thermostats control the minimum temperature.

Other than the CCD-to-heat-pipe heat path, the majority of the camera components are thermally connected to the main platform. Heaters on the platform prevent the camera optics and FPU electronics from dropping below −35°C. There are no radiators associated with the platform. The cameras and platform are isolated from the environment by MLI blankets and will not overheat due to the relatively small amount of heat dissipated in the electronics.

4 Instrument Calibration

Laboratory measurements described in the following sections were used to derive values for the terms of the calibration equation as shown in (2) for both the WAC and NAC. Both instruments measure relative light intensity in engineering units referred to as DNs. The raw engineering units are converted to the physical units of radiance, L (W m^{-2} µm^{-1} sr^{-1}), following the calibration equation:

$$
\begin{aligned}
L(x, y, f, T, \tau, b) &= \frac{[DN(x, y, f, T, \tau, b, MET) - Dk(x, y, T, \tau, b, MET)]}{Flat(x, y, f, b) * Resp(f, T, b) * \tau} \\
&\quad - \frac{Sm(x, y, \tau, b) - Scat(x, y, f, \tau, b)}{Flat(x, y, f, b) * Resp(f, T, b) * \tau},
\end{aligned} \tag{2}
$$

where:

$DN(x, y, f, T, \tau, b, MET)$ is the raw DN measured by the pixel in column x, row y, through filter f, at CCD temperature T and exposure time τ, for binning mode b and Mission Elapsed Time MET;
$Dk(x, y, T, \tau, b, MET)$ is the dark level in a given pixel, derived either from the dark strip or estimated from exposure time and CCD temperature;
$Sm(x, y, \tau, b)$ is the scene-dependent frame transfer smear for the pixel;
$Scat(x, y, f, \tau, b)$ is the contribution of scattered light from elsewhere in the scene;
$Flat(x, y, f, b)$ is the nonuniformity or "flat-field" correction at this pixel location;
$Resp(f, T, b)$ is the responsivity relative to the baseline operating temperature; and
τ is the exposure time in milliseconds.

The analysis used to derive each term of the calibration equation is described in the following sections.

4.1 Ground Calibration Facility and Test Results

All calibration tests were performed in the Optical Calibration Facility (OCF) at APL. The OCF consists of large, linked vacuum chambers with a host of support equipment. The largest of these chambers, the instrument chamber, has an internal diameter of 1.3 m and a length of 2 m and permits mounting the instrument on a three-axis motion stage. The motion stage can rotate the instrument in azimuth and elevation through a range limited only by the instrument harness and mounting hardware. Translation is possible over a limited range, but it is sufficient to center the instrument in the collimated beam of the chamber. The interior and ends of the instrument vacuum chamber are surrounded by cold walls, within which an external refrigerator circulates a cooling fluid to reduce the internal temperature to approximately $-40°C$. Most of the calibrations were performed at approximately $-30°C$. However, additional thermal configurations over the range $-34°C$ to $+25°C$ were also performed, to bracket the behavior of the instrument over the range of expected flight conditions. Figure 36 shows a schematic of the OCF test setup. Inside the instrument chamber, there are two sources of calibrated illumination. In one configuration the instrument views a 508-mm diameter integrating sphere through a quartz window. Dark cloth was draped around the window and sphere opening in order to block unwanted external light from entering the instrument optical path. Sphere spectral radiance was calibrated by the manufacturer just prior to the start of instrument calibration, for each configuration of the sphere's four bulbs: two 45-W lamps and two 150-W lamps. One photometer was coupled through a fiber optic to monitor the internal sphere brightness during all calibration runs. Two additional photometers (not shown in Fig. 36) were mounted on the camera and allowed additional monitoring of incident irradiance.

Rotating the azimuthal stage of the OCF 180° enables the instrument to look along a beam tube into an off-axis parabolic collimating mirror, which has a focal length of 1.43 m, operates at $f/7$, and is focused on a grating monochromator. The monochromator may be illuminated by an incandescent lamp with quartz optics. At the input slit, a set of neutral density filters may be used to attenuate the light by known amounts. At the exit slit, a set of long-pass filters is used to remove higher orders (i.e., shorter wavelengths) from the grating. The wavelength can be changed manually or can be set to scan under computer control. The grating may be positioned to zeroeth order, which allows the full incandescent spectrum to be passed.

A MgF_2 window partitions the monochromator from the unit under test, allowing the monochromator to be removed without breaking vacuum so that other sources may be placed at the focus of the collimator. These sources include a point source (pinhole), pinhole array, and test samples illuminated by an incandescent source and viewed off a silver fold mirror.

Four networked computers were used to acquire and monitor the calibration data. Each computer had a specialized set of tasks to perform. The first was dedicated to running most of the OCF instrumentation such as the motion stage, monochromator wavelength adjustment, and filter wheel selection. The second computer served as a frame grabber that communicated directly to the instrument DPU. The operators ran the calibration scripts and monitored the progress of the runs from the third computer. The fourth computer presented real-time images and plots on its screen, allowing monitoring of calibration results during each experiment.

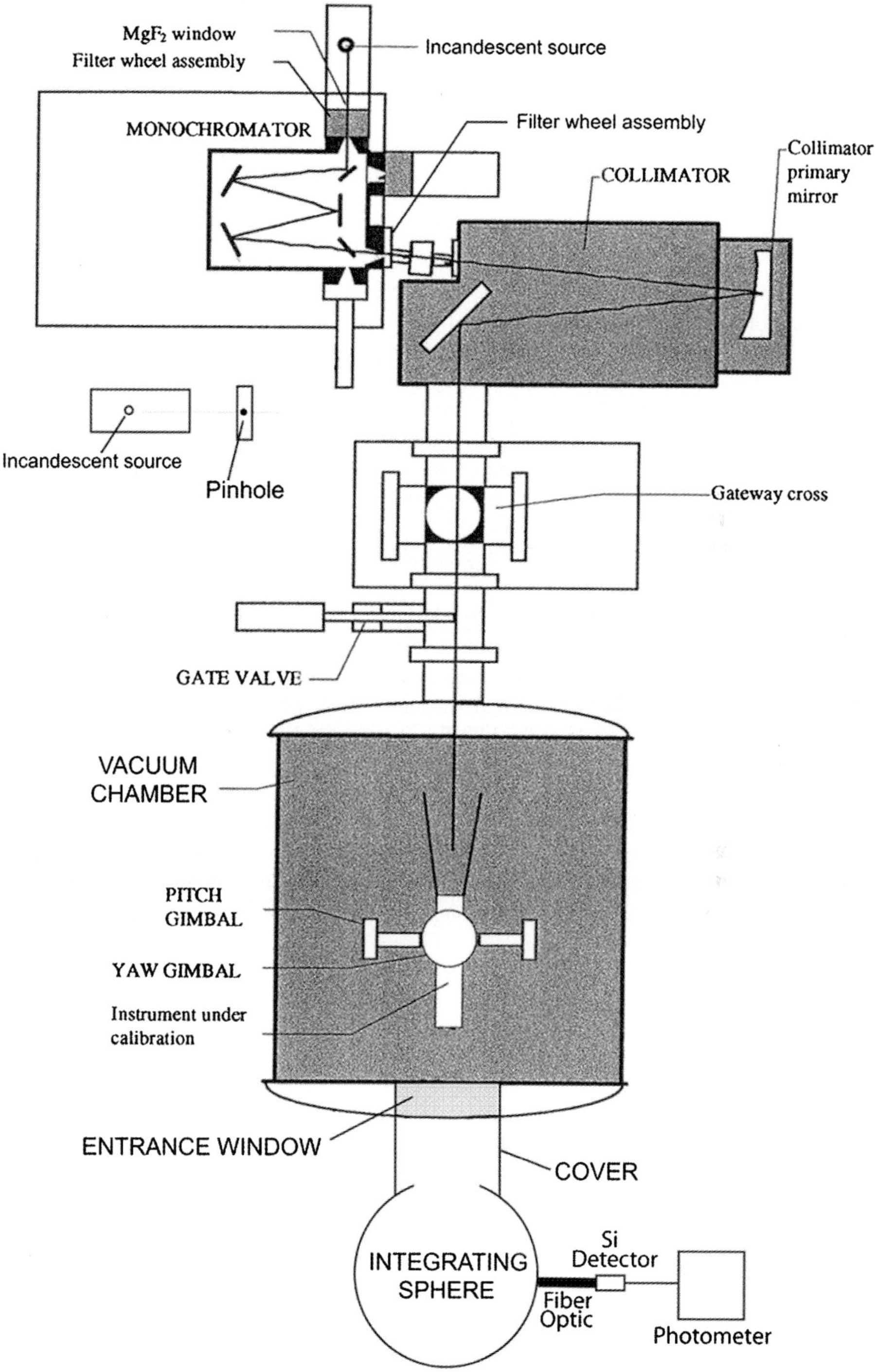

Fig. 36 Optical Calibration Facility hardware configuration used for MDIS calibration

All data were saved locally on disk and were also automatically copied to the remotely located MESSENGER Science Operations Center (SOC) within approximately 30 s of acquisition. The SOC provides a Web interface to the data, organized by test, which includes facilities for searching the database for images in particular configurations. In addition, the SOC processed the image data, performing dark-correction (using the dark strips) and providing a variety of plotting and profiling options. These features were used in near-real time to examine the data for instrumental issues and to validate test results.

4.2 Dark Level

The measured signal from the CCD, in the absence of incident photons, is the sum of three major components: (1) dark current from thermal electrons, (2) an electronic offset, or bias, of ~240 DN intentionally added to the readout to prevent occurrence of negative values that would appear as zero DN words, and (3) noise picked up by instrument electronics. Throughout calibration there has been no evidence of picked-up noise. In the following discussion, the sum of dark current and bias is treated together as "dark level." The response of pixel-dependent dark level to temperature and exposure time was investigated separately for the NAC and WAC without on-chip binning (full-frame) and with on-chip binning.

4.2.1 Response of the Sensor System to Exposure Time and Temperature

Characterization of the WAC and NAC dark level was conducted over the CCD temperature range 25°C to −40°C, as measured by a temperature sensor (AD590) bonded to the CCD heat-sink mounting plate on each camera (Fig. 34D). This temperature sensor was calibrated during instrument-level thermal-vacuum testing, where it was possible to install multiple thermocouples throughout the MDIS instrument. In particular, a thermocouple located in close proximity to the thermal link on the wax pack of each FPU was used as the best measurement to the actual CCD temperature. A linear regression was used to fit the measured response of the CCD temperature sensor with the calibrated thermocouple response. The other instrument temperature sensors, (also AD590s, see Fig. 34) located on each FPU housing and the WAC filter wheel and NAC telescope, were calibrated in the same way. The final expressions to convert the digital raw output to temperature are

$$T[\deg C] = A \times Raw[DN] + B. \tag{3}$$

The coefficients A and B are listed in Table 10.

Figure 37 shows dark-image means (in DN) for the NAC and WAC full-frame (not binned) and binned images as functions of exposure time and CCD temperature. In the

Table 10 Conversion coefficients from raw DN to °C for internal MDIS temperatures

Description	Name	*A*	*B*
WAC CCD temperature	ccd_temp	0.2718	−318.455
WAC FPU temperature	cam1_temp	0.5022	−262.258
WAC FW temperature	cam2_temp	0.553	−292.760
NAC CCD temperature	ccd_temp	0.2737	−323.367
NAC FPU temperature	cam1_temp	0.5130	−268.844
NAC FW temperature	cam2_temp	0.4861	−269.718

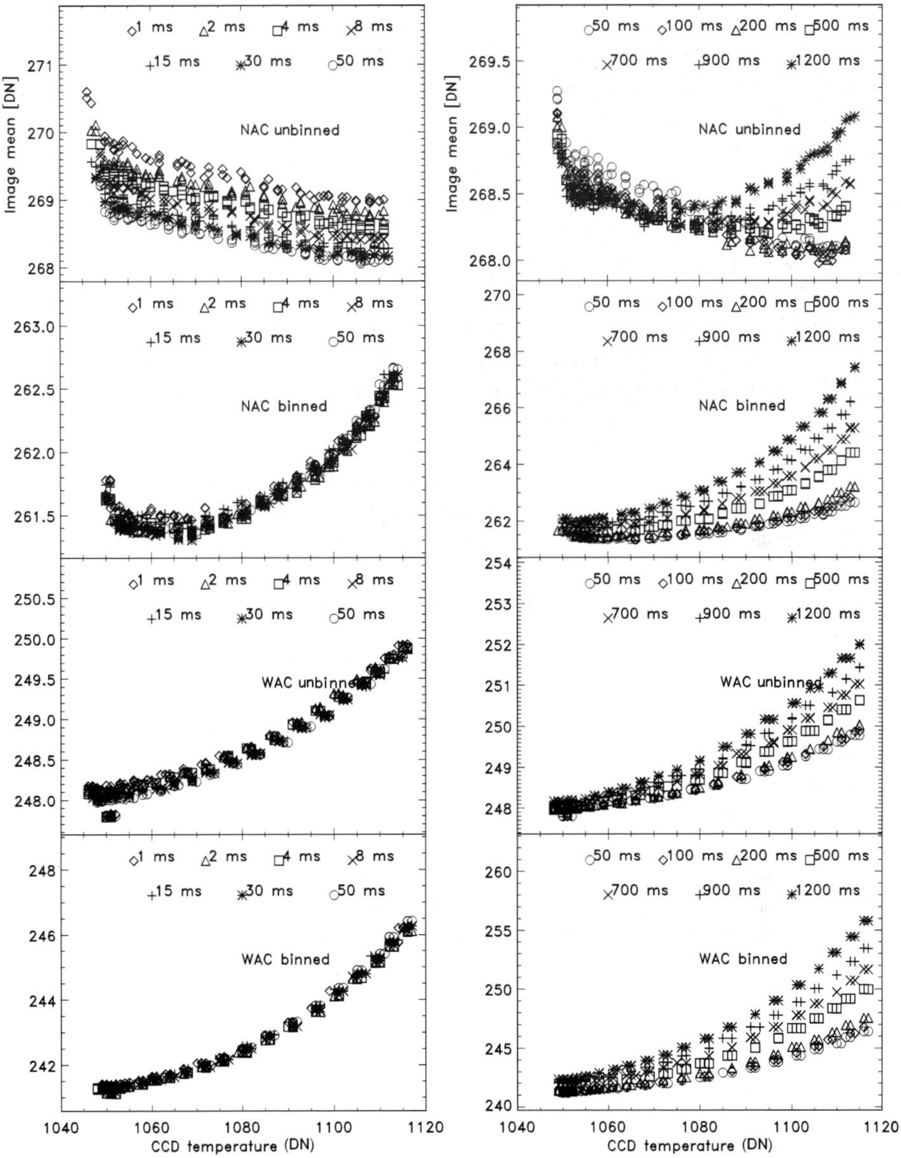

Fig. 37 Dark image mean as a function of temperature (DN) and exposure time for the WAC and NAC in each binning state

NAC full-frame data, at short exposure times and low temperatures, the dark level is dominated by the electronic bias and the image mean decreases with increasing temperature. This relationship originates in the amplifier in the read-out electronics, which has a temperature-dependent offset (not shown in this figure). After an image has been exposed, large current pulses are sent from the power supply to the clocking signals controlling the movement of charge in the CCD. These high-current clocking pulses increase the overall signal levels for a very short time, as observed in the measured DNs. At longer exposure times (>200 ms) and higher CCD temperatures ($>-30^\circ$C) the effects of dark current, which increases exponentially with temperature, dominate. The NAC full-frame data show this expected behavior of increasing dark level with increasing temperature and exposure time. Because on-chip binning adds the outputs of four pixels prior to quantization, the impact of the dark current is larger, and the onset of increasing dark level with temperature and exposure time occurs at lower temperature and shorter exposure time. WAC dark levels are dominated by the increase in dark current with higher temperature and longer exposure time.

4.2.2 Dark Models

For the purpose of modeling dark level, the WAC and NAC were treated as four different instruments: NAC full-frame, NAC binned, WAC full-frame, and WAC binned. Two dark models were derived from the OCF calibrations for each of the four modeled sensors: (1) a forward model of the dark level (in DN) as a function of column number, row number, exposure time, and raw CCD temperature; and (2) a backward model that uses the row-dependent dark levels in the dark strips and extrapolates them across the FOV using the same variables as the forward model. For both the forward and backward models, dark level is treated as a linear function of column number, row number, and exposure time. The exponential temperature dependence was approximated with a third-order polynomial for ease of calculation. It is known that the CCD dark level temperature response will change with time due to exposure to radiation in space, and this specific time dependence will be determined in flight over the lifetime of the mission.

Forward Model. An overview of the forward dark model is illustrated in Fig. 38 depicting the effect of dark level with increasing exposure and temperature. Consider the first of four variables to the dark model: column position x, row position y, exposure time τ, and temperature T. For the column-dependence at a particular row, exposure time, and temperature, dark level is modeled linearly, where an offset $\alpha(y,\tau,T)$ is the value at column $x=0$ and a slope $\beta(y,\tau,T)$ is the increase per column (per increment of x). Then, to find the dark level at a given column, the model for this row, exposure time, and temperature is:

$$Dk(x,y,\tau,T) = \alpha(y,\tau,T) + \beta(y,\tau,T)x. \tag{4}$$

However, the column-dependence will vary with row, so that in the second level of the model, the offset α and slope β are each functions of row position y. Thus, linear expansions

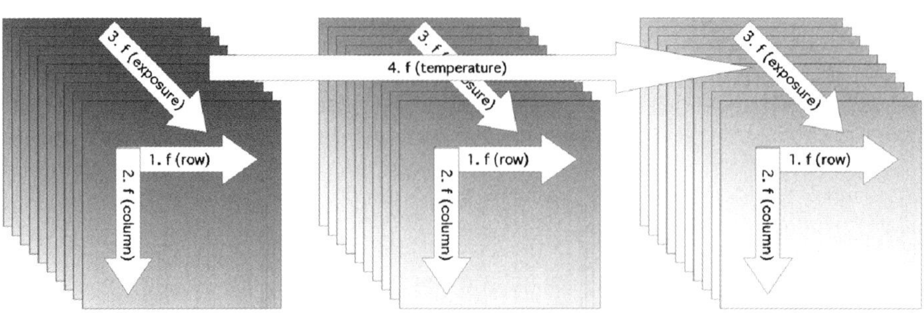

Fig. 38 Schematic representation of dark model calculation decision tree. Each key parameter is identified, and the arrow shows the direction of increase for that parameter. The *gray scale* suggests the effect of dark level in the image as a function of these parameters

of α and β are

$$\alpha(y,\tau,T)=A(\tau,T)+B(\tau,T)y, \tag{5}$$

and

$$\beta(y,\tau,T)=M(\tau,T)+N(\tau,T)y. \tag{6}$$

The offset α to the column-dependent dark level has a value $A(\tau,T)$ in row $y=0$, and in each subsequent row $y+1$ that value is incremented by slope $B(\tau,T)$. Similarly, the slope to column-dependent dark level β changes with each row by an offset $M(\tau,T)$ and slope $N(\tau,T)$.

Note that at this level, the row-dependences of column-dependences are themselves dependent on exposure time and temperature. A third level of linear regression can be added to the model, such that $A(\tau,T)$, $B(\tau,T)$, $M(\tau,T)$, and $N(\tau,T)$ can each be expressed as its own linear function of exposure time τ, using a temperature-dependent offset $C(T)$, $O(T)$, $E(T)$, or $Q(T)$, respectively, and a temperature-dependent slope $D(T)$, $P(T)$, $F(T)$, or $S(T)$, respectively:

$$A(\tau,T)=C(T)+D(T)\tau, \tag{7}$$

$$M(\tau,T)=O(T)+P(T)\tau, \tag{8}$$

$$B(\tau,T)=E(T)+F(T)\tau, \tag{9}$$

and

$$N(\tau,T)=Q(T)+S(T)\tau. \tag{10}$$

The expanded dark model at this level becomes

$$\begin{aligned} Dk(x,y,\tau,T) &= C(T)+D(T)\tau+\big[E(T)+F(T)\tau\big]y \\ &\quad +\big\{O(T)+P(T)\tau+\big[Q(T)+S(T)\tau\big]y\big\}x. \end{aligned} \tag{11}$$

The fourth and final level of the model, $C(T)$, $D(T)$, $E(T)$, $F(T)$, $O(T)$, $P(T)$, $Q(T)$, and $S(T)$, are all third-order functions of temperature, for example,

$$C(T)=H_0+H_1T+H_2T^2+H_3T^3. \tag{12}$$

The forward dark level model was derived by first finding a linear fit of dark level as a function of column number for each row in the dark calibration data (2). Second, the bias and dark DN accumulation rate per column increment were fitted linearly as functions of row (3–4). A, B, M, and N were fitted as functions of exposure time, from data binned into 15 CCD temperature bins (each bin 10 DN wide) covering the range 1,000–1150 DN (−40.8°C to −6.8°C) using (5–8). Finally, the linear fit coefficients as functions of exposure time were fitted across the temperature bins with a third-order polynomial (12).

The coefficients of these expansions are given in Table 11a for the four sensor models. After launch these coefficients are serving as the bias for the time dependence of the dark model, with the slopes to be calculated from post-launch observations.

To assess the accuracy of the model, the fitted model was differenced from the data used to create it, and the rms errors between the data and model were measured. Figure 39 shows that, for all four sensor models, over the expected range of temperatures and for exposure times of 1,200 ms or less, the difference between the dark level models and the data used to derive them is comparable to the read noise for each CCD (∼1 DN for the WAC, ∼2 DN for the NAC).

Table 11a NAC full-frame dark models

Coefficient	H_0	H_1	H_2	H_3
Forward dark level model				
C	4202.30	−10.7314	0.00974273	-2.94302×10^{-06}
D	−64.4884	0.179181	−0.000165891	5.11765×10^{-08}
E	−2.58253	0.00754599	-7.35219×10^{-06}	2.38873×10^{-09}
F	−0.000464774	1.31009×10^{-06}	-1.23407×10^{-09}	3.88515×10^{-13}
O	0.0143372	-4.21571×10^{-05}	4.12534×10^{-08}	-1.34637×10^{-11}
P	-6.86389×10^{-05}	1.80933×10^{-07}	-1.57409×10^{-10}	4.50925×10^{-14}
Q	−0.000169733	4.92790×10^{-07}	-4.77059×10^{-10}	1.53993×10^{-13}
S	1.35800×10^{-07}	-4.33913×10^{-10}	4.56621×10^{-13}	-1.58643×10^{-16}
Backward dark level model				
C	5451.46	−14.2965	0.0131305	-4.01470×10^{-06}
D	−66.9307	0.185911	−0.000172068	5.30647×10^{-08}
E	−2.42407	0.00705565	-6.84831×10^{-06}	2.21669×10^{-09}
F	0.00526574	-1.44654×10^{-05}	1.32342×10^{-08}	-4.03222×10^{-12}
O	−1.81968	0.00519216	-4.93259×10^{-06}	1.55995×10^{-09}
P	0.00351380	-9.68964×10^{-06}	8.90189×10^{-09}	-2.72453×10^{-12}
Q	−0.000402942	1.21434×10^{-06}	-1.21845×10^{-09}	4.07108×10^{-13}
S	-8.26754×10^{-06}	2.6993×10^{-08}	-2.07595×10^{-11}	6.32380×10^{-15}

Table 11b NAC binned dark models

Coefficient	H_0	H_1	H_2	H_3
Forward dark level model				
C	−5809.80	17.2831	−0.0163855	5.17322×10^{-06}
D	−18.7770	0.0535211	-5.08542×10^{-05}	1.61084×10^{-08}
E	−52.1256	0.148428	−0.00014089	4.45864×10^{-08}
F	−0.00425778	1.22892×10^{-05}	-1.18214×10^{-08}	3.78984×10^{-12}
O	0.676937	−0.00190954	1.79397×10^{-06}	-5.61987×10^{-10}
P	0.00180111	-5.16803×10^{-06}	4.94001×10^{-09}	-1.57305×10^{-12}
Q	−0.00688223	1.94574×10^{-05}	-1.83419×10^{-08}	5.76568×10^{-12}
S	9.00182×10^{-06}	-2.57628×10^{-08}	2.45929×10^{-11}	-7.83098×10^{-15}
Backward dark level model				
C	−8188.55	24.0208	−0.0227352	7.16515×10^{-06}
D	−16.5122	0.0471223	-4.48311×10^{-05}	1.42195×10^{-08}
E	−38.7242	0.110516	-1.05155×10^{-04}	3.33600×10^{-08}
F	−0.0110371	3.13365×10^{-05}	-2.96580×10^{-08}	9.35730×10^{-12}
O	7.69642	−0.0217915	2.05305×10^{-05}	-6.43961×10^{-09}
P	−0.00486634	1.36687×10^{-05}	-1.27902×10^{-08}	3.98703×10^{-12}
Q	−0.0463603	0.000131138	-1.23619×10^{-07}	3.88354×10^{-11}
S	2.89199×10^{-05}	-8.17259×10^{-08}	7.69996×10^{-11}	-2.41893×10^{-14}

Table 11c WAC full-frame dark models

Coefficient	H_0	H_1	H_2	H_3
Forward dark level model				
C	1238.24	−2.76843	0.00256473	-7.86953×10^{-07}
D	−3.48338	0.0101166	-9.79576×10^{-06}	3.16249×10^{-09}
E	−2.42999	0.00714405	-7.00585×10^{-06}	2.29185×10^{-09}
F	−0.00053432	1.49958×10^{-06}	-1.40025×10^{-09}	4.34984×10^{-13}
O	0.0206338	-6.29517×10^{-05}	6.39338×10^{-08}	-2.16318×10^{-11}
P	−0.00033310	9.70986×10^{-07}	-9.43818×10^{-10}	3.05943×10^{-13}
Q	0.000517513	-1.51006×10^{-06}	1.46957×10^{-09}	-4.77044×10^{-13}
S	1.17016×10^{-07}	-3.34717×10^{-10}	3.19031×10^{-13}	-1.01330×10^{-16}
Backward dark level model				
C	−3182.94	9.81730	−0.00936597	2.97916×10^{-06}
D	12.7966	−0.0364940	3.46355×10^{05}	-1.09384×10^{-08}
E	11.4352	−0.0325653	3.08604×10^{-05}	-9.73046×10^{-09}
F	−0.0755373	0.000216107	-2.05838×10^{-07}	6.52700×10^{-11}
O	1.26711	−0.00353596	3.28770×10^{-06}	-1.01873×10^{-09}
P	0.00127217	-3.57421×10^{-06}	3.33658×10^{-09}	-1.03455×10^{-12}
Q	0.000637356	-1.79417×10^{-06}	1.67841×10^{-09}	-5.21619×10^{-13}
S	-2.32720×10^{-06}	6.63236×10^{-09}	-6.28992×10^{-12}	1.98479×10^{-15}

Table 11d WAC binned dark models

Coefficient	H_0	H_1	H_2	H_3
Forward dark level model				
C	−484.568	2.11771	−0.00206813	6.75547×10^{-07}
D	−10.2411	0.0299233	-2.91567×10^{-05}	9.47476×10^{-09}
E	−27.9169	0.0813578	-7.90653×10^{-05}	2.56248×10^{-08}
F	0.000646762	-1.91510×10^{-06}	1.89122×10^{-09}	-6.22884×10^{-13}
O	−0.550564	0.00152559	-1.40630×10^{-06}	4.30355×10^{-10}
P	−0.00201059	5.92984×10^{-06}	-5.83228×10^{-09}	1.91308×10^{-12}
Q	0.0127947	-3.69327×10^{-05}	3.55415×10^{-08}	-1.14037×10^{-11}
S	-1.54738×10^{-06}	4.29029×10^{-09}	-3.95289×10^{-12}	1.20989×10^{-15}
Backward dark level model				
C	876.088	−1.72949	0.00155692	-4.62411×10^{-07}
D	−8.67913	0.0253792	-2.47483×10^{-05}	8.04835×10^{-09}
E	−24.2259	0.0705204	-6.84466×10^{-05}	2.21526×10^{-08}
F	−0.00286560	8.40352×10^{-06}	-8.21840×10^{-09}	2.68055×10^{-12}
O	−4.55575	0.0128502	-1.20770×10^{-05}	3.78006×10^{-09}
P	−0.00660749	1.93028×10^{-05}	-1.88063×10^{-08}	6.11108×10^{-12}
Q	0.00192523	-5.01882×10^{-06}	4.27240×10^{-09}	-1.17912×10^{-12}
S	8.80210×10^{-06}	-2.61144×10^{-08}	2.58360×10^{-11}	-8.52398×10^{-15}

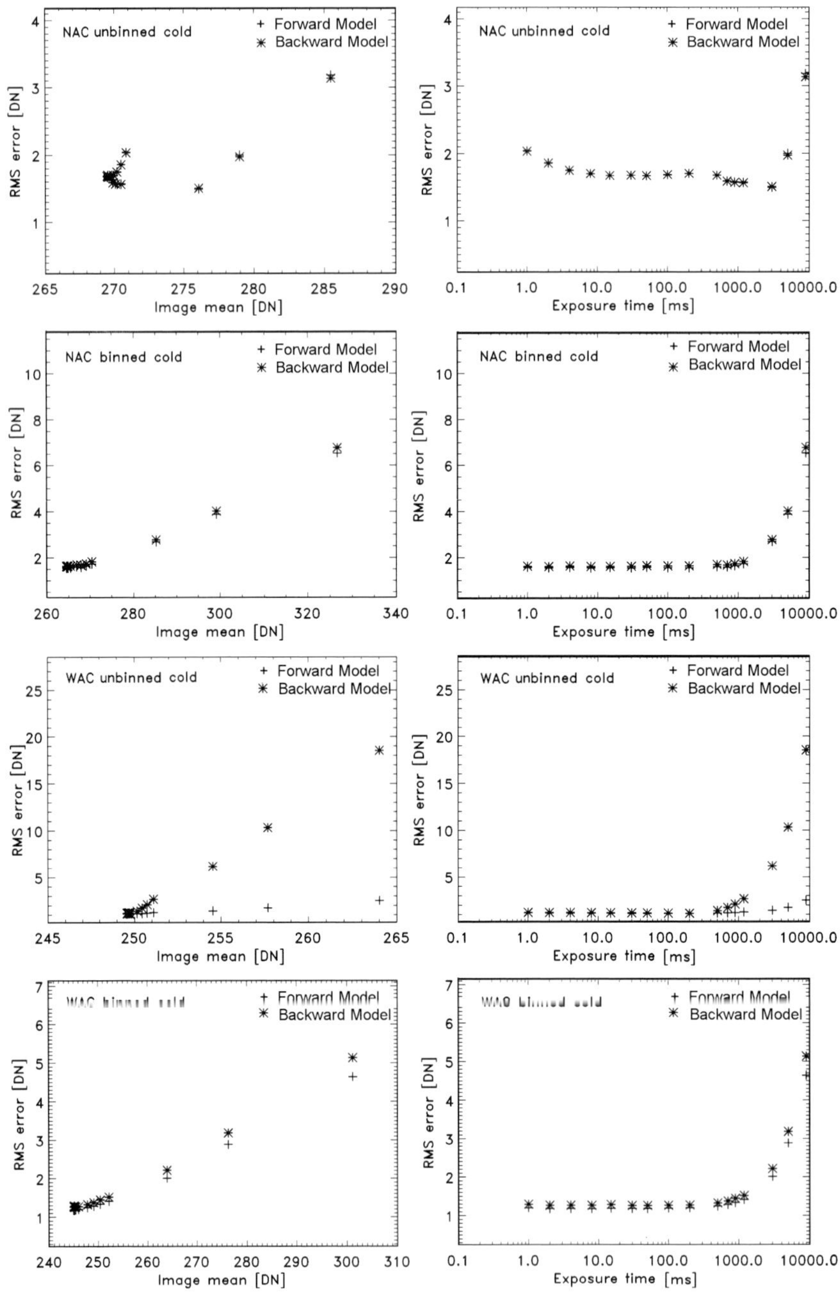

Fig. 39 RMS error of the dark models to image pixels as a function of dark image mean and exposure time for the NAC and WAC, binned and not binned, from OCF calibration data

Backward Model. In the full-frame mode for either the WAC or NAC, the four dark columns behave identically to the scene as a function of row, exposure time, and temperature to within 0.26 DN. A second method to derive the dark level from the image data is referred to as the Backward Model and uses the dark columns and a fitting procedure to determine the dark level similar to that for the dark Forward Model.

In the binned mode for both cameras, true dark columns are unavailable due to an incorrect mapping of these columns in firmware. However, the second column of a binned image provides a lower response than a column in the active image area. This lower-response column does show a temperature- and exposure-time response that can be modeled, making it a functional "dark column." Therefore, the Backward Model simply uses the second column of an image (binned or full-frame) to be representative of the dark strip properties.

The same fitting procedure used to derive the dark level for the Forward Model was used for the Backward Model with the following exception. The bias for the fit to each row as a function of column number was defined to be the value of the fit to the dark column at that row number. Then, the regression error for a range of accumulation rates assuming this fixed bias was calculated. The accumulation rate minimizing the regression error was used as the accumulation rate of each row as a function of column number. From this set of coefficients (Table 11a), the exposure time dependence and the temperature bins were derived and, as above, the rms error of fit regressions calculated. Since the bias derived from the dark columns is not the same as the bias derived directly from the dark frames, it is expected that rms errors will be slightly greater. For exposure times of less than 1,200 ms, the rms error for the dark level derived from dark column Backward Model is, on average, larger than the dark level derived from the Forward Model. Figure 39 compares the rms errors resulting from these two models. The large disparity between the Forward and Backward models at long exposures in the full-frame WAC is a consequence of a poor fit to the Backward Model at long exposures.

Given the initial better fit of the dark level Forward Model to the data, as well as the nature of the binned "dark" columns, it is tempting to rely on the Forward Model exclusively. However, the dark strips, even for binned data, serve as an indicator of the variations of each CCD's response to radiation and, as such, a means to calibrate the changes in the behavior of the CCD with time. Thus, both dark models will be periodically reevaluated en route to Mercury and during the orbital phase of the mission.

4.3 Frame Transfer Smear

Frame transfer smear corrections for MDIS follow the technique described by Murchie et al. (1999) for the NEAR MSI imager. In brief, an image is exposed for a nominal integration time and is then transferred to the memory zone of the CCD, from which the analog signal is digitized line by line. Accumulation of signal continues during the finite duration of frame transfer, inducing a streak or frame-transfer smear in the wake of an illuminated object in the field of view, parallel to the direction of frame transfer, provided pointing remains stable to the end of the frame transfer period. Quantitatively, the smear correction is:

$$Sm(x,y,T,\tau,f,b) = \sum_{1}^{y-1} \frac{t_2}{\tau} \frac{[DN(x,y,\tau,b) - Dk(x,y,T,\tau,b,MET)] - Sm(x,y,b,\tau)}{Flat(x,y,b,f)}, \quad (13)$$

where $Sm(x,y,T,\tau,f,b)$ is the smear in column x and row y at exposure time τ and temperature T in binning mode b and filter f. $Dk(x,y,T,\tau,b,MET)$ is the dark level in column

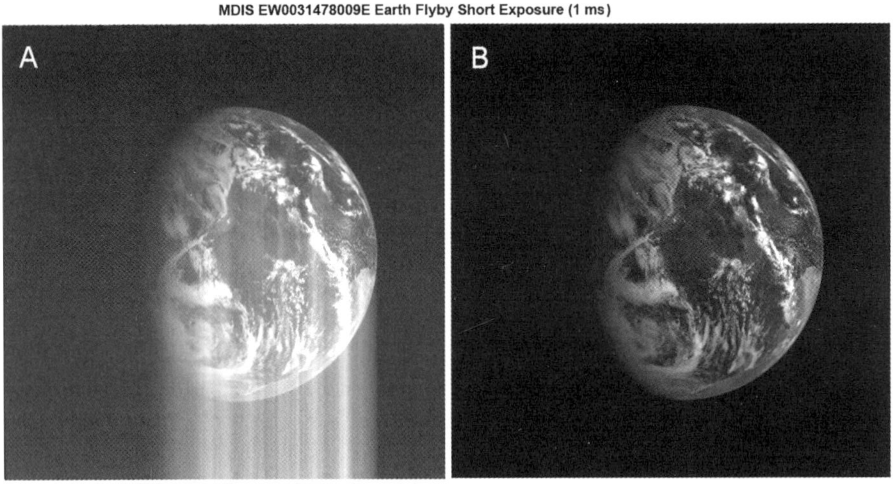

Fig. 40 (**A**) Frame-transfer smear is apparent in this unbinned NAC short-exposure (1-ms) image of Earth during the MESSENGER Earth flyby in August 2005. (**B**) Image after processing with desmear algorithm

x and row y at exposure time τ and temperature T in binning mode b at mission-elapsed time MET. $Flat(x, y, b, f)$ is the flat-field correction in column x and row y in binning mode b and filter f. τ is exposure time, and t_2 is the time for frame transfer (3.84 ms) divided by the number of lines in the image in the direction of frame transfer, i.e., 1,024 for full-frame images or 512 for binned images. Empirically, it was found that this correction removes frame transfer smear to the level of noise in typical, field-filling scenes for which exposure time is >2 times the frame transfer time. For high-contrast scenes such as well-exposed, non-field-filling extended sources imaged against a black background, artifacts are at or below the noise when exposure time is >3 times the frame transfer time.

Figure 40 (left) shows an example of frame transfer smear for a full-frame NAC image acquired during MESSENGER Earth flyby in August 2005. The image was exposed for 1 ms, less than the guideline, but taken to test the effectiveness of the desmear algorithm. In the corrected image (Fig. 40, right), although the smear is slightly overcorrected, the visual appearance of the corrected image is dramatic. The overcorrected artifact disappears in typical scenes for exposure times $\geq$8 ms. Thus, for analysis of MDIS filter passbands (Sect. 4.5) and radiometry (Sect. 4.6), only images with exposure times $\geq$8 ms were used.

4.4 Geometric Calibration

4.4.1 NAC and WAC FOV, IFOV, and Offset

To determine the fields-of-view and the relative alignment of the WAC and NAC, the output slit of the OCF monochromator was imaged using white light at different positions of the motion stage. The ends of the slit were measured from dark-corrected, desmeared images, and the pixel positions and stage positions were used to determine the IFOV, FOV, and relative pointing of each camera (Fig. 41). The angular difference per pixel position defines the IFOV, and 1,024 times that value defines the FOV. The WAC FOV is $10.54° \pm 0.02°$, and the IFOV is 179.6 ± 0.3 µrad. For a 14-µm pixel pitch (specified by the manufacturer), these figures imply a focal length of 77.96 ± 0.15 mm. The NAC FOV is $1.493° \pm 0.001°$, and the IFOV is 25.44 ± 0.02 µrad, implying a focal length of 550.3 ± 0.5 mm.

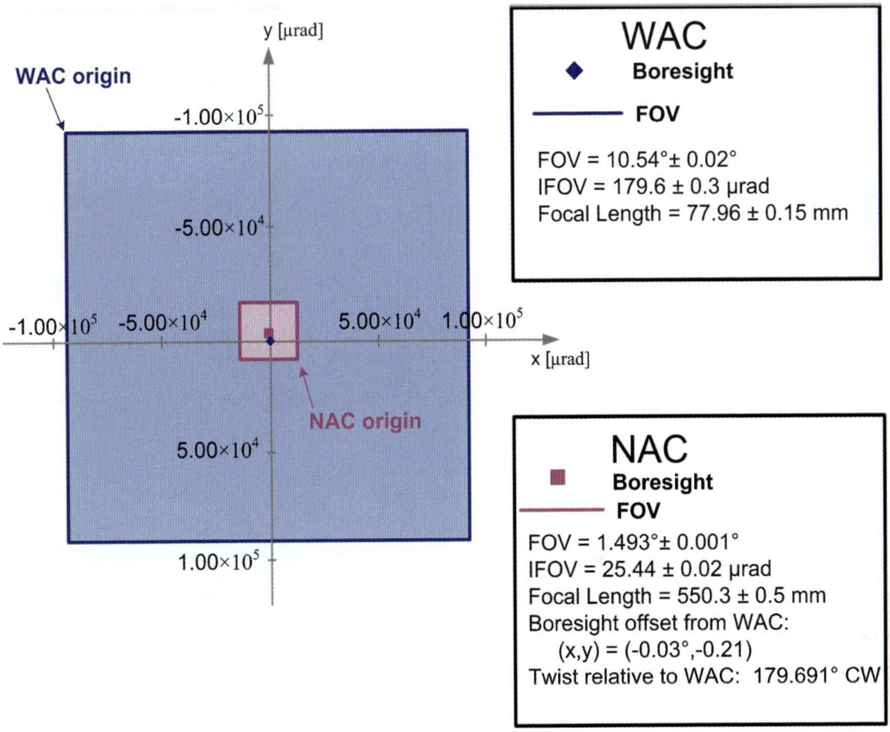

Fig. 41 WAC and NAC fields of view as determined from the ground calibration. Note that the WAC and NAC are rotated about 180° relative to one another. Scale, offset, and relative twist of NAC and WAC images are given in the inset

This measurement also provides information on the relative orientations of the two FOVs. The pixel positions measured at each stage position define a vector for each camera whose angular separation defines the twist between the two FOVs. The NAC FOV is rotated relative to the WAC's by approximately 179.69° clockwise (CW), so that a scene appears "right-side up" in one camera and "upside down" in the other. Allowing for this twist, solving for the center pixel position in each camera in angular coordinates, and differencing the two results yields the offset in boresight positions. The two boresights are offset by 0.21°.

4.4.2 Position within Pivot Plane

Through mechanical tolerances, deviation of the pivot plane from normal due to mechanical error is less than 85 μrad. Sampled measurements over the range of motion of the actuator are consistent with the nominal stepping of 0.01°/step. We attempted to characterize fully the single-step response of the pivot actuator but found systematic effects dominating the measured response.

The flight software does not routinely command the pivot actuator to the operational hard stop, but halts motion at a software stop located at +50°. To establish a second known position within the ∼240° range of motion of the pivot, detailed alignment measurements were conducted during spacecraft-level tests. These measurements were repeated at the NASA Goddard Space Flight Center facility after thermal-vacuum testing and just prior to the final close-out of the spacecraft at Astrotech in Florida. The technique employed multiple

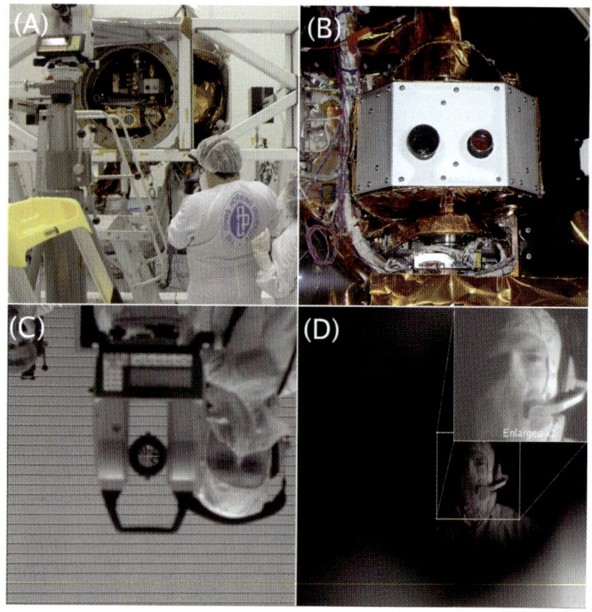

Fig. 42 Final ground-based coalignment measurement between WAC, NAC, and spacecraft. (**A**) Configuration of spacecraft on test fixture, showing theodolite in foreground. (**B**) Close-up of MDIS orientation with WAC on left and NAC on right. (**C**) WAC image and correct orientation relative to geometry shown in (**A**). Note cruciform projected into theodolite gives a particular pixel, mapped into the spacecraft frame. (**D**) NAC image, acquired by shining a bright white light on subject located behind theodolite. The reticule of the theodolite provides the vector mapped into the spacecraft frame

theodolites to reference the spacecraft master alignment cube and map the MDIS alignment cube and boresights at a given pivot position into the spacecraft system. After commanding the pivot to the 0 position of the actuator (apertures directed along spacecraft $+Z$-axis), a reference flat was placed on the nonflight protective cover of the WAC. The theodolite was autocollimated to this mirror, ensuring that the projected beam from the theodolite would at least fall in the WAC FOV. The coordinates at this setting were recorded, and without touching the spacecraft structure the red-tag cover was removed from the WAC. Several images were snapped by the WAC mapping a known field location in the WAC image into the spacecraft coordinate system.

A similar procedure was done with the NAC, but because of the bandpass filter on the NAC the green light of the theodolite could not be passed through the optical system of the NAC. Instead, after autocollimating from a reference flat on the NAC red-tag cover and recording the measurements in the spacecraft frame, a bright white light source was placed at the eye-piece of the theodolite. After passing through the theodolite, the light was collimated and successfully imaged in the NAC. Further, it was possible to identify the reticule on the theodolite that was coaligned with the cruciform, which was autocollimated on the reference mirror on the NAC before removing it. Figure 42 shows these two images. Note the WAC image is rotated 180° relative to the NAC.

4.5 Wavelength Calibration

The NAC uses a single filter, while the WAC views the scene through a filter wheel outfitted with 12 filters of varying widths. The WAC filters are labeled by their position counterclockwise around the wheel as viewed from the CCD. Using the OCF monochromator, NAC and WAC filter transmissions were measured as functions of wavelength for CCD temperatures ranging from $-35°C$ to $\sim 26°C$. The source appeared in the data as a bright rectangle in the center of each image. After subtracting the dark model from the image values and performing smear corrections, a mask was derived from each image based on bright and dark pixel distributions. The values in the mask were set to zero for the dark regions and one for the source rectangle. After multiplication, the pixel values for the resulting image were summed to achieve maximum response with varying source wavelength. A Gaussian shape was used to model most of the WAC and NAC filter centers and widths, except for WAC filters 6 (430 nm) and 11 (1,010 nm), which were modeled with a combination of Gaussian and polynomial fits. The center wavelength and FWHM passbands were calculated for filters 6 and 11 through a cumulative distribution technique, which is similar to a weighted average. Specifically, after normalization, the data were summed as

$$C(\lambda_n) = \sum_{i=0}^{n} F(\lambda_i), \tag{14}$$

where $C(\lambda_n)$ is the cumulative distribution to λ_n, $F(\lambda_i)$ is the filter response at wavelength λ_i, and n varies between 0 and M, the number of monochromator wavelength settings within the filter passband. The 50% point of the cumulative distribution was defined as the center wavelength of those two filters, and the 75% and 25% points were differenced to find the filter width.

Figure 43 shows fits calculated to WAC data acquired in the OCF at a CCD temperature of −26°C, which is about the median operating temperature expected in Mercury orbit. These compare well with the manufacturer-measured passbands (Table 6). Filter 2 is a clear fused silica filter spanning the entire passband of the CCD (395–1,040 nm) and is not shown here. The NAC filter was measured to have a center frequency of 751 nm and a passband of 88 nm.

The center wavelengths shift systematically with temperature of the filter. However, with the exception of the longest wavelength WAC filter (#11), in which the passband is affected by a temperature dependence of the response of the CCD itself, they agree with the nominal values to within 2 nm. Figure 44 shows the difference in filter nominal center wavelengths between room temperature and a CCD temperature of −26°C. The scatter in the wavelength offsets at the lower temperature (−35°C) should be considered as a measure of the precision of determining the center wavelength (±0.5 nm). All filters show the expected shift to longer wavelengths with increasing temperature. Filter 11 has a greater shift than the rest of

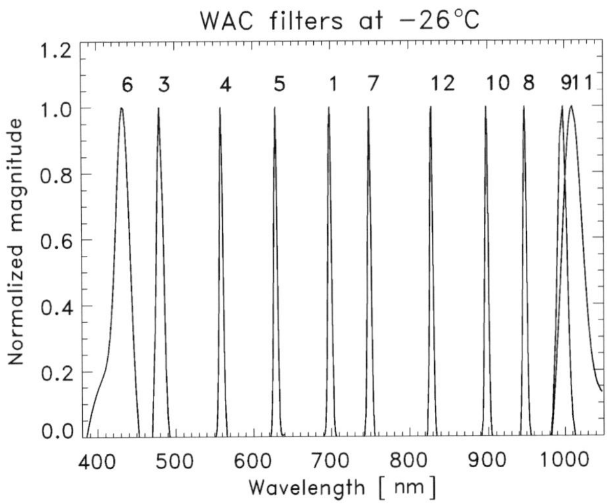

Fig. 43 WAC filter centers and passbands at −26°C

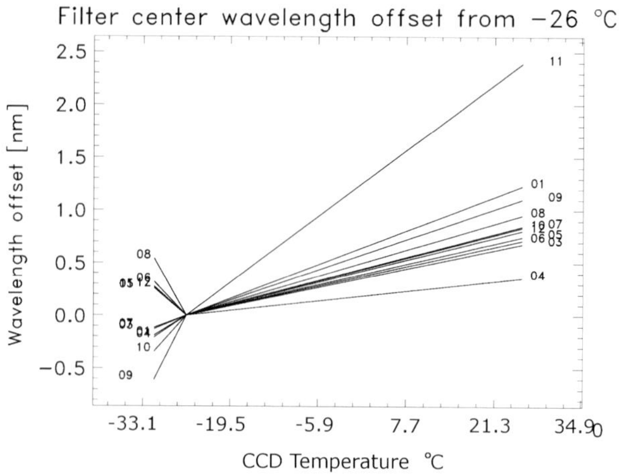

Fig. 44 Offset in WAC filter center relative to position at −26°C, as a function of temperature

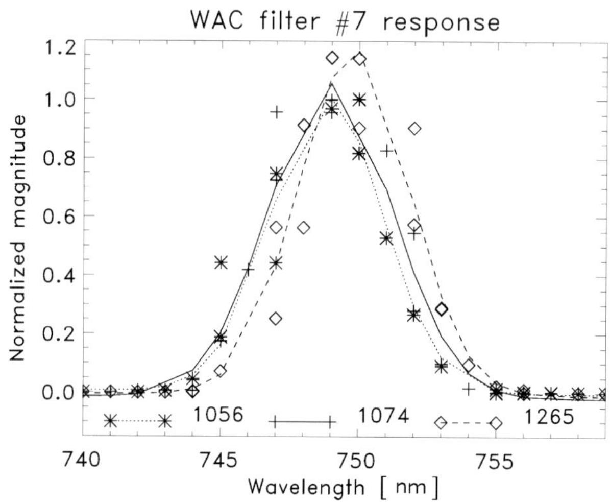

Fig. 45 Shift in WAC filter 7 (750 nm) passband between −31°C. Symbol labels give temperature in DN and room temperature

the filters, due to enhanced long-wavelength response of the CCD at higher temperatures. Figure 45 shows fitted measurements for an example passband, the WAC 750-nm filter (#7), and how the passband shifts with temperature.

4.6 Responsivity, Response Linearity, and Response Uniformity

Calibration of the signal accumulation rate per unit time per unit radiance at the sensor was conducted in the OCF for the WAC and the NAC over a broad range of exposure times and source light intensities. The exposure times to saturation were determined empirically in the initial set-up. Using a variety of temperatures (−34°C to 25°C) and source intensities, images were acquired after varying the exposure times until saturation occurred. Source intensities were stepped in a fixed decreasing-light pattern through the eight levels that could be achieved with the integrating sphere's two 150-W and two 45-W bulbs. The bulbs required about 20 minutes of warm-up before asymptotically approaching constant light output. The WAC calibration procedure began with all lamps on then subsequently turned lamps off in sequence. The exposure sequence was repeated for each filter for the WAC, but only a single exposure sequence was required for the single filter of the NAC.

With MDIS mounted in the OCF vacuum chamber with the integrating sphere set back from the door looking into the window of the door, the sphere did not fill the FOV of the WAC. It was not possible to acquire field-filling sphere measurements in vacuum and at cold temperatures. However, room-temperature, ambient-pressure measurements were acquired with the chamber door open and the sphere moved forward close to the instrument, filling the FOV.

4.6.1 Response Linearity

The relationship between the raw DN output from the camera per unit exposure time and radiance, as a function of variation in exposure time or radiance, is referred to as response linearity. Using dark-corrected, desmeared DN image (DCDSI) values taken from the center quarter of the images, response linearity was measured with both the NAC and WAC binned and full-frame. Linearity was first examined separately at room temperature and cold for (a) linearity with respect to exposure time at individual light levels, and (b) linearity with respect to radiance at individual exposure times. To first order, the results are not significantly different, so all data were merged to examine linearity with respect to measured photons (the product of radiance and exposure time). However, it was found that each detector departs from linearity at low signal levels. Nonlinearity only appears to be correlated with a specific CCD (WAC or NAC), and not correlated with either binning state or detector temperature. For each (WAC and NAC) filter, the responsivity was normalized to unity at a reference dark-corrected, desmeared DN level of 1,500. Responsivity of each detector was found to correlate with the background subtracted DN level in a linear-log relationship, as shown in Figs. 46 and 47 for the WAC and NAC, respectively.

The measurements were taken using a calibrated integrating sphere, and a matrix of exposure times and brightness levels. The clustering of the data in Figs. 46 and 47 shows that there is no difference in system responsivity at a given signal level, whether that signal level is achieved by differences in exposure times or by changes in scene brightness.

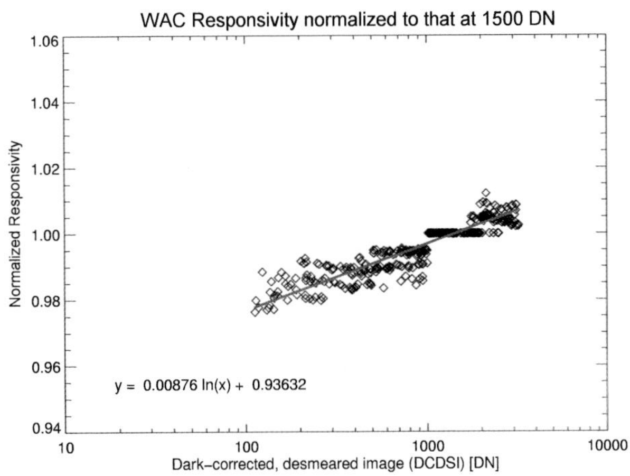

Fig. 46 WAC linearity response shows no distinction between short exposure and brightness variation. These data were acquired over all filters and a variety of exposures and brightnesses

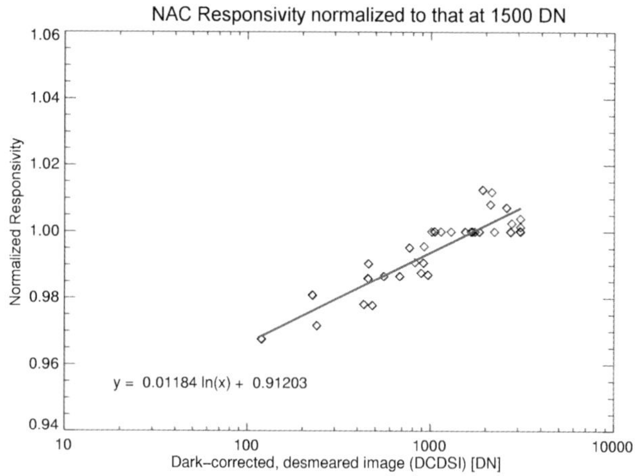

Fig. 47 NAC linearity response, although different from WAC, shows no systematic difference between varying exposure, brightness level, or temperature

4.6.2 *Responsivity*

Responsivity is the dark-corrected, desmeared DN per unit time per unit radiance at the reference DN. Integrating-sphere radiances were calibrated by Labsphere using each of the four lamps separately, over a wavelength range of 350–1,100 nm in 5-nm steps. To determine radiances through each of the WAC and NAC filters, the band passes described in Sect. 4.5 were convolved with the sphere's spectral radiance. For data taken at room temperatures, the chamber door was open, but for lower-temperature measurements the sphere was viewed through the quartz window in the OCF chamber door. Therefore, for the low-temperature data an additional correction was applied to the sphere radiance for each filter to account for the window. This correction was derived from room-temperature data, as the ratio of corrected DNs per unit time with the door closed to that with the door open, for a single configuration of sphere bulbs.

Figures 48, 49, and 50 show the responsivities measured for each filter for the full-frame WAC at a range of radiances and exposures, for three different temperatures. Data points are represented by crosses, which have widths showing 2-σ errors of the dark model and heights showing 2-σ errors in responsivity. Each line connects the responsivities measured at a single sphere radiance at different exposure times. For each filter, the data overlay to within the errors in the dark model. Table 12 gives the actual responsivities for the NAC and WAC. Figure 51 shows the temperature dependence of the full-frame WAC responsivities for each filter, normalized to the CCD temperature of -30.25°C. For these data, differences between -30°C and -34°C at wavelengths <700 nm are assumed to be a measure of systematic errors. No significant temperature dependence of responsivity is expected in these filters in that temperature range. Note that the plot shows the CCD temperature, and not the temperature of the filters. The exact temperature of the filter is uncertain, but a characteristic of these filters is a shift in wavelength toward the red with increasing temperature. Although the temperature coefficients are small (0.015 nm/° at 400 nm and 0.025 nm/° at 1,000 nm), the dispersions in the plots are largest near those filters at the edge of the quantum efficiency curve and so could have a large effect on responsivity. This effect goes in the wrong direction, unless the filter temperature lagged in temperature from the CCD. A second factor that may contribute to the large dispersion is that as the CCD temperature increases, the edges of the band gap widen, which would tend to increase the responsivity, especially at longer wavelengths. Figure 51 shows that around the operating temperature of MDIS, variation in

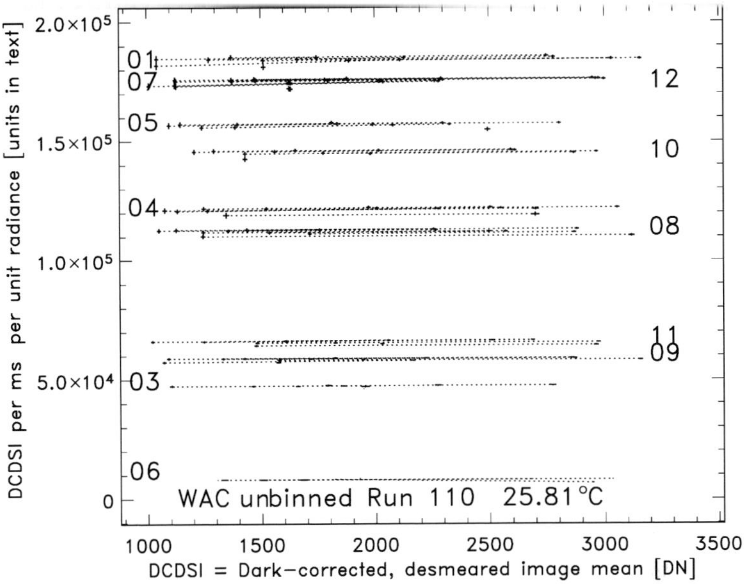

Fig. 48 Full-frame WAC responsivity at CCD temperature 1,267 DN (26°C). Radiance is in units $\mathrm{W\,m^{-2}\,nm^{-1}\,sr^{-1}}$. The lines represent different light combinations; groupings of lines correspond to different filters, whose numbers as given in Table 6 are indicated. Note that the broadband filter (02) used for optical navigation was not fully characterized and is not shown

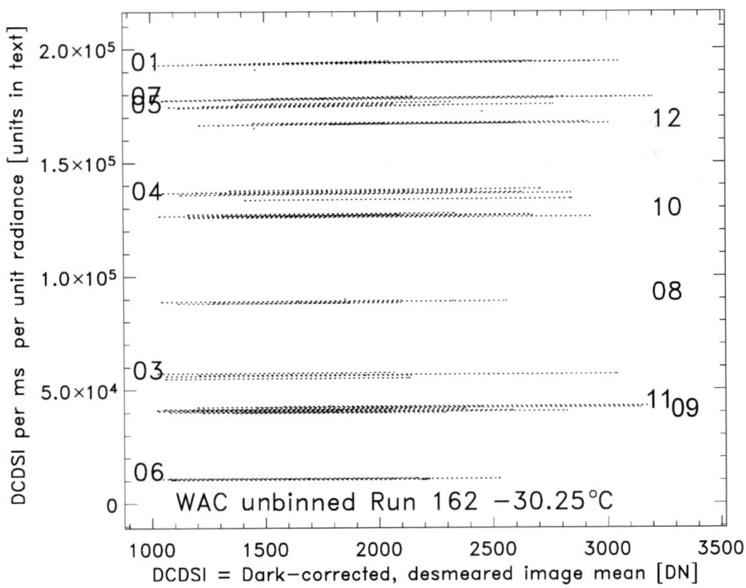

Fig. 49 Full-frame WAC responsivity at CCD temperature 1,060 DN (−30°C). Radiance is in units $\mathrm{W\,m^{-2}\,nm^{-1}\,sr^{-1}}$. The lines represent different light combinations; groupings of lines correspond to different filters, whose numbers as given in Table 6 are indicated. Note that the broadband filter (02) used for optical navigation was not fully characterized and is not shown

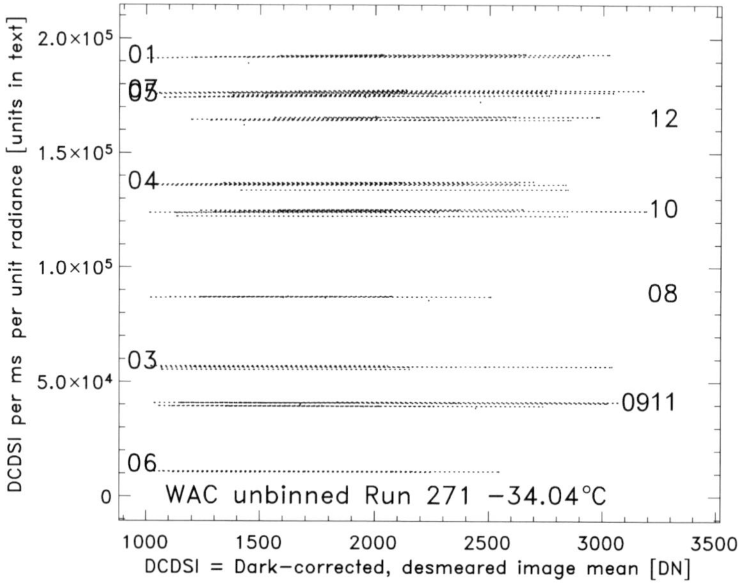

Fig. 50 Full-frame WAC responsivity at CCD temperature 1,046 DN (−34°C). Radiance is in units $\mathrm{W\,m^{-2}\,nm^{-1}\,sr^{-1}}$. The lines represent different light combinations; groupings of lines correspond to different filters, whose numbers as given in Table 6 are indicated. Note that the broadband filter (02) used for optical navigation was not fully characterized and is not shown

responsivity is a small effect. For the binned WAC, responsivity is greater than full-frame WAC responsivity by a factor of 4.

Figures 52 and 53 show the responsivities measured for the NAC at cold temperatures. Data points are represented by crosses, which have widths showing 2-σ errors of the dark model and heights showing 2-σ errors in responsivity. Each line connects the responsivities measured at a single sphere radiance through a single filter at different exposure times. Note the highly exaggerated scale; the scatter in the determinations is only about 4%, which is a measure of the absolute accuracy of the radiometric calibration.

Table 12 summarizes the responsivities of the NAC and WAC, both binned and full-frame, at a reference CCD temperature of −30.3°C. It is expected that the temperature dependence of responsivity in both cameras over the CCD operating temperature range of −45°C to −10°C will be approximately quadratic. However, ground-based measurements were acquired only at intermediate temperatures, and therefore the best correction available from these data is linear in form. A second-order correction will be determined from targets of opportunity in flight (e.g., Canopus from Mercury orbit, at which time the CCD temperature will vary over its operational range). The application of the linear correction is

$$R_{f,T,b} = R_{f,-30.3°\mathrm{C},b}\left[correction\_offset_f + T_{CCD}[DN] \times correction\_slope_f\right], \qquad (15)$$

where $R_{f,T,b}$ is responsivity in filter f at CCD temperature T_{CCD} in units of DNs, and b is the binning mode. $R_{f,-30.3°\mathrm{C},b}$ is responsivity in filter f at CCD temperature of 1060 DN (−30.3°C) as given in Table 12, and $correction\_offset_f$ and $correction\_slope_f$ are camera- and filter-dependent temperature correction offset and slope, also as given in Table 12. The temperature correction defaults to unity at the reference CCD temperature of −30.3°C.

Table 12 WAC and NAC responsivities

Filter number	Responsivity (DN ms^{-1})/(W m^{-2} nm^{-1} sr^{-1})	Temperature correction offset	Temperature correction slope
WAC full frame			
01	1.9359×10^5	1.249	-2.3460×10^{-4}
02	9.8165×10^4	0.956	4.1897×10^{-5}
03	5.6441×10^4	1.831	-7.8410×10^{-4}
04	1.3643×10^5	1.569	-5.3712×10^{-4}
05	1.7488×10^5	1.533	-5.0237×10^{-4}
06	1.0788×10^4	2.303	-1.2284×10^{-3}
07	1.7803×10^5	1.091	-8.6054×10^{-5}
08	8.8803×10^4	−0.353	1.2761×10^{-3}
09	4.0546×10^4	−1.284	2.1543×10^{-3}
10	1.2662×10^5	0.240	7.1685×10^{-4}
11	4.2273×10^4	−1.855	2.6924×10^{-3}
12	1.6698×10^5	0.746	2.3968×10^{-4}
WAC binned			
01	7.7437×10^5	1.249	-2.3460×10^{-4}
02	3.9266×10^5	0.956	4.1897×10^{-5}
03	2.2576×10^5	1.831	-7.8410×10^{-4}
04	5.4572×10^5	1.569	-5.3712×10^{-4}
05	6.9952×10^5	1.533	-5.0237×10^{-4}
06	4.3152×10^4	2.303	-1.2284×10^{-3}
07	7.1210×10^5	1.091	-8.6054×10^{-5}
08	3.5521×10^5	−0.353	1.2761×10^{-3}
09	1.6219×10^5	−1.284	2.1543×10^{-3}
10	5.0648×10^5	0.240	7.1685×10^{-4}
11	1.6909×10^5	−1.855	2.6924×10^{-3}
12	6.6792×10^5	0.746	2.3968×10^{-4}
NAC full frame			
	3.5504×10^4	1.319	-3.0184×10^{-4}
NAC binned			
	1.3574×10^5	1.133	-1.2592×10^{-4}

Slope and offset are for correction to the average responsivity at a CCD temperature value of −30.25°C. Binned responsivity in each filter is four times the full-frame value, but the temperature correction offset and slope are the same

4.6.3 Response Uniformity (Flat Field)

Response uniformity, or flat field, is a measure of pixel-to-pixel variations in responsivity. Measurements of response uniformity of the WAC and NAC were conducted in the OCF, at room temperature and while cold, by imaging the integrating sphere using two 45-W bulbs.

Figure 54 shows a 200-ms-exposure, binned WAC image acquired through the port hole window in the OCF chamber door. As noted earlier, the port hole window is smaller than

Fig. 51 Relative responsivity vs. temperature for unbinned WAC measurements, normalized to the responsivity at −30°C

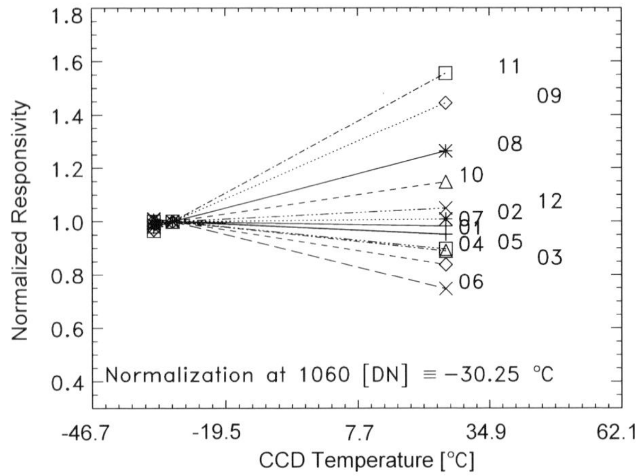

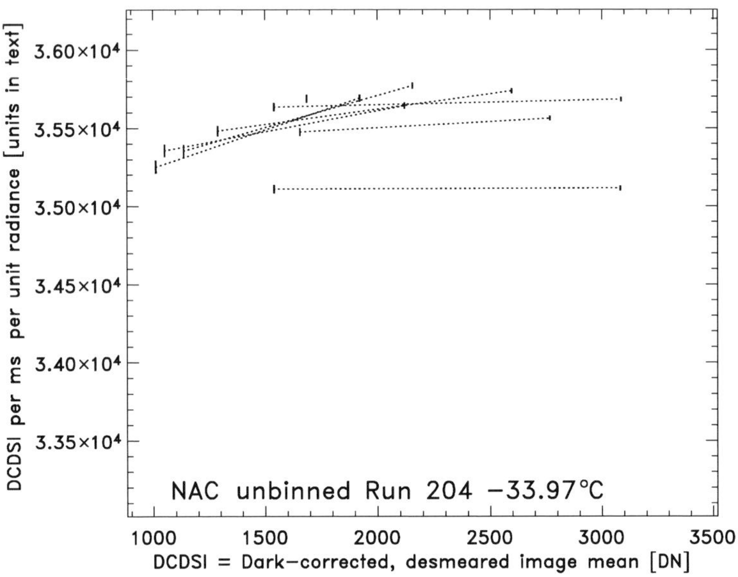

Fig. 52 Full-frame NAC responsivities for CCD temperature 1,057 DN (−34°C). Radiance is in units $\mathrm{W\,m^{-2}\,nm^{-1}\,sr^{-1}}$. The lines represent different light combinations

the WAC FOV. Four significant nonuniformities are evident in the data: (1) bright patches around the rim of the sphere opening, nicknamed "clouds," (2) a slight fall-off in brightness with field angle from near the center of the FOV, (3) slight horizontal striping, and (4) darker spots scattered across the FOV. The clouds are believed to be reflections of the integrating sphere off the radiator and thermal blankets that were mounted between the cold wall (on the door) and the chamber door. Removing the thermal blankets in later tests significantly reduced the clouds. The fall-off in brightness with angular distance from the center of the FOV (Fig. 55) is the expected $\cos^4$ fall-off in response with field angle from the optic axis. The horizontal striping is a function of the binning and is absent in the full-frame images.

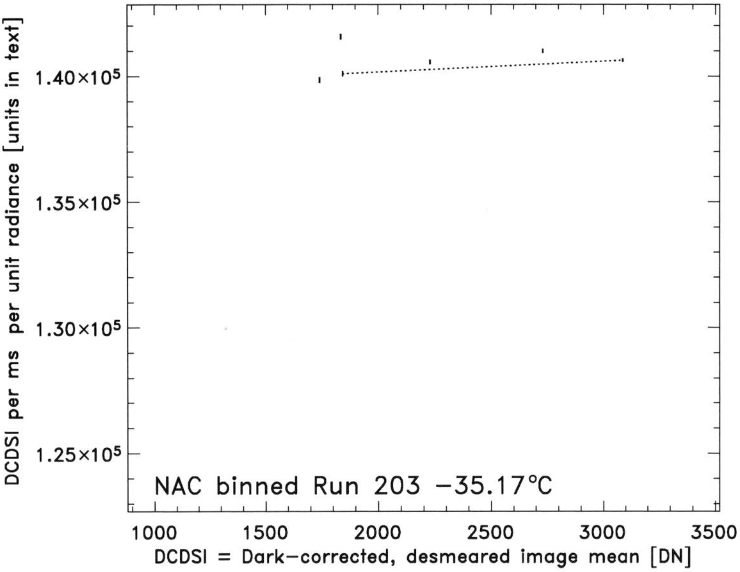

Fig. 53 Binned NAC responsivities for CCD temperature 1,053 DN (−35.2°C). Radiance is in units $\mathrm{W\,m^{-2}\,nm^{-1}\,sr^{-1}}$

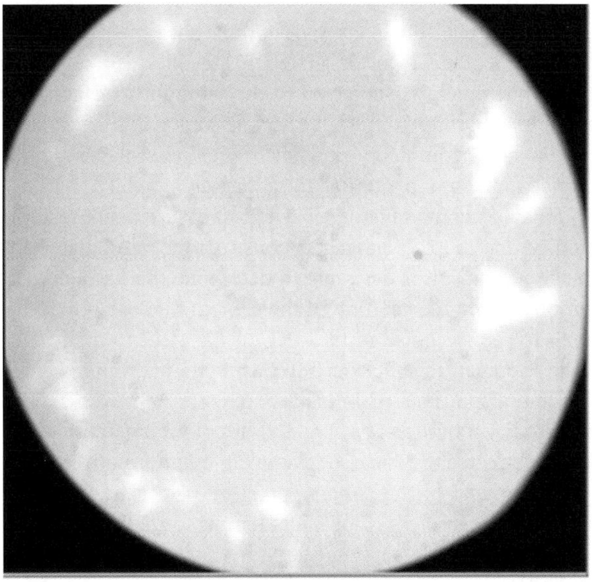

Fig. 54 Nonuniformities are visible in integrating sphere images acquired through the quartz window in the OCF chamber door of the calibration facility

The darker spots scattered across WAC images are fixed with respect to the CCD regardless of filter wheel setting, though their intensities do vary slightly with filter. The sizes of the spots are consistent with shadows of $\ll$35-μm dust on the CCD window, and their number density is consistent with the standards for a class-10,000 clean room in which the camera was assembled. Also consistent with this hypothesis, following instrument vibration during environmental testing, the locations of several spots changed. With the exception of

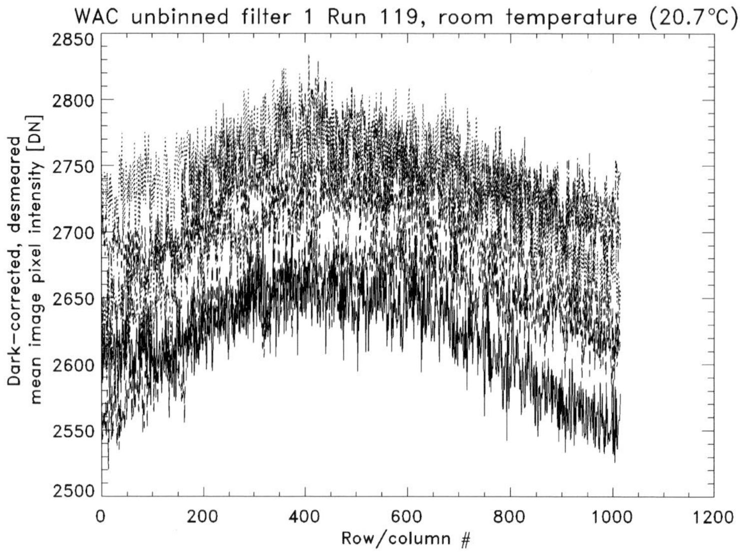

Fig. 55 Profile cuts through an unbinned 699-nm filter WAC integrating sphere image, showing the $\cos^4$ falloff in response off the optic axis. Image is average of 100 images after dark-model and desmearing correction

a single particle (the black spot in the right center of Fig. 54) the dust spots do not significantly affect the DN levels. Given this result, it is likely that the spots themselves will move as the instrument is subjected to the vibrations of launch and flight. Images of the Venus cloud tops acquired during the second Venus flyby will be used to redetermine the flat field post-launch. Preliminary analysis of the flight images acquired thus far shows no significant change in the number or location of spots.

To avoid contaminating flat-field measurements resulting from reflections in the OCF facility, and to avoid having to construct the WAC flat field from tiling image data to form image mosaics, it is desirable to determine the flat field data at room temperature with the door open and the aperture of the camera in close proximity to the integrating sphere aperture. To reduce noise in the derived flat field to approximately 10^{-3}, ~100 images have to be averaged together per filter, camera, and binning mode. Figures 56 and 57 compare the central portions of binned, corrected, and averaged WAC integrating sphere images acquired at room temperature and at cold temperature, respectively. When the images are dark-subtracted, desmeared, averaged, and normalized to the image mean, the relative DN levels are nearly identical on a pixel-by-pixel basis. Since the data are in good quantitative agreement regardless of physical temperature, the flight flat-fields are derived from room-temperature, door-open images.

Full-frame and binned flat fields are shown in Figs. 58 and 59, respectively. The values are normalized to unity in the central part of the images used for responsivity determinations, so that updates to responsivity from non-field-filling sources and to flat field from field-filling sources can be decoupled. The NAC flat field is shown in the upper right panel in both figures, and the 11 narrow-band WAC filters are shown in the remaining panels. Because the spots in the NAC images were not observed to move during environmental qualification of the instrument and they are immobile relative to the detector, redoing the NAC flat field in-flight is not a requirement for the NAC as it is with the WAC.

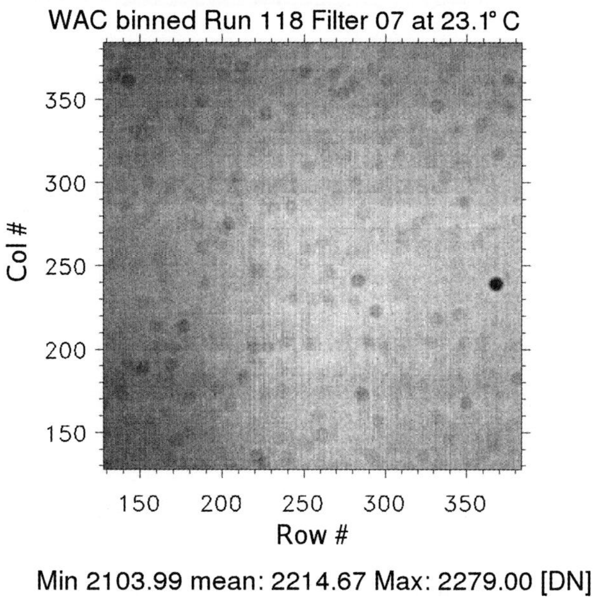

Fig. 56 Contrast-enhanced, room-temperature, binned WAC integrating sphere image with chamber door open

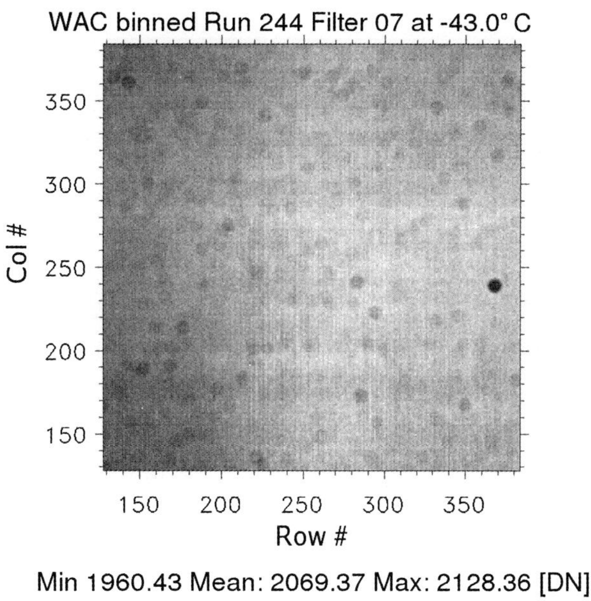

Fig. 57 Contrast-enhanced, cold, binned integrating sphere image with chamber door closed

4.7 Point Spread Function

The point spread function (PSF) is a measure of the two-dimensional distribution of the radiance measured at the detector emanating from a point source. For a conceptual but non-physical imager, light emanating from a point would all fall into a single detector element. In practice the PSF is broadened by diffraction, surface imperfections of optical elements, and scatter centers on optical surfaces. The expected size of the Airy disk (approximately,

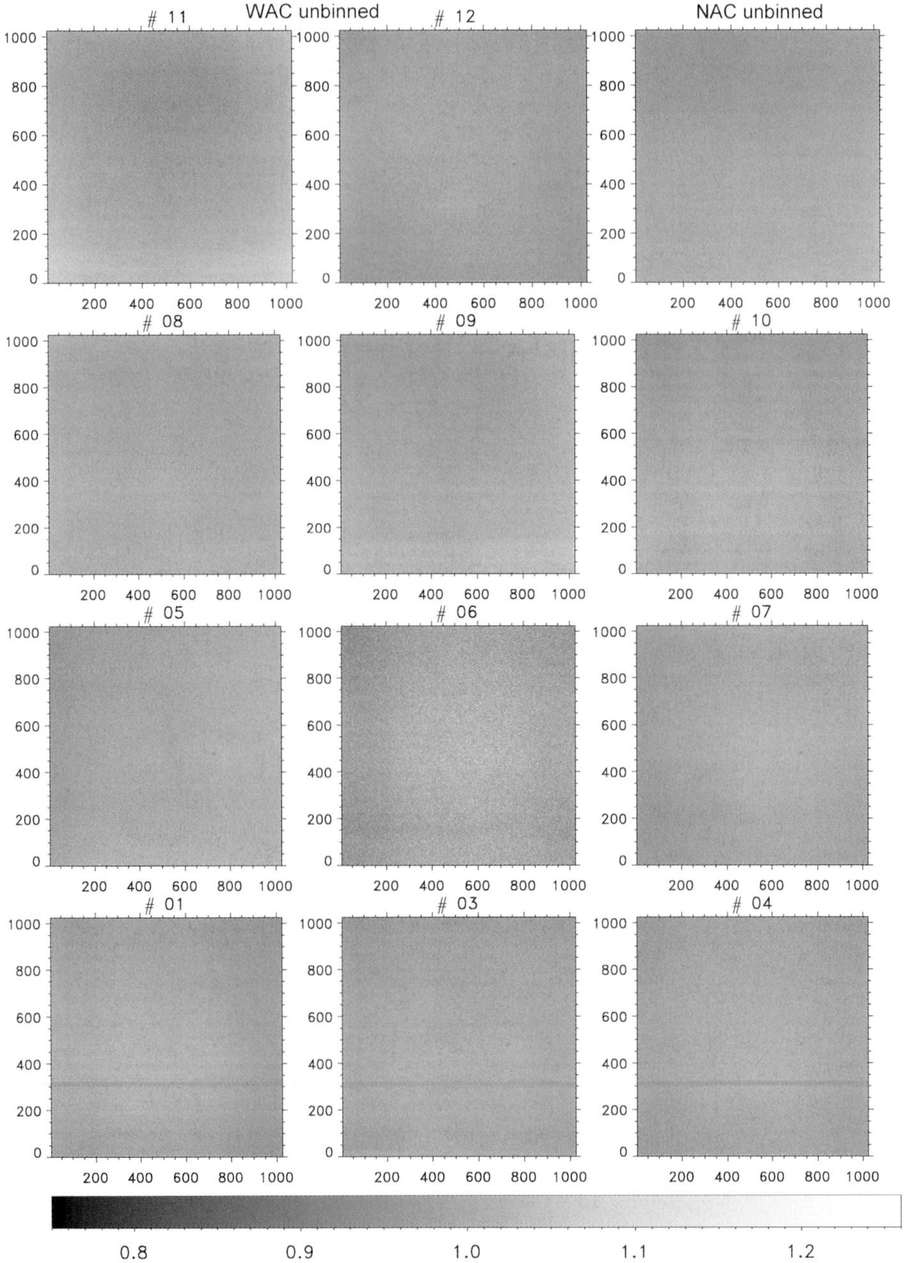

Fig. 58 Contrast-enhanced, unbinned flat fields

the FWHM of the PSF including only effects of diffraction) is >2 pixels for the NAC and ~1 pixel for the WAC. In the particular case of the NAC, the PSF is significantly broader than if the system were purely diffraction limited at the NAC focal length. For nadir imag-

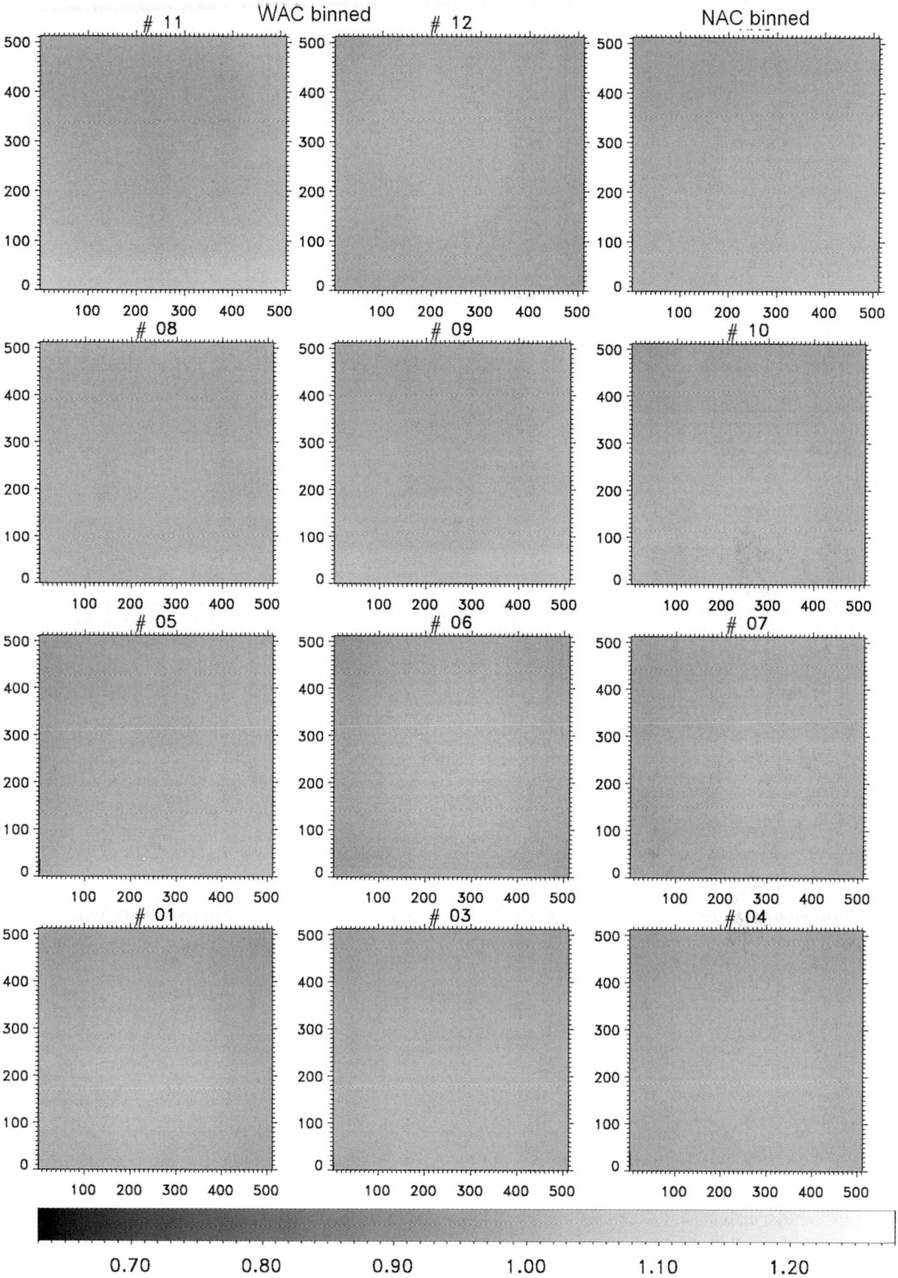

Fig. 59 Contrast-enhanced, binned flat fields

ing at high southern latitudes, this degradation will limit effective sampling of the surface to significantly worse than the desired 250 m/pixel spatial sampling.

The approach adopted for the NAC was to characterize carefully the PSF during ground testing and use Fourier image restoration (optimal filter) techniques to improve the PSF dur-

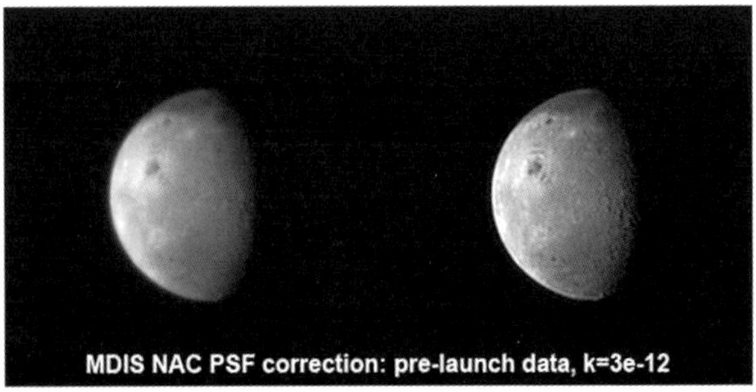

Fig. 60 Effect of image restoration using the optimal filter on an MDIS NAC image of the Moon obtained during Earth flyby. The *left* image is before PSF correction

ing calibration. A similar strategy was applied to NEAR images to remediate PSF degradation due to contaminants on the imager's optics (Li et al. 2002). The optimal filter deblurring algorithm can be presented as

$$I = CF\frac{|H|^2}{H(|H|^2 + K)}, \tag{16}$$

where I is the deblurred image, F is the original, calibrated image in units of radiance, H is the PSF, $|H|^2$ is the norm of complex number H (the square root of the sum of the squares of the real and imaginary parts of H), K is an empirically determined noise constant, and C is an empirically determined constant that maintains radiometric accuracy. K is determined as a value that provides the most sharpening of an image before "ringing" is introduced at abrupt bright-dark boundaries. The output of the correction is in arbitrary units that are directly proportional to the input units of radiance, so C is the ratio of the summed pixel values in the original and deblurred images.

For the deblurring procedure to work effectively, $C$and K must be validated across a variety of scenes, and the PSF must be both well determined and nearly constant across an image. In the case of the NEAR imager, these constraints were all satisfied. There is no requirement, however, that the PSF be radially symmetric about its central pixel. The effectiveness of the procedure is shown in Fig. 60, a NAC image of the Moon acquired during the MESSENGER Earth flyby.

For the NAC, the PSF was assembled from images of a subpixel pinhole imaged in white light at the focus of the OCF collimator. Multiple exposure times were used, the lowest of which is unsaturated, and the longer of which are progressively more saturated. At each exposure time, multiple images are acquired. After correction for dark level, frame transfer smear, and flat-field nonuniformity, the images are divided by exposure times to convert to units of DN. The multiple images at each exposure time are averaged to improve statistics. Starting with the longest exposure, saturated parts of the corrected, averaged image are zeroed out and replaced with unsaturated parts of the next longest exposure. The procedure is continued until the central pixel is reached, and finally all values are normalized to that of the central pixel. This procedure to assemble the PSF is repeated over a 3×3 grid of positions within the FOV (center, corners, and edges) to assess uniformity of the PSF across the FOV.

Figure 61 shows the 3×3 grid of NAC PSFs, displayed using a linear stretch between 0 and 1. The brightest part of the PSF exhibits its expected 2–3 pixel diameter, and the PSF is closely similar across the FOV; there is no evidence for shapes of the PSF that are related to position or distance relative to the optic axis. Figure 62 shows the PSFs similarly, except displayed using a logarithmic stretch between 10^{-6} and 1. The shape of the PSF is well

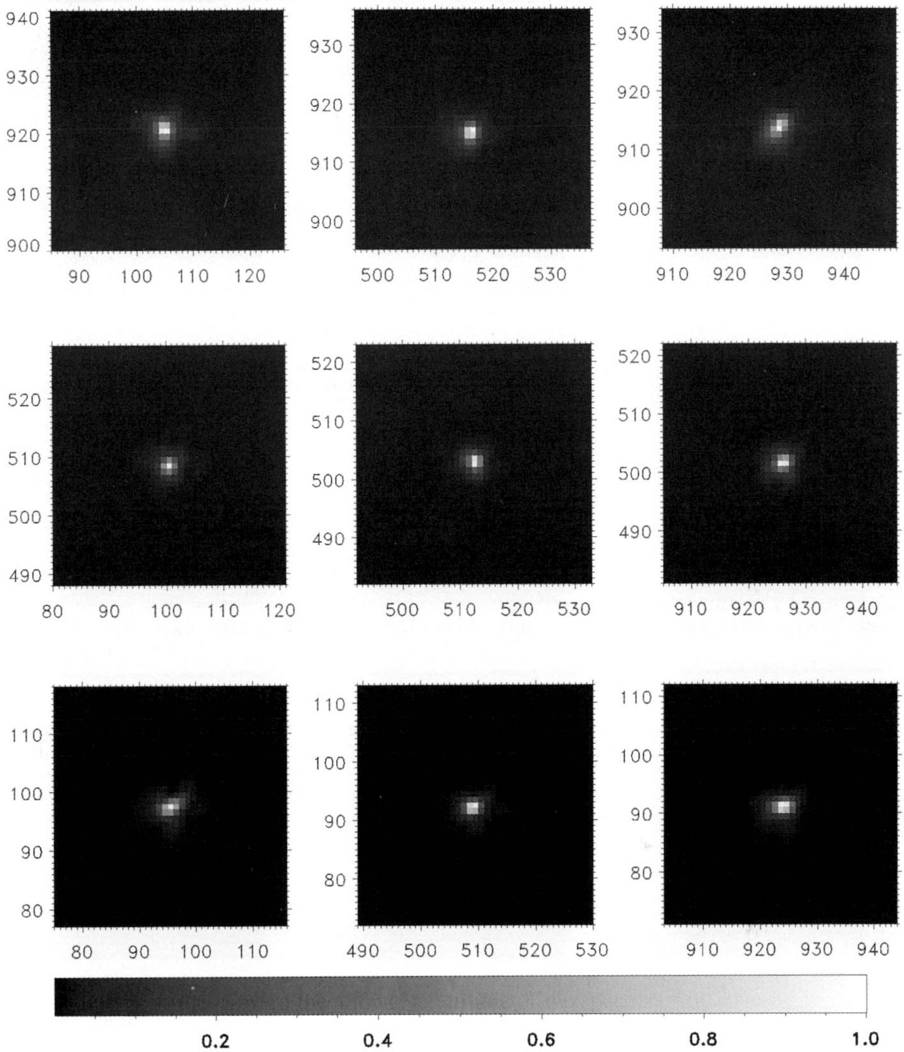

Fig. 61 Two-dimensional intensity distribution of the NAC point spread function, at a 3×3 grid of positions across the FOV. The ordinate and abscissa are labeled with pixel location. Pixel $(0, 0)$ corresponds to the lower left. The stretch is from 0 to 1

determined out to a radius of 10 pixels. The shape is not exactly radially symmetric, but it does appear nearly uniform across the FOV. Thus, the measured NAC PSF appears perfectly suited to application of the optimal filter to improve effective spatial sampling of Mercury in the NAC.

5 Mission Operations and Data Products

The MESSENGER Mission Operations Center (MOC) (Holdridge and Calloway 2007) will conduct most spacecraft activities with reusable command sequences. These activities will

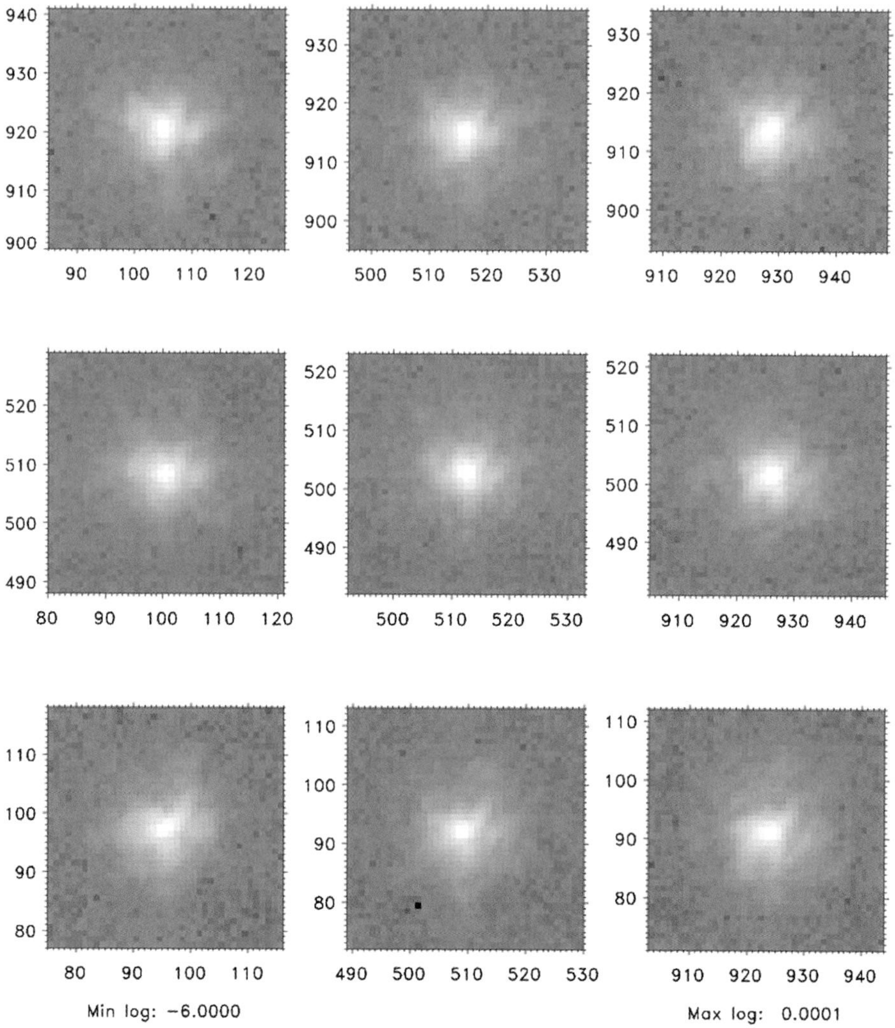

Fig. 62 Two-dimensional intensity distribution of the NAC point spread function, at a 3×3 grid of positions across the FOV. The ordinate and abscissa are labeled with pixel location. Pixel $(0, 0)$ corresponds to the lower left. The stretch is logarithmic from 10^{-6} to 1

be translated into time-tagged commands that call macros prepared, uploaded, and verified by the MOC using the MESSENGER planning and scheduling system. Typically these activities will include recorder operations, DSN track activities, and orbital and attitude maneuvers and the integration of instrument operations. Instrument commands and housekeeping commands will be merged at the MOC, and the sequences will be checked to ensure that all activities are within operational limits and within available spacecraft resources (e.g., memory, power, thermal, recorder).

These reusable command sequences are built by the MOC with input from the subsystem and instrument engineers. The sequences are built from a hierarchy of command groups. The lowest level consists of fragments, which include one or more commands (with defined relative timing) required to implement simple actions on a single instrument or subsystem. The next level consists of Canned Activity Sequences (CASs). A CAS includes one or more fragments with defined relative timing required to perform an entire activity. This activity may cross subsystem boundaries. CASs and fragments usually have input parameters to provide flexibility in use, e.g., logic, timing, and command parameter values. An example of a CAS would be an Optical Navigation Sequence, which first commands the Guidance and Control (G&C) subsystem to point the spacecraft at a star using G&C fragments. MDIS is then commanded to take images using MDIS fragments. These CASs and fragments are tested with hardware and software simulators and are configured on the planning and scheduling system. The highest command level is the request, which calls one or more of these CASs and associates a real spacecraft time and a set of input parameters defining each instantiation of each CAS. Request files are created by the MOC for spacecraft commanding and the science teams for instrument commanding. The concept of CASs and fragments is essential to the philosophy of reusing pretested command blocks to save time in the preparation and validation of spacecraft activities.

The MOC compiles spacecraft and instrument flight rules and operational constraints. This rule set is translated into software run on the ground in the planning and scheduling system. These rules assist the MOC and the science team in flagging command sequences that may put the spacecraft or instruments in improper configurations, exceed onboard resources (e.g., CPU memory, SSR space), or pass parameters to CASs and fragments that are not permitted. These flagged errors are eliminated by adjusting timing between CASs, altering CAS input parameters, or changing which CASs are used to make the command sequence.

6 Data Acquisition Strategy

6.1 Flyby Imaging

The MESSENGER trajectory provides three flyby opportunities of Mercury: January 2008, October 2008, and September 2009 (Table 2, Fig. 63). During the first flyby, approximately half of the hemisphere not viewed by Mariner 10 will be illuminated (subsolar longitude

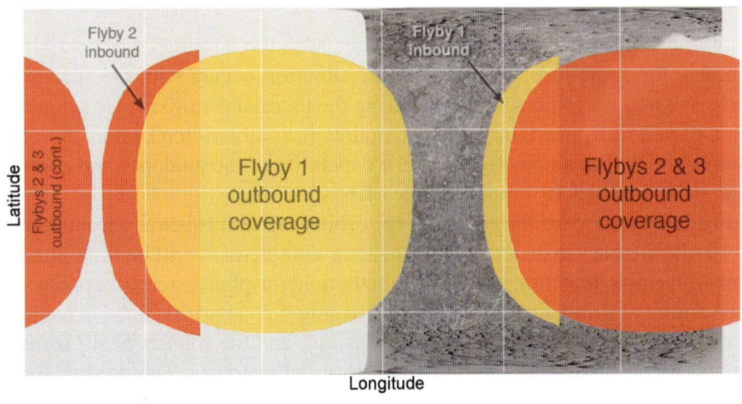

Fig. 63 Preliminary map, in simple cylindrical projection, of MDIS flyby imaging coverage at ≤500 m/pixel (at subspacecraft point) from a range of 20,000 km and emission angle ≤70°

190°E); the first Mercury data return from MESSENGER will thus observe new terrain, including the previously unseen western half of the Caloris basin and its ejecta. During the second flyby, illumination will be centered on the eastern edge of the Mariner 10 hemisphere (subsolar longitude 4°E). The lighting geometry for the third encounter will be nearly identical to that of the second encounter with the subsolar point at the prime meridian (0°E); the approach and departure phase angles will be less extreme, however, resulting in better inbound imaging. During the second and third flybys, most of the remaining unseen portion of Mercury will be imaged. Total coverage between Mariner 10 and the three flybys will exclude only the poles and a small longitudinal gap ~6°-wide, centered at ~97°E longitude. Due to the equatorial closest approach on all three flybys, resolution will diminish towards the poles and limb.

Image mosaics from the various photometric geometries obtained during the flybys and from orbit will require an accurate photometric model of the planet at the wavelengths of the NAC and WAC filters. Therefore, MESSENGER will perform a sophisticated photometric characterization of Mercury's surface from data acquired during the flybys, as well as during orbital operations, through observations of the same point on the ground acquired at various emission, incidence, and phase angles. During each flyby one area on the surface will be observed at 11 different phase angles in 10° increments. Because of Mercury's slow rotation these observations will have a fixed incidence angle—only the emission and phase angles will change.

6.2 Orbital Imaging

Global Monochrome Basemap. One of the primary goals of MDIS is to acquire a global monochrome base map at 250-m/pixel average spatial sampling, low emission angle, and moderate incidence angle. For a given area coverage is obtained at the first opportunity when local nadir is viewed at solar incidence angles of 55–75°. Selection of the combination of emission and incidence angles dictates that the spacecraft will be nearly in a dawn–dusk orbit, minimizing thermal disturbances to pointing. The choice of NAC or WAC is driven by the necessity of maintaining both cross-track overlap and near uniform spatial resolution (140 m/pixel, Fig. 64): the NAC will be used to image the southern hemisphere, whereas the WAC will be used in the northern hemisphere (Fig. 64). For monochrome imaging, the 750-nm filter is used in the WAC to match the 750-nm filter of the NAC. The global nadir-viewing basemap is planned for completion during the first Mercury solar day (i.e., during the first half of the orbital mission).

Stereo Mapping. The off-nadir stereo-complement to the basemap will consist of images taken at nearly the same local solar time. This situation occurs with the spacecraft in the same nearly dawn–dusk orbit, again minimizing the thermal disturbance to pointing and the propagation of pointing uncertainty into uncertainty in southern-hemisphere elevations. Off-nadir pointing will be accomplished using the capability of the guidance and control system to point up- or down-track by commandable offsets.

The downlink profile over the course of the orbital mission is heavily front-loaded due to favorable Earth–Mercury distance, so the off-nadir complement will be time-phased to take advantage of this high downlink rate. In the southern hemisphere, stereo mapping will be accomplished entirely during the first solar day. Imaging on one orbit will be at nadir (0° emission), then on the next orbit off-nadir (25° emission), and the sequence will be repeated so that nadir and off-nadir mapping will be built up simultaneously. The northern hemisphere will be imaged at nadir on the first solar day, then on the second solar day the same image sequence will be repeated off-nadir, covering locations at the same local solar time as in the nadir map. The vertical precision of the stereo map will typically be about 100 m.

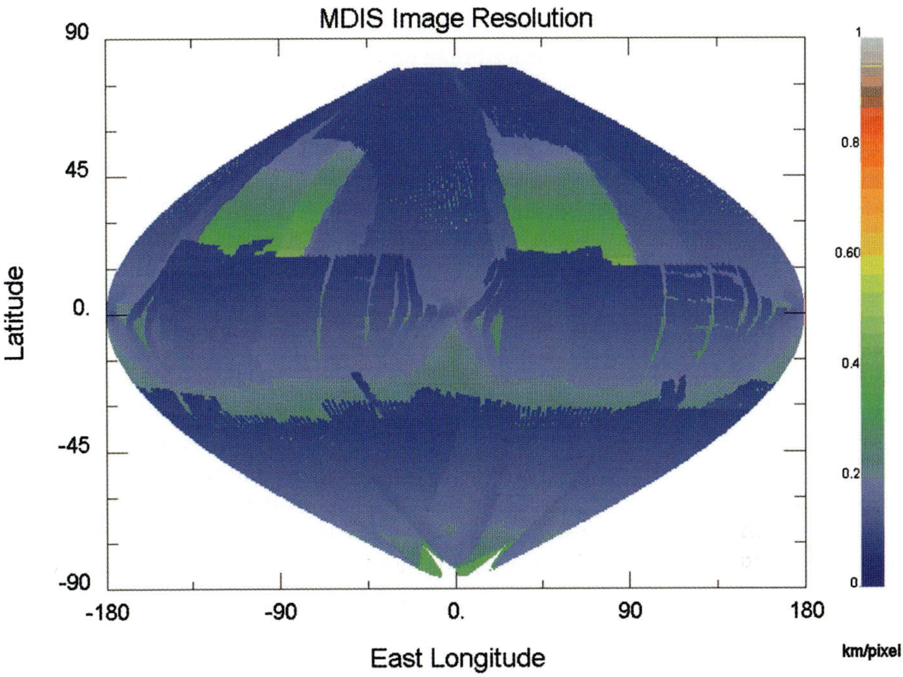

Fig. 64 Preliminary map of spatial sampling in the MDIS global base map, in simple sinusoidal projection

Gap-filling the Flyby Color Map. Once in orbit, the remaining gaps in the flyby color maps will be filled. Details of the mapping strategy are still under development, but acquisition will mesh with the global monochrome and stereo mapping. The southern hemisphere will likely be imaged as mosaics centered on given latitudes and longitudes, while the northern hemisphere will probably be imaged as strips that are pixel-binned on-chip, in order to obtain images in a greater number of filters with no gaps.

Polar Mapping. In order to identify permanently shadowed (and permanently illuminated) areas, polar regions will be imaged repeatedly throughout the Mercury solar day. Over two orbits (24 hours), the subsolar longitude will change by only $\sim 2°$. This strategy provides coverage of all areas near their minimum solar incidence angle. Coverage will extend from 85° or lower latitude to the pole on the night side, and from 80° or lower latitude to the pole on the dayside. At the south pole, the campaign will be divided between the two solar days. On the first solar day, the WAC will be used while the spacecraft is at high altitude, providing ~ 1.5 km spatial sampling and extending equatorward to approximately 70° latitude on the dayside. On the second solar day a more limited region will be covered at higher resolution, using one 2×3 NAC mosaic per day with its long axis aligned north–south. The latter provides about 200 m/pixel spatial sampling.

For the northern polar region, the entire first solar day is required for the monochrome base map, so the polar campaign will be delayed until the second solar day. The WAC will be used because its wider FOV is required for latitude coverage. The imaging strategy will parallel that used for the NAC in the southern polar region, using one 2×3 mosaic per day with its long axis aligned north–south.

High-Resolution Imaging. Selected areas of the northern hemisphere, identified for the most part in flyby imaging, will be imaged from orbit at resolutions of up to ~20 m/pixel. For this purpose the NAC has a "fast" 1-Hz mode, which will acquire images binned 2×2 on-chip in continuous strips 512 pixels wide. At low dayside altitudes (280 km), spatial sampling is 11 m/pixel but along-track motion smear is 18 m at the shortest exposure, consistent with low artifacts from frame transfer smear. Because of MESSENGER's highly eccentric orbit, high-resolution imaging is possible only in the northern hemisphere.

Targeted Color. Selected regions of the planet will be targeted on the second solar day for full-resolution color imaging with spatial sampling up to ~400 m/pixel using the ability of the guidance and control system to track a commanded latitude and longitude on the surface. The maximum spacecraft angular velocity dictates the minimum altitude and thus best spatial sampling at which this imaging can be done. Initial targets will be identified from Mariner 10 data and MESSENGER flyby results.

Color Photometry. During the orbital phase of the mission, photometric geometry complementary to that measured during the flybys is possible: incidence angle variations will be measured at nearly constant emission angle. Key features in low southern latitudes will be imaged repeatedly while the spacecraft is high (10,000 km) above high southern latitudes as the subsolar point moves from terminator to near noon-time. Several target areas will be selected: one to represent average Mercury (heavily cratered highlands), another representing intercrater plains, and a third representing smooth plains materials (probably Tolstoj plains).

Including the flyby and orbital observations, the photometric properties of Mercury will be characterized over greater than 80° of emission angle, 100° of phase angle, and 80° of incidence angle. Results from this experiment will not only characterize photometric properties of the surface but also photometric normalization to a standard geometry (30° solar incidence angle, 0° emission angle) for production of mosaicked products.

On-orbit Calibrations. In order to maintain accuracy of southern-hemisphere elevation measurements, pointing calibrations will be interleaved with nadir and off-nadir global mapping. Once or twice per orbit, depending on orbital geometry, MDIS will be pointed off the planet using the pivot at least at two different pivot angles, permitting star images to be acquired using the WAC. Subframes centered on stars will be used to minimize the data volume required. On-ground, these data will be used to solve for pivot plane orientation as a function of temperature at the base of MDIS.

The MESSENGER project will archive all MDIS Experiment Data Records (EDRs) and Reduced Data Records (RDR) with the Planetary Data System (PDS) in a timely fashion (Table 13). EDRs consist of the raw image data (DNs) fully documented in terms of geometric and radiometric variables. In addition to the raw images, the MESSENGER project will also archive radiometrically corrected versions in units of I/F or radiance, as appropriate. The information contained in this paper and prelaunch calibration files archived with the PDS will allow a user to apply any advances to the state of calibration as needed to the original raw EDRs.

Critical science observations will be obtained during the three Mercury flybys and are of four basic types:

- quadrature monochrome NAC mosaics (200–600 m/pixel),
- medium-resolution NAC quadrature stereo mosaics,

Table 13 MESSENGER MDIS EDR delivery schedule to PDS

Mission data	Product	MESSENGER planned delivery date
Prelaunch calibration	EDR	6 months after launch
Earth flyby	EDR	6 months after 2nd Venus flyby encounter
Venus flyby 1	EDR	6 months after 2nd Venus flyby encounter
Venus flyby 2	EDR	6 months after 2nd Venus flyby encounter
Mercury flyby 1	EDR	6 months after 1st Mercury flyby encounter
Mercury flyby 2	EDR	6 months after 2nd Mercury flyby encounter
Mercury flyby 3	EDR	6 months after 3rd Mercury flyby encounter
Mercury orbit	EDR	every 6 months after Mercury orbit insertion

- synoptic 11-color mosaics (1–8 km/pixel), and
- approach and departure movies (4.4–18 km/pixel).

Data products from these observations will serve as the foundation for assessing the orbital mapping strategy and planning special targeted, high-resolution sequences. RDR products associated with the flyby observations will be reviewed and archived with the PDS on a schedule similar to that for the EDRs.

During orbital operations systematic mapping with both the NAC and WAC will result in:

- a global monochrome basemap with an average resolution of 250 m/pixel,
- merged flyby and orbital multispectral cubes and mosaics in up to 11 spectral filters at $\sim$1 km/pixel,
- local stereo-based digital elevation models, and
- very-high-resolution NAC local strip mosaics in the northern hemisphere (best resolutions $\sim$20 m/pixel).

7 Conclusion

The MDIS instrument and its associated DPU form an integral part of the MESSENGER science payload. MDIS will provide critical measurements tracing Mercury's origin and evolution. The pivoting design of MDIS provides for an extra degree of freedom to map the surface of Mercury and acquire critical optical navigation images without violating the stringent Sun-keep-in rules to maintain the spacecraft sunshade in its proper orientation. The innovative thermal design ensures that the MDIS detectors remain within their operating temperature range even during periapsis at local noon.

On August 3, 2004, the MESSENGER spacecraft was launched from Cape Canaveral, FL. At the time of this writing, all aspects of the instrument have been tested in flight and verified to be working as designed. On May 11, 2005, when the spacecraft was 29.6 million km from Earth, the spacecraft was commanded to look toward the Earth to test the pointing of the MLA instrument. Figure 65 shows one of the six images snapped of the $\sim$0.43 mrad (18-pixel) Earth as observed by the MDIS NAC. In this same frame we observed another bright object—the Moon. This serendipity bodes well for the discoveries yet to be made at Mercury by MDIS and the entire MESSENGER science payload.

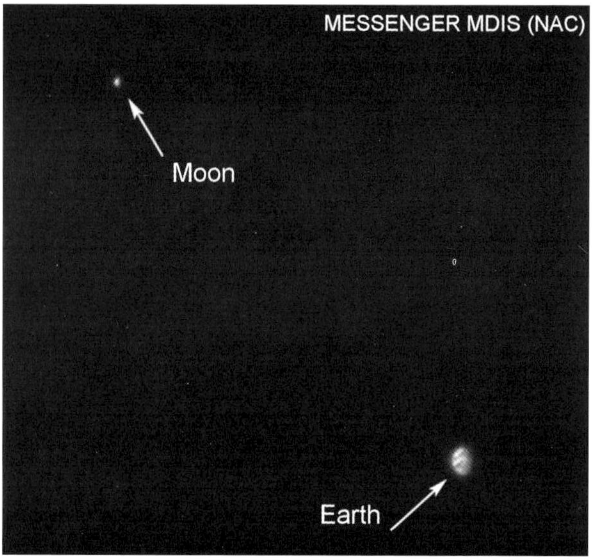

Fig. 65 Taken May 11, 2005, this processed image comes from the NAC when MESSENGER was about 29.6 million kilometers from Earth. The Moon is visible in this frame, but its contrast has been enhanced. Although the Earth only subtends about 18 pixels, bands of clouds between North and South America are apparent on Earth's sunlit side

Acknowledgements The authors would like to gratefully acknowledge the very detailed and thorough contributions of the referees in preparing this manuscript. Their comments significantly improved the quality of the paper.

Appendix 1: Radiation Effects on Filters

We deemed it necessary to characterize the degradation in transmission in the long-pass filter glasses used in the WAC by exposure to ionizing radiation. This appendix summarizes the experimental setup and the results of those measurements.

The manufacturer-provided specifications for the Schott colored glass filters were closely matched to the flight filters. The 25-mm diameter filters were 3.00 ± 0.25 mm thick. The polished surfaces had a surface quality of 80–50 scratch and dig and parallelism to 2 arcminutes. Only one filter, S8612, was 0.5 mm thick.

The equipment used in this experiment included a 0.75-m focal length triple-grating imaging monochromator/spectrograph with a sample holding cell, a stabilized light source, a silicon photodiode detector, and a computer-controlled six-slot filter wheel. All data collection was accomplished with a desktop computer to provide control of the monochromator. A ^{60}Co radiation source chamber was used to irradiate the samples.

The transmission of each test filter was measured prior to any exposure to radiation in order to establish the baseline. The test filters were irradiated in a sealed chamber containing a ^{60}Co source and received radiation at a rate of approximately 0.18 krad per minute. The error introduced in the actual dose due to the time taken to lower and raise the samples from the chamber is estimated to be $\pm 5\%$.

The experimental arrangement to measure the transmission of the test filters is shown in Fig. 66. Light from the stabilized source passes through the monochromator order-sorting filter wheel and enters the monochromator through a variable-width slit; it exits on the opposite side through a similar slit to which the sample holding cell is attached. The sample cell consists of a collimating lens, a sample holding base, and another lens to refocus the image

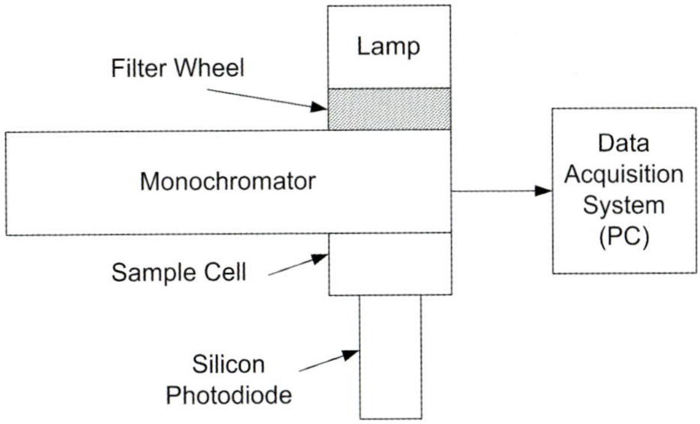

Fig. 66 Experimental test setup for measuring effect of radiation on MDIS filter glasses

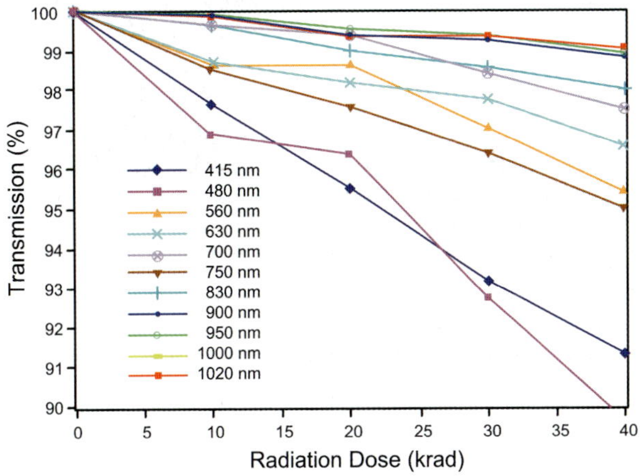

Fig. 67 Inferred transmission loss over passbands of WAC filters as a function of radiation dose

onto the silicon photodiode for measurement Every transmission test followed the same test sequence under the same environmental conditions.

A transmission test sequence included acquiring dark measurements for background subtraction to ensure that only the light passing through the filter was considered in the test. The lamp was turned on and allowed to warm up for ten minutes before the test sequence began. Once the lamp warmed up its spectrum was measured with no filter present. The transmission of the short-wavelength filters was recorded over the range 350–650 nm with no order-sorting filter present in the monochromator beam. The longer-wavelength filters were scanned over the range 650–1,100 nm with an order-sorting filter that attenuated all wavelengths below 550 nm in the beam. The spectrum of the lamp was recorded midway through the measurements and at the end of the filter measurements to verify that the stability of the lamp was below 0.2%. At the end of the filter measurements, the lamp was turned off and a spectrum of the background was taken. Finally, the filter transmissions were calculated by subtracting the dark signal from the values and dividing the filter signal by the lamp-only signal.

The measured transmissions of the test filter glasses generally show a drop of transmission with radiation dose that is wavelength dependent. The maximum transmission of the filters, roughly 90%, is limited by the reflection loss from the surfaces. The loss of transmission as a consequence of radiation is smaller at the longer wavelengths. This trend occurred for each individual filter and continued from filter to filter.

The transmission of the glass filters over the WAC filter passbands was inferred by averaging the results for the test filter glasses over the spectral passband for which they are used in each WAC filter. Figure 67 shows the transmission variation with radiation dose relative to the transmission before irradiation. The transmission loss was observed to decrease with increasing wavelength. The legend in the graph identifies the filter for each curve. Note that these curves are for 2-mm thick glass samples, whereas the actual thicknesses may be as high as 2.93 mm. At the estimated mission radiation dose of $\sim$15 krad the transmission loss will be $<4\%$ in the worst case and will average to $<2\%$ for all the filters. In-flight calibration will be used to track this change if it is measurable.

References

G.B. Andrews et al., Space Sci. Rev. (2007, this issue). doi:10.1007/s11214-007-9272-5
D.T. Blewett, P.G. Lucey, B.R. Hawke, G.G. Ling, M.S. Robinson, Icarus **129**, 217–231 (1997)
J.F. Cavanaugh et al., Space Sci. Rev. (2007, this issue). doi:10.1007/s11214-007-9273-4
B.M. Cordell, R.G. Strom, Phys. Earth Planet. Interiors **15**, 146–155 (1977)
E.H. Darlington, M.P. Grey, Proc. SPIE **4498**, 197–206 (2001)
R.E. Gold, R.L. McNutt Jr., S.C. Solomon, the MESSENGER Team, in *Proceedings of the 5th International Academy of Astronautics International Conference on Low-Cost Planetary Missions*, ed. by R.A. Harris. Special Publication SP-542 (European Space Agency, Noordwijk, 2003), pp. 399–405
O.L. Hansen, Astrophys. J. **190**, 715–717 (1974)
J.K. Harmon, M.A. Slade, Science **258**, 640–643 (1992)
J.K. Harmon, P.J. Perillat, M.A. Slade, Icarus **149**, 1–15 (2001)
S.E. Hawkins, III et al., Space Sci. Rev. **82**, 31–100 (1997)
M.E. Holdridge, A.B. Calloway, Space Sci. Rev. (2007, this issue). doi:10.1007/s11214-007-9261-8
J.R. Janesick, *Scientific Charge-Coupled Devices*. SPIE Press Monograph PM83 (SPIE, Bellingham, WA, 2001), 920 pp
W.S. Kiefer, B.C. Murray, Icarus **72**, 477–491 (1987)
J.C. Leary et al., Space Sci. Rev. (2007, this issue). doi:10.1007/s11214-007-9269-0
J.S. Lewis, Earth Planet. Sci. Lett. **15**, 286–290 (1972)
J.S. Lewis, Ann. Rev. Phys. Chem. **24**, 339–351 (1974)
H. Li, M.S. Robinson, S. Murchie, Icarus **155**, 244–252 (2002)
H.J. Melosh, D. Dzurisin, Icarus **35**, 227–236 (1978)
H.J. Melosh, W.B. McKinnon, in *Mercury*, ed. by F. Vilas, C.R. Chapman, M.S. Matthews (University of Arizona Press, Tucson, 1988), pp. 374–400
S. Murchie et al., Icarus **140**, 66–91 (1999)
S. Murchie et al., Icarus **155**, 229–243 (2002)
B.C. Murray, J. Geophys. Res. **80**, 2342–2344 (1975)
B.C. Murray, R.G. Strom, N.J. Trask, D.E. Gault, J. Geophys. Res. **80**, 2508–2514 (1975)
J.B. Pechmann, H.J. Melosh, Icarus **38**, 243–250 (1979)
A. Potter, T.H. Morgan, Science **229**, 651–653 (1985)
A. Potter, T.H. Morgan, Icarus **67**, 336–340 (1986)
B. Rava, B. Hapke, Icarus **71**, 397–429 (1987)
M.S. Robinson, P.G. Lucey, Science **275**, 197–200 (1997)
M.S. Robinson, J.G. Taylor, Meteorit. Planet. Sci. **36**, 841–847 (2001)
M.A. Slade, B.J. Butler, D.O. Muhleman, Science **258**, 635–640 (1992)
S.C. Solomon et al., Planet. Space Sci. **49**, 1445–1465 (2001)
P.D. Spudis, J.E. Guest, in *Mercury*, ed. by F. Vilas, C.R. Chapman, M.S. Matthews (University of Arizona Press, Tucson, 1988), pp. 118–164
R.G. Strom, Phys. Earth Planet. Interiors **15**, 156–172 (1977)
R.G. Strom, N.J. Trask, J.E. Guest, J. Geophys. Res. **80**, 2478–2507 (1975)
N.J. Trask, J.E. Guest, J. Geophys. Res. **80**, 2462–2477 (1975)
F. Vilas, in *Mercury*, ed. by F. Vilas, C.R. Chapman, M.S. Matthews (University of Arizona Press, Tucson, 1988), pp. 59–76.
T.R. Watters, M.S. Robinson, C.R. Bina, P.D. Spudis, Geophys. Res. Lett. **31**, L04701 (2004)
G.W. Wetherill, Geochim. Cosmochim. Acta **58**, 4513–4520 (1994)
D.E. Wilhelms, Icarus **28**, 551–558 (1976)

Space Sci Rev (2007) 131: 339–391
DOI 10.1007/s11214-007-9262-7

The MESSENGER Gamma-Ray and Neutron Spectrometer

John O. Goldsten · Edgar A. Rhodes · William V. Boynton · William C. Feldman · David J. Lawrence · Jacob I. Trombka · David M. Smith · Larry G. Evans · Jack White · Norman W. Madden · Peter C. Berg · Graham A. Murphy · Reid S. Gurnee · Kim Strohbehn · Bruce D. Williams · Edward D. Schaefer · Christopher A. Monaco · Christopher P. Cork · J. Del Eckels · Wayne O. Miller · Morgan T. Burks · Lisle B. Hagler · Steve J. DeTeresa · Monika C. Witte

Received: 22 May 2006 / Accepted: 10 August 2007 / Published online: 8 November 2007

Abstract A Gamma-Ray and Neutron Spectrometer (GRNS) instrument has been developed as part of the science payload for NASA's Discovery Program mission to the planet Mercury. Mercury Surface, Space ENvironment, GEochemistry, and Ranging (MESSENGER) launched successfully in 2004 and will journey more than six years before entering

J.O. Goldsten (✉) · E.A. Rhodes · G.A. Murphy · R.S. Gurnee · K. Strohbehn · B.D. Williams · E.D. Schaefer · C.A. Monaco
The Johns Hopkins University Applied Physics Laboratory, Laurel, MD 20723, USA
e-mail: john.goldsten@jhuapl.edu

W.V. Boynton
Lunar and Planetary Laboratory, University of Arizona, Tucson, AZ 85721, USA

W.C. Feldman · D.J. Lawrence
Los Alamos National Laboratory, Los Alamos, NM 87545, USA

J.I. Trombka
NASA Goddard Space Flight Center, Greenbelt, MD 20771, USA

D.M. Smith
Department of Physics, University of California, Santa Cruz, CA 95064, USA

L.G. Evans
Computer Sciences Corporation, Science Programs, Lanham-Seabrook, MD 20706, USA

J. White (deceased)
Scientific and Engineering Solutions, Inc., Chagrin Falls, OH 44023, USA

N.W. Madden · C.P. Cork · J. Del Eckels · W.O. Miller · M.T. Burks · L.B. Hagler · S.J. DeTeresa · M.C. Witte
Lawrence Livermore National Laboratory, Livermore, CA 94550, USA

P.C. Berg
Space Sciences Laboratory, University of California, Berkeley, CA 94720, USA

Mercury orbit to begin a one-year investigation. The GRNS instrument forms part of the geochemistry investigation and will yield maps of the elemental composition of the planet surface. Major elements include H, O, Na, Mg, Si, Ca, Ti, Fe, K, and Th. The Gamma-Ray Spectrometer (GRS) portion detects gamma-ray emissions in the 0.1- to 10-MeV energy range and achieves an energy resolution of 3.5 keV full-width at half-maximum for ^{60}Co (1332 keV). It is the first interplanetary use of a mechanically cooled Ge detector. Special construction techniques provide the necessary thermal isolation to maintain the sensor's encapsulated detector at cryogenic temperatures (90 K) despite the intense thermal environment. Given the mission constraints, the GRS sensor is necessarily body-mounted to the spacecraft, but the outer housing is equipped with an anticoincidence shield to reduce the background from charged particles. The Neutron Spectrometer (NS) sensor consists of a sandwich of three scintillation detectors working in concert to measure the flux of ejected neutrons in three energy ranges from thermal to $\sim$7 MeV. The NS is particularly sensitive to H content and will help resolve the composition of Mercury's polar deposits. This paper provides an overview of the Gamma-Ray and Neutron Spectrometer and describes its science and measurement objectives, the design and operation of the instrument, the ground calibration effort, and a look at some early in-flight data.

Keywords MESSENGER · Mercury · Gamma-ray spectrometry · X-ray spectrometry · Surface composition

1 Introduction

MErcury Surface, Space ENvironment, GEochemistry, and Ranging (MESSENGER) is part of NASA's ongoing Discovery Program, which seeks to explore the solar system with lower-cost, highly focused science investigations. The 1,100-kg MESSENGER spacecraft (more than half of which is fuel) was launched successfully aboard a Delta II rocket on August 3, 2004, and will follow a circuitous 6.6-year trajectory (McAdams 2004) involving one flyby of Earth, two of Venus, and three of Mercury before entering an elliptical orbit around Mercury on March 18, 2011. One Earth year of orbital operations are planned. Although Mariner 10 flew by Mercury three times (1974–1975), much is still unknown about Mercury's geologic history and the processes that led to its formation. As the first spacecraft to orbit Mercury, MESSENGER promises to address the outstanding science issues regarding this least explored terrestrial planet (Solomon et al. 2001).

Although much of Mercury's surface will be imaged during the three flybys, most of the detailed science observations will be carried out during the orbital phase of the mission. A one-year campaign to study the exosphere, magnetosphere, surface, and interior of Mercury will be accomplished using a carefully chosen set of miniaturized space instruments. This comprehensive science payload includes: a dual imaging system with wide-angle and narrow-angle cameras, which will provide color and monochrome imaging of the entire planet (Hawkins et al. 2007); an integrated ultraviolet, visible, and infrared spectrometer to detect atmospheric emissions and surface absorption features (McClintock and Lankton 2007); an X-ray spectrometer for remote geochemical mapping of the surface (Schlemm et al. 2007); gamma-ray and neutron spectrometers for geochemical mapping of the subsurface; a vector magnetometer to examine Mercury's active magnetic field (Andrews et al. 2007); a laser altimeter to study the surface topography and planet libration (Cavanaugh et al. 2007); and an energetic particle and plasma spectrometer to characterize ionized species in the magnetosphere (Anderson et al. 2007).

Measurements by the X-ray, gamma-ray, neutron, and infrared spectrometers are complementary, and their combined observations will enable the development of maps of Mercury's surface composition. The Gamma-Ray and Neutron Spectrometer (GRNS) is designed to identify elements within tens of centimeters of the surface by remotely detecting their characteristic gamma-ray emissions in the 0.1- to 10-MeV energy range and by characterizing the flux of escaping thermal, epithermal, and fast neutrons (to ∼7 MeV deposited energy). Taken together, the gamma-ray and neutron measurements will be used to infer composition over localized regions using established techniques most recently demonstrated on Lunar Prospector (Feldman et al. 2004) and Mars Odyssey (Feldman et al. 2002a; Boynton et al. 2004). At Mercury, the GRNS is expected to yield the abundances of the naturally occurring radioactive elements (K, Th, and U), where present at detectable concentrations, those major elements that emit gamma-rays when excited indirectly through bombardment by galactic cosmic rays (H, O, Na, Mg, Si, S, Ca, Ti, and Fe for example), and elements that strongly moderate or absorb neutrons produced by cosmic rays (H and rare earth elements).

The following sections provide an overview of the GRNS science and measurement objectives, describe the design and operation of the instrument, discuss the ground calibration effort, and present some early in-flight performance data. A more detailed review of the GRNS science investigation may be found in Solomon et al. (2001).

2 Science Objectives

The science objectives of the GRNS are to contribute to four of the six prime objectives of the Mercury mission: (1) What planetary formational processes led to the high metal/silicate ratio in Mercury? (2) What is the geological history of Mercury? (3) What are the radar-reflective materials at Mercury's poles? (4) What are the important volatile species and their sources and sinks on and near Mercury? (Solomon et al. 2001).

2.1 Gamma-Ray Spectrometer Science Objectives

Mercury is an end member of the family of terrestrial planets with respect to its mass, density, and heliocentric distance, being the smallest, densest (after correcting for self compression), and closest to the Sun (Strom 1987). Knowledge of the elemental composition of Mercury's crust would distinguish among models for the origin and evolution of Mercury and also place constraints on models for the formation of all terrestrial planets. Almost nothing is presently known concerning Mercury's chemical composition. By the process described in this paper, orbital gamma-ray spectroscopy can yield estimates of Mercury's surface elemental composition and thus provide a significant contribution to planetary science.

Because Mercury has almost no atmosphere to absorb galactic cosmic rays (GCRs), nor a strong enough magnetic field to deflect them, GCRs continuously bombard and penetrate its crust and create many secondary nuclear particles, among them energetic neutrons. These neutrons react with nuclei in crustal materials to create gamma-rays of discrete energies characteristic of an element, either through a process of inelastic scattering of fast neutrons or by capture of neutrons that have moderated to thermal energies through elastic scattering. A fraction of these gamma-rays, as well as those from the decay of radiogenic elements (such as K and Th), escape from the crustal surface to space, where they can be detected by an orbiting gamma-ray spectrometer at altitudes up to 1,000 km. Gamma-rays up to ∼10 MeV emanate from depths of ten centimeters beneath the surface, depending on the crustal density and elemental composition. Detected fluxes are generally low and require numerous orbital passes over a specific region to obtain a statistically well-defined energy spectrum (Evans et al. 2006; Boynton et al. 2007a).

If the gamma-ray spectrometer is calibrated for detection efficiency as a function of energy and incidence angle, the intensities of the gamma-ray spectrum peaks can be related to the elemental composition of the crust from knowledge of the relevant neutron reaction and gamma-ray production cross-sections, by using radiation transport codes to model the production and transport of secondary nuclear particles from GCRs for an assumed crustal composition (Brückner and Masarik 1997). If the crustal nuclear composition assumed in the transport code models is substantially different from that obtained by analysis of the orbital gamma-ray spectra, particularly for elements associated with strong cross-sections, an iterative process is required, in which the model compositions are adjusted until the changes in the resulting spectral compositions are sufficiently small.

The Mariner 10 flybys confirmed that Mercury is denser than the other terrestrial planets after correcting for self compression by internal pressure (Anderson et al. 1987). One scenario for planetary formation that attempts to explain this result is that processes in the early solar nebula changed the Fe/Si ratio in the inner part of the disk compared to the accretion zones for the other terrestrial planets and that the proximity to the Sun caused a reducing environment, so that Si behaved as a metal and contributed to the core phase. As a consequence, Mercury's crust would contain very little FeO (Wänke and Gold 1981). In a second class of models, volatility effects would lead to large chemical fractionations, so that the silicate phase would be depleted in alkalis, FeO, and SiO_2, and would be enriched in CaO, MgO, Al_2O_3, and TiO_2 (Cameron 1985; Cameron et al. 1988). Another class of models is based on the premise that the accretion of Mercury was similar to that of other terrestrial planets, but later giant impacts led to preferential loss of a large fraction of the silicate mantle (Wetherill 1988; Cameron et al. 1988). Also, the K/Th ratio could shed light on Mercury's former position in the solar nebula during accretion (Cameron et al. 1988).

Radar observations of Mercury reveal areas of anomalous reflectivity near the north and south poles that correlate with the floors of large impact craters (Slade et al. 1992; Harmon and Slade 1992; Harmon et al. 1994). Thermal modeling indicates very cold maximum surface temperatures within permanently shadowed polar crater floors (Paige et al. 1992). These observations are consistent with the presence of water ice cold trapped in areas of permanent shadow. Sulfur is another candidate for the radar-bright material (Sprague et al. 1995).

The scientific objectives of the Gamma-Ray Spectrometer (GRS) component of the GRNS instrument are to (1) provide surface abundances of major elements, particularly those elements whose abundances are thought to be indicative of planetary evolution; (2) provide surface abundances of Fe, Si, and K, infer alkali depletion from K abundances, and provide abundance limits on H (water ice) and S (if present) at the poles; (3) map surface element abundances where possible, and otherwise provide surface-averaged abundances or establish upper limits. The measurement of Si and Fe content and other elements (such as Na, Al, Ca, Ti, K, Th, and Mg) should allow classes of formation and evolutionary models for Mercury to be distinguished and address other planetary evolution issues (Boyton et al. 2007b). Of course, only elements sufficiently abundant to be above detection limits can be measured; it is not known with certainty which elements will be detectable. Detection of H and S by MESSENGER is likely only at the north pole, where low-altitude coverage will be substantially superior to that for the south pole, although at 80° inclination the orbit will pose a challenge to direct measurements of the composition of permanently shadowed regions.

2.2 Neutron Spectrometer Science Objectives

The Neutron Spectrometer (NS) sensor, like the GRS, measures spectra derived from element nuclear reactions occurring in the planetary crust, in this case thermal, epithermal, and fast neutron spectra. The neutron flux is substantially higher than the gamma-ray flux, yielding better count statistics, and the NS spectra provide progenitor input into transport computations for renormalizing gamma-ray fluxes in GRS element abundance assays. But determination of surface elemental abundances from neutron energy spectra are in general much more model dependent than from gamma-ray spectra, because there are no element-identifying peaks in neutron spectra and many elements can significantly affect the neutron energy. Very few specific elements can be identified or quantified from neutron spectra alone; an exception is hydrogen. The H nucleus mass is nearly identical to the neutron mass, so H downscatters neutrons to lower energies much more effectively than any other element expected to be present in the crust, leading to decreases in the fast and epithermal neutron fluxes and, possibly, increases in the thermal neutron flux. The flux changes depend on the H abundance and its depth distribution, particularly whether the H is buried. Generally the epithermal flux decrease is most distinctive (Feldman et al. 1997). Also, since higher-mass nuclei downscatter neutrons less effectively than lower-mass nuclei, in favorable cases a coarse map of average element mass can be made (Feldman et al. 1997).

It may also be possible to use NS data to map approximate abundances of the rare earth elements as a group, due to their high thermal neutron absorption cross-sections, if they are present in sufficient abundance to depress significantly the thermal neutron flux. To do this analysis successfully will require a map of the thermal neutron flux depression caused by other elements that also have significant thermal neutron absorption cross-sections, such as Fe and Ti, e.g., from GRS data or from MESSENGER's X-Ray Spectrometer (XRS) (Schlemm et al. 2007). Gd and Sm are the rare earth elements expected to be present in any significant abundance.

NS sensor data will be used to establish and map the abundance of hydrogen over most of the northern hemisphere of Mercury and help interpret measured gamma-ray line strengths in terms of elemental abundances. The hydrogen maps should provide significant new information regarding the presence of water ice within and near permanently shaded craters near the north pole. These maps may also outline surface domains at the base of both northern and southern cusps of the magnetosphere where the solar wind can implant hydrogen in surface material. The NS can also distinguish high-Fe and -Ti basalts from low-Fe and -Ti anorthositic lithologies.

2.2.1 Polar Water Ice on Mercury

The most widely accepted interpretation of enhanced same-sense circular polarization echoes from radar-imaging experiments using the Goldstone/VLA and Arecibo facilities of Mercury's north and south poles has been that nearly pure deposits of water ice fill the permanently shadowed portions of several impact craters (Slade et al. 1992; Harmon and Slade 1992; Butler et al. 1993; Harmon 1997; Harmon et al. 2001). Theoretical studies of the thermal state of polar terrain on Mercury confirm that these craters should be sufficiently cold to maintain such deposits for the age of the solar system (Paige et al. 1992; Ingersoll et al. 1992; Vasavada et al. 1999). An alternate hypothesis is that these deposits are composed of sulfur that has been outgassed from the crust of Mercury and/or released from cometary impacts (Sprague et al. 1995). However, simulations of the migration of volatiles on the surface of Mercury show that sulfur should be stable over both the entire northern and southern polar caps and not just within permanently shadowed craters (Butler 1997). This result makes a sulfur interpretation of the enhanced radar-reflective echoes less likely.

Differences between the results of radar-return experiments at Mercury and the Moon (Harmon 1997; Campbell et al. 2003, 2006) reveal the uncertainties in all interpretations of the origin of polar deposits on these bodies. The size of the polar zone that can retain water ice is predicted to be larger on the Moon than on Mercury (Ingersoll et al. 1992; Salvail and Fanale 1994), and the retention of water vapor released by comets and meteorites after impact should be greater at the Moon than on Mercury because average impact speeds are smaller at the Moon (Moses et al. 1999), but the evidence for polar deposits from radar is stronger on Mercury than the Moon and the number of sites of such deposits is also larger. Potential resolutions of these differences from expectations include (1) the source of water ice on Mercury is endogenic rather than exogenic (Butler et al. 1993), which would imply differences in crustal chemistry and/or outgassing histories between the two bodies; (2) the history of obliquities for the Moon and Mercury differ because of the proximity to the Earth and Sun, respectively; (3) chance has resulted in a very recent cometary or meteoroid impact with Mercury but not the Moon; and (4) the Lyman-alpha resonance-florescence glow of the terrestrial hydrogen exosphere has caused the breakup and subsequent loss of lunar surface ice by photo-dissociation. Mercury may have additional deposits of polar subsurface dirty ice that cannot be seen by the polarized radar echo technique.

Neutron spectroscopy from orbit will be the most sensitive technique to differentiate between water ice and sulfur, as well as to identify the presence of dirty water on Mercury (Feldman et al. 1997; Lawrence et al. 2006).

2.2.2 Search for Terrains that Contain High-Fe and -Ti Basalts and/or High Gd and Sm

Mercury's high density implies a very high ratio of Fe metal to silicate. However, observations of the composition of its surface provide little evidence for Fe^{2+}. The bulk of the iron in Mercury must therefore reside in a large core. Most formation models consistent with a large core fraction involve early melting of the planet. If a differentiated crust solidified from a magma ocean on Mercury, there may be evidence of localized deposits of basalts rich in Fe and Ti or similar to lunar KREEP basalts (Lawrence et al. 2002, 2003). Deposits that are similar to lunar KREEP basalts should be rich in Th, Gd, and Sm. A combination of GRS, XRS, and NS data can provide the abundance of Gd and Sm, as was accomplished for the Moon with the Lunar Prospector Neutron Spectrometer (Elphic et al. 2000; Maurice et al. 2004).

3 Measurement Objectives

3.1 Gamma-Ray Spectrometer Measurement Objectives

Measurement objectives for GRS are based on gamma-ray fluxes obtained from radiation transport computations for three Mercury compositional models (Brückner and Masarik 1997). These three models, "refractory," "preferred," and "volatile," span the range of crustal compositions from refractory-rich to volatile-rich suggested by the different formation models (Goettel 1988) and are shown in Table 1. The Na, Al, Ca, Ti, and Fe contents vary substantially among the models and are good discriminants if their signals can be measured accurately. Si and Fe content are needed to determine the Fe/Si ratio. O content is basically constant and serves as a data normalizer (in particular, its strong lines from neutron inelastic scattering can be used to gauge variation in GCR flux).

Brückner and Masarik (1997) made the simplifying assumption that the crustal compositions are the same as Goettel (1988) model compositions for the bulk silicate portion of Mercury. Because the crust probably formed from melt products of the mantle, this assumption is not likely to hold. Their models are nonetheless sufficient for a sensitivity analysis, but measurement results from Mercury orbit will warrant a more definitive analysis.

Table 1 Model compositions (weight percent) for Mercury's crust (Brückner and Masarik 1997)

Element	Refractory model	Preferred model	Volatile model
O	43.81	45.34	42.89
Na	~0	0.28	1.04
Mg	20.85	22.59	19.33
Al	8.80	2.96	1.73
Si	15.23	21.51	21.05
Ca	10.88	4.02	2.17
Ti	0.43	1.01	0.08
Fe	~0	2.29	11.71

In addition to the gamma-ray fluxes at discrete energies from crustal elements (Brückner and Masarik 1997), which are the signal source, substantial gamma-ray detector background from other sources is present and must be considered in determining the sensitivity of the GRS. GCRs react with elements of the GRS sensor to produce gamma-rays of the same discrete energies as those of the same elements in Mercury's crust. The GRS sensor housing and structure are constructed mainly of Al, with some Fe and Mg, and the sensor Al background may be too great to measure Al crustal content successfully. Estimates of the Fe and Mg sensor background line fluxes were made on the basis of Mars Odyssey neutron flux during flight cruise (Feldman et al. 2002a) and neutron cross-sections. GCRs also react with the spacecraft and with the detector to create discrete and continuous energy background in the sensor; much of the continuous energy background comes from scattered gamma-rays in structural materials and in the detector itself. This background was approximated by the Mars Observer GRS background measured during cruise when the sensor boom was retracted (L.G. Evans, private communication, 2001), scaled to the MESSENGER GRS. The MESSENGER and Mars Observer GRS sensor designs are quite similar; both include high-purity Ge detectors with plastic anticoincidence shields, and they have cylindrical crystals of length and diameter 5.0 cm and 5.5 cm, respectively. The continuous energy background caused by the GCRs reacting with the crust, much of it from gamma-ray scattering, was approximated by adding the Apollo GRS lunar continuum (Bielefeld et al. 1976), scaled to the MESSENGER GRS detector effective area divided by that of the Apollo GRS detector.

The total continuous background under a gamma-ray peak determined in such a manner substantially exceeds the peak signal, even for the narrow energy resolution of a Ge detector. In this case, the minimum gamma-ray flux required for a three-standard-deviation confidence level is well approximated by $f = 3(B\Delta/t)^{1.2}/\varepsilon A$ in units of photons/cm^2/s, where t is the measurement time, A is the detector frontal area in cm^2, ε is the detector intrinsic efficiency, B is the total background in counts/s/keV, and Δ is the energy peak width in keV. B, ε, and Δ vary with gamma-ray energy. ε is approximated by a GEANT4 radiation transport code computation (D.M. Smith, private communication, 2002). Δ varies slowly with energy, so a fixed value of 5 keV full-width at half-maximum (FWHM) was chosen; this conservative value is larger than the measured value to account for broadening by radiation damage. MESSENGER orbits are highly eccentric, with an apoapsis altitude of 15,000 km and a periapsis altitude of 200 km. As a result, only a small fraction of each orbit is spent at altitudes low enough to effectively fill the field of view (FOV) of the GRS, a necessary compromise given the challenging thermal environment. The total accumulated measurement time for altitudes less than 1,000 km and for one year of orbital operations is approximately 137 hours, and a measurement time of 8 hours is considered necessary to carry out coarse mapping of elemental composition over the surface, so these values were

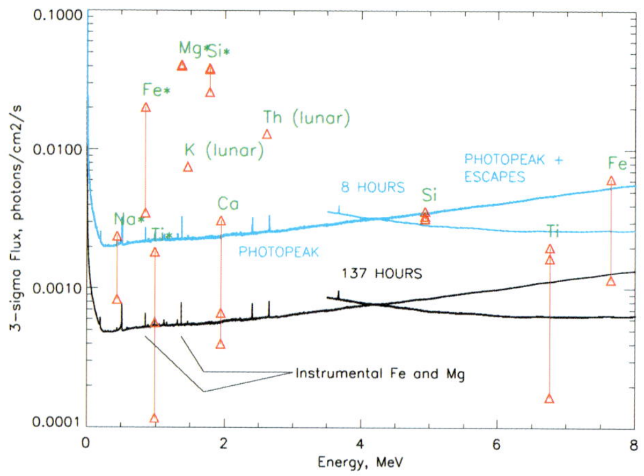

Fig. 1 Calculated sensitivities to elements of MESSENGER's GRS for three Mercury crustal composition models (*diamonds*). Sensitivity baselines are for measurement times of 137 hours (full mission) and 8 hours (coarse mapping)

chosen to calculate the sensitivity of the GRS to the major discriminating elements (with perhaps only tens of pixels or so up to more than a hundred, depending on GRS element sensitivity).

Shown in Fig. 1 is f for 137-hour (full mission) and 8-hour (coarse mapping) measurement times, along with flux ranges at energies for specific elements for the three crustal composition models, for uncollided gamma-rays of discrete energy (Brückner and Masarik 1997). Lunar fluxes are plotted for the radioactive elements K and Th (Bielefeld et al. 1976), for want of known Mercury abundances. Where a flux range appears above the corresponding f, sufficient signal definitely exists for elemental content detection, and also for quantitative measurement to some precision level. Figure 1 indicates that Fe, Si, K, Th, and Mg can be mapped for all model compositions, that Ca can be mapped for one composition model but measured only for the average surface for another, that Ti can be measured only for the average surface for two composition models, and that Na can be mapped for one composition model and measured only for the average surface for another. If the null result is interpreted as no significant content, then each of the elements shown can act as a discriminant for the compositional models.

The detection of water ice at the poles depends on the detection of H, which can be accomplished by several methods, all based on the H moderation of fast neutrons to epithermal and thermal energies: measurement of the H-capture gamma-ray 2,223-keV peak intensity, measurement of the fast, thermal, and epithermal neutron fluxes by the NS (Feldman et al. 1997), and measurement of the ratios of the gamma-ray capture peak intensities to the inelastic peak intensities at available energies for the same element (Evans and Squyres 1987). Radiation transport model computations indicate increases in the ratio of capture to inelastic peak averaging ~25% for elements such as Fe, Si, and Mg with as little as 0.5 weight percent water uniformly added to the crust (Brückner and Masarik 1997). Although water ice would be distributed nonuniformly as patches in permanent shadows in unknown amounts, this computation gives some indication of GRS sensitivity to water ice. Because Fe, Si, and Mg were found mappable in Fig. 1, significant deposits of water ice at the north pole might be detected by the MESSENGER GRS. The GRS is expected to be substantially less sensitive to H than the NS, since the gamma-ray flux is much lower than the neutron flux. In principle, S at the north pole should be detectable through its 2,230-keV inelastic gamma-ray or 5,424-keV capture gamma-ray, if present in sufficient amounts.

3.2 Neutron Spectrometer Measurement Objectives

The scientific objectives addressed by the Neutron Spectrometer sensor can be met by measuring separately the flux of thermal neutrons (having energies between 0 and 1 eV), epithermal neutrons (having energies between 1 eV and about 500 keV), and fast neutrons (having energies between 500 keV and about 7 MeV). The counting rates should be sufficiently high to allow the mapping of all three neutron energy bands over all sub-satellite positions for which the spacecraft altitude is less than about 1,000 km. Also required will be (1) a continuous monitor of the gains of all three detector channels to provide for in-flight gain calibration; (2) measurements of all counting rates as functions of orientation of the detector relative to the spacecraft velocity vector while in Mercury orbit; (3) monitoring the flux of high-energy protons using coincidences between all three neutron spectrometer scintillators, as well as using some of the energetic particle sensors aboard MESSENGER; (4) determining spacecraft backgrounds during the full range of solar energetic particle environments in all three neutron energy ranges during cruise in transit to Mercury; and (5) calibrating the sensitivities of all three energy ranges by measuring counting rates during the second Venus flyby and all three Mercury flybys as functions of spacecraft altitude.

The science goal to identify and map the abundance of hydrogen over most of the northern hemisphere of Mercury will be accomplished by mapping the epithermal counting rates over all subsatellite positions for which the spacecraft altitude is less than about 1,000 km. Hydrogen is signaled by a reduction in epithermal counting rates relative to that measured above dry soil. Based on the fourth periapsis pass of Mars Odyssey (which, during that pass, had about the same orientation relative to Mars that the MESSENGER NS will have relative to Mercury at periapsis), the epithermal counting rate of the MESSENGER NS should then be between about 1.5 and 4.7 counts per second. In other words, the epithermal sensor of NS should register about 800 counts in 300 s and obtain 1% statistics (10,000 counts) in 12 passes over the same regolith area.

Distinguishing between high-Fe and -Ti basalts and low-Fe and -Ti anorthositic lithologies could be important for MESSENGER because it would extend the knowledge of Mercury's surface composition provided by the GRS. Elemental composition will be determined by GRS only as an average over relatively large horizontal scales because of expected low counting rates. The thermal neutron counting rates measured aboard Lunar Prospector (which provided a basalt-anorthosite discrimination) varied by a factor of three between mare and highland terrains. From the thermal neutron counting rates measured by Mars Odyssey (which has a similar front- and back-facing scintillator arrangement and can be used to provide an estimate of the counting rate of the front-facing minus the counting rate of the back-facing Li-glass scintillator elements for the MESSENGER NS), we should expect between 1 and 6 counts per second for this difference at Mercury if high-Fe and -Ti units are present.

Fe and Ti compositional information can also be inferred from a map of the flux of fast neutrons. Because both the thermal and fast neutron fluxes are independent, their combination provides an ability to account and correct for contributions to both counting rates from neutrons processed by the spacecraft.

A last measurement requirement is to support the analysis of gamma-ray spectra by providing the surface neutron number density and flux of fast neutrons. This task should be much more straightforward for Mercury (as it was for the Moon) than for Mars because of the expectation that hydrogen will not be abundant within nonpolar terrains on Mercury. Measurement of gamma-ray line strengths are proportional to the respective elemental abundances through the neutron number density for (n, γ) transitions and through the flux of fast neutrons for $(n, n'\gamma)$ transitions. Whereas the neutron number density was found from the analysis of Lunar Prospector data to depend monotonically on the ratio of thermal to epithermal neutron counting rates, the multiplicative factor for the $(n, n'\gamma)$ reactions has been shown to be directly proportional to average atomic mass (Lawrence et al. 2002; Prettyman et al. 2006).

4 Instrument Design

4.1 Instrument Overview

The MESSENGER mission was conceived to answer outstanding science questions about the formation and evolution of Mercury while adhering to the schedule and cost constraints of the Discovery Program. By its nature, an orbital mission to Mercury is extremely challenging and must necessarily include tight restrictions on the resources available to the payload. Mass limitations precluded the use of a boom to distance the sensors from the locally generated neutron/gamma background of the spacecraft as was done on Lunar Prospector and Mars Odyssey (Feldman et al. 1999; Boynton et al. 2004). The high-purity germanium (HPGe) detector size is smaller than that of Mars Odyssey to ensure an adequate cooling margin while at Mercury. Because the gamma-ray spectrometer is actively cooled and therefore relatively compact, it was possible to incorporate an anticoincidence shield to reduce the unwanted background due to charged particles. The instrument thermal design in turn makes good use of the extra heat capacity of the shield housing to lessen the spike-like temperature transients that will be experienced during some of the worst-case orbits. But it was not until the shield mass was minimized that the neutron spectrometer detector size could be increased to match that on Mars Odyssey. These are the types of systems issues and trades that often drove the design of the instrument—and the payload—as there was always an awareness of the immediate impact on the other instruments and the spacecraft.

There are seven instruments that make up the MESSENGER science payload. The GRNS is one of those seven from the viewpoint of representing a single science investigation, but it is partitioned as a GRS sensor and an NS sensor that are physically separate and operate independently. Each sensor has its own associated electronics box with a dedicated microprocessor, power supply, and telemetry interface. Mass was, of course, at a premium, but it is unlikely that any significant mass savings could have been realized with shared electronics, as the sensors are located on different deck surfaces and the additional cabling required would have absorbed most of the difference. Four major assemblies make up the GRNS: the GRS sensor, the GRS electronics, the NS sensor, and the NS electronics.

The payload instruments and their locations are shown in Fig. 2, in a view looking up from the planet. False colors are used to highlight instrumentation. The narrower FOV instruments are clustered inside the payload adapter ring on an isolated deck. The GRS is located on a deck just outside the adapter ring, which allows its passive radiator a view of cold space and also keeps any vibrations produced by the mechanical cryocooler from reaching the optical instruments. As a result, a portion of the aluminum adapter ring falls within the FOV of the GRS, although when the spacecraft is nadir-pointing its overall attenuation is expected to be less than 8%. The GRS sensor is mounted atop a set of standoffs ("flexures") that elevate the GRS detector to the allowed height limit of the clamp band (which attached the spacecraft to the third stage of the launch vehicle). The lower portion of the sensor, which mainly consists of the photomultiplier tube (PMT) assembly for the shield, extends down through a hole in the deck. The NS is located on the deck opposite

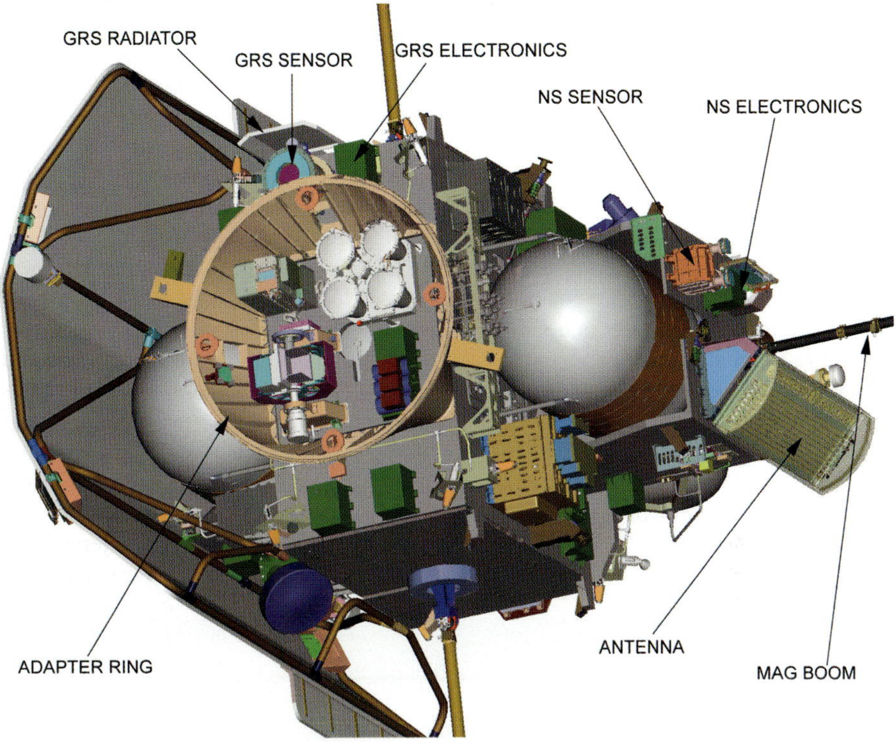

Fig. 2 View of the MESSENGER spacecraft showing the locations of the GRS and NS sensors. The GRS is located outside the adapter ring, which provides mechanical isolation and allows its passive radiator a view to cold space. The NS is located near the base of the MAG boom and detects thermal neutrons along and opposite the orbital velocity direction of the spacecraft, while detecting fast neutrons towards nadir

the sunshade near the base of the Magnetometer boom. The spacecraft mounting surface for the NS was extended downward slightly the better to clear the FOV for the NS low-energy neutron detectors, but a completely clear FOV is not possible for detectors with hemispherical response. The two low-energy neutron detectors view directions along and opposite to the spacecraft orbital velocity with respect to the planet (when in a "dawn/dusk" orbit), with the major asymmetry being the Magnetometer boom and part of the rear-facing phased-array antenna (Leary et al. 2007). The detector background may also differ slightly for the two detectors, as the inboard detector has an increased view of the spacecraft propulsion tanks. The fast-neutron spectrometer has an omnidirectional response, so a portion of its view toward the planet will always contain a major cut through the spacecraft (primarily the propulsion tanks). Background subtraction is a major challenge for both the GRS and NS experiments, and their science investigations must rely on in-flight calibration measurements carried out during the cruise phase, during Venus and Mercury flybys, and at apoapsis during orbital operations. Details of the design of the sensors, electronics, and flight software are given in the following sections.

4.2 Gamma-Ray Spectrometer

The GRS detector is a high-resolution coaxial germanium crystal 50 mm in diameter and 50 mm in length, N-type, chosen for its resistance to radiation damage and its annealing capabilities. The detector is rigidly clamped in a hermetically sealed Al capsule pressurized with clean, dry nitrogen. The capsule is thermally isolated by nested thin, low-emissivity shields suspended on Kevlar strings and is cooled to an operating temperature in the 85–95 K range by a mechanical cryocooler attached to an external passive radiator that rejects its heat to cold space. The capsule is equipped with two Zener diodes capable of annealing the Ge crystal up to 358 K (85°C) or higher. A plastic scintillator anticoincidence shield (coupled to a photomultiplier tube) surrounds the Ge detector on its sides and back, for rejection of cosmic-ray background. In addition to survival heaters on the sensor housing having fixed on and off temperatures, the cooler compressor is outfitted with a heater with its temperature regulated under instrument control. Electronics includes low-voltage and high-voltage power supplies, cooler power and Ge temperature regulation, annealing power and temperature regulation, uncooled preamplifiers, shaper amplifiers, high-speed analog-to-digital converters (ADCs), field-programmable gate arrays (FPGAs), and a microprocessor.

Figure 3 shows a photograph of the complete GRS on a test stand just prior to installation on the spacecraft. The electronics box is seen in the foreground with the sensor housing behind it. The Ge preamplifier box is seen attached to the sensor housing in front in the photo, with the Ge high-voltage (HV) filter box attached to the sensor housing in back. The passive radiator (which is always oriented facing outer space, away from the planet and

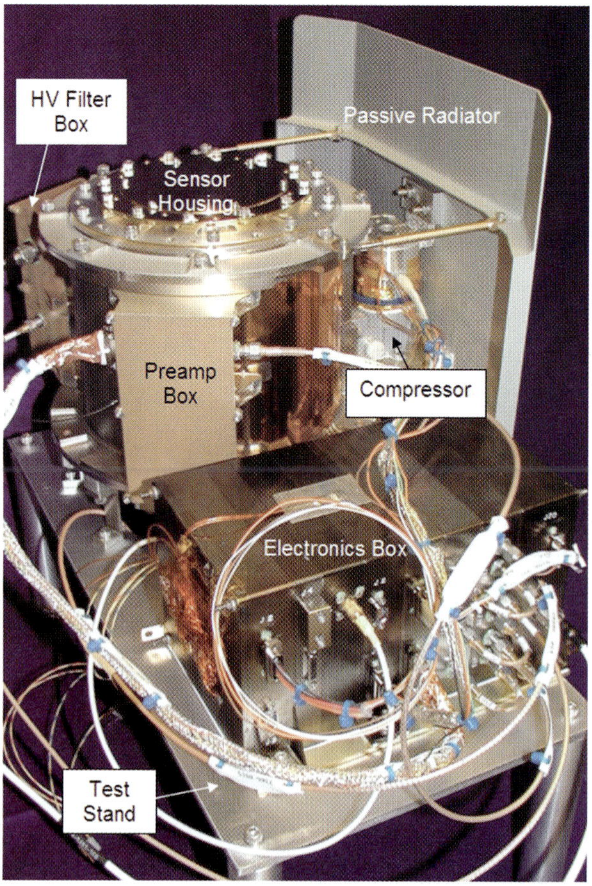

Fig. 3 Photograph of GRS flight unit on a test stand, prior to mounting on the spacecraft. Major visible exterior parts of the instrument are labeled. For scale reference, the electronics box dimensions are approximately 10 cm × 10 cm × 18 cm

the Sun) is to the right of the sensor housing, with the cryocooler compressor between the radiator and the housing.

Table 2 gives a summary of GRS specifications and performance characteristics. The energy resolution exceeds the target assumed for science measurements. The electronics are designed to reduce the effects of nonlinearities, noise (including that from cryocooler vibration), and temperature change on energy resolution. The preamplifier gain drift with temperature, together with the largest sensor housing temperature change in orbit near periapsis, yields the largest potential smearing in energy resolution during data collection. Correction for this gain drift, however, coupled with small data collection intervals near periapsis, can

Table 2 GRS summary specifications

Primary sensor	High-purity germanium (HPGe) coaxial detector, 5 cm × 5 cm beveled cylinder
Anticoincidence shield	Borated plastic scintillator (BC454) cup coupled to a 76-mm PMT
Effective FOV	Limb-to-limb (omnidirectional, with some shadowing by anti-coincidence shield and adapter ring)
Cryocooler	Miniature rotary Stirling cycle, 0.5-W heat lift
HPGe operating voltage	3,200 V, leakage current <10 pA
HPGe energy range	60 keV to 9 MeV
No. of spectral channels	15,000 HPGe; 1,024 shield
Energy resolution (FWHM)	3.34 keV @ 121 keV, 3.49 keV @ 1,332 keV, 4.77 keV @ 6,130 keV
Pulser resolution (FWHM)	3.07 keV @ 7985 keV, rate = 7.6 Hz
Intrinsic efficiency	~0.236 @ 511 keV, ~0.104 @ 1,332 keV, ~0.018 @ 6,130 keV
Maximum throughput	>5,000 events per second
Background rejection ratio	2:1 @ 1 MeV, 7:1 @ 8 MeV
Differential nonlinearity (DNL)	< count statistics (less than 0.3% at 10^5 counts/channel)
Integral nonlinearity (INL)	±1 channel (~0.6 keV) over full energy range, stable over temperature
Electronics offset	±0.2 channels over temperature
Electronics gain drift	<20 ppm/°C
Preamplifier gain drift	<95 ppm/°C
Pulser drift	<50 ppm/°C
Sensor thermal environment	$-25°C < T_{radiator} < 40°C$
Cool-down time to 90 K	< 36 hrs, $T_{radiator}$ @ 25°C
HPGe crystal temperature regulation	<0.25°C, $-25°C < T_{radiator} < 65°C$ (transient conditions)
HPGe crystal annealing temperature	85°C, regulated
GRS sensor mass	7.3 kg (incl. passive radiator)
GRS electronics mass	1.9 kg (incl. harness + adapter plate)
Total instrument mass	9.2 kg
Power dissipation	23 W peak, 16.5 W operational
Data rate	2500 bps peak, <175 bps orbital average

reduce this smearing to a negligible fraction of the available energy resolution. (See Burks et al. (2004) for a discussion of GRS thermal design and performance.)

4.2.1 GRS Sensor

A computer-aided design (CAD) model cutaway view of the GRS sensor in Fig. 4 shows internal parts. In the center of the housing is shown the Ge crystal (pink) inside its sealed Al capsule, which is thermally isolated inside three nested, thin, Al cylindrical thermal shields. These elements are contained within an open cage. Low-conductivity pegs on this cage and on the capsule form anchors for a crisscrossing annular network of thin, low-conductivity, strong Kevlar strings under tension that suspend the capsule. Surrounding this cage is a vented Al cup and cap.

These parts form the Ge detector cryostat, which is cooled by the cryocooler (silver-blue) cold finger, through a flexible copper braid attached to the capsule. The Al shell thermal shields are suspended by low-conductivity standoffs having minimal surface contact area that are attached to the crystal capsule. The thermal radiation shields, capsule, cage, cup, and cap (nearly all parts of the cryostat) are coated with a highly reflective gold film that has very low emissivity (Burks et al. 2004) to minimize infrared radiation reaching the capsule. Small holes are provided in the thermal shields for high-voltage, signal, and other leads and for the cold finger. The calculated accumulated conductive heat load is small compared with the measured total heat load. The intended direction for the GRS to view the planet is in the upward direction in Fig. 4, along the vertical centerline of the Ge crystal, although the detector is sensitive to gamma-rays from all directions, so the view field is a cone that extends from limb to limb on the planet surface.

The mechanical cryocooler is the Ricor K508 rotary Stirling-cycle unit, which has space mission heritage from the Comet Nucleus Tour (CONTOUR) and Rosetta programs, and most recently the Mars Reconnaissance Orbiter (an earlier K506 model was used in the Clementine program). In order to provide lifetime margin for the one Earth-year orbit of Mercury, a He fill pressure was chosen to yield a nominal mean time to failure (MTTF) of $\sim$12,000 hr and a nominal heat lift of 0.5 W at a cold tip temperature of 85 K for a 20°C ambient temperature and 13 W input power. Cooler mass is 450 g, including motor drive electronics. Flight-grade coolers were chosen from a qualification program of selection, testing, and customization at the manufacturer, based on heat lift, cool-down time, He leak rate, motor speed, thermal cycling, and consistency trends, according to specifications appropriate for the MESSENGER mission. The top performing few percent of production coolers were selected and sent to The Johns Hopkins University Applied Physics Laboratory, where final environmental testing and selection were completed. The intent of the qualification program was to eliminate any "infant mortality" issue and further improve the chances for an expected lifetime of the flight cooler beyond the nominal MTTF.

The passive radiator is an Al plate clear-anodized on its front surface for high infrared emissivity but low solar radiation absorptivity. It is designed to face in the direction of cold space when the spacecraft is orbiting Mercury, in order to reject waste heat from the cooler compressor. It is conductively coupled to the cooler compressor by its mount and also by a copper strap attached to the cooler stator and the sensor housing, visible in Figs. 3 and 4. The passive radiator allows the cooler to operate at higher efficiency and lower power, which helps extend the cooler lifetime. To protect the sensor housing from the radiated heat of Mercury, a tent of multilayer insulation surrounds the housing and is attached around the inner edges of the passive radiator (not shown in the figures). Through its conductive couplings to the cooler, the overall heat capacity of the sensor greatly reduces cryocooler temperature transients near orbit periapsis.

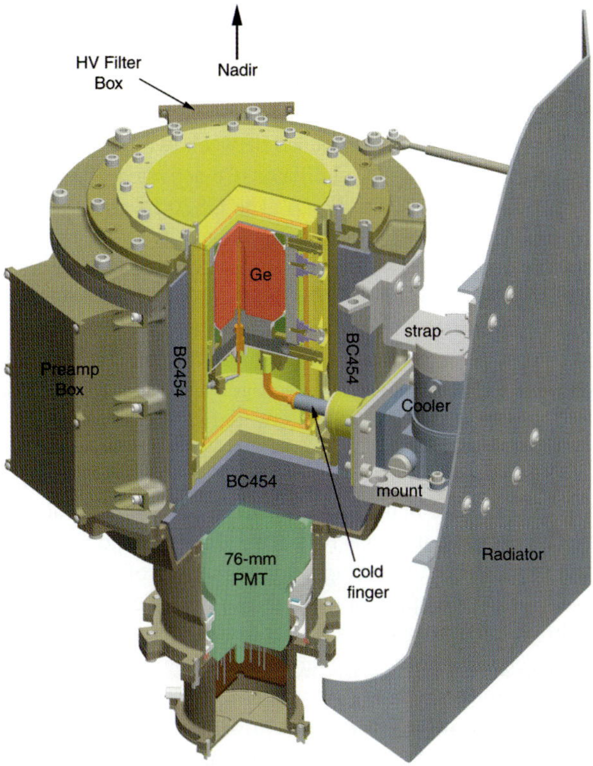

Fig. 4 CAD model cutaway view of GRS sensor. The *upper left part* of the passive radiator is cut away to show instrument internal parts. The Ge crystal, *highlighted in pink*, is 5 cm in diameter and 5 cm in length. Intended view direction of the planet is upward along the Ge detector centerline (same as the cut axis)

The cosmic-ray anticoincidence shield is formed by an annular cylinder of BC454-type borated plastic scintillator (blue) around the cryostat and another thick disk of BC454 below the cryostat. The scintillator blocks are wrapped in ultraviolet (UV) reflective materials and are optically coupled together and to a photomultiplier tube (green) by pads (cyan) under compression. BC454 consists of 5-weight-percent boron in polyvinyl toluene. The boron preferentially absorbs neutrons that would otherwise be captured by hydrogen in the shield, which would produce gamma-rays that could enter the Ge detector and give false indications of hydrogen in the Mercury crust. The scintillator blocks are enclosed by an outer Mg housing and the PMT (with a 70-mm-diameter active surface) has an Al housing, which contains a spring-loaded PMT mount with elastomeric potting to protect the PMT during launch. The bottom part of the sensor outer housing is attached to the GRS deck by ten short Al feet (some visible in Fig. 3). The photomultiplier housing protrudes down through a circular hole in the deck (not shown in the figures).

As shown in Fig. 4 for the cooler cold finger, each of the penetrations into the Ge detector cryostat has an O-ring vacuum seal. The outer cup is vented at the top (into the vacuum of outer space in flight) and is designed to accept an O-ring-sealed top hat that screws into the ring of holes visible in Fig. 4. The top hat has been fabricated with a side-on vacuum hose connector. This pump hat is made of thin Al so as to minimally attenuate gamma-rays of interest and allows full operation of the GRNS without a large vacuum chamber. Since the detector parts can be assembled and disassembled nondestructively, this arrangement was helpful in laboratory development of the GRNS and in spatial energy-efficiency ground calibration at a nuclear reactor site. It also allowed full GRNS operation and testing after mounting on the spacecraft to complete the comprehensive prelaunch checkout.

4.3 Neutron Spectrometer

The NS is based on a configuration of three scintillators, each wrapped separately in UV-reflective materials and optically coupled to a separate PMT. A beveled cube of BC454 100 mm on a side is sandwiched between two beveled square plates (100 mm on a side) of 4-mm-thick GS20-type scintillator (Ce-activated glass loaded with ~7 weight percent ^{6}Li, as ~95% enriched Li in Li_2O). The GS20 Li-glass scintillators are primarily sensitive to thermal neutrons, with decreasing sensitivity to epithermal neutrons. When the spacecraft has a velocity vector with a component normal to the Li-glass surfaces (relative to Mercury), one Li-glass exposed plate face moves toward the planetary neutrons and the opposite exposed plate face moves away from the planetary neutrons. A Doppler filter technique can be used to obtain a coarse spectrum of thermal neutrons (~0.025 eV to ~1 eV), providing a sensitive indicator of surface hydrogen and water (Feldman and Drake 1986).

The open BC454 faces are covered by Gd 0.25 mm thick, so the BC454 does not see neutrons below ~1 eV in energy. The BC454 detector is sensitive to epithermal neutrons and fast neutrons up to ~7 MeV deposited energy, with decreasing sensitivity at higher neutron energies. In the capture-gated mode (Drake et al. 1986), this detector acts as a spectrometer for fast neutrons between ~0.5 MeV and ~7 MeV, which allows an approximate determination of the average atomic number of surface material (Gasnault et al. 2001). In this mode, a prompt light pulse due to multiple neutron elastic scatterings (primarily off of hydrogen atoms) occurs as the neutron moderates in energy, followed by a second, delayed-coincidence light pulse that arises when the moderated neutron is absorbed by a ^{10}B atom in a reaction that results in an alpha particle of known energy. This two-pulse sequence indicates that the incident neutron deposited all of its energy in the BC454 and that the amplitude of the first pulse relates to its initial energy. This two-pulse sequence is also a powerful discriminator against background events.

Electronics include low-voltage and high-voltage power supplies, preamplifiers, high-speed analog-to-digital converters, field-programmable gate arrays, and a microprocessor. Figure 5 shows a photograph of the complete NS on a test stand just prior to installation on the spacecraft. The PMTs are labeled according to their faceplate diameters. The electronics box is positioned with respect to the sensor housing as it is on the spacecraft, with the electronics mounted above the sensor when the BC454 detector is viewing the planet along the axis of the 51-mm PMT and to the right. The two 19-mm PMTs are optically coupled to each of the Li-glass scintillators on either side of the BC454 scintillator and are labeled as outboard and inboard, indicating which Li-glass plate detector is farthest from the spacecraft centerline along the back "hatbox" (Leary et al. 2007), where the NS is mounted.

Table 3 gives a summary of NS specifications and performance characteristics. The coarse energy resolution of the NS is adequate to meet the science objectives. Unlike the GRS, there are no sharply defined emission peaks in the neutron energy spectrum that directly indicate elements at the planetary surface.

4.3.1 NS Sensor

A CAD model cutaway view of the NS sensor in Fig. 6 shows internal parts. The scintillator housing is Mg, and each PMT housing, which contains a spring-loaded PMT mount with elastomeric potting to protect the PMT during launch, is Al. The inboard Li-glass scintillator (yellow) and the BC454 scintillator (blue) are shown, as are the PMTs (green) and optical coupling pads (cyan) for these scintillators. In the upper right of the figure are shown the intended view directions of the three detectors, with the BC454 pointed at the planet and the

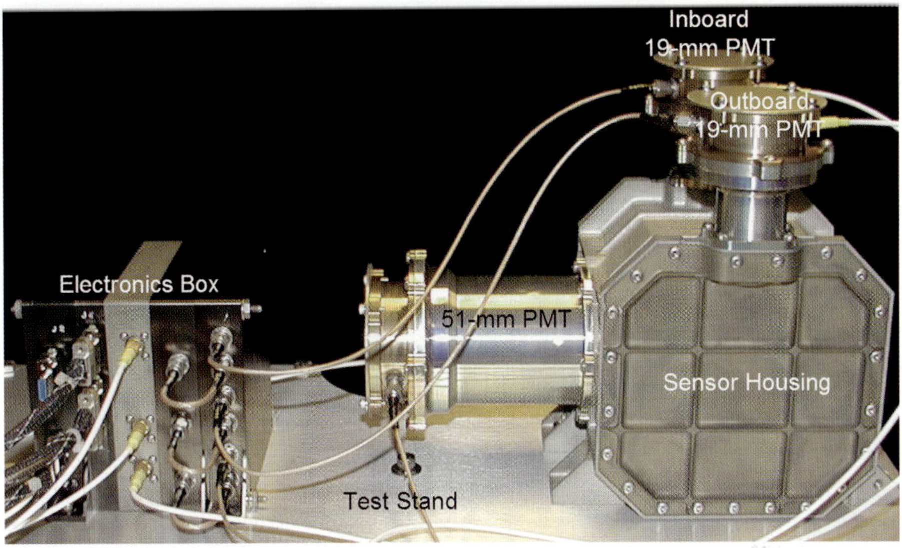

Fig. 5 Photograph of NS flight unit on a test stand for calibration experiments. Each major external part of the instrument is labeled. For scale reference, the electronics box dimensions are approximately 10 cm × 10 cm × 10 cm

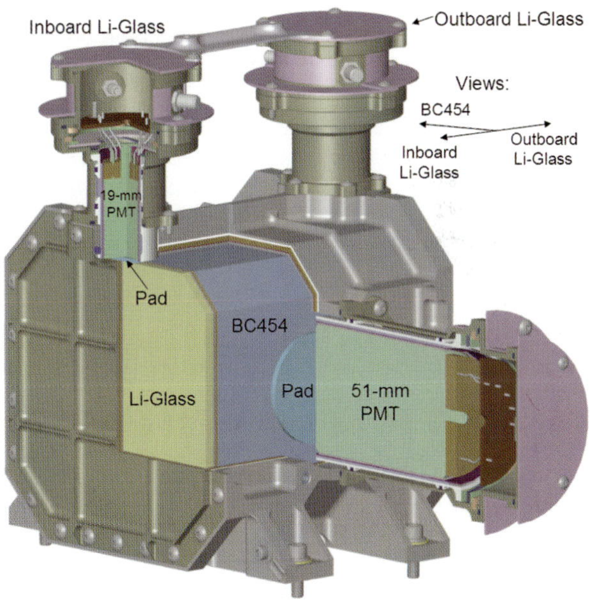

Fig. 6 CAD model cutaway view of NS sensor. The housing is cut open to show the internal scintillator and PMT arrangements. The intended view is given for each scintillator. The direction given for BC454 is nadir, towards the planet

two Li-glass detectors facing in the direction (and opposite direction) of spacecraft velocity relative to the planet, in the "ideal" case where the spacecraft velocity is normal to the Li-glass plates (a "dawn-to-dusk" orbit). The BC454 is sensitive to neutrons from all directions, so its actual view of the planet is from limb to limb.

Table 3 NS summary specifications

Thermal neutron detector	GS20 Li-glass scintillators (2), 10 cm × 10 cm × 0.4 cm, coupled to 19-mm PMTs
Epithermal and fast neutron detector	BC454 borated plastic scintillator, 10 cm × 10 cm × 10 cm beveled cube, coupled to a 51-mm PMT
Energy range	Thermal (∼0.025 eV to ∼1 eV)
	Epithermal (∼1 eV to ∼500 keV)
	Fast (∼500 keV to ∼7 MeV)
Energy resolution (FWHM/mean)	Li-glass: ∼50%, neutron capture
	BC454: ∼30%, neutron capture (93 keV electron-equivalent)
Effective FOV	Li-glass: Hemispherical response to thermal neutrons along and opposite spacecraft velocity direction
	BC454: Omnidirectional response to epithermal and fast neutrons towards nadir.
No. of spectral channels	Li-glass: 64; BC454: 256
Lower-level discriminators	Separate adjustable discriminator settings for each Li-glass detector and separate settings for BC454 prompt and capture pulses
Maximum throughput	>10,000 events per second
Differential nonlinearity (DNL)	±0.1 channels
Integral nonlinearity (INL)	±0.5 channels
Operating voltage	19-mm PMTs: 820 V
	51-mm PMT: 675 V
High-voltage stability	<20 ppm/°C
Electronics offset	±0.2 channels
Sensor gain drift	Li-glass: <2%/°C;
	BC454: <0.25%/°C
NS sensor mass	2.8 kg
NS electronics mass	1.1 kg (incl. harness)
Total mass	3.9 kg
Power dissipation	6 W
Data rate	400 bps peak, ∼70 bps orbital average

The large PMT is located on the opposite side of the BC454 view direction, to avoid absorbing or scattering neutrons entering the scintillator from the planet. The small PMTs are placed along the edges of the Li-glass plates, so as not to absorb or scatter neutrons along the spacecraft velocity vector, at the edge location that yields the maximum and most uniform light pulse response (which happens to be the edge center). For a "noon-to-midnight" orbit, the spacecraft velocity vector is at a right angle to the Li-glass plate normals, in the plane of the Li-glass plates, so there is no Doppler energy shift between the two plates. For most orbits, the spacecraft velocity vector will be at an angle between 0 and 90° with the plate normals, so the Doppler shift will be present but reduced.

The purple-tinted parts in Fig. 6 are gadolinium pieces 0.25 mm thick, placed to absorb thermal neutrons (by capture reactions) before they reach the BC454. (The external Gd pieces are not seen in Fig. 5, because this photo was taken prior to their installation.) Visible as a thin purple line is the Gd wrapped around the BC454 faces not adjacent to Li-glass plates, which also absorbs thermal neutrons. Also visible are cylindrical Gd shells around the PMTs. External Gd pieces around the PMT bases act as baffles that require thermal neutrons to go through several scatterings in small solid-angle ranges in order to get by the Gd, greatly reducing the probability that thermal neutrons entering the PMT base areas will reach the BC454. The Gd pieces were applied externally around the PMT bases (rather than internally) to keep Gd (which is electrically conductive) away from high-voltage electrical leads and "bleeder" (resistor bias string) boards (shown as dark brown in Fig. 6).

4.4 Electronics

The Johns Hopkins University Applied Physics Laboratory has developed an electronics packaging scheme that minimizes mass and allows quick reuse of standard designs. The scheme uses a series of modular "slices" that can couple together to form a compact electronics stack without the need or added mass of a backplane. Slices developed for use across the MESSENGER payload include: a low-voltage power supply (LVPS), a microprocessor board (called an Event Processing Unit or EPU), and a dual high-voltage power supply (HVPS). Standard slices are 10 cm × 10 cm and are either 1.4 or 2.1 cm thick. Each slice has its own magnesium frame, and a typical mass for a completed assembly is 150 g (power slices with integral heat sinks are about 240 g).

Equally important to the common hardware is the common Command and Data Handling (C&DH) software that provides all of the basic functionality to test and develop the customized electronics for an instrument. Also convenient is the common Ground Support Equipment (GSE) that was produced to mimic the spacecraft interface and provide real-time displays of instrument telemetry data. This approach created a standardized development environment that allowed the instruments to proceed quickly from bench testing to spacecraft integration without incident.

To minimize the development time of the electronics unique to the GRNS, the design approach made use, where possible, of functional elements common to both the GRS and NS portions of the instrument. Where needed, the design was then augmented or adapted to implement the unique parts of each experiment. Examples include the 1.5-kV high-voltage power supplies for the PMTs, the PMT resistive-divider ("bleeder") boards, and the ADC boards. The last example is particulary noteworthy because, although the pulse processing requirements are markedly different for the GRS and the NS, the same ADC board design was able to work for both by simply changing the programming of the FPGA.

4.4.1 Common Electronics

As we have discussed, the GRS and NS electronics incorporate electronic slices used across the instrument payload. This architecture reduced development time, and the standardized interfaces simplified integration and test.

The LVPS slice provides multiple secondary voltages to power the EPU and the low-power portion of the instrument electronics (the GRS cryocooler has its own power supply). The board has built-in housekeeping functions to measure voltages, currents, and temperatures. It also provides two switched primary outputs. The GRS makes use of one of these outputs to control the external warm-up heater on the cryocooler to prevent cold starts and to keep the cryocooler above −20°C when possible during cruise.

The EPU board is based on a radiation-hardened, reduced instruction set computer (RISC) architecture designed for the direct execution of Forth. The processor board contains

32,000 words of programmable read-only memory (PROM), 128,000 words of random access memory (RAM), and 128,000 words of electrically erasable PROM (EEPROM). The processor first boots from PROM, copies the operational software from EEPROM to RAM, and then executes out of RAM to maximize speed. The processor clock rate is 6 MHz. The EPU communicates with the main Data Processing Unit (DPU) over a 38.4-kbps serial link, although data transfers are limited to 10 kbps for each of the nonimaging instruments.

The dual HVPS slice was developed to support the GRS, NS, and XRS sensors. The slice contains two completely independent high-voltage supplies with an output range of 0 to 1.5 kV. Negative high voltages are used to bias the GRS and NS PMTs. The board provides its own housekeeping and control functions, i.e., it accepts digital control values in and provides digitized voltage monitor readings out. For the GRS, only one PMT supply is required and so the other supply was left unpopulated.

4.4.2 GRS Electronics

Figure 7 shows a functional block diagram of the GRS. The HPGe detector is shown suspended inside a thermal enclosure by Kevlar strings. The cyrocooler cold finger penetrates the cryostat and cools the detector via a short, flexible copper strap. A small electrically insulating block at the connection site prevents unwanted ground loops that could disrupt the detector signal. Redundant silicon diodes, mounted to the underside of the detector capsule, accurately monitor the temperature of the detector crystal. Closed-loop regulation under software control maintains the crystal temperature to within a fraction of a degree.

When detector annealing is required, up to 2 W of controllable input power can be applied to the redundant power Zener diodes mounted to the underside of the detector capsule.

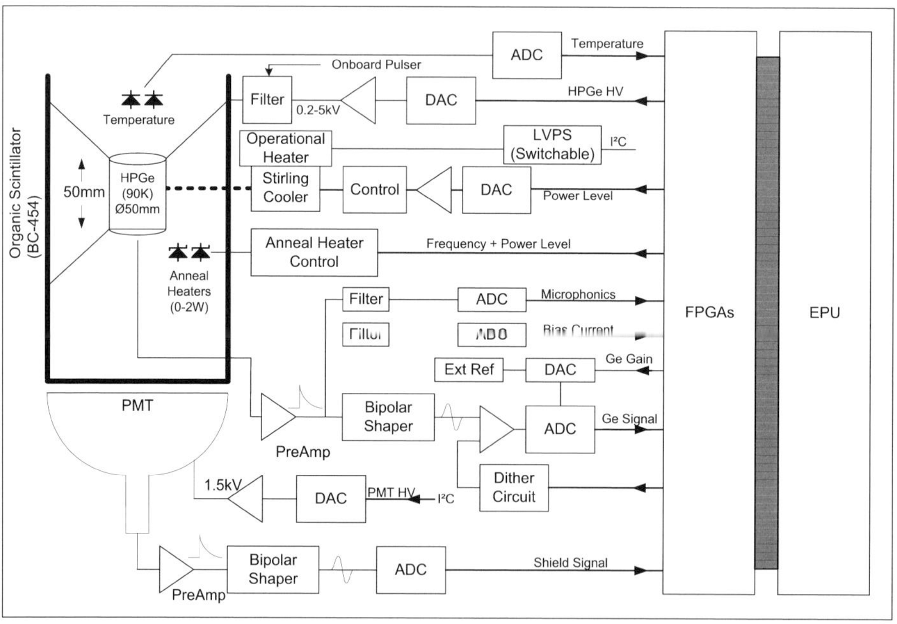

Fig. 7 GRS functional block diagram. The HPGe preamplifier and high-voltage filter modules are mounted on the GRS sensor housing. All signal processing and power electronics are located in the GRS electronics box. The GRS has its own microprocessor, power supplies, and telemetry interface. I^2C denotes an inter-integrated circuit serial interface. See text for further details

Tests show that less than 0.5 W is needed to maintain the crystal at the maximum annealing temperature of 85°C. High-voltage Zener diodes (∼90 V) are used to keep drive currents low, which allows the use of thin stainless steel wires for better thermal isolation. The annealing temperature is set by software command, but the crystal temperature is regulated directly in hardware by an FPGA using simple thermostatic control. This circuitry contains many fault-protection features.

The HPGe detector itself is essentially a large coaxial diode operated under reverse bias. The capacitance of the detector reaches an irreducible minimum of about 25 pF when the detector is fully depleted. The depletion voltage is a function of the impurities and varies from detector to detector. The GRS flight unit fully depletes at approximately 2,400 V, but it is operated at 3,200 V to guarantee good charge collection throughout the entire active volume of the detector (the depletion voltage can also shift with exposure to radiation). An added benefit is that in an over-biased condition the detector signal becomes relatively insensitive to changes in the bias voltage.

The detector must be operated cold, at less than about 140 K, to keep the leakage current in check (to a few pA); otherwise excessive noise will result in severely degraded energy resolution. More important is the detector's susceptibility to radiation damage, which can be greatly reduced by maintaining the detector at much colder temperatures (Brückner et al. 1991). As a suitable compromise, an operating point of 90 K was chosen to slow the accumulation of radiation damage while ensuring the long-term operation of the mechanical cryocooler. When not operating, the detector comes into thermal equilibrium with the outer housing and typically rests at a temperature of around −10°C, which is sufficiently warm to avoid radiation damage issues, although in general an annealing cycle is planned prior to each cooldown of the detector.

The HPGe detector measures the deposited energies of individual photons. When a gamma-ray interacts with the detector, electron-hole pairs are created that are then swept to the appropriate electrodes under the influence of a strong electric field. The resulting charge is collected by a charge-sensitive preamplifier that produces a voltage step whose amplitude is proportional to the amount of energy deposited in the crystal. For an HPGe detector of this size, the charge collection time is ∼100 ns and varies slightly depending on the site of the interaction. The preamplifier output signal is then amplified and filtered ("shaped"), and its peak value is measured using an ADC. The EPU collects the measurements and sorts them into bins according to "pulse height" or energy absorbed. After an appropriate accumulation interval, within which time many photons have been processed, an energy spectrum (histogram) develops that can be sent to the ground for analysis.

The detector preamplifier is mounted in an enclosure on one side of the sensor housing. The front-end transistor, an N-channel junction field effect transistor (JFET), is located outside of the cryostat along with the rest of the preamplifier circuitry and operates near room temperature. Admittedly, a cooled JFET offers better noise performance, but the added heat load and development time were considered unnecessary risks. Similarly, a large-geometry JFET was chosen for its known robustness as opposed to qualifying for space some of the newer dual-gate devices. The preamplifier is direct-coupled to the detector and employs resistive feedback to restore in-range pulses and to provide an output monitor of the detector leakage current. For large overloads (>100 MeV), a pulsed-reset circuit, based on a design developed at Lawrence Berkeley National Laboratory (Landis et al. 1982), provides quick recovery so that no line-broadening or significant dead time results from energetic cosmic rays.

The signal developed from the anticoincidence shield is much larger in amplitude than the signal from the HPGe detector because of the electron gain in the PMT. This characteristic allows its charge preamplifier to be located conveniently on the same board as the shaper circuits. The preamplifier for the shield is slightly different in design than for the HPGe detector, with more emphasis on low power dissipation rather than low noise, but it retains the important pulsed reset feature.

The analog shaping circuits occupy a slice in the GRS electronics box. The shapers produce bipolar pulses with a 6-μs peaking time for the HPGe channel and a 4-μs peaking time for the shield channel. Bipolar shaping is generally worse than unipolar shaping at reducing preamplifier noise, but it provides better rejection of low-frequency interference, such as microphonic pickup from the cryocooler (the Ricor cooler contains a rotary-driven piston that generates unbalanced vibrations with harmonics extending up into the kilohertz range).

The bipolar-shaped signals are then passed to the adjacent ADC board where they are simultaneously sampled at 10 megasamples per second. The digitized samples feed into an FPGA, which then performs all pulse height analyzer functions in the digital domain, such as lower-level discrimination and triggering, peak detection, coincidence timing, and pile-up rejection. Many modern laboratory pulse height analyzers now work this way and are even able to perform the pole-zeroing and pulse shaping functions in an all-digital implementation, but the current generation of radiation-hardened FPGAs contain far fewer gates than those commercially available. However, the hybrid approach presented here is well-suited to the unique requirements of this application and retains most of the performance features of the laboratory units.

There are three separate ADC channels on the ADC board: one dedicated to the HPGe channel, the second for the anticoincidence shield, and a third that samples the output of the HPGe preamplifier more directly to look for cryocooler mechanical vibrations that have been transformed into undesired electrical signals (microphonics). The type of ADC used is a successive approximation converter with 14-bit resolution. The HPGe channel has an external reference for improved stability and a Digital-to-Analog Converter (DAC) to provide an adjustable range. Another DAC is used to add a small amount of dither to the input signal to improve linearity, particularly the differential nonlinearity (DNL), which is a measure of how evenly spaced the channel transitions occur. The dither voltage is updated on a pulse-by-pulse basis using a look-up table of pseudo-random values. The inclusion of dither reduces the usable energy range slightly and in this case has been restricted to the top 5% of full scale, which sits well above the highest energies of interest.

The pulse processing algorithm is somewhat unconventional in that it captures both the positive and negative peaks of the bipolar pulse. Their difference is the reported pulse height. This approach offers several unique advantages: differencing, a form of high-pass filtering, provides additional rejection of low-frequency microphonics; it removes transient baseline errors due to pulse pile-up or pole-zero mismatch, which can distort the line-shape of energy spectra; it conveniently removes the added dither value without having to rely on a priori knowledge, which is likely to change with temperature and radiation; and, finally, it cancels all the electronics offsets, including those internal to the ADC itself, thereby providing a very stable and accurate "energy zero" that simplifies energy scale calibration.

The randomly occurring analog pulses entering the pulse height analyzer are asynchronous with the ADC sampling clock, so the digital peak detectors capture the sample nearest to the peak and not the actual peak value itself, although on average the pulse height will be correct. This technique, however, with no additional processing, would introduce a small noise term that would tend to broaden the energy resolution as a function of energy (~5%). To mitigate this effect, the digitized samples are first passed through an eight-sample moving average filter. This filter not only provides smoothed peak values, but it reduces the wideband noise component internal to the converter itself. It also helps to smooth out local variations due to DNL and integral nonlinearity (INL).

Pulse pile-up is detected by measuring the time (clock ticks) between the captured positive and negative peaks. For a valid event, this time should be approximately a fixed value determined by the shaping circuitry. Events with times that fall outside a window around the expected time are rejected. Except during a solar particle event, the counting rates at Mercury will in general be very low, where pile-up is not a major issue. Coincidence between the shield and HPGe detectors is detected in a similar fashion by measuring the time between the two channel waveforms. Events with a coincidence time less than 1 μs are flagged for rejection; otherwise they are processed as independent events.

A block of hardware counters, gated by the master one-pulse-per-second (PPS) input, provide scalar information such as the number of trigger events or the number of charge resets for each channel. A dead-time counter accumulates 1-μs ticks during each 1-s interval whenever the channels are busy processing events, settling, or inhibited by charge resets, or the event buffer is full. In practice, the onboard pulser has proved more useful in accurately determining the true system dead time.

All of the circuitry was designed to exhibit low temperature sensitivity to minimize smearing effects (resolution broadening) for cases when the instrument temperature is rapidly changing. Temperature sensors are strategically located all along the signal chain to allow detailed post-correction, if needed. Critical temperatures include the HPGe crystal, the front-end preamplifier, the cryocooler body, the shaper board, and the ADC board.

The onboard pulser may provide the best measurement of end-to-end gain shifts. The electronics can inject a small, stable pulse onto the detector HV line. This pulse couples across the detector capacitance and appears as a photon event to the electronics. The pulse energy is fixed near the upper end of the energy scale at about 8 MeV. The pulse rate is set to $\sim$8 Hz, which should provide sufficient count statistics for even the shortest integration periods. The pulser itself has a nonzero temperature coefficient that depends on components located on the shaper board and inside the HV filter box, and it also depends on the detector capacitance, which is itself a function of crystal temperature and high voltage. Fortunately, the combined effects appear to produce an overall temperature characteristic that is two to three times better than the charge-preamplifier and so will be useful in providing first-order corrections to spectral data.

The controller and cooler power boards perform the main housekeeping functions of the GRS instrument. Power for the cryocooler and annealing heaters is provided by the cooler power board. The controller board performs cooler and annealing heater control, detector temperature reading, voltage and current monitoring, and control of the 5-kV high-voltage power supply.

The Ge detector temperature must be tightly maintained to have a negligible effect on the energy resolution. Closed-loop control is needed to achieve the required stability. The detector temperature is measured using two silicon temperature diodes. If one of the diodes fails as an open or short circuit, the electronics detect this failure and use only the operational diode. Each diode output voltage is buffered and channeled into a multiplexed voltage-to-frequency (V to F) converter. For each temperature reading, the output of the V to F converter is accumulated for $\sim$200 ms by a counter in the controller FPGA. The extended measurement interval results in excellent noise averaging and an effective resolution of $\sim$15 bits. A stable voltage reference is used to calibrate continuously the V to F system. The onboard calibration corrects for temperature drift and long-term stability of the V to F circuit and the oscillator used to set the gate period in the FPGA. Using this technique, the detector temperature is known to $\sim$0.1°C over the complete operating range of the electronics.

The Ricor cooler power level is controlled via a control voltage provided by a DAC on the controller board. A software algorithm implements a proportional plus integral (PI) control loop that updates the control voltage each second. The integral term guarantees settling to the temperature set point by eliminating any gain errors. Because the Ricor cooler and DAC are both inside the control loop, the detector temperature is dependent only on the accuracy of the temperature readings and not the stability characteristics of the other electronics.

The 5-kV HVPS slice was supplied by the Space Sciences Laboratory (SSL) at the University of California, Berkeley. Their existing design was repackaged to accommodate the slice packaging format but was placed in a self-contained enclosure to control radiated emissions and protect the adjacent electronics. The supply is capable of outputting –5,000 V and dissipates about 250 mW when operating. Output ripple is very low and goes through a final reduction by a two-stage network inside the HV filter box, also supplied by SSL. The HV filter box mounts directly to one side of the GRS sensor housing. A HV coaxial connector and cable system rated to 18 kV connects the two units.

4.4.3 NS Electronics

Figure 8 shows a functional block diagram of the NS. The NS is a sandwich of three scintillation detectors. The light output from each scintillator is collected by a separate PMT. The two Li-glass detectors share a HVPS; the BC454 detector, which operates at a different voltage, has its own supply. The signals from the NS detectors are relatively large in amplitude, which allows all of the processing electronics, including the preamplifiers, to be located in the NS electronics box. The NS sensor is completely passive; the only electrical circuits are the PMT bleeder boards.

The charge-sensitive preamplifiers are identical in design to that used for the GRS anticoincidence shield. The three preamplifiers are packaged together on one slice with copper barriers between them to prevent crosstalk. All include pulsed-reset capability to recover quickly from overload pulses due to energetic cosmic rays. The signature of a fast neutron is a prompt pulse followed by a delayed capture pulse. The NS preamplifier was specifically designed to provide a fast rise time (~20 ns) without appreciable ringing, as ringing could produce a trigger that could be misinterpreted as a delayed pulse.

The preamplifier outputs feed directly into the high-speed ADCs. This arrangement is different from the signal chain for the GRS, in which there is an intervening shaping circuit

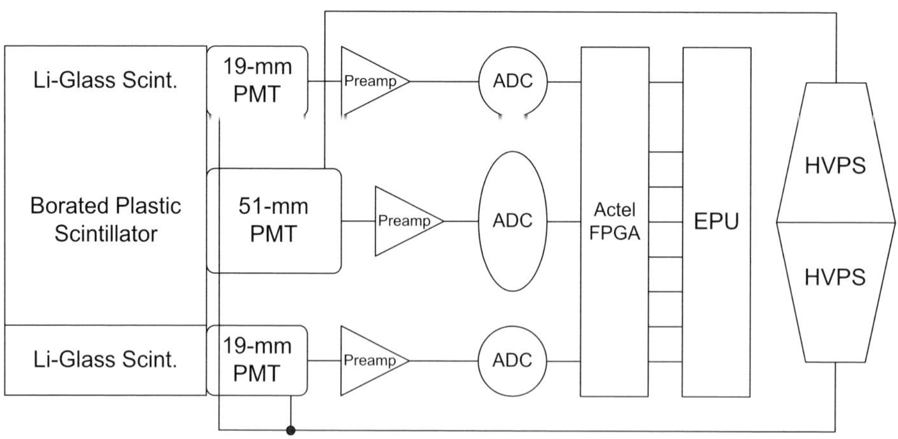

Fig. 8 NS functional block diagram. The light output from each scintillator is collected by a separate PMT. The two Li-glass detectors share a HVPS; the borated plastic scintillator (BC454) channel has its own supply. All signal processing electronics, including the preamplifiers, are located in the NS electronics box

to reduce noise. For the NS, the PMTs provide ample gain, such that the energy resolution of the detectors is dictated more by the number of photons collected than by the electronic noise in the preamplifiers. Here the emphasis is on speed and timing to detect closely spaced events. The buffer amplifiers on the ADC board were removed to preserve the fast response of the preamplifiers.

The energy resolution for the NS detectors is very broad (30–50% FWHM relative to the mean), so using a high-resolution ADC may seem like overdesign. The ADC was initially chosen to satisfy the more exacting requirements of the GRS, but the extra bits of precision allow the NS to fine tune and match detector gains in software, which is especially important for the two Li-glass detectors, which share a common HVPS. The high-precision ADC also results in improved DNL in all of the NS spectra.

The pulse processing FPGA detects trigger events by a simple differencing of successive samples. Complex logic then looks for coincident events between detectors as well as delayed events falling within a 25-μs window following the start of an event. Case-by-case algorithms are used to capture accurately the pulse height information from each detector. Care is taken to adjust for the longer light decay time of the Li-glass scintillators and also the alignment of the pulses relative to the sample clock as all arrival times are random. For the BC454 channel, the time-to-second-pulse (TTSP) is also measured. At the end of the timing window, the complete status of the event (who fired and when) is reported to the EPU for further event validation, classification, binning, etc.

Detection of coincidence between adjacent detectors (or between all three) serves to reject background events due to cosmic rays. This capability is important for the Li-glass channels, which cannot easily discriminate between thermal neutrons and minimum-ionizing charged particles, and also for the detection of epithermal neutrons in the BC454, which is subject to a high background due to "corner-clipping" charged particles. Coincidence windows are about 200 ns wide and are used to validate both prompt and delayed pulses.

The two-pulse sequence signature for fast neutrons provides a natural rejection scheme of not only the cosmic-ray background, but also the local gamma-ray background. Those non-neutron events that make it through are mostly limited to chance coincidences. However, PMTs can exhibit afterpulsing, which is the result of occasional ion feedback inside the PMT (mostly due to He and Ar ions), and unfortunately these afterpulses come out with delay times similar to fast neutrons. A unique feature of the NS trigger circuitry is that it provides separate thresholds for prompt and delayed pulses, and this feature helps to reject afterpulse events by applying separate criteria. The amplitude of an afterpulse is generally less than 5% of the prompt pulse, so most afterpulses will fall below the level produced by a typical neutron capture. This technique allows a lower trigger threshold to be set for prompt pulses, thereby extending the instrument response to lower energies and gaining sensitivity, while maintaining a higher threshold for the delayed pulses to minimize event corruption by PMT afterpulses. A further restriction in software rejects those events with delayed-pulse energies that are above the expected capture energy. This restriction further reduces the acceptance of events with chance coincidence timing. All of the threshold and gain values are adjustable. The NS instrument maintains a block of hardware and software counters to record the occurrences of the most relevant types of coincidence possibilities.

4.5 Software

4.5.1 Common Software

The MESSENGER Instrument Common Software forms the core software component of six of the seven instruments on MESSENGER. It supplies most of the necessary functionality that is not directly related to the gathering and processing of the scientific data. Much of the functionality in the common software is provided as customizable services that can be adapted to meet the needs of the unique software for each instrument. The common software provides the command and telemetry interface to the DPU and handles all command parsing, macro execution, data compression algorithms, telemetry packet generation, housekeeping interface functions, monitoring and alarming, and instrument status reporting.

4.5.2 GRS Software

The GRS software is designed to configure and operate the GRS, keep it safe from damage at all times (autonomy), process and classify gamma-ray events, and collect the science and housekeeping data at regular intervals.

Autonomy is an essential function and is always active (under extraordinary circumstances certain overrides are provided). Major autonomy functions protect the HPGe crystal, Ricor cooler, PMT, and high-voltage supplies. For example, the application of high voltage to the HPGe detector is not permitted unless the detector temperature is in the target cooling range. Likewise, the HV will be removed in the event that the detector temperature rises above 105 K. Other checks include ensuring that the two temperature sensors are comparable and reasonable, the detector leakage current is within limits, and the annealing circuits are operating correctly. In the event of a spacecraft emergency, the software performs an orderly shutdown of the instrument.

High-voltage ramping is handled entirely under software control. For the HPGe detector, the ramping algorithm contains six distinct phases with constant checking to see if it is safe to proceed either to the next voltage level or to the next phase. If the leakage current rises beyond acceptable limits, then further ramping of the high voltage is immediately stopped. In extreme cases (very high leakage currents and the detector voltage close to operational levels), the high voltage is ramped back down to the depletion voltage (zero voltage is not optimum for resistance to radiation damage, C.P. Cork, private communication, 2002). Although the shield PMT is much more robust than the HPGe detector, the PMT voltage is also applied as a controlled ramp and is immediately brought down to zero volts in the event the count rate exceeds a safing threshold. Safing is likely to occur during strong solar particle events, and so the safing algorithm is designed to attempt autonomously to restore operation after a commanded time interval. If multiple retries all fail, then the instrument simply awaits a ground command to restore high voltage.

The software is structured such that all time-critical operations are tied to the master 1-PPS interrupt. Examples include the annealing heater control, the warm-up heater, and the implementation of the cryocooler PI control loop. Voltage ramping, safing, and general alarming are also performed during the 1-PPS process.

All of the remaining background operation is devoted to event processing. The EPU continuously polls the status of the pulse-processing FPGA for new events. When a new event is available, the EPU reads in all the necessary data registers and quickly rearms the hardware. The event is then validated, classified, and binned into the appropriate histogram. The pulse-processing hardware rejects events too closely spaced for proper analysis (pile-up rejection), but at high count rates, events may arrive faster than the EPU can process them, in which case a portion of these events are lost. For the low counting rates expected at Mercury, the system throughput is expected to be $\sim$97%.

At the end of a reporting period, the software stops accumulating data and generates the various data records, usually in a compressed format, and transfers the data to the DPU. The different types of science telemetry may be individually enabled or disabled. The reporting period is commandable and is the primary means for controlling the overall data volume.

4.5.3 NS Software

The NS software works in much the same way as that for the GRS but is in general less complex because of its simplified autonomy needs. The PMT voltage ramping and safing algorithms are the same, but there are no temperature-critical or closed-loop operations. The NS event processing is also similar to the GRS but has many more types of events to classify and bin at varying degrees of resolution. The NS throughput is similar to the GRS with an average event requiring $\sim$100 μs to process. In addition, the NS has the capability to capture raw event-mode data for fast neutrons and place them in a variable-length buffer. This capability provides a high-resolution subset of events. A separate diagnostic mode captures 96 raw events (all types) per reporting period and is useful for characterizing the in-flight operation of the FPGA pulse processor. This mode was particularly useful in developing a new method to track the gain of the Li-glass detectors. In the original design, events that were coincident between the Li-glass and BC454 detectors were simply discarded as cosmic rays or undetermined interactions, but after launch it was discovered that these events could be exploited to help track the gain of the Li-glass detectors. The energy deposited in a Li-glass detector for a minimum-ionizing proton (cosmic ray) is only about 20% higher than for a neutron capture. A triple coincidence condition selects those cosmic rays that pass through the detector sandwich nearly normal to the Li-glass surfaces and thus produce an energy spectrum with a clearly defined peak and width comparable to thermal neutrons. The measured flux ($\sim$40 cps) should be sufficient to track the gains of the Li-glass detectors throughout the mission.

One feature that is unique to the NS is its Gamma-Ray Burst (GRB) mode. When enabled, this mode continuously monitors the 1-s background rate of the fast neutron detector and triggers on sudden jumps in the rate. If a burst is detected, then a time series of pre-trigger and post-trigger samples is captured in a buffer and added to the science telemetry. This is the only science mode that can generate autonomous data, although GRBs occur infrequently with a rate of detection of only a few per day. A side benefit is that GRBs with known positions can be used to check the accuracy of the MESSENGER onboard clock.

5 Ground Performance and Calibration

Detailed GRNS ground performance and calibration will be described elsewhere. Here only critical performance parameters are discussed and calibration experiments are described in general terms, with some results given that have been analyzed to date.

5.1 GRS Ground Performance Testing and Calibration

This section describes the performance of the completed instrument and its calibration on the ground. Measurements include DNL, INL, gain and offset, energy resolution, and spatial energy efficiency. Descriptions of functional performance include detector depletion, cryocooler cooldown, cryocooler microphonics, detector annealing, and anticoincidence shield effectiveness.

5.1.1 GRS Electronics Measurements

A broad measure of DNL is the standard deviation of the differences in channel width across the energy spectrum, measured using many pulses spaced uniformly over the spectrum, such that the statistical fluctuations in each channel are small compared with measurement accuracy. A pulse generator, with its amplitude modulated by a slowly varying ramp generator, was fed into the flight electronics coupled to a gamma-ray detector simulator and was operated at ∼5 kHz to accumulate quickly ∼120,000 counts in each channel. Some minor nonlinearities were introduced but were removed by a low-order polynomial fit. The flight ADC has a dither circuit, which slightly reduces the number of available channels from 16,384 to ∼15,000. No overall trend or significant periodicity effects were seen over the spectrum, verifying the dither circuit operation, which should not result in any DNL patterns. The measured value of the DNL ($<0.3\%$) was consistent with that expected from the counting statistics alone, which suggests that the DNL contribution is smaller than the counting statistics and is therefore negligible.

The INL is the deviation from an assumed linear fit between energy and channel number measurements when using radioisotope sources or between the calculated and measured channel positions when using a linear pulse generator. A pulser was used to complete the test in a reasonable amount of time. During this test the flight electronics (not including the preamp) were placed in a climate chamber. A precision pulser was used to characterize the INL by manually adjusting its precision attenuator to approximately 50 different equally spaced settings. For each setting, 500 pulse height samples were captured and averaged to reduce noise. The pulser rate was kept low (∼50 Hz) to maintain linearity. The chamber was set to four different temperatures over the temperature extremes predicted to occur for the GRS electronics box, with added temperature to compensate for the lack of a vacuum, from −30°C to 64°C. After subtracting a linear fit for each temperature and a small estimated sublinear INL due to the laboratory equipment, the total INL spread over the spectrum was only ∼1.5 channels over all temperatures. Near the ^{60}Co 1,332 keV line, the INL change over the full temperature range was ∼0.4 channel, only 2.4% of the ∼17 channels covering this peak. The largest full-temperature INL changes were 1.2 channels at ∼channel 15,000 and 1.0 channels at ∼channel 7,200, but the energy peaks cover substantially more channels at these points. The INL is negligible over the spectrum and temperature range.

The gain and offset of a system give the energy-channel relation as a function of other variables, such as temperature or voltage settings. The gain is the energy per channel, and the offset is the energy at channel zero. Each electronic component displays a unique behavior in response to a change in temperature. In each case, this behavior can be characterized in terms of a gain and an offset with respect to the gamma spectra. For each component, then, a correction based on temperature is derived. Comprehensive end-to-end data were taken during the instrument thermal vacuum (TV) test using a ^{60}Co radioisotope source. The bulk temperature of the GRS sensor and the GRS electronics components were varied independently. These data will include any subtle effects due to thermal gradients but have not yet been analyzed. The GRS contains internal temperature sensors at each stage of the signal processing chain, and all of these temperatures are available on-orbit, if such second-order corrections become necessary.

Coarser gain characteristic measurements were also made using a thermal climate chamber and an external pulser and have been analyzed. The GRS preamplifier and the GRS electronics were measured separately. The GRS electronics measurement includes contributions from the shaper, programmable gain stage, ADC, and voltage reference, as well as any parametric influences due to the temperature dependence of the power supplies. Measurements were made at discrete temperature plateaus and during temperature ramping. Electronics gain drift was 20 ppm/°C or less and will not have much effect on the energy calibration, since one channel represents 60 ppm of full scale. The offset shows negligible dependence on temperature, being held to within 0.5 channels over the full 100°C temperature range.

This outcome is a result of the design of the pulse processing system, which captures both the positive and negative peaks of a bipolar pulse and digitally computes the difference. The gain characteristic of the GRS preamplifier shows a more dominant dependence on temperature and reaches 95 ppm/°C when hot. Small gain corrections for the GRS preamplifier temperature will likely be required.

There is an internal pulser built into the Ge detector electronics to provide an internal pulse height reference for the energy spectrum. It is set at a specific energy slightly less than 8 MeV, but its position changes with varying shaping amplifier gain. The internal pulser is comprised of electronic components that reside in both the GRS sensor and in the GRS electronics (as well as the cable between them), so each contribution must be separately characterized. For calibration, end-to-end data were taken during instrument TV tests using a ^{60}Co radioisotope source during which the bulk temperature of the GRS sensor and the GRS electronics components were independently varied. Preliminary analysis of a few data points suggests that the internal pulser correctly tracks variations in the sensor preamplifier and that its overall sensor temperature coefficient is substantially smaller that of the preamplifier. Pulser variations due to temperature changes of the components inside the GRS electronics have yet to be analyzed but in general should be less critical, because the electronics box, completely blanketed with multilayer insulation, does not experience the dramatic temperature swings of the sensor as it traverses the hot planet. Noise width of the internal pulser line consistently measures 3.0 keV FWHM, which provides a good diagnostic for the state of the preamplifier, especially for cases in which the energy resolution of the detector has degraded due to radiation damage.

A lower-level discriminator (LLD) is used to exclude electronic noise at the low end of the spectral range for each detector and has a digital value that can be set from 0 to 16,384. LLD settings have been calibrated with respect to channel positions. The HPGe LLD is currently set to a value of 10 ($\sim$30 keV), which is above the noise floor but also well below the 300-keV minimum energy requirement. Such a large margin makes detailed characterization of the LLD less important. However, a series of LLD tests was performed during the final GRS comprehensive performance test to determine the LLD conversion factor more precisely and to characterize any effects on the system throughput. The GRS pulse processor does not contain an analog comparator, but rather implements level discrimination inside the FPGA on the basis of differences in successive ADC samples. As a result, the LLD function is exceedingly stable and does not exhibit problems normally associated with analog components, such as sensitivities to voltage offset, temperature, crosstalk, or hysteresis.

5.1.2 Germanium Detector Energy Resolution Measurements

Energy resolution is the most important parameter for the GRS. It determines basic sensitivity (through the signal-to-background ratio) and the ability to resolve closely spaced peaks associated with the surface elemental composition. Ground energy resolution measurements were performed using various radioactive sources to cover the energy range of interest, including ^{60}Co (1,173, 1,332 keV), ^{226}Ra (352, 609, 1,120, 1,764 keV), ^{228}Th (239, 727 keV), ^{56}Co (847, 1,038, 1,238, 1,771, 2,035, 2,598, 3,253 keV), and Pu–^{13}C, the highest energy being 6,130 keV from Pu–^{13}C.

Galactic cosmic rays can have a profound effect on the energy resolution of a Ge-based system because of the large overloads created by these energetic particles. The design of any high-resolution gamma-ray spectrometer for use in space must address this issue and ensure it will adequately perform in the cosmic ray environment. The GRS design contains several features that mitigate the effects of large overloads. The charge-sensitive preamplifier has charge-reset capability and can recover from overloads greater than 100 MeV within ~10 μs. Under the same conditions, a standard preamp might remain in a saturated condition for several milliseconds. Not only does such a condition introduce additional dead time, but while the preamp is saturated, a serious pole-zero mismatch results in baseline errors that can severely degrade energy resolution and take a significant amount of time to settle out. With charge-reset active, the GRS recovers from most overloads in nearly the same amount of time as for in-range pulses. Even with charge-reset disabled (a commandable feature), the "tri-polar" shaping technique employed in the GRS dynamically eliminates baseline errors due to pole-zero mismatch. These features were tested using a high-level square-wave pulser capable of simulating >1 GeV pulses. Most cosmic rays will deposit less than 60 MeV in the detector and will therefore stay within the linear range of the preamp. The shaper networks will of course saturate, as they are designed for a 10-MeV full scale, but they recover relatively quickly and the tri-polar shaping technique reduces any residual errors. Finally, the pile-up rejection logic enforces a settling window for each event and is commandable in 1-μs increments. Optimized values were determined using the overload pulser. As will be shown in Sect. 6, results from the first flight test show near identical performance to ground measurements, proving the efficacy of this approach.

The MESSENGER GRS has an additional source of noise from the mechanical cooler that may contribute to the energy resolution. A series of energy resolution measurements was made using ^{60}Co with both the cooler running and with the cooler temporarily off. When the cooler was turned off, the detector temperature would rise quickly such that integration times had to be limited to 120 s to prevent significant energy smearing. Interestingly, there was no significant difference in energy resolution with the cooler on versus the cooler off. This result suggests that the vibrations generated by the cooler are sufficiently isolated so as not to affect adversely the preamp, or that the GRS tri-polar shaping technique is so effective in rejecting microphonics that one doesn't see a difference, or possibly a combination of both. This question can be answered only by taking additional measurements with standard laboratory electronics to characterize the effects of microphonics for different pulse shapes (unipolar and bipolar) as a function of shaping time. No further tests in which the cooler is turned on and off were conducted on the flight detector, because it was noted that turning the cooler off while high voltage is still applied to the detector may lead to a possible coronal discharge from deposits boiling off the cold finger.

Traditionally, performance baselines are established for Ge detector energy resolution and line shape using the ^{60}Co 1,332-keV line. The flight detector end-to-end peak shape for this line is shown in Fig. 9. The figure illustrates the type of analysis provided by the Aptec MCArd software, which has been integrated into the GRNS ground support equipment. The log scale shows little tailing on the low energy end, indicating minimal pole-zero mismatch, as expected for this electronics design, and the 3.49-keV FWHM energy resolution exceeds the design goal of 4 keV.

The best simulation of flight-like conditions for temperatures and operating with other instruments was with the detector mounted on the spacecraft during TV tests. Data from these tests will be analyzed as part of the final calibration effort. Using radioisotope sources, a room-temperature performance test of the full end-to-end (ETE) flight unit on the spacecraft with all other instruments operating has been conducted and analyzed, using Aptec MCArd software to determine FWHMs. The results are shown in Table 4. No escape peaks (from pair production) are included. In order to conserve cooler life, the tests have limited measurement times and thus limited counts accumulated in some peaks, particularly at high energy. Accumulated counts (minus background) are given in the table, to indicate which peaks have FWHMs with more measurement uncertainty. The pulser peak in Table 4 of course is not

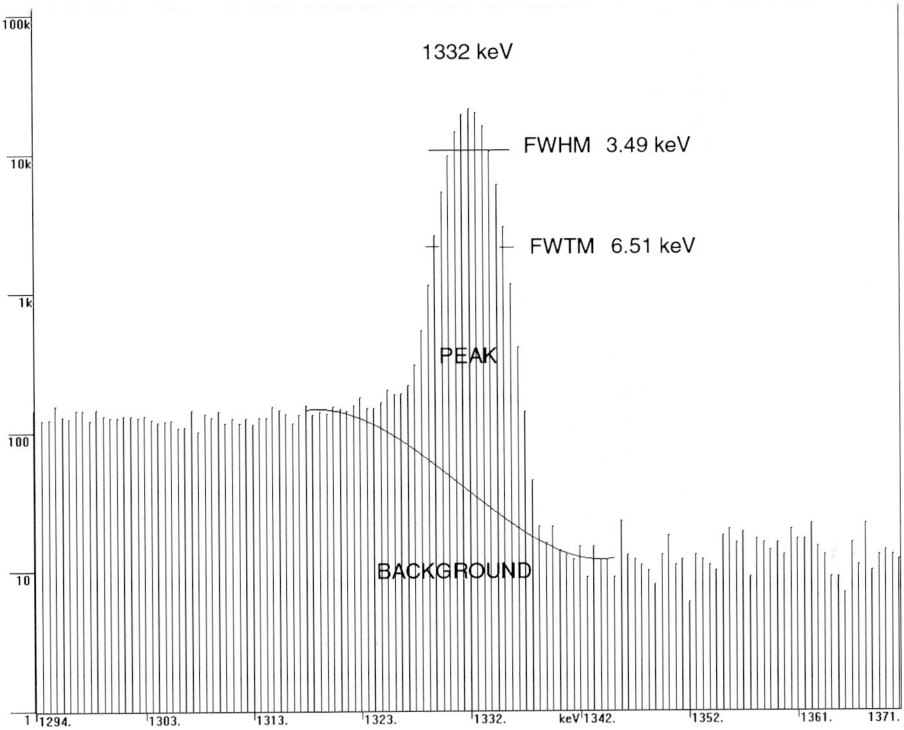

Fig. 9 Aptec MCArd screen analysis of GRS ^{60}Co 1332-keV peak. *Vertical line* heights are counts accumulated in energy channels. The *line shape* shows little tailing, and the 3.49-keV FWHM energy resolution exceeds the design goal of 4 keV

indicative of ETE resolution. The radioisotope-source gamma peaks shown span nearly all gamma-ray energies of interest, except for the Fe capture doublet at ~7,646 keV. The peak widths are all less than the 5-keV FWHM resolution used to calculate the GRS sensitivity baselines for the major elements (see Sect. 3.1).

5.1.3 GRS Spatial Efficiency Calibration

The purpose of the spatial calibration is to map out the spatial (angular) response of the GRS sensor head to photons incident on the detector from a distance large compared with detector dimensions, at all energies and angles of interest for detection and mapping of elements at Mercury's surface. This objective amounts to determination of the effective area or intrinsic efficiency of the GRS over a wide range of energies and angles, corrected for energy and angle-dependent attenuation of the radioisotope sources in the experimental setup and associated with the position of the GRS on the spacecraft. A sufficient number of counts must be accumulated for each energy peak at each angle to yield count statistics that will allow an accurate assay of surface elemental composition. This is the most time-consuming and complex portion of the calibration. Measurements taken at Mercury must be converted from counts in the detector to photon flux incident on the detector. Since this instrument is not on a boom, blocking due to spacecraft structural members (including the Al adapter ring) will modify an otherwise smoothly varying response function, in addition to structures within the instrument itself. Both the intrinsic detector angular response and

Table 4 ETE energy resolution measurements

Energy, keV	FWHM, keV	Counts in peak	Source
7,985	3.07	26,721	Pulser
6,130	4.77	974	Pu–^{13}C
3,548	3.97	217	^{56}Co
3,451	4.24	978	^{56}Co
3,273	4.37	2,060	^{56}Co
3,253	4.44	9,213	^{56}Co
3,202	4.07	3,679	^{56}Co
3,010	3.66	1,166	^{56}Co
2,598	4.05	24,191	^{56}Co
2,035	3.77	13,961	^{56}Co
2,015	3.64	5,422	^{56}Co
1,771	3.73	31,130	^{56}Co
1,360	3.53	10,214	^{56}Co
1,332	3.49	130,741	^{60}Co
1,238	3.59	170,942	^{56}Co
1,173	3.44	140,235	^{60}Co
1,038	3.48	39,016	^{56}Co
847	3.47	319,479	^{56}Co
121	3.34	24,037	background

"shadowing" by structural members are functions of the energy at which the response is being measured.

The response is a convolution of these effects with the "point source" measurements of the calibrations. To maximize mapping resolution, a scheme to deconvolve these effects to some approximate level needs to be devised. The GRS sensor detects gamma-rays from all directions. For mapping in orbit around Mercury, the GRS has an effective viewfield defined by the solid angle subtended by Mercury, which will change as a function of time throughout each orbit due to spacecraft altitude changes. Centered on the nadir direction, this angle is described by a half-cone angle given by $\Theta = \sin^{-1}[1 + h/R)]^{-1}$, where R is the radius of Mercury (2,440 km) and h is the spacecraft altitude. The orbits will be highly eccentric, with the closest distance ~200 km defining the largest cone (and the most sensitive measurements, having the highest gamma-ray flux). Thus GRS response to the planet is limited to a half-cone angle of 67.55°. It is desirable to make efficiency measurements at a number of angles distributed within this cone along the nadir direction to account for detector asymmetries and at a lesser number of angles over the full unit sphere, particularly in the forward direction, in order to help determine and correct for background and for non-nadir-pointing orbital measurements.

The gamma-rays of interest extend from the Na inelastic scattering line at 440 keV to at least the 7,646-keV Fe capture doublet, and measurements at lower and higher energies are desirable to improve efficiency accuracy near the ends of the spectrum. Previous spatial calibrations of Ge gamma-ray spectrometers flown in space have relied on radioisotope sources, which are available with reasonably high source strengths, reasonably long half-lives, and reasonably accurate calibrations with gamma-rays only up to ~3,500 keV, with weak approximately calibrated sources at 6,130 keV. Efficiencies at higher gamma-ray energies were generally extrapolated using Monte Carlo computations.

Instead the spatial calibration of the GRS flight unit for MESSENGER was conducted at a reactor at the National Institute of Standards and Technology (NIST), where intense beams of cold neutrons are available to irradiate neutron-capture targets and excite strong fluxes of gamma-rays across the entire energy range of interest. This set-up allowed sufficient data rates at high gamma-ray energies to complete the calibration in only 5.5 days of 24-hr measurements, while providing superior efficiency accuracy across the desired energy range. Targets of Cr and Cl were chosen, which provide many strong gamma peaks in the range from 517 to 9,717 keV, with intensities well characterized by recent International Atomic Energy Agency standards. These targets provide only relative efficiencies because the neutron fluxes and cross-sections are not known accurately enough to yield absolute efficiencies. To normalize to absolute efficiencies, fairly strong sources of ^{56}Co (100 μCi), ^{226}Ra (10 μCi), and ^{228}Th (10 μCi) were procured and calibrated by NIST. These sources provide many strong gamma-ray peaks between 239 and 3,451 keV and complement the capture target peaks by having more lower-energy peaks, as well as many peaks in the same range as the capture targets, to obtain proper normalization. The detailed results of this calibration will be published separately.

5.1.4 GRS Functional Performance

In addition to calibration measurements, there are several basic GRS functions that are required to operate the sensor. The voltage applied to the flight Ge detector must be sufficiently high to deplete fully the semiconductor crystal. The cryocooler must cool the Ge detector to the required operating temperature, reaching temperature stability in a reasonable length of time. The annealing heaters must also be able to bring the Ge detector up to the annealing temperature in a reasonable length of time. The plastic anticoincidence shield should reliably detect and reject cosmic rays. A comprehensive performance test has been developed to determine that the spectrometer is functioning properly.

The Ge crystal depletion voltage is determined by plotting crystal capacitance as a function of high voltage applied. Crystal capacitance is measured by injecting pulses on the high-voltage line using either the internal or an external pulser. For the flight detector, the capacitance curve begins to flatten out at 2,200 V, becoming nearly completely flat at 3,000 V, which indicates the point of full depletion. Since the flight detector is operated at 3,200 V, it is beyond full depletion and therefore insensitive to small changes in the high voltage. The high-voltage supply does not need to be regulated to close tolerance, such as is required for photomultipliers.

When the cryocooler begins cooling down, the initial temperature change is relatively fast, but it then progressively slows because the heat lift of the cooler (related to its efficiency) drops as a function of the temperature difference between the cold tip and the ambient environment. At the same time the cryostat heat load is increasing due to radiative coupling. Remarkably, the slope of the cooling curve as a function of the crystal temperature is fairly linear and is expected to be proportional to the difference between the cooler heat lift and the total heat load, and inversely proportional to the heat capacity of the detector capsule. The time required to reach regulation at 85 K from a 20°C ambient temperature is 37.5 hours. If the cooler is turned off, the Ge crystal temperature drifts up to 100 K within ~1.3 hr, leaving the crystal outside of the desired operating range. If the cooler is then turned back on, about ~24 hr is required to bring the Ge capsule back into regulation, so there is a substantial penalty for removing cooler power for even short periods. Although data could be taken throughout this period, the peaks would be highly smeared due to the 185 ppm/°C temperature coefficient of the Ge detector (unless corrected over very short collection intervals, which is impractical given the limited downlink), but more importantly, the cooler will incur additional stress (reducing its lifetime) and the detector will be subject to increased radiation damage. It is therefore preferable that the detector operating temperature be maintained continuously in orbit, except for preplanned outages, such as during the long eclipse season or when annealing the detector is necessary to repair radiation damage.

There are two Zener diodes attached to the bottom of the detector capsule that are used to heat the Ge crystal to anneal radiation damage from cosmic rays. They each consume up to 1 W and can be used individually or in tandem. During an annealing cycle, the cooler is turned off and the crystal is heated up to 85°C, held at a constant temperature for three Earth days, and then cooled back down to operating temperature by the cooler. If both diodes are used at full power, the crystal will reach the annealing temperature in ~14 hr. Maintaining 85°C requires only 0.5 W, so the heater circuits provide ample margin. The cooldown from 85°C takes ~40 hr, which is, interestingly, only a few hours longer than cooldown from ambient due to the tremendous heat lift available from the cooler for the case when the cold tip is actually hotter than ambient. A full annealing cycle can therefore be accomplished in five days. It is expected that an annealing cycle will need to be performed about once every Mercury year (88 Earth days). During the long eclipse season, the spacecraft will be in a low-power mode for about seven days, and there will be insufficient power available to operate the GRS. However, the annealing function requires relatively little power, so this will be an ideal time to accomplish annealing without having to sacrifice science data collection.

The purpose of the anticoincidence shield is to veto cosmic rays. Proper operation requires detection of penetrating particles over all parts of the shield. The light path from the top part of shield to the PMT at the bottom center is particularly tortuous, and efficient detection requires that enough light be received to exceed the lower-level discriminator setting. A collimated ^{137}Cs source was used to examine the relative light output from different portions of the shield. Positions along the side of the shield exhibited Compton edges around channel 50, while illumination of the bottom portion of the shield produced Compton edges around channel 90, nearly a factor of two higher. Actually, a factor of two variation in light collection is surprisingly good given the complex geometry of the anticoincidence cup with its thin walls, optical coupling pads, and cryostat feedthroughs. The shield LLD cuts off below channel 5, which roughly converts to ~50 keV. The thickness of the shield walls were sized (18 mm) so that minimum-ionizing particles will deposit significantly more energy than this amount, providing very high overall detection efficiency for penetrating cosmic rays.

An unusual function of the GRS is the "microphonics" channel, which is designed to monitor the mechanical vibrations produced by the cooler and possibly infer the level of wear. A buffered and filtered version of the HPGe preamplifier output signal is sent to a sampling ADC that acts as the cryocooler "microphone." The data are sampled at a maximum rate of 5 kHz for 6 s. The fundamental frequency of the cooler is ~60 Hz. The stator contains 22 poles, so the cooler also exhibits a strong "electronic" frequency of ~1,320 Hz. The total vibration spectrum is rich in harmonic content and extends nearly unabated to beyond 2 kHz. When the sampled signal was played back through a loudspeaker, the sound appeared to the human ear to have good fidelity compared with the sound heard directly from the cooler and was judged adequate to detect some sound changes that might indicate bearing noise, presaging cooler failure. However, the background pops and clicks that will be generated by cosmic rays striking the detector in space may make it difficult to detect cooler sound nuances in flight (more sophisticated signal processing may be required).

5.2 NS Ground Calibration

NS preflight ground-calibration data were taken at two facilities in order to make measurements for the full range of energies from thermal to fast neutrons. Specifically, NS data were taken at the Columbia University Radiological Research Accelerator Facility (RARAF) using monoenergetic fast neutrons having energies from 0.5 to 14 MeV. NS data were also taken at the Free-air Neutron Source Facility at Los Alamos National Laboratory for measurements of thermal, epithermal, and fast neutrons. Summaries for both of these sets of measurements are given in the following. The experimental details and analysis will be described elsewhere.

5.2.1 Fast Neutron Measurements at RARAF

The RARAF facility beamline target area is a relatively small, concrete-lined room. The NS, with full flight electronics, was set up at the RARAF facility and measurements were made at eight different neutron energies and at various incident angles in order to calibrate the energy scale and to measure the neutron efficiency as a function of energy for fast neutrons. However, due to the large background from room-scattered neutrons, a better estimate of fast neutron efficiency will be obtained using data taken with a bare ^{252}Cf source (discussed later).

Typical fast-neutron measurements are shown in Fig. 10. Figure 10a shows prompt pulse data for 1.5-MeV neutrons that exhibit delayed coincidences within the acceptance time window. There is a well-defined peak around channel 15 due to the monoenergetic neutrons at 1.5 MeV. The energy resolution is $\sim$50% after subtracting background, which is normal for these types of scintillator measurements. The tail at lower energies is due mostly to room-scattered neutrons at lower energies. Figure 10b shows the delayed pulse for neutrons at 1.5 MeV. The peak around channel 50 is due to the ^{10}B(n,α)^{7}Li* reaction at an equivalent electron energy of 93 keV. The counts at high channels are due to the Compton continuum of 478-keV gamma-rays that are emitted by excited lithium nuclei in coincidence with the charged-particle recoils. Figure 10c shows a TTSP spectrum for 1.5 MeV neutrons. The exponential portion at low channels is the expected spectrum for BC454 scintillator plastic. The flat portion at high channels is due to chance coincidences. After subtracting a background of 50 counts due to chance coincidences, the remaining spectrum (gray curve) has an e-folding time τ of 1.98 μs, which is close to the expected time of 2 μs based on previous BC454 sensors (Feldman et al. 1991, 2004).

Figures 11 and 12 show the behavior of the NS for various monoenergetic neutron energies ranging from 0.5 to 14 MeV. Figure 11a shows data for the prompt pulse where the measured counts are now plotted versus energy (see Fig. 12 for how the energy scale is derived). Note that the counts for each spectrum in Fig. 11a have been arbitrarily scaled to show all the data on one plot. This plot, therefore, cannot be used to determine the absolute efficiency. In addition, for the highest energy (14 MeV), the high voltage on the BC454 PMT was lowered to keep the prompt pulse on scale. This gain change is reflected in Fig. 11b where the peak position of the delayed pulse is shifted to a lower channel. Finally, we note that for increasingly higher energies, the background from lower-energy, room-scattered neutrons becomes greater so that for the two highest energies (5.8 and 14 MeV), most of the measured neutrons are lower energy than the main beam energy.

Figure 12 shows an energy calibration using prompt pulse data, where the neutron energy is plotted versus gain-corrected channel number for the centroid of the prompt peak. Since

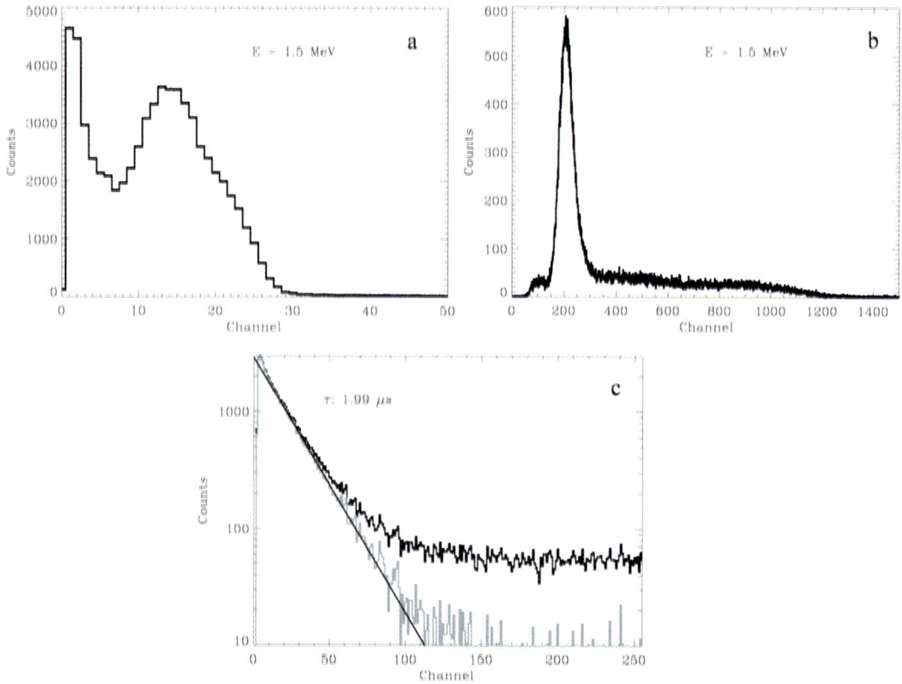

Fig. 10 (**a**) Histogram of NS prompt pulses in coincidence with a delayed pulse for monoenergetic fast neutrons of 1.5 MeV. (**b**) Histogram of NS delayed pulses in coincidence with a prompt pulse for monoenergetic fast neutrons of 1.5 MeV. (**c**) Histogram of time-to-second pulse for the same data shown in (**a**) and (**b**). Each channel corresponds to 100 ns. Shown in *gray* is the resulting spectrum when the background due to chance coincidences is subtracted from the original spectrum

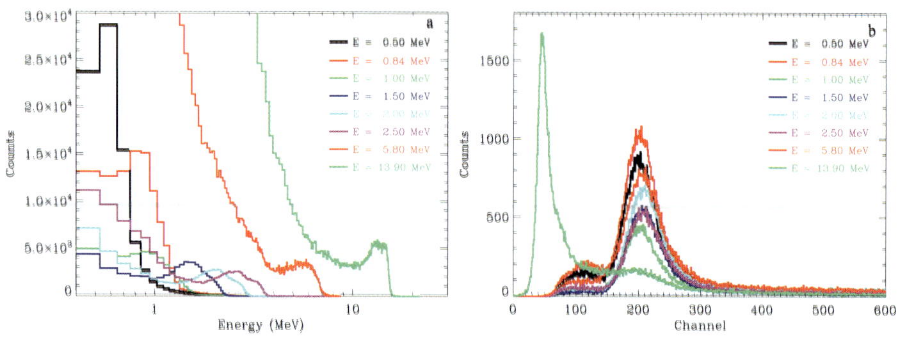

Fig. 11 (**a**) Histogram of NS prompt pulses in coincidence with a delayed pulse for monoenergetic fast neutrons having energies from 0.5 to 14 MeV. The different colors indicate the different energies. (**b**) Histogram of NS delayed pulses in coincidence with a prompt pulse for monoenergetic fast neutrons having energies from 0.5 to 14 MeV

the data points lie on a straight line in log–log space, a power law has been fit to the data (black line in Fig. 12):

$$C = 6.7E^{1.69}, \tag{1}$$

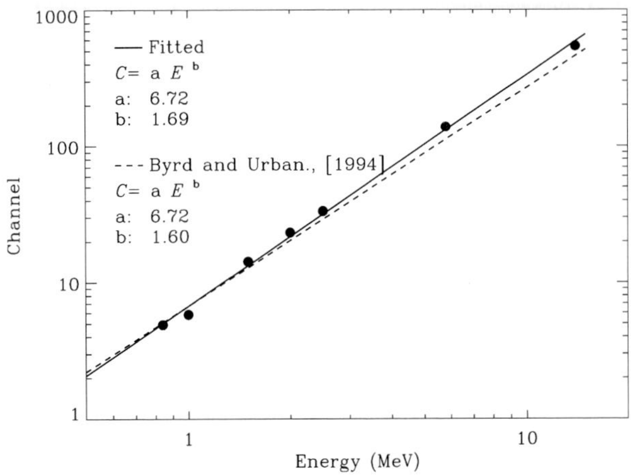

Fig. 12 Plot of the channel position centroids from Fig. 11a versus the neutron energy (*circles*). The *black solid line* is a power law fit to the data. The *gray dashed line* shows the fit of Byrd and Urban (1994) for the same type of scintillator material

where C is the channel number and E is the neutron energy. The power law exponent of 1.69 compares very favorably to the exponent of 1.6 that was determined (Byrd and Urban 1994) for BC454 using a model of BC454 plastic (dashed line in Fig. 12).

5.2.2 Neutron Measurements at Los Alamos National Laboratory

For thermal and epithermal neutrons, a ^{252}Cf source having a neutron count rate of 6.504×10^7 n/s was used in a test facility at the Los Alamos National Laboratory. The Cf source was surrounded by a 22-cm diameter D_2O sphere in order to moderate the higher energy (>1 MeV) neutrons coming from the source. In addition, the D_2O sphere had a removable Cd shield to block out thermal neutrons (Cd is a highly effective absorber of neutrons with energies less than 0.4 eV). Therefore, epithermal neutrons are measured with the Cd on the D_2O sphere, and thermal plus epithermal neutrons are measured when the Cd cover is removed. Thermal neutrons are measured by taking the difference between the measured counts of the Cd and no-Cd configurations. The entire source/detector configuration was placed on a ~3.5-m-high platform that is far away from any dense wall material in order to reduce room-scattered background neutrons. A benchmarked neutron transport model of the room exists to determine remaining backgrounds so that absolute neutron efficiencies will be able to be determined with as little error as possible. We carried out a long (~12 hour) run where we illuminated the NS with the bare ^{252}Cf source. This was done to obtain measurements from which we will be able to determine the fast neutron efficiency with good accuracy.

Neutron measurements were made with both the Cd and no-Cd configurations for nine different azimuthal neutron beam angles ranging from −90° (facing normal to outboard Li-glass detector LG2) to 90° (facing normal to inboard Li-glass detector LG1) around the axis of the Li-glass (LG) PMTs, and seven different polar angles ranging from −45° to +45° about an axis perpendicular to both axes of the PMTs. Typical measurements for both LG detectors, shown in Fig. 13, were made where each LG sensor was pointing normal to the ^{252}Cf source direction. The black lines show thermal plus epithermal neutrons (i.e., no Cd on the D_2O sphere), and the gray lines show epithermal neutrons (i.e., Cd was on the D_2O sphere). Very clear peaks due to the 4.78-MeV ^{6}Li(n,α)^{3}H reaction are seen for both sensors.

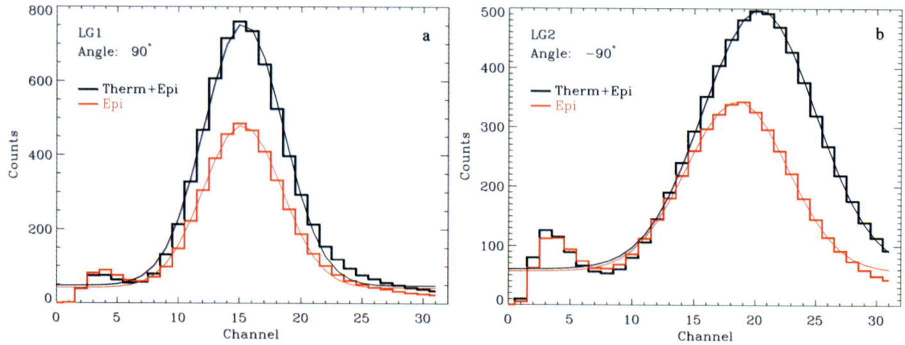

Fig. 13 (**a**) Histogram of LG1 counts when the LG1 detector was directly illuminated by the ^{252}Cf source. The *black line* shows thermal plus epithermal neutrons, and the *red line* shows epithermal neutrons. (**b**) Same type of data as (**a**), but now the LG2 detector is illuminated directly by the ^{252}Cf source

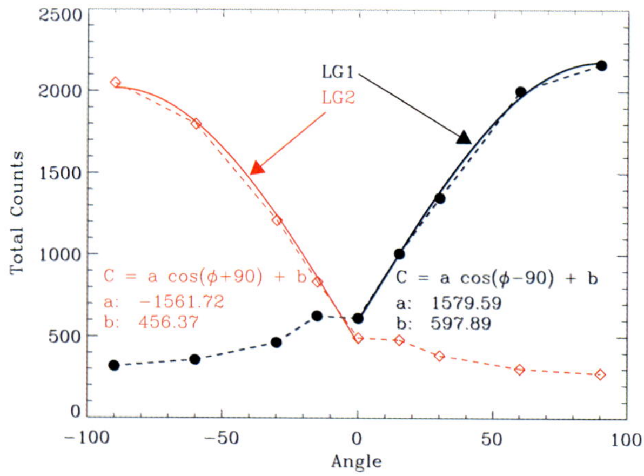

Fig. 14 The *circles* and *diamonds* show LG count rates as a function of azimuthal angles for both LG1 (*black circles*) and LG2 (*red diamonds*). Both sets of data are fit to (2) and show that when the detectors are facing the source they are well fit by a cosine function, which is expected for planar detectors

The azimuthal angular dependence is seen in Fig. 14. Since these are uncollimated, planar detectors, their angular response will have a $\cos\theta$ dependence of:

$$C = a\cos\theta + b, \tag{2}$$

where θ is the azimuthal angle and a and b are fitted constants. Figure 14 shows that for both the LG1 and LG2 sensors, the angular dependence of the sensors when they are facing the source is well fit by (2). In order to understand the behavior of the sensors for greater angles, a detailed modeling of the NS sensor must be carried out to account for the NS material and the background neutrons at the free-air facility.

6 In-flight Performance

6.1 First GRS Flight Performance Test

During the first two weeks of November 2004, the GRS was turned on and collected approximately five days of measurements during cruise. The purpose of these measurements was to determine the energy and intensity of background gamma-ray peaks, to identify the sources of background gamma-rays, to measure the effectiveness of the anticoincidence shield, to measure the energy resolution of the detector as a reference point for future measurements, and to ascertain the overall flight functionality of the instrument.

The processes that produce gamma-rays from the surface also produce gamma-rays from the spacecraft, from the local material surrounding the detector, and from the detector itself. Some of these background gamma-ray lines are at the same energies as lines from the planet as they come from the same elements (e.g., Al, O, K, and Th). Other background gamma-rays, while not coming from the same elements, occur at energies that are at the same or close to gamma-rays from the planet. Besides the peaks that can interfere with the planetary measurements, it is important to identify and characterize all background peaks. There might be other weaker background peaks, from the same isotopes, that can be identified in these background measurements, which may cause interference in future long accumulations. Thus, it is important to determine the count rates and identification of as many background peaks as possible, before the spacecraft goes into orbit around Mercury.

During the first week of November 2004, the Ge crystal was annealed for three days at 85°C, in order to clear out any contaminants that might collect inside the cryostat and increase the cryocooler heat load. (Little radiation damage would have accumulated since the launch on August 3, 2004, but it had been found in ground testing that annealing aided cooldown after a long period at ambient temperatures and decreased the cooler heat load.) By the time the detector had cooled down to operating temperature, a solar particle event (SPE) was in progress. High-energy protons from the Sun caused a large increase in the detector count rate, both peaks and continuum. In fact, so many particles were impacting the crystal when high voltage was initially applied that the detector leakage current measured 2–3 nA, nearly an order of magnitude higher than expected under quiescent conditions. Measurements taken over subsequent days showed a decrease in this continuum and eventually

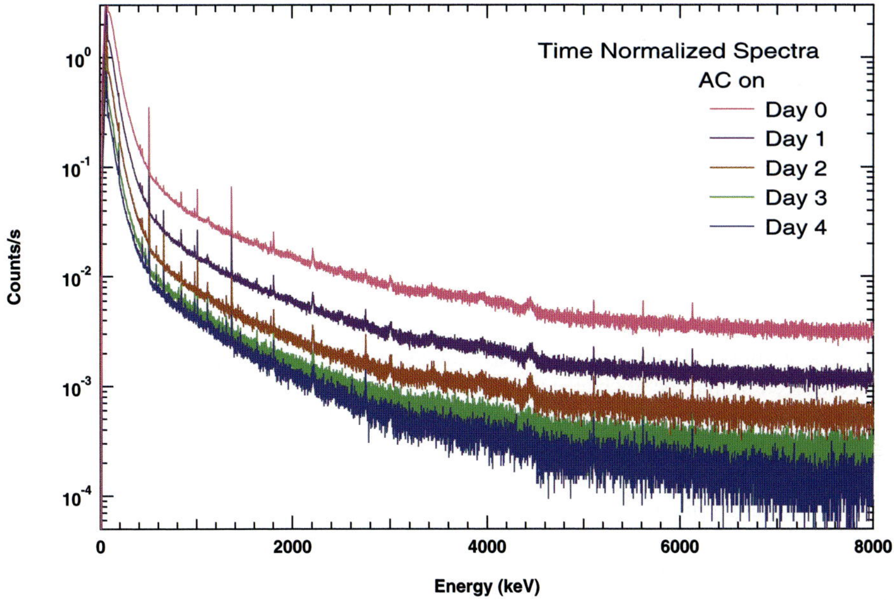

Fig. 15 Normalized anticoincidence Ge spectra for five time intervals. The increase in count rate caused by the SPE decreased after Day 1 and was essentially gone by Day 4

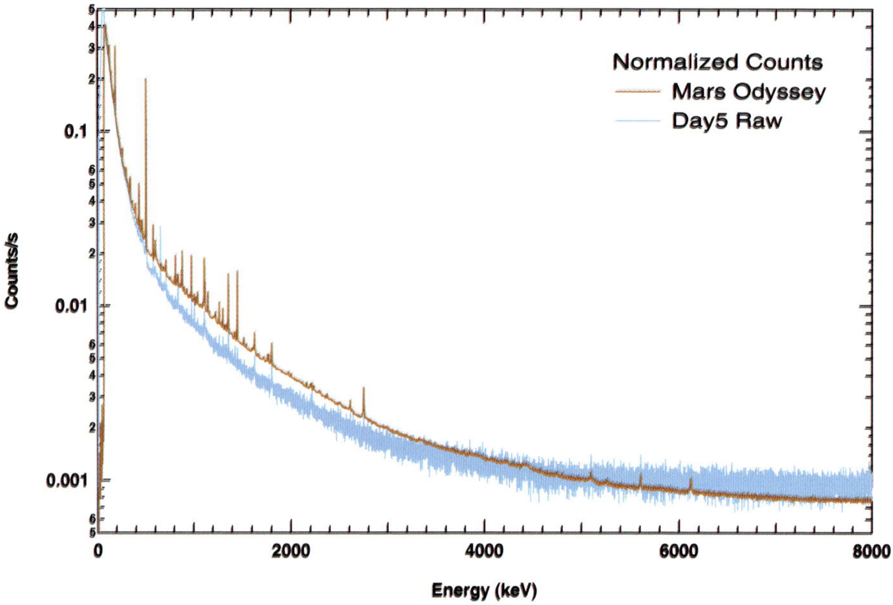

Fig. 16 Comparison between MESSENGER Day 5 (after the SPE) and Mars Odyssey cruise spectra shows slightly different shape but similar intensity

by Days 4 and 5 (November 14 and 15) the continuum levels seemed to reach the pre-SPE levels. Shown in Fig. 15 are the anticoincidence spectra normalized to counts per second per bin for five of the six time intervals. "Day 0" is the first 14 hours of accumulation during the SPE. The spectrum for Day 5 lies on top of that shown for "Day 4," indicating that count rates were back to normal for the last two days. Figure 16 shows a comparison between the cruise spectra for MESSENGER GRS for Day 5 (raw data, anticoincidence off) and that for Mars Odyssey GRS (Evans et al. 2002) for no SPE. The continua have a somewhat different shape but similar intensity.

The GRS detector system anticoincidence shield is used to reject charged particles that interact in the shield and then the central detector. This behavior helps to reduce the background continuum, particularly at higher energies. The efficiency of the background reduction can be determined by comparing the raw germanium spectrum (all events) with the simultaneous anticoincidence spectrum. This ratio is shown in Fig. 17 during the SPE and during the last day of measurement. The ratio has been smoothed by co-adding several adjacent energy bins to illustrate more clearly the measured result. The efficiency of charged particle rejection may be unchanged for the two cases. The lower ratio during the SPE is likely due to protons streaming into the front face of the detector, where anticoincidence is not involved.

Gamma-ray peak results were obtained by analyzing the Day 4 plus the Day 5 anticoincidence spectrum. This sum was used to increase the statistics and reduce the uncertainties. The spectrum was analyzed using GANYMED, an interactive software analysis tool developed for Ge detectors (Brückner et al. 1991), for all the peaks except the region around the 2,223-keV region. The fit for that complex region was done with gamma-ray analysis software developed in the Igor code. The fits are shown in Fig. 18 for the Day 4+5 spectrum and Fig. 19 for the Day 1 spectrum. Note the increased intensity of the 2,211 keV peak due to

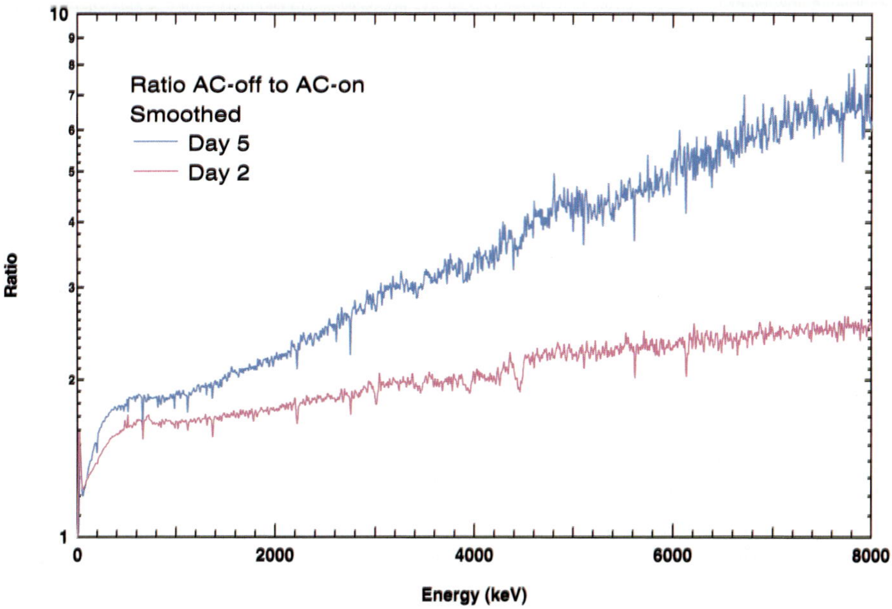

Fig. 17 Ratio of raw Ge count rate to that with anticoincidence (AC) on Day 2 is during the SPE and Day 5 is after the SPE. The lower ratio during the SPE is likely due to solar protons that stream into the front face of the detector and do not produce veto signals in the shield

the interactions of the high-energy protons with the aluminum in the spacecraft and material around the detector.

The peak analysis results are given in Table 5. For each source identified, the spectrum channel number and corresponding energy are listed. The area under each peak is given in counts, where the column Error (%) is the statistical uncertainty, and the column CPM is the counts per minute. A nuclide with an asterisk indicates that the gamma-ray is from an excited state of that nucleus that can be made by several processes. Table listings with the +K notation include the K binding energy of the captured electron. Single and double escape peaks are noted by SE and DE, respectively. For comparison, the count rates from the measured Mars Odyssey (MO) cruise spectrum are also shown (note that the Mars Odyssey detector is larger by a factor of more than two in volume than the MESSENGER detector, being a 67 mm $\times$ 67 mm cylinder).

As can be seen, all of the gamma-rays listed here were also detected on Mars Odyssey (except ^{137}Cs). The sources fall into three basic categories: natural radioactivity (e.g., ^{40}K, Th, and U), lines from germanium and from spallation products of germanium (e.g., ^{75m}Ge, ^{67}Ga, and ^{65}Zn), and lines from the spacecraft and other materials (e.g., Al, Mg, Ti, H, O). There are some gamma-rays that can be measured in the early spectra, but not in the Day 4+5 spectrum (e.g., 1,778.9 keV, probably from ^{28}Al). As with Mars Odyssey measurements, the only neutron capture line detected in the spectrum is from hydrogen. Not listed in the table is a (broad) ^{10}B(n,α) peak at 478 keV in the raw spectrum, present at reduced intensity in the anticoincidence spectrum. Ground testing showed that this peak from the BC454 anticoincidence shield (gamma-rays entering the Ge detector from moderated neutrons being absorbed by the boron in the plastic) was essentially eliminated in anticoincidence mode as expected. So this peak in the anticoincidence spectrum is from boron elsewhere on the spacecraft, probably from boron in the PMT glass of the instrument and in the spacecraft

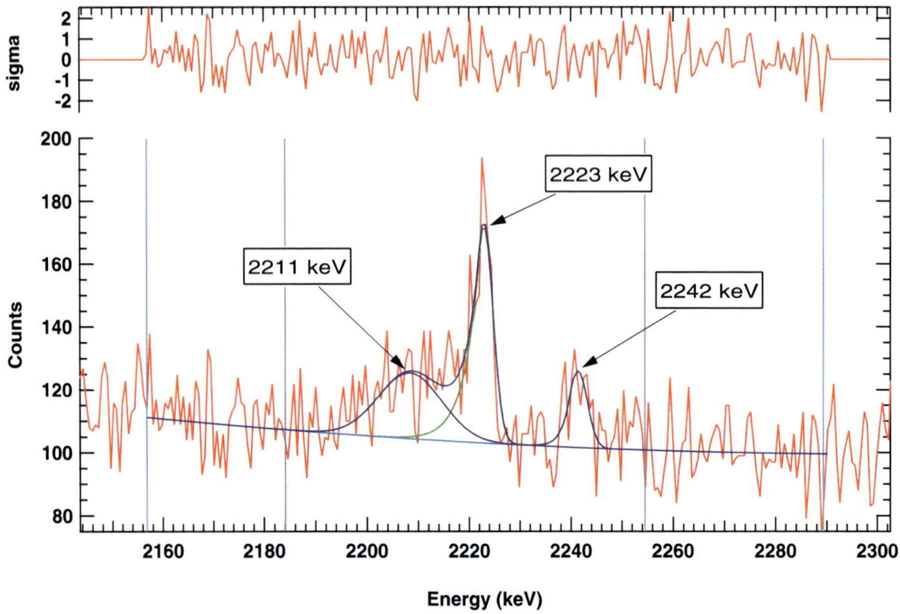

Fig. 18 Fit to the 2,223 keV region for Day 4 plus Day 5 (after the SPE) with narrow peaks at 2,223 keV from H(n,g) and 2,242 keV from ^{24}Na single escape (SE) and a broad peak at 2,211 keV from ^{27}Al*. *Sigma graph* shows goodness of fit in standard deviations

sunshade. Neither of these boron sources are present for Mars Odyssey. We suspect that the ^{137}Cs is a contamination of some metal parts (it is sometimes found in steel, for example).

The energy resolution determined during these cruise measurements will be the reference point for future decisions regarding annealing the detector. The peak shape fit to the MESSENGER cruise spectra was a pure Gaussian with no low-energy tailing for all peaks. This is in contrast to the Mars Odyssey detector that showed significant tailing due to radiation damage (possibly also from residual pole-zero mismatch), even after it had been annealed in orbit around Mars. A comparison of the peak shape fit to the 1,368-keV peak for spectra from both instruments is shown in Fig. 20. The MESSENGER GRS peak has a FWHM of 3.70 keV and a full width at tenth-maximum (FWTM) of 6.74 keV. The peak in the Mars Odyssey GRS spectrum has a FWHM of 4.69 keV and a FWTM of 8.77 keV. A pulser peak shift for Days 4 and 5 (the days used for summing in Fig. 20) indicates a smearing of ~0.2 keV for the MESSENGER GRS, consistent with the measured temperature change and temperature coefficient for the preamplifier. The temperature-corrected flight FWHM is ~3.5 keV, the same as seen in Fig. 9 for the ground measurements.

Sometimes the time dependence of peak intensities can be used to help identify the particular isotope that was activated by the solar particles in the case where two isotopes decay with emission at the same gamma-ray energy. However, with the limited time resolution of these measurements, it was not possible to determine accurately any half-lives of source isotopes and resolve any ambiguity in isotope identification. Limited time was allocated to these cruise measurements in order to conserve cryocooler lifetime and ensure measurement of the Mercury surface composition from orbit. It was verified that intensities of the natural radioactivity peaks resulting from potassium, uranium, and thorium were time independent,

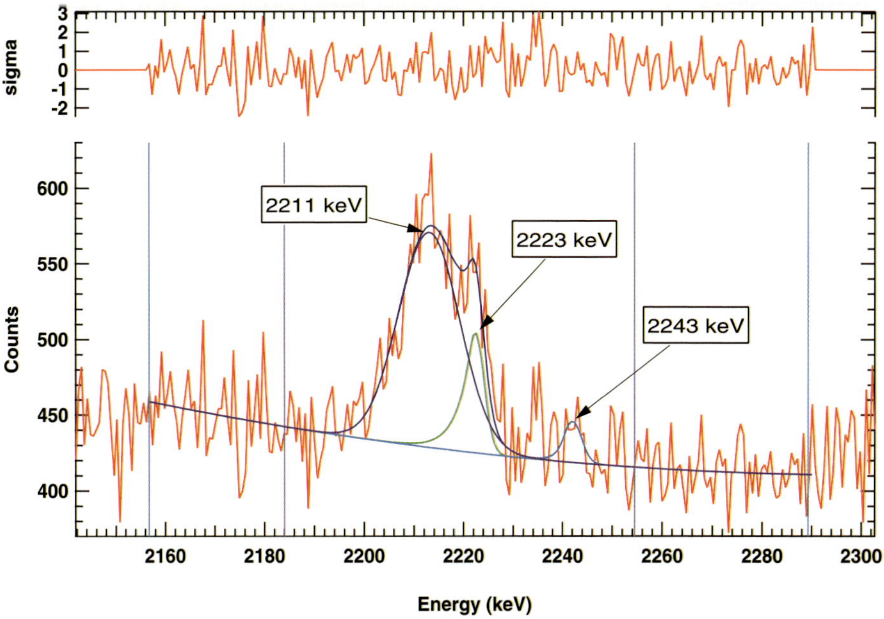

Fig. 19 Fit to the 2,223 keV region for Day 1 (during the SPE) with narrow peaks at 2,223 keV from H(n,g) and 2,242 keV from ^{24}Na SE and a broad peak at 2,211 keV from ^{27}Al*. Note the relative differences in peak intensities between this figure and Fig. 18

while the intensities of the total count rate, the annihilation peak, the ^{27}Al peaks, and the oxygen inelastic peaks decreased with time similar to the SPE fast-particle rates.

6.2 NS In-flight Performance

The NS instrument was turned on and operated for extended testing from April 18 to May 23, 2005. Following a flight software revision in June 2006, the NS was turned on for the duration of the mission on January 31, 2007. All data modes have been exercised, and substantial long-term cruise science telemetry has been collected. Some preliminary results are presented in the following.

6.2.1 Li-glass Detectors

Figure 21 overlays NS spectra for the two Li-glass detectors. Both detectors measure thermal neutrons but are facing in opposite directions. During cruise, the thermal neutron flux is expected to be very low because the few fast neutrons that are created in the spacecraft structure tend to escape without being moderated, and those that are successfully moderated within the fuel tanks are then likely to be absorbed by the large amount of nitrogen (Feldman et al. 2002b). If detected, thermal neutrons will appear as a broad Gaussian line centered near channel 24. The outboard-facing detector, LG2, shows only the gamma-ray background due to Compton scattering in the glass. The inboard-facing detector appears to show a hint of a thermal peak.

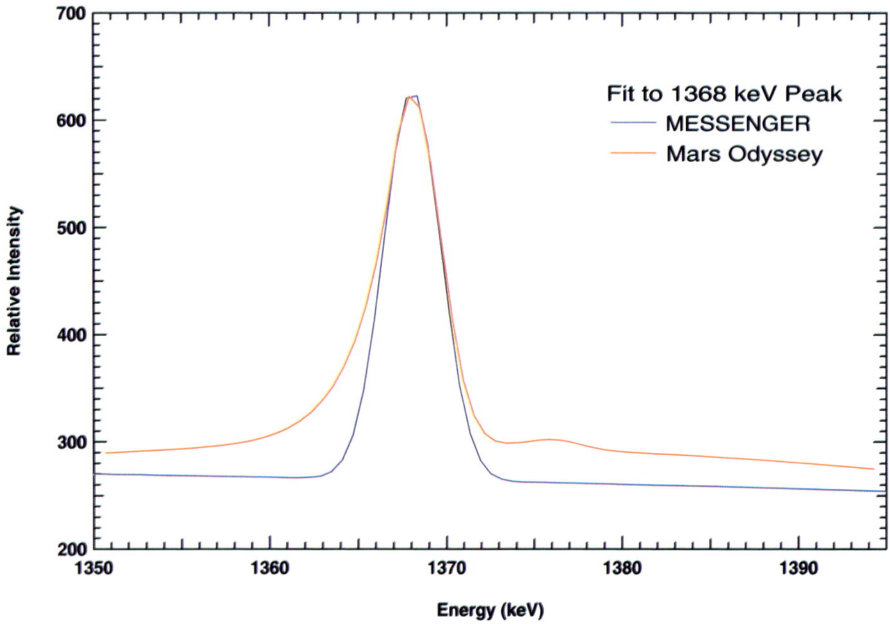

Fig. 20 Peak shape comparison between the MESSENGER GRS (after the SPE) and the Mars Odyssey GRS. The MESSENGER instrument has a narrower peak and less low-energy tailing

Fig. 21 NS Li-glass in-flight spectra (48-hr accumulation). The two Li-glass detectors both detect thermal neutrons but are facing in opposite directions. Thermal neutrons will appear as a broad Gaussian line centered near channel 24. The spacecraft background exhibits very few thermal neutrons, but the inboard detector appears to show a hint of a thermal peak

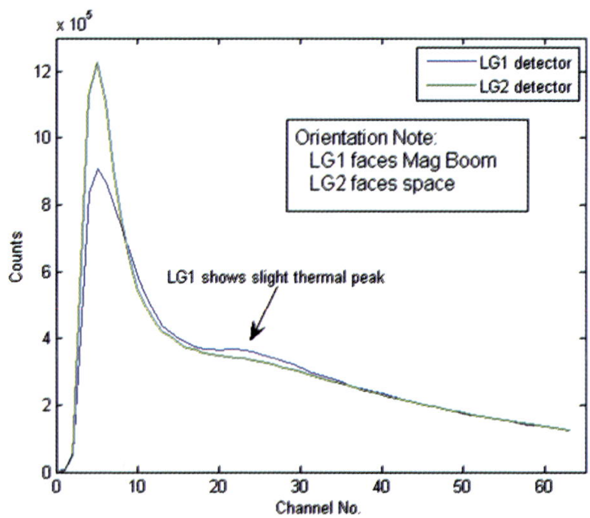

6.2.2 *Fast Neutron Detector*

The BC454 boron-loaded plastic scintillator contains its own mixture of moderator and absorber and is capable of detecting fast neutrons directly with energies up to ∼7 MeV deposited energy. The NS counts fast neutrons that exhibit a prompt light pulse, due to scattering and moderating in the BC454, followed by a delayed coincidence light pulse that arises when the moderated neutron is captured (absorbed) by a ^{10}B nucleus. While the prompt

Table 5 GRS results for the Day 4+5 spectrum

Source	Channel	Energy (keV)	Area (cnts)	Error (%)	CPM	FWHM (keV)	MO CPM	Ratio to MO
^{75m}Ge	232.9	139.9	21095	5.7	6.89	3.01	18.42	0.37
U–^{235}U	309.14	185.9	11356	9.0	3.71	3.04	5.72	0.65
^{67}Ga+K	323.77	194.7	12439	9.7	4.06	3.05	7.38	0.55
^{71m}Ge	329.69	198.3	67714	1.9	22.13	3.05	59.22	0.37
Th–^{212}Pb	396.58	238.7	4126	21.2	1.35	3.07	1.40	0.96
U–^{214}Pb	583.0	351.1	3119	19.1	1.02	3.14	4.63	0.22
^{69m}Zn	729.02	439.2	6372	9.6	2.08	3.20	8.04	0.26
^{24m}Na	783.89	472.3	3793	21.0	1.24	3.22	1.83	0.68
annihilation	848.01	511.0	102241	0.5	33.41	4.17	68.34	0.49
Th–^{208}Tl	969.06	584.0	3175	12.8	1.04	3.28	4.43	0.24
U–^{214}Bi	1008.95	608.1	1133	46.7	0.37	3.30	1.84	0.20
^{137}Cs	1097.74	661.7	16999	3.0	5.55	3.33	–	–
^{58}Co	1356.31	817.7	1348	22.2	0.44	3.43	2.42	0.18
^{27}Al*	1398.75	843.3	2176	19.2	0.71	3.44	0.94	0.76
^{56}Fe*	1404.04	846.5	1671	25.7	0.55	3.45	1.04	0.53
^{69}Ge+K	1462.98	882.0	1034	21.6	0.34	3.47	1.29	0.26
^{46}Ti*	1474.85	889.2	726	36.5	0.24	3.47	3.41	0.070
^{48}Ti*	1630.86	983.3	2480	13.1	0.81	3.53	3.71	0.22
^{27}Al*	1681.96	1014.1	2306	18.6	0.75	3.55	0.64	1.17
^{69}Ge+K	1852.76	1117.2	2634	10.5	0.86	3.61	3.07	0.28
^{65}Zn+K	1864.78	1124.4	897	33.4	0.29	3.62	1.37	0.21
^{56}Fe*, U–^{214}Bi	2051.98	1237.4	530	51.7	0.17	3.65	0.46	0.40
^{22}Na	2112.54	1273.9	631	42.0	0.21	3.68	1.19	0.18
^{48}Ti*	2175.16	1311.7	1460	23.4	0.48	3.69	1.00	0.48
^{24}Na, ^{24}Mg*	2268.59	1368.0	4840	5.6	1.58	3.70	3.95	0.40
^{52}Mn	2377.35	1433.7	839	24.6	0.27	3.81	0.30	0.90
^{40}K	2421.10	1460.0	831	33.9	0.27	3.82	4.31	0.063
^{20}Ne*	2707.31	1632.7	919	25.4	0.30	3.93	0.79	0.38
^{26}Mg*	2997.85	1808.0	1223	19.7	0.40	4.04	0.83	0.48
Al^{27*}	3668.24	2211.3	1300	7.3	0.42	14.53	0.25	1.68
H(n,γ)	3684.27	2223.0	1051	6.9	0.34	4.22	0.16	2.12
^{24}Na SE	3716.47	2243.0	416	13.3	0.13	4.23	0.20	0.65
Th– ^{208}Tl	4333.48	2613.8	487	31.6	0.16	4.54	0.26	0.62
^{24}Na	4564.92	2753.4	1283	10.1	0.42	4.62	0.79	0.53
O(n,n$'\gamma$) DE	8463.42	5105.4	344	28.7	0.11	6.07	0.12	0.92
O(n,n$'\gamma$) SE	9310.32	5616.3	379	21.6	0.12	6.39	0.19	0.63
O(n,n$'\gamma$)	10158.38	6127.9	307	25.0	0.10	6.70	0.18	0.56

spectrum shown in Fig. 22b is mostly featureless (a power law), the capture spectrum shown in Fig. 22a clearly shows one of the characteristic signatures for fast neutrons. Another telltale sign that the detector is indeed measuring fast neutrons is to plot the distribution of times between the prompt and capture pulses as shown in Fig. 23. Background subtraction

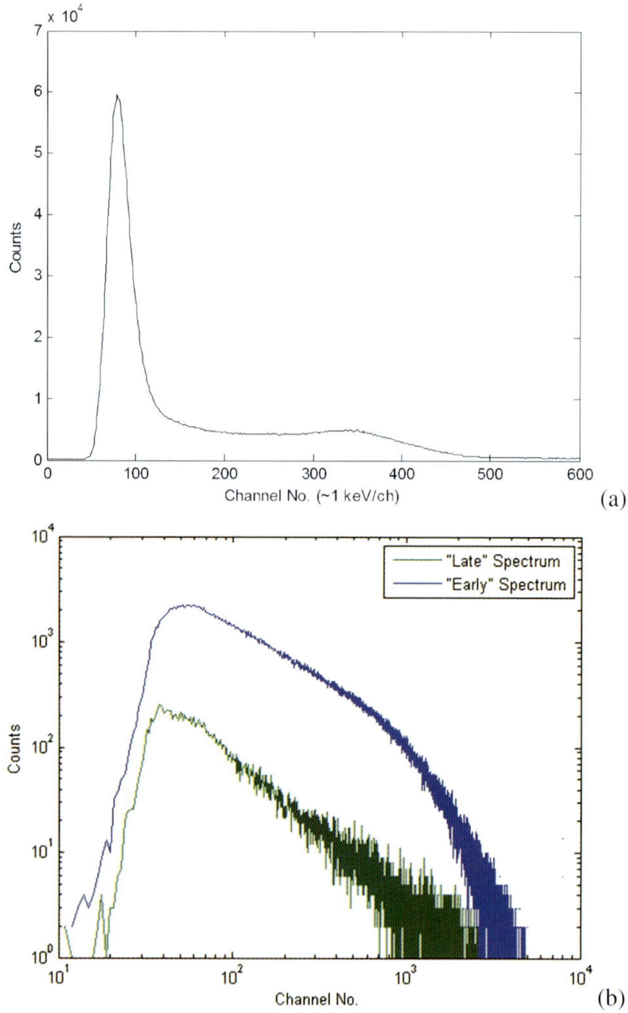

Fig. 22 NS fast neutron spectra. (**a**) Fast neutrons that are moderated in the boron-loaded plastic scintillator and that are eventually captured by a ^{10}B atom yield a light pulse equivalent to that of a 93-keV electron, due to emission of an alpha particle in the reaction. This capture peak is clearly observed. The reaction simultaneously releases a 478-keV gamma-ray (with 94% probability) that usually escapes the detector but occasionally scatters on the way out, producing the characteristic Compton edge and continuum to the right of the capture peak. (**b**) Histogram of prompt events with early time-to-second-pulse values (*blue*), which are dominated by fast neutrons, and for late events (*green*) that are mostly background

removes the effects of chance coincidences, which here are approximated as uniformly distributed in time given the low input singles rate, leaving an exponential distribution with a characteristic time of $\sim$2 μs, which is expected for a scintillator having 5 weight percent loading of boron. It should be noted that Figs. 22a and 23 are in close agreement with corresponding Figs. 10b and 10c for fast neutrons measured during ground calibration at the RARAF facility.

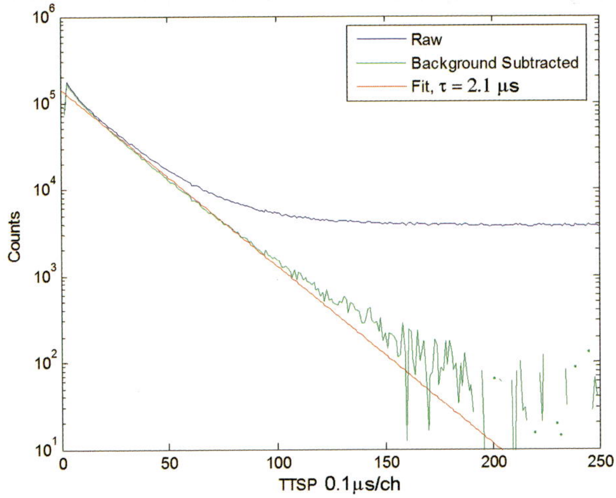

Fig. 23 NS fast neutron time-to-second-pulse (TTSP) distribution. Times beyond ~10 μs are mostly due to chance coincidences. Subtracting this background leaves an exponential distribution with a characteristic time of 2.1 μs, which is expected for a 5 weight percent loading of boron. The departure at very short times is a geometric effect of a finite-size detector. The absence of spurious features at specific times shows the effectiveness of the dual-threshold technique in eliminating the effects of afterpulsing (see Sect. 4.4.3)

The BC454 is enclosed by materials in all directions that strongly absorb thermal and epithermal neutrons. Energetic neutrons that can penetrate this shielding but have insufficient energy to produce a detectable prompt light pulse ($<\sim$500 keV) can still moderate and produce a capture pulse. These single-interaction events are recorded separately from the delayed-coincidence events and provide a measure of the neutron flux at intermediate energies; however, without coincidence, measurements in this energy region are subject to higher overall background.

6.2.3 Gamma-Ray Burst Mode

The NS GRB mode was configured for operation from April 23 to May 23, 2005. Bursts are detected as sudden jumps in the background of the BC454 detector. The nominal background count rate is ~75 counts s^{-1}. A 164-s moving average of the background rate is continuously computed in the flight software, and each new 1-s sample is compared against this value. For this test, the trigger threshold was set to approximately 4-σ above the background, so that one should not expect more than a few false alarms per day.

There was a significant solar flare that occurred on May 13, 2005, with an associated Coronal Mass Ejection (CME) that subsequently caused very high count rates in the detector due to energetic particles. No GRB events could reliably be detected during this time. However, during the quiescent periods, a total of 23 bursts were detected. Of these, 10 had times close to those reported by other sources (K.C. Hurley, private communication, 2005). Figures 24 and 25 show the intensity profiles for two representative bursts. From these figures, the NS appears capable of detecting both short- and long-duration bursts. For the longer-duration bursts (~30 s), the NS appears sufficiently sensitive to provide information on the structure of the burst profile. Confirmed bursts from multiple sources in the Interplanetary Network can also be used in reverse to verify the accuracy of the onboard spacecraft clock. A test of this technique showed the absolute timing accuracy of the MESSENGER precision clock to be within 200 ms.

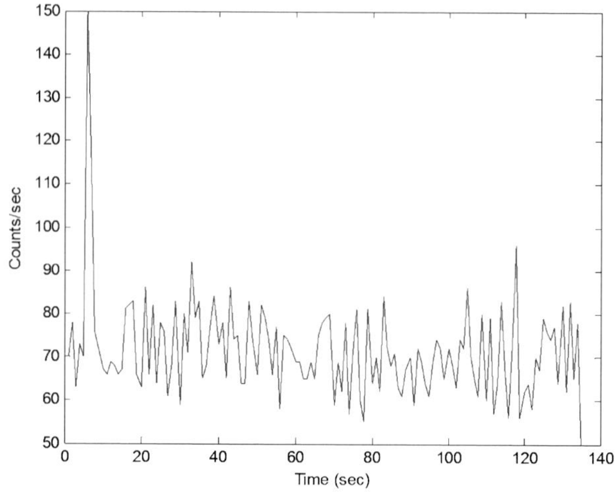

Fig. 24 Time profile of a short-duration gamma-ray burst detected by the NS. This 10-σ burst returned to background levels within seconds

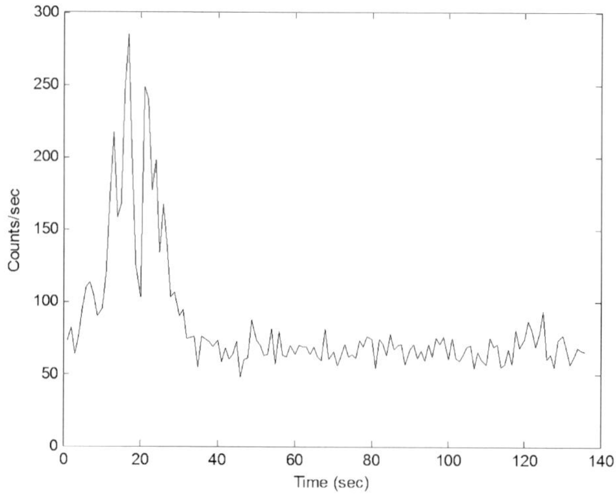

Fig. 25 Time profile of a long-duration gamma-ray burst detected by the NS. This large burst, measuring about 23-σ at its peak, continued for nearly 30 s. No two bursts are alike, and the NS appears to have sufficient sensitivity to provide information on the structure of the burst profile

7 Telemetry Data Products

The GRS and NS each produce a range of telemetry packets for science and engineering use. The raw science telemetry is collected by the MESSENGER Mission Operations Center and then converted to Experimental Data Record (EDR) format by the Science Operations Center. Status and housekeeping telemetry is also provided in EDR format. All GRNS EDR data, higher-level derived data products, and reduced instrument calibration data will be submitted to the Planetary Data System (PDS) for archiving and general use. Specialized engineering telemetry that is intended only for diagnostic use is typically not provided to the science community.

Data on gamma-ray bursts are derived from the NS rather than the GRS. Gamma-ray burst detection capability (primarily intended for astrophysics use) was implemented in the NS for several reasons: (1) The NS is likely to be on during a greater portion of the mission; GRS operation during cruise is minimal in order to conserve cryocooler lifetime and is curtailed by the occasional need for annealing during orbit, while the NS is a low-power instrument that can be operated even during long eclipses at Mercury. (2) The relatively large block of plastic scintillator in the NS is sufficiently sensitive to the gamma-rays of interest. (3) Solar neutron bursts (rarely observed in Earth orbit due to the short lifetime of a free neutron) are preceded by solar gamma-ray bursts and benefit in detection probability by having gamma-ray bursts detected in the same instrument.

7.1 GRS Data Products

There are seven GRS EDR data products: the Ge detector raw spectrum, the Ge detector anticoincidence spectrum, the shield spectrum, the software counters, the status packet, the FPGA ADC block, and the microphonics time series. Spectral bins contain 16-bit unsigned data, and scalar counts are 32 bits. The Ge detector raw spectrum has 16,384 bins (due to the 14-bit ADC). Each science packet can accommodate 1,024 bins, so the Ge detector raw spectrum is created from 16 science packets. Each science packet header contains the mission elapsed time (MET), accumulation time, start and stop bins, and application identifier (ApID). The Ge detector anticoincidence spectrum has the same format as the raw spectrum, but the data contain only those events not in coincidence with the shield events. The shield spectrum has 1,024 bins and fits in a single science packet. The cryocooler microphonics time series is of variable length up to 32 science packets. Normal collection consists of 32,768 raw ADC samples taken at 5,000 samples per second ($\sim$6.6 s duration). All science packets can be compressed by the lossless Fast algorithm. The software counter packet contains sums of various valid and invalid events, such as Ge detector raw and anticoincident events, shield events, charge resets, and pile-up rejections. The status packet contains all instrument temperatures, voltages, and currents, Ge detector leakage current, and other analog data, as well as digital and software status parameters. The FPGA ADC packet contains electronic event processing parameters, including trigger levels, settling times, and raw event data.

7.2 NS Data Products

Most of the data of geochemical interest are contained in the NS science data packet, which contains the time tag, accumulation period, neutron energy spectra, software counters, event mode data, and some instrument mode parameters. The energy spectra consist of 64 channels for each of the two Li-glass scintillators (thermal neutrons), 64 channels for the borated plastic non-time-correlated (single) events (epithermal neutrons), 256 channels for the borated plastic time-correlated events falling within the "early" time window (fast neutrons), and 256 channels for the borated plastic time-correlated events falling within the "late" time window (mostly chance coincidences). Spectral channels contain 16 bits. There are 18 32-bit counters that record raw events, charge resets, over-range and valid events for each scintillator, dead time, and interscintillator coincidences. The event mode data provide high-resolution samples of time-correlated events occurring in the borated plastic. The event mode data buffer is variable in length and can hold up to 256 events per accumulation period. The buffer is organized as three arrays: 16-bit prompt energies, 16-bit capture energies, and 8-bit TTSP values.

The NS status packet contains critical data on the instrument state, including command and alarm counts, voltages and temperatures, voltage settings, modes, and parameter settings for event processing. A special diagnostic data packet, present only when enabled, captures all event types (both Li-glass and borated plastic) and places the first 96 events of each accumulation period into a record that is organized in an event-by-event format. All energies have full ADC precision, and each event is tagged with a data quality value. This mode is useful when adjusting event processing parameters.

The NS GCR packet contains the specialized spectra used to track the gain of the Li-glass detectors and can be enabled or disabled independently of the NS science data packet. The GCR data accumulation interval is synchronized with the science collection interval. The GCR spectra consist of 64 channels for each of the two Li-glass detectors and use the same scaling parameters that apply to the data in the corresponding science spectra.

The gamma-ray burst and neutron short science data packets are of astrophysical interest and are present when the corresponding mode is activated. Gamma-ray bursts are detected as counts that exceed a settable threshold over background as determined by a running average of a counter, either the borated plastic hardware trigger counter or a "processed" counter consisting of all borated plastic counts in anticoincidence below a settable energy threshold (to exclude cosmic rays). The gamma-ray burst packet contains a time series of 164 1-s counts starting and ending at settable pre- and posttrigger times, along with the trigger time and other parameters. Neutron short science packets contain a subset of the normal neutron science packets (fast neutrons only) that allows determination of solar neutron bursts as a reduced data set, enabling detection of neutron bursts on much shorter time scales with little increase in data rate. Neutron short science data are further divided into two packet types that can be independently enabled: the neutron short science histograms, which contain spectra for the early and late time-correlated events; and a counters packet, which contains only the corresponding accumulated counts.

8 Operations Plans

The GRS and NS will have different operations plans because of different sensor and science needs, but common to both are the MESSENGER spacecraft solid-state recorders and downlink, which are limited resources whose use and allocation will be negotiated as part of the detailed science planning effort. The telemetry data rates for the GRS and NS are independently adjustable and may be raised or lowered by commanding shorter or longer integration periods. The shortest useful integration periods are about 20 s for the NS and about 300 s for the GRS, on the basis of count statistics, although the NS is capable of producing spectra every second and the GRS can produce full spectra about every 60 s. A short integration period for the GRS may be needed to avoid smearing of energy resolution due to temperature transients near periapsis. The MESSENGER orbit is highly eccentric, and less than 30 minutes out of each 12-hr orbit is spent at low altitude. Most of the time is spent at very high altitude, which allows the spacecraft a "cooling off" period, and where much longer integration periods can be used without penalty. The orbit-averaged bit rate for the GRNS is expected to be about 200 bps.

8.1 Gamma-Ray Spectrometer Operations Plan

The most important GRS instrument operation issue is the preservation of cryocooler lifetime to yield maximum GRS science data. Flight operation of the cooler outside of science operations will be restricted to that absolutely necessary for instrument testing, maintenance, and spacecraft background data collection, since the cruise phase is much longer than the orbital phase of the mission. The cooler manufacturer recommends that the cooler be run for a short period at least every six months, so this maintenance activity will be scheduled every six months to include a 30-minute run, but no such activity will be performed if the cooler is run for some other operation, such as a Mercury flyby, within a given six-month interval. The first full functional test has been completed, which established the baseline performance of the GRS and characterized the early-cruise spacecraft background. Further cruise science data collection periods will be restricted to one to three more tests of one-week duration or less coincident with the three scheduled flybys of Mercury. To guard against any possible loss of helium gas in the cooler, and also minimize any contamination of the Ge detector cryostat, the cooler and sensor will be kept relatively warm and at a relatively constant temperature by continual operation of the instrument warm-up heater during cruise (and possibly other warm-up operations involving spacecraft orientation). (Warmer temperatures will also promote low-level annealing of the Ge detector, protecting it against radiation damage, and helping to preserve energy resolution.)

In the orbital phase of the mission, the overall GRS strategy will be to maximize operational hours at low altitudes because the eccentric orbit results in only a small fraction of time below 1,000 km altitude where count statistics are good and surface elemental mapping is feasible. Since the GRS is not on a boom and has a fixed spacecraft mount, the spacecraft will generally be oriented as nearly nadir-looking as possible when the GRS is taking science data, except for a few occasions near passage over the north pole when the spacecraft may point the lower spacecraft deck toward the pole. The Ge detector cannot be operated during about seven Earth days of the long eclipse season because of power restrictions, a period sufficiently long to perform an annealing cycle. The long eclipse season occurs every 88 Earth days, and it is thought that annealing will not likely be required more often to maintain acceptable energy resolution. Except for the cooldown time, annealing can take place with no loss of science operation time. The annealing cycle would not be observable in near real time, but a sufficient number of annealing cycles have been performed with the flight Ge detector that this is no longer felt to be an issue. Should an increase in cooler microphonics, loss of heat lift, or other symptoms indicate a probable shortening of cooler lifetime, then the option of raising the Ge crystal operating temperature to lengthen cooler lifetime at the expense of increased sensitivity to energy resolution loss due to radiation damage will be considered.

Data volume will be managed by increasing the data collection interval progressively for altitudes above 1,000 km. It may be desirable to have data collection intervals as short as $\sim$60 s near periapsis in order to avoid smearing of energy resolution due to rapid temperature changes. Fortunately, the Fast algorithm compression ratio will increase dramatically for shorter integration intervals because there are many fewer counts in each spectral bin. Using this approach, the orbit-averaged data rate is expected to be approximately 150 bps.

8.2 Neutron Spectrometer Operations Plan

There are three phases of the mission for the NS. The first is during cruise after permanent turn-on (after the second Venus flyby on June 5, 2007). Once the instrument is properly configured, the only operating parameter is the data-accumulation time. The present plan is to maintain a constant accumulation period of 50 s, which will provide necessary information of time-variable backgrounds induced by solar variability. One part of this background is the response of the NS detector to energetic particles incident on both the detector and the spacecraft. A second part has to do with neutrons generated in the solar atmosphere, which can reach the spacecraft before they decay if they have sufficiently high energies. The energies of these neutrons can be measured using their transit time from the Sun given by the time difference between their arrival at the spacecraft and the time of occurrence of a flare-generated gamma-ray burst. A 50-s accumulation period should be sufficient to provide better than 10% energy resolution for all heliocentric locations between the orbits of Venus and Mercury. Full neutron science mode is preferred (all detector spectra), but if data rate becomes an issue, the neutron short science subset will suffice.

The next phases in the MESSENGER mission for the NS are the planetary flybys. Here an accumulation period of 20 s would be needed for about plus and minus six hours about the times of closest approach in order to define counting rates as a function of planetary distance. Short accumulation times also provide some insurance against possible occurrences of time-variable solar-particle events during the flybys.

The last phase is in Mercury orbit. Operation here should be broken into three zones that depend on time from periapsis. Within plus and minus 15 min of periapsis, the plan is to have an accumulation time of 20 s (or about 60-km surface track). For about plus and minus 75 min before and after the first time zone, the accumulation time will be increased to 30 s. And for the remaining part of the orbit, consisting of plus and minus 4.5 hours about apoapsis, the accumulation can be increased to 600 s. If a gamma-ray burst is detected while in the outermost zone, it is desirable to switch the accumulation time quickly to 40 s for a time period of about two hours. Using this approach, the NS maximizes spatial and temporal resolution while minimizing data volume.

Acknowledgements The authors thank all of the GRNS team members for their dedicated efforts in developing this instrument and making the Gamma-Ray and Neutron Spectrometer a success. We also wish to express our gratitude to the Technical Services Department of The Johns Hopkins University Applied Physics Laboratory for meeting a very demanding schedule, the MESSENGER Mission Operations team for assisting in the collection of many hours of in-flight data, and payload manager R.E. Gold and former MESSENGER program manager M. Peterson for their courage to allow us to move ahead with this advanced sensor. Many thanks also go to J. Boldt, G. Marcus, P. Wilson, J. Rossano, J. Connelly, R. Rumpf, W. Bradley, D. Landis, and E.H. Darlington.

A special acknowledgment goes to J. White who, sadly, will not be able to witness the end result of his superb craftsmanship. Jack, we couldn't have done it without you.

References

B.J. Anderson et al., Space Sci. Rev. (2007, this issue). doi:10.1007/s11214-007-9246-7
J.D. Anderson, G. Colombo, P.B. Esposito, L.E. Lau, G.B. Trager, Icarus **71**, 337–349 (1987)
G.B. Andrews et al., Space Sci. Rev. (2007, this issue). doi:10.1007/s11214-007-9272-5
M.J. Bielefeld, R.C. Reedy, A.E. Metzger, J.I. Trombka, J.A. Arnold, Proc. lunar sci. conf. 7th. Geochim. Cosmochim. Acta, Suppl. **7**, 2662–2676 (1976)
W.V. Boynton et al., Space Sci. Rev. **110**, 37–83 (2004)
W.V. Boynton et al., J. Geophys. Res. (2007a, in press)
W.V. Boynton et al., Space Sci. Rev. (2007b, this issue). doi:10.1007/s11214-007-9258-3
J. Brückner, J. Masarik, Planet. Space Sci. **45**, 39–48 (1997)
J. Brückner et al., IEEE Trans. Nucl. Sci. **38**, 209–217 (1991)
M. Burks et al., *Proc. IEEE Nucl. Sci. Symposium Conference Record*, vol. 1 (2004), pp. 390–394
B.J. Butler, J. Geophys. Res. **102**, 19283–19291 (1997)
B.J. Butler, D.O. Muhleman, M.A. Slade, J. Geophys. Res. **98**, 15003–15023 (1993)
R.C. Byrd, W.T. Urban, Technical Report LA-12833-MS, Los Alamos National Laboratory, Los Alamos, NM, 1994, 52 pp
A.G. Cameron, Icarus **64**, 285–294 (1985)
A.G.W. Cameron, W. Benz, B. Fegley Jr., W.L. Slattery, in *Mercury*, ed. by F. Vilas, C.R. Chapman, M.S. Matthews (University of Arizona Press, Tucson, 1988), pp. 692–708
B.A. Campbell, D.B. Campbell, J.F. Chandler, A.A. Hine, M.C. Nolan, P.J. Perillat, Nature **426**, 137–138 (2003)
D.B. Campbell, B.A. Campbell, L.M. Carter, J.-L. Margot, N.J.S. Stacy, Nature **443**, 835–837 (2006)
J.F. Cavanaugh et al., Space Sci. Rev. (2007, this issue). doi:10.1007/s11214-007-9273-4
D.M. Drake, W.C. Feldman, C. Hurlbut, Nucl. Instr. Meth. Phys. Res. **A247**, 576–582 (1986)
R.C. Elphic et al., J. Geophys. Res. **105**, 20333–20345 (2000)

L.G. Evans, S.W. Squyres, J. Geophys. Res. **92**, 9153–9167 (1987)
L.G. Evans, W.V. Boynton, R.C. Reedy, R.D. Starr, J.I. Trombka, in *Proceedings of SPIE 4784, X-Ray and Gamma-Ray Detectors and Applications IV*, ed. by R.B. James, L.A. Franks, A. Burger, E.M. Westbrook, R.D. Durst (SPIE, Seattle, 2002), pp. 31–44
L.G. Evans, R.C. Reedy, R.D. Starr, K.E. Kerry, W.V. Boynton, J. Geophys. Res. **111**, E03S04 (2006). doi:10.1029/2005JE002657
W.C. Feldman, D.M. Drake, Nucl. Instrum. Meth. Phys. Res. **A245**, 182–190 (1986)
W.C. Feldman, G.F. Auchampaugh, R.C. Byrd, Nucl. Instrum. Meth. Phys. Res. **A306**, 350–365 (1991)
W.C. Feldman, B.L. Barraclough, B.L. Hansen, A.L. Sprague, J. Geophys. Res. **102**, 25565–25574 (1997)
W.C. Feldman et al., Nucl. Instrum. Meth. Phys. Res. **A422**, 562–566 (1999)
W.C. Feldman et al., J. Geophys. Res. **107**, 5016 (2002a). doi:10.1029/2001JE001506
W.C. Feldman et al., J. Geophys. Res **107**, 1083 (2002b). doi:10.1029/2001JA000295
W.C. Feldman et al., J. Geophys. Res. **109**, E07S06 (2004). doi:10.1029/2003JE002207
O. Gasnault et al., Geophys. Res. Lett. **28**, 3797–3800 (2001)
K.A. Goettel, in *Mercury*, ed. by F. Vilas, C.R. Chapman, M.S. Matthews (University of Arizona Press, Tucson, 1988), pp. 613–621
J.K. Harmon, Adv. Space Res. **19**, 1487–1496 (1997)
J.K. Harmon, M.A. Slade, Science **258**, 640–643 (1992)
J.K. Harmon, M.A. Slade, R.A. Vélez, A. Crespo, M.J. Dryer, J.M. Johnson, Nature **369**, 213–215 (1994)
J.K. Harmon, P.J. Perillat, M.A. Slade, Icarus **149**, 1–15 (2001)
S.E. Hawkins, III et al., Space Sci. Rev. (2007, this issue). doi:10.1007/s11214-007-9266-3
A.P. Ingersoll, T. Svitek, B.C. Murray, Icarus **100**, 40–47 (1992)
D.A. Landis, C.P. Cork, N.W. Madden, F.S. Goulding, IEEE Trans. Nucl. Sci. **29**, 619–624 (1982)
D.J. Lawrence et al., J. Geophys. Res. **107**, 5130 (2002). doi:10.1029/2002JE001530
D.J. Lawrence, R.C. Elphic, W.C. Feldman, T.H. Prettyman, O. Gasnault, S. Maurice, J. Geophys. Res. **108**, 5102 (2003). doi:1010.1029/2003JE002050
D.J. Lawrence et al., J. Geophys. Res **111**, E08001 (2006). doi:10.1029/2005JE002637
J.C. Leary et al., Space Sci. Rev. (2007, this issue). doi:10.1007/s11214-007-9269-0
S. Maurice, D.J. Lawrence, W.C. Feldman, R.C. Elphic, O. Gasnault, J. Geophys. Res. **109**, E07S04 (2004). doi:10.1029/2003JE002208
J.V. McAdams, Astrodynamics 2003, Adv. Astronaut. Sci. **116** (part III), 643–662 (2004)
W.E. McClintock, M.R. Lankton, Space Sci. Rev. (2007, this issue). doi:10.1007/s11214-007-9264-5
J.I. Moses, K. Rawlins, K. Zahnle, L. Dones, Icarus **137**, 197–221 (1999)
D.A. Paige, S.E. Wood, A.R. Vasavada, Science **258**, 643–646 (1992)
T.H. Prettyman et al., J. Geophys. Res. **111**, E12007 (2006). doi:10.1029/2005JE002656
J.R. Salvail, F.P. Fanale, Icarus **111**, 441–455 (1994)
C.E. Schlemm II et al. (2007, this issue). doi:10.1007/s11214-007-9248-5
M.A. Slade, B.J. Butler, D.O. Muhleman, Science **258**, 635–640 (1992)
S.C. Solomon et al., Planet. Space Sci. **49**, 1445–1465 (2001)
A.L. Sprague, D.M. Hunten, K. Lodders, Icarus **118**, 211–215 (1995)
R.G. Strom, *Mercury: The Elusive Planet* (Smithsonian Institution Press, Washington, 1987), 197 pp
A.R. Vasavada, D.A. Paige, S.E. Wood, Icarus **141**, 179–193 (1999)
H. Wänke, T. Gold, Phil. Trans. R. Soc. Lond. **303**, 287–302 (1981)
G.W. Wetherill, in *Mercury*, ed. by F. Vilas, C.R. Chapman, M.S. Matthews (University of Arizona Press, Tucson, 1988), pp. 670–691

Space Sci Rev (2007) 131: 393–415
DOI 10.1007/s11214-007-9248-5

The X-Ray Spectrometer on the MESSENGER Spacecraft

Charles E. Schlemm II · Richard D. Starr · George C. Ho · Kathryn E. Bechtold · Sarah A. Hamilton · John D. Boldt · William V. Boynton · Walter Bradley · Martin E. Fraeman · Robert E. Gold · John O. Goldsten · John R. Hayes · Stephen E. Jaskulek · Egidio Rossano · Robert A. Rumpf · Edward D. Schaefer · Kim Strohbehn · Richard G. Shelton · Raymond E. Thompson · Jacob I. Trombka · Bruce D. Williams

Received: 22 May 2006 / Accepted: 19 July 2007 / Published online: 24 October 2007

Abstract NASA's MESSENGER (MErcury Surface, Space ENvironment, GEochemistry, and Ranging) mission will further the understanding of the formation of the planets by examining the least studied of the terrestrial planets, Mercury. During the one-year orbital phase (beginning in 2011) and three earlier flybys (2008 and 2009), the X-Ray Spectrometer (XRS) onboard the MESSENGER spacecraft will measure the surface elemental composition. XRS will measure the characteristic X-ray emissions induced on the surface of Mercury by the incident solar flux. The Kα lines for the elements Mg, Al, Si, S, Ca, Ti, and Fe will be detected. The 12° field-of-view of the instrument will allow a spatial resolution that ranges from 42 km at periapsis to 3200 km at apoapsis due to the spacecraft's highly elliptical orbit. XRS will provide elemental composition measurements covering the majority of Mercury's surface, as well as potential high-spatial-resolution measurements of features of interest. This paper summarizes XRS's science objectives, technical design, calibration, and mission observation strategy.

Keywords Mercury · MESSENGER · X-ray spectrometry · Surface composition · X-ray emissions · Elemental composition

C.E. Schlemm II (✉) · G.C. Ho · K.E. Bechtold · S.A. Hamilton · J.D. Boldt · W. Bradley · M.E. Fraeman · R.E. Gold · J.O. Goldsten · J.R. Hayes · S.E. Jaskulek · E. Rossano · R.A. Rumpf · E.D. Schaefer · K. Strohbehn · R.G. Shelton · R.E. Thompson · B.D. Williams
The Johns Hopkins University Applied Physics Laboratory, Laurel, MD, USA
e-mail: chuck.schlemm@jhuapl.edu

R.D. Starr
Department of Physics, The Catholic University of America, Washington, DC, USA

W.V. Boynton
Department of Planetary Science, Space Sciences Building, University of Arizona, Tucson, AZ, USA

J.I. Trombka
Goddard Space Flight Center, Code 691, Greenbelt, MD, USA

1 Introduction

Mercury has been difficult to study because of its close proximity to the Sun, but its chemical composition is important for the clues it offers to the formation of the planets (Solomon et al. 2001). MESSENGER (MErcury Surface, Space ENvironment, GEochemistry, and Ranging) is a NASA Discovery Program mission that will study Mercury's exosphere, magnetosphere, surface, and interior. One of the instruments in the MESSENGER payload is the X-Ray Spectrometer (XRS) that will measure the elemental composition of the surface.

Surface elements fluoresce in response primarily to the solar X-ray flux. These elemental X-ray fingerprints can be measured from orbit at Mercury because the planet lacks an appreciable atmosphere to cause attenuation. XRS will make these measurements during the one-Earth-year orbit mission at Mercury. These measurements will provide quantitative information on elemental composition, a key measurement in understanding the planetary formation process.

The XRS instrument owes much of its heritage to the X-ray/gamma-ray spectrometer (XGRS) instrument on the Near Earth Asteroid Rendezvous (NEAR) Shoemaker mission (Starr et al. 2000). Three planet-viewing gas proportional counters (GPCs), including balanced filters and energetic particle detection, are used to maximize the active detector area and resolve the abundances of the lighter elements from Mercury. A small solid-state detector is used as the solar monitor, which looks directly at the Sun through thin Be foils to limit the heat flow, and has a small aperture to reduce the potentially high rates during an intense solar flare. X-ray measurement will be taken over the energy region from 1 to 10 keV for a major portion of the planet surface, as well as for the Sun. The Kα lines for the elements Mg (1.254 keV), Al (1.487 keV), Si (1.740 keV), S (2.308 keV), Ca (3.691 keV), Ti (4.508 keV), and Fe (6.403 keV) will be detected in this energy range. Solar flares will be used to detect the concentrations of the heavier elements Fe, Ti, Ca, and S, while quiet-Sun periods will yield the abundances of Mg, Al, and Si. The XRS field-of-view (FOV) provides spatial resolutions ranging from 42 km at periapsis and 3200 km at apoapsis when counting statistics are not a limiting factor.

2 Science Background/Objectives and Requirements

Mercury is the innermost planet in our solar system. Its proximity to the Sun makes it difficult to study from Earth-based observatories, and the number of spacecraft to visit this second smallest planet in the solar system has been limited to just one, Mariner 10, which flew by Mercury twice in 1974 and once in 1975. Mariner 10 provided a wealth of new information about Mercury, yet much still remains unknown about Mercury's geologic history and the processes that led to its formation.

Mercury's chemical composition offers clues to how the planets formed. For example, several theories have been developed to explain the unusually high metal-to-silicate ratio in Mercury inferred from its high bulk density. In one hypothesis Mercury formed in the hottest regions near the Sun, where silicates were vaporized by solar radiation early in the Sun's evolution (Bullen 1952; Ringwood 1966), depleting FeO and volatiles while enriching refractory elements. An alternative hypothesis, due mostly to Wetherill (1988), is that a large impact, occurring after differentiation, ejected much of Mercury's silicate crust and mantle. By this scenario, FeO and volatiles would be less depleted than in the vaporization model, and refractories less enriched.

Important aspects of Mercury's geologic history can be understood by the identification and quantification of major elemental chemistry. With spatial resolution sufficient to discern principal geological units, such a study can distinguish material excavated and ejected by young impact craters from volcanic deposits or a possible veneer of cometary and meteoritic material.

These and other fundamental geologic questions are awaiting more and improved data to differentiate between competing hypotheses. Mercury plays an important role in comparative planetology, and the data returned by the MESSENGER mission will go a long way towards illuminating Mercury's history as well as that of Earth and the other terrestrial planets.

3 Measurement Objectives

The MESSENGER XRS will measure characteristic X-ray emissions induced at the surface of Mercury by the incident solar flux. The Kα lines for the elements Mg, Al, Si, S, Ca, Ti, and Fe will be detected. Because of the very fast drop-off slope of the incident solar spectrum at high energy, the fluorescence from the heavier elements (Fe, Ti, Ca, and S) will be detected only during solar flares. Fluorescence from Mg, Al, and Si will be best detected and analyzed even during quiescent solar conditions.

The resolution of the X-ray measurements will depend upon the distance to the planet surface, intensity and shape of the exciting solar spectrum, the elemental abundances, and the background rejection efficiency. The XRS will attain its best spatial resolution ($\sim$40 km) at periapsis, which occurs over Mercury's northern hemisphere. However, because of the highly elliptical orbit of the MESSENGER spacecraft, such resolution will be obtained only for $\sim$15 minutes out of every 12-hour orbit, or 180 hours during the one-year orbital phase of the mission. In the southern hemisphere, spatial resolution will be $\sim$3000 km.

3.1 Measurement Technique

The X-ray spectrum of a planetary surface measured from orbit is dominated by a combination of the fluorescence excited by incident solar X-rays and scattered solar X-rays. The sampling depth is dependent on energy but is always less than 100 μm. The most prominent fluorescent lines are the Kα lines (1–10 keV) from the major elements Mg (1.254 keV), Al (1.487 keV), Si (1.740 keV), S (2.308 keV), Ca (3.691 keV), Ti (4.508 keV), and Fe (6.403 keV). The strength of these emissions from a planetary surface is strongly dependent on the chemical composition of the surface as well as on the incident solar spectrum, but it is of sufficient intensity for Mercury to allow orbital measurement by detectors such as that on the MESSENGER spacecraft.

The coherently and incoherently scattered solar X-rays are one source of background signal. Astronomical X-ray sources, which could also be sources of background, are eliminated at Mercury, because the XRS is collimated to a 12° field of view and the planet completely fills this field of view even when the spacecraft is at apoapsis.

Because incident solar X-rays are the excitation source for X-rays generated from a planetary surface, knowledge of the solar spectrum is necessary for quantitative analyses. The solar flux from 1 to 10 keV is composed of a continuum and discrete lines, both of which vary with solar activity. This process is well understood, and theoretical models accurately predict the solar spectrum. The solar intensity decreases by three to four orders of magnitude over the range from 1 to 10 keV. Fluorescence lines as well as the scatter-induced background, therefore, have greater intensity at lower energies. As the level of solar activity increases, relatively more output occurs at higher energies, the slope of the spectrum becomes less steep, and the overall magnitude of the X-ray flux increases. This process is called hardening. The hardened X-ray solar illumination causes a greater amount of fluorescence from the higher-mass elements at the surface, allowing their relative abundances to be measured.

3.2 Planetary Sensors

Remote sensing of soft X-ray emission from the Moon (Adler et al. 1972) and the asteroid 433 Eros (Trombka et al. 2000; Nittler et al. 2001) by orbiting spacecraft has been conducted using gas-filled proportional counters. This type of sensor provides a large geometric factor that is currently unmatched by any newer X-ray fluorescence measurement technique (i.e., solid-state sensors). For this reason, the MESSENGER XRS planetary viewing sensor consists of three identical gas-filled proportional counters. Designed and built by Oxford Instruments Analytical Oy of Finland (formerly Metorex International Oy), the counters are an improved version of those flown on the NEAR mission (Starr et al. 2000). Gas-proportional counters measure individual X-ray photons that enter the gas-filled chamber through a thin beryllium window and produce an ion electron pair (by the photoelectric effect). The relatively heavy ion achieves very little average energy between collisions, but the free electron is easily accelerated by the electric field towards the single anode wire within the gas-filled chamber causing further ionization processes. Electrons liberated by this secondary ionization process will also undergo acceleration by the electric field and ultimately lead to an avalanche process. This multiplication process terminates when all free electrons have been collected at the anode, producing a signal pulse. Under normal operation, the number of secondary ionization events will be proportional to the number of primary ion pairs formed. Furthermore, the photoelectron energy is directly related to the X-ray energy, so that photon energies can be identified from the position of corresponding peaks in the measured pulse height spectrum. Because the proportionality is related to the applied electric field inside the counter, the voltage being applied in the counter has to be carefully monitored and held constant to within fairly tight tolerances (see Sect. 4.3.1). A more detailed discussion of gas proportional counters can be found in Knoll (1999).

The XRS planetary viewing sensor must be able to resolve all the low energy X-ray lines in the 1–10 keV energy range. However, the energy resolution of a gas proportional counter is not sufficient to resolve some of the low-energy X-ray lines from Mg (1.25 keV), Al (1.49 keV), and Si (1.74 keV). Hence, the XRS uses balanced filters to make the required measurement. This technique uses a thin Mg foil over one GPC entrance window and a thin Al filter over another; this combination exploits the small energy difference between the K-absorption edges in Mg and Al to separate the lower energy lines. The same technique was also employed in the NEAR XGRS instrument with success (Starr et al. 2000).

One of the disadvantages of using gas proportional counters is that they are susceptible to large background levels caused by penetrating ions within their large collection volume. Two different techniques are used on XRS to mitigate the expected high background. First, similar to the NEAR XGRS gas proportional counter, MESSENGER XRS counters use rise-time discrimination electronics to reduce the background induced by cosmic rays and gamma rays. The rise time of the electronic signals produced in a proportional counter due to particle and gamma-ray interaction is longer than that for X-rays. Rise-time discrimination can eliminate up to $\sim$80% of these background events. In addition to the rise-time discrimination circuitry, the MESSENGER XRS also has anti-coincidence wires around the periphery of the gas chamber (except by the entrance window) and parallel to the center anode. When a penetrating particle enters the proportional counter through any area that has anti-coincidence coverage, two signals will be generated by the event, one by the anti-coincident wires and another by the center anode. Events such as these will be eliminated by coincidence logic, a procedure that provides up to an additional ~80% reduction in background.

3.3 Solar Monitor

The Mercury X-ray emissions are related to the solar X-ray input, so in order to interpret the planetary data accurately, a solar X-ray monitor is required. Direct viewing of the Sun at 0.3 AU provides a substantial challenge for selecting an appropriate solar monitor. Ultimately, a small silicon-PIN (Si-PIN) solid-state detector was chosen because of its compactness and consequent relaxed thermal exclusion requirement. A similar Si-PIN detector was flown as a back-up solar monitor on the NEAR mission (Starr et al. 2000). However, this detector failed repeatedly during cruise due to charge build-up in the SiO_2 surface layer (Starr et al. 1999). The MESSENGER Si-PIN was designed with this failure mode in mind, and ground tests indicate that it is not sensitive to charge build-up.

The Si-PIN detector was placed on the spacecraft sunshade (Leary et al. 2007) to provide direct viewing of the Sun. Unlike the gas proportional counters, the PIN detector will experience a very high intensity of solar X-rays. Hence, it has a small (0.03 mm^2) aperture that views through 75-μm-thick beryllium to limit and harden the solar X-ray input to a manageable level for most solar flares. The 75-μm beryllium foil is actually three separate 25-μm foils staged to limit the solar thermal input to the detector.

One disadvantage of the Si-PIN type detector is that it is susceptible to radiation damage from the high-energy particle environment in space. Fortunately, such damage in the PIN detector chosen is largely reversible when the detector is heated to 100°C for 24 hours. Hence, the Si-PIN diode is mounted on a miniature two-stage thermoelectric cooler (TEC) that can operate in a heater mode to provide in-flight annealing.

4 Instrument Design

The XRS measures X-rays emitted from the planet's surface and from the Sun and accumulates these measurements into energy histograms. The energy histograms along with instrument status and housekeeping data are then telemetered to the payload Data Processing Unit (DPU) and from there to the spacecraft Main Processor (MP) for storage and transmission to the ground. The telemetry is used for science analysis and instrument health monitoring, calibration, and debugging.

There are three physical units to the XRS (Fig. 1). The planetary detectors and measurement circuitry comprise the Mercury X-ray Unit (MXU), which is located within the payload adapter ring along with the Mercury surface imaging instruments. The MXU is co-located with the power converter, the processor board, and the high-voltage power supplies that collectively constitute the Main Electronics for X-rays (MEX) unit. The power and communications for the XRS are through the MEX unit, which receives power from a cable to the Power Distribution Unit (PDU) and communications from a cable to the DPU. The MEX and the MXU are connected through five approximately 30-cm cables, three of which are high-voltage coaxial cables. The remaining two low-voltage cables are divided between power and signals. The third physical unit, which measures the solar X-rays, is called the Solar Assembly for X-rays (SAX) and is mounted on the spacecraft sunshade. It connects through two approximately 4-m-long cables to the MEX and MXU. The cable to the MXU is for power and signals, and the cable to the MEX is for annealing heater control.

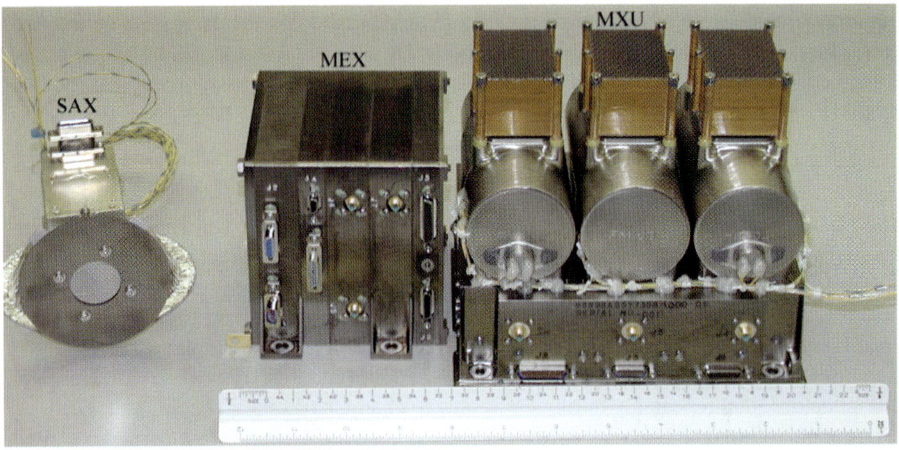

Fig. 1 The XRS flight instrument

Table 1 XRS characteristics

Measured elements	Mg, Al, Si, S, Ca, Ti, Fe	
Energy range	1 to 10 keV	
Integration period	20 s for flare; 450 s @ apoapsis	
Raw science data	10 kbits per integration period (instrument state + 4 spectra)	
Mass	3.4 kg (including inter-box cables)	
Power	6.85 W nominal, 11.7 W during annealing	
Monitors	Mercury	Solar
Detector	Gas Proportional Counter	Si-PIN, 500 μm thick
Active area	30 cm^2 (3 × 10 cm^2)	0.03 mm^2 (aperture)
Field of view	12° (hexagonal)	42° (±15° × ±15°)
Window	25 μm	25 μm + 2 × 25 μm = 75 μm
Filters	None, 4.5 μm Mg, or 6.3 μm Al	None

The photograph in Fig. 1 shows the Sun-facing section of the staged beryllium filter of the SAX unit, and the MEX and MXU arranged as they are on the spacecraft. The MXU electronics box is at the bottom right with the three GPC tubes above the electronics and the collimators over the entrance apertures of the GPCs.

The primary characteristics of the XRS instrument are listed in Table 1, including the science specifications, the 3.4-kg mass, the 6.85-W nominal power, the science telemetry requirements, and the detector specifics.

4.1 Block Diagram and Design Basics

The XRS functional block diagram is shown in Fig. 2. The X-rays are detected, and the signals are shaped, measured, validated, and accumulated into energy histograms.

The GPC detectors are provided with a 12° (full-angle) hexagonal field of view using beryllium-copper collimators. The material was chosen to minimize mass and X-ray conta-

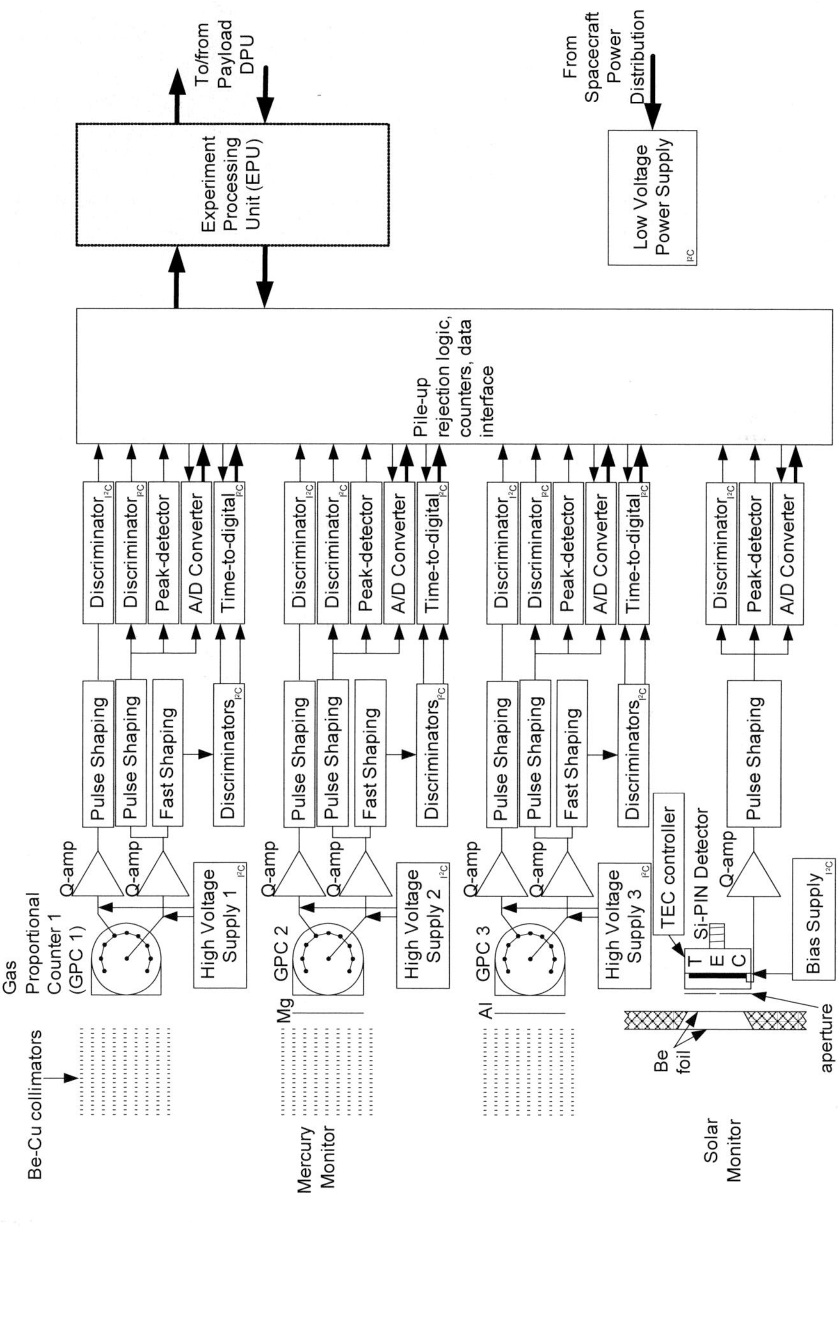

Fig. 2 Block diagram of the XRS instrument

mination generated from the collimator surfaces. Additionally, two of the three GPCs have a thin magnesium or aluminum filter, which allows the low-energy X-ray lines to be separated.

The solar monitor Si-PIN detector is located behind three 25-μm-thick layers of beryllium foil, which harden the solar spectra being measured and restrict the heat into the detector. The solar monitor is passively cooled to below -20°C using a radiator pointed nominally to deep space. The field of view is a 42° cone, to keep the Sun on the detector for all allowed spacecraft pointing angles.

Each GPC has its own dedicated high-voltage power supply. This arrangement allows the gas gains of each GPC to be adjusted individually by command and limits a supply failure to the loss of data from a single GPC, providing graceful degradation. The solar monitor detector also has its own dedicated bias voltage supply, which is a variant of the same supply used by the GPCs.

The GPC Q-amps are charge-sensitive amplifiers that feed into semi-Gaussian pulse-shaping circuitry to maximize the signal-to-noise ratio and prepare the signal for conversion to a digital value. The discriminators have programmable references and produce a logic-level signal when the analog signal crosses the reference point. The peak detector determines the time for the analog-to-digital converter (ADC) to sample the signal amplitude by differentiating the shaped signal and triggering at the zero-crossing. The time-to-digital converter (TDC) counts the time between logic-level start and stop signals associated with the GPC energy signal rise-time.

Programmable logic is used to screen incoming events to remove significant amounts of background from the X-ray events. The logic includes the commandable ability to accept only those events without a simultaneous veto signal (since most background events cause a veto), or events that have a valid rise-time measurement (the rise time of some background events is sufficiently slow that the fast shaping attenuates the signal, preventing a rise-time measurement from being possible), or events without rise time or energy pileup. All the events from a sensor can be turned off, which is useful for debugging or to disable sensor circuitry that fails in a noisy state. The events from the three GPCs are placed into a single three-event-deep, first-in first-out (FIFO) buffer, which permits new events to be accumulated while the data system is reading and processing previous events. The solar monitor has a dedicated FIFO that is 10 events deep, due to the much higher maximum event rate. The programmable logic also contains a number of counters for each sensor, to measure the number of events at several stages of processing, an arrangement that helps to determine the rejection efficiency and total X-ray event rates.

The TEC controller provides a commandable-level, constant-current source to power the TEC integrated to the solar monitor Si-PIN detector. This current source provides up to 600 mA and is relay-switched to allow cooling during ground testing as well as heating during flight to anneal radiation damage to the detector.

The Experiment Processing Unit (EPU) is the data system used by the XRS. It contains the flight code, the memory, the processor, and the telemetry interface to the rest of the instrument as well as to the spacecraft. The EPU is a common design to most of the MESSENGER instruments.

The Low-Voltage Power Supply (LVPS) takes the spacecraft primary power and produces the $+5$, -5, $+12$, and -12 V supply voltages used by the rest of the instrument. It also provides switched primary power for the SAX heater. The LVPS is also a common design for most of the MESSENGER instruments.

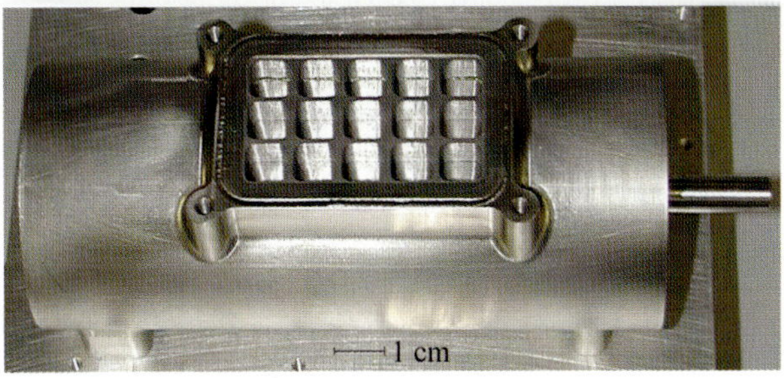

Fig. 3 Engineering model of the GPC detector

4.2 Detectors

The X-ray detectors were chosen on the basis of active area, operating temperature, and mass. A large active area and a wide operating temperature range were needed for the Mercury sensors, and a small active area and low mass were needed for the solar sensor.

4.2.1 Planetary GPC Detectors

GPCs are used to achieve the 30-cm^2 active area needed to accumulate the relatively low X-ray fluorescence emissions from the planet surface. Figure 3 shows the GPC engineering model detector, which is identical to the flight unit. Each GPC is cylindrical in shape with an approximate overall diameter of 60 mm and a length of 150 mm. The collimator (not shown in the figure) is attached directly over the X-ray entrance aperture, which is a 25-μm-thick Be foil bonded to a Be support matrix. The collimator defines the field-of-view of the instrument. Each GPC is filled with a standard mixture of 90% Ar and 10% CH_4 (P10) and pressurized to 0.15 MPa (1.5 bar). The units were designed and qualified to operate from −30°C to 90°C. During normal operation the center anode is biased at nearly 1.5 kV, which translates into a gas gain of about 1500. GPCs rely on gas multiplication to produce a measurable signal that is proportional to the energy of the incident X-ray photon. We employed 1.5-kV high-voltage power supplies with excellent stability over temperature and space environments to provide the multiplication potential (see Sect. 4.3.1).

The GPCs are enhanced with 14 veto anode wires to detect the passage of highly energetic charged particles so that this source of background contamination can be reduced from the spectra. The veto wires enclose the majority of the circumference of each GPC except for the collimator and both ends. These veto wires are biased down slightly from the potential on the center anode and powered from the same high-voltage supply.

Mass reduction was another improvement over previous missions. Electron-beam-welded titanium was used instead of stainless steel, reducing the mass from about 1 kg to 0.3 kg per detector.

Field enhancement structures, associated with the boron nitride insulators at the ends of the tubes holding the anode wires, are also new to this design and eliminated charging effects experienced with the NEAR detectors (Floyd et al. 1999; Starr et al. 2000) and improved gain uniformity over the entrance aperture. This change allowed the end insulators to be located closer to the aperture window edge, reducing the overall length of the tube, which in turn reduced the collection volume where the background is accumulated.

4.2.2 Solar Silicon-PIN Diode Detector

The requirements for the XRS solar monitor were challenging because the detector must make measurements of the solar X-ray spectra over a large dynamic range in photon rate

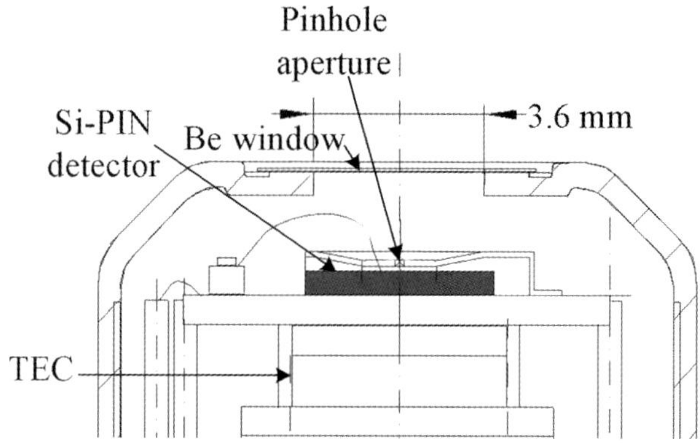

Fig. 4 Cutaway drawing of the solar monitor detector

while at the same time withstand the harsh thermal environment looking at the Sun through the spacecraft sunshade. A Si-PIN detector with a 200-μm-diameter pinhole aperture is used for the solar monitor (Fig. 4) because of its small (3 mm^2) detector active area, low mass, and ability to cool passively. The detector also has an integral TEC element that allowed cooling during ground testing and heating in flight to anneal much of the accumulated radiation damage. This detector showed the least amount of damage from proton bombardment of the candidates tested and recovered substantially after a 24-hour annealing at 100°C.

The detector was configured with resistor charge replenishment. This arrangement simplified the electronics design over pulsed replenishment and reduced the heat produced near the detector, allowing for passive cooling. The resolution is comparable to that of the GPCs at the Fe Kα line energy. The detector is biased by 110 V provided by a dedicated power supply.

4.3 Electronics

Components of the XRS electronics are located in each of the three physical units. The MEX contains the LVPS, the EPU, two High-Voltage Power Supply (HVPS) boards, and an interface board to connect to the MXU. The MXU contains the GPC preamplifiers, the shaping channels for the GPCs and solar monitor, the preamplifier power switches, the analog-to-digital converters, the event selection logic, and the TEC control. The MXU also provides the X-ray event data to the EPU and interfaces to the SAX unit. The SAX houses the detector, preamplifier, some signal shaping, and temperature measurement circuitry.

4.3.1 Main Electronics for X-Rays (MEX)

The LVPS, on one end of the MEX stack, operates from the spacecraft primary power and produces the +5, −5, +12, −12 V, and switched primary voltage used by the XRS electronics. Voltages and signals are passed from board to board in the stack by a sequence of stacking connectors. The EPU uses +5 V, along with the logic circuitry in the MXU. The preamps and shapers use +5 and −5 V, and the HVPSs use the +12-V supply and a small amount from the −12-V supply. The switched primary power is used to preheat the SAX unit during the detector annealing process.

The next in the MEX stack is the EPU, the single-board instrument computer that interfaces the XRS sensor electronics to the two redundant payload Data Processing Units

(DPUs). The EPU provides the telemetry interface through an RS422 signal interface to the two DPUs. Commands in, telemetry out, and synchronization signals are communicated across the three signal groups. The communication to the XRS sensor is through a parallel interface and an I^2C (Inter-Integrated Circuit) serial interface.

Next are the two HVPS assemblies. The HVPSs provide the acceleration potentials to the three GPCs up to 1.5 kV DC and the bias voltage to the solar monitor up to 150 V DC. There is one independent supply for each detector, allowing each GPC voltage to be separately controlled, and providing graceful degradation in the event of a supply failure. The precision and temperature stability of the output voltage are stringent requirements for XRS, because the GPC gain changes nearly 0.5% for each 1-V change of the supply. The output voltage is set using a 12-bit digital-to-analog converter, resulting in a 366-mV step size. The thermal stability was addressed by a 10 ppm/°C specification on the output voltage, which amounts to about 0.5 V over the expected temperature range at operating voltages. Two HVPSs are arranged on a single MEX circuit board (slice). This supply design is also common to the MESSENGER Gamma-Ray and Neutron Spectrometer instrument (Goldsten et al. 2007).

The last board in the stack is the Interface board, which is simply a connector interface between the MEX and the MXU. Supply voltages from the LVPS and digital signals from the EPU pass through this board to two connectors that connect to the MXU through cabling.

4.3.2 Mercury X-Ray Unit (MXU)

The MXU electronics consist of a multi-purpose logic board, a GPC shaper board, and three GPC preamplifier boards. All five boards have their own electromagnetic interference (EMI) enclosure and make use of chassis ground to minimize signal noise.

The preamplifier boards contain high-voltage dividers and filters and charge amplifiers for the center anode and the veto anode; there is one board for each GPC. There is a separate EMI enclosure for each preamp board to eliminate crosstalk between detectors and noise from other boards in the MXU. The anode connections from the GPC tube penetrate the EMI enclosure through two holes in the MXU chassis and are connected to the preamplifier through a 4-cm lead with a pin-and-socket connection to aid assembly. This design eliminated the need for the signal-attenuating high-voltage coaxial connection that had been used in previous instruments. The single-inline package (SIP) version of the AMPTEK A250-charge-sensitive amplifier, containing an internal junction field-effect transistor, was used to minimize the board size needed. The center anode amplifier and the veto amplifier were separated as much as possible on the board to minimize further the chance of crosstalk. A 250-μs charge-restoration time constant was used. The signals were then passed to the GPC shaper board within the MXU.

The GPC shaper board provides the signal shaping for both GPC anode signals, as well as the signal-level discriminator for the veto anode and the individual power supply filtering for all three GPC detectors. Approximately 1-μs, bipolar pulse shaping was used for the slower pulse shaping. Bipolar shaping was used, despite the increased noise, to reduce the telephonic problem experienced when the NEAR sensor was initially integrated near the spacecraft reaction wheels. A GPC anode produces a signal during mechanical vibration from the resulting capacitance changes between the high-voltage anode wire and the surrounding ground-potential inside the detector. Although the MESSENGER reaction wheels are distant from the XRS MXU, there are articulated sensors on the payload deck near the MXU that could be a source of mechanical noise during the mission, and the ground test environment frequently included cryogenic pump noise. The shaped veto signal is then threshold discriminated on the basis of a commanded level. The fast shaping channel uses approximately 100-ns shaping designed to emphasize the rise time of the anode signal. The circuitry for all three GPC detectors is on the same board, but the individual signal pathways are kept as separate as possible to reduce crosstalk. The primary side of the board was filled with ground plane and contained no sensitive signal traces to act as an integral EMI shield to the adjacent and electrically noisy ADC logic board. The logic board is where the shaped center anode signal, the fast-shaped center anode signal, and the logic-level veto signal are all passed.

The logic board takes up nearly the entire MXU footprint and provides the interface between the sensors and the data system. The board handles the analog-to-digital conversion and the independent event validity determination for all four sensors. It also performs the fast-shaped signal discrimination and time-to-digital conversion, the preamplifier power switching, the solar monitor TEC control power, many of the housekeeping analog measurements, and communication to and from the XRS EPU data system. The final low-pass filter pole for the fast-shaped signal is also located on the ADC logic board to help control the noise at the input of the fast discriminators. The rise-time measurement is made by measuring the time between the positive-going start-threshold crossing and the negative-going stop-threshold crossing. The thresholds can be changed by command and are adjusted to compensate for pulse-height variations between the positive and negative regions of the bipolar pulse. An Application Specific Integrated Circuit (ASIC), designed at The Johns Hopkins University Applied Physics Laboratory (APL), was employed to make the 11-bit rise-time measurement. A 14-bit ADC was used to make the X-ray energy measurement, with the conversion triggered by a peak detector, after it passes through the Field Programmable Gate Array (FPGA) that was used to provide all the logic circuitry on the board. The FPGA included independent event-logic determination (indicating the signals needed for a valid X-ray measurement) for each sensor, event buffering in the FIFO registers, rate accumulation in counters, and various logic-level signals. Another APL ASIC, the Temperature Remote Input/Output (TRIO), was used to measure voltages, currents, and temperatures, and communicated to the EPU over the I^2C bus. There is also a TEC control circuit, which provides current (taken from the -5 V supply, to help balance the load to the ± 5-V supply) to the TEC attached to the solar monitor detector, to provide heating for radiation annealing or cooling for ground testing. Two non-latching mechanical relays were used, with one to determine the current flow, and therefore heating or cooling, and the other to supply power when on, or ground when off, to reduce noise to the detector.

4.3.3 Solar Assembly for X-Rays (SAX)

The SAX electronics consist of the silicon-PIN X-ray detector, the preamplifier and shaping circuitry, bias voltage filtering, and temperature measurements. Two small boards are used: one containing the detector and charge preamplifier and the other the shaper and temperature measurement circuitry. Power dissipation needs to be minimized because the SAX box is cooled using a small radiator. The SAX box is electrically isolated from the sunshade support structure, so the box is grounded through the shielded cabling to the single-point ground in the MEX.

The preamplifier uses resistive feedback, and although the energy resolution is poorer than pulsed reset, the circuitry is much simpler and lower in power. The detector mounts to the outside of the box, is in good thermal contact with the box wall through the integral copper mounting stud (so the TEC can be used), and is socketed onto the preamplifier board. The preamplifier board is attached at a right angle to the shaper board using strain-relieved electrical connections. Both boards are hard-mounted to the SAX chassis.

The shaper provides a differential output to the electronics in the MXU for improved noise immunity, because the detector signal traverses nearly 4 m across the spacecraft to get to the MXU for the final shaping stages and analog-to-digital conversion. The detector and electronics box temperatures are also measured and sent to the MXU for analog-to-digital conversion. The detector temperature circuit employs a dual slope scheme to maintain better than 1° resolution both in the extended operational range of −55°C to 30°C and at the annealing temperature of 100°C.

4.4 Software Description

The XRS EPU shares a significant amount of the software across the payload as well as a common hardware design. XRS flight software uses the FORTH interpretive programming language. The interpretive characteristics of the language permitted efficient debugging and code verification. The flight code is comprised of the software common for the science instruments, called Common Code, and the custom XRS software.

The Common Code includes the software for the telemetry interface with the DPU, the alarm monitoring and reaction service, common commands and command handling service, and a macro command service. The Common Software section of the Mercury Dual Imaging System (MDIS) paper (Hawkins et al. 2007) describes the Common Code in greater detail.

The custom XRS software handles the XRS-specific commands, the data collection and accumulation, and telemetry formatting. The XRS-specific commands control data collection intervals, high-voltage power supply limits and settings, preamplifier power, event logic, threshold settings, and telemetry modes.

4.4.1 Data Collection

The event data collection process starts with either an X-ray or background radiation event triggering the electronics. The event is converted to digital words, placed in a hardware FIFO, and a signal set for the processor to read. The processor then continuously checks this signal until the signal is asserted. When the event is pulled from the event queue and processed, it is added to the appropriate histogram counter. For those events detected by the solar monitor, the event is simply added to the accumulating solar histogram. However, for the GPCs, the event energy and rise-time are first analyzed and, if found valid and indicating an X-ray, are added to the accumulating histogram.

There are five main software processes running, with the highest priority being the telemetry process, which runs once per second and is initiated by the Common Code software. The second priority is the I^2C process, which also runs once per second and handles housekeeping data collection. The third priority is the 1-Hz process, and the fourth priority is the command process, which is triggered by the receipt of a command. The lowest priority is the background process, which continuously polls the sensor for event data when no other processes are running.

The rates from the sensors are read once per second, and the housekeeping analog values, such as voltages, currents, and temperatures, are measured once per second as well. These values are used for instrument mode and alarm determination and are made available for use by the telemetry process.

4.4.2 Telemetry Formatting

The routine XRS telemetry takes two forms: the instrument Status packet and the Science Record. Both use commandable reporting periods so the telemetry can be tailored to fit the available instrument telemetry volume and desired temporal and spatial resolution. The Status packet is smaller (132 bytes) and reports only the instrument health, including analog measurements for trending, and event rates over a 1-s interval.

The Science Record contains a snapshot of the data contained in the Status packet, as well as additional instrument state information for interpreting the included science data. The science data consist primarily of the four energy histograms and the event counts summed during the accumulation period. The energy histograms are separately fast-compressed to minimize the telemetry volume (nominally about 1216 bytes). The Science Record covers the events and rates accumulated over a commanded accumulation period and reported every commanded report-period interval. Typically these intervals are the same, but shortening the accumulation period could allow high-spatial resolution measurement with reduced telemetry production.

4.5 Operational Modes

The XRS has two nominal operational modes: normal (quiet Sun) mode and flare mode. The difference between the two modes is mainly the data accumulation period. During the flare mode, the accumulation times are shortened to a pre-commanded level, typically 20 s, and an additional flare-mode solar stability data array is added to the science data. The added array consists of the solar monitor counting rate for each of 10 sections of the science record accumulation period and provides a measure of the solar input stability during the accumulation period.

The counting rate of the solar monitor determines whether flare mode can be entered and is based on a commandable threshold. Flare mode produces a higher rate of telemetry and therefore is limited in duration and in frequency by uploadable parameters. The default setting is one 60-min or shorter flare allowed per 24-hour period.

An additional special operational mode is thermal annealing for the solar monitor. The solar detector PIN is subject to particle radiation damage, which increases the noise of the detector and can be partially repaired by annealing at an elevated temperature. The XRS PIN detector is heated to 100°C for about 24 hours at a time, as needed, to reduce the effects of radiation damage.

4.6 Mechanical Design

The XRS consists of three physical units (see Fig. 1), two of which are mounted to the payload deck and the third mounted to the spacecraft sunshade frame. The MXU and MEX are attached to the payload deck using slipwashers and bolts, and the SAX is attached to the sunshade tubing by a mounting bracket.

The MEX was constructed using the "slice" technology, where 10 cm × 10 cm boards are housed individually and stacked to make an assembly. Through-bolts are used to hold the stack together, and connections between boards are made through stacking connectors that continue throughout the stack. The five-slice MEX is approximately 10 cm in length as well.

The MXU consists of a chassis housing the electronics boards, with the GPC tubes mounted on top. The GPC preamplifier boards are mounted within enclosed cavities in the chassis to reduce EMI and crosstalk.

The SAX is simply the foil assembly mounted to a small box containing the detector board and preamplifier board. The two staged beryllium foils are contained by the titanium foil assembly to harden the solar X-ray spectrum and limit heat into the detector. The foil assembly emerges from an opening in the sunshade but is not mechanically connected to the sunshade itself. The SAX assembly mounts to a bracket attached to the sunshade support tubing.

4.7 Thermal Design

Each of the three physical units of the XRS uses a different method for thermal control. The MEX has no apertures and is thermally mounted to the payload deck, so no heaters are required and only blankets are needed to reduce the heat leak, because the payload deck is maintained within a reasonable range for the electronics.

The base of the MXU is thermally mounted to the payload deck and the entire unit thermally blanketed except for the collimators. The planet-viewing GPCs have thermostatically controlled heaters because the apertures present a radiative heat leak. The collimators are thermally isolated from the rest of the MXU unit since they are mounted to the GPC tubes, which are made of titanium, a relatively poor thermal conductor. The collimators are exposed outside the thermal blankets so they can radiate the planetary heat load acquired during periapsis passage through the remainder of the orbit.

The SAX is mounted to the spacecraft sunshade support structure using thermal insulating standoffs, but the sunshade structure can reach temperatures as high as 100°C, so thermal blanketing and a radiator are needed to keep the temperature below −10°C in order to minimize detector noise. Staged beryllium foils are also used to reduce the heat from the Sun into the solar monitor Si-PIN detector. Thermostatically controlled heaters are needed to maintain the SAX above −55°C during portions of cruise and during eclipses.

5 Performance Testing, Analysis, and Calibration

To meet the mission's scientific goals, the XRS must be able to process a significant subset of the X-ray events detected, resolve the X-ray energy, maintain a stability of energy measurement over temperature and time, and reject most of the background cosmic events.

A special telemetry item was added for XRS to provide raw event data for diagnostic purposes. The energy measurement, the rise-time measurement, and the trigger status are provided for a selected maximum (0 to 255 GPC and 0 to 255 solar monitor) number of individual events, which are reported once per second. The telemetry volume produced by this structure is fairly large, so this item was enabled for ground testing and calibration and is being used sparingly in flight to tailor the rise-time rejection efficiency.

Small radioactive sources were used for most of the performance and ground calibration operations. An activity-calibrated ^{55}Fe source was used for much of the testing. Sources were also employed to illuminate target elements that we expect to see at Mercury. A gamma-ray source was employed over a multi-day period to verify that there were no insulator-charging gain effects (see Sect. 4.2.1) in the GPC.

5.1 Event Rate

The GPCs are expected to experience approximately 200 background events per second and only a few foreground X-ray events per second during orbital operation at Mercury, so XRS was designed to process greater than 1000 events per second from each of the three GPCs. The solar monitor is expected to produce only a few counts per second during normal Sun conditions but up to 50,000 events per second during a solar flare, so the electronics were

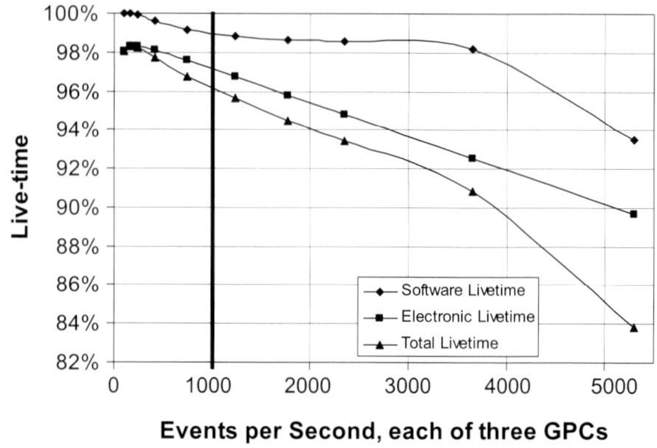

Fig. 5 GPC rate performance, expressed as fractional live-time versus X-ray flux

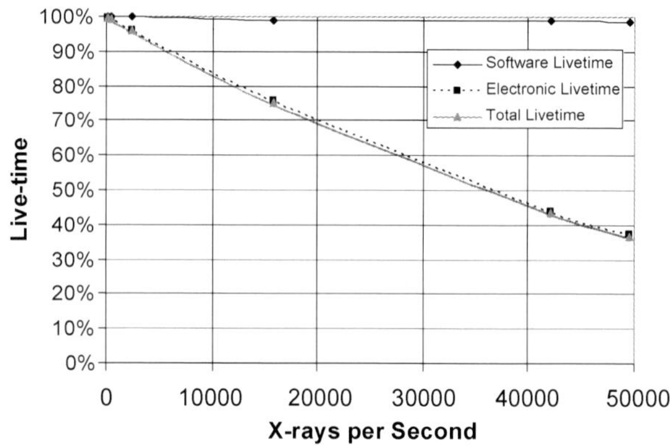

Fig. 6 Solar monitor rate performance, expressed as fractional live-time versus X-ray flux

designed to balance energy resolution and total event throughput. FIFO event buffers, three-deep for the GPCs and 10-deep for the solar monitor, were provided to minimize event loss due to the random nature of the X-ray events and the periodic nature of the data processing.

The goal for the event rate performance was to provide a 90% live time for GPC rates up to 1000 per second per detector and 50% live time for solar monitor Si-PIN rates up to 50,000 per second. Figure 5 shows the rate performance of the GPCs is better than 95% at a total rate of 1000 events per second per detector, and Fig. 6 shows that the rate performance of the solar monitor is only 36% at 50,000 events per second and is 50% at about 38,000 events per second. This performance was a result of the shaping and recovery time for the solar monitor signal but will still provide reasonable statistics.

5.2 Energy Resolution

The GPC energy resolution at full-width half maximum (FWHM) was specified to be 15% or less at the ^{55}Fe 5.90-keV emission, and the solar monitor Si-PIN detector system was required to display a resolution equivalent to or better than the GPC resolution. Figure 7 shows the measured spectrum of the GPC. The GPC resolution of the ^{55}Fe peak, without a chromium filter to suppress the Kβ line, is $880/5900 = 14.9\%$ at 5.90 keV, which meets the specification. Figure 8 shows the measured spectrum for the solar monitor, again without the chromium filter. The solar monitor energy resolution is $598/5900 = 10.1\%$ at 5.90 keV, which is better than the GPC resolution.

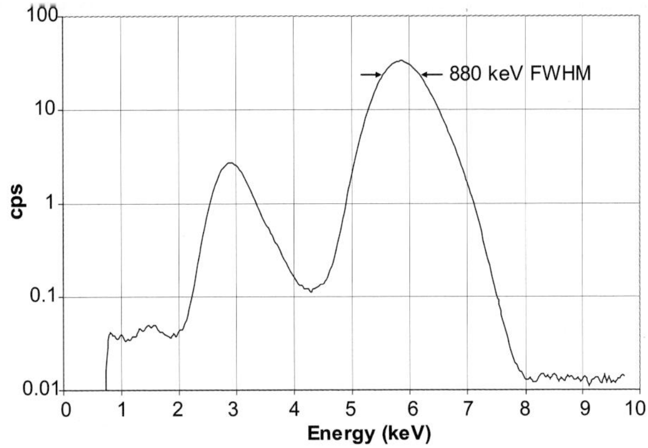

Fig. 7 Resolution of ^{55}Fe measured by the GPC

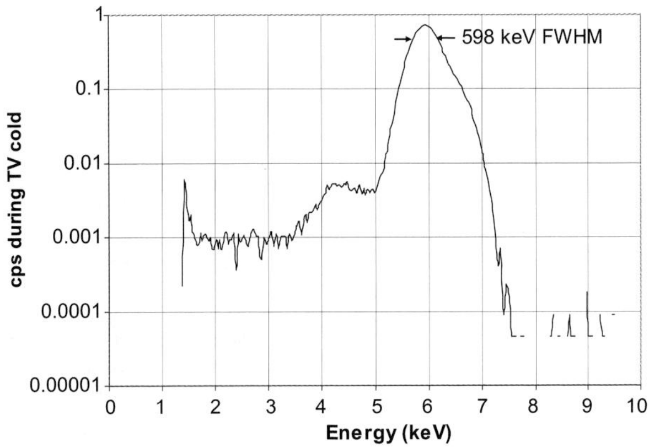

Fig. 8 Resolution of ^{55}Fe measured by the solar monitor during thermal vacuum (TV) cold testing

5.3 GPC Gain Stability

The sensor gain needs to be stable (to a fraction of a channel) over the measurement time. Temperature variations are the largest contributor to gain changes, so the XRS system needed to limit the thermal effects to the measurement system.

The GPCs are sensitive to internal pressure differences, but the pressure change due to the temperature range was small. Also, the GPCs were designed, constructed, and tested to have a leak rate small enough not to significantly affect the gain over the mission lifetime.

A variation of 1 V in the supply voltage to the GPCs causes about a 0.5% change in the gas gain. Since there are 256 measurement channels, the gain change needs to be less than 0.4%. To minimize this effect, the GPC high-voltage power supplies were designed to be stable to a fraction of a volt, at up to 1.5-kV supply voltage, over the expected temperature range and mission lifetime (see Sect. 4.3.1).

The analog electronics uses low-temperature-coefficient components to minimize the thermal effects. This was needed for the solar monitor electronics as well as the GPC electronics because neither have in-flight calibrators.

The XRS instrument was subjected to thermal vacuum testing as part of the flight qualification process, which included cycling between the flight temperature extremes. ^{55}Fe X-ray sources were included in the test, and the GPC gain was confirmed to be within one measurement channel.

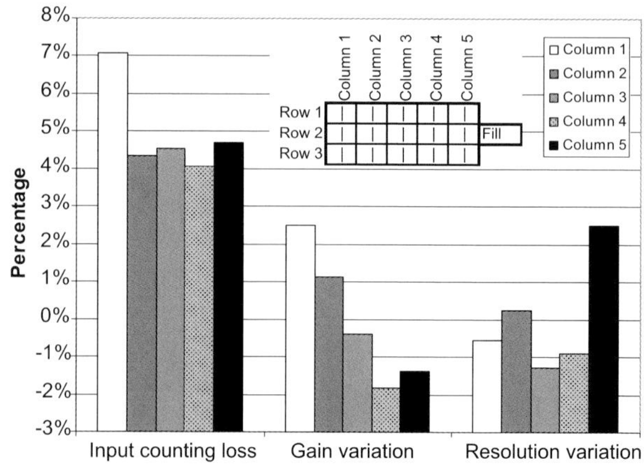

Fig. 9 Variations in response across aperture columns for one of the flight GPCs

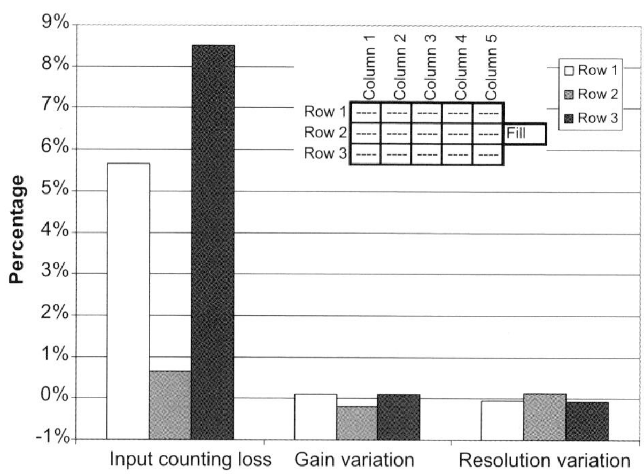

Fig. 10 Variations in response across GPC aperture rows

5.4 GPC Uniformity over the Window

Each XRS GPC has a 10-cm^2 total aperture opening that is divided into 15 sub-windows arranged in a 3 by 5 matrix (Fig. 3). Due to the internal geometries of the GPC detector, the field strengths tend to vary slightly over the X-ray window. On NEAR, the gain at the ends was affected by charging the insulators supporting the anode wire (Floyd et al. 1999; Starr et al. 2000), so the design was changed for XRS with structures designed to make the field more uniform at the ends as well as eliminate the charging effect due to high-energy particles. This goal was successfully achieved (see Fig. 9, where the variations at the ends are only a few percent). Additionally, no gain changes were observed during extended exposure to a gamma source.

However, when the veto anode voltage was raised to increase the veto efficiency, it was discovered that the X-ray efficiency at the sides was reduced. This was because the field from the veto wires extended into the region near the entrance aperture, but the effect could have been reduced or eliminated by iterating the GPC design if the science were at risk. Figure 10 illustrates this effect (see the "Input counting loss" bars).

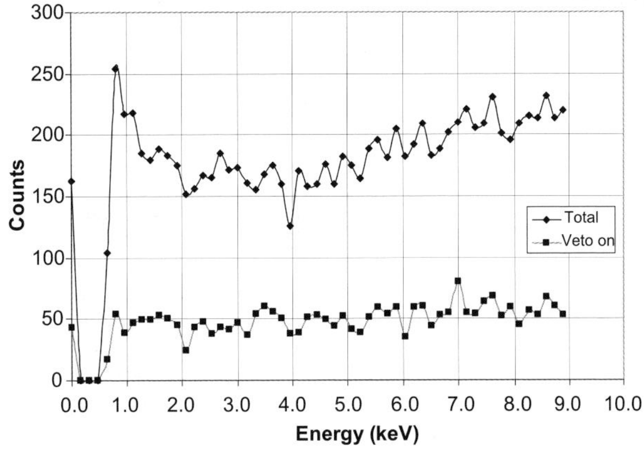

Fig. 11 GPC background rejection

5.5 GPC Background Rejection

GPC background rejection is a critical element of the MESSENGER mission because the fraction of fluorescent X-rays from Mercury to charged particle background is small. The goal was to reject 80% of charged particles while accepting 95% of measured X-rays. A two-pronged approach was taken to ensure that the XRS meets the goal. Unlike the NEAR GPC, the XRS GPC X-ray measurement volume is surrounded by veto wires to cover more than 80% of the full sensor geometry. Background caused by cosmic rays penetrating the GPC titanium housing will be vetoed out by this method. The second approach for background reduction is a more advanced implementation of the rise-time rejection technique employed by the NEAR XGRS unit. A particle interacts differently than a photon inside the GPC: it usually interacts at the walls of the detector and leaves long ionization trails that exhibit a characteristically longer charge-collection time. Particle background rejection can therefore be based on the measured rise time of the signal detected in the GPC. The rise rejection is performed by the data system with a threshold that is linear with energy, rather than a fixed threshold such as used on NEAR, to maximize background rejection while minimizing foreground rejection across the measured energy spectrum.

Laboratory background measurements with the veto rejection on and off are shown in Fig. 11. The rejection rate is between 60% and 75%, but this rate included gamma-ray background, which would not be removed by this GPC veto design.

The combined effect of the veto and rise rejection in flight is shown in Fig. 12. In this example the veto system eliminates between 50% and 75% of the background, and the rise rejection reduces the background by another 50% to 70%. The result is about an 80% reduction at 1 keV to about 90% at 8 keV. The foreground rejected was expected to be between 5% and 10% with this single flat threshold. Additional observations of astronomical X-ray sources are planned to optimize further the threshold to maximize the foreground accepted and the background rejected.

5.6 In-flight Checkout and Calibration

The supernova remnant X-ray source Cassiopeia-A (Cas-A) is being used for periodic in-flight calibration because of its relatively high intensity in the XRS energy range. Once or twice a year XRS is pointed to Cas-A for about 48 hours and then away from that source for another 48 hours to accumulate background measurements. The background is subtracted

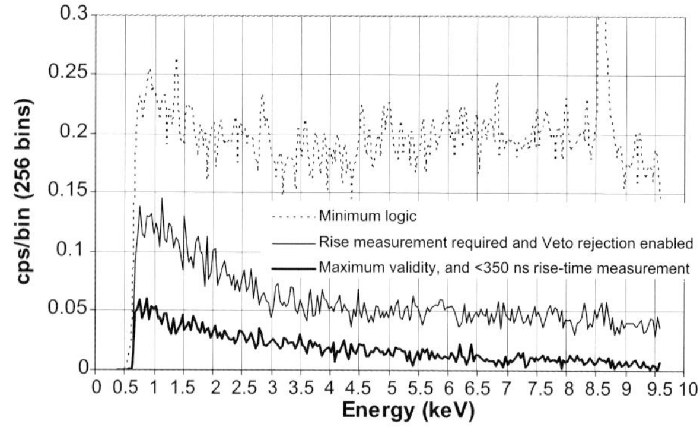

Fig. 12 GPC background rejection with no filter

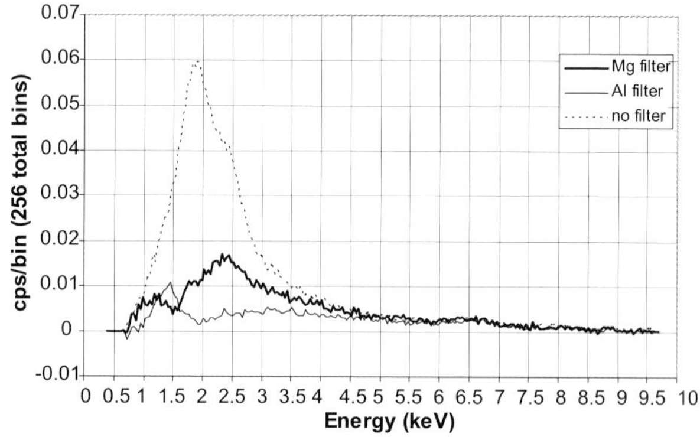

Fig. 13 X-ray accumulations (background subtracted) from the Cas-A X-ray source taken on February 6, 2005

from the foreground; results for the first observation (February 6, 2005) are shown in Fig. 13. These results were with sub-optimal background-rejection settings for the GPCs, so subsequent observations will be useful for further adjustment and calibration of XRS in preparation for later phases of the mission.

6 Data Products

An uncompressed XRS science data record is 2171 bytes long. Each record contains four energy spectra, housekeeping, status, and rate data for one integration period. The spectra correspond to the Mg-filtered detector (GPC1), the Al-filtered detector (GPC2), the unfiltered detector (GPC3), and the solar monitor (SAX). All the spectra have 256 bins, but the highest 244 bins for the GPC and the highest 231 bins for the SAX are telemetered. All bins are 16 bits deep. There are five 32-bit rate counters for SAX and nine 24-bit rate counters for each of the three gas proportional counters. Rate counters for all four detectors include raw, valid event, analyzed, pileup, and high-energy rates. The gas detectors also have counters for the veto-anode, veto, rise-pileup, and rise-rejection rates. Housekeeping data (43 bytes) provide voltage, current, and temperature readings measured at the end of each integration period. An XRS science record also reports record and software status (34 bytes) as well as hardware settings (28 bytes) and integration and reporting times (6 bytes).

Integration periods will vary depending on distance to the planet. Every 12-hour orbit is divided into red (2000 s), yellow (10,000 s), and green (31,200 s) zones, corresponding to closest approach (<1000 km), mid-range (1000–3000 km), and greatest distance

(>3000 km) to the planet, respectively. Integration periods are adjustable over a wide range, but default values are 40 s (red), 200 s (yellow), and 450 s (green). The XRS also has a flare mode, which is triggered if the SAX detects the onset of a solar flare. During flare mode integration periods are automatically set to 20 s. Flare mode is enabled for up to 60 minutes. The typical uncompressed data volume (excluding any flare data) is 4.6×10^6 bits per day. Data volume for compressed records, which is the normal mode of operation, is 3.4×10^6 bits per day. Data volume can be adjusted by increasing or decreasing the duration of integration periods.

7 Operation Plans/Observing Strategy

MESSENGER will take 6.6 years between the August 2004 launch and Mercury orbit insertion (McAdams et al. 2007). The planned trajectory will take MESSENGER past Earth once, Venus twice, and Mercury three times before the Mercury orbital phase. During the long interplanetary cruise to Mercury, there are only limited XRS operations planned. This cruise duty cycle is to ensure that the instrument can perform as expected when MESSENGER enters the prime orbital phase at Mercury.

7.1 Cruise Phase Operation

Unlike the NEAR XGRS instrument, there is no onboard X-ray calibration source for the MESSENGER XRS instrument to provide a stable reference measurement. Routine star observations are therefore required to ensure XRS performance. Currently, a minimum of one star calibration is scheduled for XRS per year. Typically, each star calibration will last four days with two days of constant pointing at the observed star and two days of background rate accumulation. Initial results indicate that XRS is functioning at an expected performance level (Fig. 13).

In addition to routine star calibration, XRS will normally be turned on during the flybys to collect X-ray emission data from each planet encountered. XRS was not turned on during the Earth flyby, however, because of spacecraft power constraints.

7.2 Orbital Operation

MESSENGER orbital operations at Mercury will nominally last one Earth year. During this phase of operation, the instruments on MESSENGER will use different orbit pointing and data acquisition strategies to address the multiple scientific questions at Mercury.

7.2.1 XRS Observation Strategy

The XRS instrument detects characteristic fluorescent X-ray emission from Mercury. These characteristic X-ray emissions are diagnostic of elemental composition within 1 mm of the surface and are best observed when there is an ongoing solar X-ray event (flare) to illuminate the planet, the solar X-ray to surface normal is less than 80°, and the view angle to surface normal is less than 60°. In the absence of active solar activity, long periods of data accumulation over a given region of interest are required to acquire the necessary statistics. Hence, the XRS observation strategy at Mercury is naturally divided into quiet periods and active periods. Table 2 (Solomon et al. 2001) shows the expected elemental sensitivities at Mercury.

Table 2 XRS sensitivity

Element	Assumed abundances	Proportional counters with 95% background rejection		Proportional counters with 75% background rejection	
		Normal	Flare	Normal	Flare
Fe	2.3%	–	80 s	–	2 min
Ti	1.0%	–	3 min	–	4 min
Ca	4.0%	–	10 s	–	30 s
Si	21.5%	7 min	2 min	18 min	3 min
Al	3.0%	2 hr	22 min	8 hr	50 min
Mg	22.5%	5 min	2 min	11 min	3 min

Counting times are those required to reproduce the assumed composition at the 10% uncertainty level. Assumed abundances in weight percent are from Brückner and Masarik (1997). Count rates are scaled from actual measurements made by NEAR at Eros in July 2000 (1.78 AU) to the assumed composition at Mercury (0.387 AU)

7.2.1.1 Quiet Period The MESSENGER orbit has an initial periapsis altitude of 200 km and an initial latitude of periapsis of 60° N; the orbit is inclined 80° to the equatorial plane of the planet and has a 12-hour period. Three observation zones are identified according to the spacecraft altitude relative to the planet surface. XRS will change the spectrum accumulation period for each zone, by a timed command, to maximize statistics reflecting the varying distance to the planet. At periapsis, the XRS 12° FOV gives 42 km resolution for 40-s integration with spacecraft velocity of 3.3 km/s. Even at apoapsis (15,200 km altitude), XRS has a 3200-km resolution while at a 450-s integration period.

7.2.1.2 Active Period During active solar activity, abundant solar X-rays will illuminate Mercury, and the characteristic fluorescent X-ray emission can readily be detected by an orbiting spacecraft. The best measurements made by NEAR XRGS were during moderate to intense solar flares, and we expect that the MESSENGER XRS will perform similarly. As discussed in Sect. 5.1, XRS is designed to operate during moderate to intense solar flares (up to X-1 class) without going into count rate saturation. Most intense solar flares have a relatively fast onset and decay profile. During a solar flare, X-ray emissions from the planet can increase by several orders of magnitude in less than an hour. Therefore instead of accumulating several hours of data, the same spectra can be deduced in minutes, which could increase the spatial resolution of the measurement (see Table 2). Hence, XRS monitors the solar monitor count rate to detect the solar X-ray activity. When a commanded threshold is crossed, XRS will automatically enter into solar flare mode. When in this mode, the data accumulation interval shortens to a nominal 20 s for up to one hour, a transition that will help XRS achieve a finer spatial resolution. Both the integration interval and duration are commandable, and the solar flare mode can be deactivated by ground command.

8 Summary

The XRS, along with the other scientific instruments that make up the MESSENGER payload, is poised to provide the data upon which breakthrough scientific understanding can be achieved regarding the composition and formation of the planet Mercury, which in turn will yield insight into planetary formation for all of the inner solar system. The XRS will provide the crucial surface elemental composition using X-ray fluorescence measurements of individual regions of Mercury's surface, which will eventually extend to most of the planet surface. The enhanced GPC and electronic design will reduce much of the background and permit shorter accumulation periods and an improved science return. The narrow field-of-view at periapsis will also allow measurements of selected features of interest discovered throughout the mission.

Acknowledgements The authors gratefully acknowledge many individuals who spent considerable time and care in building, qualifying, and calibrating the XRS instrument, with special thanks to the many people at Oxford Instruments who designed, built, and tested the detectors that are the foundation of the XRS instrument.

References

I. Adler et al., Science **175**, 436–440 (1972)
J. Brückner, J. Masarik, Planet. Space Sci. **45**, 39–48 (1997)
K.E. Bullen, Nature **170**, 363–364 (1952)
S.R. Floyd et al., Nucl. Instrum. Method. Phys. Res. A **422**, 577–581 (1999)
J.O. Goldsten et al., Space Sci. Rev. (2007, this issue). doi:10.1007/s11214-007-9262-7
S.E. Hawkins et al., Space Sci. Rev. (2007, this issue). doi:10.1007/s11214-007-9266-3
G.F. Knoll, *Radiation Detection and Measurement* (Wiley, New York, 1999), Chap. 6
J.C. Leary et al., Space Sci. Rev. (2007, this issue). doi:10.1007/s11214-007-9269-0
J.V. McAdams, R.W. Farquhar, A.H. Taylor, B.G. Williams, Space Sci. Rev. (2007, this issue). doi:10.1007/s11214-007-9162-x
L.R. Nittler et al., Meteorit. Planet. Sci. **36**, 1673–1695 (2001)
A.E. Ringwood, Geochim. Cosmochim. Acta **30**, 41–104 (1966)
S.C. Solomon et al., Planet. Space Sci. **49**, 1445–1465 (2001)
R. Starr et al., Nucl. Instrum. Method. Phys. Res. A **428**, 209–215 (1999)
R. Starr et al., Icarus **147**, 498–519 (2000)
J.I. Trombka et al., Science **289**, 2101–2105 (2000)
G.W. Wetherill, in *Mercury*, ed. by F. Vilas, C.R. Chapman, M.S. Matthews (University of Arizona Press, Tucson, 1988), pp. 670–691.

Space Sci Rev (2007) 131: 417–450
DOI 10.1007/s11214-007-9246-7

The Magnetometer Instrument on MESSENGER

Brian J. Anderson · Mario H. Acuña · David A. Lohr · John Scheifele · Asseem Raval · Haje Korth · James A. Slavin

Received: 22 May 2006 / Accepted: 19 July 2007 / Published online: 30 October 2007

Abstract The Magnetometer (MAG) on the MErcury Surface, Space ENvironment, GEochemistry, and Ranging (MESSENGER) mission is a low-noise, tri-axial, fluxgate instrument with its sensor mounted on a 3.6-m-long boom. The boom was deployed on March 8, 2005. The primary MAG science objectives are to determine the structure of Mercury's intrinsic magnetic field and infer its origin. Mariner 10 observations indicate a planetary moment in the range 170 to 350 nT R_{M}^3 (where R_{M} is Mercury's mean radius). The uncertainties in the dipole moment are associated with the Mariner 10 trajectory and variability of the measured field. By orbiting Mercury, MESSENGER will significantly improve the determination of dipole and higher-order moments. The latter are essential to understanding the thermal history of the planet. MAG has a coarse range, ±51,300 nT full scale (1.6-nT resolution), for pre-flight testing, and a fine range, ±1,530 nT full scale (0.047-nT resolution), for Mercury operation. A magnetic cleanliness program was followed to minimize variable and static spacecraft-generated fields at the sensor. Observations during and after boom deployment indicate that the fixed residual field is less than a few nT at the location of the sensor, and initial observations indicate that the variable field is below 0.05 nT at least above about 3 Hz. Analog signals from the three axes are low-pass filtered (10-Hz cutoff) and sampled simultaneously by three 20-bit analog-to-digital converters every 50 ms. To accommodate variable telemetry rates, MAG provides 11 output rates from 0.01 s^{-1} to 20 s^{-1}. Continuous measurement of fluctuations is provided with a digital 1–10 Hz bandpass filter. This fluctuation level is used to trigger high-time-resolution sampling in eight-minute segments to record events of interest when continuous high-rate sampling is not possible. The MAG instrument will provide accurate characterization of the intrinsic planetary field, magnetospheric structure, and dynamics of Mercury's solar wind interaction.

Keywords Mercury · MESSENGER · Magnetometer · Magnetic field · Magnetosphere

B.J. Anderson (✉) · D.A. Lohr · A. Raval · H. Korth
The Johns Hopkins University Applied Physics Laboratory, 11100 Johns Hopkins Road, Laurel, MD 20723, USA
e-mail: brian.anderson@jhuapl.edu

M.H. Acuña · J. Scheifele · J.A. Slavin
NASA Goddard Space Flight Center, Greenbelt, MD 20771, USA

1 Introduction

Mercury's magnetic field was first measured in 1974 by Mariner 10 (Ness et al. 1974), and the presence of an intrinsic planetary magnetic field was surprising because the planet's size, expected thermal state, and angular momentum had seemed to rule out the possibility of an active dynamo (Solomon 1976; Srnka 1976; Jackson and Beard 1977). Additional encounters of Mercury by the Mariner 10 spacecraft in 1975 (Ness et al. 1975, 1976) confirmed the initial results and allowed the estimation of the planetary magnetic dipole moment to within perhaps a factor of two (Connerney and Ness 1988). The discovery prompted a variety of suggestions for the source of the intrinsic field, including a hydromagnetic dynamo, a non-traditional dynamo, and a remanent field. The presence of sufficient sulfur in the outer core would allow a thin outer fluid core to persist to the present and perhaps to support a shell dynamo sustained by thermal energy released by continued solidification of core material and growth of an inner core (Stevenson et al. 1983; Stanley et al. 2005; Heimpel et al. 2005). An alternative possibility is that the seed magnetic field is produced by thermoelectric currents generated along a topographically rough core-mantle boundary (Stevenson 1987; Giampieri and Balogh 2001, 2002). Remanent magnetization of the crust and mantle was an early suggestion (Stephenson 1976), but because Runcorn (1975a, 1975b) had shown that under idealized assumptions a uniform spherical shell magnetized with a centrally located field does not exhibit an externally detectable field, this idea had been discounted (Stevenson et al. 1983; Stevenson 1983). Relaxing the idealized assumptions in Runcorn's analysis has subsequently indicated that remanent magnetization may nonetheless be important. The remanent magnetization is not simply proportional to the magnetizing field (Merrill 1981; Merrill and McElhinny 1983), and this non-linearity will yield an external field even for uniform spherical shells (Lesur and Jackson 2000). In addition, Aharonson et al. (2004) demonstrated that non-uniformity in the magnetized crustal shell thickness produced by latitudinal and longitudinal thermal variations at Mercury could result in a detectable external field even when the susceptibility is linear. The observations of intense remanent magnetization at Mars by Mars Global Surveyor (Acuña et al. 1998, 1999) provide additional observational support for the possibility of crustal magnetization at Mercury.

The detailed structure of the planetary field is one key to determining how Mercury's field is generated. In an effort to constrain early hypotheses for the origin of the internal field, more sophisticated models of the Mariner 10 data were developed (Whang 1977; Engle 1997). Connerney and Ness (1988) pointed out, however, that it is not possible to derive more precise estimates of the dipole and higher-order terms from the Mariner data because the flyby trajectories do not adequately constrain the spherical harmonic representation of the field (Connerney and Ness 1988; Korth et al. 2004). A major improvement to the estimation of higher-order terms can be obtained from data acquired by a low-altitude, high-inclination, orbiting spacecraft. Conducting these measurements is one of the primary objectives of the MErcury Surface, Space ENvironment, GEochemistry, and Ranging (MESSENGER) mission (Solomon et al. 2001; Gold et al. 2001).

This paper describes the Magnetometer (MAG) instrument on the MESSENGER spacecraft. Topics covered include performance requirements, instrument design, sensor boom, electronics, software, spacecraft magnetic cleanliness program, pre-launch testing, and initial performance in space. Post-launch instrument checkout activities are ongoing and include observations during the Earth flyby on August 2, 2005.

2 Science Objectives and Instrument Requirements

2.1 Intrinsic Magnetic Field

The primary science objectives of the MAG investigation are the structure of Mercury's magnetic field and its interaction with the solar wind. Analyses of the Mariner 10 flyby observations (Ness et al. 1975, 1976; Connerney and Ness 1988; Korth et al. 2004) yield estimates for the planetary dipole magnetic moment in the range 170 to 350 nT R_M^3, where R_M is the planet radius (2,440 km). The inferred dipole axis is aligned within 11° of the planet's rotation axis. On the basis of these results, the surface magnetic field magnitude is expected to be ~600 nT over the poles and ~300 nT at the equator. The orbital coverage provided by the MESSENGER mission will allow a much improved determination of the planetary dipole moment and an estimation of higher-order terms, including corrections due to external fields associated with the solar wind interaction (Korth et al. 2004).

The required accuracy and time resolution of the MAG measurements are determined by the expected intensity of Mercury's magnetic field in orbit and the variable contributions due to external currents. These currents contribute fields as large as two thirds of the intrinsic field at the surface near the night-side equator (e.g., Ness et al. 1974) and respond directly to the imposed solar wind and interplanetary magnetic field (IMF) (Luhmann et al. 1998; Glassmeier 2000). By contrast, at Earth the external currents generate fields that are typically less than 1% of the intrinsic field at the surface. Relative to the planetary radius Mercury's magnetosphere is ~15% as large in linear dimension as Earth's (e.g., Slavin and Holzer 1979). The external currents contribute a large, variable background signal and, in addition to the spacecraft orbits, are the primary factors limiting the accuracy and resolution of the models used to represent the planetary field and its geometry. These factors in turn will limit the constraints that magnetic field observations will place on Mercury's interior structure, thermal evolution, and the effects of early large impacts if crustal fields are detected (Korth et al. 2004; Connerney et al. 1989; Acuña et al. 1999).

2.2 Magnetospheric Magnetic Fields

2.2.1 Fields Due to External Currents

Characterizing the geometry and time variability of the magnetospheric field is an essential MESSENGER science objective, for two reasons. First, accurate determination of dipole and higher-order terms in the multipole expansion of the intrinsic field requires specification of and correction for the external magnetic field. Second, such characterization is necessary to understand possible induction effects in Mercury's interior as well as particle acceleration, magnetic substorms, and reconnection processes. For planetary field analysis, the statistical uncertainty in the estimated dipole moment, including the effective noise contribution due to variable solar wind and IMF conditions at Mercury, will be reduced significantly for MESSENGER compared with that for Mariner 10. MESSENGER observations will also allow unambiguous identification of quadrupole and octupole components contributing more than about 1% of the field intensity at the surface (Korth et al. 2004). To achieve these objectives the magnetometer needs to have a full scale greater than about 600 nT, a full-scale accuracy of roughly 1%, and a sensitivity of 1 nT or better.

2.2.2 Magnetospheric Boundaries, Currents and Waves

There are several additional science objectives that the magnetic field measurements address (Russell et al. 1988; Slavin 2004). These include: (1) wave-particle interactions, both those associated with field-line resonances that should have periods comparable to ion gyroperiods (~1 to 7 Hz for protons) (Glassmeier et al. 2003) and those associated with heavy ions that may play a role in the formation in of Mercury's exosphere (Potter and Morgan 1990); (2) magnetotail dynamics, including phenomena possibly analogous to substorms in the Earth's magnetosphere (Ogilvie et al. 1975; Christon 1987); (3) magnetopause structure and dynamics for Mercury's small magnetosphere (relative to the planet's radius and to energetic ion gyroradii) under the solar wind conditions that generally prevail at Mercury (Siscoe and Christopher 1975; Slavin and Holzer 1979; Burlaga 2001) and that are very different from the average conditions prevailing at Earth and the outer planets; and (4) characterization of field-aligned currents linking the planet with the magnetosphere (Slavin et al. 1997). One unique feature of Mercury is the absence of an ionosphere that is directly related to the flow and closure of field-aligned currents. Magnetopause turbulence is expected to be several nT in amplitude and should extend above ion gyrofrequencies (~2 Hz). Fluctuation amplitudes could be significantly higher if instabilities associated with ion pickup by charge exchange or electron impact ionization are present (Mazelle et al. 2004; Crider et al. 2004; Bertucci et al. 2005). Magnetotail dynamics may have signatures of a few tens of nT and occur on time scales of seconds to minutes. Regions where field-aligned currents may be present will be traversed in a few minutes and may result in signatures of a few tens of nT amplitude. The MAG instrument should therefore have a frequency coverage and amplitude resolution of up to a few Hz and ~0.1 nT, respectively.

2.3 Performance Requirements

The MAG performance capabilities were tailored to address the scientific objectives and to make efficient use of spacecraft resources (Santo et al. 2001; Gold et al. 2001). To support measurements over the range of expected fields in orbit at Mercury, the instrument provides a "low" or fine range covering ±1,530 nT for each of three orthogonal axes. A "high" or coarse range covering ±51,300 nT is provided to simplify ground testing and operations in Earth's field. The absolute accuracy in the fine range needed to meet the science objectives is 3 nT or 0.2% of full scale. This goal imposes requirements that the spacecraft magnetic field at the MAG sensor vary by less than ~1 nT. To achieve 1% accuracy in the field measurement, the maximum attitude uncertainty is ±0.6° with respect to inertial coordinates. The magnetic measurements are digitized to a resolution of 20 bits internally for digital processing and telemetered with 16-bit resolution. The telemetered resolution is 1.6 nT in coarse range and 0.047 nT in fine range. The scale factors were determined to an accuracy of 0.01% and 0.08% in coarse and fine ranges, respectively.

2.4 System and Operational Constraints

To meet the science objectives within mission resource constraints, MAG can be operated in a variety of modes adapted to the mission and orbital phases. For example, at the minimum periapsis altitude of 200 km, the spacecraft velocity relative to the surface is ~3 $\mathrm{km\,s^{-1}}$. The shortest scale lengths of the intrinsic planetary field that could be detected from orbit

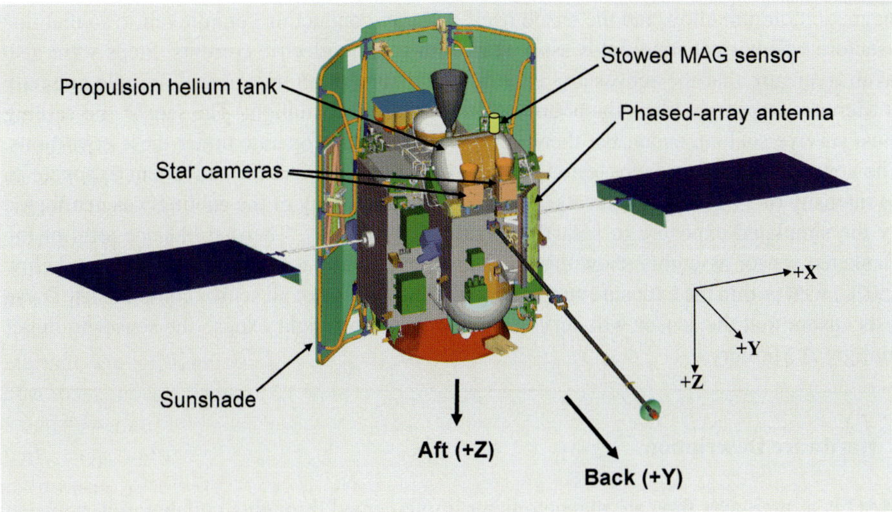

Fig. 1 View of the MESSENGER spacecraft showing the instrument coordinates and the Magnetometer boom deployed in the $+Y$ direction, which is the anti-sunward direction during orbital operations. The MAG sensor stowed location is indicated (*yellow cylinder*) as are the propulsion system helium tank, the star cameras, and the phased-array antenna nearest the stowed Magnetometer sensor. The spacecraft sunshade is on the $-Y$ side of the vehicle as shown in *light green*

are approximately half the periapsis altitude, and for a minimum altitude of 200 km a vector sample every 30 s would be sufficient to achieve this along-track resolution. However, the high time-variability of Mercury's magnetosphere, from a few seconds to fractions of a second at the boundaries, plus the short characteristic times associated with wave-particle interactions (the proton gyrofrequency at 500 nT is about 7.5 Hz), require sampling commensurate with a Nyquist frequency higher than a few Hz.

The magnetic field observations are also required to interpret data acquired by the MESSENGER Energetic Particle Spectrometer (EPS) and Fast Imaging Plasma Spectrometer (FIPS) sensors on the Energetic Particle and Plasma Spectrometer (EPPS) instrument (Andrews et al. 2007). The EPS sensor has a minimum sample interval of 0.5 s for the electron channels, and the FIPS minimum integration time at a given energy step is 1 s. It is therefore desirable to be able to sample MAG at 1 s^{-1} and faster. During the mission cruise phase, assessment of the static spacecraft magnetic field requires long-term observations of the IMF only at low data rates. MAG therefore uses selectable filter and sub-sampling techniques providing output sample rates from 0.01 s^{-1} to 20 s^{-1}.

To minimize the effects of the spacecraft field on the measurements, the MAG sensor is mounted at the end of a deployable boom that extends 3.6 m from the spacecraft (Fig. 1). The spacecraft coordinate system is also indicated. The MAG boom extends in the $+Y$ direction, which will nominally be anti-sunward during the mission's orbital phase. The payload adapter ring, which houses remote sensing instruments, is located at the $+Z$ end of the spacecraft. The $+X$ direction completes the right-hand system. The MAG axes were chosen to correspond to those of the spacecraft, and only small corrections are required to align the two systems.

During orbital operations the spacecraft can be pointed such that $+Y$ is not strictly anti-sunward. To protect the MAG sensor from direct exposure to solar illumination at 0.3 to 0.4 AU, a small conical shade is mounted near the sensor. The shade frame is made of non-magnetic titanium alloy, and the shade itself is a non-conducting ceramic fabric so that this structure will not generate fields associated with thermo-electric currents. Steps were also taken to ensure that the sensor and cabling could survive an unexpected attitude anomaly at Mercury that might place the boom and sensor in direct sunlight. The sensor and cabling must survive such an event, but there is no requirement to operate under these conditions. The sensor and cabling were tested on the ground for survival to inadvertent exposure to an intensity of 11 Suns. The structural and electrical integrity of the cabling was unaffected by the simulated exposure to solar illumination at Mercury. Thermal balance tests of the blanketed sensor assembly show that in such a situation its temperature will remain below 150°C, well within the exposure survival temperature rating of all sensor components. These tests ensure that the sensor will survive an anomaly that might expose the sensor to direct sunlight at Mercury.

3 Hardware Description

MAG is a three-axis fluxgate magnetometer implemented through a collaboration between NASA Goddard Space Flight Center (GSFC) and The Johns Hopkins University Applied Physics Laboratory (JHU/APL). The instrument has extensive flight heritage derived from more than 50 space missions (Acuña 1974; Behannon et al. 1977; Neubauer et al. 1987; Acuña et al. 1992, 1997; Acuña 2002). The design also benefits from experience gained in previous collaborative efforts, including the Active Magnetosphere Particle Tracers Explorers/Charge Composition Explorer spacecraft (Potemra et al. 1985), the Viking auroral mission (Zanetti et al. 1994), the Upper Atmospheric Research Satellite (Winningham et al. 1993), the Advanced Composition Explorer (ACE) (Smith et al. 1998), and the Near Earth Asteroid Rendezvous (NEAR)-Shoemaker mission (Lohr et al. 1997). For MESSENGER, MAG employs a low-mass electronics design. Relevant instrument characteristics are summarized in Table 1. A complete list of acronyms and abbreviations used here is given in Table 2. The instrument consists of an electronics box and a sensor assembly mounted at the end of the deployable boom. The MAG electronics (Fig. 2) include the following subsystems: sensor drive and signal processing circuits, analog-to-digital (A/D) conversion control logic and programmable gate arrays, event processing unit (EPU) and onboard memory, power converter, sensor heater proportional controller, and interface to the payload data processing unit (DPU).

3.1 Sensor and Boom

The sensor assembly (Fig. 3) uses three orthogonally mounted ring-core fluxgate sensors. Small mechanical deviations from true orthogonality are corrected in ground processing using an "alignment matrix" determined in pre-flight calibration. To protect the magnetometer from exposure to direct sunlight during normal operations and ensure operation within the qualified temperature range, a conical sunshade, shown in Fig. 4, is mounted on the boom just inward of the sensor. Exposure to radiative heating from Mercury near periapsis is not anticipated to heat the sensor above 20°C since it will normally operate in complete shadow near −40°C and it will take several hours over the hot planet to warm to 20°C. The most temperature-restricted components in the sensor are capacitors rated down to −55°C for repeated temperature cycling. To ensure operation in this range, a non-magnetic foil-element heater is included that operates in proportional mode between −15°C and −40°C. To eliminate magnetic contamination fields at the sensor, the circuit drives current to the heating

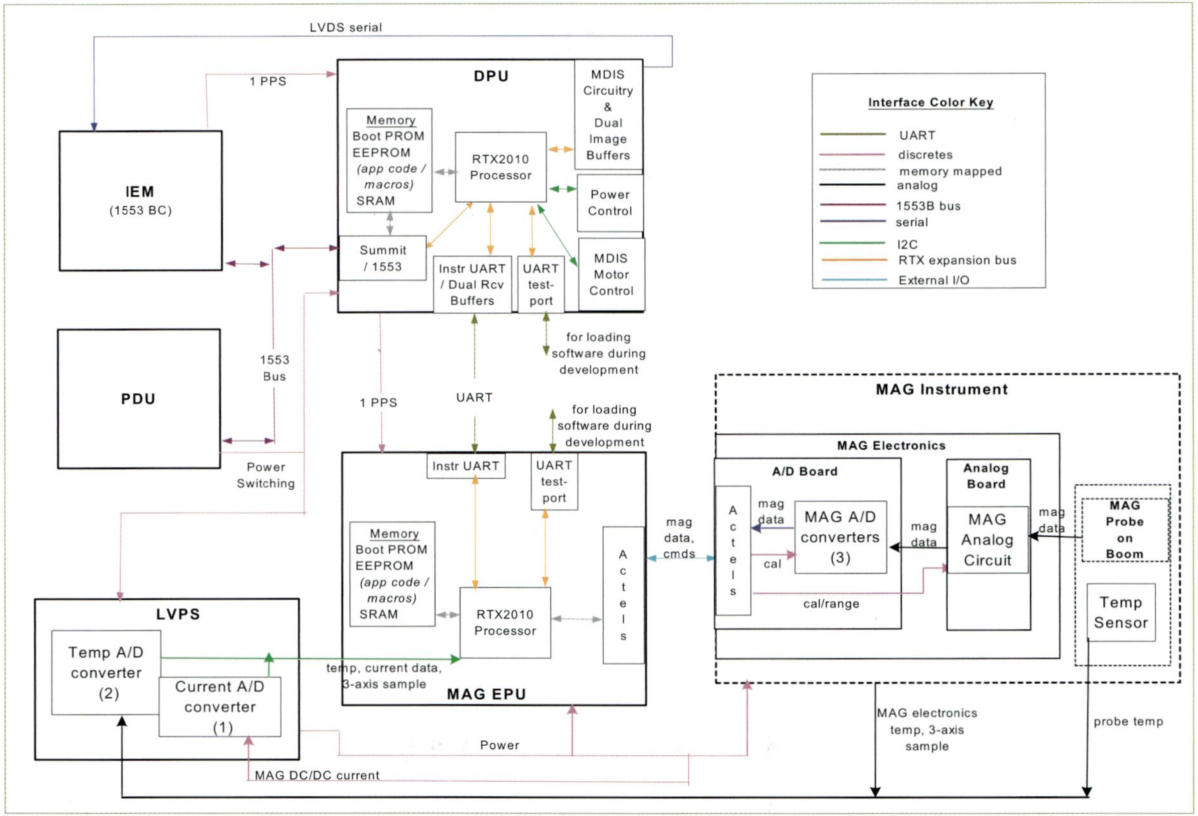

Fig. 2 MESSENGER Magnetometer block diagram showing the DPU, EPU, digital and analog electronics, LVPS, and their interfaces. Acronym and abbreviation definitions are given in Table 2

Table 1 MESSENGER Magnetometer resources and performance characteristics

Dimensions:	Sensor	8.1 cm × 4.8 cm × 4.6 cm	
	Electronics	13.0 cm × 10.4 cm × 8.6 cm	
	Boom	3.6 m long	
Mass:	Sensor	184 g	
	Electronics	835 g	
	Boom	2.66 kg	
	Cable	408 g	
	Total	4.09 kg	
Power:	Instrument	4.2 W	
	Probe heater	0.93 W (2 W limited)	
Type:	Low-noise tri-axial fluxgate (<20 pT/$\sqrt{}$Hz at 1-Hz intrinsic noise level)		
A/D:	20-bit, 20 conversions per second		
	Self calibrating on command		
	Three independent units, one dedicated for each axis		
Ranges:	Coarse	±51,300 nT full scale	1.56-nT resolution (16 bits out)
	Fine	±1,530 nT full scale	0.047-nT resolution (16 bits out)
Output rates:	Maximum	20 s^{-1} (10-Hz analog filter)—internal A/D	
	Filtered	10 s^{-1}, 5 s^{-1}, 2 s^{-1}, 1 s^{-1} (digital filter at Nyquist frequency)	
	Sub-sampled	0.5 s^{-1}, 0.2 s^{-1}, 0.1 s^{-1}, 0.05 s^{-1}, 0.02 s^{-1}, 0.01 s^{-1} (0.5-Hz filter)	
Output data volume:	Vector field	48 bits per three-axis sample	
	AC channel	8 bits log-compressed per sample (one axis)	
	Packet average	56.5 bits per sample (before compression)	
	Compression	Lossless differential pulse code modulation (DPCM)	

The total instrument mass does not include 0.3 kg for pyrotechnic releases, temperature sensors, and associated cabling added to the boom assembly during integration

element via a transformer at a frequency of 54 kHz. The instrument gain and offset stability are determined principally by the core materials whose behavior has been characterized for previous missions, e.g., Voyager (Behannon et al. 1977; Acuña 2002; Burlaga et al. 2003) and yield less than 0.25% change in gain over the range −40°C to +50°C and a change in offset of less than 0.04 nT/°C. The instrument will therefore meet the accuracy requirements for the mission throughout the expected temperature range.

The boom was built using two sections of carbon composite tubing joined by spring-loaded hinges. The inboard hinge is attached to the spacecraft, while the outboard hinge is located near the mid-point. The latter is shown in Fig. 4 in the deployed configuration including a cable service loop, prior to the installation of thermal blankets. All materials, including the spring, locking pin, frame, and damper are non-magnetic. A twist damper dissipates energy released on deployment. Repeatability was shown to be within <0.05° in pre-launch tests. Locking pins secure the hinges after deployment. Cabling between the sensor and electronics is threaded inside the blanketed boom, and service loops at both hinges allow free hinge motion (cf. Fig. 4).

Table 2 Acronyms and abbreviations

Acronym or abbreviation	Definition
AC	Alternating Current
ACE	Advanced Composition Explorer
A/D	Analog to Digital
CCSDS	Consultative Committee on Space Data Systems
CDR	Calibrated Data Record
CME	Coronal Mass Ejection
DC/DC	Direct Current to Direct Current (converter)
DPCM	Differential Pulse Code Modulation
DPU	Data Processing Unit
EDR	Experiment Data Record
EEPROM	Electrically Erasable Programmable Read Only Memory
EPPS	Energetic Particle and Plasma Spectrometer
EPS	Energetic Particle Spectrometer
EPU	Event Processing Unit
FIPS	Fast Imaging Plasma Spectrometer
FPGA	Field Programmable Gate Array
GSFC	Goddard Space Flight Center
Hz	Hertz
I^2C	Inter-Integrated Circuit
IEM	Integrated Electronics Modules
IIR	Infinite Impulse Response
IMF	Interplanetary Magnetic Field
IMP	Interplanetary Monitoring Platform
I/O	Input/Output
JHU/APL	Johns Hopkins University Applied Physics Laboratory
kBaud	Kilobaud
kbytes	Kilobytes
kHz	Kilohertz
LH	Local Heliospheric (coordinates)
LVDS	Low Voltage Differential Signaling
LVPS	Low-Voltage Power Supply
MAG	Magnetometer (MESSENGER)
MDIS	Mercury Dual Imaging System
MESSENGER	MErcury Surface, Space ENvironment, GEochemistry, and Ranging
MET	Mission Elapsed Time
MHz	Megahertz
MLA	Mercury Laser Altimeter
MP	Main Processor
ms	Millisecond
NASA	National Aeronautics and Space Administration
NEAR	Near Earth Asteroid Rendezvous
nT	Nanotesla
PDU	Power Distribution Unit
PPS	Pulse Per Second

Table 2 (Continued)

Acronym or abbreviation	Definition
PROM	Programmable Read Only Memory
RTX	Real Time Express
SRAM	Static Random Access Memory
UART	Universal Asynchronous Receiver Transmitter
Temp	Temperature
UT	Universal Time
UTC	Universal Coordinated Time

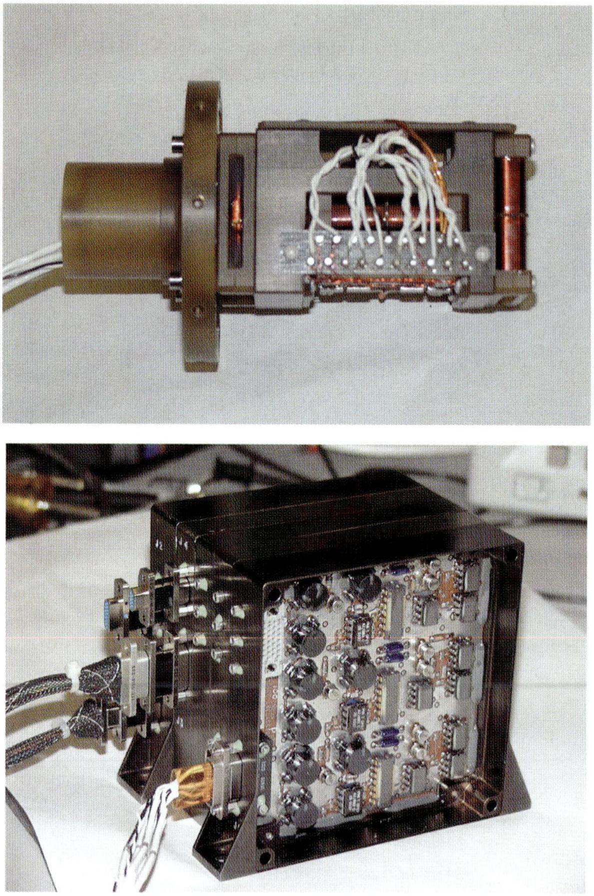

Fig. 3 Flight MESSENGER MAG sensor (*top*) and electronics (*bottom*) with view of analog slice. The sensor is shown mounted to a test boom adapter flange. Sensor dimensions not including the flange are 8.1 cm × 4.8 cm × 4.6 cm. The electronics dimensions are 10.4 cm and 8.6 cm in cross section as shown and 13.0 cm deep

Fig. 4 Flight MESSENGER MAG sensor and sensor sunshade integrated to the boom and spacecraft prior to final thermal blanketing (*top*) and MAG boom midpoint hinge (*bottom*) during deployment testing prior to installation of thermal blankets

3.2 Electronics Enclosure

The electronics are housed in a chassis composed of a stack of five 4″ × 4″ "slices" or boards integrated with a frame and connectors. One slice is devoted entirely to the fluxgate analog drive/sense electronics and amplifiers. The three science A/D converters, one for each magnetometer axis, and digital electronics to control the A/D sampling and the analog board make up the second slice. The EPU and interface to the DPU are housed on the third slice. The Low-Voltage Power Supply (LVPS), which includes lower-resolution A/D channels for instrument housekeeping data, is housed on the fourth slice. The fifth slice is devoted to the sensor survival heater, which is electronically isolated from the other four slices and powered by command from the spacecraft main processor rather than the payload DPU to ensure that this heater operates when the payload is turned off. The LVPS provides power to the EPU and MAG electronics under spacecraft control. The MAG EPU interfaces with the LVPS via an inter-integrated circuit (I^2C) bus to collect MAG sensor temperature, electronics temperature, and MAG input current together with three redundant magnetic field sampling channels with 14-bit resolution. A view of the electronics chassis with the end cover removed (Fig. 3) shows the analog electronics slice. The radiation exposure for MESSENGER is principally due to solar energetic particles and is not severe (Gold et al. 2001; Santo et al. 2001). The electronics were designed to meet a 20-kRad total dose behind 2 mm of aluminum. The complete electronics package volume is $<1{,}200\ \text{cm}^3$, and its mass is $\sim$850 g.

3.2.1 Analog Electronics

A functional schematic of the analog signal-processing electronics is shown in Fig. 5. The triaxial ring cores are driven at 15 kHz. This drive signal is derived from a 30-kHz clock, which is also used as a reference to synchronously detect the signals from the fluxgate sensors. The output from each synchronous detector is applied to a high-gain integrator, which in turn controls a precision voltage-to-current converter that supplies the feedback current to null the field seen by the sensor, thus completing the feedback loop. The output voltage of the integrators is directly proportional to the component of the magnetic field along each orthogonal axis. The intrinsic root mean square noise of the fluxgate sensor is <10 pT (10-Hz bandwidth). The magnetometer design heritage spans more than 30 years (Acuña 2002), and the MESSENGER implementation takes advantage of the latest developments in low-mass electronics design and performance optimization.

As mentioned earlier, the instrument has two dynamic ranges giving full scales of $\pm1{,}530$ nT and $\pm51{,}300$ nT per axis. The range is selected either automatically or by command, and all three axes always operate in the same range. The output voltages corresponding to the three axes are digitized internally to 20-bit resolution. Internal self-calibrations are provided for the A/D converters, and a precision step-function input signal can be applied to the voltage-controlled current sources to provide end-to-end verification of the electronics performance in flight.

The sensor heater 54-kHz controller is powered independently of the rest of the MAG electronics to allow operation when the main MAG electronics are powered off, and spacecraft protection is provided by a floodback current limiter. The sensor assembly and foil heater are covered by multilayer insulating blankets to reduce radiative losses.

3.2.2 A/D Conversion

The three analog outputs of the magnetometer are sampled simultaneously with three independent 20-bit sigma-delta A/D converters (Crystal 5508) at a rate of 20 conversions per second (Fig. 6). To ensure uniformly distributed time sampling, the conversions are synchronized to a hardware clock internal to MAG. A single-pole 10-Hz anti-alias low-pass filter is used on each channel to limit the input bandwidth. The A/D converters introduce additional filtering with a -3 dB point at 17 Hz. A self-calibration feature is integral to the sigma-delta A/D converters and updates the conversion scale factor against a voltage reference when invoked by command. On the basis of experience gained from NEAR-Shoemaker (Lohr et al. 1997; Anderson et al. 2001; Acuña 2002) with similar sigma-delta devices, only passive latch-up protection is provided in the MAG instrument. No latch-up events occurred in 11 months of interplanetary observations including the Earth flyby transit of the radiation belts.

Timing of magnetometer samples is accomplished as follows. A time-synchronization interrupt is issued by the spacecraft clock at 1-s intervals, and the precision of the timing

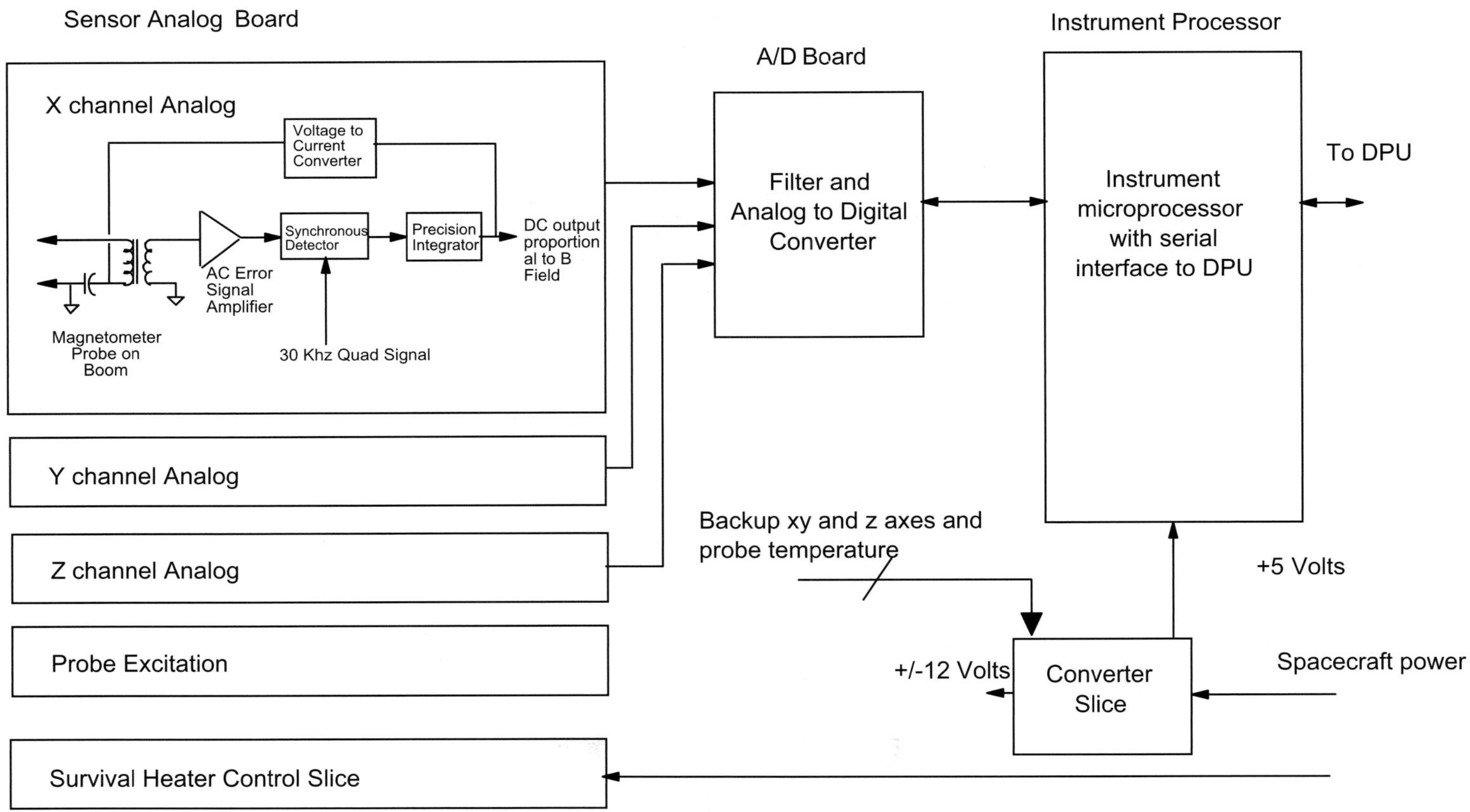

Fig. 5 Magnetometer electronics functional schematic showing the three independent synchronous detection sense circuits, one for each axis, and the functional interfaces between the A/D converter, MAG EPU, power supply, and thermostatic sensor heater control

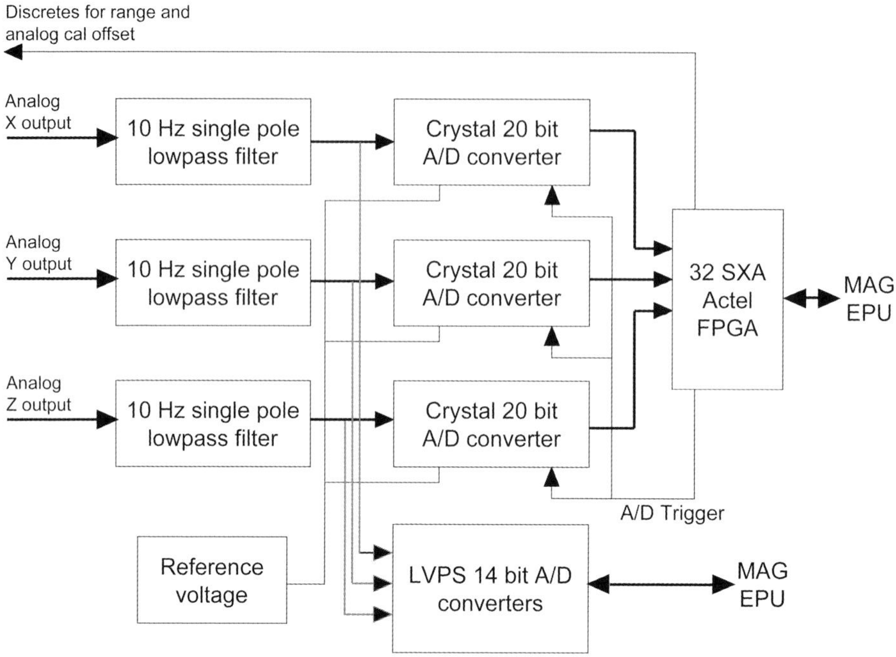

Fig. 6 A/D conversion electronics functional schematic showing the three independent anti-alias analog filters, three 20-bit A/D converters, parallel 14-bit LVPS channels, and the conversion trigger/data buffer FPGA interfaces

it provides to MAG is limited by the EPU clock interval, or 0.7 μs. This signal is used to time-tag magnetometer samples relative to Mission Elapsed Time (MET). The three A/D converters are triggered simultaneously every 50 ms to ensure simultaneous sampling. To ensure that the A/D samples are evenly spaced in time the converter trigger is derived from an oscillator internal to the MAG EPU and is not synchronized to the spacecraft clock. By driving the A/D sampling from the MAG EPU clock, the spacing between samples is guaranteed to be uniform to 0.7 μs. Absolute timing is provided for each data block, i.e., packet, by reporting the time for the A/D trigger of the first sample in the packet. This absolute time tag is provided with a resolution of 50 ms. Because the MAG EPU and spacecraft clocks can drift relative to each other, an error of up to 50 ms in the absolute time tag can occur, but this is sufficiently small to ensure that MAG can accomplish its science objectives. At a maximum relative velocity of 3 $\mathrm{km\,s^{-1}}$ relative to the planet at periapsis, the maximum uncertainty in spacecraft position knowledge for a given MAG sample is just 150 m, far smaller than the maximum spatial resolution possible for the reconstruction of the planetary intrinsic magnetic field structure (Korth et al. 2004).

The LVPS provides eight 14-bit A/D channels that are converted once per second. These are used to monitor the sensor and electronics temperatures and the MAG electronics input current, and to provide a backup sample of the three analog outputs to recover primary science data should a problem occur with the science A/D converters. These parameters are forwarded to the EPU. The resolution of the LVPS converters corresponds to 0.4 nT in fine range, sufficient to achieve primary science objectives for MAG.

3.2.3 Digital Logic and Processor

Instrument operation and modes are controlled by the EPU using a 16-bit RTX2010 processor running at a clock speed of 6 MHz. An Actel RT54SX32S Field Programmable Gate Array (FPGA) (Fig. 6) is used as an interface to the MAG analog electronics for data collection and instrument control and for many logic functions including watchdog timer and Universal Asynchronous Receiver/Transmitter (UART). The MAG software interfaces with the MAG electronics via memory-mapped input/output (I/O) for data collection, range control, and electronics calibration. The MAG EPU communicates with the DPU via a 38.4-kBaud RS422/UART interface. Under normal operations the EPU software resets a 5.6-s watchdog timer once per second. If the watchdog circuit times out, indicating an anomaly, an EPU reset is triggered and an alarm is sent to the DPU, which reports to the spacecraft main processor (MP) where error counters are updated and a decision to turn the Magnetometer off may be made under autonomy control. Range control is performed in the EPU on the basis of data from all three science A/D converters as described in the following.

The interfaces between the DPU and MAG electronics are controlled by interrupts that the EPU software services. UART interrupts are used to read a command from the DPU and to indicate that the DPU is ready to receive data from the MAG. The MAG EPU interfaces with the LVPS via an I^2C counter-timer interrupt to collect MAG sensor temperature, electronics temperature, MAG input current, and vector magnetic field samples. This interrupt occurs once per second. An internal MAG clock drives the A/D conversion at 20 s^{-1}, and the A/D conversion data ready signal is sent to the EPU as an interrupt every 50 ms. Each sample sequence consists of an echo of the commanded state of the Magnetometer (analog calibration on/off, A/D calibration on/off, and range setting) and the counts for each axis.

4 Instrument Software

4.1 Software Overview

The MAG software manages the data and command interface between the DPU and MAG, performs several data processing tasks, and controls autonomous instrument operation. The command and spacecraft interface functions include reception of MAG commands from the DPU, implementation of these actions in the instrument, receipt of spacecraft time, and calculation of associated time tags. The software also handles transmission of the instrument status, health and safety data, science data telemetry, and turn-off requests to the DPU. Data processing tasks performed by the software consist of digital filtering to implement a range of sampling rates, 1-to-10-Hz bandpass filter output amplitude calculation, data compression, and science data formatting including header information following the Consultative Committee for Space Data Standards (CCSDS) telemetry packet standards. Autonomous instrument operation controlled by the software includes range control, burst data triggering, and fault detection.

MAG uses the FORTH language. A set of common routines is used for the EPU-DPU interface, LVPS interface, CCSDS packet generation, and data compression. Many elements of the MAG code, including the digital filtering, range control, and variable sample rate implementation were originally written for the NEAR-Shoemaker magnetometer (Lohr et al. 1997). New features used for MESSENGER include data compression, burst data collection including pre-burst buffering, and LVPS A/D backup data acquisition. All features of the code were subjected to extensive testing to verify proper functionality.

The MAG EPU memory is organized in 13 blocks or pages of 64 kbytes and is allocated as follows. One page is reserved for boot Programmable Read-Only Memory (PROM); four pages of Electrically Erasable Progammable Read-Only Memory (EEPROM) are available for code, macros, and tables; and eight pages of Static Random Access Memory (SRAM) are available for run-time code, macros, and data buffers. The SRAM is allocated as follows: one page each for run-time code, macros, data buffer, telemetry buffer, and burst data buffer.

4.2 Instrument Control

4.2.1 Commanding

The MAG software controls the MAG electronics via memory-mapped I/O. The MAG software sends a command from the EPU to the magnetometer electronics by writing a 16-bit command word into an output control register via the FPGA. Other functions including the output sample rate, autonomous range control logic, and various parameter settings for EPU processing are performed by the EPU. Any command that changes the range or sampling rate or interrupts the sampling spacing (e.g., the A/D calibration) is acted on only after the instrument finishes collecting the current science packet. A "terminate science packet" command is provided to truncate the current packet, enabling prompt execution of commands that would otherwise be delayed.

4.2.2 Range Control Strategy

The instrument range can be controlled in automatic or manual modes. In manual mode, the instrument range is set by command. In automatic range mode, the software compares the largest absolute value of the signed count values against the current range's full scale and adjusts the range if needed to maximize resolution while keeping the readings on scale. On startup, the range is controlled in automatic mode and the default range is fine ($\pm$1,530 nT full scale). If the automatic control algorithm changes the range, the current science record packet is closed and a new packet with a new header reflecting the new range is started with the first sample in the new range. The number of samples in a packet is given in the packet header.

Range control processing is the first action taken when a new sample is received by the EPU. The software evaluates the absolute value of the signed X, Y, and Z counts. The maximum absolute value of the signed X, Y, and Z counts is compared against the full-scale count. If the current range is 0 or fine ($\pm$1,530 nT full scale) and the maximum absolute value is greater than 95% of full scale for $N_{\mathrm{to\_coarse}}$ consecutive samples, then the range is changed to 1 or coarse ($\pm$51,300 nT full scale). The default value for $N_{\mathrm{to\_coarse}}$ is 1 so that the instrument immediately shifts to coarse range if needed. If the current range is 1 (coarse) and the maximum is less than 2.5% of full scale for $N_{\mathrm{to\_fine}}$ consecutive samples, the range is changed to 0. The default value for $N_{\mathrm{to\_fine}}$ is 1,200. Both $N_{\mathrm{to\_fine}}$ and $N_{\mathrm{to\_coarse}}$ can be set by command.

4.2.3 Internal Calibrations

Two internal calibrations are available and are exercised by command. The Magnetometer calibration circuit injects a known current into each of the fluxgate sensors simulating a magnetic field offset of one-quarter full scale. This calibration is controlled on and off by setting an output register bit. The sigma-delta A/D converters include a self-calibration capability that is invoked by setting an output register bit high for two conversion cycles. This calibration lasts two conversion cycles (100 ms), and the A/D outputs no data during this time. The A/D calibration is turned off automatically after calibration is complete. A/D calibrations are allowed only between science data packets and are not allowed during bursts (Sect. 4.4) to preserve uniform sampling in each packet. The digital filters are also reset after calibration and require a few seconds to settle. For this reason, A/D calibrations are planned only when they do not impact primary science data acquisition (e.g., near apoapsis and in the solar wind).

4.3 Data Processing

Magnetic field data are processed as follows. Every 50 ms the EPU retrieves a 20-bit word for each magnetometer axis from the digital electronics. If digital filtering is enabled, the EPU software converts the data to signed 32-bit values, scales the values to physical units, and removes fixed offsets before applying digital filtering and sub-sampling commensurate with the commanded data output rate. The scaling corrections are then removed and the data are truncated to 16-bit resolution and assembled into packets of 200 vector samples. Data compression may be enabled using differential pulse code modulation (DPCM). For the MAG implementation of DPCM, the first full-word vector sample is retained and the remaining 199 samples are recorded as differences from the previous value. The number of bits used for the difference values is given in the packet header and is chosen to be the smallest that will accommodate the largest difference. Similar compression algorithms have been used in previous space missions (Behannon et al. 1977; Acuña et al. 1992; Neubauer et al. 1987). The time tag for the first sample in the packet is the MET time in seconds for the first sample together with a counter giving the number of 50-ms intervals elapsed from the first sample in the packet to the immediately previous MET second boundary. The header (instrument status), up to 200 vector data, and 1–10 Hz filter output values (discussed later) are assembled in a CCSDS telemetry packet and forwarded to the DPU. This data stream is the MAG standard science packet. An uncompressed science packet is 11.3 kbits long, corresponding to data throughput of 56.5 bits per vector sample so that the packet overhead and header amount to 8.5 bits per sample. Data acquired in the interplanetary medium to date typically require 5-bit deltas, which corresponds to a factor of 2.7 compression or 20.7 bits per sample.

A combination of digital filtering and sub-sampling is used to provide a wide range of sample output rates to provide maximum flexibility in output bit rate (cf. Table 1). For rates of 1, 2, 5, or 10 s^{-1} the data are filtered using an infinite impulse response (IIR) Butterworth low-pass filter with a corresponding Nyquist frequency of 0.5, 1, 2.5, and 5 Hz, respectively. Section 5.3 describes the frequency responses of the digital filters. Filtering can be commanded on or off for diagnostic purposes but is normally on. For the 20 s^{-1} rate no digital filtering is applied. If digital filtering is enabled and the commanded rate is 0.5 s^{-1} or lower, the software applies the 0.5-Hz low-pass filter and sub-samples the data to the commanded rate to a maximum sampling interval of 100 s. It is anticipated that sample intervals coarser than 1 s^{-1} would be used only if the telemetry rates or recorder space are severely limited. Because the spacecraft will be moving rapidly relative to the planet at low altitudes, sub-sampling at low sample rates was adopted to ensure that the data represent a sample at a well-defined position rather than an average over a range of spacecraft locations. For the present mission design we do not anticipate using sample rates coarser than 1 s^{-1} once in orbit at Mercury.

In addition to the vector samples, the software evaluates the average output amplitude of a 1-to-10 Hz bandpass filter for a single magnetometer axis selected by command. This parameter is evaluated once per second, and the result is digitally scaled to match the binary vector data. Its representation is a logarithmic "logAC" value with a four-bit mantissa formed by the four most significant non-zero bits and a four-bit power-of-two exponent of the fourth (least significant) bit of the mantissa. This parameter provides a continuous monitor of the 1-to-10-Hz ambient field fluctuations regardless of the commanded vector sample rate. It is included in the MAG high-priority housekeeping data to enhance the detection of transient events and magnetospheric boundaries as well as to assist in orbital operations planning. It is used internally to trigger burst data collection. The logAC samples are included in the standard science packet at a rate of 1 s^{-1} or at the commanded science rate, whichever is lower.

4.4 Additional Onboard Data Products

In addition to the standard science data, three other types of data are generated: low-rate housekeeping, instrument status, and burst. The low-rate housekeeping data provide continuous monitoring of instrument health and low-resolution science data. This data stream was designed to provide prompt, low-data-volume access to essential instrument health and science data to facilitate operations planning during periods of limited communications link margin. Status data are contingency data that are normally not downlinked but are stored in the spacecraft recorder if needed.

A burst feature was included because the downlink rate from MESSENGER at Mercury will not allow routine collection of 20 s^{-1} data for long periods. Burst data consist of eight contiguous minutes of 20 s^{-1} data, and this mode is triggered when the logAC value exceeds a commanded level. To provide context, 32 s of data prior to burst triggering are included in the burst interval. Burst sampling is enabled or disabled via time-tagged command and is disabled on startup. Burst data collection occurs in parallel with standard science and other data collection and therefore provides a convenient means to acquire the 20 s^{-1} data that are input to the digital filtering and logAC calculations. This feature was used extensively in pre-flight checkout and will be used in flight to verify that all filters are working properly.

5 Pre-flight Calibration and Testing

5.1 Scale Factors and Alignment

Scale-factor calibration and sensor alignment testing were conducted over several phases of instrument development. Initial parameters were established at GSFC's Laboratory for Extraterrestrial Physics during instrument tests. Absolute coarse-range scale factor calibration and sensor alignment measurements were conducted at the magnetics test facility at the NASA Wallops Flight Facility. This facility consists of a set of 20-foot coils that null the geomagnetic field and generate precise artificial fields in any desired direction. A proton precession magnetometer with 1-nT absolute accuracy was used to measure the applied fields. Optical cubes on the sensor test fixture provided precise alignment knowledge of the sensor relative to reference theodolites attached to the facility structure. During these tests, scale factors and relative alignment coefficients were determined independently using a variant of the technique developed by Merayo et al. (2000) using 27 sensor positions in a constant ambient field. The coarse-range absolute calibration was transferred to the fine range using the JHU/APL magnetics test facility.

5.1.1 Absolute Calibration and Relative Alignment

The technique used to provide scale factor calibration and precise relative orientation of the three axes consisted of applying a constant ~50,000-nT field in an arbitrary fixed direction and then rotating the sensor in steps of ~45° about three orthogonal axes. A total of 27 vector measurements was obtained. Precise positioning of the sensor is not required in this technique. The resulting alignment determination is relative to an arbitrarily chosen magnetic axis. This procedure was repeated for a field of ~10,000 nT. Because the sensor linearity is better than 1 part in 50,000 and the ambient field magnitude is constant for all sensor positions, the gain and orientation parameters can be obtained using least-squares minimization. Denoting the readings (counts) in the X, Y, and Z axes by c_x, c_y, and c_z and the fixed offsets as c_{x0}, $c_{y0,}$ and c_{z0}, the magnetic field may be expressed in sensor coordinates as

$$\begin{aligned} B_x &= k_x(c_x - c_{x0}),\\ B_y &= \alpha k_x(c_x - c_{x0}) + k_y(c_y - c_{y0}),\\ B_z &= \beta k_x(c_x - c_{x0}) + \gamma k_y(c_y - c_{y0}) + k_z(c_z - c_{z0}), \end{aligned} \quad (1)$$

where k_x, k_y, and k_z are the gain coefficients for each axis and α, β, and γ measure the contributions of X in the Y axis (α), X in the Z axis (β), and Y in the Z axis (γ), respectively, as off-diagonal elements of an alignment correction matrix $[M]$

$$[M] = \begin{bmatrix} 1 & 0 & 0 \\ \alpha & 1 & 0 \\ \beta & \gamma & 1 \end{bmatrix}. \quad (2)$$

Because the applied field, B_{appl}, is the same for all orientations of the sensor, the least-squares estimate minimizes the residual

$$\Delta^2 = \frac{1}{N}\sum_i \left(B_{x,i}^2 + B_{y,i}^2 + B_{z,i}^2\right) - B_{\text{appl}}^2. \quad (3)$$

The offsets c_{i0} were determined in three independent ways: by measuring the residual field with the sensor placed in a magnetic shield; by flipping the sensor 180° about the three axes in a small ambient field during the absolute calibration testing at Wallops; and by treating the offsets as free parameters in (3). The three techniques gave results differing by as much as 37 nT. The Wallops facility flip test was judged to give the most reliable measurement, so these values were used in pre-flight software. The offsets in the coarse and fine ranges are given in Table 3. Final offset determinations, including corrections for residual fixed spacecraft fields, will be determined from data acquired in flight (see Sect. 6.3).

Initial values for the fine-range scale factors were determined in bench-level tests at GSFC by measuring the relative response in fine and coarse range to the internal analog calibration signal. The fine-range scale factors were verified using the 2-m coil system at JHU/APL. The JHU/APL magnetics facility provides a zone 60 cm in diameter in which the field is uniform to 0.01%. The ratios of the scale factors between the two ranges were determined by measuring the response per unit facility coil current in both ranges. The gain factors for both ranges are given in Table 3. Coarse and fine-range scale factors are accurate to 0.01% and 0.08%, respectively.

Table 3 Absolute gain, sensor alignment calibration values, and internal instrument offsets

Parameter	Coarse range (±51,300 nT)	Fine range (±1,530 nT)
k_x: nT/count	1.56513	0.046769
k_y: nT/count	1.56419	0.046673
k_z: nT/count	1.61029	0.047997
α	–0.00462	–0.00462
β	0.00053	0.00053
γ	–0.00736	–0.00736
X offset: counts (nT)	–30.4 (–48)	–1017 (–48)
Y offset: counts (nT)	–75.2 (–118)	–2520 (–118)
Z offset: counts (nT)	–16.2 (–26)	–544 (–26)

All values are pre-flight calibration results. The offset determinations do not account for spacecraft fields and are the initial values used in the onboard software. Final offsets were determined following boom deployment

5.1.2 Sensor Orientation

Calibration of the sensor orientation relative to spacecraft coordinates was accomplished in two separate measurements. The first was conducted prior to launch and measured the orientation in azimuth about the Y axis, i.e., the axis of the boom, using optical cubes on the sensor and spacecraft as follows. At the Wallops magnetics test facility a magnetic field was applied in the X–Z sensor plane, and the sensor was rotated to align it precisely with an optical reference. The ratio of X and Z fields then yields the angle of the sensor axes relative to vertical (or horizontal). After integration to the boom and with the boom deployed (vertically), the orientation of the same optical cube was measured relative to a cube mounted on the boom root hinge. In final spacecraft alignment tests, the root hinge cube orientation was determined with respect to the spacecraft reference axes. To account for the relative azimuth orientation the conversion from sensor (mag) coordinates to spacecraft (SC) coordinates is

$$\begin{aligned} B_{x-\mathrm{sc}} &= 0.99635 B_{x-\mathrm{mag}} + 0.08542 B_{z-\mathrm{mag}} \\ B_{z-\mathrm{sc}} &= -0.08542 B_{x-\mathrm{mag}} + 0.99635 B_{z-\mathrm{mag}}. \end{aligned} \tag{4}$$

The second measurement—to determine the deviation of the Y-axis sensor direction from the spacecraft Y-axis—is difficult to make prior to launch but can be assessed in flight as described in the following. In addition, the Earth flyby of August 2, 2005, provided an opportunity to verify the sensor alignment, both the azimuth about and alignment to the spacecraft Y axis, by comparison against geomagnetic field reference models. Detailed comparison with the predicted field for the Earth flyby should yield alignment estimates accurate to 0.1° (Anderson et al. 2001), and initial analysis confirms the pre-launch alignment to within 0.5°.

5.2 Frequency Response and Timing

The frequency response of the Magnetometer is determined by the output low-pass filters and the additional low-pass filter in the A/D converters. The left panel of Fig. 7 shows the attenuation of the combined filters that together yield a −3 dB point at 11.3 Hz and attenuation of 17 dB/octave at 15 Hz (26 dB/octave at 20 Hz). The filter characteristics for the analog and the digital filters are given in Table 4. The right panel of Fig. 7 shows the 1-Hz low-pass IIR filter response (solid circles) and the 1-Hz high-pass IIR filter (open circles) used to implement the 1-to-10-Hz bandpass filter. The high-pass filter has a −3 dB point at 0.87 Hz and an attenuation of −75 dB/octave below 0.8 Hz.

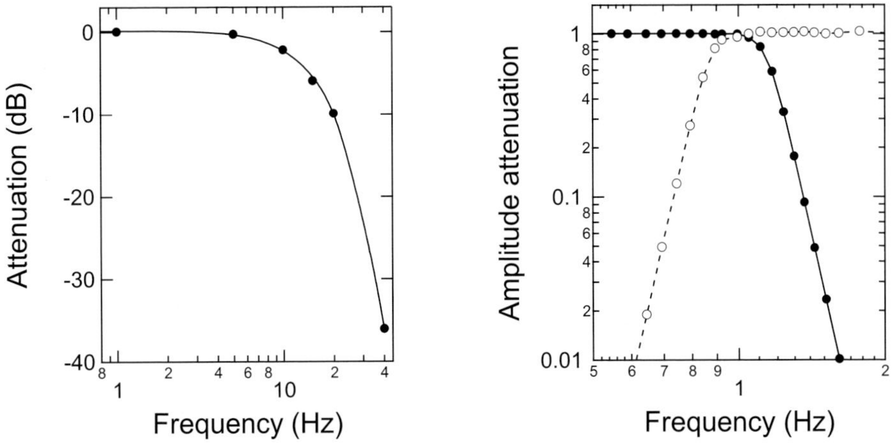

Fig. 7 Frequency response of the attenuation of analog electronics and A/D low-pass anti-aliasing filter (*left*) and digital filter (*right*). The analog anti-alias filter has its −3 dB point at 11.3 Hz, −17 dB/octave at 15 Hz. The 1-Hz low-pass (*solid symbols*) and 1-Hz high-pass (*open symbols*) IIR digital filters (*right*) characteristics are −3 dB at 1.14 Hz, −73 dB/octave above 1.25 Hz, and −3 dB at 0.87 Hz, −75 dB/octave below 0.8 Hz, respectively

Table 4 MESSENGER Magnetometer sample rates, digital IIR filter −3 dB points, filter attenuation characteristics, IIR time lags and net time lags

Rate setting	Sample rate (s^{-1})	Filter: −3 dB (Hz)	Attenuation (dB/octave)	IIR lag (s)	Net lag (s)
0	0.01	0.567*	–72*	2.316	2.358
1	0.02	"	"	"	"
2	0.05	"	"	"	"
3	0.10	"	"	"	"
4	0.20	"	"	"	"
5	0.50	"	"	"	"
6	1.00	"	"	"	"
7	2.0	1.141*	–73*	1.144	1.186
8	5.0	2.83*	–97*	0.435	0.477
9	10.0	5.38*	–147*	0.181	0.223
10	20.0	11.3	–17	0.0	0.042

*Characteristics of IIR digital filter

Digital filtering introduces additional delays relative to the high-rate data. The digital filters are Butterworth IIR filters with their cutoff frequency adjusted for the corresponding sampling rate as discussed earlier and shown in Table 4. The time delays are given by the properties of the digital filters and were verified during testing by acquiring simultaneous burst and digitally filtered data. Comparison of these two data sets allows direct measurement and confirmation to within 1 ms of the delay introduced by the digital filter. The total delays for all filters, including both the instrument and digital filter time lags, are given in Table 4.

5.3 Environmental Testing

The sensor, cabling, and electronics were tested in vacuum over a range of temperatures prior to spacecraft integration to ensure functionality in Mercury's environment. The sensor and electronics were cycled over the temperature range $-106°C$ to $+47°C$ and $-44°C$ to $+63°C$, respectively, in all possible combinations of extremes, spanning a range broader than the expected temperature range at Mercury for normal operating conditions. Instrument performance was monitored throughout by exercising the A/D self-calibration and the Magnetometer step calibration feature in both ranges. Fine-range operation was achieved using cancellation coils located near the sensor in the test chamber that reduced the field at the sensor to less than 1% of the ambient field. The relative axis alignment and the scale factors were tested at the JHU/APL facility before and after environmental testing and were identical to within the accuracy of the measurements, 0.08% for the scale factor and 0.03° for the alignment.

5.4 Spacecraft Magnetics Control Program

All components of the spacecraft employing magnets or magnetic materials were screened for magnetic signatures. Current loops were avoided in the design of the battery, solar array, power distribution system, and spacecraft harnesses. The three spacecraft systems that employed magnets or magnetic materials that could generate large signatures at the MAG sensor were, in decreasing order of significance: the propulsion latch valves, the momentum reaction wheels, and the battery. The steps used to mitigate the effects of these magnetic field sources on the measurements are summarized in the following.

5.4.1 Static Spacecraft Field

The magnetic moment of each latch valve in the propulsion system was characterized at the JHU/APL magnetics facility prior to propulsion system assembly. A mathematical model of the field from the propulsion valves was generated using the geometry of the propulsion system and used to estimate the field from the propulsion system at the deployed Magnetometer sensor location. The same model was used to design the intensity and distribution of magnets to cancel the field at the sensor location. Two permanent magnets were attached to the spacecraft structure to cancel the field of the latch valves at the MAG sensor location.

The battery cells use nickel in both the casing and in the active elements and therefore can exhibit "soft" (Acuña 2002) magnetization when exposed to external fields. The cells were demagnetized in an ambient field of <0.5% of Earth's field at the JHU/APL magnetics facility. This step reduced the cells' "as received" residual magnetic moment by an order of magnitude. Other spacecraft components exhibiting "hard" behavior were demagnetized to minimize their contribution to any spacecraft residual field. These included: stainless steel hardware, solar array cell interconnects, and stainless steel fastening material for the spacecraft sunshade. A spacecraft control process was followed to ensure that none of the magnetically permeable components was exposed to magnetizing fields (e.g., due to magnetized tools) during spacecraft integration.

The spacecraft static moment was assessed near the end of integration by means of a "swing" test in which the spacecraft was suspended from an overhead crane and allowed to swing like a pendulum with a lateral displacement of ~60 cm. The magnetic signature of the spacecraft was monitored simultaneously with three test magnetometers. The swing motion modulates the detected signature and allows the estimation of the total moment and approximate source location. This test confirmed that there were no magnetic field sources other than those identified above and that no "soft" components had been remagnetized. In-flight determination of the spacecraft field is discussed in Sect. 6.

5.4.2 Variable Spacecraft Field

The momentum wheels of the attitude control system were found to produce variable magnetic signatures that would have been larger than 0.3 nT at the deployed sensor location. The sources were identified as steel and magnets in the rotor assembly. Demagnetization of the rotors and bearings reduced the magnetic signature by more than a factor of three. Additional reductions, about another factor of three, were obtained using a single layer of 20-μm-thick amorphous magnetic shield. The shielding mass for each wheel was 33 g. Tests during integration confirmed that the field from the shielded wheels was negligible at the location of the deployed MAG sensor.

The magnetic fields generated by the power system were minimized using compensation loops and twisted pairs for all power distribution lines. The solar arrays were backwired, that is, the return for each string of cells was routed along the back of the panel to retrace the string on the front face. In addition, adjacent strings were arranged with reversed polarities so as to approximately cancel the moment due to the small remaining loop created by the panel thickness. Special attention was also paid to the Mercury Laser Altimeter (MLA) instrument (Cavanaught et al. 2007), which includes high-current electronics to drive the laser pulses. Cancellation loops were added in the design of the power distribution electronics. The battery cells were wired in a figure-eight pattern and the orientations of the cells optimized to minimize the total magnetic moment of the battery. The charge/discharge fields were verified to be primarily quadrupolar out to a distance of $\sim$1 m. Tests and design calculations indicate that the field at the deployed sensor arising from the solar arrays and battery are less than about 20 pT.

Variable spacecraft fields were monitored with the flight sensor in its stowed configuration during spacecraft integration and environmental testing. As for MAG environmental testing, coils were used to cancel out the ambient field to allow operation in the fine range. The background noise levels measured by MAG in fine range during these tests ranged from one to several nT, and no signals from the power system or other instruments above this level were detected. Measurements with a portable test instrument were carried out in the integrated configuration of the reaction wheels, power distribution system, solar array drive motors, and other systems and revealed no magnetic signatures at a level greater than 10 nT at a distance of 10 to 20 cm from each component, consistent with a negligible field at the deployed location of the MAG sensor.

6 Flight Checkout and Performance

6.1 Turn-on and Initial Checkout

MESSENGER was launched at 2:16 AM EDT on August 3, 2004, and nine days later the MAG instrument was turned on. The first data returned after MAG turn-on including an initial checkout are shown in Fig. 8. The data are not converted to physical units but plotted in counts the better to show the range control and calibration operations. At turn-on, the instrument remained in its sensitive range, because the baseline field in the stowed position was well within fine-range full scale. The baseline field in the stowed configuration was approximately 7,000 counts (300 nT) in X, –5,000 counts (–235 nT) in Y, and 2,000 counts (95 nT)

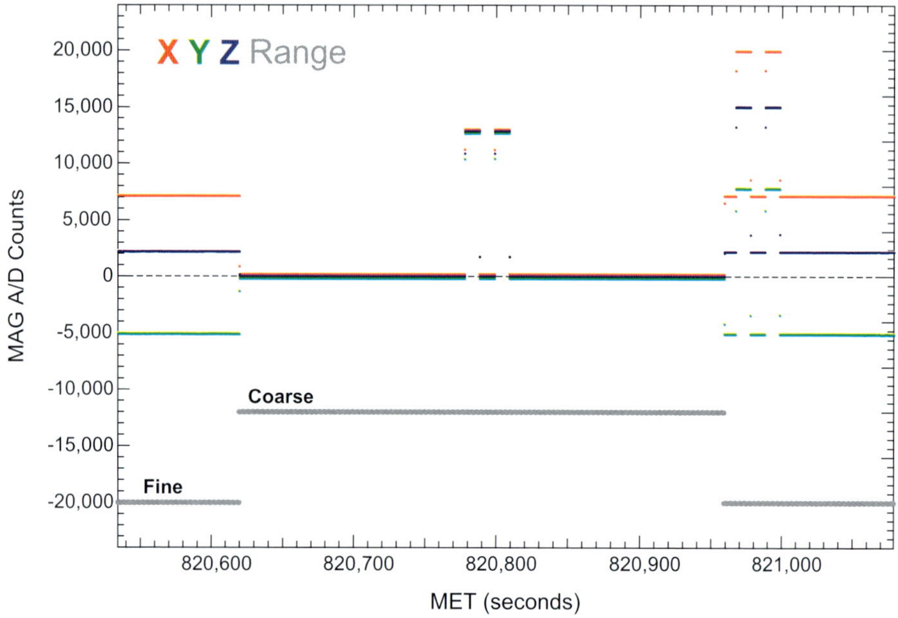

Fig. 8 First in-flight data from the MESSENGER MAG on August 12, 2004, nine days after launch. Plot shows initial checkout operations including analog calibration and range commanding. The two square-wave pulses near 820,800 MET and 821,000 MET are activations of the analog calibration. One major division is 235 nT. The Magnetometer sensor remained in its stowed configuration until March 8, 2005

in Z, consistent with expectations for the stowed sensor location inasmuch as the sensor was close to propulsion latch valves and thrusters mounted on the spacecraft $-Z$ deck. The initial checkout consisted of commanding the instrument to coarse range, activating the analog calibration twice, commanding to fine range, again activating the analog calibration twice, and returning to auto-ranging mode. The instrument performed as commanded, and the results were consistent with pre-launch testing. The analog calibration steps were within 0.1% of pre-launch values on average (0.26% worst case). Instrument performance was nominal and indicated no changes from pre-launch values.

6.2 Boom Deployment

The Magnetometer boom was deployed on March 8, 2005, and high-time-resolution ($20\ \mathrm{s}^{-1}$) Magnetometer data were recorded throughout the deployment (Fig. 9). Deployment was performed in two steps, first by releasing the mid-point hinge and allowing hinge oscillations to damp out, then by releasing the root hinge located at the base of the boom. Both hinges are spring loaded and have an equilibrium point defined by flat faces with orientation reproducible to within 0.05°. The hinges allow rotation beyond equilibrium in both directions. The magnetic field shows two step-like decreases in the field, as expected from the spacecraft magnetic field model determined from pre-flight measurements. Offsets characterized subsequently (see Sect. 6.3) were used in processing these data. The midpoint hinge overshoots and executes five swings of successively smaller amplitude. The field decreased from 400 nT to less than 20 nT as the result of the mid-point hinge release. This change is consistent with the spacecraft field model, including compensation magnets in their designed

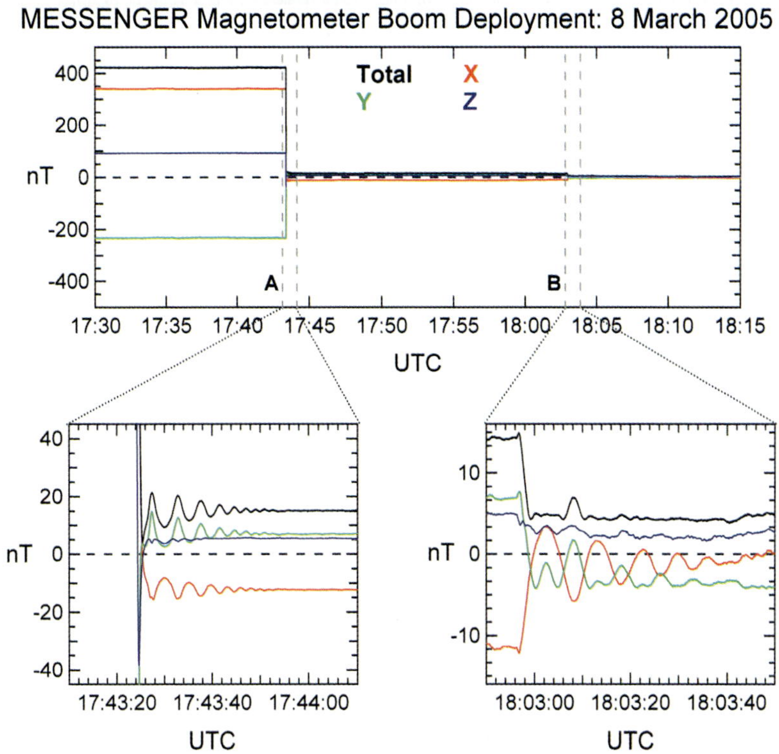

Fig. 9 Magnetic field records during the boom deployment on March 8, 2005. Top panel shows data spanning the entire boom deployment sequence. *Lower panels* are expanded views of the data for each hinge deployment: (*A*) the mid-point hinge deployment at 17:43:25 UTC; (*B*) the root hinge deployment at 18:02:58 UTC. Data are in sensor coordinates

orientation (polarity). The field observed following the root hinge release also overshoots and oscillates five times before damping sufficiently to settle at the hinge flat faces. The field decreased to less than 10 nT following the root hinge release. This magnitude field is consistent with expectations for a nominal solar wind field. The magnetic field oscillations allowed quantitative estimates of the amplitude of the hinge overshoot rotations and were consistent with mechanical design expectations.

Data obtained at high sample rate shortly after boom deployment allowed initial assessment of possible residual spacecraft noise at the deployed sensor location. In the stowed position, noise from components within 30 cm of the sensor was detected in association with star camera image acquisition, but no other signals (for example, from spacecraft reaction wheels) could be identified. Data following boom deployment on April 1, 2005 (day 091), are plotted in Fig. 10. The fluctuations are typical of interplanetary magnetic field variations showing greater variation transverse to the field direction than in the total field magnitude. Note the relatively large variations in Y and Z compared with X and the total field. The expanded-scale plots show that the solar wind variations are observed down to the digitization level of the data, 0.047 nT.

Figure 11 shows power spectra of these data (left panel) together with spectra of synthetic data (right panel) to indicate the effects of amplitude digitization. The left panel shows power spectra of the three components of the magnetic field in sensor coordinates. The black line

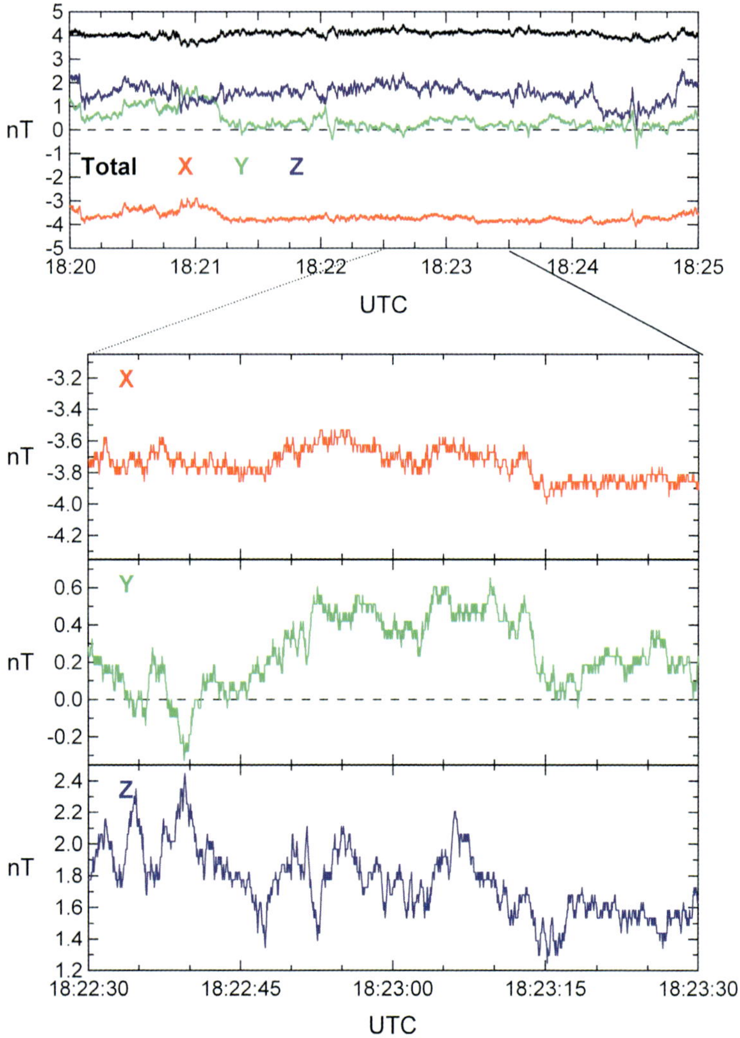

Fig. 10 Data in sensor coordinates on April 1, 2005 (following boom deployment), taken at 20 samples per second. Signals show nominal solar wind fluctuations and reveal no signals indicative of spacecraft noise above the digitization resolution of 0.05 nT

is a power-law fit to the X-component power spectrum below 3 Hz. The spectrum below 3 Hz follows a power law, $P \sim f^{-\alpha}$, dependence typical of interplanetary magnetic field fluctuations (e.g., Denskat et al. 1983), where $\alpha = -1.37$. There is also a broad enhancement in the power from 0.1 to 1 Hz typical of Aflvénic turbulence (e.g., Leamon et al. 1998; Tsurutani et al. 2001).

The spectra flatten above about 3 Hz (Fig. 11) giving an average spectral density of $\sim 2.5 \times 10^{-4}$ nT^2/Hz. This is higher than the lowest detectable power associated with the digitization step size, $P_{\Delta B} > \Delta B^2/(12F_{\text{Nyq}})$ which for these data is 1.8×10^{-5} nT^2/Hz (e.g., Bendat and Piersol 1986). This outcome is to be expected because the lower detection limit, $P_{\Delta B}$, is not a comprehensive measure of high-frequency noise due to digitization. In fact,

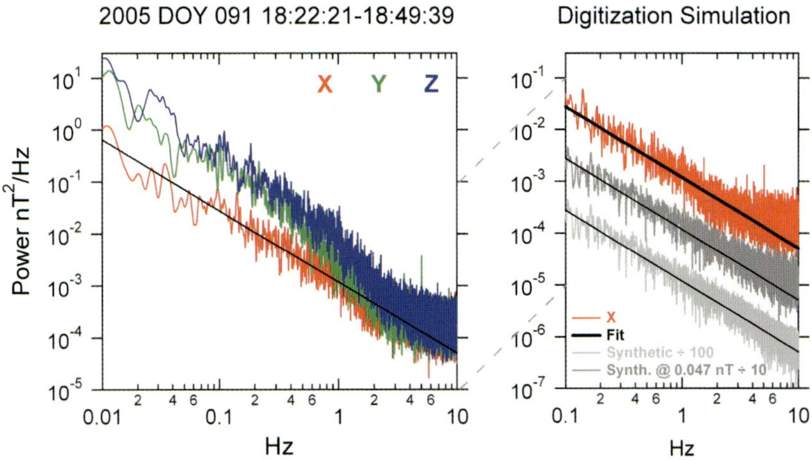

Fig. 11 Power spectra of 20-samples-per-second data shortly after boom deployment on April 1, 2005, showing flattening of power spectra as expected for 0.047-nT digitization and 0.05-s sampling

digitized data processed with discrete Fourier transforms will yield artificial high-frequency power above $P_{\Delta B}$ that depends on the low-frequency power (Welch 1961). Artificial higher-frequency power is introduced so the transform can reproduce the discrete steps in lower-frequency signals. To verify this effect for these data, we constructed an artificial time series with the same spectral intensity below 1 Hz as the X-component data and verified that digitizing these artificial data at 0.047 nT yielded a flattened spectrum above a few Hz. The right-hand panel of Fig. 11 shows the power spectrum of the synthetic data (lowest trace), the spectrum of the synthetic data digitized with a 0.047-nT step size (middle trace), and the spectrum of the X-component data. The results demonstrate that the spectra are consistent with the expected instrument response to the observed turbulent interplanetary magnetic field. One feature indicating a possible contamination field from the spacecraft is the narrow peak at 5.04 Hz reaching 4×10^{-3} nT^2/Hz. Integrating this peak yields an amplitude estimate for this signal of ~0.02 nT. This result confirms that at this time and at least for frequencies between 3 and 10 Hz, the spacecraft noise was below 0.05 nT.

6.3 Initial In-flight Offset Determination

After boom deployment, two spacecraft maneuvers were performed to determine the residual offsets. The data from these maneuvers are shown in Fig. 12. For the offset analysis, the data were processed by removing pre-flight offset estimates and applying the calibrations to convert to engineering units. In the first maneuver, the spacecraft was rolled about the Y-axis, parallel to the spacecraft–Sun line, for three complete revolutions. In the second maneuver, the spacecraft was rolled once about the X-axis, parallel to the solar panel axis. The X-axis roll was executed once rather than three times because this rotation stresses the spacecraft systems in that the solar panels turn away from the Sun and the star cameras are exposed to scattered sunlight. Offsets were determined by fitting sine waves to the time spans of the rolls as shown by the black traces. The offset estimate is given by the shift of the fit sine wave from zero, and differences in the sine fit amplitudes between the two roll-plane components provide a measure of uncertainty due to variations in the interplanetary field during the maneuver. The offset values and corresponding uncertainties are given in Table 5. The two Z-axis offsets agreed within the uncertainties, and only the lower uncertainty

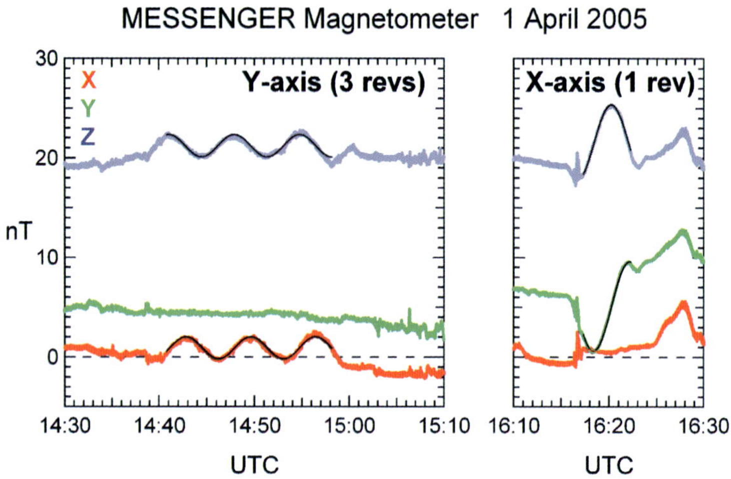

Fig. 12 Magnetic field data during spacecraft roll maneuvers on April 1, 2005. Left panel shows data for the three rolls about the spacecraft Y-axis while the right panel shows data for the single roll about the spacecraft X-axis. Black sinusoidal traces are fits to the components displaying a roll signal, X and Z for the Y-axis rolls and Y and Z for the X-axis roll. The zeros of these sinusoid fits correspond to the fixed field offsets

Table 5 Offsets determined with the instrument in fine range following boom deployment from data taken during the spacecraft maneuvers and as shown in Fig. 12

Sensor axis	Counts	nT
X	19.6 ±1.6	0.92 ±0.08
Y	92 ±13	4.29 ±0.61
Z	442.0 ±1.6	21.20 ±0.08

value is reported. Errors for the X and Z axes are smaller because they were derived from three spacecraft rolls, whereas the Y-axis offset is determined from only one spacecraft roll about the X axis.

The offsets determined in flight are considered to be corrections to the pre-launch estimates for the instrument offsets (Table 3) rather than spacecraft residual fields. The Y and Z offsets in Fig. 12 are unlikely to be unexpectedly large spacecraft fields because the near-field measurements agree with the pre-launch estimate for the spacecraft field and errors in the spacecraft field yielding a 20-nT residual field at the deployed sensor location would correspond to large discrepancies, hundreds of nT, between the model and observed fields near the spacecraft. Several techniques were used in pre-flight testing to determine the internal instrument offsets, but they did not all agree and differed by as much as 37 nT in Y and 34 nT in Z. Although a best judgment was made as to which pre-launch values were most reliable, the flight results are considered definitive. The flight offset corrections deviate by less than the variation in pre-launch offset estimates, so we regard the flight results as a refinement to the instrument offsets.

The roll maneuvers provide evidence that the deployed sensor orientation is consistent with the mechanical design. The roll signals in the roll-axis directions were both zero to within the uncertainties associated with variations in the IMF. Analysis of data acquired during the Earth flyby of August 2, 2005, provides verification of the sensor orientation. Final results of the flyby observations will be reported in an in-flight instrument calibration paper, but preliminary results indicate that the rotation about the Y-axis is ~0.5° less than the pre-flight alignment result and that the tilt of the Y-axis is $\sim 0.3^\circ$ in the Y–Z plane and less than 0.05° in the X–Y plane.

7 Data Acquisition Strategy and Plans

7.1 In-flight Calibration Activities

The science objectives for MAG require that the spacecraft residual field be known and stable to $\sim$1 nT and 0.05 nT, respectively. There are additional activities that will allow refinement of the offset determinations, verify the sensor orientation, and validate coordinate transformation software. The Y-axis spacecraft roll maneuver established the offsets in the X and Z axes to within this requirement. A second method to determine the offsets relies on the known properties of the IMF and will be used to complement the spacecraft rolls. On time scales of hours and days, variations in the IMF are predominantly rotational, and long-term observations can be used to determine the offsets. These techniques have been used on many interplanetary missions including Interplanetary Monitoring Platform (IMP) 8, Wind, Voyager, ACE, and NEAR-Shoemaker (Ness 1970; Smith et al. 1998; Acuña 2002; Anderson et al. 2001). In addition, the Earth flyby closest approach altitude of 2,300 km ensures a sufficiently strong maximum magnetic field to provide alignment and coordinate transformation validation checks accurate to $<0.5^\circ$ (Anderson et al. 2001). Initial analysis confirms that the sensor orientation and transformation code are correct.

7.2 Mercury Flyby Observations

The three Mercury flybys by MESSENGER will provide unique opportunities to detect magnetic signatures of the equatorial regions of the planet. In the primary mission orbit, the minimum altitude at Mercury's equator varies from $\sim$900 km to $\sim$1,700 km, so any small-scale crustal features or higher-order internal-field terms near the equator will not be resolvable from orbital data. Low-altitude data near the equator provide important information to constrain the spherical harmonic representation of the field (e.g., Korth et al. 2004; cf. Connerney and Ness 1988), and the flybys are the only opportunities in the mission to sample the equatorial regions at low altitude. The closest approach of all flybys to the surface will be $\sim$200 km, closer than any of the Mariner 10 Mercury encounters, and the trajectories will be distributed in planetary longitude. The data from the Mercury flybys will therefore improve our knowledge of the planet's magnetic field significantly prior to Mercury orbit insertion.

7.3 Mercury Orbital Observations

The MESSENGER orbit at Mercury will be 12 hours in period and highly eccentric, reaching a periapsis altitude of $\sim$200 km at $\sim 60^\circ$ N latitude and an apoapsis distance 7.6 R_M. Depending on the local time of the orbital plane, the spacecraft will spend from one third to one fifth of each orbit within Mercury's magnetosphere. These magnetosphere transits are important because characterization of the intrinsic magnetic field geometry depends on a quantitative understanding of the magnetospheric structure and currents. In addition, the dependence of the magnetospheric currents on the IMF implies that measurements acquired when MESSENGER is in the solar wind are also essential to the achievement of science objectives and will be used in at least two ways. First, they provide a basis to estimate the distribution of IMF conditions imposed on Mercury's magnetosphere during the mission's orbital phase. This statistical data set will provide a basis for estimating the contribution of the external currents as was done, for example, by Korth et al. (2004). Second, although the IMF will often be highly variable, there are likely to be some periods of prolonged stability, identified by a paucity of time variations within the magnetosphere together with comparable IMF values prior to and after transit through the magnetosphere. These events will provide opportunities to check the IMF-driven external field model and may form a high-fidelity data subset for analysis of the planet's intrinsic field.

The observation strategy is also influenced by the available data downlink. Because the total telemetry volume for MESSENGER available to MAG is not sufficiently large to allow continuous high-rate sampling, MAG operations will be tailored to focus on the regions of greatest interest while still providing full-orbit coverage at lower sampling rates. The telemetry downlink rates from MESSENGER will also vary during orbital operations. In particular, when Mercury is close to the Sun as viewed from Earth, the downlink rates will be so low that standard science data will be stored onboard for a month or longer before playback and only low-volume data will be available daily for health and safety monitoring. Because observations during the orbital phase will be planned no less than two weeks prior to their execution, the low-volume data must provide sufficient information to allow observation planning.

7.3.1 Low-Rate High-Priority Data

The low-rate housekeeping data described in Sect. 4.4 will be telemetered daily even during periods of low link margin. During orbital operations these data will provide instrument temperature and current data every 500 s, and vector magnetic field and logAC fluctuation level samples every 50 s (860 samples per orbit), thus giving insight into both instrument performance and recent magnetic field observations. The burst capability is triggered by the logAC fluctuation level and will be used to allow high-time-resolution observations of intermittent waves, substorm occurrence, or boundaries whose locations or timing cannot be accurately predicted. Because the fluctuation levels associated with these phenomena are not presently known, the logAC values are included in the continuous science data stream to aid in identifying the burst trigger levels appropriate to the local time, phenomenon, or physical region of interest. The full orbit vector data will provide information on the locations of boundaries and their variability, both to identify the appropriate time windows during which bursts should be enabled and to identify time ranges during the orbit to be used for enhanced sampling rates in the standard science data.

7.3.2 Standard Science Data

The primary science observations are provided in the standard science data format, which consists of vector and logAC samples at the commanded rate as described in Sect. 4.3. During orbital operations the rates used will be varied according to spacecraft position in Mercury's magnetosphere and relative to its boundaries. Early in the orbital phase, the link margins will be sufficiently high to support high-time-resolution sampling, 20 s^{-1}, throughout the orbit. At other times, the sampling rate will have to be tailored to the specific regions of interest. In general, solar wind observations will be made at lower sampling rates while higher rates will be used closer to the planet. Sampling rates less than 2 s^{-1} are not anticipated. The logAC channel provides fluctuation levels down to 1 Hz, so 2 s^{-1} vector sampling is required to achieve complete frequency coverage. A minimum sampling interval of 2 s^{-1} in the magnetosphere ensures full frequency coverage and provides sufficiently rapid pitch angle information to the Energetic Particle and Plasma Spectrometer (Gold et al. 2001; Andrews et al. 2007). Periods when the orbit crosses the sub-solar magnetopause will be targeted for higher-time-resolution sampling, as will magnetotail neutral sheet crossings.

7.3.3 Burst Data Collection

The burst mode will be used to collect high-rate vector data during portions of the orbit and/or during times of the mission when high-rate sampling is limited. This mode will be implemented by selecting a time window during which burst collection is enabled. The time window would be selected to span the region of interest and the logAC trigger level selected appropriate to the target region. Within the specified time window, a burst consisting of an eight-minute (480 s) span of data taken at 20 s^{-1} is taken if the logAC value exceeds the trigger level. To provide context for the trigger event, the first sample in the burst corresponds to a time 32 s prior to the logAC value that triggered the burst. The number of bursts allowed within a given time window is commandable, and bursts are acquired until the commanded limit is reached. Bursts may be contiguous but do not overlap. Present plans allow for taking an average of one burst per day.

8 Data Products

Timely access to data is important for instrument monitoring, observations planning, and science analysis. To support planning and science analysis, several MAG data products will be generated. These include Experiment Data Records (EDRs), quick look displays, and Calibrated Data Records (CDRs). Data packets are first converted to EDRs consisting of time-ordered records of all data of a given type received for each day. For burst data, however, all 15 packets for a given burst are converted into a single EDR file regardless of possible day boundaries during a burst. The EDRs are used for subsequent processing and to generate quick-look plots allowing rapid assessment of instrument status and performance. Calibrated data for science analysis, CDRs, are obtained from the EDRs by converting to engineering units and transforming to orthogonal coordinates and then into spacecraft coordinates. This intermediate product will consist of the vector samples in physical units, transformed to spacecraft coordinates together with time in both MET and UTC. In addition, the housekeeping data, instrument state (e.g., range, sample rate, AC axis), and burst status will be collected in a MAG activity/state log.

Because MESSENGER will transit a variety of regions and will address science appropriate to different coordinate systems, a next-level product will be generated that will include spacecraft location and data converted to all relevant coordinate systems. These systems will include an inertial system (e.g., J2000), Mercury-centered body-fixed, Mercury-centered magnetospheric, Mercury-centered solar wind, and heliocentric (appropriate for cruise). Magnetospheric coordinates depend on the orientation of Mercury's magnetic dipole as estimated by derived analysis products and will, therefore, be updated during the orbital phase of the mission. Corresponding coordinates for Earth and Venus will also be provided to support analysis of data acquired during the gravity-assist flybys of these planets. All subsequent analysis products and displays will be generated using these calibrated data. One of the principal objectives of the Earth flyby observations has been validation of the coordinate transformation algorithms used at this stage in the processing.

An example of interplanetary observations is shown in Fig. 13 for an interplanetary magnetic cloud interval observed by MESSENGER beginning on May 14, 2005. At that time

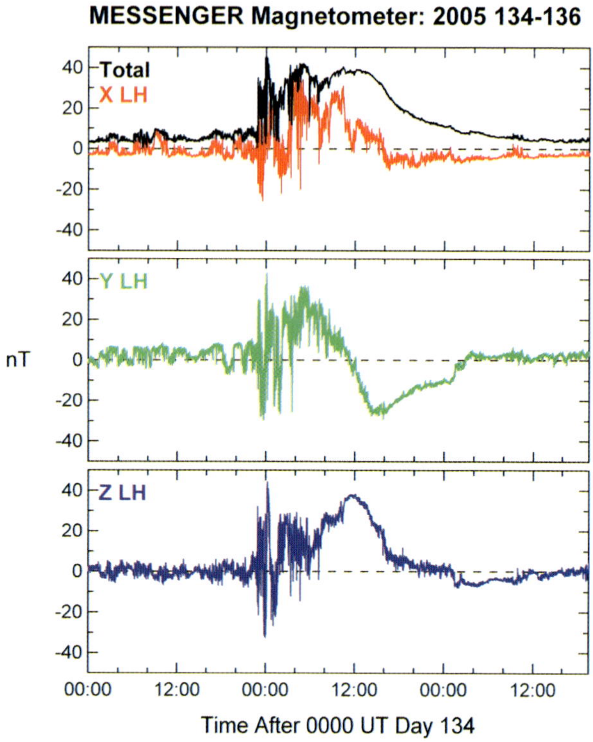

Fig. 13 Magnetometer data for an interplanetary magnetic cloud observed on May 14, 15, and 16 2005, associated with a coronal mass ejection. Data shown are processed science data at 1 sample/s in heliocentric coordinates, where X is radially inward, Y is normal to the ecliptic and X, and Z completes the right-hand system

MESSENGER was located 10° east of the Earth–Sun line and at a heliocentric distance of 0.93 AU. These data are in local heliospheric (LH) coordinates with X toward the Sun, Y positive eastward, parallel to the ecliptic plane and perpendicular to X, and Z completing the right-handed system, close to the ecliptic normal. The sheath of magnetized solar wind plasma compressed ahead of the coronal mass ejection (CME) arrives at MESSENGER near 2330 UT on May 14, and the magnetic cloud of the CME begins at about 0330 UT. The same CME was observed by the ACE spacecraft positioned in a halo orbit at the Earth's first Lagrangian point. Magnetic field data from ACE (Smith et al. 1998) show that the sheath arrived at ACE near 0200 UT on May 15 and the magnetic cloud began near 0530 UT. These data demonstrate that in addition to providing calibration of the MESSENGER offsets, interplanetary observations can be used to characterize the structure of magnetic cloud characteristics using multi-point and coordinated coronal observations (cf. Mulligan et al. 1999, 2001; Rust et al. 2005).

9 Summary

The MESSENGER Magnetometer will provide accurate magnetic field measurements for the characterization of Mercury's planetary and magnetospheric magnetic field. The MAG instrument provides a measurement capability up to ±51,300 nT in the coarse range, resolution of 0.047 nT in fine range, and full vector sampling up to 20 s^{-1}. Pre-flight calibrations have established an absolute measurement accuracy of 0.08% in the sensitive range (±1,530 nT). The instrument has been adapted to MESSENGER mission telemetry constraints to provide a wide range of sampling rates, from 100 s between samples (0.56 bits per second) to 20 samples per second (1,130 bits per second). A bandpass filter, 1 to 10 Hz, is applied to acquire continuous sampling of ambient field fluctuations useful for boundary and transient detection and for observation planning. Burst sampling is triggered by the bandpass amplitude when the telemetry rate does not allow continuous high-rate sampling. MAG therefore provides maximum science return within the mission constraints.

Success in meeting the observational objectives depends also on ensuring that magnetic field contamination effects are minimized. Through a cost-effective magnetics control program, sources of fixed and variable magnetic fields were identified early in the spacecraft development process, design recommendations were developed, and appropriate mitigation steps were taken. Measurements during spacecraft integration and in flight following boom deployment indicate that the variable field contribution from the spacecraft is below the MAG resolution in fine range, 0.047 nT. The fixed spacecraft field is expected to be below a few nT, and the measurements during boom deployment confirmed the spacecraft magnetic field model, indicating that this objective was met. The MESSENGER MAG investigation will therefore provide accurate characterization of the intrinsic planetary field, the magnetosphere, and the structure and dynamic processes accompanying Mercury's interaction with the solar wind.

Acknowledgements This work was supported under NASA Contract NAS5-97271 to JHU/APL. The authors are grateful for the efforts of the spacecraft engineers of the JHU/APL Space Department for the successful design, integration and test, and commissioning of the MESSENGER spacecraft. Additional thanks are offered to the mechanical, power system, propulsion system, and attitude control system engineering teams for supporting the magnetics mitigation effort. The authors are also grateful to the mission operations team for their efforts in the successful MESSENGER spacecraft commissioning and operations.

References

M.H. Acuña, IEEE Trans. Magn. **MAG-10**, 519–523 (1974)
M.H. Acuña, Rev. Sci. Instr. **73**, 3717–3736 (2002)
M.H. Acuña et al., J. Geophys. Res. **97**, 7799–7814 (1992)
M.H. Acuña, C.T. Russell, L.J. Zanetti, B.J. Anderson, J. Geophys. Res. **102**, 23751–23759 (1997)
M.H. Acuña et al., Science, **279**, 1676–1680 (1998)
M.H. Acuña et al., Science, **284**, 790–793 (1999)
O. Aharonson, M.T. Zuber, S.C. Solomon, Earth Planet. Sci. Lett. **218**, 261–268 (2004)
B.J. Anderson et al., IEEE Trans. Geosci. Rem. Sens. **39**, 907–917 (2001)
G.B. Andrews et al., Space Sci. Rev. (2007, this issue). doi:10.1007/s11214-007-9272-5
K.W. Behannon, M.H. Acuña, L.F. Burlaga, R.P. Lepping, F.M. Neubauer, Space Sci. Rev. **21**, 235–257 (1977)
J.S. Bendat, A.G. Piersol, *Random Data: Analysis and Measurement Procedures* (Wiley, New York, 1986), 566 pp
C. Bertucci, C. Mazelle, M.H. Acuña, C.T. Russell, J.A. Slavin, J. Geophys. Res. **110**, A01209 (2005). doi:10.1029/2004JA010592
L.F. Burlaga, Planet. Space Sci. **49**, 1619–1627 (2001)
L.F. Burlaga et al., Geophys. Res. Lett. **30**, 2072 (2003). doi:10.1029/2003GL018291
J.F. Cavanaugh et al., Space Sci. Rev. (2007, this issue). doi:10.1007/s11214-007-9273-4
S.P. Christon, Icarus **71**, 448–471 (1987)
J.E.P. Connerney, N.F. Ness, in *Mercury*, ed. by F. Vilas, C.R. Chapman, M.S. Matthews (University of Arizona Press, Tucson, 1988), pp. 494–513
J.E.P. Connerney et al., Science **284**, 794–798 (1989)
D.H. Crider, D.A. Brain, M.H. Acuña, D. Vignes, C. Mazelle, C. Bertucci, Space Sci. Rev. **111**, 203–221 (2004)
K.U. Denskat, H.J. Beinroth, F.M. Neubauer, J. Geophys. Res. **54**, 60–67 (1983)
I.M. Engle, Planet. Space Sci. **45**, 127–132 (1997)
G. Giampieri, A. Balogh, Planet. Space Sci. **49**, 1637–1642 (2001)
G. Giampieri, A. Balogh, Planet. Space Sci. **50**, 757–762 (2002)

K.-H. Glassmeier, in *Magnetospheric Current Systems*, ed. by S. Ohtani, R. Fujii, M. Hesse, R.L. Lysak, Geophysical Monograph, vol. 118 (American Geophysical Union, Washington, 2000), pp. 371–380
K.-H. Glassmeier, P.N. Mager, D.Yu. Klimushkin, Geophys. Res. Lett. **30**, 1928 (2003). doi:10.1029/2003GL017175
R.E. Gold et al., Planet. Space Sci. **49**, 1467–1479 (2001)
M.H. Heimpel, J.M. Aurnou, F.M. Al-Shamali, N. Gómez Pérez, Earth Planet. Sci. Lett. **236**, 542–557 (2005)
D.J. Jackson, D.B. Beard, J. Geophys. Res. **82**, 2828–2836 (1977)
H. Korth et al., Planet. Space Sci. **54**, 733–746 (2004)
R.J. Leamon, C.W. Smith, N.F. Ness, Geophys. Res. Lett. **25**, 2505–2508 (1998)
V. Lesur, A. Jackson, Geophys. J. Int. **140**, 453–459 (2000)
D.A. Lohr et al., Space Sci. Rev. **82**, 255–281 (1997)
J.G. Luhmann, C.T. Russell, N.A. Tsyganenko, J. Geophys. Res. **103**, 9113–9119 (1998)
C. Mazelle et al., Space Sci. Rev. **111**, 115–181 (2004)
R.T. Merrill, J. Geophys. Res. **86**, 937–949 (1981)
R.T. Merrill, M.W. McElhinny, *The Earth's Magnetic Field, Its History, Origin and Planetary Perspective*. International Geophysics Series, vol. 32 (Academic, London, 1983), 401 pp
J.M.G. Merayo, P. Brauer, F. Primdahl, J.R. Petersen, O.V. Nielsen, Meas. Sci. Tech. **11**, 120–132 (2000)
T. Mulligan et al., J. Geophys. Res. **104**, 28217–28223 (1999)
T. Mulligan, C.T. Russell, B.J. Anderson, M.H. Acuña, Geophys. Res. Lett. **28**, 4417–4420 (2001)
N.F. Ness, Space Sci. Rev. **11**, 459–554 (1970)
N.F. Ness, K.W. Behannon, R.P. Lepping, Y.C. Whang, K.H. Schatten, Science **185**, 151–160 (1974)
N.F. Ness, K.W. Behannon, R.P. Lepping, Y.C. Whang, J. Geophys. Res. **80**, 2708–2716 (1975)
N.F. Ness, K.W. Behannon, R.P. Lepping, Y.C. Whang, Icarus **28**, 479–488 (1976)
F.M. Neubauer, M.H. Acuña, L.F. Burlaga, B. Franke, B. Gramkow, J. Phys. E **20**, 714–720 (1987)
K.W. Ogilvie et al., Science **185**, 145–151 (1975)
T.A. Potemra, L.J. Zanetti, M.H. Acuña, IEEE Trans. Geosci. Remote Sens. **GE-23**, 246–249 (1985)
A.E. Potter, T.H. Morgan, Science **248**, 835–838 (1990)
S.K. Runcorn, Nature **253**, 701–703 (1975a)
S.K. Runcorn, Phys. Earth Planet. Inter. **10**, 327–335 (1975b)
C.T. Russell, D.N. Baker, J.A. Slavin, in *Mercury*, ed. by F. Vilas, C.R. Chapman, M.S. Matthews (University of Arizona Press, Tucson, 1988), pp. 494–513
D.M. Rust et al., Astrophys. J. **621**, 524–536 (2005)
A.G. Santo et al., Planet. Space Sci. **49**, 1481–1500 (2001)
G. Siscoe, L. Christopher, Geophys. Res. Lett. **2**, 158–160 (1975)
J.A. Slavin, Adv. Space Res. **33**, 1587–1874 (2004)
J.A. Slavin, R.E. Holzer, J. Geophys. Res. **84**, 2076–2082 (1979)
J.A. Slavin, J.C.J. Owen, J.E.P. Connerney, S.P. Christon, Planet. Space Sci. **45**, 133–141 (1997)
C.W. Smith, M.H. Acuña, L.F. Burlaga, J. L'Heureux, Space Sci. Rev. **86**, 613–632 (1998)
S.C. Solomon, Icarus **28**, 509–521 (1976)
S.C. Solomon et al., Planet. Space Sci. **49**, 1445–1465 (2001)
L.J. Srnka, Phys. Earth Planet. Inter. **11**, 184–190 (1976)
S. Stanley, H. Bloxham, W.E. Hutchison, M.T. Zuber, Earth Planet. Sci. Lett. **234**, 27–38 (2005)
A. Stephenson, Earth Planet. Sci. Lett. **28**, 454–458 (1976)
D.J. Stevenson, Rep. Prog. Phys. **46**, 555–620 (1983)
D.J. Stevenson, Earth Planet. Sci. Lett. **82**, 114–120 (1987)
D.J. Stevenson, T. Spohn, G. Schubert, Icarus **54**, 466–489 (1983)
B.T. Tsurutani et al., J. Geophys. Res. **106**, 30223–30238 (2001)
P.D. Welch, IBM J. Res. Dev. **5**, 141–156 (1961)
Y.C. Whang, J. Geophys. Res. **82**, 1024–1030 (1977)
J.D. Winningham et al., J. Geophys. Res. **98**, 10649–10666 (1993)
L.J. Zanetti et al., Space Sci. Rev. **70**, 465–482 (1994)

Space Sci Rev (2007) 131: 451–479
DOI 10.1007/s11214-007-9273-4

The Mercury Laser Altimeter Instrument for the MESSENGER Mission

John F. Cavanaugh · James C. Smith · Xiaoli Sun · Arlin E. Bartels · Luis Ramos-Izquierdo · Danny J. Krebs · Jan F. McGarry · Raymond Trunzo · Anne Marie Novo-Gradac · Jamie L. Britt · Jerry Karsh · Richard B. Katz · Alan T. Lukemire · Richard Szymkiewicz · Daniel L. Berry · Joseph P. Swinski · Gregory A. Neumann · Maria T. Zuber · David E. Smith

Received: 28 August 2006 / Accepted: 24 August 2007 / Published online: 9 November 2007

Abstract The Mercury Laser Altimeter (MLA) is one of the payload science instruments on the MErcury Surface, Space ENvironment, GEochemistry, and Ranging (MESSENGER) mission, which launched on August 3, 2004. The altimeter will measure the round-trip time of flight of transmitted laser pulses reflected from the surface of the planet that, in combination with the spacecraft orbit position and pointing data, gives a high-precision measurement of surface topography referenced to Mercury's center of mass. MLA will sample the planet's surface to within a 1-m range error when the line-of-sight range to Mercury is less than 1,200 km under spacecraft nadir pointing or the slant range is less than 800 km. The altimeter measurements will be used to determine the planet's forced physical librations by tracking the motion of large-scale topographic features as a function of time. MLA's laser pulse energy monitor and the echo pulse energy estimate will provide an active measurement of the surface reflectivity at 1,064 nm. This paper describes the instrument design, prelaunch testing, calibration, and results of postlaunch testing.

Keywords Mercury · MESSENGER · Topography · Laser altimeter

J.F. Cavanaugh (✉) · J.C. Smith · X. Sun · A.E. Bartels · L. Ramos-Izquierdo · D.J. Krebs · J. McGarry · A.M. Novo-Gradac · J.L. Britt · J. Karsh · R.B. Katz · D.L. Berry · J.P. Swinski · G.A. Neumann · D.E. Smith
NASA Goddard Space Flight Center, Greenbelt, MD 20771, USA
e-mail: john.f.cavanaugh@nasa.gov

R. Trunzo
Swales Aerospace, Beltsville, MD 20705, USA

R. Szymkiewicz
Orbital Sciences, Greenbelt, MD 20770, USA

A.T. Lukemire
Space Power Electronics, Kathleen, GA 31047, USA

M.T. Zuber
Massachusetts Institute of Technology, Cambridge, MA 02129, USA

1 Introduction

1.1 Mission Overview

The Mercury Laser Altimeter (MLA) is one of seven scientific instruments on the MErcury Surface, Space ENvironment, GEochemistry, and Ranging (MESSENGER) spacecraft. MESSENGER was launched on August 3, 2004, and will enter Mercury orbit in 2011 to perform scientific measurements for a period of one Earth year, equivalent to four Mercury years. MESSENGER will be in a highly elliptical and near-polar orbit around Mercury with a periapsis altitude of 200 to 400 km, an apoapsis altitude of ~15,000 km, and a period of 12 hours. Figure 1 shows the MESSENGER orbits and MLA measurement geometry. The spacecraft altitude will come within the MLA ranging capability near periapsis for 15 to 45 minutes during each orbit, depending on the time of the year. The periapsis latitude is at 60–70°N to increase the measurement coverage for low-latitude regions. The intense heat from the Sun will require the spacecraft to have its sunshade facing the Sun at all times, which during noon–midnight orbits will confine the payload deck to point within 10° about the normal to Mercury's orbital plane with MLA ranging at a slant angle as high as 50° (Solomon et al. 2001).

1.2 MLA Instrument Overview

MLA is a time-of-flight laser rangefinder that uses direct pulse detection and pulse edge timing to determine range to the surface. Its laser transmitter emits 5-ns pulses at an 8-Hz rate with 20 mJ of energy at a wavelength of 1,064 nm. Return echoes are collected by an

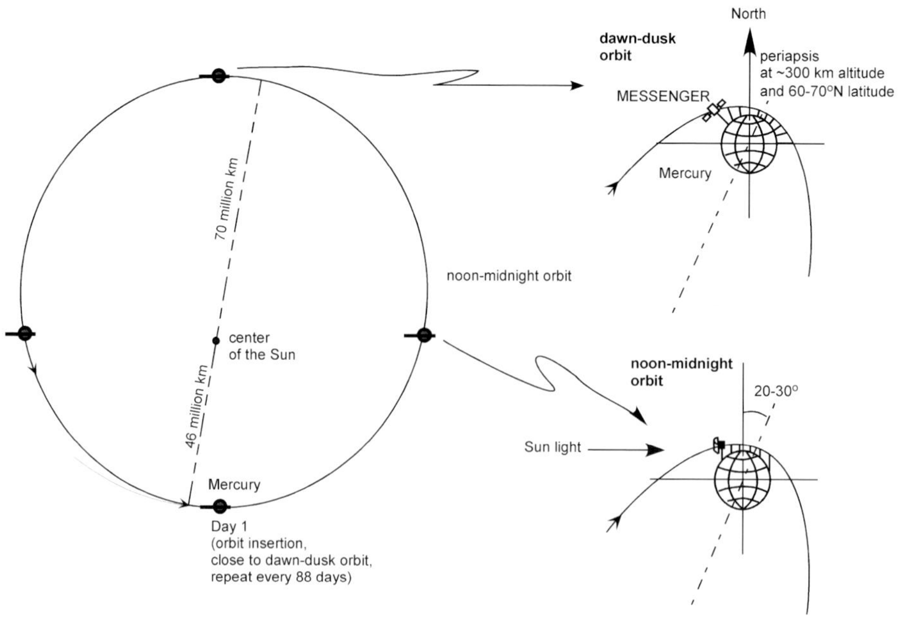

Fig. 1 MESSENGER spacecraft orbit around Mercury. MLA will measure range to the surface when the spacecraft is less than 800 km from the surface and the instrument's optical axis is within 50° of nadir. For optical axis angles approaching 0° (nadir pointing) MLA can range to the surface from 1,200 km

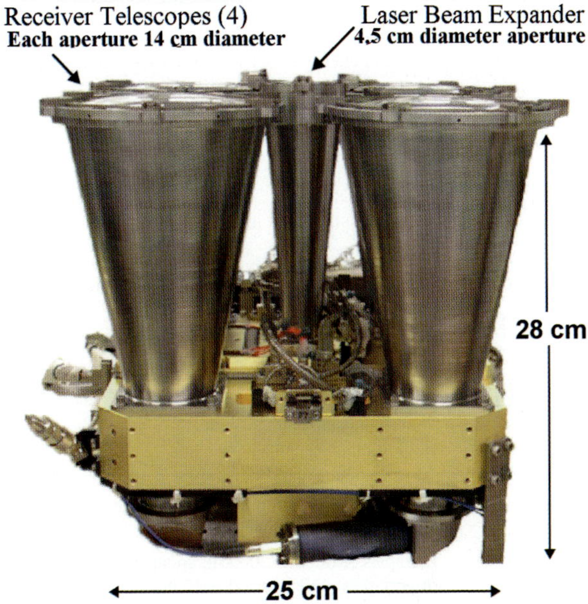

Fig. 2 MLA instrument shown without thermal protection blankets

array of four refractive telescopes and detected using a single silicon avalanche photodiode and three matched low-pass electronic filters. Pulse timing is measured using a combination of crystal-oscillator-based counters and high-resolution time-of-flight application-specific integrated circuits (ASICs). The MLA instrument is shown in Fig. 2.

2 Science and Measurement Objectives

2.1 MLA Science Objectives

The primary science measurement objectives for MLA are to provide a high-precision topographic map of the high northern latitude regions; to measure the long-wavelength topographic features at mid-to-low northern latitudes; to determine topographic profiles across major geologic features in the northern hemisphere; and to detect and quantify the planet's forced physical librations by tracking the motion of large-scale topographic features as a function of time. An additional goal of the MLA instrument is to measure the surface reflectivity of Mercury at the MLA operating wavelength of 1,064 nm (Sun et al. 2004).

2.2 MLA Instrument Objectives

MLA will measure the topography of the Mercury northern hemisphere via laser pulse time-of-flight data and spacecraft orbit position data in an approach similar to the Mars Orbiter Laser Altimeter (MOLA) (Abshire et al. 2000). MLA is designed to perform range timing measurements up to 1,800 km from the planet's surface. Since the single-pulse signal link margin is close to 0 dB for altitudes above 800 km, the MLA data acquisition scheme allows collection and downlink of up to 15 returns per shot in order to allow the use of correlation techniques to process the range signals on the ground. Measurement of both the outgoing pulse energy and received pulse shape enable the surface reflectivity measurement. The MLA functional requirements are enumerated in Table 1; allocations and constraints are listed in Table 2.

Table 1 Summary of MLA functional requirements

Time-stamp laser pulses to better than ±1 ms with respect to the ephemeris
Support determination of pointing stability to ±50 μrad peak-to-peak
Measure laser pulse time of flight from 1 to 8 ms with return pulse widths from 6 to 1,000 ns
Total ranging error <1 m with probability of detection $P_d > 95\%$ at 200 km altitude nadir pointing
Laser pulse repetition rate of 8 Hz
Produce laser footprint ≤16 m diameter at 200 km altitude, nadir pointing
$P_d > 10\%$ at 800 km slant range and 50° angle
Maintain long-term ranging bias error to ≤0.50 m over the mission lifetime
Measure the reflectivity of the ranging targets

Table 2 MLA design allocations and constraints

Parameter	Requirement
Instrument design life	6.6-year cruise with power off, followed by 1 year of operation
Orbit definition	Periapsis altitude = 200 km, apoapsis altitude = 15,193 km, inclination = 80°, latitude of periapsis ≤60°, period = 12 hours
Flight environment	Infrared irradiance from Mercury not to exceed 1 W/cm^2 onto the instrument for a 60 minute pass centered about periapsis; cold space viewing for the remaining 11 hours of the orbit
Testing environment	Protoflight verification program per MESSENGER Component Environmental Specification, drawing #7384-9101
Mass	7.4 kg
Power	<25 W for 60 minutes centered on periapsis and <16 W for the remaining 11 hours of the orbit. <17 W orbit average
Telemetry	2.0 Mbit average per 12-hour orbit about Mercury
Time correlation	1 pulse per second time tick from the spacecraft along with announcement with <50 ms accuracy in real time and <1 ms following navigation data processing
Relative pointing	Determine laser output pointing direction with respect to the spacecraft axes over temperature to ±50 μrad peak to peak
Envelope	28 cm × 28 cm × 26 cm

3 MLA Instrument Design

3.1 Background

MLA builds on the experiences gained from the development of several space-based and airborne laser altimeter systems flown over the past two decades including the Mars Observer Laser Altimeter (MOLA-1; Zuber et al. 1992), the Mars Orbiter Laser Altimeter (MOLA-2; Smith et al. 1998), two Shuttle Laser Altimeters (SLA-1 and SLA-2; Garvin et al. 1998), the Geoscience Laser Altimeter System on the Ice Cloud and Land Elevation Satellite (GLAS/ICESAT; Zwally et al. 2002), and the airborne Microchip Laser Altimeter System Microaltimeter (Degnan 2002). The measurement scheme for MLA is similar to that of MOLA, in which the outgoing pulse starts a counter that is stopped by the detection of a received echo pulse. A significant change implemented in MLA is the ability to record several return pulses, process them onboard, and send the "most likely" signal returns to the ground for correlation analysis. This technique is similar to and derived from experience with the Microaltimeter (Degnan 2002).

3.2 MLA Functional Overview

An overview of MLA operation may be gained from the MLA functional block diagram (Fig. 3). A trigger pulse initiated by the range measurement electronics starts the optical pumping of the MLA laser. This trigger pulse also initiates a 5-MHz 23-bit counter, 16 bits of which are used for coarse range timing. After optically pumping the laser for approximately 150 µs the laser emits a 5-ns, 20-mJ pulse, which propagates along the instrument line-of-sight to the planet's surface. A small fraction ($<10^{-6}$) of the emitted pulse impinges on a four-segment photodiode that performs three functions. One output of this photodiode is fed to the laser electronics where it is used to terminate the laser diode pump array drive current. The second output goes to a pulse integrator used to measure emitted energy, and a third output is fed to a comparator, which generates a logic pulse synchronous with the transmitted optical pulse. The leading edge of this latter pulse starts the first of six fine-resolution time-of-flight counters, which is subsequently stopped by the next edge of the 5-MHz clock. These counters provide a relative time within each 200-ns coarse clock cycle. The coarse counter value is latched at this point, and the two counters provide the leading-edge start timing value (with respect to the trigger pulse) for the range measurement. The trailing edge of the start pulse is timed in an identical manner. These first pair of counts establish the start time of the range timing cycle. The pulse width is computed by subtracting the leading edge measurement from the trailing edge measurement.

A portion of the pulse energy reflected off the planet's surface is collected by MLA's four receiver telescopes and focused onto four optical fibers that relay the signal through an opti-

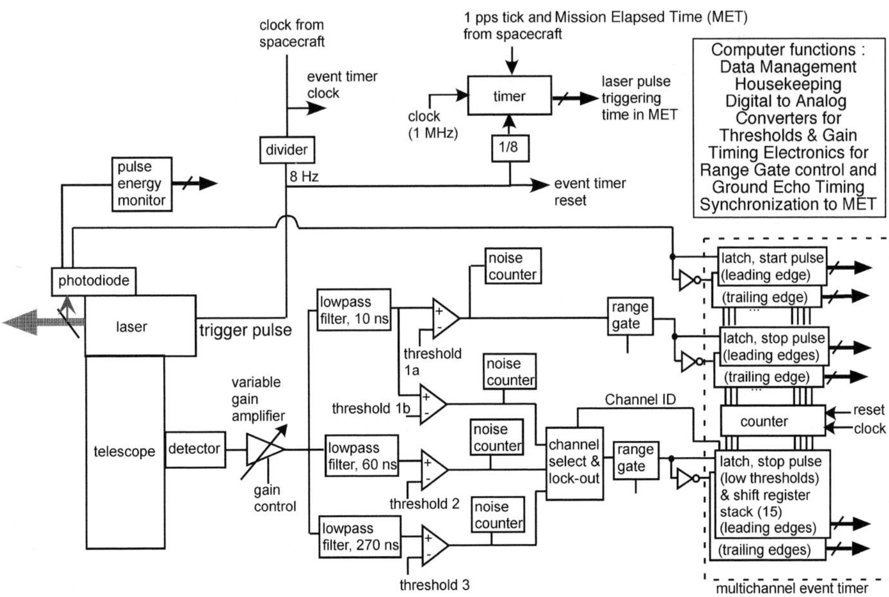

Fig. 3 MLA functional block diagram

cal bandpass filter to a silicon avalanche photodiode (SiAPD), which preserves to the extent of its bandwidth the deformation of the incident pulse caused by interaction with surface features. Slant angle ranging, surface roughness, or large-scale features on the surface can cause temporal pulse spreading. To maximize the probability of detecting spread returns, three matched filters are used after the detector amplifier. The first, a high-bandwidth filter output with a 10-ns impulse response, is split and fed to two comparators with independently programmable thresholds referred to as channel 1 high and low, respectively. The channel 1 high comparator signal is fed to the second pair of time-of-flight counters, which record the leading and trailing edge times of this pulse. The channel 1 low comparator and the other filter outputs with 60-ns and 270-ns impulse response, referred to as channels 2 and 3, respectively, are fed to the third pair of counters. For this last pair of counters the first received pulse leading edge to cross its corresponding threshold will stop the counter and lock out the other channels for approximately 1 μs in order to prevent a different channel from stopping the trailing edge counter. The MLA Range Measurement Unit (RMU) can collect and store up to 15 signal pulse returns from this last set of counters.

The nominal operational timebase for MLA event sequencing and range measurement is the MESSENGER spacecraft's oven-controlled crystal oscillator. The 5-MHz signal from either of the spacecraft's redundant oscillators can be used. The laser trigger operating at 8 Hz is also synchronized to the spacecraft 1-pulse-per-second (1-PPS) timing reference. A third backup 5-MHz crystal oscillator on the MLA RMU board can also be used.

To maximize the probability of return signal detection in MESSENGER's dynamic operating environment, MLA uses a real-time embedded algorithm to monitor background noise and adjust comparator threshold levels to minimize false signals. The MESSENGER spacecraft also provides MLA with once-per-second updates of estimated line-of-sight range to Mercury based on spacecraft attitude and orbit data. This information is used to dynamically set the range-gate time delays and gain levels for the detector amplifier.

For altitudes greater than 800 km on the inbound and outbound orbit trajectories the MLA signal-to-noise ratio becomes sufficiently weak that the "high" threshold channel is no longer detecting the return. In this situation the software collects and downlinks the multiple return pulses from the "low" channel timers.

The signal-tracking algorithm and the MLA command and data-handling software are implemented on a radiation-hardened 80C196 processor that, with a control and data communications field-programmable gate array (FPGA) and memory devices, make up the central-processing unit (CPU) electronics board. The RMU board is housed with this board in a box underneath the main instrument housing (shown in Fig. 2). The detector, amplifiers, and comparators are mounted on a separate printed wiring board and mechanically mated to the aft optics, which filter and focus the signal from the receiver fiber optics.

Housekeeping signal conditioning and data conversion functions are performed on another circuit board referred to as the Analog Electronics Module (AEM). Laser control and drive functions are implemented on two assemblies beneath the laser bench referred to as the Laser Electronics Assembly (LEA). All of the secondary voltages are converted in the Power Converter Assembly (PCA) located in the magnesium box bolted directly to the spacecraft deck adjacent to the main MLA housing.

Command and data communications to the MESSENGER payload Data Processing Unit (DPU) are through a low-voltage differential signal (LVDS) serial link using the RS-422 signal standard. An additional LVDS signal carrying the 5-MHz spacecraft reference clock is also provided to MLA and connects directly to the RMU board. The MLA instrument has single-string components but can operate from either one of the redundant DPU, clock, and power connections.

MLA has two defined software modes, Boot and Application. Boot mode is entered immediately after power up. MLA will remain in Boot mode until commanded to execute its onboard code residing in electrically erasable programmable read-only memory (EEPROM), at which point it enters Application mode. Within Application mode there are three defined operational modes through which MLA will cycle every orbit. The lowest power mode is Keep Alive, in which only the CPU, AEM, and laser diode's thermo-electric cooler (TEC) are powered. Upon transition to Standby mode the RMU is powered on. Finally, in Science mode, the laser power supply is turned on and the laser fires.

3.3 Technology Advances Implemented in MLA

MLA has several new technologies which provide a compact and efficient laser altimeter to the MESSENGER mission. The most prominent is the miniaturized laser transmitter shown in Fig. 4, delivering 20 mJ of energy in a 75-μrad beam width from a compact 14 cm × 9 cm × 3 cm package to facilitate high-resolution sampling of Mercury's surface. Its unique breathable filter allows for simpler testing in vacuum environments, providing significant test time in a flight-like environment.

MLA also incorporates a multiple aperture refractive telescope receiver, which has proven easier and less costly to manufacture and test than reflector telescopes used on previous altimeters and is capable of performing at the extreme temperatures expected during this mission. The modular design can be scaled to match the aperture requirements for a variety of missions. Fiber coupling makes the alignment and integration process simpler and decouples the detector assembly from the optomechanical integration and test (I&T) flow.

The MLA Range Measurement Unit is another significant advancement, using a low-speed 5-MHz counter and time-of-flight Complementary Metal Oxide Semiconductor (CMOS) ASIC developed by The Johns Hopkins University Applied Physics Laboratory (APL) to achieve timing resolution equivalent to a 2-GHz counter with significantly less power. The overall timing accuracy of MLA and component contributions to timing errors are shown in Table 3.

The detection scheme on MLA also makes possible operation and signal retrieval beyond the theoretical range limit of single-pulse detection electronics by acquiring and downlink-

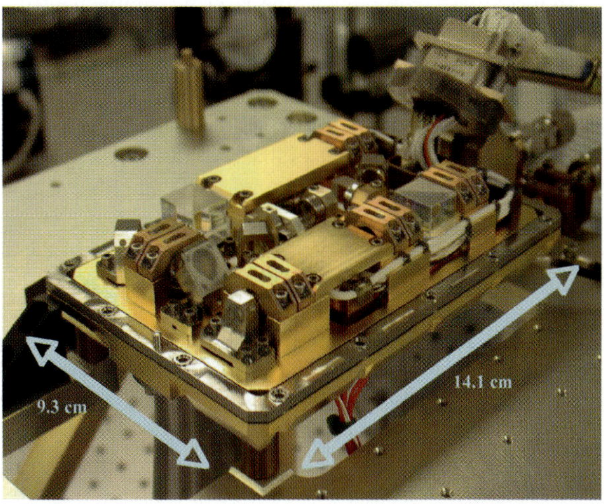

Fig. 4 MLA laser optical bench assembly

Table 3 Component contributions to MLA range error

Error source	Contribution
Leading edge timing	0.06 m
Clock frequency error (0.1 parts per million)	0.20 m
Measurement quantization (2.5 ns)	0.11 m
Pointing angle uncertainty (0.13 mrad)	0.68 m
Spacecraft orbit knowledge error	0.75 m
Total (root sum squared)	1 m

ing up to 15 returns per shot. Correlation processing on the ground will be used to extract valid range signals from these additional data points.

3.4 Laser Transmitter

The MLA laser transmitter is an evolutionary design building on the past 15 years of spaceflight laser designs from MOLA to GLAS. There are common features shared among these and other space-based lasers, which include a semiconductor laser-pumped solid-state approach to the laser, "zigzag" slabs made of neodymium-doped yttrium-aluminum garnet that is chromium co-doped for radiation tolerance (Nd:Cr:YAG) with ends cut at Brewster's angle, and porro prism and mirror or crossed-porro prism laser resonators for stability against vibration and thermal stresses. The laser parameter requirements are enumerated in Table 4. The total number of shots required to complete the mission is on the order of 30 million, a relatively small number when compared with other laser altimeter experiments such as MOLA-2 (660 million shots) and GLAS. Salient parameters such as pulse energy and pulse repetition rate are also lower.

The MLA laser transmitter is also required to survive and operate occasionally (once or twice per year for a few hours) during the 6.6-year MESSENGER cruise phase as well as in the extreme thermal environment encountered in Mercury orbit. During the latter phase of the mission the laser will operate for 15- to 45-minute periods every 12 hours. During this operating time the laser bench temperature will climb from its initial heater-controlled temperature of 15°C at rates up to 0.5°C per minute. The thermal design of the instrument is described in greater detail in Sect. 3.12.

To meet these requirements the laser was implemented with an oscillator and amplifier design as shown in Fig. 4 and Fig. 5. The oscillator is a miniaturized version of the GLAS laser oscillator incorporating a zigzag Nd:Cr:YAG slab, GaInAsP laser diode pump arrays, a passive Q-switch, crossed porro prisms, and polarization output coupling. The laser diode arrays are temperature controlled with a TEC to maintain the optimum pump wavelength. The oscillator emits 3-mJ pulses that are then fed through a 2X beam expander and a second Nd:Cr:YAG laser diode-pumped single-pass amplifier slab, which provides approximately 7 to 9 dB of amplification resulting in an output pulse energy of 15 to 22 mJ. The gain of the amplifier stage is dependent on the temperature of the amplifier pump diodes and slab, which are not actively controlled (Krebs et al. 2005).

The output of the amplifier stage is coupled into a 15X Galilean beam expander, which reduces the final beam divergence to approximately 75 microradians. Since the beam expander output face is exposed to the surface of Mercury during operations, a sapphire window is used to minimize IR coupling through its glass elements.

The MLA laser electronics assembly delivers a 100-A current pulse to the semiconductor laser diode pump arrays. The oscillator and amplifier diode arrays are connected in series and are driven from a capacitor bank charged to approximately 35 volts direct-current

Table 4 MLA laser transmitter design requirements

Parameter	Requirement
Wavelength	1,064.5 nm ± 0.2 nm
Pulse energy	20 mJ ± 2 mJ
Pulse width	6 ns ± 2 ns
Pulse repetition rate	8 Hz
Beam divergence ($1/e^2$)	less than 80 μrad

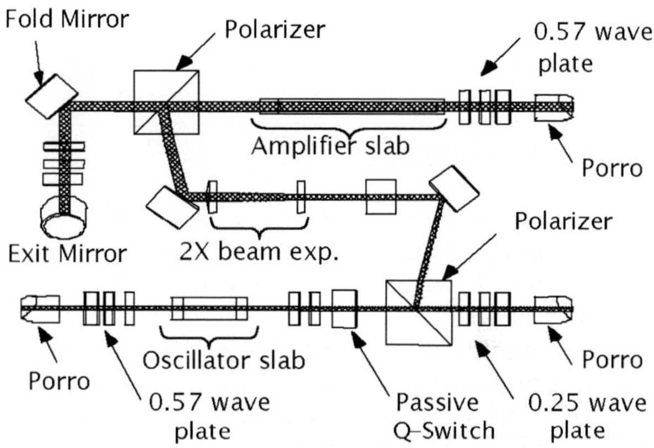

Fig. 5 MLA laser optical layout. The *bottom portion* shows the passively Q-switched crossed-porro master oscillator, and the *top portion* shows the amplifier path

(VDC) generated as a regulated secondary supply by the MLA Power Converter. When triggered from the MLA timing electronics, current is switched through the diode arrays by redundant metal-oxide-semiconductor field-effect transistors (MOSFETs). Current flowing through the laser diode arrays generates the 808-nm optical pump pulses for both the oscillator and amplifier Nd:YAG slabs. When the Q-switch, a saturable absorber in the oscillator, becomes transparent or "bleached," the 5-ns, 1,064-nm laser pulse is emitted. A photodiode in the laser detects the emitted pulse, and this signal is used to terminate the current pulse. The diode pump pulse is nominally 150 μs in duration and is limited by the electronics to 255 μs. As the diode lasers age and emit less power the pump pulse will last longer, allowing constant 1,064-nm pulse energy output for the duration of the mission.

The laser electronics also implement a control loop to maintain the oscillator pump diode lasers at 17 ± 0.1°C via thermoelectric cooler. The photodiode used to terminate the pump pulse is a quadrant detector that is placed behind a diffuser and views residual energy transmitted through the final turning mirror prior to the laser beam expander. Three segments of this detector are used for timing functions, and the fourth segment output is used to monitor the energy.

One breadboard, one engineering model, and one flight model laser were fabricated at the Space Lidar Technology Center operated by the NASA Goddard Space Flight Center (GSFC). Laser fabrication was performed in a Class-100 clean room. The engineering and flight units underwent vibration and thermal-vacuum cycling tests at the subassembly level during which all parameters were verified.

3.5 MLA Receiver Optics

The MLA receiver telescope design is a significant departure from prior telescopes used for MOLA, SLA, and GLAS. These altimeters all used Cassegrain reflector telescopes made

Table 5 MLA receiver optics design requirements

Parameter	Requirement
Aperture	417 cm^2
Field of view	400 μrad full angle, circular
Bandpass filter	0.7 nm pass band centered at 1,064.5 ± 0.2 nm
Detector	0.7 mm diameter silicon avalanche photodiode

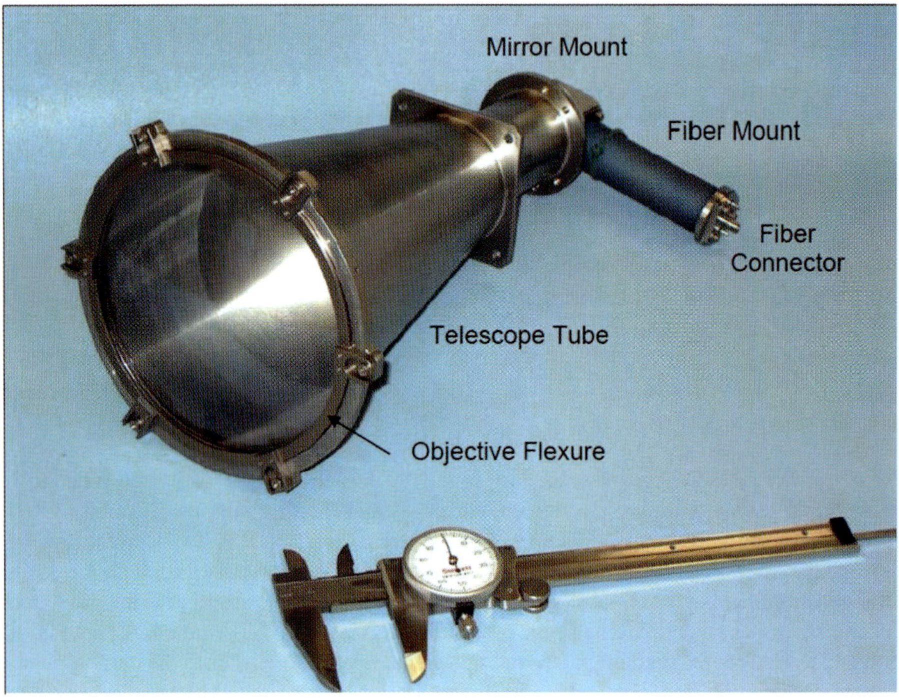

Fig. 6 One of the four MLA receiver telescope assemblies

with beryllium in the cases of MOLA and GLAS. The original design for MLA was a beryllium reflector similar to these, but once the MESSENGER thermal environment was understood it became apparent that the Cassegrain reflector would not perform adequately with the significant thermal gradients resulting from viewing Mercury's surface. The requirements for the MLA receiver are listed in Table 5.

The ensuing design shown in Fig. 6 is a set of four refractive telescopes with an aperture equivalent to a single 0.25-m diameter reflector with a 15% secondary obscuration. The objective lenses in each telescope are sapphire, chosen for its ability to withstand thermal shock, its lower absorption in the infrared, and its resistance to radiation darkening. At the back of each telescope is a dielectric fold mirror which reduces the total height of the assembly and reflects only a narrow band about 1,064 nm, allowing protection against accidental direct solar viewing by passing most visible solar radiation through its frosted back. At the focal plane of each telescope is a 200-mm core diameter fiber, which limits the field of view to 400 μrad. Alignment of each telescope's field of view (FOV) to the MLA laser is accomplished solely by translating each fiber at the focal plane (Ramos-Izquierdo et al. 2005).

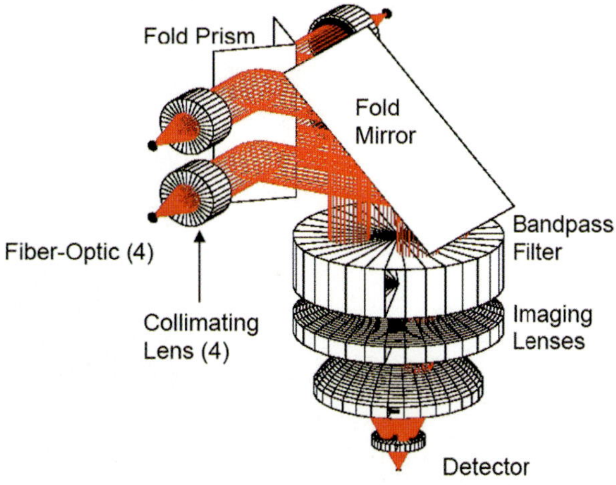

Fig. 7 MLA aft optics layout showing the four fiber optics from the receiver telescopes coupled through a common bandpass filter to one detector

Each fiber optic couples the received power to a single aft optics assembly (Fig. 7), which collimates the fiber output through a single optical bandpass filter with a 0.7-nm bandwidth. The center wavelength of the bandpass filter can be adjusted by tilting it to match the center wavelength of the laser pulse, which for MLA is 1,064.5 nm. After the bandpass filter the light from all four fibers is then focused onto a single silicon avalanche photodiode detector with a 0.7-mm diameter.

Five engineering model (EM) telescopes and an EM aft optics assembly were fabricated out of aluminum to validate optical, mechanical, and thermal performance models. These EM units also served to develop the integration procedures and ground test equipment needed for the flight model telescopes. Four of the telescopes and the aft optics with fibers were also integrated with an aluminum MLA EM housing and laser beam expander assembly. This approach provided a complete optomechanical EM that also served to validate models and develop instrument-level alignment verification tests, which were used successfully on the MLA flight instrument.

3.6 Receiver Electronics

The detector hybrid, shown in Fig. 8 was first developed during in the 1980s by then EG&G Optoelectronics Canada for optical communication programs. It consists of a SiAPD chip, a low-noise preamplifier, and a high-voltage bias circuit, all contained in a hybrid circuit housed in a one-inch-diameter hermetically sealed package. These devices were used in SLA, MOLA-1, and MOLA-2, the latter of which operated in space from 1996 until the end of the Mars Global Surveyor mission in 2006 with little performance degradation. Additional development of this device by the manufacturer in the 1990s improved the bandwidth from 50 MHz to more than 100 MHz. The higher bandwidth version was used for GLAS and MLA.

The electronics for received pulse detection duplicates in concept subsets of the circuits used for MOLA and GLAS and includes the same SiAPD detector hybrid and variable gain amplifier (VGA) stage as GLAS. The output of the VGA is split three ways that define the MLA "channels" 1 through 3. Channel 1 is the highest bandwidth version of the signal and has a 10-ns matched filter. The next two splits feed into the 60-ns and 270-ns matched filters. These latter two filter outputs are gain compensated for the filter loss. Channel 1 has two comparators, each with separately programmable thresholds (channel 1 "high" and

Fig. 8 MLA detector hybrid package

channel 1 "low"). The remaining channels have single comparators. When the signal level crosses a programmed threshold voltage, the high-speed comparator switches a logic level, which is transmitted to the RMU via low-voltage differential signal (LVDS) interface for signal timing.

3.7 Range Measurement Electronics

Conceptually the MLA range timing circuitry builds on the basic pulse edge timing technique used for MOLA and SLA. The timing cycle for each laser pulse is shown in Fig. 9. Rather than using the laser pulse emission to start the counter, as was done with MOLA, the trigger pulse to the laser is itself used as the start time and the laser pulse is the first event in each cycle. The important change is the implementation of a low-frequency counter operating at 5 MHz coupled with a tapped delay line ASIC to determine the intra-cycle timing within the coarse counter resolution, providing much better resolution without the need for a high-frequency counter.

The receiver event timers consist of a set of time-to-digital converters (TDCs). The TDCs are based on the tapped delay line technique, and each channel is implemented in a silicon ASIC specially designed for space applications, designated TOF-A by their APL developers (Paschalidis et al. 2002). The tapped delay lines consist of a series of logic gates with precise and uniform propagation delay times. An on-chip delay-lock-loop is used to self-calibrate the delay time against an external reference clock signal. The TOF-A can perform subnanosecond timing without the need for high-frequency logic circuits and clock oscillators yet has limited full-scale range; thus a coarse-resolution 5-MHz counter is synchronized with each TOF-A to provide a full-scale pulse time-of-flight. Digital logic circuits are implemented in an FPGA, an Actel RT54SX72S. The combined circuit can time the leading and trailing edges of the transmitted laser pulses and the received echo pulses to better than 500-ps accuracy with 13-ms dynamic range.

Six of these circuits make up the MLA RMU. A signal edge designated T0 starts the coarse counters. Each edge of the LVDS comparator signals from the detector board starts a corresponding TOF-A, and this circuit counts until the next leading edge of the 5-MHz clock stops it, providing subcycle timing. The start pulse and channel 1 "high" pulse TDCs record one event per edge (leading and trailing). The "low" threshold pulse TDCs can store multiple events from each shot; therefore a lockout circuit (as described in the overview) and a means to associate the source of the pulse (channel 1, 2, or 3) are implemented. This circuit records the times of up to 15 events for every transmitted pulse with a dead-time of

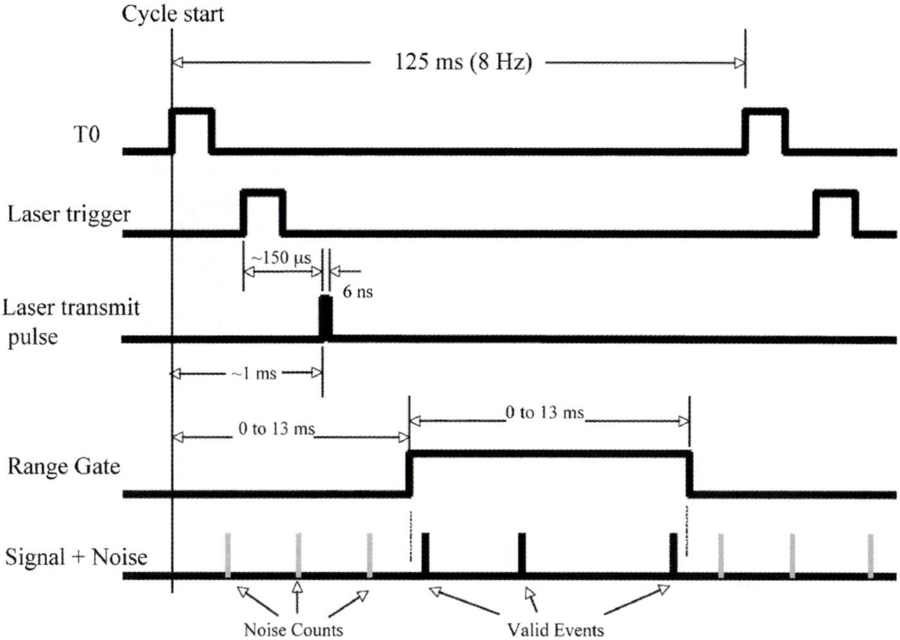

Fig. 9 MLA range measurement timing diagram. Events are referenced to an internal signal (labeled T0) that is synchronized to the spacecraft's 1-Hz mission elapsed time (MET) reference signal

Fig. 10 MLA Range Measurement Unit (RMU) assembly shown mounted in its beryllium housing

several hundred nanoseconds. A separate set of event counters totalizes pulses received on each channel inside the range gate. These counters are read and reset after every shot.

The flight RMU is shown in Fig. 10. An engineering unit is integrated into the instrument EM. Subassembly testing of the RMU demonstrated range precision capability under 1 ns. Standard deviation of the error signal was approximately 250 ps with a mean error of 480 ps.

3.8 Command and Data Handling Electronics

The MLA command and data-handling (C&DH) electronics are assembled on one printed wiring board referred to as the MLA CPU board. The unit was custom built for MLA but uses components with extensive heritage. The CPU itself is a 80CRH196KD device from United Technologies Microelectronics Corp. operated at a 16-MHz clock frequency. All control and interface functions are implemented in an Actel RT54SX72S FPGA. Two each 64-kB programmable read-only memories (PROMs), 256-kB EEPROMs, 512-kB static random-access memories (SRAMs), an oscillator, and LVDS transceivers round out the board. A block diagram of the CPU is shown in Fig. 11. The CPU board is shown in Fig. 12.

All command and data interfaces flow through the FPGA on the CPU board. Each MLA subsystem's control and data lines tie into this device. The FPGA therefore has a functional block for each MLA subsystem, two universal asynchronous receiver/transmitter (UART) blocks for communications, and one block for address control and reset functions. The subsystem functional blocks are mapped to 80196 data memory space and defined for the PCA, AEM, and RMU. An interrupt controller is also implemented in the FPGA for interrupts

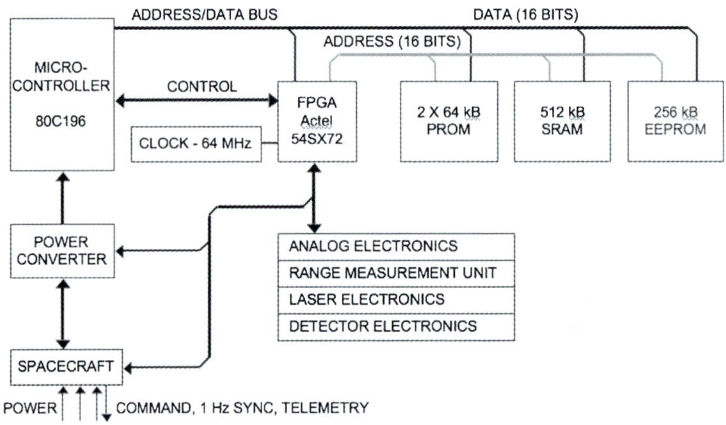

Fig. 11 MLA CPU block diagram

Fig. 12 MLA CPU board

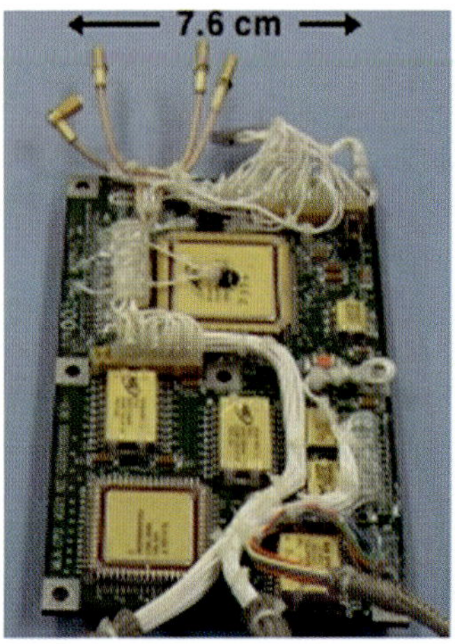

from the RMU, AEM, address decode, and UART. All transfers to and from the subsystem and UART blocks are byte-wide only. Each UART is connected to one of the redundant MESSENGER DPUs (A and B). All command and data transfer to the MESSENGER DPU passes through the UARTs. An additional UART integrated into the 80196 CPU was used for development. The 80196 timer function is also used to provide a reference to the phase of the spacecraft 1-PPS signal and for flight software timing.

MLA operational modes (Keep Alive, Standby and Science, described in Sect. 3.2) are defined by which subsystems are powered, so all mode changes are implemented by setting bits in the PCA block to enable secondary voltages. The RMU block transfers all range data and control functions. This block must function when the RMU is powered off as is the case in Keep Alive mode. The AEM block executes an autoconversion sequence to acquire laser energy and diode current monitor samples each time the laser fires without software intervention. All other analog-to-digital and digital-to-analog conversions are done under software control.

PROM memory contains the bootstrap code. One EEPROM device is permanently write protected and contains the last fully ground-tested version of the flight software. The second EEPROM can be overwritten via ground commands to allow updates to the flight software during the mission.

One breadboard, two engineering models, and one flight unit were fabricated for MLA. The breadboard CPU was used for flight software development and testing, and one EM was used in the instrument EM.

3.9 Power Converter Electronics

The MLA PCA is shown in Fig. 13.

The PCA generates all the secondary voltages on MLA. Requirements for secondary voltages and the subsystems that use them are enumerated in Table 6. The 2.5-V, 35-V, and 550-V converter outputs are all switched on or off by control bits from the CPU FPGA to control power consumed by MLA. The 12-V, 5-V, and 2.5-V sources are forward converter topology DC-DC converters. The negative 5-V source is produced from the positive 5-V

Fig. 13 MLA Power Converter Assembly, shown mounted on the MESSENGER instrument deck adjacent to the MLA main housing

Table 6 PCA power requirements for each subassembly and secondary supply

Subsystem	Power required (W)					
	+2.5 V	+5.0 V	−5.0 V	+12.0 V	35 V	+550 V
CPU	0.275	0.87	0	0	0	0
RMU	0.088	0.25	0	0	0	0
RMU heater	2					
Detector	0	0.775	0.47	0	0	0.01
Analog board	0	0.157	0.073	0.019	0	0
Laser electronics	0	0.26	0.11	0.324	5.54	0
Laser thermo-electric cooler	0	2.5	0	0	0	0
Laser amplifier heater	0	0	0	2	0	0
Load totals	2.36	4.81	0.65	2.34	5.44	0.01
	1.11	1.44	0.35	0.91	0.74	0.20
Power requirement				20.37		
Schottky diode loss				0.370		
Average prime power required				**20.75**		

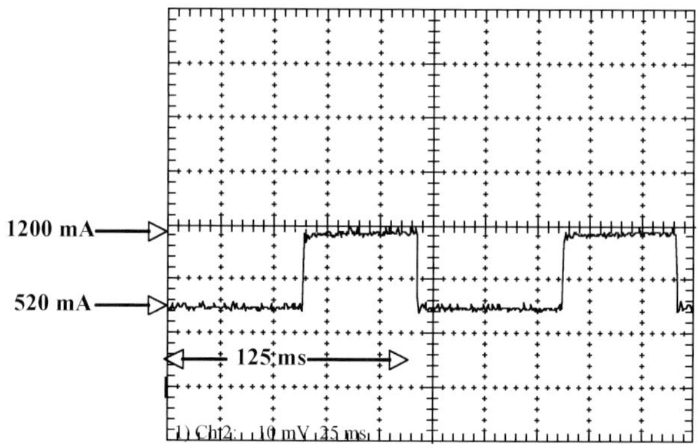

Fig. 14 Oscilloscope trace showing MLA PCA current waveform captured with a current probe placed on the 28-V input power line

supply by a flyback converter. The 550-V SiAPD bias voltage is also converted from the positive 5-V supply by a resonant converter. A 35-V source used solely for charging the capacitor bank that powers the laser pump diodes is implemented in a flyback converter.

The laser, operating at an 8-Hz pulse rate, discharges the capacitor bank by switching 100 A through the laser pump diodes in a 150-μs current pulse. The capacitor bank must then be recharged in time for the next pulse. The resultant charge current, drawn through the 35-V converter, is reflected on the spacecraft power bus current in the form of an 8-Hz square wave with approximately 700-mA amplitude and duty cycle from 25% at maximum bus voltage to 80% at minimum bus voltage as shown in Fig. 14. Since the MESSENGER general electromagnetic compatibility (EMC) requirements specified that current ripple be less than 500 mA, a waiver was granted for MLA allowing up to 800 mA ripple. Subsequent cross-compatibility tests showed that this ripple did not affect other spacecraft subsystems.

All secondary voltages are monitored, scaled, and sampled once per second with an 8-bit analog-to-digital converter, and these values are reported in the MLA Status telemetry packet. To reduce circuit board area, secondary supply current monitors were not implemented. Only the laser pump diode current is sampled during every shot 50 μs after being triggered. This value, along with a counter value indicating the duration of each pump pulse, is stored in the MLA Science telemetry packet.

The 8-Hz laser trigger is synchronized to the spacecraft 1-Hz timing reference. Since the spacecraft C&DH samples the power bus current synchronously with this reference the MLA current is always sampled at the peak of 8-Hz square wave described earlier.

The PCA is housed in a magnesium box, which is bolted directly to the spacecraft deck. A thermal gasket allows conduction of dissipated power to the deck. The spacecraft power and communications connectors are on the PCA box. Command and data signals are simply routed through the PCA to the MLA CPU board.

One engineering model and one flight model PCA subassembly were delivered to GSFC by Space Power, Inc., the contractor for the PCA design and fabrication. The units as delivered were functionally tested over the specified temperature range.

3.10 Software

The MLA flight software consists of multiple tasks running under a real-time executive. The two main groups of tasks are the C&DH set comprised of timing, communications, and system maintenance, and the Science tasks that acquire, process, and compress the critical science data. Code for MLA was developed in the C programming language and operates under a Real Time Operating System (RTOS) from CMX Systems. The MLA Flight Software Tasks are depicted in Fig. 15.

The science algorithm builds on heritage primarily from MOLA and Microaltimeter experience. Like the MOLA software, the MLA algorithm integrates background noise counts (detector events outside the range gate) and sets comparator thresholds to minimize false detections within the range window. MLA also uses line-of-sight range information from the spacecraft attitude control system (ACS) to update the range gate delay, width, and detector gain based on altitude and descent or ascent rate. More significantly the MLA algorithm takes in the multiple returns from the RMU low channel and maintains a histogram of range

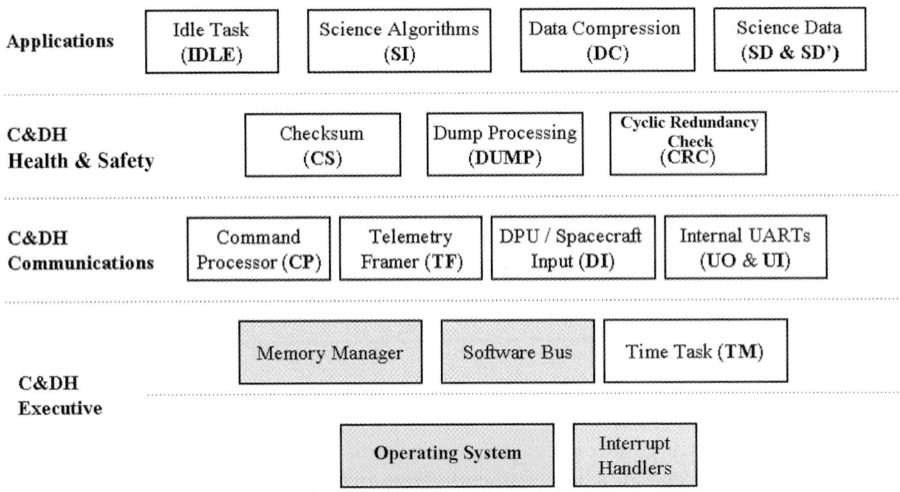

Fig. 15 MLA flight software task hierarchy. Tasks in *gray boxes* comprise the kernel; tasks in *white boxes* are interrupt-driven. Within the latter category the Time Task has highest priority and schedules execution of all other tasks

measurements within the window. If a valid return is detected on the high threshold channel, then the three low returns closest to it are downlinked to the ground. If there is no high channel return, then all the low returns (up to a preset limit) are downlinked to the ground to be processed using correlation techniques.

The flight software timing is handled by the RTOS and interrupt service routines (ISRs). The RTOS is a preemptive, priority-based scheduler with semaphore and delay functionality. The processor immediately branches to the ISRs when their associated interrupt is generated, except when they are masked off, in which case the interrupt is pending but no action is taken by the processor.

The base cycle period of the flight software is 1 s. The spacecraft demarcates this period with the 1-PPS signal. All telemetry for that second is required to be packaged in an Instrument Transfer Frame (ITF) and sent to the spacecraft. During each second the flight software runs an entire pass through all of its code. There are some tasks that have counters that enable them to do different things on different seconds, but those counters are always maintained locally to the task.

Three scheduled sequences occur in each second. The first is the internal flight software timing sequence, the second is the RMU laser firing sequence, and the third is the purely data-driven, low-priority tasks that are scheduled either from command packets or the time message. The first two sequences are independent and can run asynchronously, but they are both configured to synchronize with the 1-PPS and therefore, in effect, synchronize with each other. The execution of the last sequence is dependent on the content of the command input and the spacecraft time message.

Every time the CPU board is powered on or reset, the Boot Loader executes. A command is required to signal the software that the Boot Loader should load the flight software out of EEPROM and into SRAM for execution.

There are three ways to reset the MLA instrument: a processor reset, an FPGA watchdog timeout, and a spacecraft power cycle. All three ways can be commanded. There is, on the other hand, only one way to go from the PROM Boot Loader code to the EEPROM flight software code, and this is by processing a load directive. A load directive is a structure in memory used by the Boot Loader that contains instructions on where and how to initialize SRAM, as well as where to copy sections of EEPROM code and data into SRAM to be executed. There are only two commands that cause the Boot Loader to process a load directive.

Bench testing was performed on three separate platforms. First, the MLA flight software was executed on a PC-based simulator. Second, the MLA CPU breadboard was used as a stand-alone test bed. Finally, before loading the software to the flight instrument EEPROM it was tested on the MLA engineering model instrument. Once software was loaded to the flight instrument, additional orbit simulation testing was performed during spacecraft I&T that included inputs from the spacecraft attitude control system and simulated optical signals to mimic the expected return signal from the Mercury surface.

3.11 Mechanical Design

MLA's optomechanical structural elements, like those of MOLA and GLAS, are made primarily of beryllium, which was selected for its superior stiffness-to-weight ratio and high thermal conductivity. The PCA box, fabricated from magnesium to minimize mass, is located off the optomechanical structure. This was done to reduce power dissipation and thermal gradients in the alignment-sensitive portion of the instrument. A significant departure from previous altimeters is the use of four refractive telescopes, also made with beryllium.

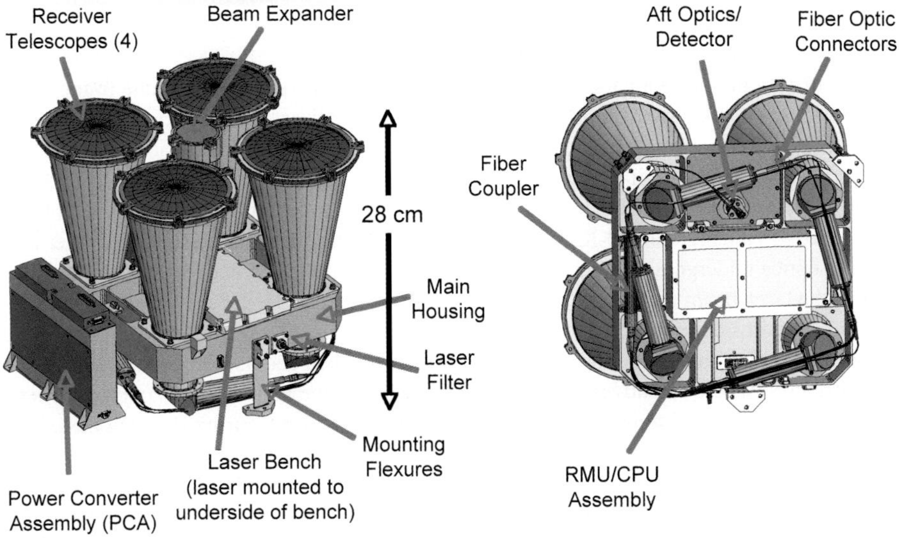

Fig. 16 MLA mechanical subassemblies

Previous altimeters have used reflective telescopes, but in this case a refractive telescope design was deemed less costly and better suited to the expected thermal environment. The main mechanical elements of MLA are depicted in Fig. 16.

The primary metering structure that maintains the boresight alignment is the MLA main housing. This beryllium structure supports the four telescope tubes, the beam expander, and the laser bench. The detector, aft optics, and housekeeping electronics also occupy the housing, and a separate box with the RMU and CPU is bolted to the housing. The main challenge in the design of this structure is to keep the telescope tubes coaligned with the beam expander over the extreme temperature swings to be seen on orbit. Tight assembly tolerances were required to bring all four telescope tubes to within 2 mrad of the beam expander output at initial assembly. The fiber optic couplers could then be adjusted to align the receiver and transmitters to within 10 μrad. The main housing bolts to the spacecraft deck via kinematic mounts using three titanium flexures to minimize thermal conduction to the deck and avoid mechanical distortion of the housing.

The laser bench is a 14-cm by 9.3-cm slab of beryllium to which the laser resonator and amplifier components are mounted. The beam expander bolts directly to this bench and comprises the laser subassembly. The housing has a cavity to contain the laser, which was precision cleaned and kept sealed until laser integration. This cavity has a breathable filter that allows the laser to vent during vacuum testing and after launch. The laser bench is bolted to the housing with 14 bolts around the perimeter. The four beryllium telescope tubes also bolt directly to the housing with four bolts each. The lens mounts for the beam expander and telescope tubes are titanium and hold the BK7G18 glass optics for the beam expander and the sapphire objective lenses in the telescopes.

The main housing, laser bench, and electronics housing were all gold plated for thermal performance. Most of the beryllium used is instrument grade (I-220-F) save for the telescope flanges, which are structural grade. All beryllium parts are nickel plated to prevent oxidation and personnel exposure. Phosphor-bronze helicoils were used for all bolt holes. Aluminum engineering models of all components were fabricated to develop assembly and integration procedures and to validate structural models.

3.12 Thermal Design

MLA will operate under a harsh and highly dynamic thermal environment due to the large variation in heat flux from the Mercury surface from daytime, nighttime, and deep space views. Figure 17 shows the predicted temperatures during the hottest mission orbit. As the spacecraft approaches orbit periapsis the MLA laser is turned on, increasing power dissipation by approximately 10 W, and the instrument view transitions from seeing the cold of deep space to viewing the 700-K surface of Mercury. During this period the transmitter and receiver optics undergo a rapid and uneven temperature rise at a rate of tens of degrees per hour in the laser-beam expander and the receiver telescope.

MLA acquires data for 15 to 45 minutes of each 12-hour orbit during which time it fires the laser, resulting in maximum power dissipation. This data acquisition period takes place near periapsis, where radiative input from Mercury's surface is also at a maximum. These factors result in a rapid temperature rise through the instrument during the data pass. The remaining 11 hours of the orbit are used to radiate the absorbed heat into deep space. The primary radiators are the telescope tubes and the laser beam expander. MLA, as with the rest of the spacecraft, never reaches thermal equilibrium during operations and will see temperature excursions similar to those shown in Fig. 17 during every 12-hour orbit, resulting in more than 700 thermal cycles during the 12-month mission.

The entire instrument is covered with a ten-layer multilayer insulation (MLI) blanket as shown in Fig. 18. The outer layer of the blanket is vapor-deposited gold (VDG), and the inner layers are vapor-deposited aluminum (VDA). During spacecraft I&T a layer of

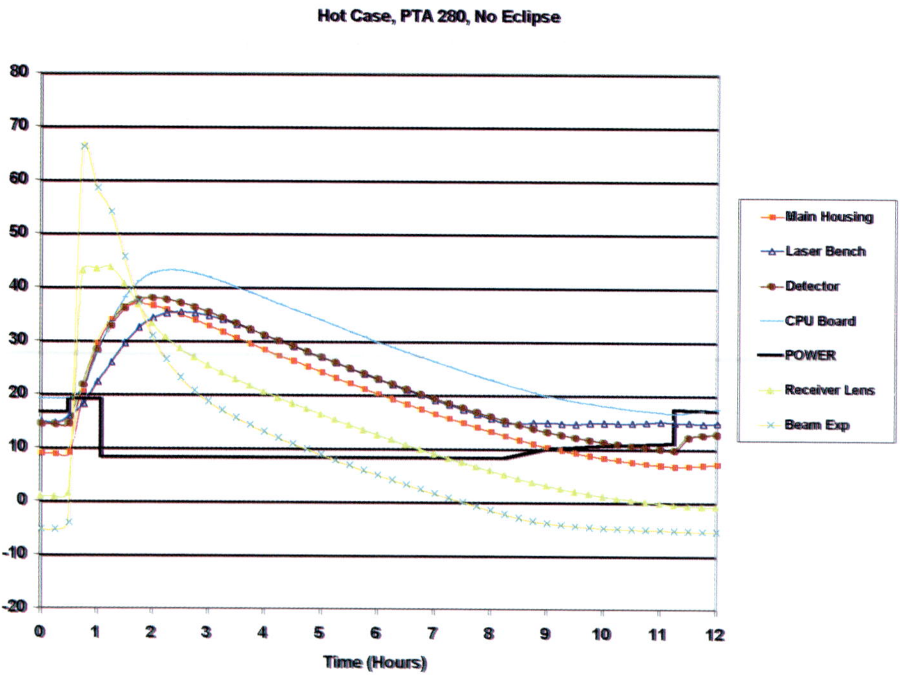

Fig. 17 Predicted temperatures of MLA subsystems during the hottest expected orbit. The plot time starts approximately 30 minutes before firing the laser. The laser beam expander output window temperature (*orange line*) swings from −5°C to +65°C in less than 30 minutes

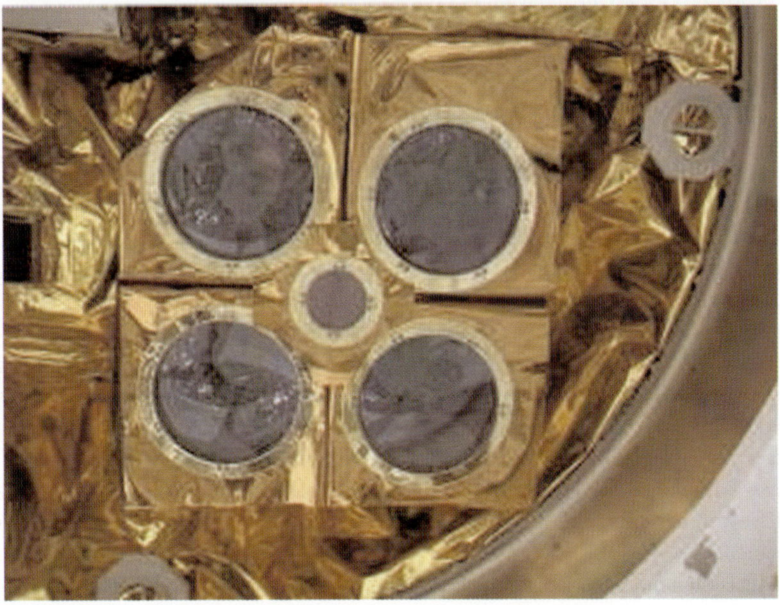

Fig. 18 The MLA instrument shown mounted to the MESSENGER payload deck with thermal blankets installed

aluminized Kapton was added to the nadir-facing surface to alleviate concerns about solar illumination of the VDG layer. A thin strip of black Kapton was also added to minimize reflections on the side facing the MESSENGER Dual Imaging System, MDIS (Hawkins, III et al. 2007). The PCA box conducts its dissipated power directly to the spacecraft deck and is enclosed in another MLI blanket to avoid radiative coupling to the rest of the instrument. Titanium flexures minimize conduction from the main housing to the deck, and another MLI blanket underneath the instrument provides radiative isolation from the spacecraft. The sapphire objective lenses have excellent thermal conductivity compared with other transparent materials, and while they absorb about half of the infrared radiation from the planet's surface they can withstand the resultant thermal shock. A sapphire flat is also used on the beam expander output to improve its thermal properties.

Two sets of 17-W survival heaters powered by separate spacecraft circuits are used. The "B" side circuit has a −20°C thermostat, and the "A" side trips at −15°C. During operations the "A"-side heaters will be used. The "B"-side heaters are needed during cruise and orbit eclipse cases.

Laser transmitter performance is dependent on the laser bench temperature. Optimal operating temperatures for the laser bench range from 15 to 25°C, so to maximize laser energy during the data acquisition part of the orbit the laser bench should be at 15°C just prior to entering Science mode. To achieve this objective an operational heater is located on the laser bench and is powered by the 12-V supply when MLA enters Standby mode. The duration in Standby mode may be modulated through the mission to achieve this starting temperature under varying orbit conditions. Currently the predicted time in Standby mode ranges from 15 minutes for the hot case to 480 minutes for the cold case.

4 MLA System Integration and Test

4.1 Instrument Assembly

The MLA assembly sequence was driven primarily by the cleanliness requirements for the laser. Prior to any assembly activity the main housing was precision cleaned, and the cavity that enclosed the laser was sealed during integration of the electronics with the housing. Laser bench integration was performed in a Class-100 clean room. Once the laser was sealed in the housing, a ground-support equipment (GSE) air filter was attached in series with the built-in flight filter to provide additional protection during I&T.

The receiver telescope tubes and aft-optics were assembled on a Class-100 laminar flow bench in a Class-1000 clean room. The MLA housing with the laser and electronics was delivered to this area, and the telescope tubes were attached. The assembled instrument was then attached to its handling fixture. Before the fiber optics were connected, a series of free-space tests with the laser were performed to establish baseline performance and to calibrate the GSE power measurements that would be used throughout ground testing to monitor laser performance.

MLA was then mounted to an optical bench that contained the alignment optics for boresighting. The optical axis of the alignment fixture was coaligned with the MLA laser's optical axis. Each receiver's fiber optic was then aligned to the fixture's axis. As a final post-assembly check each fiber was back-illuminated while the laser was firing, producing an image at the alignment fixture's focal plane showing the laser image with the receiver's illuminated FOV. Since the alignment was done in the lab at one atmosphere, shims on the fiber optic connectors positioned each fiber at the vacuum focus of each telescope, causing the image to blur but allowing for accurate evaluation of the centered images. Subsequent evaluation of the alignment in vacuum was performed during environmental testing. The fibers were then attached to the aft optics, completing the assembly sequence prior to instrument-level testing.

4.2 Contamination

After assembly of MLA the particulate contamination requirements were relaxed to Class-10 000 level since the most sensitive optics were sealed. All optical surfaces, particularly the beam expander output window, were visually inspected prior to firing the laser. MLA GSE that came in close contact with the instrument was cleaned and inspected prior to each use. GSE covers were made for each telescope objective and the beam expander window. These were attached when the instrument was not under test. After MLA was installed on the spacecraft with the thermal blankets, a sheet of lumalloy was taped over the receiver objectives when not testing. Final inspection and cleaning of all optical surfaces was performed prior to encapsulation using bright white and ultraviolet light sources.

The GSE air filter remained attached to the laser housing until final closeout at the spacecraft level save for mass properties testing during which a small plug was installed. A gaseous nitrogen purge was applied to the spacecraft payload area during integration up to launch. This purge line was filtered, and a manifold distributed the purge gas to each instrument. The purge line for MLA was attached near the laser filter snout at final closeout after the GSE air filter was removed.

Table 7 MLA optical alignment allocations and margins. These parameters could be verified to within 10 μrad during testing with the MLA alignment GSE

	Specification
Integration margins	
Laser beam axis parallel to receiver telescope axis	<2 mrad
Laser beam axis perpendicular to mounting plane	<5 mrad
Alignment margins	
Receiver telescopes to laser beam axis (boresight)	±50 μrad
Laser beam axis to MLA reference cube (knowledge)	±50 μrad
Stability	
Laser beam axis to MLA mounting plane	±50 μrad
Receiver telescopes to laser beam axis (boresight)	±100 μrad

4.3 Performance Testing

Table 3 and Table 7 show the MLA ranging and optical alignment error budgets associated with the performance requirements listed in Table 1. A suite of tests was designed to verify MLA performance at the instrument level and during spacecraft testing. Subsets of this suite were used for Aliveness and Functional testing, and the entire suite comprised the Comprehensive Performance Test. Ancillary tests such as boresight alignment verification and timing tests were used at key points during integration and test as calibration points and to demonstrate compliance with the MESSENGER Component Environmental Specification.

4.3.1 Bench Checkout Equipment

The set of test equipment used to verify performance was in itself a system of electronic, electro-optic, and optomechanical subsystems referred to as the Bench Checkout Equipment (BCE). The BCE hardware is composed of three subsystems. One equipment rack contains the main data acquisition system, spacecraft interface simulator, timing references, and electro-optic sources. A second rack contains the alignment data acquisition and control system. An optomechanical target assembly is used for functional and alignment testing. Several hundred meters of fiber optic cable were also used to couple signals and provide constant time-delay paths.

During instrument-level testing the BCE used the MESSENGER-supplied spacecraft interface simulator to provide power and transfer commands and data to and from MLA. After integration with the spacecraft the BCE acquired real-time telemetry via the MESSENGER ground data system network. During all phases of testing the BCE captured and time-stamped each and every laser pulse emitted by MLA and provided simulated return signals and noise at the MLA receiver telescope. Laser pulse energy was continuously monitored.

Simulated return signals were generated by a diode-pumped Nd:YAG laser housed within the rack that produced 5-μJ 1-ns pulses when triggered by the timing electronics. These pulses were fed through an optical attenuator and fiber coupled to a holographic diffuser, which distributed the signal power in a 0.006-steradian cone to be collected by one of the MLA receiver telescope apertures. Optical background noise was simultaneously coupled in a similar manner using a continuous-wave (CW) Nd:YAG laser source.

The BCE used a Stanford Research Systems SR620 time-interval analyzer and two InGaAs photodiodes to independently measure the delay between each emitted MLA laser pulse and the corresponding simulated return pulse. The timebase for the SR620 is a rubidium-based oscillator. The drift of this oscillator is measured by recording the phase between its 1-Hz output to a 1-Hz tick generated by a Global Positioning System (GPS) receiver.

4.3.2 Laser Performance Testing

Two different laser beam termination schemes were used for monitoring the MLA laser. Both were designed to provide a "light-tight" seal to the MLA beam expander to prevent damaging levels of power from leaking into the receivers and afford an additional level of personnel safety. In addition both schemes were required to minimize back reflections into the laser itself.

One assembly is illustrated in Fig. 19. This scheme, known as the beam dump, was an integral part of the alignment test fixture and used a beam splitter to direct 90% of the laser pulse power into a holographic diffuser and then focused a portion thereof into a 0.22 numerical-aperture optical fiber for transmission to a Molectron JD2000 Joulemeter energy monitor and an InGaAs photodiode used for timing and temporal pulse width measurement. The remainder of the energy was transmitted to a charge-coupled device (CCD) camera used for beam diagnostics and redirected via retroreflector into one of the four MLA telescopes. Filter holders allowed for placement of volume-absorbing neutral-density filters to attenuate optical power to usable levels.

The second assembly, referred to as the beam stop, was used primarily for spacecraft-level testing and incorporated a Macor ceramic insert to diffusely reflect the incident laser power. The Macor transmits approximately 1% of the laser pulse, allowing for placement of an optical fiber on the back of the beam stop used with the BCE in the same manner as the beam dump fiber.

4.3.3 Boresight Alignment Verification

Two different methods were used to verify the MLA boresight alignment, or angular offset between the transmitted laser beam and each receiver telescope's field of view.

The initial boresight alignment was performed on a laboratory optical bench using a 2.5-m-focal-length collimator with a 400-mm-diameter off-axis parabola. The initial alignment described in Sect. 4.2 used this fixture. To verify the alignment with this fixture the MLA laser beam axis was aligned to the optical axis of the fixture. With the MLA laser turned off, a 1,064-nm continuous-wave point source at the focus of the fixture was mechanically swept across the MLA field of view while monitoring noise counts in MLA telemetry. MLA noise counts were then plotted against the point source offset to assess the receiver alignment. This method also enabled simultaneous alignment testing of all four receiver telescopes.

During MLA instrument environmental testing and spacecraft-level testing the laser beam alignment to the receiver telescopes was measured by redirecting a portion of the transmitted laser pulse back into one of the four receiver tubes using a lateral transfer hollow retroreflector (LTHR). At the output of the LTHR a pair of motorized Risley prisms deflected the beam at a programmable angular offset into the telescope. By sweeping the angular offset in orthogonal directions across each telescope's field of view and monitoring the power received at the MLA detector, a centroid was computed that indicated the relative

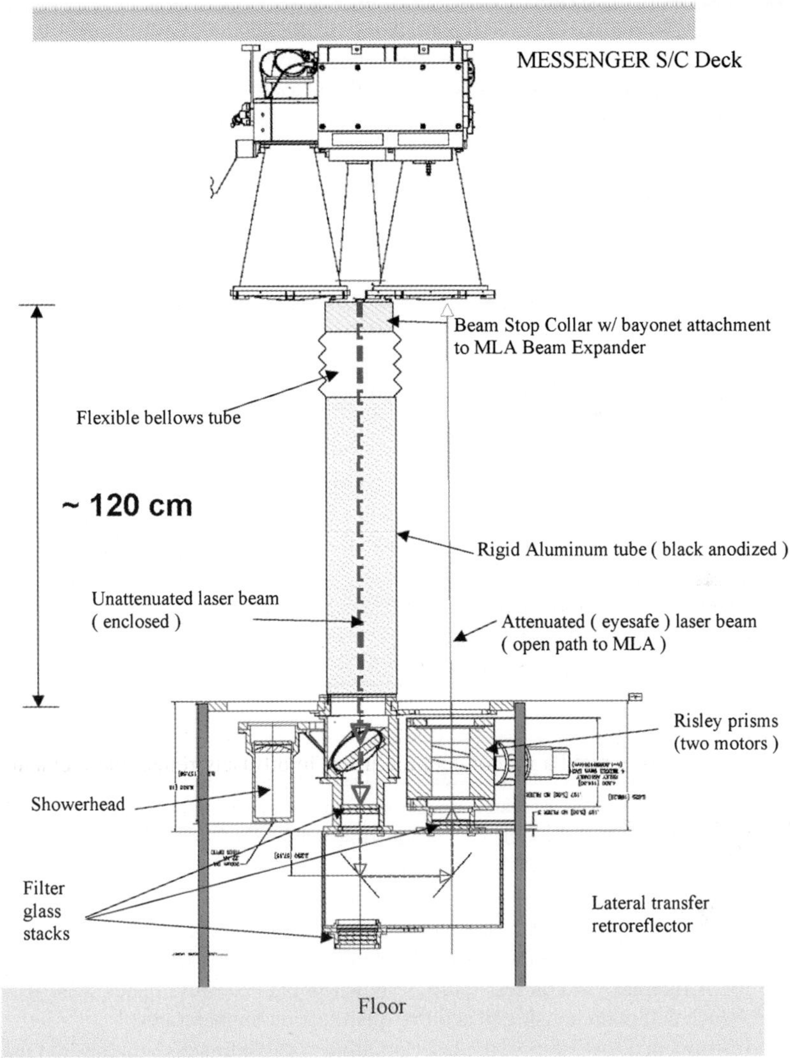

Fig. 19 MLA laser beam dump and alignment fixture shown in the configuration used to verify alignment after MLA was installed on the MESSENGER spacecraft S/C

angular offset of the laser beam to each telescope. Relative pulse power received was measured by computing the change in detected pulse width and normalizing for each tube. The setup used during spacecraft-level testing is depicted in Fig. 19.

4.4 Calibration and Characterization

4.4.1 Instrument Calibration

MLA provides three measurements: the laser-pulse time-of-flight, the echo-pulse width, and the echo-pulse energy, along with a precise epoch time. To calibrate these measurements the MLA BCE continuously monitored laser pulse energy and range timing over the entire dynamic range of the instrument at all detector gain settings and comparator thresholds.

Three methods were used to calibrate the range measurement. First, MLA and a calibrated time interval analyzer simultaneously measured a sequence of simulated range signals generated by the BCE. These simulated optical signals varied in time delay from approximately 1 ms to 13 ms after the transmitted MLA laser pulse, corresponding to the expected time delays to be seen in Mercury orbit from 200 km to 1,900 km.

Second, a series of closed-loop delay tests was performed by transmitting the MLA laser pulse via the beam dump described in Sect. 4.3.2 through five different lengths of fiber optic cable coupled back to the receiver using the holographic diffuser described in Sect. 4.3.1. The fiber lengths ranged from approximately 140 m to 270 m. Finally, the instrument range bias was evaluated using the boresight alignment fixture LTHR described in Sect. 4.3.3. This setup reflected the attenuated MLA laser pulse directly into the receiver telescope over an optical path length of 1 m.

To locate the measurement point on Mercury's surface a precise knowledge of the MLA boresight axis relative to the spacecraft frame is also needed. Pointing verification tests were performed during instrument thermal-vacuum (TVAC) testing, and on the spacecraft after integration, during TVAC, and prior to launch at the Astrotech integration facility in Titusville, FL.

Timing tests with the MESSENGER spacecraft were also performed after integration to verify that each laser shot was correctly time-stamped. All timing measurements were referenced to GPS-based Universal Time Coordinated (UTC) epoch time.

4.4.2 Environmental Testing

All engineering models of electronic, optical, and laser subassemblies were thermally cycled at ambient pressure. Optics were tested under vacuum to verify focus shift. The flight laser subassembly underwent full thermal vacuum testing while monitoring beam quality, pulse shape, and energy (Krebs et al. 2005).

A planet simulator target was also used during instrument-level thermal balance testing to provide a radiative input to MLA and measure the instrument's response to thermal transients. These tests provided valuable insight to the thermal performance during orbit conditions. The MLA thermal vacuum test setup is shown in Fig. 20. Cold plates were attached to the laser bench and main housing to achieve qualification temperatures.

During instrument TVAC testing the laser pointing was continuously monitored using a telescope and charge-coupled device (CCD) camera outside the chamber that imaged the

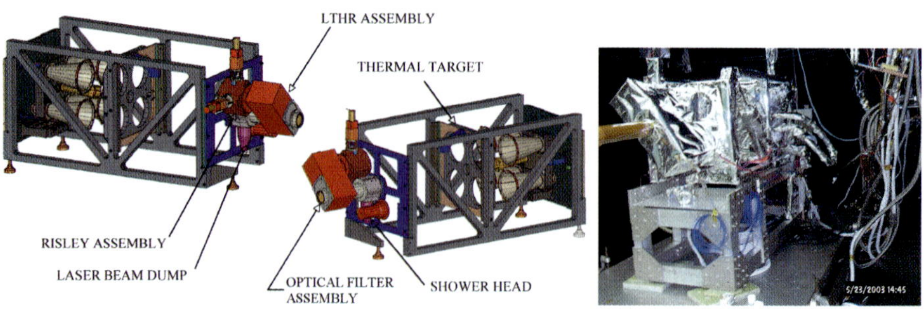

Fig. 20 MLA thermal vacuum test fixture used for instrument-level thermal vacuum testing. The fixture provided a stable platform to monitor alignment and inject optical signals during testing

MLA laser spot along with a reference cube on the instrument. Boresight verification was performed using the LTHR and Risley setup at each TVAC test plateau as well. After instrument TVAC testing the boresight was again verified using the optical bench setup.

During spacecraft environmental testing cold plates were again attached to the MLA laser bench and main housing to provide additional temperature cycles. MLA testing on the spacecraft was limited to range timing with simulated signals. Pointing verification was done before and after spacecraft TVAC.

4.5 Postlaunch Checkout Results

MLA was first powered on August 19, 2004, 16 days after launch (L+16). It remained in Standby mode for approximately 24 hours to allow time for the laser bench to warm up. This warm up also allowed any contaminants (our primary concern was water condensate left as ice on the optics) to sublime. A detector noise characterization test was performed shortly after turn on. Noise levels were observed to be nearly identical to the values seen during ground testing. Another noise characterization was run on L+17 with a similar result.

On August 20, 2004 (Launch + 17 days), MLA was commanded to Science mode, causing the laser to fire. Laser performance was normal with the laser energy monitor reporting 19 to 20 mJ of energy. Figure 21 shows a plot of postlaunch laser energy with an equivalent cold turn-on during vacuum testing. The slightly lower energy (~0.4 mJ, or 2%) can be attributed to the lack of beam termination at the end of the beam expander. During testing it was observed that the laser energy monitor was sensitive to back reflection from either the beam stop or the beam dump. Since it was impossible to fire the laser unterminated on the ground this condition could not be verified, but the drop in the monitor reading was expected. The matching upward slope after several minutes of operation and the laser diode current pulse width indicate a healthy laser.

The need for additional power from the operational laser bench heater during this first cruise operation of the laser was unanticipated. As this heater was switching on for four

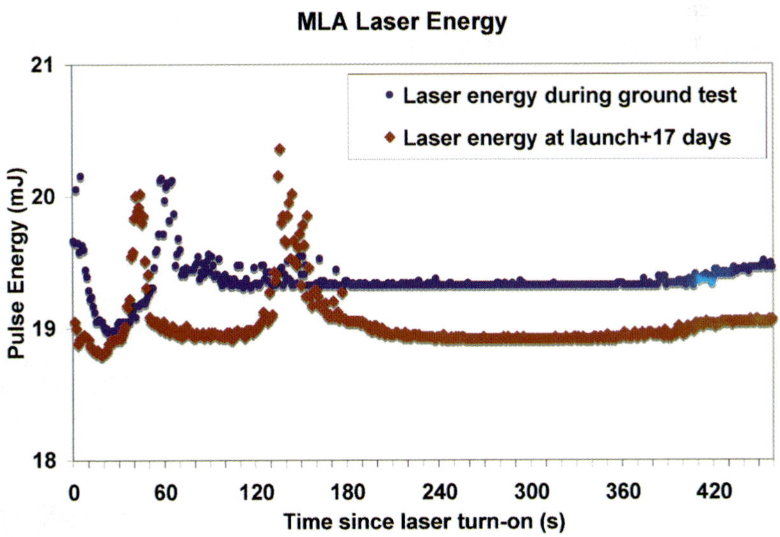

Fig. 21 MLA laser energy comparison showing pulse energy over a seven-minute window measured before launch and 17 days after launch

minutes during a nine-minute period it added approximately 250 mA to the total MLA current. Since the spacecraft C&DH computes power by multiplying the current monitor value times the bus voltage, and the MLA current is only sampled at the 35 ms peak (see Fig. 14), the computed power with the heater was 40.8 W. This apparent power draw violated a MESSENGER spacecraft autonomy rule that automatically shuts down any instrument drawing excessive current. This condition was realized just prior to the planned laser turn-on and the test proceeded with the agreement of the payload systems and spacecraft systems engineers. The autonomy rule executed as expected and turned off the instrument after approximately eight minutes. This rule was subsequently updated to account for this condition.

4.6 Cruise Calibration

In May 2005 MESSENGER successfully performed a sequence of Earth scan maneuvers to calibrate MLA pointing, radiometry, and laser pulse timing during cruise (Smith et al. 2006). Additional radiometric calibrations were performed during the two Venus flybys.

4.7 Data Products

The time-of-flight range data from MLA will be combined with MESSENGER spacecraft pointing knowledge and Radio Science range and range-rate tracking data obtained from the Deep Space Network to produce Digital Elevation Models (DEMs) of the planet's surface referenced to Mercury's center of mass.

MESSENGER's inertial measurement unit and star tracker provide the required pointing data. The radio frequency data link signals are used to determine the precise trajectory of the spacecraft with respect to the planet's center of mass. Once the spacecraft orbit and attitude determinations are initially made, the MLA time-of-flight measurements are used to determine the surface altitude. These surface measurements sample the planet's shape and are fed back into the precision orbit and pointing determination to improve these data sets. Crossover analysis identifies and correlates multiple range measurements of proximate surface features made during different orbits. Reiteration of this process produces the final data product in the form of a DEM. This DEM will be used to determine and track the planet's shape to refine the measurement of the amplitude of the forced physical libration.

Simultaneous pulse width data from the high- and low-threshold channels will be used to determine received pulse energy, which along with the transmitted energy measurement will indicate surface reflectivity at 1,064 nm. Pulse width measurements are also used to assess laser spot-scale, surface slope, and roughness.

4.8 Operation Plans and Observing Strategy

MLA will operate in Science mode, sampling the planet's surface, when the line-of-sight range to Mercury is less than 1,200 km under spacecraft nadir pointing or the slant range is less than 800 km. These limits result in an operating time of approximately 20 to 40 minutes during each 12-hour orbit. Over the life of the mission or approximately 700 orbits, MLA will make over 8 million range measurements at its laser pulse repetition rate of 8 Hz.

For the remainder of each orbit MLA will be commanded to Keep Alive mode to conserve power. In this mode the laser and range measurement electronics are powered off and the instrument dissipates the heat generated by the laser and absorbed from the planet during the brief measurement period. Depending on the orbit, MLA will then be commanded to Standby mode 15 to 480 minutes prior to the next Science mode pass. In Standby mode the laser and range measurement electronics are again powered without actually firing the laser to increase overall instrument power and ensure that the laser amplifier is at its minimum operating temperature of 15°C prior to entering Science mode. During eclipse periods when Mercury blocks the Sun from the MESSENGER solar panels the instrument is powered off.

5 Summary

The successful development, integration, testing, and deployment of MLA have demonstrated the practicability of its scalable, miniaturized laser transmitter, low-cost scalable receiver architecture, and low-power time-of-flight measurement electronics for future space flight missions. Scaling of laser transmitter power is achievable by either removing the laser amplifier (using just the oscillator section) to generate a laser pulse approximately one order of magnitude less powerful or by adding an amplifier to achieve higher laser pulse power. Scaling of the receiver aperture area may be accomplished by adding or removing individual telescopes. A scaled version of the MLA architecture has been used in the design of the Lunar Reconnaissance Orbiter (LRO) Lunar Orbiter Laser Altimeter (LOLA) instrument (Chin et al. 2007). LOLA will use the MLA laser oscillator design to generate 2.7-mJ laser pulses and a single receiver telescope to profile the lunar surface from a nominal altitude of 50 km above the Moon. LRO is scheduled to launch in October 2008.

MLA will provide the first precise laser pulse time-of-flight measurements of the surface of Mercury. This unique data set along with MESSENGER spacecraft pointing and Deep Space Network tracking data will be used to accurately determine the detailed topography of Mercury's northern hemisphere, measure topographic profiles across major geologic structures, track large-scale planetary shape to measure the planet's libration, and measure the surface reflectivity of Mercury at 1,064 nm.

Acknowledgements We wish to thank the following individuals for their significant contributions to the successful completion of the MLA instrument effort: Edward Amatucci, Adrienne Beamer, Pete Dogoda, Tom Feild, Ron Follas, Ame Fox, Jeff Guzek, Randy Hedgeland, Sid Johnson, Igor Kleyner, Steve Li, Steve Lindauer, Billy Mamakos, Dave McComas, Roger Miller, Tony Miller, Lou Nagao, Karen Pham, Steve Schmidt, Stan Scott, Nancy Stafford, Jon Vermillion, Ken Waggoner, Tony Yu, Ron Zellar, and the MESSENGER team at APL, in particular Steve Jaskulek, Eliot Rodberg, Chuck Schlemm, Stan Kozuch, Jack Ercol, and Ted Hartka.

References

J.B. Abshire, X. Sun, R.S. Afzal, Appl. Optics **39**, 2449–2460 (2000)
G. Chin et al., Space Sci. Rev. (2007). doi:10.1007/s11214-007-9153-y
J.J. Degnan, J. Geodyn. **34**, 503–549 (2002)
J. Garvin et al., Phys. Chem. Earth **23**, 1053–1068 (1998)
S.E. Hawkins, III et al., Space Sci. Rev. (2007, this issue). doi:10.1007/s11214-007-9266-3
D.J. Krebs, A.M. Novo-Gradac, S.X. Li, S.J. Lindauer, R.S. Afzal, A.W. Yu, Appl. Optics **44**, 1715–1718 (2005)
N. Paschalidis et al., IEEE Trans. Nucl. Sci. **49**, 1156–1163 (2002)
L. Ramos-Izquierdo et al., Appl. Optics **44**, 1748–1760 (2005)
D.E. Smith et al., Science **279**, 1686–1692 (1998)
D.E. Smith et al., Science **311**, 53 (2006)
S.C. Solomon et al., Planet. Space Sci. **49**, 1445–1465 (2001)
X. Sun, J.F. Cavanaugh, J.C. Smith, A.E. Bartels, *Proceedings of the 22nd International Laser Radar Conference (ILRC 2004)*. Special Publication SP-561, European Space Agency, Noordwijk, The Netherlands, 2004, pp. 961–964
M.T. Zuber et al., J. Geophys. Res. **97**, 7781–7797 (1992)
H.J. Zwally et al., J. Geodyn. **34**, 405–446 (2002)

Space Sci Rev (2007) 131: 481–521
DOI 10.1007/s11214-007-9264-5

The Mercury Atmospheric and Surface Composition Spectrometer for the MESSENGER Mission

William E. McClintock · Mark R. Lankton

Received: 22 May 2006 / Accepted: 10 August 2007 / Published online: 25 October 2007

Abstract The Mercury Atmospheric and Surface Composition Spectrometer (MASCS) is one of seven science instruments onboard the MErcury Surface, Space ENvironment, GEochemistry, and Ranging (MESSENGER) spacecraft en route to the planet Mercury. MASCS consists of a small Cassegrain telescope with 257-mm effective focal length and a 50-mm aperture that simultaneously feeds an UltraViolet and Visible Spectrometer (UVVS) and a Visible and InfraRed Spectrograph (VIRS). UVVS is a 125-mm focal length, scanning grating, Ebert-Fastie monochromator equipped with three photomultiplier tube detectors that cover far ultraviolet (115–180 nm), middle ultraviolet (160–320 nm), and visible (250–600 nm) wavelengths with an average 0.6-nm spectral resolution. It will measure altitude profiles of known species in order to determine the composition and structure of Mercury's exosphere and its variability and will search for previously undetected exospheric species. VIRS is a 210-mm focal length, fixed concave grating spectrograph equipped with a beam splitter that simultaneously disperses the spectrum onto a 512-element silicon visible photodiode array (300–1050 nm) and a 256-element indium-gallium-arsenide infrared photodiode array 850–1,450 nm. It will obtain maps of surface reflectance spectra with a 5-nm resolution in the 300–1,450 nm wavelength range that will be used to investigate mineralogical composition on spatial scales of 5 km. UVVS will also observe the surface in the far and middle ultraviolet at a 10-km or smaller spatial scale. This paper summarizes the science rationale and measurement objectives for MASCS, discusses its detailed design and its calibration requirements, and briefly outlines observation strategies for its use during MESSENGER orbital operations around Mercury.

Keywords Atmosphere · Exosphere · Mercury · MESSENGER · Spectrometer · Surface

1 Introduction

The Mercury Atmospheric and Surface Composition Spectrometer (MASCS) is one of seven science instruments aboard the MErcury Surface, Space ENvironment, GEochemistry, and

W.E. McClintock (✉) · M.R. Lankton
Laboratory for Atmospheric and Space Physics, University of Colorado, Boulder, USA
e-mail: william.mcclintock@lasp.colorado.edu

Ranging (MESSENGER) spacecraft. It is designed to provide measurements that address four of the six science questions that frame the MESSENGER mission:

1. What planetary formation processes led to the high metal/silicate ratio in Mercury?
2. What is the geological history of Mercury?
3. What are the radar reflective materials at Mercury's poles?
4. What are the important volatile species and their sources and sinks on and near Mercury?

MASCS consists of a small Cassegrain telescope that simultaneously feeds a Visible and InfraRed Spectrograph (VIRS) and an UltraViolet and Visible Spectrometer (UVVS). It will obtain maps of surface reflectance spectra that will be used to investigate the mineralogical composition of the surface on spatial scales of 5 km or less. UVVS will measure altitude profiles of known species in order to determine composition and structure of the exosphere and its variability. It will also search for signatures of possible volatile polar deposits and previously undetected exospheric constituents.

2 Science Objectives

2.1 Exosphere

The atmosphere of Mercury is a surface-bounded exosphere, which acts as an interface between the surface and external stimuli impinging upon it (Killen and Ip 1999; Domingue et al. 2007). Its composition and behavior are controlled by its interactions with the magnetosphere and the surface. The MESSENGER exospheric science investigation is based upon four key questions related to the nature of the exosphere and its relationship to the external processes that modify the surface:

1. What are the composition, structure, and temporal behavior of the exosphere?
2. What are the processes that generate and maintain the exosphere?
3. What is the relationship between exospheric and surface composition?
4. Are there polar deposits of volatile material, and how are the accumulation of these deposits related to exospheric processes?

1. What are the composition, structure, and temporal behavior of the exosphere?
Mercury's exosphere is known to contain six elements (H, He, O, Na, Ca, and K) that taken together have a surface density at the subsolar point of approximately 10^4 atoms cm^{-3} (Hunten et al. 1988). The airglow spectrometer aboard the Mariner 10 flyby mission detected H, He, and O (Broadfoot et al. 1974a, 1974b). Potter and Morgan discovered Na and K using high-resolution ground-based spectroscopic observations (Potter and Morgan 1985, 1986). Bida et al. (2000) detected high-temperature Ca using the Keck I telescope. Searches for additional constituents have not been successful (Hunten et al. 1988).

Ground-based studies of Na indicate that the exosphere is spatially and temporally variable. There are orderly changes in Na surface density related to changes in solar radiation pressure and distance (Killen et al. 1990), but there are also chaotic variations in the exosphere. The subsolar Na density has been observed to change by a factor of two on time scales of less than a week, and bright Na emission spots have been observed at high northern latitudes and over the Caloris basin.

With the existing data our inventory of Mercury's exospheric composition is incomplete. Current understanding of the source processes (discussed later) suggests the presence of as yet undetected species including Ar, Si, Al, Mg, Fe, S, and OH (from impact vaporization of

H_2O). With the exception of Ar, all of the species listed above have strong ground-state emission lines (predicted intensities in the range 10–1,000 Rayleighs, e.g., Morgan and Killen 1996) in the spectral range 130–600 nm, but observational constraints have prevented us from detecting these species from the ground or from Earth-orbiting spacecraft.

The measurements required to support the investigation of Mercury's exospheric composition and structure are: (1) altitude profiles of known species (H, O, Na, Ca, and K) measured with a vertical resolution comparable to an exospheric scale height (25–50 km) and a latitude/longitude resolution of 10–20°, and (2) a sensitive search for predicted species that have not been previously observed (e.g., Si, Al, Mg, Fe, S, OH).

The UVVS channel of MASCS is specifically designed to make these measurements. It provides broad spectral coverage (115–600 nm), moderate spectral resolution (0.6 nm), and high sensitivity (detection limit ~100 Rayleighs), enabling it to produce a detailed inventory of the species in the exosphere of Mercury (or strong upper limits) together with the vertical and horizontal distributions of the most abundant species. In addition to determining composition and structure, these data will provide the basis for determining exospheric processes, studying the relationship between surface and exospheric composition, and studying surface–exosphere–magnetosphere interactions.

2. What are the processes that generate and maintain the exosphere?

The processes that supply and remove exospheric material have been identified, but their relative importance is poorly understood. Hydrogen and helium are likely derived from neutralized solar wind ions, although photodissociation of meteoritic water and crustal outgassing should supply a portion of these two species. Proposed sources for Na, K, Ca, and O include impact vaporization, ion sputtering, photon stimulated desorption, thermal desorption, and crustal degassing. Currently, there is strong disagreement about the relative importance of these four mechanisms (McGrath et al. 1986; Sprague 1990; Morgan and Killen 1996). Determining a comprehensive inventory of exospheric species and measuring their spatial and temporal distributions will allow us to quantify the dominant source mechanisms for the various exospheric species.

The principal loss mechanisms are thermal escape and photoionization with subsequent loss through transport along open magnetic field lines. Although thermal escape appears to be the dominant loss mechanism for both hydrogen and helium, it is probably unimportant for Na and K (Hunten et al. 1988). Photoionization rates for Na and K are relatively well known; however, the total loss rates from the exosphere are uncertain by a factor of ten because the efficiency with which ions are swept away by the convecting electric field is highly uncertain. Based on lunar studies (Mendillo et al. 1991), gas–surface interactions may also be an important sink for sodium and potassium.

UVVS measurements of composition and structure will provide the data required to characterize exospheric processes. Distributions for sodium and other species with strong emission lines will permit a definitive determination of their surface interaction. By understanding the interactions of the major species with the surface we can use the inventory to determine the relative importance of the key source processes.

Correlating UVVS data with Energetic Particle and Plasma Spectrometer (EPPS) (Andrews et al. 2007) and Magnetometer (Anderson et al. 2007) data will provide an additional tool for understanding exospheric processes. The magnetic field controls the location and strength of sputtering ion flux as well as the efficiency of photo-ion loss from the planet. By relating local variations in exospheric composition with the loci of charged particle precipitation on the surface, we can isolate sputtering as a source. Similarly, production rates from impact vaporization, which depend on the magnitude of impact velocity, should peak in the ram direction of Mercury's orbital motion.

We can also examine the importance of crustal outgassing and search for surface features (e.g., the Caloris basin, Sprague 1990; Sprague et al. 1998) that may be associated with enhanced concentrations of exospheric sodium or potassium. Establishing a convincing exosphere–basin correlation requires that we remove the contributions from all other effects, most of which are related to the magnetosphere, the solar wind, and changes in solar ultraviolet (UV) and extreme ultraviolet (EUV) flux.

In addition to neutral species, the UVVS should detect such prominent ions along lines of sight as Ca^+ and Mg^+, which pass over the poles and extend along the magnetotail. Strong resonance lines for S^+, Si^+, and Al^+ also fall in the spectral range of the UVVS. A measurement of an element in the exosphere and the corresponding ion in the tail will help us determine the rate at which ions are permanently lost from the planet.

3. What is the relationship between exospheric and surface composition?

Except for the noble gases, hydrogen, and a few volatile species such as sulfur, which are abundant in the micrometeoroid population, species found in the exosphere are thought to be derived from the regolith and crust. If we can quantify the sources, sinks, and gas–surface interactions of the exosphere, then measuring other regolith-derived elements (Ca, Mg, Al, Fe) will allow us to estimate the relative abundances of these species on the surface. For example, Morgan and Killen (1996) showed that if sputtering is the dominant source, then the relative abundances of Ca, Al, Fe, Mg, and Si in the exosphere are related to surface composition.

Exospheric detection, together with detection by the Gamma-Ray and Neutron Spectrometer (Goldsten et al. 2007) and by color imaging (Hawkins et al. 2007) and surface reflectance spectroscopy, provide the data required to support the correlation between exospheric and surface composition.

4. Are there polar deposits of volatile material, and how are the accumulation of these deposits related to exospheric processes?

One of the most exciting recent discoveries about Mercury comes from ground-based radar backscatter measurements that have been interpreted as arising from polar deposits of water ice on the surface or at shallow depths (Slade et al. 1992; Butler et al. 1993). An alternative interpretation was offered by Sprague et al. (1995), who concluded that the observations indicate deposits of elemental sulfur. In either case, one mechanism that may explain the accumulation of volatile species at the poles is impact vaporization of infalling material followed by exospheric transport to the poles and trapping in permanent shadows.

The UVVS will search for the presence of water ice deposits at the poles by measuring the distribution of the daughter product of water dissociation, the hydroxyl molecule (OH). Impact vaporization, sublimation, and degassing are expected to contribute water to the exosphere, where photodissociation by sunlight produces OH. Similar source mechanisms will also supply sulfur to the exosphere. Killen et al. (1997) calculated the expected exospheric concentration of OH from both buried and exposed ice deposits for a wide variety of surface conditions. They concluded that the OH surface density should exceed 10^3 cm^{-3} and that the OH emission rate at wavelength $\lambda = 306$ nm should be greater than 100 Rayleighs if the surface temperature is greater than about 110 K. This intensity is easily detectable by the UVVS. (Similar calculations for the sublimation and degassing rates for sulfur have not yet been completed; nonetheless, we expect that a surface density of 8×10^3 cm^{-3} would produce an emission strength of about 200 Rayleighs at $\lambda = 181.3$ nm, which would also be easily detected by the UVVS.)

Specifically, the MASCS exosphere measurement objectives are to provide exospheric composition over the polar regions and to make spectral measurements of exospheric composition and density as functions of latitude, time of solar day, and time of solar year.

2.2 Formation and Geological History

The MASCS instrument will also provide data to address MESSENGER science questions regarding the formation process of Mercury and its geological history:

1. What is the mineralogy of Mercury's surface and its variation with geological unit?
2. What is the rate of space weathering on Mercury, and how does it affect spectral interpretations?

1. What is the mineralogy of Mercury's surface and its variation with geological unit?

The mineralogical composition of Mercury's surface material is largely unknown. Although Earth-based observations indicate that the reflectance spectra of Mercury show similarity in shape and slope with those of the lunar highlands, it is recognized that the spectra of the two bodies also show a marked difference. Whereas the Moon's spectrum shows an absorption band near 1,000 nm caused by ferrous iron in pyroxenes and olivines (Adams and McCord 1970), Mercury's spectrum shows, at most, a weak, distorted feature near 1 μm, and evidence is accumulating that Mercury's surface is low in Fe^{2+} (Vilas 1985), limiting the average FeO content to less than about 3–6% by weight (Blewett et al. 1997). Further, Warell et al. (2006) argued that a broad feature near 1,000 nm recently observed at mid latitudes and 190°E longitude results from the presence of iron-poor, high-calcium clinopyroxene. Clementine UV–visible multispectral measurements have led to the identification of large regions on the lunar farside that are extremely low in FeO (<3% by weight) (Lucey et al., 1998, 2000a). Such areas are believed to be composed of >90% plagioclase feldspar. Furthermore, recent observations of the Moon have revealed small areas of the nearside that lack the otherwise ubiquitous 1-μm ferrous iron absorption band (Blewett et al. 2002, and references therein), leading to the possibility that Mercury's surface composition can be classified as pure anorthosite. In addition to anorthosite, there are alternative mineralogies for Mercury with featureless reflectance spectra. Earth-based mid-infrared observations show emission features consistent with the presence of both calcic plagioclase feldspar containing some sodium and very-low-FeO pyroxene (Sprague et al. 2002).

The current paradigm for Mercury, derived from Earth-based observations, suggests a crust composed predominantly of plagioclase feldspar (anorthitic to albitic) with little basaltic materials, showing virtually no large-scale compositional variation. However, Mariner 10 color data indicate the presence of several units, too small to be detected from Earth, thought to be enriched in opaque minerals (relatively low albedo and blue color) (Rava and Hapke 1987; Robinson and Lucey 1997; Robinson and Taylor 2001). Mercury's surface also contains an unknown proportion of meteoritic material, perhaps as much as 5–20% (Noble and Pieters 2003), which can provide nontypical constituents, such as sulfur or carbon. Spectral reflectance information obtained in the range 400–1,500 nm can identify key minerals and estimate their abundances. On the basis of results from Clementine lunar observations (Lucey 2004) the anticipated accuracy at Mercury should be on the order of 10% or better; although Warell and Blewett (2004) argued that differences in composition and rates of space weathering on Mercury and the Moon (Cintala 1992) may nullify many of the detailed retrieval algorithms for lunar materials (e.g. Lucey et al. 1998, 2000a, 2000b). Mariner 10 images also show regions with albedo differences that may indicate small variations in mafic minerals. Characterizing the major mineral constituents of Mercury's crust places important constraints on formation models of the planet.

Mariner 10 color images show that, despite the heavily agglutinized nature of the surface, many color differences exist that must be due to compositional rather than maturity differences. These units fall into three broad categories: first is the "average" Mercury surface, composed predominantly of the ancient heavily cratered terrain and intercrater plains.

Second, there is a relatively low-albedo, blue material that generally appears insensitive to local topographic undulations and exhibits feathered or indistinct margins. This material is possibly relatively rich in opaque mineral content and may have been emplaced in explosive eruptions (Robinson and Lucey 1997). Interestingly, several smooth plains units appear to embay and therefore postdate this unit. The color properties of the smooth plains are similar to the global average. Detailed spectral characterization of the smooth plains is a high priority for MASCS. Determining compositional variations within and between smooth plains deposits will give the best chance of understanding the evolution of the crust and the composition of Mercury's mantle (Robinson and Taylor 2001). Third, one of the more mysterious units seen in the Mariner 10 color images is a class of materials exhibiting high albedo and red color. These materials were noted in early papers as anomalous high-albedo patches in the floors of several craters (Dzurisin 1977). Similar to the Tycho crater on the Moon, a low-albedo annulus is present around the crater Basho.

2. What is the rate of space weathering on Mercury, and how does it affect spectral interpretation?

A serious obstacle to the interpretation of any spectral data for Mercury is the intense space weathering environment at Mercury's orbit, which results in a high percentage of the surface being converted to agglutinates or glass (Cintala 1992). Agglutinates tend to shift band centers, but more importantly they suppress band depths, thus making the identification of iron bands difficult—especially with FeO contents below 10% by weight. To overcome this problem a key MESSENGER strategy is the acquisition of spectra of as many Kuiperian (immature) ejecta blankets as possible. When an impact crater is formed, material from depth is excavated and deposited near the rim. Initially this material is nearly agglutinate free, but over time the space weathering process continually converts crystalline material to agglutinitic material, thus lowering the albedo, reddening the spectral slope, and suppressing absorption bands. Even if the true mineralogic properties of fully mature material cannot be recovered, the fresh ejecta blankets will provide a random sample of the local bulk composition. Lucey (2004) followed such a strategy with Clementine spectral data to produce an interpolated 1-km-per-pixel major mineralogy map of the Moon (clinopyroxene, orthopyroxene, olivine, plagioclase feldspar).

Specifically, the MASCS surface composition science measurement objectives will contribute to the goals of the MESSENGER geology investigation (Head et al. 2007) by providing spectral measurement of the surface mineralogy associated with Fe-bearing, Ti-bearing, and other units via visible and near-infrared absorption bands.

3 Measurement Requirements and Implementation

MASCS measurement requirements are summarized in Table 1. In general the MESSENGER mission design and launch vehicle selection severely limit both payload mass and data downlink volume. These considerations led to a design for an exospheric experiment that employs a scanning grating monochromator with photomultiplier tube detectors rather than one using fixed gratings and multielement detectors. The primary science objectives require a small, low-mass instrument with moderate resolution and very high sensitivity, which can measure a small number (10–20) of isolated emissions at known wavelengths spread over a very broad range. With the exception of the noble gasses, the known and predicted exospheric species have observable emission lines in the far ultraviolet to visible (115–600 nm). Noble gasses were excluded from the exosphere measurement objectives

Table 1 MASCS measurement requirements

Parameter	Atmospheric composition	Surface composition
Wavelength range	115–600 nm	115–1,450 nm
Spectral resolution	0.6 nm	5 nm
Spatial resolution	25-km height on limb 15–20° latitude/longitude	<500 μrad
Sensitivity	Detection of 10 Rayleighs in 100 s; $130 < \lambda < 425$ nm	Signal-to-noise >100

because their resonance lines lie in the extreme ultraviolet (<100 nm), requiring a separate optical train. Although most of these emissions are optically thin, the notable exception is the important sodium resonance doublet at 589.0 nm and 589.6 nm, which is optically thick for many observational geometries. Accurate retrieval of sodium column abundances requires that MASCS have sufficient spectral resolution to determine individual emission strengths of these two lines.

When the spectrum consists of a few isolated features, a scanner has a sensitivity advantage over a fixed grating design because it has very large light-gathering power per spectral resolution element. On the other hand, the requirement for simultaneous and continuous spectral coverage dictates the need for a fixed grating and multielement design for the visible and near-infrared surface reflectance measurements. Here, downlink limitations led to a point spectrograph rather than a mapping spectrograph design for surface composition studies. Obtaining visible and infrared spectra on a fixed latitude and longitude grid of 5 to 10 km instead of over a continuous swath will provide context for the full-coverage color images obtained with Mercury Dual Imaging System (Hawkins et al. 2007). Mercury thermal infrared emission limits the spectral coverage for reflectance spectroscopy to <1.5 μm, which is still adequate to measure signatures of ferrous oxide near 1 μm.

4 Instrument Description

4.1 Instrument Overview

Figure 1 is an image of the MASCS taken during instrument testing. The instrument consists of a small Cassegrain-style telescope simultaneously feeding both an Ultraviolet-Visible Spectrometer (UVVS) and a Visible-Infrared Spectrograph (VIRS). The telescope-UVVS configuration is nearly identical to the UltraViolet Spectrometer (UVS) aboard the Galileo spacecraft (Hord et al. 1992). It uses a plane-grating monochromator equipped with three photomultiplier tube (PMT) detectors to measure far ultraviolet (FUV-PMT, 115–190 nm), middle ultraviolet (MUV-PMT, 160–320 nm), and visible (VIS-PMT, 250–600 nm) wavelengths. Far ultraviolet wavelengths are observed in the second order of the grating with a typical resolution of 0.3 nm, while middle ultraviolet and visible wavelengths are observed in first order with resolutions that vary from 0.7 nm at 200 nm to 0.45 nm at 600 nm. UVVS is optimized for measuring the composition and structure of the exosphere and for measuring ultraviolet (wavelength $\lambda < 300$ nm) surface reflectance. VIRS is mounted on top of the UVVS and is coupled to the telescope focal plane with a short fiber optics bundle. It uses a fixed concave grating and two solid-state array detectors to measure visible (VIS, 300–1,050 nm) and near-infrared (NIR, 850–1,450 nm) wavelengths, both with a resolution

Fig. 1 MASCS instrument image taken during instrument testing at the Laboratory for Atmospheric and Space Physics (LASP) at the University of Colorado

of 4.7 nm (a dispersion of 2.33 nm per pixel). It is optimized for measuring visible and near-infrared ($0.3 \le \lambda \le 1.45$ μm) surface reflectance. A contamination door equipped with a MgF_2 window covers the telescope aperture. The MASCS control and interface electronics module, which uses a microprocessor-based logic system to manage instrument configuration, control UVVS spectral scanning, and provide spacecraft-instrument communications, is mounted on the side of the UVVS spectrometer housing. Figure 2 shows a functional block diagram that illustrates the instrument optical paths and electronic signal flow.

The UVVS will measure altitude profiles of known species (H, O, Na, K, and Ca), which will be used to determine exospheric temperature and density and to map the extended distributions of these species. It will make observations at a series of fixed wavelengths to search for predicted species that have not been previously detected (e.g., Si, Al, Mg, Fe, S, OH)

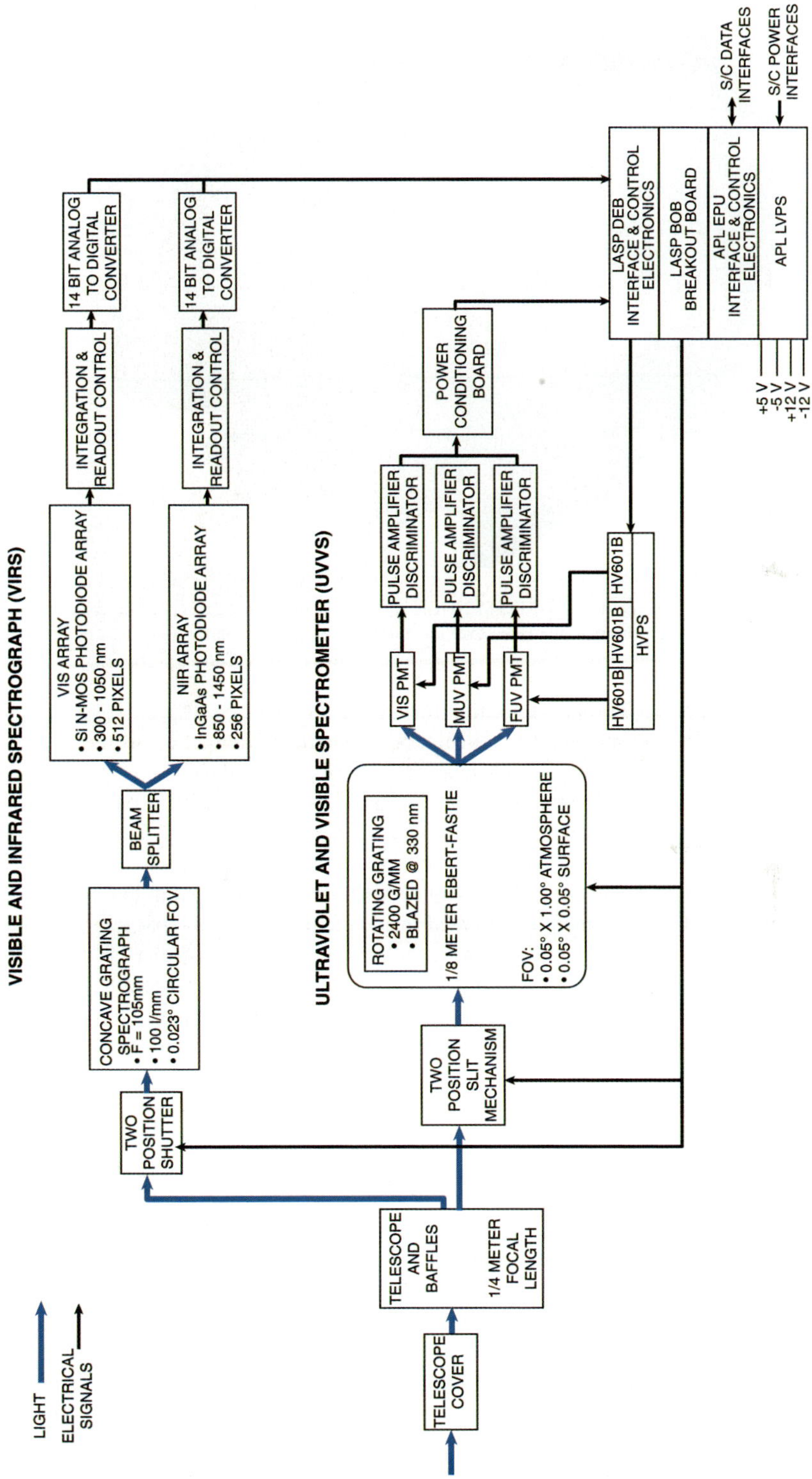

Fig. 2 MASCS functional block diagram showing optical paths and electronic signal flow

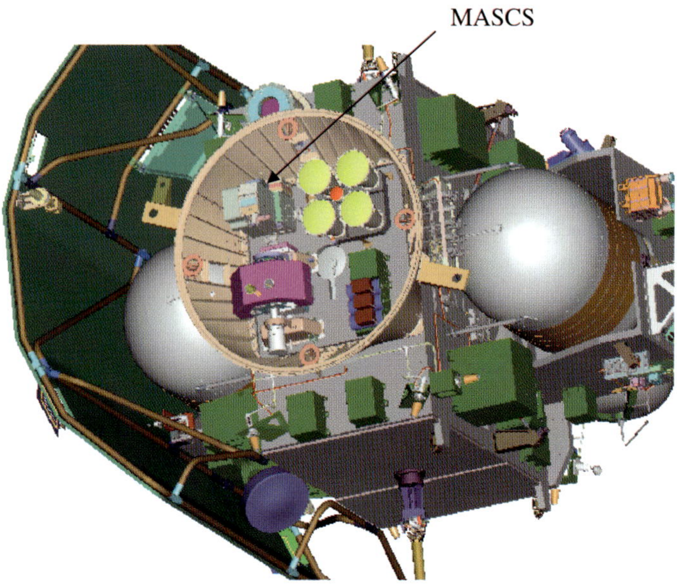

Fig. 3 MASCS (*arrow*) placement on the MESSENGER spacecraft aligns the telescope optic axis with the spacecraft $+Z$ axis. For flight, thermal blankets cover the entire instrument except for the telescope aperture

and in spectral scans (115–600 nm) to search for new species (see Sect. 6.1). Together, the UVVS and VIRS will measure surface reflectance at middle ultraviolet to visible to near-infrared wavelengths to search for ferrous-bearing minerals (spectral signatures near 1 μm), Fe-Ti-bearing glasses (spectral signatures near 340 nm), and ferrous iron (strong band near 250 nm). These measurements will be made with a spatial resolution of 5 km or better. Table 2 summarizes the key performance characteristics for the two channels.

Figure 3 shows the placement of MASCS on the MESSENGER spacecraft before thermal blanket installation. The instrument is mounted inside the third-stage-to-spacecraft adapter ring with its telescope optic axis aligned with the spacecraft $+Z$ axis (Leary et al. 2007) and with the long axis of the UVVS entrance slit aligned with the spacecraft (S/C) $\pm$Y axis. In this configuration the UVVS slit "points at" the spacecraft sunshade. Except for the telescope aperture, the entire instrument is covered with thermal blankets for flight.

4.2 Optical Design and Performance

Figure 4 is a schematic diagram of the MASCS optical system showing the major components, detectors, and telescope baffles. The design is derived from the Galileo UVS in which a Cassegrain-style, concentric-mirror telescope and a plane grating monochromator equipped with three PMT detectors form an integral optical-mechanical system. It was modified for MASCS by adding a spectrograph, which is located on an external optical bench and coupled to the telescope focal plane by an optical fiber.

4.2.1 Telescope-Ultraviolet and Visible Spectrometer

The telescope optical design is a Dall-Kirkham configuration, which has a concave elliptical primary mirror with a 254.0-mm radius of curvature and a conic constant equal to 0.8786.

Table 2 MASCS instrument summary

Telescope	
Focal length	257.6 mm
Aperture	50.3 × 52.8 mm
Ultraviolet and visible spectrometer	
Focal length	125 mm
Grating	2400 groove/mm—blazed at 330 nm
Spectral resolution	
FUV channel	0.3 nm
MUV channel	0.7 nm
VIS channel	0.6 nm
Wavelength range	
FUV channel	115–190 nm
MUV channel	160–320 nm
VIS channel	250–600 nm
Field of view	
Exosphere	1° × 0.04°
Surface	0.05° × 0.04°
Visible and infrared spectrograph	
Focal length	210 mm
Grating	100 groove/mm—sinusoidal profile
Dispersion	46.6 nm/mm, 2.33 nm/pixel
Spectral resolution	4.7 nm
Wavelength range	
VIS channel	300–1,050 nm
NIR channel	850–1,450 nm
Field of view	0.023° circular
Instrument	
Mass	3.1 kg
Average power	6.7 W
Dimensions	195 × 205 × 310 mm

A convex spherical secondary mirror focuses the final image ~17 mm behind the primary mirror vertex, producing a system with a 257.6-mm effective focal length and an $f/5$ focal ratio. Baffles located directly behind the secondary mirror and in the center of the primary mirror shield the telescope focal plane from direct illumination by the external scene.

For exosphere observations, the UVVS is equipped with a 0.175-mm-wide × 4.5-mm-long slit, corresponding to a 0.04° × 1.0° field of view (FOV). A two-position mechanism, located behind the telescope focal plane, inserts an opaque mask over the top and bottom of the exosphere slit, reducing the aperture height to 0.23 mm, to provide a 0.04° × 0.05° FOV for surface observations. The VIRS 0.023°-circular FOV is determined by the 0.1-mm diameter of the quartz optical fiber that feeds the spectrograph input. It is located 1.71 mm (0.38°) from the center of the UVVS surface slit.

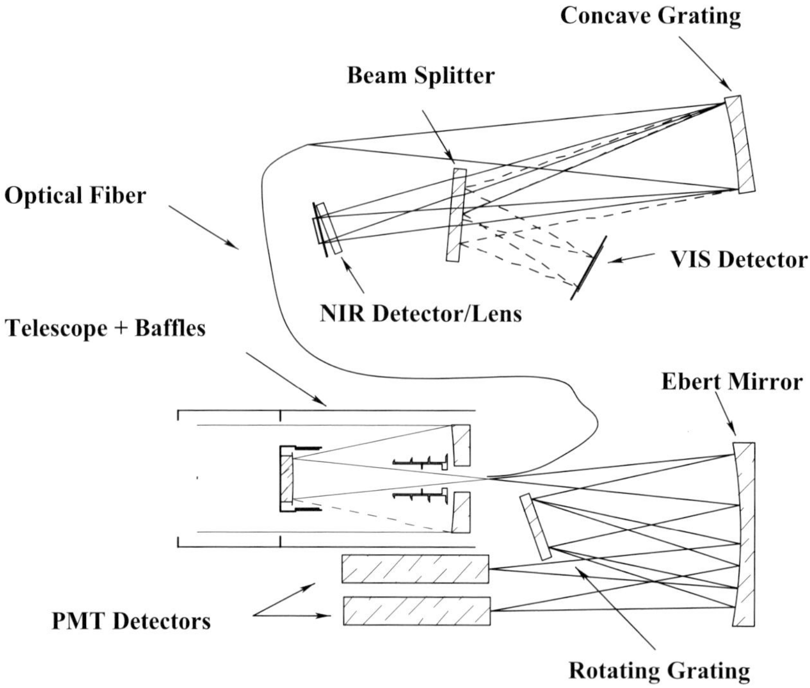

Fig. 4 MASCS optical system schematic. For clarity, only two of the three UVVS photomultiplier tube detectors are shown

In a Dall-Kirkham telescope, spherical aberration in the primary and secondary mirrors cancel and the images near the optic axis are only blurred by coma. Spot diagrams computed from ray tracing a circular object with a 0.023° angular diameter located 100 km from the telescope aperture were used to evaluate imaging performance for the VIRS FOV. The results are shown in Fig. 5 as intensity contours, normalized to a peak value of 1. These contours represent the irradiance in the focal plane that results from convolving the telescope imaging point spread function with the angular FOV defined by the spectrograph input fiber. The shape of the contour plot is consistent with analytic estimates from third-order aberration theory, which predict that the point source comatic blur at the field location of the VIRS fiber is 8 μm compared to its 100-μm diameter. Similar results were obtained for the UVVS FOV, and the imaging performance can be accurately approximated as the convolution of an 8-μm telescope point spread function and a 0.175-mm × 4.5-mm rectangular aperture.

The UVVS spectrometer is a standard, 125-mm focal length, Ebert-Fastie design, which uses a single spherical mirror as both collimator and camera. Light from the telescope enters the spectrometer through the entrance aperture and is collimated and reflected toward the diffraction grating by one side of the Ebert mirror. After diffraction a narrow band of wavelengths is imaged onto the spectrometer focal plane by the other half of the Ebert mirror. Here three Hamamatsu photomultiplier tubes (FUV-PMT, MUV-PMT, and VIS-PMT), behind separate exit slits, each separated by 10.2 mm in the spectrometer focal plane, record the spectrum in pulse counting mode, providing maximum sensitivity for the exospheric observations. Their photocathodes (CsI for FUV, CsTe for MUV, and Bi-alkali for VIS) are chosen to optimize measurements in narrow spectral ranges and minimize the effects of

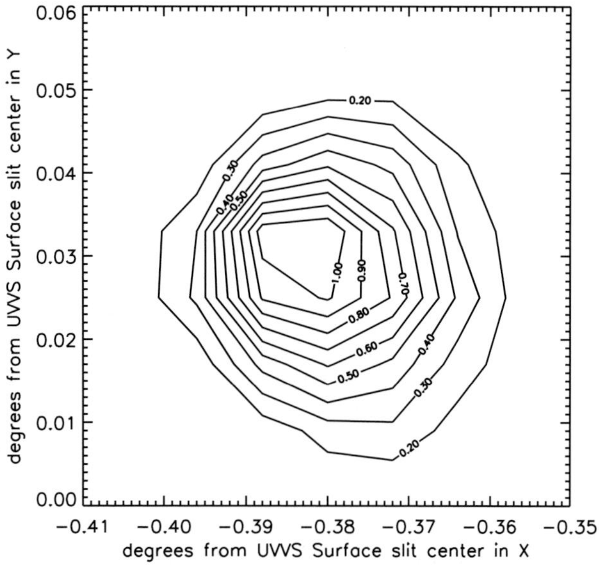

Fig. 5 Modeled imaging performance for the VIRS FOV. Contours are irradiance values normalized to 1

wavelength-dependent scattered light within the spectrometer (Hord et al. 1992). Spectral scanning is accomplished by rotating the diffraction grating in discrete steps.

Analytic calculations and ray tracing were used to select the spectrometer parameters (entrance and exit slit widths, grating spacing, and grating drive angular range and step size) required to meet the MESSENGER measurement objectives summarized in Table 1. The wavelength and imaging equations for an Ebert-Fastie spectrometer provided the point of departure for this analysis:

$$\begin{aligned} m\lambda &= d\big[\sin(\alpha) + \sin(\beta)\big], \\ m\lambda &= 2d\sin(\theta)\cos(\phi), \\ \theta &= \frac{\beta + \alpha}{2}, \\ \phi &= \frac{\beta - \alpha}{2}, \end{aligned} \tag{1}$$

where λ is the wavelength, d is the grating spacing, m is the diffraction order number, α is the grating angle of incidence, and β is the grating angle of diffraction. $\theta = \alpha + \phi$ is the grating rotation angle, and ϕ is the half angle difference between β and α, which is fixed by the geometry of the spectrometer and is unique for each channel. Rotating the grating through angle $\Delta\theta_S$ changes the wavelength by $\Delta\lambda_S$ given by

$$\Delta\lambda_S = \frac{2d\cos(\theta)\cos(\phi)}{m}\Delta\theta_S. \tag{2}$$

The minimum step size consistent with minimum Nyquist sampling is 1/2-bandpass-per-step. A goal for the MASCS design was to reduce that step size by 33%, providing about three steps per bandpass.

The UVVS imaging function for a monochromatic line is the convolution of the image of the entrance slit with the physical width w_{en} of the exit slit. In the absence of image aberration, the entrance slit image width is given by:

$$w'_{en} = w_{en} \cdot \cos(\alpha)/\cos(\beta). \tag{3}$$

The ratio of cosines is referred to as anamorphic magnification.

For an Ebert-Fastie configuration, the image of the entrance slit is always larger than the slit itself, and the resulting imaging function is a trapezoid with a full-width half maximum (FWHM) equal to the greater of the entrance slit image or the exit slit width. Differentiating the standard grating equation with respect to the diffraction angle, β, and multiplying by the angular width of the imaging function gives an expression for the spectrometer bandpass

$$\begin{aligned} \Delta\lambda &= \frac{d}{m \cdot F_l} \cdot \cos(\beta) \cdot w_{ex} \quad \text{when } \cos(\beta) \cdot w_{ex} > \cos(\alpha) \cdot w_{en}, \\ \Delta\lambda &= \frac{d}{m \cdot F_l} \cdot \cos(\alpha) \cdot w_{en} \quad \text{when } \cos(\alpha) \cdot w_{en} > \cos(\beta) \cdot w_{ex}, \end{aligned} \tag{4}$$

where $F_l = 125$ mm is the spectrometer focal length and w_{en} and w_{ex} are the entrance and exit slit widths, respectively. If the spectrometer imaging aberration is negligible, then the instrument spectral resolution is equal to the bandpass. Equation (4) shows that for fixed bandpass, smaller values of grating spacing lead to wider slits and therefore greater instrument sensitivity (see (6) in Sect. 5). An assumed maximum grating rotation angle of ~45° limits the smallest spacing from commercially available gratings to 416.67 nm (2,400 grooves mm^{-1}).

In order to maintain high efficiency, it is common practice to restrict the operating wavelengths for a diffraction grating to a range specified by the blaze wavelength, λ_B: $2\lambda_B/3 < \lambda < 2\lambda_B$ (James and Sternberg 1969). The blaze wavelength is related to the blaze angle, which is the angle that the groove facets make with respect to the grating normal, through (1) when $\phi = 0$ [$\lambda_B = 2d\sin(\theta_B)$]. A 600-nm maximum operating wavelength suggests $\lambda_B \geq 300$ nm and a 200-nm minimum operating wavelength. Since the blaze range is actually a restriction on incidence and diffraction angles, high efficiency for the FUV wavelength range is accommodated in MASCS by using the grating in second order ($m = 2$), resulting in an effective first-order range 230–380 nm. These considerations led to the selection of a standard catalog Jobin Yvon grating with a 2,400-g/mm ruling density and a 330-nm blaze wavelength for the UVVS.

Once the grating was selected, ray-tracing trade studies were used to maximize entrance and exit slit widths and to place the VIS detector in the focal plane, consistent with the requirement that the instrument resolve the sodium resonance doublet, i.e., the D1 and D2 lines. Although (4) suggests that placing the VIS detector at the outboard position in the focal plane to maximize β will produce the best spectral resolution, the strong variation in instrument point spread function across the focal plane causes the inboard position to be the best choice of the three. The values selected from the trade study for entrance and exit slits are 0.175 mm and 0.23 mm, respectively. Spectral bandpasses using these slits and a 2,400-g/mm grating are shown in Fig. 6 as solid lines. Dashed curves show the spectral sampling for the three channels obtained with a 1-arcmin grating step size ($\Delta\theta_S = 0.0167°$), providing the required $\sim 1/3$ bandpass interval.

The adopted UVVS design meets its requirement to resolve the sodium doublet over the entire range of permissible D1/D2 emission ratios. This is demonstrated by the two ray-tracing model profiles shown in Fig. 7 constructed for the limiting cases where the column

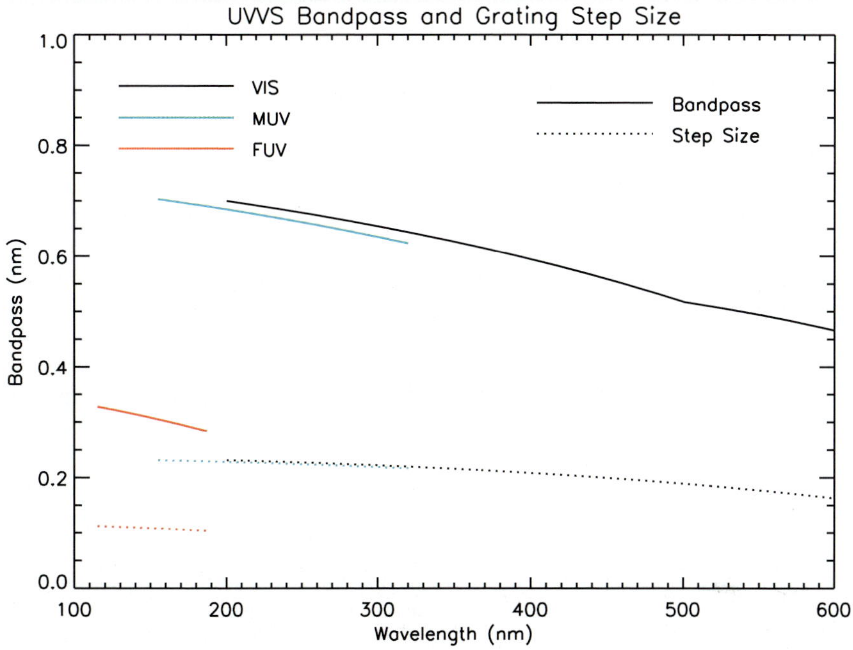

Fig. 6 UVVS channel bandpasses and spectral sampling intervals

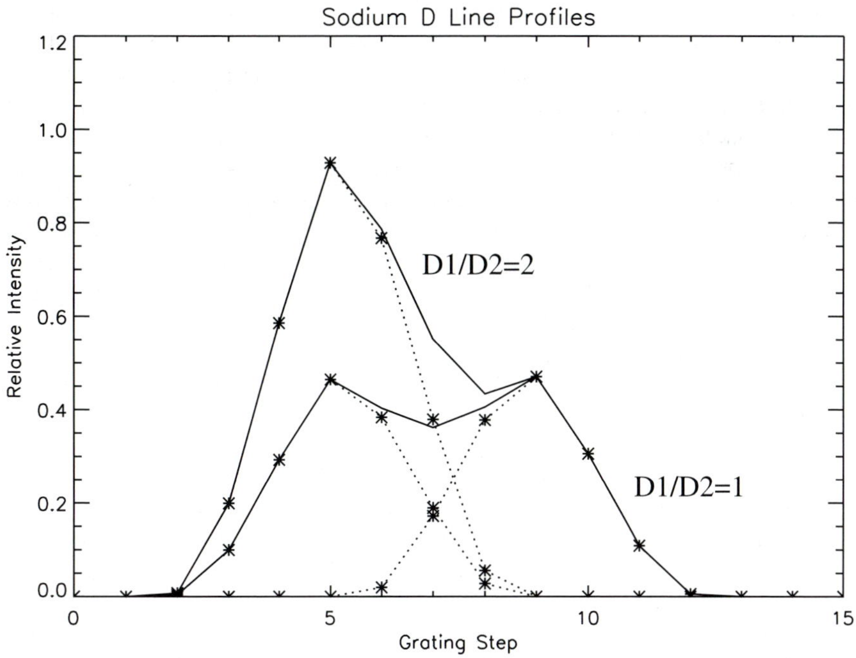

Fig. 7 Ray-trace models for UVVS observations of the sodium D lines in the two limiting cases D1/D2 = 1 and D1/D2 = 2. *Solid curves* show the observed profiles, which are the sum of the individual emission lines (*dotted lines*). *Stars* mark the "observed values" at the individual grating positions

emission in both lines is optically thick (D1/D2 = 1) and optically thin (D1/D2 = 2). Although the spectrometer point spread function causes the spectral resolution to be approximately 15% larger than the bandpass shown in Fig. 6, the two emission lines are separable in both cases. Additional ray-tracing analysis demonstrated that this resolution performance is maintained over the entire operating temperature range of the instrument (−30°C to +30°C).

4.2.2 Visible and Near-Infrared Spectrograph

A single-mode optical fiber, which InnovaQuartz, Inc., of Phoenix, AZ, contributed to the MASCS project, transmits light from the telescope focal plane to the VIRS spectrograph. It was fabricated from fused silica with low water content, which minimizes the light loss caused by strong OH absorption at 1.2 μ that is often encountered in commercial fibers, and was drawn to maintain an $f/5$ beam divergence compatible with the telescope. The polished output surface of the fiber, which is 100 μm in diameter, provides the entrance aperture to the spectrograph.

The VIRS uses a concave grating to both disperse and image the spectrum onto an extended focal plane. Light diverges from the fiber output and is diffracted and imaged by the grating toward a beam splitter, which reflects visible wavelengths onto a silicon (Si) photodiode array detector, referred to as the VIS detector. Infrared wavelengths are transmitted by the beam splitter and are imaged onto an array of indium-gallium-arsenide (InGaAs) photodiodes, after passing through a condensing lens located directly in front of the NIR detector.

In addition to meeting the measurement requirements for spectral coverage (300–1,450 nm) and spectral resolution (5 nm), the VIRS optical design was constrained to use existing, commercially available detectors and grating and to fit within the UVVS mechanical footprint. MASCS uses self-scanning array detectors, developed by Hamamatsu Corporation, to record simultaneously the visible (300–1,050 nm) and near-infrared (880–1,450 nm) spectrum. Hamamatsu produces Si arrays in a variety of formats up to 512 in length with pixel dimensions that are either 50-μm wide × 500-μm tall or 50-μm wide × 2,500-μm tall. InGaAs pixels are restricted to a single length, which is 500-μm tall. A 50-μm pixel pitch is well matched to the input fiber diameter, satisfying the standard minimum Nyquist sampling requirement (two pixels per resolution element). If two pixels span a 5-nm spectral resolution element over a 100-μm displacement in the focal plane, then the grating has to provide 50-nm-per-mm dispersion. This had to fit within an instrument with an overall length of 250 mm, which limited the VIRS grating focal length to ~200 mm. The dispersion equation for a fixed grating spectrograph defines the relationship between ruling density and focal length

$$\frac{d\lambda}{dx} = \frac{d\cos(\beta)}{m \cdot F}, \tag{5}$$

where $d\lambda/dx$ is the linear dispersion at the focal plane and F is the spectrograph focal length. If $d\lambda/dx = 50$ nm per mm and $F \sim 200$ mm, then $d \sim 10^4$ nm, assuming β is small. Only a small number of off-the-shelf gratings exist that have parameters close to those required for VIRS. Of these, the Jobin Yvon grating 523-00-060 provides the best match for the VIRS application. This grating is concave with a 210.6-mm radius of curvature and an average of 100-grooves-per-mm ruling density, providing a nominal dispersion of 46.6 nm per mm (2.33 nm per 50-μm pixel). It is manufactured using holographic recording as a Type II grating that produces a flat focal plane and good spectral imaging over the entire 250–1450 nm wavelength range with a point spread function that has a 10-μm or less FWHM. In the cross-dispersion direction the output image height for the 0.1 mm entrance fiber grows to as much as 1.5 mm at the shortest wavelengths in the VIRS spectrum (near 300 nm) and up to 0.75 mm at the longest wavelengths (near 1,450 nm) because the 523-00-060 is corrected for astigmatism at only a single wavelength (1,050 nm).

The VIS detector pixels are tall enough to accommodate the astigmatic images produced by the grating. On the other hand, fore optics are required to compress the spectrum in the cross-dispersion direction before it is imaged onto the NIR array. This is accomplished by fabricating a beam splitter with a cylindrical second surface and placing a fused silica cylindrical, convex-plano lens directly in front of the array. The beam splitter second surface, which has a 250-mm radius of curvature, causes the rays from the grating to focus behind the detector in the vertical dimension. They are subsequently compressed by the condensing lens, which has a 6.6-mm radius on its first surface, and reimaged on the detector with an average demagnification of 2. Because the magnitude of astigmatism introduced by the grating changes with wavelength, the lens must be tilted to increase the degree of demagnification toward longer wavelengths. Figure 8 compares the imaging performance for the two configurations at $\lambda = 1{,}450$ nm.

The left column shows a spot diagram and intensity profile for the grating and a 6.2-mm-thick beam splitter with plane-parallel faces. The right column shows the corresponding plots for the VIRS flight design, which includes a 6.1-mm-thick beam splitter with plano-concave surfaces and the tilted condensing lens. Although adding the condensing lens slightly degrades spectral imaging performance (shown in the lower two panels), the design still exceeds the 5-nm performance requirement.

Once the imaging performance is determined from ray tracing, the detector array coverage and beam splitter coatings can be defined. At a 50-μm pitch approximately 365 pixels of the VIS array are required to capture the spectral range 200–1,050 nm. (This somewhat broader range provided additional pixels below and above the nominal instrument sensitivity cutoff, which can be used to measure backgrounds and scattered light.) For wavelengths greater than 550 nm (pixels greater than 153), the detector input window was coated with a long-wavelength-pass filter to suppress contamination of the spectral signal by short-wavelength light diffracted into second order by the grating (Maymon et al. 1988). Wavelengths 895–1,490 nm cover the 256 pixels of the NIR array. No order-sorting filter is required for NIR because InGaAS photodiode sensitivity has a short-wavelength cutoff at 750 nm. The initial beam splitter concept employed a multilayer dichroic coating for its first surface and an infrared antireflection coating for its output surface. This design was later abandoned for the flight instrument in favor of a more robust approach in which the top half of the first surface was coated with Al and its bottom half with an infrared antireflection coating.

4.3 Design Approach

An exploded view of MASCS is shown in Fig. 9. Its design was strongly driven by the MESSENGER mission mass constraints. As originally proposed, MASCS was intended to follow closely the mechanical design of the Galileo UVS with additional structure to support the VIRS optical elements and detectors. The MASCS UVVS section is optically very similar to the Galileo UVS, and since the VIRS section is coupled to the UVVS telescope by a flexible fiber optic its location was not critical as long as its elements were correctly located in relation to each other. However, a detailed study showed that a one-piece instrument case could result in a lighter instrument than the assembly-of-boxes heritage design from the Galileo UVS.

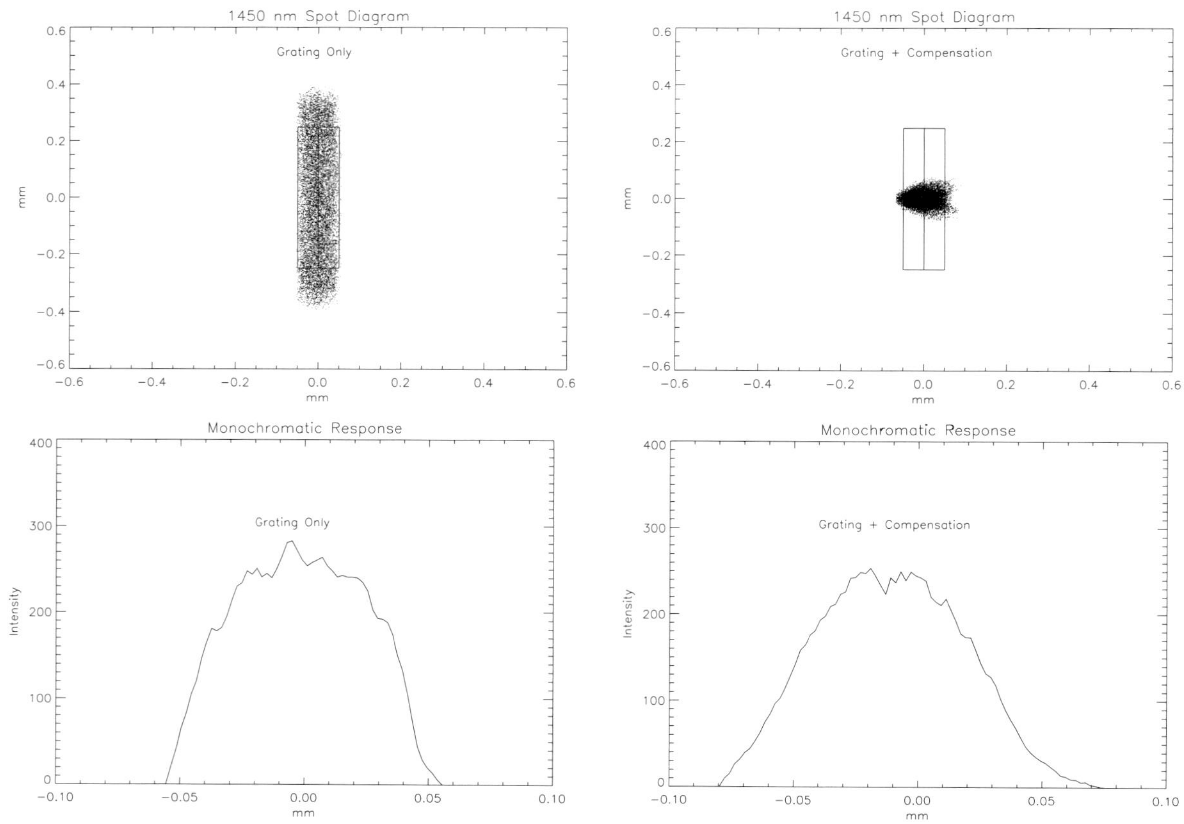

Fig. 8 Imaging performance comparison at $\lambda = 1{,}450$ nm. The *left* and *right* columns show spot diagrams and intensity profiles for the grating + plane-parallel beam splitter and for the grating + beam splitter + condensing lens, respectively. In both cases the spectrograph input is the 100-μm diameter fiber aperture. The condensing lens is required to match the spectral image to the size of an NIR detector pixel pair (*solid lines* in the *upper panels*), which defines a spectral resolution element for the spectrograph

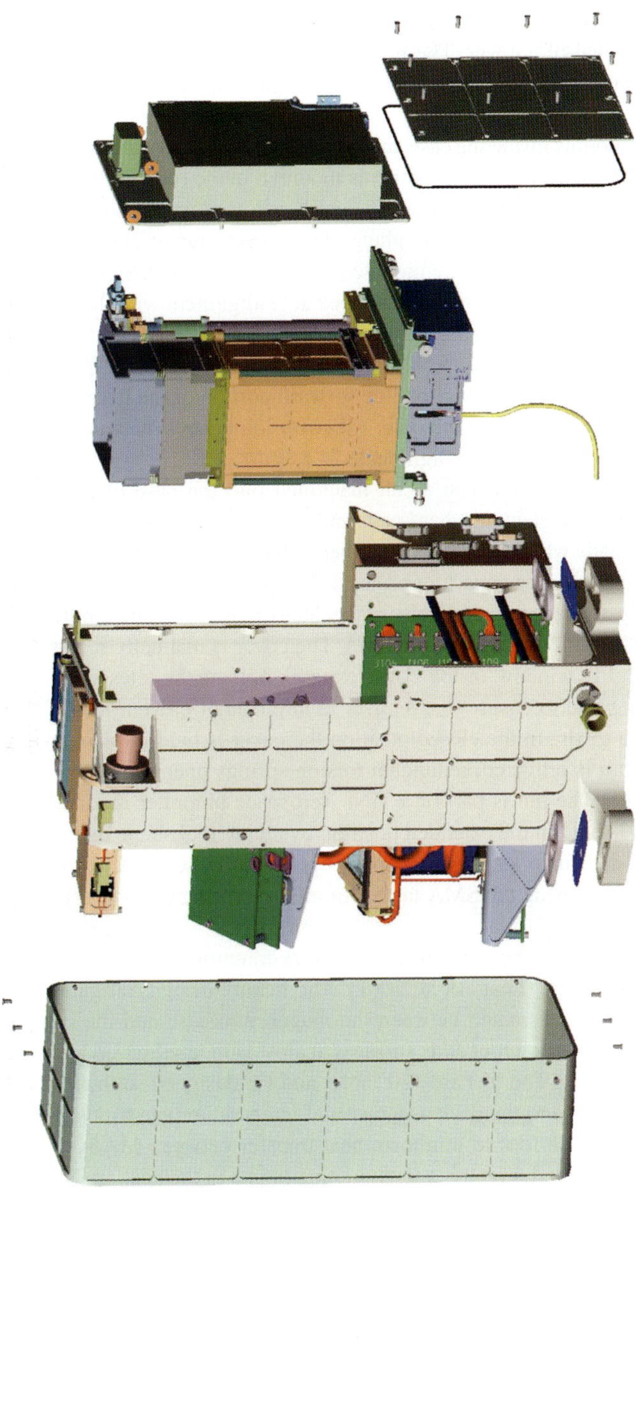

Fig. 9 An exploded view of MASCS. The *right side* of the figure shows the two major assemblies that comprise the UVVS component. The *left side* shows the VIRS cover

The case provides five sides of the UVVS enclosure. The back face of the case serves as an optical bench for the VIRS section of the instrument. The VIRS components are completely accessible by removal of a cover. The fiber optic input and shutter mount, grating, beamsplitter, and both detectors could be adjusted during alignment without disturbing other parts of the instrument.

The UVVS assembly mounts inside the case and is designed so that it can be removed as a unit. With the exception of the Ebert mirror, which is mounted in the base of the instrument case, all UVVS components are mounted to an optical bench that is perpendicular to the optic axis. The optical bench is located at the plane of the spectrometer entrance and exit slits, so the telescope and PMT assembly are mounted on one face while the grating drive is mounted on the other. The optical bench maintains repeatable alignment with the instrument case with a combination of dowel pins and precision shoulder screws. The high-voltage power supply is mounted directly above the PMT assembly. The base of the high-voltage power supply enclosure acts as a lid to close out the forward portion of the instrument case. A separate cover seals the rear portion of the case. All cover joints have machined channels and small-diameter O-ring seals to prevent light leaks.

The UVVS Ebert mirror is mounted in a cell machined into the base of the case. As with most other MASCS optical elements, hard pads and opposing springs secure the mirror against launch vibration loads while allowing adjustments during spectrometer alignment.

A hinged cover over the telescope aperture protects the optics from external contamination. The cover is a spring-loaded, one-time release mechanism that is opened in flight after spacecraft outgassing has reached an acceptable level. The cover is manually reclosable for ground operations. A $MgFl_2$ window in the cover exposes half of the telescope aperture to facilitate ground testing and to ensure that a cover failure cannot prevent MASCS from accomplishing its science goals. In the closed position the cover is held in place by a pin-in-hole latch. When the pin is retracted, redundant torsion springs open the cover, rotating it by 190° about the hinge. The pin is part of a TiNi Aerospace pinpuller mounted on the instrument case. The pin is held in the extended position by an internal shape memory alloy (SMA) latch. To actuate the pinpuller MASCS applies power to either of a pair of redundant heaters in the device, releasing the SMA latch. An internal spring then retracts the pin, allowing the cover to open.

The base of the case includes the three integral flexure-mounting feet that mount the instrument to the spacecraft deck (Leary et al. 2007). The flexure design mitigates the coefficient of thermal expansion mismatch between the magnesium case and the composite deck. A titanium cap plate over each foot adds stiffness against launch vibration loads while allowing the feet to flex radially. The feet are 120° apart and are designed so the instrument boresight does not move relative to the deck as a result of temperature variations. Titanium washers are installed under each foot to minimize heat transfer between MASCS and the instrument deck. MASCS is bolted directly to its mounting points in the deck; no shimming was required to meet alignment requirements.

The instrument case and most other structural pieces are machined from magnesium and aggressively weight-relieved. Some small parts are machined from aluminum where the difference in mass compared to magnesium is small. The magnesium parts are electroless nickel plated for corrosion resistance and conductivity. Black paint is applied over much of the case surface to enhance optical and thermal properties. The extensive use of magnesium reduced the mass of the instrument significantly, though the cost of manufacturing the parts was somewhat higher than it would have been for an all-aluminum structure. Overall, the use of magnesium saved approximately 0.4 kg, a significant savings in a 3.1-kg instrument.

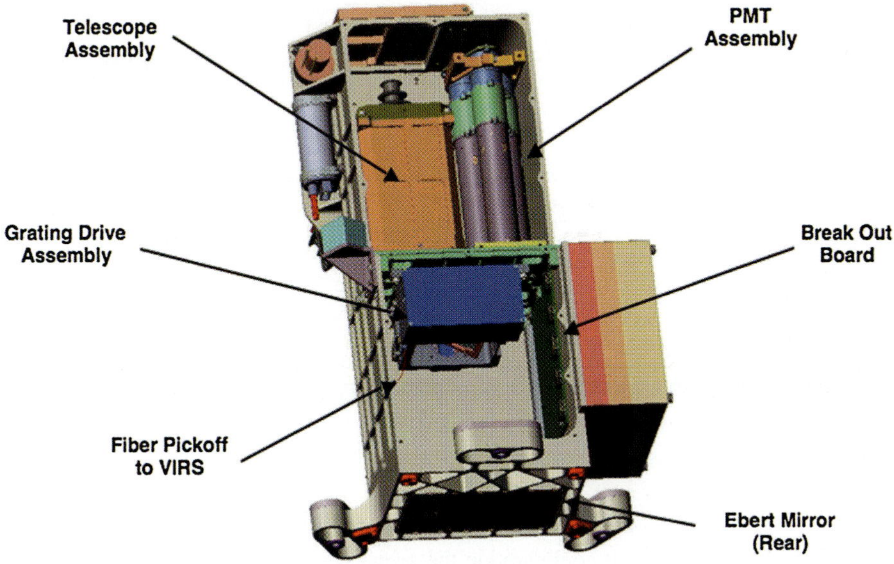

Fig. 10 Solid-model drawing showing the UVVS components

4.4 Telescope-Ultraviolet and Visible Spectrometer

Figure 10 is a solid-model drawing that shows the details of the Telescope-Ultraviolet and Visible Spectrometer mechanical assembly and its five major components: telescope, entrance aperture plate, grating drive, detector head, and Ebert mirror. Four of these components mount directly to a central plate that is oriented perpendicular to the optic axis. In this drawing two instrument covers, as well as a telescoping light shield between the front of the telescope housing and the door on the front of the instrument case, are not shown.

4.4.1 Telescope

Figure 11 shows a cross-section view of the MASCS telescope, which was adapted from the Galileo UVS telescope design with only minor modification. In this design, the telescope housing, which attaches directly to the central mounting plate with screws and locating pins, acts as the primary structural element. The primary mirror is registered in the housing laterally by Cu–Be leaf springs that push the mirror against Vespel® side pads. A back plate with a central hub, which contains the mounting surface for the telescope central baffle, and additional leaf springs register the optical surface of the primary against gold-alloy pads inserted into the housing. The secondary is retained by springs and pads located in a separate mirror cell that is attached to the ends of three invar metering rods, rather than to the main housing. The other end of each rod is attached to the telescope housing near the location of the primary mirror optical surface. As the aluminum housing expands and contracts with changes in temperature, the invar rods, which have a thermal expansion coefficient that is matched to that of the fused silica mirrors, maintain a nearly fixed primary-to-secondary separation. This allows the telescope to image over the instrument operational temperature range without significant change in focus.

The MASCS metering system is an improvement over the Galileo design in which the secondary cell was mounted to the telescope housing and invar rods contacted the primary

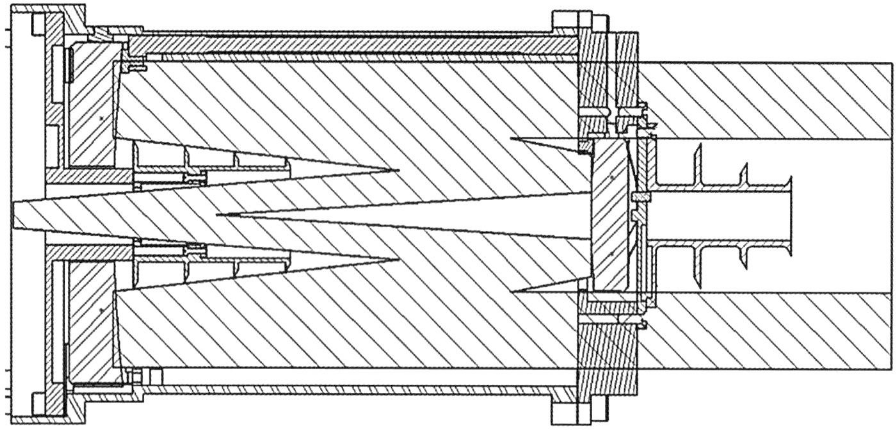

Fig. 11 Telescope cross-section view

mirror optical surface. In that configuration the primary-to-secondary distance was maintained, but both mirrors were free to move relative to the telescope focal plane, causing a small shift in focus. This was acceptable for Galileo because it operated over a much smaller temperature range than MASCS will encounter in Mercury orbit.

Suppression of stray and scattered light within the MASCS telescope is important because the instrument must observe faint exospheric emissions above the sunlit limb of Mercury. Primary and secondary mirror baffles shield the telescope focal plane from direct illumination by the external scene. The primary baffle was fabricated as a series of concentric rings to minimize grazing incidence reflections from its exterior. In addition, the walls of the main housing are coated with black anodizing, and the primary and secondary baffles are plated with black-nickel to suppress secondary light scattering from those surfaces. Finally a light shield is installed between the front of the telescope housing and the instrument door to seal the telescope cavity against leaks.

4.4.2 Telescope Focal Plane Assembly

Light from the telescope is focused onto an aperture plate assembly, which holds both the UVVS entrance slit and the VIRS input fiber. Figure 12 shows a view of this assembly as seen from inside the spectrometer. The UVVS slit, which is 4.4 mm long $\times$ 0.175 mm wide, is photoetched into a 1.3-mm-thick nickel substrate and is located on the main assembly plate with steel dowel pins. A ferrule that holds the fiber optic feed for the VIRS is mounted next to the UVVS slit.

The aperture plate also supports the mechanism that selects either the exosphere or surface FOV for the UVVS. For the surface mode, an Aeroflex 16305 stepper motor, which has a 10-mm diameter, rotates a balanced arm counter clockwise through 90° in order to cover the UVVS slit. This places a 0.22-mm-wide horizontal slot across the center of the slit defining a 0.04°-wide by 0.05°-tall field-of-view. For exosphere mode the motor rotates the arm clockwise to the position shown in Fig. 12. In this orientation, a small knife edge located on the back of the arm breaks the light path between a light emitting and receiving diode pair, producing an electrical signal that indicates its location.

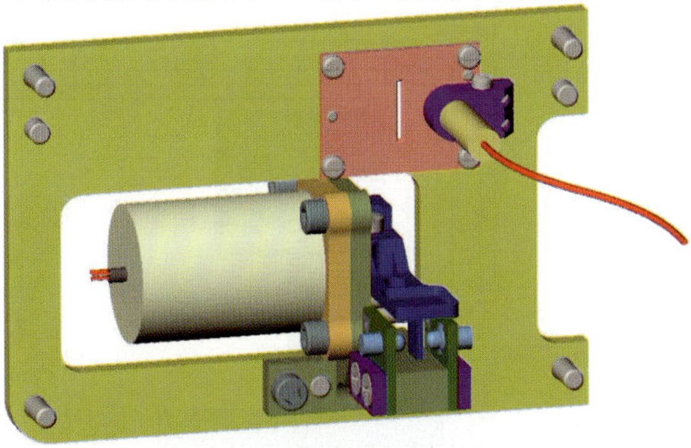

Fig. 12 Telescope focal plane assembly. The arm that defines the UVVS surface/exosphere FOV is shown in the "exosphere" position

Fig. 13 Solid-model and cut-away views of the UVVS grating drive mechanical assembly

4.4.3 Grating Drive Assembly

The UVVS incorporates a grating drive assembly that provides the required precise rotation of the spectrometer grating. The full range of the grating rotation is limited by hard stops to approximately 45°, with a minimum step size of 1 arcmin. The required accuracy is ±15 arcsec, with a settling time of 1 ms per step. The maximum rate of the grating drive is 200 steps per second. The grating drive assembly, which is shown in Fig. 13, consists of a magnesium housing supporting a grating rotor, drive motor, and optical position encoder. The grating drive assembly is removable from the instrument as a unit for ease of testing.

The grating rotor is a single machined part that supports the grating and provides concentric shafts at opposite ends for the drive motor and the position encoder. The grating itself is approximately 33-mm wide by 30-mm tall. It is mounted in the rotor with adjustable hard pads that allow alignment in tip, tilt, and focus. Opposing springs ensure that the grating is firmly located against the hard pads. The drive motor is a brushless DC torque motor procured from Aeroflex.

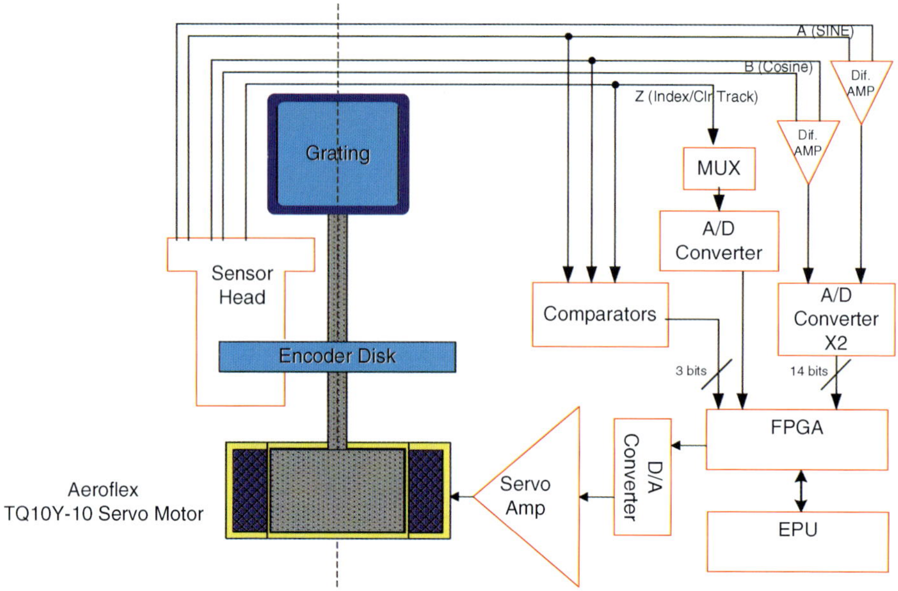

Fig. 14 UVVS grating drive control system block diagram. MUX denotes a mutiplexer

The optical encoder is a semi-custom unit procured from Dynamics Research Corporation. The design uses a rotating glass encoder disk and fixed reticules with light-emitting-diode (LED)/phototransistor pairs to generate and sense fringe patterns. The encoder glass produces 4,096 cycles per full revolution. This allows coarse position knowledge to about 5-arcmin resolution. The LED/phototransistor pairs produce two sinusoidal signals (A and B) 90° out of phase. Interpolation by a factor of 32 allows fine position knowledge at better than 10-arcsec resolution, which is sufficient for the grating drive control system to meet the rotational position requirement. A separate index track on the encoder disk provides a zero reference point. Since most of the index track is clear, the index signal is also used to monitor signal level degradation over the life of the mission. The encoder indicates relative rather than absolute rotation, so the zero reference index must be found after each power cycle.

Figure 14 shows the block diagram for the grating drive position closed-loop control system, which is implemented in a combination of hardware and software. Encoder signals are processed in a field-programmable gate array (FPGA) on the Digital Electronics Board (DEB, see Sect. 4.5).

The grating drive is initialized by a two-step process when the instrument is powered on. The first step generates a look-up table that is used for efficient position calculations. Software, which is resident within the instrument Event Processing Unit (EPU), commands the drive motor to sweep from one hard stop to the other in an open-loop mode and monitors the A and B encoder signals to establish their full dynamic range, then calculates and stores in memory the interpolated arctan look-up table that allows quick position determination within an encoder cycle. Since the table is calculated using the measured dynamic range of the encoder signals, small variations over the life of the mission are compensated automatically. Larger variations can be compensated by changing the power levels of the LEDs by software command. Lookup table generation takes about 20 s to complete.

The second step in grating drive initialization finds the index mark of the encoder and establishes the zero reference position. This zero reference position is near the center of travel. It is used internally by the software as an offset so that from the user's point of view all commanded grating step positions are positive, with step 0 near the short-wavelength hard stop.

The position control algorithm is a proportional-integral-differential (PID) calculation with an added feed-forward element to improve step performance. The control loop runs at 3 kHz. A hardware timer in the FPGA on the DEB generates a 3-kHz interrupt to the microprocessor. The FPGA also controls the output signals to the drive motor and the input signals from the encoder at 3 kHz, providing precise timing for the control loop. The FPGA also maintains a coarse position counter by monitoring and comparing the encoder A and B signals. This hardware "fringe counter" ensures that the control system has positive knowledge of each encoder cycle and relaxes software real-time performance requirements.

At each 3-kHz interrupt the software calculates the current position by reading FPGA registers to retrieve the latest values of the A and B signals, then uses those values as indices into the look-up table to determine the position within the current encoder cycle. This position is added to the coarse position counter value, also retrieved from an FPGA register, to give the current position. The current position is compared with the desired position to derive the position error, which becomes the input to the PID calculation. The output of that calculation is written to an FPGA register. At the start of the next control cycle the FPGA writes this value to the digital-to-analog (D/A) converter that controls the grating drive motor current to reposition the grating. The feed-forward parameter of the control system takes effect when the commanded step position changes. It provides an additional boost to the drive motor during the first three control cycles of a step to maximize the initial acceleration of the mechanism without affecting the stability of the PID control when the grating settles at the new position.

In operation the UVVS scans a wavelength region of interest by stepping the grating so that a series of discrete wavelengths are presented to the spectrometer exit slits and thus to the PMTs. Commonly commanded modes include short back-and-forth or "zigzag" scans across a spectral feature and longer one-direction scans to explore a wider spectral range. At each point of the scan the grating drive steps to the next desired position, then accumulates data from one or two of the three photomultiplier tubes for the commanded integration time. At the end of each integration time the grating drive moves to the next position until the commanded scan is complete. The step size is commandable from 0 to 15 arcmin in 1-arcmin steps. If the step size is 0 the grating drive controls at a single position and the integrations are performed back to back.

Two commandable parameters help ensure that the grating position will be stable during the integration times. First, the time allowed for the grating drive to step and settle is configurable in units of $1/3{,}000$ s. Second, the timing between the end of an integration period and the start of the next step can be modified so that the step begins just before the integration time ends. This phase offset parameter takes advantage of the fact that the grating does not actually move out of the position tolerance band for a step at the instant the step is initiated.

Grating drive health and performance are monitored in flight. Each UVVS science telemetry packet includes a counter value indicating the number of control cycles in which the grating drive position exceeded its ± 15 arcsec tolerance band. This provides a measure of performance during each UVVS observation. The grating drive can also be commanded to perform a step response test during which a configurable set of control system parameters are sampled at 3 kHz. This test generates diagnostic telemetry packets, which are downlinked and analyzed to determine grating drive step performance.

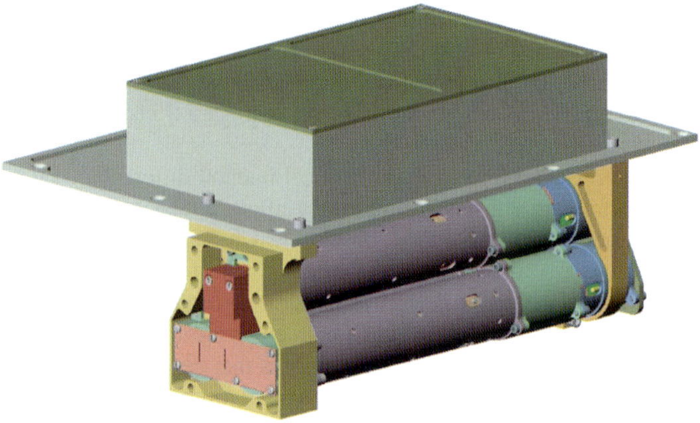

Fig. 15 Detector head assembly

4.4.4 UVVS Detector Assembly

The instrument detector head assembly is shown in Fig. 15. It consists of a pair of tri-brackets that support the three individual photomultiplier tube assemblies and the spectrometer exit slit assembly.

Each exit slit pair is separated by 10.2 mm. This spacing, which is smaller than the 13.5-mm diameter of an individual tube, is accommodated by placing the MUV detector above the spectrometer dispersion plane. A pair of mirrors that are oriented at 45° with respect to the light path form a periscope, which directs light from the middle exit slit to the tube (see Fig. 16). Spherical surfaces on the mirrors funnel the diverging beam toward the center of the detector input window. The bottom of the high-voltage power supply (HVPS) housing, which is oversized to form a cover for the detector assembly after it is mounted in the main instrument case (see Fig. 9), provides lateral support for the tri-brackets. High-voltage connections to the detectors are hard-wired to feed-through connectors that protrude through the bottom of the HVPS. External electrical connections to the detector assembly are made through a single 25-pin Airborn micro-connector. This design allows the entire assembly to be easily removed from the instrument without disturbing the sensitive high-voltage connections, which are encapsulated for flight.

Hamamatsu photomultipliers, from the 13-mm diameter head-on family, were selected for MASCS, replacing the EMR Photoelectric 510-series detectors used in the Galileo UVS. The UVVS detectors operate in pulse-counting mode with grounded photocathodes and output pulses capacitively coupled into Amptek A-111F pulse-amplifier-discriminators (PADs). A-111F PADs are high-sensitivity ($\sim 5 \times 10^4$ e^-), hybrid devices with a ~500-ns pulse-pair resolution.

Although the Hamamatsu tubes' borosilicate glass envelopes make them less rugged than 510 detectors, their larger active areas significantly increase the instrument FOV uniformity. Their multiplication sections have ten-stage linear-focused dynodes, which produce stable, saturated pulse height distributions. Multiplier modal gain is photocathode dependent. Table 3 summarizes the operating characteristics for three UVVS wavelength ranges.

The detectors were subjected to a comprehensive characterization and qualification program because only the R1081 had been previously used for a long-duration spaceflight application (Esposito et al. 2004). Characterization measurements for all tubes included quantum

Table 3 UVVS detector characteristics

Parameter	Value		
	FUV channel	MUV channel	VIS channel
Type number	R1081	R759	R760P
Photocathode	CsI	CsTe	BiAlkali
Window material	MgF_2	Fused silica	Fused silica
Dynode material	BeCu	Cs_3Sb	Cs_3Sb
Operating voltage	1,800–2,250 V	800–1,250 V	800–1,250 V
Modal gain	1×10^6	2×10^6	3×10^6
Photocathode diameter	6 mm	10 mm	10 mm

Fig. 16 UVVS photomultiplier tube assembly cross-section view

efficiency, photocathode spatial uniformity, and pulse height distribution, measured as functions of accumulated counts ranging from an initial value of 10^8 to a final value of 10^{10}. No degradation in the modal gain was detected for this level of accumulated count. There was no detectable loss in responsivity with amplifier thresholds set at flight values (1×10^5 e^- for CsTe, 5×10^4 e^- for CsI, and 1×10^5 e^- for BiAlkali), and no detectable change in spatial uniformity. Just before final assembly, the operating voltage for each tube was determined by illuminating it with a constant-intensity light source and identifying the plateau in a plot of output count versus high voltage.

Figure 16 shows a cut-away drawing with a cross-section view of the MUV detector and the periscope assembly. The 13.5-mm-diameter × 71-mm-long tube is registered inside a Delrin sleeve by an O-ring at the front and a slip fit at the rear. Twelve feed-through pins, which provide the electrical connections to the dynodes and photocathode, exit the detector at the rear.

A 13-mm-diameter three-board circuit stack, which provides the resistor divider chain for the high voltage, is soldered directly to dynode pins. The entire area around the output pins and divider circuit is encapsulated to provide mechanical rigidity and protection from high-voltage arcing. Two additional 13-mm diameter boards carry the A-111F amplifier and its output-pulse driver electronics. A μ-metal housing with a 10-mm-diameter aperture, located 5 mm in front of the input window, surrounds the entire photomultiplier tube to protect it from magnetic fields. Laboratory tests show that this shield geometry is adequate to eliminate magnetic field effects for strengths at the detector of up to 15 gauss, which is a factor of 5 greater than ambient fields produced by torque motors, located in the exit slit housing (~2.5 gauss at the tube).

Two effects cause the UVVS detector outputs to exhibit a small but nonnegligible dependence on temperature. The smaller of the two arises from small variations in the high-voltage power supply output and divider string resistance, which cause the multiplier gain to fluctuate with temperature. For the UVVS detectors, the measured value for the relative change in output is $\sim 5 \times 10^{-5}$/°C. The larger of the two arises from the fact that the quantum efficiencies of all photocathodes are temperature dependent. Temperature coefficients depend upon both photocathode material and the wavelength of the incoming light. The relative changes for the UVVS detectors can be as large as 0.008/°C for CsTe and BiAlkali at the longest observed wavelengths. Accurate preflight calibrations and in-flight measurements of detector head temperature allow these effects to be corrected to ~2%.

A single HVPS, equipped with three Amptek HV601 high-voltage opto-isolators, provides 2,040, 900, and 900 V to the FUV, MUV, and VIS tubes, respectively. The primary supply, which operates at a nominal 2,250 V, uses a 50-kHz oscillator to drive a six-stage Cockroft-Walton voltage multiplier to produce unregulated high voltage for the 601s. Custom-regulated high voltage for each tube is produced by using a 601 in a closed-loop control system that compares output voltage, which has been divided by 10^3, to a fixed reference voltage, which was determined from that detector's output count versus high-voltage plot. The entire supply is mounted on three printed circuit boards and enclosed in a magnesium housing, which measures $7.0 \times 11.1 \times 2.5$ cm. Its mass, after encapsulation, is 250 g and its nominal operating power, referenced to its input, is 370 mW with all three detectors operating. Flight operations limit the maximum number of operating tubes to two, for which the output power is 320 mW. Bench tests, performed before instrument assembly, showed that the high-voltage output was stable to ±10 V over an operating temperature range from −45°C to +45°C.

4.5 Visible and Near Infrared Spectrograph

Figure 17 is a solid-model drawing that shows the details of the Visible and Near Infrared Spectrograph mechanical assembly and its five major components: entrance aperture assembly, grating, beam splitter, and two detector heads. Unlike the majority of the UVVS components, which are tightly integrated into a single mechanical package, the VIRS assemblies mount in individual holders that attach to an optical bench, which is one surface of the MASCS instrument case. Pairs of dowel pins pressed into each optical mount allow for easy removal and replacement of elements without the loss of precise optical alignment. The optical elements themselves are mounted in cells that use leaf springs and pads to locate their optical surfaces except for a condensing lens, which is bonded into a mounting flange that attaches to the front of the NIR detector housing. A lightweight cover, which is sealed by an O-ring that is captured in the optical bench to enclose the spectrograph, is not shown.

Stray and scattered light within the spectrograph is suppressed using surface treatments and by masking all unused surface area on both grating and beam splitter. Exterior surfaces of all the component holders are coated with black anodizing. The bench and cover interior are covered with a black polyimide-based paint. Finally, a light baffle, which is not shown in the figure, is mounted between the inboard side of the entrance assembly and the beam splitter to shield the detectors from direct illumination by light from the spectrograph input fiber.

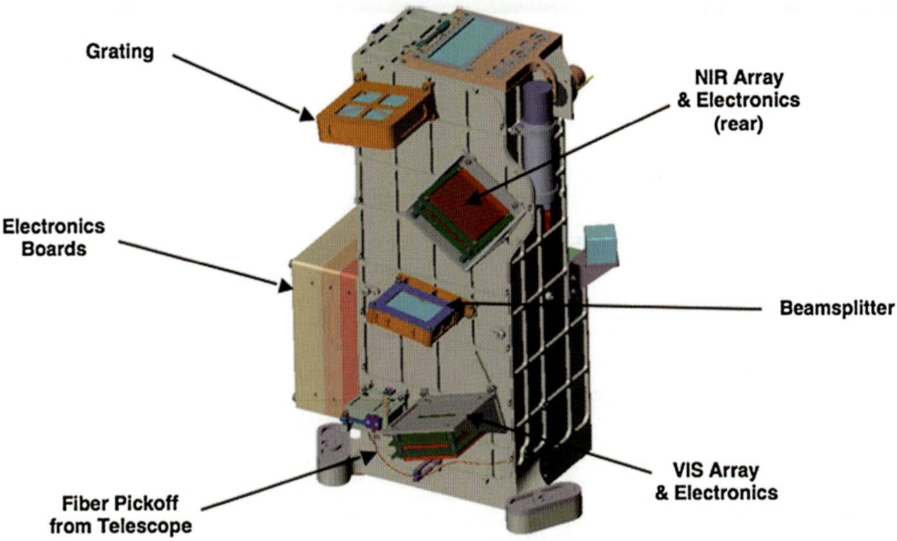

Fig. 17 Solid-model drawing showing the VIRS components

4.5.1 Optical Mechanical Design

Figure 18 shows a cross-section view of the VIRS channel. The fiber feed from the telescope enters through a hole in the floor of the optical bench and is routed to the entrance assembly, which is shown in Fig. 19. Its output end is cemented into a ferrule and clamped to the top of the assembly using a "V" block. The "slip fit" provided by loosening the "V" block retaining screws provides a convenient method for adjusting the focus of the instrument before cover installation. Two grain-of-wheat lamps (not shown in Fig. 19), which are located on either side of the ferule, provide flat-field illumination for the detectors during calibration. The entrance assembly also supports a shutter mechanism that is used to interleave dark current spectra with normal observations. This is accomplished using an Aeroflex 16305 stepper motor to rotate an arm in front of the fiber input to block the incoming beam. A light-emitting and receiving diode pair provides an electrical signal that indicates when the shutter is closed.

Light diverging from the fiber travels 210.9 mm to the grating vertex, which is rotated 9.08° with respect to the incoming beam. The grating diffracts wavelengths 200–1,500 nm into a 7.5° fan of angles projected toward a flat focal plane, which is located 206 mm from the grating pole and is rotated by 9.1° with respect to its normal.

A 6.1-mm-thick, fused-silica beam splitter, which is rotated by 19° relative to the incident fan, intercepts the rays before they reach the focal plane. The lower half of the first surface reflects all wavelengths toward the visible focal plane where they are imaged onto a Hamamatsu S3901-512SPL negative-channel metal-oxide semiconductor (N-MOS) silicon diode array. The upper half of the first surface transmits light to the second surface of the beam splitter, which has a concave cylindrical surface with a 250-mm vertical radius of curvature. Both the upper half of the first surface and the entire second surface have infrared anti-reflection coatings optimized for wavelengths greater than 900 nm. After light exits the beam splitter it travels to the infrared detector head (NIR) where it traverses a fused silica convex-plano cylindrical lens, which has a 6.6-mm vertical radius of curvature, before being

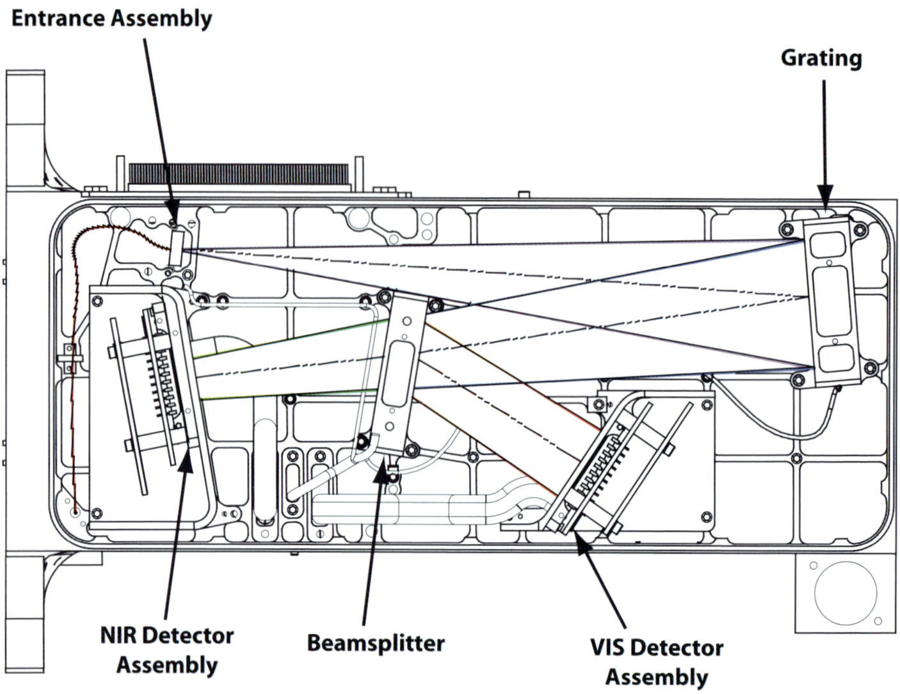

Fig. 18 Cross-section view of the VIRS showing the locations of major components

Fig. 19 VIRS entrance assembly

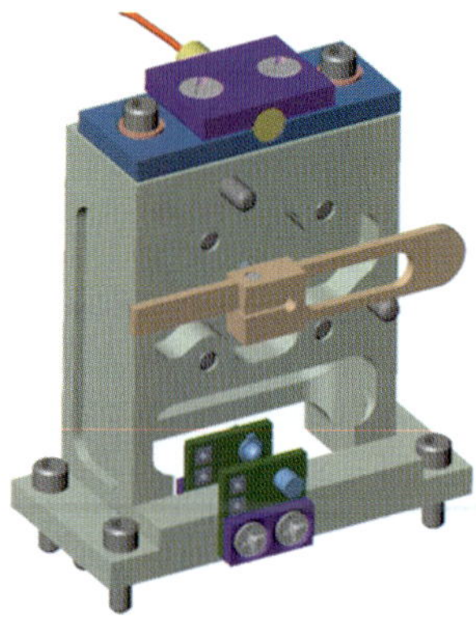

imaged onto a Hamamatsu G8162-256S InGaAs diode array. The two opposing cylindrical surfaces remove astigmatism from the spectral images in order to focus all the light from the grating onto the detector pixels (see Fig. 8).

4.5.2 Detector Head Assemblies

Figure 20 shows a solid-model view of the VIS detector head. The diode array and its control and readout electronics are mounted on a pair of 75-mm-square printed circuit boards, which are inter-wired and supported in the corners by four aluminum spacers. This arrangement produces a lightweight, compact (75 mm × 75 mm × 50 mm tall) integrated sensor package

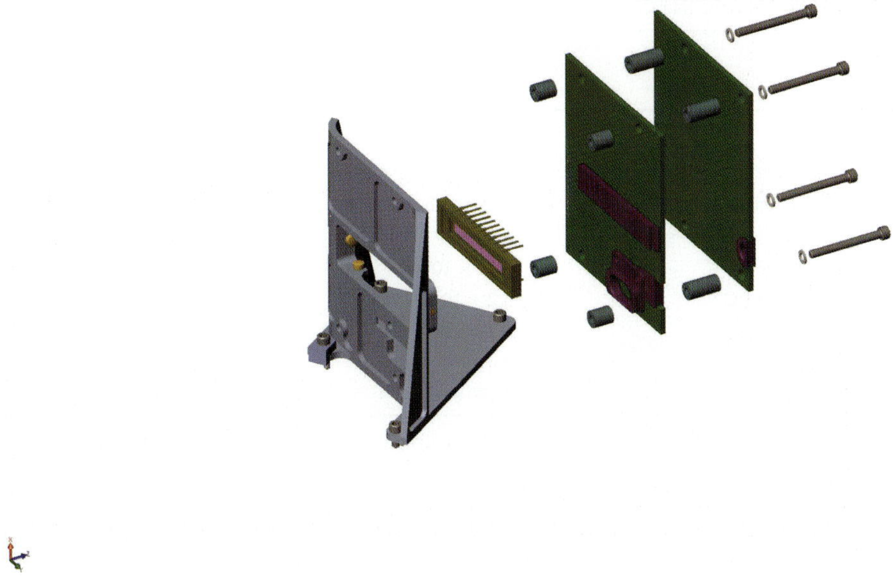

Fig. 20 Exploded view of the VIS detector assembly

with a single 15-pin Airborn micro-connector for both power and signal. It is aligned to the detector housing by inserting the diode array package, which is soldered to the front side of the first printed circuit board, into a pocket fitted with Vespel® side pads that contact the diode array package to provide horizontal and vertical registration. Custom-length spacers located between the front surface of the first printed circuit board and the housing determine the distance from the grating to the detector. Socket cap screws, which pass through the centers of the spacers, hold the sensor package in place. This approach, which uses the rigidity of the printed circuit board to maintain focus, is adequate for the VIRS application because the $f/5$ focal ratio of the spectrograph and 0.1-mm-wide imaging footprint allow for up to ± 0.1 mm change in focus with negligible loss of spectral resolution. The mechanical configuration for the NIR detector head is nearly identical to that of VIS except that the cylindrical condensing lens and mount are attached to the front surface of its housing and its mounting footprint modified for placement on the optical bench.

Table 4 summarizes characteristics for both VIS and NIR detectors. Although the mechanical configurations of their two assemblies are nearly identical, there are substantial differences in their architecture and the design of their control and readout electronics. The VIS has a linear array of 512 P–N junction photodiodes that consist of an N-type silicon diffusion layer on a P-type silicon substrate. Photons impinging on a diode are converted into electrons, which are stored in the junction. A simple linear shift register sequentially connects the diodes to an output video line. When a diode is connected to the video line, its charge is transferred to a charge-sensitive amplifier, which is located on board 1 of the sensor assembly along with the array package, and converted to an output voltage that is proportional to the charge accumulated since its last readout. This voltage is transformed into a digital number in a 16-bit analog-to-digital (A/D) converter (ADC), which is located on board 2 of the sensor assembly. After the charge is read out from the diode, it is disconnected from the video line. The readout process automatically resets the diode for its next integration. External clock signals, generated in the FPGA on the Digital Electronics Board

Table 4 VIRS detector characteristics

Parameter	Value	
	VIS channel	NIR channel
Type number	S3901-512SPL	G8162-256S
Photodiode material / substrate	N-silicon on P-silicon	P-InGaAs on N-InGaAs
Format – pixel pitch	512–50 μm	256–50 μm
Pixel height	2,500 μm	500 μm
Saturation charge	3.12×10^8 e^-	1.87×10^8 e^-
Dark current @ 10°C	3.2×10^5 e^-/pixel	9.0×10^6 e^-/pixel
Readout noise	5,500 e^-	8,900 e^-
Readout time	8 ms	4 ms
Digital scale factor	1,190 e^-/DN	1,190 e^-/DN

(Sect. 4.6), sequence the array readout. For data acquisition the array is first read out twice at a 66.67-kHz per diode rate (7.7 ms total readout time). It is then allowed to collect photocharge for some integration period before it is sequentially read out at the same 66.67 kHz. This produces a spectrum with a sliding integration time because the first pixel start and stop times are displaced from those of the last by ~7.7 ms.

The NIR detector has a linear array of 256 P–N junction photodiodes that consist of a P-type InGaAs diffusion layer on an N-type InGaAs substrate. In addition to the diode, each pixel has a sample-and-hold circuit and a charge-to-voltage converter associated with it. This architecture simplifies the design of the external readout electronics and supports simultaneous integration times because the entire array can be reset with a single control pulse from the FPGA. At the end of the integration period, another control pulse initiates a parallel transfer of accumulated charges to each pixel sample-and-hold. The voltages from the individual pixels are then read out using a shift register to connect sequentially the outputs of the pixel amplifiers to a video line at a 66.67-kHz rate. Output voltages are transformed to digital numbers using an analog-to-digital converter identical to VIS.

The external timing in the FPGA is designed to maximize the simultaneity with which the two detectors acquire a VIRS spectrum. First the VIS detector is read out twice in succession and the output discarded to reset its pixels. At the end of the second readout, the FPGA issues reset-start integration pulses to the NIR detector. After charge collection is complete, the FPGA issues a stop integration pulse to NIR and initiates a simultaneous readout from both detectors. Because the integrations on the VIS pixels start and stop at slightly different times, they each sample slightly different scenes as the spacecraft moves. During Mercury orbital observations typical integration times are either 1 or 2 s. Simulations done using the MESSENGER spacecraft orbital geometry show that the ~8 ms time skew between the first and last VIS pixel never introduces more than a 0.5% change in the area of the scene.

4.6 Interface and Control Electronics

The MASCS overall electronics design is shown in Fig. 21. The Low-Voltage Power Supply (LVPS), Event Processing Unit, Digital Electronics Board (DEB), and Breakout Board (BOB) are installed as a stack mounted on the side of the MASCS instrument case. The LVPS, EPU, and DEB are external to the case and are mounted in individual stacking frames. The BOB is mounted to the inside of the case, with its stacking connector passing through

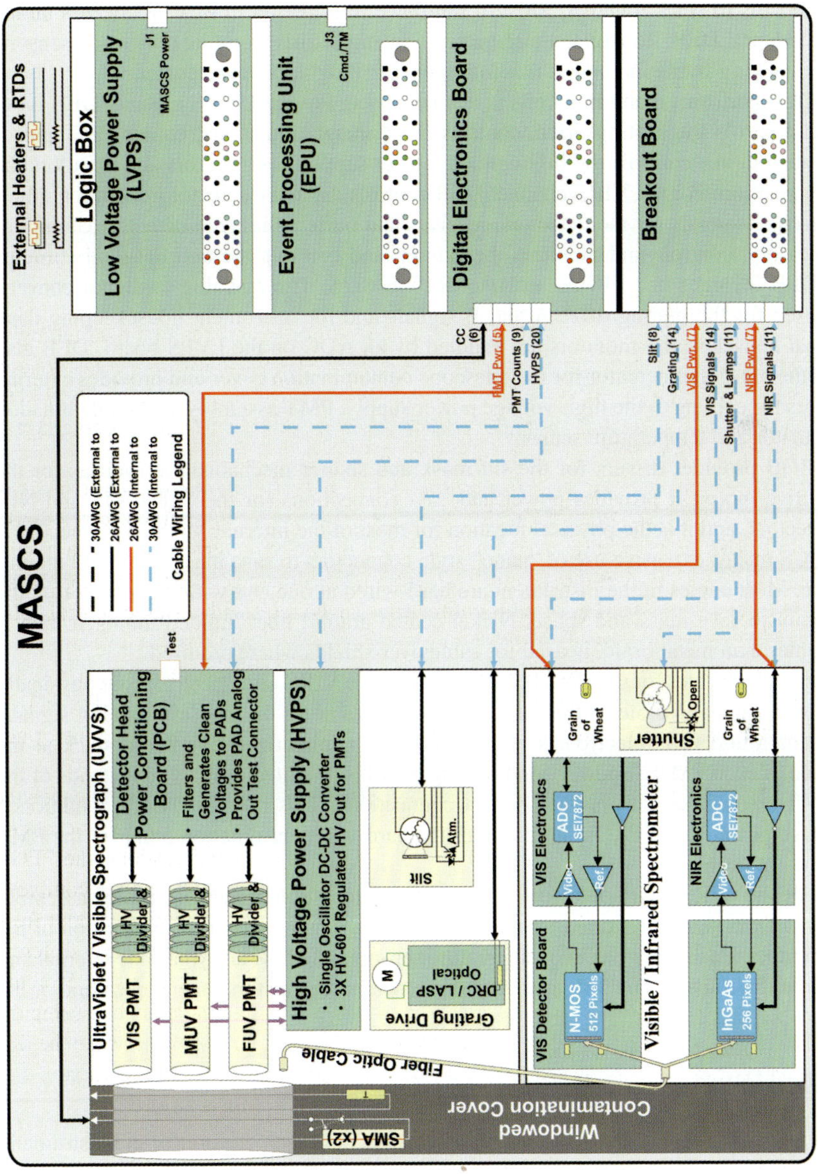

Fig. 21 MASCS electrical block diagram

an opening in the case wall. All board-to-board connections in the stack pass through the 96-pin connectors.

The LVPS and EPU assemblies (Hawkins et al. 2007) were designed and built by The Johns Hopkins University Applied Physics Laboratory (APL). The MASCS team designed the DEB and BOB assemblies to match the APL mechanical and electrical designs, allowing the four boards to stack together. Most of the power dissipation in MASCS occurs on the LVPS, EPU, and DEB. To isolate these heat-producing boards from the optics and sensors, the DEB stacking frame includes a titanium spacer that reduces heat transfer.

The DEB performs many functions in a small space. For the UVVS channel DEB monitors and controls the grating drive, controls PMT integration times, counts PMT pulses during integrations, controls the high-voltage power supply, and controls and monitors the slit mask position. For the VIRS channel DEB controls the array detector integration times, provides clock signals for the detectors, receives and buffers digitized detector data in on-board memory, controls and monitors the shutter, and controls the two flat-field stimulus lamps. DEB includes one FPGA, an Actel RT54SX32S. Three analog-to-digital converters are used for the grating drive A and B signals and for instrument housekeeping data. Additional housekeeping monitors are handled by an ADC on the LVPS board. DEB also controls the pin-puller actuator for the telescope contamination cover and provides external connectors for cabling to the high-voltage power supply, PMT assembly, and contamination cover actuator and temperature sensor.

The BOB provides drivers for the slit mask and shutter mechanism as well as for the grating drive motor. It provides power and data connections for the VIRS VIS and NIR array detectors, and it is the physical location for most of the internal MASCS connectors.

MASCS makes extensive use of "nano" style connectors to save mass and board mounting space. Most cables in the instrument are hard-wired at one end with a connector on the other, again to save mass and space. Nickel-coated aramid fiber braided tubing, which is much lighter than metal braid, is used for cable over-shields where required.

Detector electronics for the UVVS PMTs and the two VIRS array detectors are physically located as close as possible to the sensors to minimize noise on low-level analog signals. Divider chain and PAD electronics are located in a miniaturized five-board stack at the output end of each PMT. A power conditioning board is mounted directly to the ends of the three PMT assemblies with pin-and-socket connections. This board provides conditioned low-voltage power to the PADs and provides the hard-wired termination point for the PMT power and data cable.

Each of the VIRS array detectors is integrated into a two-board electronics assembly that provides the signals required to operate the detector and immediate digitization of the detector output. Each array has two temperature sensors. The NIR array has one internal and one external thermistor. The VIS array has two external thermistors, one at each end of the array.

4.7 Thermal Design

The goal of the thermal design is to maintain a stable, cool temperature for the instrument, particularly the detectors. The only active thermal control components are the redundant survival heaters, which are powered by the spacecraft and controlled by thermostats located near the MASCS center of mass. Titanium spacers under each mounting foot of the instrument thermally isolate MASCS from the spacecraft instrument deck. A titanium spacer also isolates the main electronics stack from the rest of the instrument, minimizing heat transfer into the optics and detectors. Most of the external surfaces of the instrument are treated with

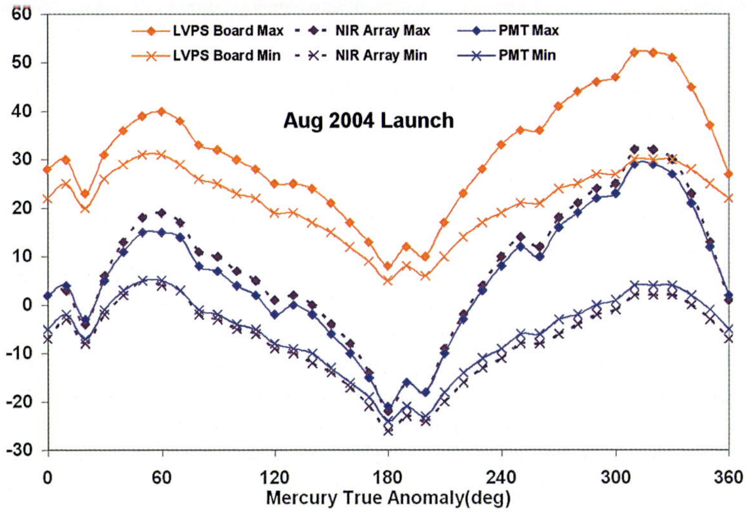

Fig. 22 Predicted MASCS component temperatures during one Mercury year

high-emissivity black paint and are radiatively coupled to the spacecraft payload adapter ring. Protection from the highly varying thermal environment when the spacecraft is in Mercury orbit is provided by a multilayer insulation (MLI) tent that exposes only the end of the instrument that contains the MASCS telescope aperture and contamination cover.

Figure 22 shows the temperatures of critical MASCS components from the instrument thermal model during a single 88-day Mercury year. Two curves for each component show the maximum and minimum temperature over a single 12-hour orbit. For example, at 60° Mercury true anomaly, the LVPS board varies between 42°C and 30°C over an orbit. The lower two curves are the predictions for the detectors indicating that they will operate at 10°C or lower during most of the mission, minimizing their dark current.

4.8 Instrument Software

The MASCS instrument flight software controls all instrument functions, accepts and executes commands, and formats and transmits telemetry for eventual downlink. The software executes on the RTX-2010 processor on the APL-provided EPU that serves as the instrument computer. As with most of the MESSENGER instruments, MASCS makes extensive use of the APL common code for the EPU. Common code functions make up about half of the 56.6-kbyte total size of the program code; MASCS-specific functions make up the other half.

At power-on MASCS runs in boot mode, executing a boot program that was created by APL and delivered in a programmable read-only memory (PROM) with the EPU. The boot program provides basic communications and limited state-of-health monitoring but no science capabilities. From boot mode MASCS is commanded into application mode, in which it has full capabilities. The application-mode program is built as an executable image using source code for both MASCS-specific and common code functions. The application mode program is stored in an electrically erasable programmable read-only memory (EEPROM). New application code can be uploaded and stored in the instrument if in-flight modifications to the program become necessary.

The MASCS software can receive and execute 35 common-code commands and 52 MASCS-specific commands. In normal science operation, most incoming commands are used to configure the UVVS or VIRS for the next observation. Once a UVVS or VIRS scan has started, no further commanding is required until the scan is complete. In general the order in which commands are sent to the instrument is not critical, except that all parameters should be set to their desired values before an observation is started. For actions where the command sequence is important, the software enforces the correct sequence and will reject commands that are sent out of order.

The MASCS software can generate 20 different Consultative Committee on Space Data Systems (CCSDS) telemetry packets. Fifteen of these are status or diagnostic packets. The other five contain science data with a distinct packet type for each MASCS detector. Each MASCS science packet contains sufficient instrument configuration information for identification and analysis of that packet's data. Missing packets from an observation would leave gaps in the data record but would not preclude analysis of the packets that are received.

UVVS science packets contain two data fields in addition to the required CCSDS primary and secondary headers. A configuration data structure contains information that completely describes the instrument's state at the time an observation was performed. A detector data field contains all of the PMT data gathered during the observation. Nominal observation planning must take into account the restriction that all of the data from one observation should fit into a single CCSDS packet. Planned science observations for MASCS fit well within this restriction.

UVVS PMT data are optionally compressed with an algorithm that reduces the 16-bit raw values to 9-bit compressed values, which are then packed end-to-end in the CCSDS packet data field. Compression allows more data values, that is, longer observations, to fit within the one-packet-per-observation constraint. The compression algorithm is a sqrt($2N$) calculation, where N is the raw data number of counts per PMT integration time, originally developed for the Cassini UltraViolet Imaging Spectrograph (UVIS) (Esposito et al. 2004). The square root calculation is a bit-wise successive approximation algorithm that returns a rounded integer result in a deterministic time. It is well suited to a real-time program running on a processor without floating point hardware. Raw data values less than 128 are not compressed, avoiding the relatively large errors that would result from computing their integer square roots. The algorithm is somewhat lossy (less so at higher values) but is a good solution for a photon counting detector.

VIRS science packets also contain two data fields in addition to the required CCSDS primary and secondary headers. As with UVVS packets, a configuration data structure describes the instrument's state at the time an observation was performed. VIRS, however, does not have the constraint that all observation data must fit within a single CCSDS packet. VIRS observation data may span many telemetry packets, though each packet contains enough information to allow successful data analysis even if neighboring packets are missing.

VIRS detector data are optionally compressed in several ways. The VIS and NIR array detectors overlap somewhat in wavelength, and in normal operations MASCS is configured to downlink only the nonoverlapping data from the VIS detector to minimize redundant data. Both VIRS detectors have a physical pixel resolution that exceeds MASCS spectral resolution requirements, so in normal operations MASCS is commanded to bin VIS and NIR data to reduce data volume without degrading the science return. Both of these methods of reducing data volume involve some losses. A third option, usually invoked in addition to the preceding methods, provides a further lossless compression by computing the successive difference of each spectrum within a packet and packing each differenced result in the minimum number of bits.

In general the MASCS instrument design does not require that the software perform fault detection and correction. However, if the VIS PMT views the illuminated surface of Mercury when its high voltage is on, the incoming light level is above its working dynamic range. If this occurs the VIS PMT noise temporarily increases above acceptable levels and will not return to normal until the detector views a dark target for some time. Full recovery can take many minutes. To avoid this problem, software monitors VIS counts whenever the photomultiplier tube high voltage is on. The result is compared against a commandable threshold value; if the value exceeds the threshold software turns off the VIS high voltage.

5 Radiance Conversion and Calibration

Both the UVVS and VIRS measure the radiance arriving at the input aperture of the instrument, and a single equation defines the conversion of instrument output (photomultiplier tube counts, solid-state detector data numbers, and ancillary engineering values) to geophysical data (radiance):

$$L(\lambda_j) = \frac{[C(\lambda_j)\cdot N(C) - D(j) - S_l(\lambda_j)]/\Delta t}{A\cdot \Delta\lambda \cdot R_c(\lambda_j)\cdot FF_j \cdot \overline{\Omega}}. \tag{6}$$

Here, C is either the photomultiplier counts at grating position j or VIRS array data number for pixel j obtained during integration time, Δt. N is the detector linearity correction, which includes analog-to-digital converter errors for the VIRS arrays, and D is the dark count or dark current data number correction. A is the area of the telescope, $\Delta\lambda$ is the spectral bandpass, R_c is the instrument responsivity at the center of the FOV for a detector array with uniform pixel sensitivity, FF is the flat-field correction for the VIRS arrays ($FF = 1$ for the photomultipliers), S_l is the stray plus scattered light correction ($S_l = S_l' + S_{\text{Stray}}$). S_l' is negative for bright spectral features (more light is scattered out of position/pixel j than is scattered in by all other wavelengths), and positive for faint spectral features (less light is scattered out of position/pixel j than is scattered in from all other wavelengths). Stray represents the light that is diffusely scattered from internal instrument surfaces. $\overline{\Omega}(\lambda_j)$ is the instrument-effective FOV (the responsivity averaged over the geometrical FOV divided by R_c).

The signals from UVVS are counts from the photon-counting detectors, and the dark counts are expected to depend weakly on temperature over a MESSENGER orbit. There will be a background signal from telescope scattered light for some limb-drift observations made near apoapsis. This background will be determined by measuring the spectrum for 2–3 steps beyond the parent emission line. VIRS dark current will be measured by periodically closing a shutter located at the entrance to the spectrograph. UVVS detector nonlinearity is caused by the processing dead time in the detector pulse-amplifier-discriminator and is typically 2.5% at 50-kHz count rates. VIRS nonlinearity is caused by well saturation in the array detectors and by differential nonlinearity in the analog-to-digital converter. It is expected to be less than 5%. The scattered light correction appears in the numerator as a value that must be subtracted from observed signal (C) to correct for light scattered from all other wavelengths into position/pixel j: $S_l'(\lambda_j) = A\cdot\Delta\lambda\cdot FF_j\cdot\overline{\Omega}\cdot\int L(\lambda')\cdot R_c(\lambda')\cdot G(\lambda_j - \lambda')d\lambda'$, where G is the grating scatter distribution function, which is often approximated as a Lorentzian profile plus constant background (Woods et al. 1994). Responsivity is the quantum throughput (QT) of the optics and detectors at the center of the FOV, and $\overline{\Omega}(\lambda_j)$ is the effective FOV: $\overline{\Omega}(\lambda_j) = \int R(\lambda_j\theta,\phi)d\Omega/R_c(\lambda_j)$. Radiometric calibrations using radiance standards (irradiance standards and reflectance screens) that fill the instrument aperture and FOV measure the quantity $\overline{R} = A \cdot \Delta\lambda \cdot R_c(\lambda_j) \cdot FF_j \cdot \overline{\Omega}$. Calibrations using standard detectors and star sources that fill only the aperture measure $A \cdot R_c(\lambda_j) \cdot FF_j$, and $\overline{\Omega}(\lambda_j)$ and $\Delta\lambda$ must be determined separately by scanning the FOV and measuring monochromatic lines, respectively. Prelaunch values for the various parameters appearing in (6) were determined during instrument characterization and calibration using both radiance and source standards (McClintock et al. 2004).

Conversion of exospheric radiance to column emission rate requires knowledge of atomic cross-sections and solar flux, both of which are well known. Surface reflectance is typically calculated by dividing measured radiance at the instrument aperture by solar irradiance that is measured by a separate instrument. This approach can introduce artifacts in the derived reflectance. To mitigate these artifacts, the MESSENGER project made careful measurements of the lunar reflectance with MASCS during an Earth flyby, which occurred in August 2005 (Holsclaw et al. 2005; Bradley et al. 2006). Comparing these data to the Mercury observations will provide an invaluable method for validating MASCS reflectance values derived from the Mercury surface radiances. In-flight measurements of stellar irradiance also provide a technique for routinely tracking values of $A \cdot R_c(\lambda_j) \cdot FF_j$.

6 Observation Scenarios

6.1 Exosphere Measurements

UVVS measures exospheric composition and structure by observing solar radiation that is scattered by atoms above the sunlit surface of Mercury. Its wavelength coverage includes five of six known species (H, O, Na, K, and Ca), excluding only He. Altitude profiles obtained from sequences of mini-scans (~16 step scans at a cadence of 0.5 s/step) centered on the resonance wavelengths will be used to determine temperature and density. Broad system scans that are executed near apoapsis will be used to map the extended distributions of these species. Observations will also include both mini-scans at resonance wavelengths to search for predicted species that have not been previously detected (e.g., Si, Al, Mg, Fe, S, OH) and broad spectral scans (115–600 nm) to search for new species. Candidate UVVS measurement wavelengths for the various species are summarized in Table 5.

Figure 23 illustrates two examples of UVVS observational geometry. The left panel shows the spacecraft in a noon–midnight obit. In this configuration the UVVS performs a classic limb drift experiment in which the instrument FOV executes an altitude scan above the equatorial limb with a vertical drift rate of ~2 km/sec. Restrictions on spacecraft pointing limit the latitude of the limb tangent point to ±15°. The right panel shows the UVVS

Table 5 Candidate UVVS measurement wavelengths

Species	Wavelength (nm)	Species	Wavelength (nm)
H (O)	121.6	Na (O)	330.4
O (O)	130.4	Fe	371.9
S	181.3	Al	394.4
Si	252.6	Al	396.2
Mg	285.1	K (O)	404.4
OH	308.5	Ca (O)	422.7
Al	309.2	Na (O)	589.3

O: previously observed species

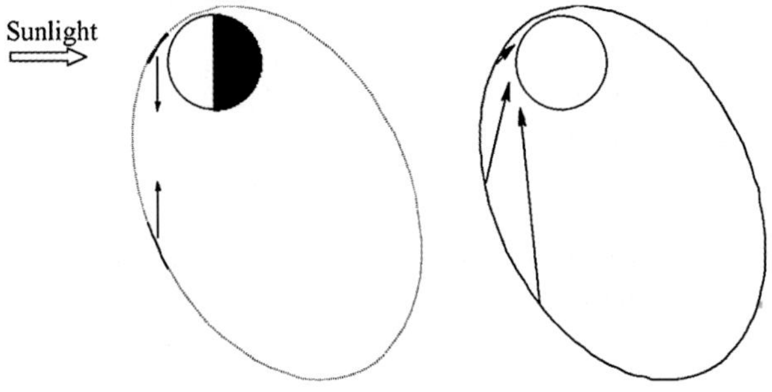

Fig. 23 UVVS observational geometries for a noon–midnight (*left*) and for a dawn–dusk orbit (*right*)

viewing geometry when the spacecraft is in a dawn–dusk orbit. As the spacecraft rolls about the Mercury–Sun line the UVVS FOV scans the limb. This orbit can be used search for exosphere dawn–dusk asymmetries and to probe high latitudes, particularly in the southern hemisphere.

MASCS is oriented on the spacecraft so that the long axis of the entrance slit is parallel to the limb in the dawn–dusk orbits; therefore, altitude resolution for exospheric observations is $\Delta H = 1.75 \times 10^{-2} * R$ and $\Delta H = 8.73 \times 10^{-4} * R$ for noon–midnight and dawn–dusk orbits respectively, where R is the distance from the spacecraft to the limb tangent point at altitude H.

6.2 Surface Reflectance Measurements

VIRS will measure surface reflectance at visible and near-infrared wavelengths to search for ferrous-bearing minerals (spectral signatures near 1 μm) and Fe-Ti-bearing glasses (spectral signatures near 340 nm). These measurements will be made with a spectral resolution of ~5 nm per pixel pair and can be binned to lower resolution to reduce data volume. Spacecraft pointing constraints restrict the phase angle ($\Theta = i + e$, where i and e are the incidence and emission angles, respectively) of these to ~$90 \pm 15°$. The VIRS field of view footprint is $\Delta X = 3.88 \times 10^{-4} * R$. The nominal integration time for VIRS is 1 s; thus the spatial footprint near apoapsis is a 5–6 km rectangle. Near periapsis, the spacecraft orbital motion during an integration period smears the field of view along the spacecraft track by 2–3 km while the cross-track resolution is better than 0.5 km.

UVVS will measure surface reflectance in the 115–300 nm wavelength range and is sensitive to a strong ferrous iron band near 250 nm. It is equipped with a mechanism that replaces the $1° \times 0.04°$ slit by a $0.05° \times 0.04°$ aperture. This aperture is offset from the VIRS entrance aperture by 0.38° in the plane of the UVVS grating dispersion. This arrangement allows VIRS and UVVS to observe the same location on the surface during dawn–dusk orbits by displacing the start of their respective integration times in order to correct for spacecraft ground-track motion.

7 Conclusion

MASCS is one of seven science instruments aboard the MESSENGER spacecraft, which was launched on August 3, 2004, on its way to a one Earth-year orbital mission around Mercury. The instrument contains two spectroscopic channels. UVVS is designed to measure exospheric composition, structure, and temporal variability. VIRS will measure mineralogical composition of the surface. MASCS incorporates innovative approaches to package proven optical designs and detector technologies into a compact, low-mass instrument that provides a wide range of measurement capability. It will provide important measurements that will address MESSENGER scientific objectives related to Mercury's formation and geologic history and to the composition and transport of volatile species on and near the planet.

Acknowledgements More than a dozen engineers, technicians, and instrument makers at the Laboratory for Atmospheric and Space Physics (LASP) contributed to the design, fabrication, and test of MASCS. Their expertise, dedication, and hard work transformed the MASCS concept into a world-class scientific instrument. We extend special thanks to James Johnson for the craftsmanship and wisdom that he brought to the assembly and test process and to Jim York at InnovaQuartz, who contributed the VIRS fiber feed to the MASCS project. We also thank Sean Solomon and an anonymous referee for valuable comments. Mark Robinson made many contributions to the discussion of Mercury's geology. This work was supported by the National Aeronautics and Space Administration's Discovery Program through contracts to the University of Colorado from the Carnegie Institution of Washington and from APL.

References

J.B. Adams, T.B. McCord, *Proc. Apollo 11 Lunar Sci. Conf.*, 1970, pp. 1937–1945
B.A. Anderson et al., Space Sci. Rev. (2007, this issue). doi:10.1007/s11214-007-9246-7
G.B. Andrews et al., Space Sci. Rev. (2007, this issue). doi:10.1007/s11214-007-9272-5
T.A. Bida, R.M. Killen, T.H. Morgan, Nature **404**, 159–161 (2000)
D.T. Blewett, P.G. Lucey, B.R. Hawke, G.G. Ling, M.S. Robinson, Icarus **129**, 217–231 (1997)
D.T. Blewett, P.G. Lucey, B.R. Hawke, Meteorit. Planet. Sci. **37**, 1245–1254 (2002)
E.T. Bradley, G.M. Holsclaw, W.E. McClintock, N.R. Izenberg, Eos Trans. Am. Geophys. Union **87** (Jt. Assem. Suppl.) (2006), abstract P41B-01
A.L. Broadfoot, S. Kumar, M.J.S. Belton, M.B. McElroy, Science **185**, 166–169 (1974a)
A.L. Broadfoot, D.E. Shemansky, S. Kumar, Geophys. Res. Lett. **3**, 577–580 (1974b)
B.J. Butler, D.O. Muhleman, M.A. Slade, J. Geophys. Res. **98**, 15003–15023 (1993)
M.J. Cintala, J. Geophys. Res. **97**, 947–973 (1992)
D.L. Domingue, P.L. Koehn, R.M. Killen, A.L. Sprague, S. Menelaos, A.F. Cheng, E.T. Bradley, W.E. McClintock, Space Sci. Rev. (2007, this issue). doi:10.1007/s11214-007-9260-9
D. Dzurisin, Geophys. Res. Lett. **4**, 383–386 (1977)
L.W. Esposito et al., Space Sci. Rev. **115**, 299–361 (2004)
J.O. Goldsten et al., Space Sci. Rev. (2007, this issue). doi:10.1007/s11214-007-9262-7
S.E. Hawkins III et al., Space Sci. Rev. (2007, this issue). doi:10.1007/s11214-007-9266-3
J.W. Head, C.R. Chapman, D.L. Domingue, S.E. Hawkins, W.E. McClintock, S.L. Murchie, M.S. Robinson, R.G. Strom, T.R. Watters, Space Sci. Rev. (2007, this issue). doi:10.1007/s11214-007-9263-6
G. Holsclaw, T. Bradley, W.E. McClintock, N. Izenberg, R. Vaughan, M.S. Robinson, Eos Trans. Am. Geophys. Union **86** (Fall Meeting Suppl.) (2005), abstract P51A-0896, p. F1198
C.W. Hord et al., Space Sci. Rev. **60**, 503–530 (1992)
D.M. Hunten, T.H. Morgan, D.E. Shemansky, in *Mercury*, ed. by F. Vilas, C.R. Chapman, M.S. Matthews (University of Arizona Press, Tucson, 1988), pp. 562–612
J.F. James, R.S. Sternberg, *The Design of Optical Spectrometers* (Chapman and Hall, London, 1969)
R.M. Killen, A.E. Potter, T.H. Morgan, Icarus **85**, 145–167 (1990)
R.M. Killen, J. Benkoff, T.H. Morgan, Icarus **125**, 195–211 (1997)
R.M. Killen, W.-H. Ip, Rev. Geophys. **37**, 361–406 (1999)
J.C. Leary et al., Space Sci. Rev. (2007, this issue). doi:10.1007/s11214-007-9269-0
P.G. Lucey, Geophys. Res. Lett. **31**, L080701 (2004). doi:10.1029/2003GL019406
P.G. Lucey, D.T. Blewett, B.R. Hawke, J. Geophys. Res. **103**, 3679–3699 (1998)
P.G. Lucey, D.T. Blewett, B.L. Jolliff, J. Geophys. Res. **105**, 20297–20306 (2000a)
P.G. Lucey, D.T. Blewett, G.J. Taylor, B.R. Hawke, J. Geophys. Res. **105**, 20297–20305 (2000b)
S.W. Maymon, S.P. Neeck, J.C. Moody Sr., Soc. Photo-Optical Instr. Eng. **924**, 10–22 (1988)
W.E. McClintock, E.T. Bradley, G.M. Holsclaw, Applied Physics Laboratory Report No. 7384-9470, 2004
M.A. McGrath, R.E. Johnson, L.J. Lanzerotti, Nature **323**, 694–696 (1986)
M. Mendillo, J. Baumgardner, B. Flynn, Geophys. Res. Lett. **18**, 2097–2100 (1991)
T.H. Morgan, R.M. Killen, Planet. Space Sci. **45**, 81–94 (1996)
S.K. Noble, C.M. Pieters, Sol. Syst. Res. **37**, 31–35 (2003)

A.E. Potter, T.H. Morgan, Science **229**, 651–653 (1985)
A.E. Potter, T.H. Morgan, Icarus **67**, 336–340 (1986)
B. Rava, B. Hapke, Icarus **71**, 397–429 (1987)
M.S. Robinson, P.G. Lucey, Science **275**, 197–200 (1997)
M.S. Robinson, G.J. Taylor, Meteorit. Planet. Sci. **36**, 841–847 (2001)
M.A. Slade, B.J. Butler, D.O. Muhleman, Science **258**, 635–640 (1992)
A.L. Sprague, Icarus **84**, 93–105 (1990)
A.L. Sprague, D.M. Hunten, K. Lodders, Icarus **118**, 211–215 (1995)
A.L. Sprague, W.J. Schmitt, R.E. Hill, Icarus **135**, 60–68 (1998)
A.L. Sprague, J.P. Emery, K.L. Donaldson, R.W. Russell, D.K. Lynch, A.L. Mazuk, Meteorit. Planet. Sci. **37**, 1255–1268 (2002)
F. Vilas, Icarus **64**, 133–138 (1985)
J. Warell, D.T. Blewett, Icarus **168**, 257–276 (2004)
J. Warell, A.L. Sprague, J.P. Emery, R.W.H. Kozlowski, A. Long, Icarus **180**, 281–291 (2006)
T.N. Woods, R.T. Wrigley III, G.J. Rottman, R.E. Haring, Appl. Optics **33**, 4273–4285 (1994)

Space Sci Rev (2007) 131: 523–556
DOI 10.1007/s11214-007-9272-5

The Energetic Particle and Plasma Spectrometer Instrument on the MESSENGER Spacecraft

G. Bruce Andrews · Thomas H. Zurbuchen · Barry H. Mauk · Horace Malcom · Lennard A. Fisk · George Gloeckler · George C. Ho · Jeffrey S. Kelley · Patrick L. Koehn · Thomas W. LeFevere · Stefano S. Livi · Robert A. Lundgren · Jim M. Raines

Received: 22 May 2006 / Accepted: 23 August 2007 / Published online: 31 October 2007

Abstract The Energetic Particle and Plasma Spectrometer (EPPS) package on the MErcury Surface, Space ENvironment, GEochemistry, and Ranging (MESSENGER) mission to Mercury is composed of two sensors, the Energetic Particle Spectrometer (EPS) and the Fast Imaging Plasma Spectrometer (FIPS). EPS measures the energy, angular, and compositional distributions of the high-energy components of the in situ electrons (>20 keV) and ions (>5 keV/nucleon), while FIPS measures the energy, angular, and compositional distributions of the low-energy components of the ion distributions (<50 eV/charge to 20 keV/charge). Both EPS and FIPS have very small footprints, and their combined mass (∼3 kg) is significantly lower than that of comparable instruments.

Keywords Plasma · Spectrometer · Ions · Electrons · Energy · Time of flight · MESSENGER · Mercury

1 Scientific Motivation

The Mariner 10 close flybys of Mercury in 1974–1975 revealed two surprises (Russell et al. 1988). The first was that Mercury has a substantial internal magnetic field, sufficiently robust to stand off the dynamic pressure of the solar wind emanating from the Sun. Because of that internal magnetic field, Mercury's space environment supports a small magnetosphere that looks superficially like that of Earth, with a "bow shock," "magnetopause," nightside "magnetotail" structure, and a variety of internal plasma structures (Fig. 1). Prior to the Mariner 10 encounters it was presumed that Mercury had long ago cooled to a sufficient extent that

G.B. Andrews (✉) · B.H. Mauk · H. Malcom · G.C. Ho · J.S. Kelley · T.W. LeFevere · S.S. Livi
The Johns Hopkins University Applied Physics Laboratory, Laurel, MD, USA
e-mail: Bruce.Andrews@jhuapl.edu

T.H. Zurbuchen · L.A. Fisk · G. Gloeckler · P.L. Koehn · R.A. Lundgren · J.M. Raines
Department of Atmospheric, Oceanic and Space Sciences, University of Michigan, Ann Arbor, MI, USA

G. Gloeckler
University of Maryland, College Park, MD, USA

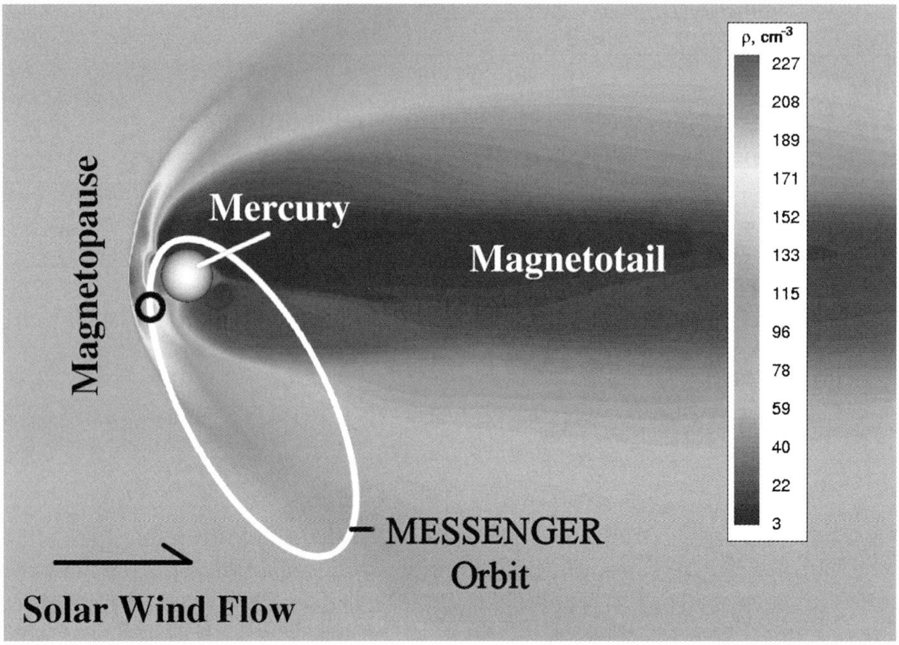

Fig. 1 The structure of Mercury's magnetosphere, with MESSENGER's orbit superposed; ρ is the number density. During each orbit FIPS will measure the solar wind, Mercury's bow shock and magnetosheath, and other magnetospheric components. The *black circle* marks a time in the orbit at which MESSENGER is expected to cross from closed fields into cusp fields with direct access to the polar regions of Mercury (Kabin et al. 2000). A major source of pick-up ions accelerated by the convective field of the magnetospheric plasma is expected in these polar regions (Koehn 2002)

the internal fluid motions necessary to sustain a magnetic dynamo had ceased (Schubert et al. 1988). The second surprise was that Mercury's magnetosphere hosts dynamical features with substantial similarities to "magnetospheric substorms" that occur within the Earth's magnetosphere in association with enhancements in the northern and southern auroral displays (Fig. 2). Not only did the Mariner encounters reveal energetic particle bursts, but also magnetic and plasma signatures consistent with "depolarization" events that, at Earth, signal the partial collapse of the magnetotail structure (Christon et al. 1987). Prior to these encounters it was presumed that substorm-like dynamics require the presence of a collisional, conducting ionosphere in order to short out sporadically the electric currents that support the magnetic field configuration that has been inflated by energy input from the solar wind. What little is known of Mercury's atmosphere (Hunten et al. 1988) reveals it to be a sparse, collisionless "exosphere."

Major questions arising from these surprises include: (1) What is the detailed configuration of Mercury's internal magnetic field, and how is it generated? (2) What is the nature of Mercury's sparse atmosphere, how is it generated, and can it carry substantial electric currents by means of active ionization or surface processes? (3) What is the configuration of Mercury's extended magnetospheric environment, what are its modes of dynamical change, and how does it energize particles? These and other questions that motivate the Energetic Particle and Plasma Spectrometer (EPPS) instrument on the MErcury Surface, Space ENvironment, GEochemistry, and Ranging (MESSENGER) spacecraft are addressed in the following.

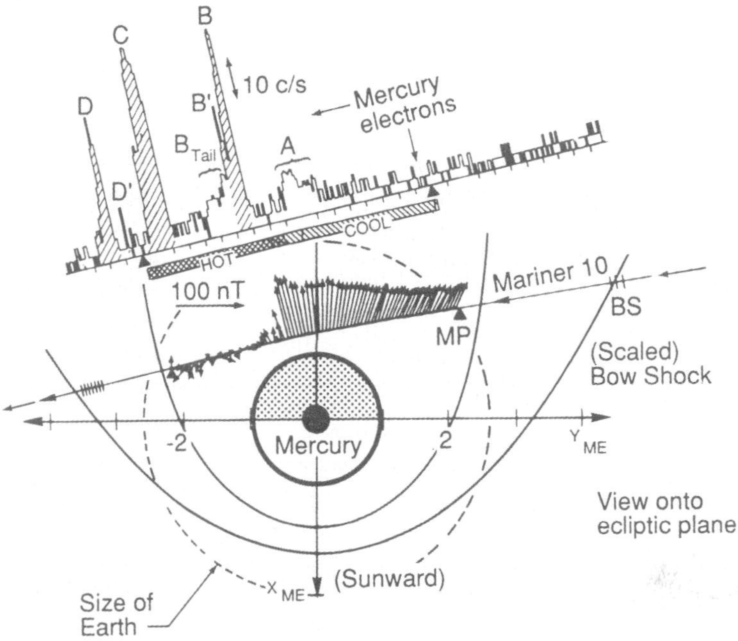

Fig. 2 Trajectory of the first Mariner 10 flyby of Mercury with representations of the energetic particle intensities (showing burst structures) and of the temperature of plasma electron components (Christon 1992)

All of Mariner 10's findings on Mercury's space environment are based on two of the three flyby encounters of the near-planet environment. The particle instrumentation on Mariner 10, analogous to the EPPS sensors, included a plasma electron sensor covering mono-directional energies from 13 to 715 eV and an energetic ion and electron particle sensor, which nominally measured mono-directional protons from 500 keV to 68 MeV and electrons from 175 keV to 30 MeV. It appears that the energetic particle sensor was at times saturated within Mercury's magnetosphere, particularly during burst events, so the species and intensities of the energetic particles are highly uncertain (Christon et al. 1979).

The MESSENGER EPPS investigation is designed to overcome the limitations of the Mariner 10 instrumentation. MESSENGER will, of course, revisit the near-planet space environment hundreds of times rather than just twice, allowing the nominal configuration to be distinguished from the special, and allowing the greater part of the space environment to be mapped out. The EPPS Energetic Particle Spectrometer (EPS) sensor covers broader ranges of energy than did the instrumentation on Mariner 10, determines angular distributions, and has sensitivities that incorporate knowledge of intensities from the Mariner mission. The EPPS Fast Imaging Plasma Spectrometer (FIPS) sensor will make the very first measurements of plasma ion distributions within Mercury's space environment.

1.1 EPPS Science Objectives

The science objectives for EPPS are described in Tables 1 and 2 from the broad science questions that motivate them and the broad observation objectives that address them (column 1), to the specific measurements needed (column 2), and finally to the EPPS measurement capabilities that were designed into the EPPS sensors in order to meet the measurement objectives (column 3). The first three science questions and objectives (Table 1) are among

Table 1 EPPS traceability—planetary science

Science questions/ MESSENGER objectives	EPPS observations	EPPS measurements
What is the nature and origin of Mercury's magnetic field?/ Determine the structure of the planet's magnetic field	1. Ion energy densities (pressures) to assess external contributions to magnetic field.	Ions 20 eV–300 keV $dE/E < 50\%$ H, He, CNO, Na, K (major species) Wide pitch-angle coverage 30° angular resolution 36-s time resolution (∼100 km)
	2. Ion and electron energy spectra vs. time for high sensitivity to contaminating temporal events.	Ions 15–300 keV Electrons 15–300 keV $dE/E < 50\%$ Broad angular coverage 30° angular resolution 2-s time resolution, selected channels
	3. Electron pitch angle distributions to determine Mercury connectivity; field configuration geometry.	Electrons: 15–300 keV $dE/E < 50\%$ FOV toward / away from planet 30° resolution with full post-acceleration 36-s time resolution, selected channels
What are the important volatile species and their sources and sinks on and near Mercury?/ Characterize exosphere neutrals and accelerated magnetospheric ions	1. Masses, moments, and angle distributions of cold, but accelerated, ions streaming away from Mercury along connected field lines. Polar materials: H or S enrichment.	Ions 5 eV–20 keV $dE/E < 25\%$ H, ^{3}He, ^{4}He, C, N, O, Ne, Na, K, Ca, Si, S, KAr, Fe Hemisphere to view along B on connected field lines 15° angular resolution 1-minute time resolution
What are the radar-reflective materials at Mercury's poles?/ Determine the composition of the radar-reflective materials at Mercury's poles	2. Masses and intensity distributions of pick-up ions borne in flowing solar wind or magnetospheric plasmas. Polar materials: H or S enrichment.	Ions 20 eV–20 keV $dE/E < 25\%$ H, ^{3}He, ^{4}He, C, N, O, Ne, Na, K, Ca, Si, S, KAr, Fe Hemisphere with edge towards aberrated solar wind direction 15° angular resolution 1-minute time resolution

Table 2 EPPS traceability—space science

Science questions/ MESSENGER objectives	EPPS observations	EPPS measurements
What is the electrical conductivity of the atmosphere/crust system?/ Determine the electrical properties of the crust/atmosphere/space-environment interface	Acceleration properties of ions close to Mercury's surface. Probe of electric field.	Ions 20 eV–20 keV $dE/E < 25\%$ H, ^{4}He, CNO, Na, K (major species) Hemisphere to view planet and aberrated solar wind 15° angular resolution 1-minute time resolution
What global energy conversion processes operate in the magnetosphere, and are they analogous to Earth magnetospheric processes?/ Determine characteristics of the dynamics of Mercury's magnetosphere and their relationships to external drivers and internal conditions	1. Energy spectra and angle distributions of particle energization events.	Ions 15–1000 keV $dE/E < 50\%$ H, He, CNO, Na, K (major species) Wide FOV normal to spacecraft-planet line 30° angular resolution 2-s time resolution, selected channels
	2. Plasma density, temperature, and composition versus region. Plasma heating/spectral changes versus boundaries, regions, and dynamic events.	Ions and electrons 20 eV–20 keV $dE/E < 25\%$ H, He^+, He^{2+}, O^+, $O^{6+,7+}$, Na, K Hemisphere normal to spacecraft–planet line 30° angular resolution 36-s time resolution
How does the interplanetary medium evolve between 0.3 and 1.0 AU?/ Measure interplanetary plasma properties in cruise and in Mercury vicinity	Plasma temperature, density, non-thermal components, and composition near Mercury and in cruise.	Ions 50 eV–10 keV $dE/E < 12\%$ Mass: H, He^+, He^{2+}, $O^{5+,8+}$, $Fe^{16+,6+}$ FOV hemisphere view near aberrated solar wind 30° angular resolution 1-minute time resolution

the major planetary science objectives identified at the time the MESSENGER mission concept was developed (Solomon et al. 2001). The second three science objectives (Table 2) are derived from the first three and focus on the space–planet interactions at Mercury. Here we discuss the implications of each of the science objectives in turn.

1.1.1 Determine the Structure of Mercury's Internal Magnetic Field

The major problem in using the MESSENGER magnetometer to determine Mercury's intrinsic field is that external electric currents flowing within Mercury's magnetosphere generate magnetic fields that are comparable to the magnetic fields generated by internal currents. EPPS contributes to separating internal from external sources of magnetic fields in three separate ways:

1. External plasmas contribute substantially to the ambient magnetic fields when the energy densities of the plasmas are comparable to the magnetic field energy densities. During the Mariner 10 encounters, for example, certain regions were identified as probably associated with high-pressure plasma populations (Connerney and Ness 1988). EPPS pressure measurements will identify regions that must be excluded from the analyses used to characterize the internal magnetic field configuration.
2. Experience within other planetary magnetospheres has shown that energetic particles often provide the most sensitive signaling of the presence of external transient events leading to changes within the magnetosphere (Mauk et al. 1999).
3. Energetic particles can also at times remotely sense magnetic topology in a fashion that cannot be ascertained by measuring local fields alone. For example, the energetic-electron angular measurements made by Galileo at the icy satellite Ganymede were highly instrumental in confirming the existence of, and determining the configuration of, Ganymede's internal magnetic field (Williams et al. 1997).

1.1.2 Characterize Exosphere Neutrals and Accelerated Magnetospheric Ions

EPPS composition measurements will identify materials emanating from Mercury's exosphere and ultimately its surface, in concert with measurements of ultraviolet emissions from exospheric neutrals made by the Mercury Atmospheric and Surface Composition Spectrometer (MASCS), another MESSENGER instrument (McClintock and Lankton 2007). Mercury's space environment, to be characterized with the help of EPPS measurements, participates in the generation of the exosphere, but it also represents a challenge in that the space environment processes the materials before they can be sensed by EPPS. In particular, the space environment must ionize and then accelerate the now-charged particle before it can be sensed (Fig. 3). Determination of configuration, sources, and sinks of exospheric neutrals and magnetospheric ions must go hand-in-hand with the characterization and understanding of the structure and dynamics of the space environment.

For example, sodium, potassium, and calcium emissions have been detected in Mercury's atmosphere (Bida et al. 2000). The Na density is observed to be highly variable in space and time. These changes appear to be associated with changes in solar radiation (Potter and Morgan 1987; Ip 1990), but they may also be a direct result of magnetosphere–planet interactions (Koehn 2002). In general, only a comprehensive inventory of atomic species in Mercury's environment, taken by in situ measurements, will provide critical clues about planetary and solar wind sources of the tenuous exosphere associated with Mercury.

1.1.3 Determine the Composition of the Radar-Reflective Materials Near Mercury's Poles

One crucial observation of the post-Mariner era is that of radar-reflective materials at Mercury's poles (Slade et al. 1992). The reflective materials are found in the floors of many craters in Mercury's polar regions. The high polarization ratios observed from these regions

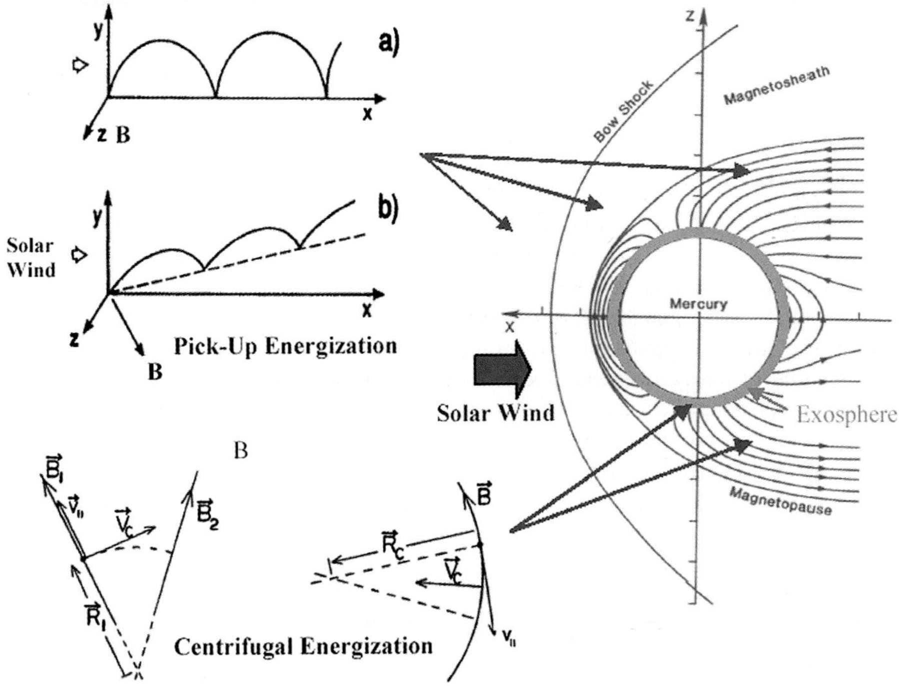

Fig. 3 Two mechanisms of particle acceleration that can occur in the vicinity of Mercury. "Pick-up" of energized particles perpendicular to the magnetic field results from ionization in plasma that is flowing perpendicular to the straight field lines. "Centrifugal" acceleration of energized particles parallel to the magnetic field lines occurs as a result of the flow of highly curved field lines, or flow that rapidly changes the direction of frozen-in field lines (Russell et al. 1988; Mauk and Meng 1989). Not represented here are the very-high-energy energization processes that can occur within the narrow confines of the neutral sheet of the more distant magnetotail

are similar to those of icy satellites of the outer planets. These reflecting areas are therefore widely thought to contain water ice. Because of the near-zero obliquity, floors of polar craters are in permanent shadow. Also, this dynamic configuration is sufficiently stable that the ice in these craters may survive for billions of years (Vasavada et al. 1999). Sprague et al. (1995) proposed the alternative hypothesis that the polar deposits are composed of elemental sulfur, cold-trapped from sulfur emitted elsewhere on Mercury's surface and redeposited in a neutral or ionized state.

EPPS will look near the poles for high concentrations of either low-energy H ions (signaling a water source) or S ions. Again, the space environment processing of these materials, including acceleration processes, must be folded into any interpretation of such signals.

1.1.4 Determine the Electrical Properties of the Crust, Atmosphere, and Space Environment Interface

The electrical properties of Mercury's near-surface region are central to the issue of magnetospheric dynamics and the role of the space–planet interface in closing electric currents that support the solar-wind-inflated magnetic configuration. To examine the near-planet electrical properties, EPPS will seek evidence of ion acceleration very close to the planet. If

Mercury behaves as a highly conducting object, the magnetic field lines will essentially be "frozen-in" to the planet. In this situation, electric fields needed to accelerate particles cannot penetrate to the vicinity of the solid planet. If Mercury's surface–space interface behaves like an insulator, then particle acceleration is expected to occur very close to the planet's surface.

1.1.5 Determine the Characteristics of the Dynamics of Mercury's Magnetosphere and Their Relationships to External Drivers and Internal Conditions

Characterization of Mercury's magnetospheric dynamic will be carried out by simultaneous measurement by EPPS and the MESSENGER Magnetometer (MAG) instrument (Anderson et al. 2007). These instruments will determine external magnetic field conditions; characterize the plasma, energetic particle, and magnetic field signatures of boundary locations, the internal magnetospheric configuration, and transient events; provide diagnostics of internal atmospheric sources of plasma; and give low-altitude ion acceleration signatures of the electrical conductivity of Mercury's surface, atmosphere, and space environment.

Most of the Mercury magnetospheric ions presumably originate from the solar wind. These magnetospheric ions will be distinguished from neutral species sputtered into the exosphere of Mercury. These particles are ionized within hours and then produce Mercury's pick-up ions. The entire system is being driven by the solar wind. This system is the best known particle source in the inner heliosphere, based on Helios data at the same heliocentric radius as Mercury (Schwenn 1990). The anticipated magnetospheric particle population is expected to be highly variable and highly dependent on the location of MESSENGER with respect to Mercury (Fig. 1).

1.1.6 Measure Interplanetary Plasma Properties in Cruise and Near Mercury

Any comparison of Mercury's response and Earth's response must take into account their different space environments. Energetic particle populations, solar wind structures, and magnetic configuration will differ between the environment near Earth and the environment near 0.3–0.4 AU, at Mercury's orbit. The evolution of the interplanetary environment between Earth and Mercury is also of intrinsic value in and of itself. EPPS will measure aspects of the high-energy ion and electron components of the interplanetary environment (the part moving towards the Sun) and will sense the medium-energy pick-up ions from interstellar and local gas sources. Nominally, the FIPS sensor is not angled to view the solar wind. The solar wind will be viewed occasionally near Mercury, however, when the spacecraft is positioned where the solar wind is deflected around the planet.

2 The Fast Imaging Plasma Spectrometer (FIPS) Sensor

FIPS is a completely new sensor, designed for the MESSENGER mission to Mercury (Zurbuchen et al. 1998; Koehn 2002). The sensor (Fig. 4) is designed to characterize ionized species in a range of energy-per-charge (E/Q) from <50 eV/Q up to 20 keV/Q. One of the innovations in this sensor is a new electrostatic analyzer (ESA) system geometry that enables a large instantaneous ($\sim 1.4\pi$ steradians) field of view. This idea was first put forth in a proposal for the *Solar Probe* mission (G. Gloeckler, personal communication, 1994), but MESSENGER carries the first fully integrated sensor in flight with a comparably large instantaneous field of view.

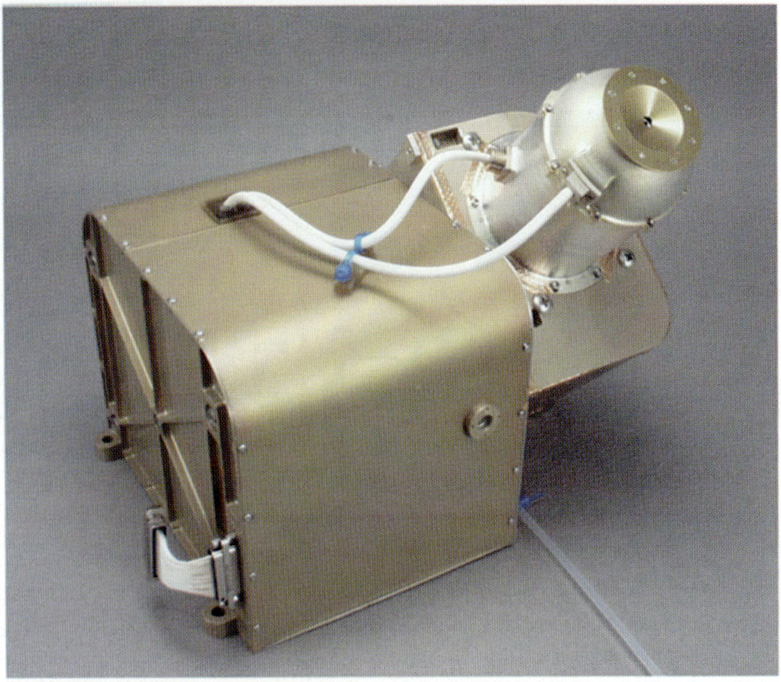

Fig. 4 The FIPS sensor

2.1 FIPS Design Parameters

The following requirements and constraints most affected the design of the FIPS sensor:

1. Low mass, volume, and power
2. Suppression of ultraviolet (UV) to permit operation in full sunlight near Mercury
3. Wide field of view
4. Large dynamic range
5. Good mass resolution
6. High voltages required for the electrostatic analyzer, and for post-acceleration, which enables low-energy ions to penetrate the carbon foil
7. Good dynamic range for energy-per-charge, fast transitions between steps
8. Fast time-resolution capability
9. Low data rate

2.2 FIPS General Description

Particles that pass through the ESA have a known E/Q, proportional to the stepped deflection voltage. They are then post-accelerated by a fixed voltage, before passing through a very thin carbon foil. The ions travel a known distance and hit the stop micro-channel plate (MCP), while forward-scattered electrons from the carbon foil are focused onto the start MCP. Position sensing with a wedge-and-strip anode in the start MCP assembly determines the initial incidence angle. The mass-per-charge (M/Q) of a given ion follows from the known E/Q and the measured time of flight, τ, allowing reconstruction of distribution

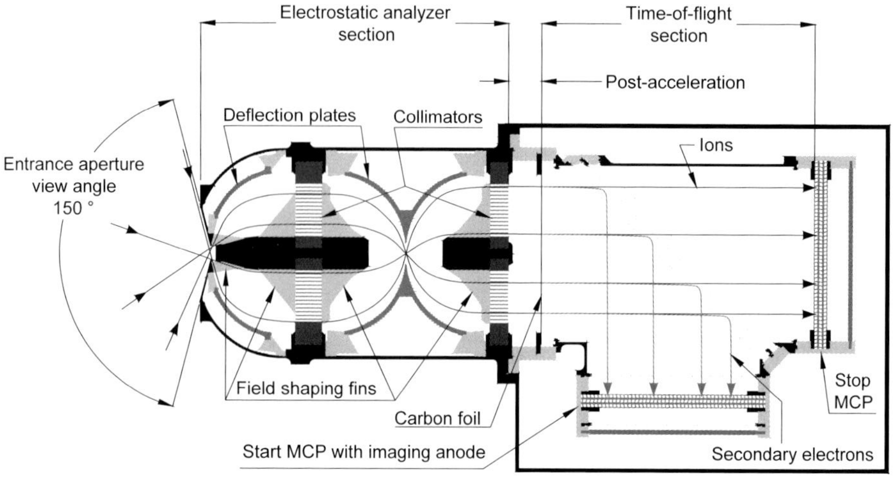

Fig. 5 Cross-section of the FIPS sensor showing major functional components. Ions are analyzed by their energy per charge, two-dimensional position, and total time of flight

functions for different M/Q species. In its Normal (nominal) scan mode, the ESA system covers the E/Q range in 64 logarithmically spaced steps every 65 s. It can also run in a fast "Burst mode" that allows highly focused sweeps in only 2 s.

FIPS is composed of three major subsystems: an electrostatic analyzer system, a time-of-flight (TOF) system, and the sensor electronics. Fig. 5 illustrates the FIPS sensor in cross-section.

2.3 FIPS Electrostatic Analyzer (ESA)

The ESA serves both as a UV trap and an energy-per-charge filter, allowing only ions of a very specific energy-per-charge interval (determined by a stepped deflection voltage) to pass through the system. Ions enter through a small annular aperture, seen on the left side of Fig. 5. Within this aperture there is a set of 24 radial openings that the ions must pass through. These slots limit both the position and the velocity of the ions to a small band along the symmetry axis of the ESA, eliminating ions that might spiral within the cylindrically symmetric ESA. The ions then enter a region where their trajectories are curved by the electrostatic field between the positive outer deflection plate and the grounded inner cylinder and field-shaping fins. Ions must then pass through the collimator. The collimator slits, an array of small circular arcs, are aligned with the axis of symmetry of the ESA so that ions that are not traveling parallel to the axis are eliminated. The E/Q resolution and geometric factor of the ESA are determined by the angular acceptance of the collimator, the area of its openings, and the size of the entrance aperture. After ions have passed through the first deflection region and first collimator, the ESA has performed its purpose as an E/Q filter. However, at this point UV attenuation is not sufficient, so an hourglass-shaped deflection region between the first and second collimators was added to provide the needed UV suppression.

Ions are post-accelerated by a potential drop of up to -15 kV just before entering the TOF section. The energy they gain enables the lowest-energy ions to pass through the carbon foil and reduces energy straggling for all ions.

2.4 FIPS Time-of-Flight (TOF) System

The TOF system is a so-called "mirror-harp" design that relies on wire harps to establish equipotential surfaces yet allows particles to pass between the wires. The speed of an ion is determined by measuring its travel time between the carbon foil and the stop MCP, separated by a distance of approximately 7 cm. As an ion passes through the carbon foil, it loses a small fraction of its energy (ε), is subjected to some directional scattering as a result of interactions with the foil, and emerges with a charge state that is predominantly neutral or singly charged. Low-energy secondary electrons are also released from the foil, due to internal ion scattering interactions. These secondary electrons are accelerated and then reflected by the diagonal mirror harp onto the start MCP, which produces a start signal. After leaving the foil, the ion travels straight through the TOF system until it strikes the second MCP, which produces a stop signal. The time difference between the start and stop signals is measured by the electronics. The time of flight of the ion is known once a correction is made for the travel time of secondary electron from foil to start MCP (a fixed offset,typically 5 ns, depending on potentials within the TOF system).

In addition to E/Q filtering and TOF measurement, FIPS measures the incident direction of the ion. Within the cylindrically symmetric ESA, an ion's trajectory is curved toward the axis by the electric field, until its path is normal to the surface of the collimator. The azimuth angle of the incoming ion is not changed by the ESA. Radial fins every 15° prevent ions from spiraling within the ESA (these fins are also used for field shaping of the ion optics). The greater the polar angle (the angle between the direction of the incident ion and the symmetry axis of the ESA), the greater the radial position of the ion where it passes through the collimator. The TOF system has a circular cross-section, with the same axis of symmetry as the ESA. Ions that exit the collimator are moving parallel to this axis, so their location on the carbon foil is the same as in the collimator. The electron optics of the TOF system map the position of the secondary electrons that are released from the carbon foil onto the start MCP (in a mirror image). A wedge-strip-zigzag (WSZ) anode behind this MCP allows two-dimensional position sensing of the event on the MCP (Martin et al. 1981). The wedge-strip-zigzag anode consists of a pattern of three interleaved conductive electrodes whose area varies according to the x and y position on the anode. When the charge output of an MCP falls on the anode and covers an area wider than the wedge-strip-zigzag pattern repetition period the centroid position of the event may be deduced from the proportion of the charge deposited on each of the three electrodes with the following formulas,

$$x = \frac{2Q_w}{Q_w + Q_s + Q_z}, \qquad y = \frac{2Q_s}{Q_w + Q_s + Q_z},$$

where Q_w, Q_s, and Q_z are the amounts of charge falling on the wedge, strip, and zigzag electrodes, respectively.

In summary, the position information recorded by the time-of-flight system can be mapped back to the ion's angle of entry into the ESA. The typical angular resolution achieved is 15° full-width half maximum (FWHM) in the angle about the FIPS symmetry axis and 15° (FWHM) perpendicular to it.

2.5 FIPS Electronics

The FIPS electronics consist of five board-level subsystems packaged with the sensor:

Table 3 Summary of FIPS characteristics

Characteristic	Value
Mass	1.41 kg
Volume	$17.0 \times 20.5 \times 18.8$ cm^3
Power (average/maximum)	1.9/2.1 W
Bit rate (average)	80 bps
Field of view	1.4π steradian
Energy range	50 eV–20 keV
M/Q range	1–40 amu/e
Scan speed (nominal/burst)	65/2 s

1. The "Digital" board includes fast-pulse amplifiers, constant-fraction timing discriminators, and a time-to-digital converter in order to measure time of flight. It also includes the interface electronics as well as a field-programmable gate array (FPGA) that handles instrument control and takes some of the load off the shared event processor housed in EPS.
2. The "Analog" board contains three channels of pulse-shaping electronics to measure position from the WSZ anode. A single analog-to-digital converter (ADC) is used in conjunction with an analog multiplexer to digitize the amplitude of these pulses and to measure various housekeeping parameters. A precision digital-to-analog converter (DAC) sets the ESA deflection voltage.
3. The deflection system high-voltage power supply provides the voltage for the ESA. Its output covers the range from +35 V to +15 kV, and it is designed to transition rapidly between output levels.
4. The post-acceleration power supply provides an output voltage up to -15 kV for the TOF system.
5. The MCP bias power supply supplies both MCP assemblies with several output voltage taps for each. Its highest output voltage is -3.6 kV.

2.6 FIPS Calibration

The sensor calibrations fall into six distinct and complementary parts. All of these parts directly affect the end to end performance. They form a complete set of calibrations. These calibrations include:

1. Electronics calibration: The time-of-flight electronics have to be calibrated to provide absolute sensitivity. This calibration affects both speed (V) and mass-per-charge (M/Q) measurements in the device. The analog circuit also has to be calibrated. This calibration affects the position sensing and hence angular resolution of the program.
2. Electrostatic analyzer (ESA): The ESA calibration is a set of end-to-end tests of the electrostatic system. The instrument performance is judged by comparisons with simulation results. This calibration affects the total geometric factor (or sensitivity), the range and resolution in energy per charge (E/Q), and the angular range and resolution.
3. System efficiency: System efficiency is calibrated from end-to-end tests of the entire sensor, including flight electronics. The results of these calibrations are test cases of mass-dependent efficiencies, as well as the dependence of efficiency on particle incidence direction.

4. Mass resolution: Finally, from a combination of tests 1–3, the mass resolution can be evaluated using the usual process. The mass resolution calibration is described in more detail in the following.
5. Input rate vs. output rate characterization and proton rejection test: The FIPS instrument is tested to characterize its processing capabilities as a function of increasing input flux, and the proton rejection function of the instrument is tested.
6. In-flight analysis tools: There are a number of electronic tools that will allow in-flight testing of the FIPS instrument. These in-flight calibration tools were calibrated and include a test-pulser designed to calibrate the position-sensing function of the instrument.
7. Background suppression: There are two aspects to the measurement of background suppression. First, the instrument's ESA has to suppress UV photons that can lead to spurious background. Second, secondary electron emission from electrostatic mirrors can also produce such background.

2.6.1 FIPS Mass Resolution

The mass resolution of a TOF instrument is directly tied to the measured time of flight of the incident particle. We know the time of travel, τ, between the carbon foil and the anode and the distance, d, traveled in that time. The ESA voltage step V and analyzer constant k give us the particle's E/Q from

$$\frac{E}{Q} = Vk. \tag{1}$$

This energy-per-charge is modified by the post-acceleration voltage, U, and the energy loss from the foil, ε, so the velocity $(d/\tau)^2$ and E/Q information from the ion's impact location give us the mass-per-charge (Gloeckler and Hsieh 1979; Gloeckler 1990),

$$\frac{M}{Q} = \frac{2(Vk+U)\varepsilon\tau^2}{d^2}. \tag{2}$$

The mass-per-charge resolution is then

$$\left(\frac{\Delta(m/Q)}{m/Q}\right)^2 = \left(\frac{\Delta\alpha}{\alpha}\right)^2 + \left(2\frac{\Delta\tau}{\tau}\right)^2 + \left(2\frac{\Delta d}{d}\right)^2 + \left(\frac{\Delta kV}{kV}\right)^2 \left(\frac{1}{1+U/(kV)}\right)^2, \tag{3}$$

where $\Delta\alpha/\alpha$ is the energy straggling of the ion from its passage through the carbon foil, $\Delta\tau/\tau$ is the uncertainty in the ion's time of flight, $\Delta d/d$ is the uncertainty in the path length of the ion's trajectory through the TOF telescope, and $\Delta(E/Q)/(E/Q)$ is the energy resolution of the electrostatic analyzer.

As an experimental measure of the mass resolution, we fold the energy straggling, path length, and timing uncertainties into a single number, measured by taking the width of the TOF spectrum for each species-energy test run from the efficiency calibration series. The calculated M/Q is then

$$\frac{M}{Q} = \frac{2(Vk+U)\tau'^2}{D^2} \tag{4}$$

where D is now the physical distance between the carbon foil and the Stop MCP, and τ' is the measured time of flight of an incident particle. The new formula for the mass resolution is

$$\left(\frac{\Delta(m/Q)}{m/Q}\right)^2 = \left(2\frac{\Delta\tau'}{\tau'}\right)^2 + \left(\frac{\Delta kV}{kV}\right)^2\left(\frac{1}{1+U/(kV)}\right)^2. \quad (5)$$

The mass resolution is then just the TOF and energy resolutions added in quadrature. Table 4 shows the resulting mass resolution trends.

Table 4 FIPS mass resolution by species

Species	Beam KE (keV)	$\Delta t/t$ (%)	$\Delta E/E$ (%)	$[\Delta(M/Q)]/(M/Q)$ (FWHM[1]) (%)
H	10	2.60	5.00	7.21
	15	6.70	5.00	14.30
	17	3.60	5.00	8.77
	20	3.90	5.00	9.26
	25	4.60	5.00	10.47
N	10	9.80	5.00	20.23
	15	6.20	5.00	13.37
	17	4.90	5.00	11.00
	20	3.60	5.00	8.77
	25	2.00	5.00	6.40
H_2O	10	12.80	5.00	26.08
	15	9.30	5.00	19.26
	17	8.60	5.00	17.91
	20	7.90	5.00	16.57
	25	10.90	5.00	22.37
N_2	10	18.40	5.00	37.14
	15	13.40	5.00	27.26
	17	12.00	5.00	24.52
	20	9.80	5.00	20.23
	25	7.00	5.00	14.87
O_2	10	28.20	5.00	56.62
	15	16.80	5.00	33.97
	17	13.60	5.00	27.66
	20	13.10	5.00	26.67
	25	N/A	N/A	N/A
Ar	10	55.30	5.00	110.71
	15	23.50	5.00	47.27
	17	19.50	5.00	39.32
	20	19.20	5.00	38.72
	25	13.80	5.00	28.05

[1]FWHM = full width half maximum

Table 5 Summary of calibration results

Requirement	Value	Measured
Energy range	50 eV–20 keV	Sufficient
Energy resolution	10%	4.5%
Angular range (sr)	1.4π	1.4π
Angular resolution (FWHM)	20°	15°
Instantaneous dynamic range	10^5	10^5
Mass/charge range	1–30	1–40
Mass/charge resolution $\{[\Delta(M/Q)]/(M/Q)\}$	Sufficient to separate C, N, O	Sufficient
Geometric factor	0.1 mm^2 sr eV/eV	0.1 mm^2 sr eV/eV
Maximum allowable valid count rate	10^5	10^5
Maximum allowable background count rate	1 kHz	18 Hz

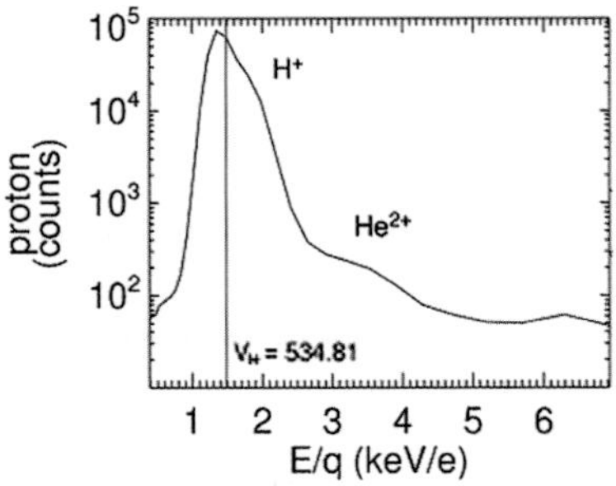

Fig. 6 FIPS proton rate spectrum acquired during the solar wind observation of April 15, 2005, and accumulated over a one-hour observation. The proton velocity calculated from these data is in good agreement with proton velocities from other spacecraft at that time

It is important to note here that energies listed in Table 4 are the particle energies after post-acceleration. Given that the maximum energy per charge that will be seen by the carbon foil is 35 keV, we can expect that the mass resolution for magnetospheric and solar wind heavy ions will improve. Additionally, the 10 keV energy values will not be seen at all, given a flight post-acceleration value of 15 keV.

FIPS mass resolution must be such that C, N, and O can be separated. The FWHM mass resolution for N in the above table is sufficient for the peak separation.

2.6.2 Calibration Summary

Summary of calibration results is shown in Table 5.

2.7 FIPS Early Flight Data

When the MESSENGER spacecraft is near Mercury, the temperatures will be much too high to expose the FIPS aperture to direct solar irradiance. The spacecraft's sunshade (Leary et al. 2007) is always placed between the Sun and the spacecraft. In this state, the FIPS instrument will be unable to sense the solar wind directly. However, early in the cruise phase of the mission the spacecraft is still at a substantial distance from the Sun, and it is safe to expose the FIPS aperture. A spacecraft maneuver was commanded to tilt the spacecraft about its Z axis by approximately 20°, allowing the FIPS sensor to measure solar wind parameters for approximately four hours. Figure 6 shows a proton rate spectrum acquired during the solar wind observation, confirming the instrument performance characteristics provided in Table 5.

3 The Energetic Particle Spectrometer (EPS) Sensor

EPS is a hockey-puck-sized, energy by time-of-flight (E x TOF) spectrometer that measures ions and electrons over a broad range of energies and pitch angles. Particle composition and energy spectra will be measured for H to Fe from $\sim$15 keV/nucleon to $\sim$3 MeV/nucleon and for electrons from 15 keV to 1 MeV. The EPS concept was developed with the support of a NASA Planetary Instrument Definition and Development (PIDDP) grant aimed at designing a low-mass, low-power sensor that can measure energetic particles, including pickup ions produced near planets and comets.

3.1 EPS Design Parameters

The following requirements and constraints most affected the design of the EPS sensor:

1. Low mass, volume, and power
2. High rejection of background
3. Good pitch-angle resolution for both electrons and ions
4. Clean electron measurements
5. Wide event-rate dynamic range
6. Good mass resolution
7. Fast time-resolution capability
8. Low data rate

3.2 EPS General Description

The sensor is housed in the same structure as the instrument's digital processing system and consists of three boards: an energy board with integral sensor module, a high-voltage (HV)/bias power supply board, and a time-of-flight board. The other three boards in the six-board EPS module are used by both EPS and FIPS. These additional three boards are the traditional components of what historically was called an instrument's digital processing unit (DPU). They consist of a common low-voltage power supply (LVPS), a common RTX2010-based event processor board (which contains the flight software used to analyze EPS and FIPS events), and a common FPGA digital input/output (I/O) and logic board. The LVPS and EPU boards are used in a number of MESSENGER instruments and are discussed elsewhere in this issue. The digital board and the flight software are designed specifically for EPPS and are discussed in the following.

A photograph of the EPS sensor (Fig. 7) plainly shows the aperture cover and collimator. The aperture cover is used to protect the thin foils in the sensor from acoustic damage, as well as from contamination and particulates. The cover is released by a set of redundant pyrotechnic retractable actuators. The semi-circular collimator consists of four nested plates, each with an array of holes. This type of collimator design is necessary for sensors that measure both ions and electrons because it traps electrons that scatter off of the collimator surfaces, minimizing the uncertainties in determining the incoming electron's pitch angle. A cross-sectional view of the electron-optics portion of the sensor (Fig. 8) shows the path of the primary and secondary particles as they pass through the collimator and enter the time-of-flight chamber prior to striking the solid-state detector (SSD) array. Figure 9 is a photograph of the SSD array and primary components of the TOF chamber.

The EPS sensor consists of a number of different electronic subsystems. These are the energy, TOF, HV, and digital processing subsystems. These subsystems are described in more detail later. A summary of EPS characteristics is shown in Table 6.

Fig. 7 Photograph of EPS mounted in the accelerator's test chamber. The saloon door cover is shown in the fully open position, revealing the collimator. The TOF chamber is located just behind the collimator, and just below the collimator are the six electronics modules (or slices) that are each 10 cm on a side

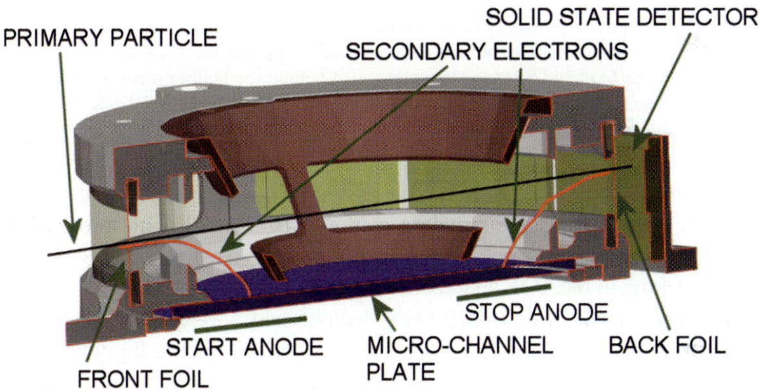

Fig. 8 Cross-sectional view of the sensor, showing how the primary particle traverses the TOF chamber, producing secondary electrons when passing through the thin foils. The secondary electrons are used to produce timing signals. The distance from front foil to back foil is approximately 6 cm

3.3 EPS Energy Subsystem

The EPS energy system consists of 24 SSDs mounted on six individual board substrates. Each substrate contains a 500-μm-thick p-on-n ion-implanted silicon detector crystal that has been pixelated into four individual detector regions. The overall dimensions for the four pixels are approximately 15 by 7 mm. The two pixels on the left side of the crystal have thin windows (<50 nm) and are used for detecting ions. The two pixels on the right are coated with a thin layer of aluminum and are used for detecting electrons. The active area of the large-ion pixel is roughly 20 times larger than the small-ion pixel, and similarly for the electron pixels. The electronics have been designed to be able to activate either the large or small pixel for the ion or electron detectors, providing a 20-to-1 dynamic range adjustment. In periods of high flux the small pixels are selected. During quiet times the large pixels are selected in order to improve the geometric factor.

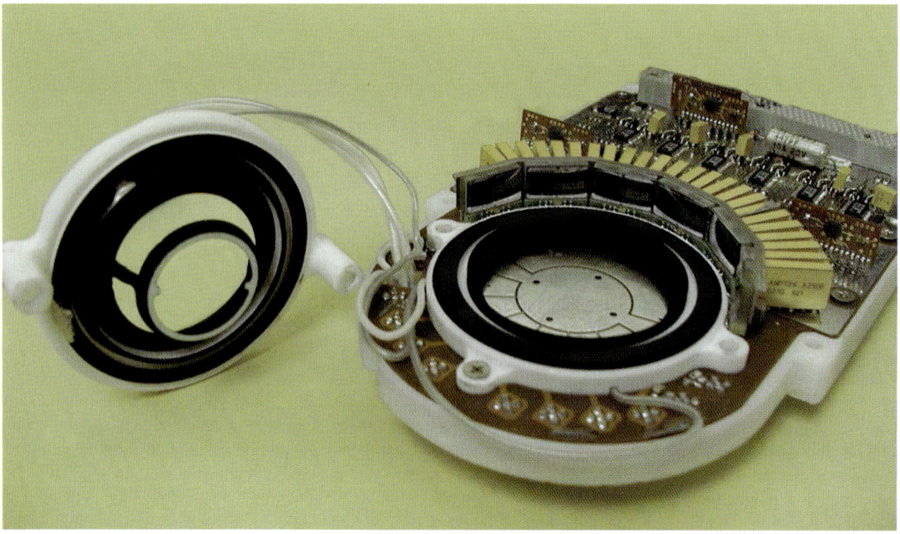

Fig. 9 Photograph of the prototype energy board with sensor components partially mounted. The TOF chamber consists of two ceramic pieces. The bottom ceramic piece is part of the MCP assembly. The MCP has been removed to show the Start and Stop anode pattern. Just behind the bottom piece is the SSD array, consisting of 24 detectors on six substrates. The top ceramic piece, which is lying to the side of the board, has a gridless conical electro-static surface that bends the secondary electrons emitted from the foils (not shown) into the MCP

The six detector substrates are mounted side-by-side along the perimeter of the TOF chamber and are positioned just behind the exit (Stop foil). Particles that pass through the entrance and exit foils in the TOF chamber will strike one of the detectors, depositing their energy. A 500-μm detector will stop 9-MeV protons and 440-keV electrons completely. Electrons with energies up to 850 keV will also stop in the detector, but with reduced efficiency.

The detectors share a common bias supply, which provides a programmable negative voltage to the entrance/junction side of the detectors. Typical operating voltage is −180 V. Each detector has its own bias resistor. If one detector were to short, the supply would be pulled down to roughly half voltage, which is still sufficient to deplete most of the detectors and operate at an acceptably low noise threshold.

The thin window detectors will detect both ions and electrons, but they are able to discriminate against electrons by requiring that the detector signal be in coincidence with the timing signal from the TOF system. Because electrons do not trigger the TOF system, any electrons striking the ion detectors are rejected by the valid event logic. Because protons have a relatively low efficiency for triggering the TOF system, many protons that strike the ion detectors are also rejected by the coincidence logic.

The flashing of aluminum on the electron detectors will stop protons up to 110 keV. This prevents the majority of the protons in Mercury's magnetosphere from triggering the electron detectors. Protons above 110 keV will produce a signal in the electron detector, but the valid event logic checks to see if the particle produced a start or stop pulse when passing through the TOF system. If so, the event is not an electron and will be rejected. Protons that produce neither a start nor stop pulse in the TOF system can be mistaken for electrons in the electron detectors. In this case, information from an ion detector can be used to determine how much proton contamination is occurring in the adjoining electron detector.

Table 6 Summary of EPS characteristics

Characteristic	Specification
Measured species	H, He, CNO, Fe, electrons
Field of view	160° by 12°
Geometrical factor	~0.03 cm^2 sr
Foils	Front foil (Poly/Al — 10 µg/cm^2), back foil (Poly/Pd — 19 µg/cm^2)
TOF range	0–200 ns, ±200 ps (1σ)
Peak input rate	1 MHz singles, 10 kHz triple coincidence
Detectors	6 Si 1,000-µm detectors with 4 pixels/detector. Pixel description:
	40 mm^2 uncoated (high-sensitivity ion detector)
	2 mm^2 uncoated (high-rate ion detector)
	40 mm^2 with 1-µm Al flashing (high-sensitivity electron detector)
	2 mm^2 with 1-µm Al flashing (for high-rate electrons)
	40.0 mm MCP, length/diameter: 60 : 1,
	channel: 10 µm; bias: 8°, high current
Energy range	Protons: 25 keV to 3 MeV total energy
	Alpha particles: 35 keV to 3 MeV total energy
	CNO: 70 keV to 3 MeV total energy
	Fe: 120 keV to 3 MeV total energy
	Electrons: 25 keV to 1.0 MeV
Integration period	Fully programmable

There are six electronics signal processing chains dedicated to the ion detectors and six electronics signal chains dedicated to the electron detectors. On the basis of ground command, either the small or large pixels will be connected to the signal processing chains. It is possible to have the small electron pixels selected at the same time as the large ion pixels and vice versa.

Each signal processing chain consists of an Amptek charge amplifier followed by a DC-coupled unipolar shaping amplifier. The shaping time is 2.4 µs. Each shaping chain has a dedicated high-speed ADC. A settable discriminator detects the output level of the first stage of a shaping chain, which is the pole-zero compensation stage. This discriminator is called the "fast" discriminator because the rise time of the first stage pulse is very fast and occurs within 30 ns of the particle entering the SSD. Another settable discriminator detects the output level of the last stage of the shaping chain. This discriminator is called the "shaped" discriminator because it senses the level of the Gaussian-shaped energy pulse. The full scale range of an ion channel is 3 MeV, and the range of an electron channel is 1 MeV.

The digitization of the peak amplitude of the energy pulse is accomplished using the time-to-peak method. A timer is started when the fast discriminator fires and the ADC is commanded to convert after 2.4 µs, provided that the shaped discriminator has also fired. All 12 electronics channels are processed independently, and the digital electronics can collect energy values at an aggregate rate of 250,000 events per second. However, each individual channel is subject to pile-up due to the finite shaping time of the analog electronics, so an individual channel is limited to a rate of $\sim$50, 000 particles per second before pile-up becomes significant. The ADC also digitizes the shaped pulse at its baseline value, immediately after the fast discriminator fires, and this pulse shape provides an indication of whether pile-up is occurring.

3.4 EPS Time-of-Flight Subsystem

The EPS time-of-flight (TOF) system consists of the TOF chamber, the anode array, and the TOF electronics. A separate HV subsystem provides the required voltages to the TOF system. A description of the various components of the TOF system follows.

The TOF chamber is a cylindrical housing roughly in the shape of a hockey puck. It consists of a front entrance foil, a back exit foil, secondary electron electrostatic deflection surfaces, and a micro-channel plate assembly. The flight path is 6 cm long.

The foils serve two functions: (1) secondary electrons are typically produced when an incoming ion penetrates a foil. The secondary electrons are steered into the micro-channel plates by the deflection system and produce a timing pulse. (2) The foils filter visible and UV light, which can cause background counts in the SSDs and MCPs. The front foil consists of a polyimide substrate coated with aluminum on each side. The substrate is supported by a 40-line-per-inch stainless steel mesh. Secondary electrons from the entrance foil are used to produce the Start timing signal. The back foil is similar in construction to the front foil, except that the metallization on each side of the polyimide is palladium rather than aluminum. Secondary electrons from the exit foil are used to produce the Stop timing signal.

The electrostatic system consists of a focusing lens placed along the perimeter of the TOF chamber, just inside of the foils that rim the chamber. The lens serves to collect and accelerate the low-energy secondary electrons towards the center of the chamber. As the secondary electrons approach the center of the chamber, they come under the influence of a deflection cone, which bends the electrons by approximately 90° into the MCP.

The micro-channel plate assembly contains a 40-mm active area matched pair of micro-channel plates in a chevron configuration. The plates have 10-μm pores and a 60-to-1 thickness-to-channel diameter ratio. The plates operate in the 2,000–2,400 V range and nominally draw about 20 μA. The gain of the plates is $\sim 5 \times 10^6$ when operated at 2,200 V.

Located directly behind the MCP assembly is an anode array that collects the electron clouds exiting the back of the MCP. Because a single MCP stack is used for both Start and Stop timing signals, a split anode configuration is required. A segmented anode array with six wedge-shaped anodes placed along the portion of the MCP perimeter proximal to the entrance foil collects the electrons that produce the Start signal. A single arc-shaped anode placed along the portion of the MCP perimeter proximal to the exit foil collects the electrons that produce the Stop signal. The Start anode is segmented in order to obtain the direction of the primary particle.

The signals from the single Start anode and the six Stop anodes are connected to the TOF board by coaxial cables. The six Start signals first pass through 50-Ω radio frequency (RF) splitters. One output from each splitter is fed directly to a charge amplifier and discriminator stage. If the signal is above the discriminator threshold, it will fire, indicating which one of the six directional sectors the particle has entered. The other output of each of the splitters is fed to a six-to-one RF combiner. The combiner produces a single Start signal, which is then amplified and input to a constant fraction discriminator (CFD).

The Stop signal also passes through a 50-Ω RF splitter. One output of the splitter is amplified and stretched and fed to two leading-edge discriminators. The thresholds of the two discriminators are set so that crude particle identification can be accomplished based on MCP pulse height. The other output of the splitter is amplified and input to a constant fraction discriminator.

A TOF chip designed at The Johns Hopkins University Applied Physics Laboratory (APL) accepts the Start and Stop pulses from the CFDs and produces a digital output representing the time interval between the Start and Stop pulses. The TOF is adjustable in terms of the time quantization level; typically 325 ps is used. The TOF range is also programmable; typically 330 ns is used.

3.5 EPS High-Voltage/Bias Subsystem

The high-voltage and solid-state detector bias supplies reside on a common circuit board. The high-voltage power supply (HVPS) provides voltage to the secondary electron deflection system and to the MCP assembly. The supply has four taps: (1) −2,900 V for the deflection cone that bends the secondary electrons into the MCP, (2) −2,600 V for the front and back foils, (3) −2,100 V for the front of the MCP, and (4) −100 V for the back of the MCP. The first two taps draw essentially no current. The load of the MCP is placed across the third and fourth taps and is nominally 20 μA, although this load varies somewhat as a function of particle flux, MCP gain, and temperature. As the gain declines on the MCP as charge is extracted over the course of the mission, it will be necessary to raise the voltage across the plates. The power supply has the capability to raise the voltage across the plates, to a maximum of 2,600 V, in 15-V increments. The voltage on the first and second taps is regulated by high-voltage zener diodes so that the second tap is always maintained at 500 V above the MCP front voltage, and the first tap is maintained at 800 V above the MCP voltage.

The high voltage is developed using a standard high-voltage transformer and Cockroft-Walton multiplier. The transformer is driven by two metal-oxide semiconductor field-effect transistors (mosfets), each operating for half of a cycle. The bi-phase drive clocks for the mosfets are generated by an Actel FPGA and are adjustable in frequency. This adjustability allows the drive frequency to be modified in-flight to compensate for any variations in the resonance in the circuit due to aging of components. The supply has an efficiency of about 25%.

Filtering of the approximately 30 V of ripple on the high-voltage tap connected to the front and back foils is necessary because this AC component capacitively couples to the SSDs as a result of the close proximity of the back foil to the SSD array. When unfiltered, the coupling produces a false signal in the detectors equivalent to 600 keV and at the same frequency as the high-voltage drive frequency. Two stages of filtering are required in order to reduce the coupling to a negligible level.

The bias supply provides voltage to the SSDs. Approximately −180 V is required to deplete the detectors fully. The detector array typically draws about 200 nA in a low-flux environment. The supply is capable of providing up to −240 V at 4 μA. This voltage is fully programmable between −6 and −240 V in 1-V increments. A single stage of filtering is provided at the output of the bias supply. Each of the 24 detectors has its own bias resistor. Individual bias resistors provide a certain amount of isolation between the detectors, so that if a particular detector draws excessive current (or completely shorts) the other detectors will not be severely affected. A fully shorted detector will cause the bias voltage to droop to about −100 V, but the detectors are oriented so that particles strike the junction side of the detector, so the detector does not need to be fully biased to operate. The noise level will increase somewhat and the energy range for electrons will decrease, but otherwise energy system operation will be unaffected.

The bias supply voltage is developed using a standard Cockroft-Walton multiplier, but a resonant inductor circuit is used in place of a transformer. The supply has an efficiency of about 10%. The inductor is driven by a transistor, which is driven by a clock developed within the gate array. This arrangement allows the drive frequency to be modified in-flight to compensate for any variations in the resonance in the circuit due to aging of components.

Both the HV and bias supplies have housekeeping (HSK) circuits to measure voltage and current. In the case of the bias supply, voltage is not measured directly. The HSK circuit actually measures current and power. The voltage can be derived from these two readings. The four parameters are read by a high-speed ADC 1,000 times per second. The readings are passed to the software over the bus for subsequent reporting in the telemetry. Additionally, the gate array has the capability to compare the HV voltage reading to a preloaded limit value and can shut down the supply instantaneously if the limit is exceeded. This feature is useful if a single event upset (bit flip) occurs in the DAC that sets the operating voltage of the supply. A bit flip could potentially cause the voltage to be set at a dangerously high level. The hardware protection circuit can detect this condition immediately and prevent damage to the supply and sensor.

3.6 EPS Digital Processing Subsystem

The digital processing subsystem is responsible for all the low-level command and control functions performed within the instrument. It has interfaces to all subsystems, including the FIPS and EPS sensors and front-end circuitry, as well as the Event Processor Unit (EPU) and flight software. It is based on two triple-voted radiation-hardened Actel SX72 gate arrays, along with associated external memories and peripheral components.

One of the major functions of this subsystem is to accept and act on commands from the flight software. The hardware command registers are made available to the software via memory-mapped I/O. Table 7 shows a subset of the commands supported by the digital processing subsystem. The software sets and clears the command bit as necessary, and commands in Command Words A and B are acted on immediately. Commands in Parameter Words A, B, and C are acted upon simultaneously, but only when a separate Sent Parameter command is executed.

Another major function is serving as the interface to the FIPS sensor. FIPS connects to the Digital Processing board via a single 37-pin cable that carries power, command, data, and temperature lines. The digital processing board contains field-effect transistor (FET) switches that allow DC power to be switched on or off to the sensor. All commands to control operation within FIPS are sent over an inter-integrated circuit (I^2C) serial interface. Additionally, housekeeping values collected by an ADC and gate array in FIPS are read over the I^2C interface. When FIPS senses a low-energy (plasma) particle, it gathers the pertinent information relative to the event, such as TOF and position, and forms the information into an event packet. A high-speed low-voltage differential interface supports transfers of up to 20,000 event packets per second. The digital processing subsystem stores these packets in a deep First In, First Out (FIFO) buffer. The output of the FIFO is mapped to the address space of the EPU, so that the software can access the event packets directly.

The Digital Processor controls the EPS sensor, which is contained in the same housing, via the command registers already described. Additionally, the digital processing electronics manage all tasks related to EPS valid event logic and data collection. A unique event packet format has been defined for each type of event detected, i.e., electron, high-energy ion, low-energy ion (TOF only), and diagnostic (energy only). All event packets, regardless of type, are loaded into a deep FIFO to await further processing by the flight software.

A number of valid event modes and features special to EPS are supported by the digital processor:

1. The width and position of the timing window used to determine the energy and time-of-flight coincidence is adjustable. This adjustment allows the window to be narrowed in high-flux environments to minimize the detection of accidental events.

2. One mode allows only high-energy ion events, those that satisfy the full triple coincidence criteria (energy, Start pulse, Stop pulse) to be accepted. Ion events with only a Start and Stop pulse (i.e., low-energy ion events) are rejected. This discrimination is useful in high-flux regions when the number of low-energy ion events can dominate and may cause the more important high-energy ion events to be excluded from processing.
3. When more than one energy channel fires within a coincidence window, the event packet is flagged to indicate that there may be ambiguity in the classification of the event.
4. The occurrence of multiple energy events on an individual channel within 12 μs can result in pile-up on the channel. The digital processing subsystem can be commanded to detect this condition and reject the event if this condition occurs.
5. Particles striking the SSDs that have been coated with aluminum are treated as electrons if the energy signal is in anti-coincidence with a Stop pulse.
6. Because electrons are expected to dominate the particle population at Mercury, the digital processor can discard a programmable fraction of electron events that occur, ensuring that ion events are not excluded from processing.

3.7 EPS Mechanical and Thermal Design

As with other instruments on MESSENGER, the APL-developed stackable card slice technology was used in the design and construction of the EPS sensor components. A unique characteristic of the EPS design was that the sensor head (i.e., the collimator, TOF chamber, and SSD array) was incorporated into a single card slice. Figure 10 (left) shows the card stack and the thermal radiator. The radiator is attached to the card stack with NuSil CV2943. Figure 10 (right) shows the sensor's mounting location on the spacecraft.

The EPS instrument thermal design required that electronics be maintained between −35° and +35°C operating and −45° to +45°C nonoperating temperature. Additionally, the temperature of the SSDs was required to be kept below −5°C for all portions of the orbit except at the subsolar point. Because the spacecraft could not support these temperature

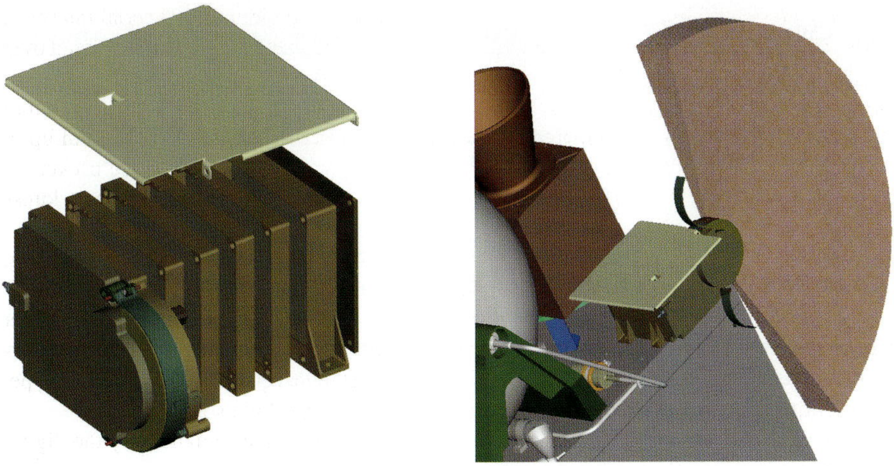

Fig. 10 (*Left*) Exploded view of the sensor, showing the six boards in the electronics stack and the radiator attached to one side of the stack. Each board is 10 cm on a side. (*Right*) The EPS sensor mounted on the spacecraft top deck next to the star trackers. The look direction is anti-sunward. The FOV is 160° in elevation and 12° in azimuth

Table 7 The command bits in hardware command register A

ID	Name	Description Command Word A	State = 1 Default state	State = 0
0	CMD-0 Ion Power Switch	Controls whether the large or small ion pixel detector preamps are powered	small pixels	large pixels
1	CMD-1 Electron Power Switch	Controls whether the large or small ion pixel detector preamps are powered	small pixels	large pixels
2	CMD-2 nReset Ion Latchup	If any of the 6 ion ADCs latch up, they will be powered down automatically.	inactive	active
3	CMD-3 nReset Electron Latchup	If any of the 6 electron ADCs latch up, they will be powered down automatically.	inactive	active
4	CMD-4 nTOF Master Reset	Master reset for TOP chip	inactive	active
5	CMD-5 Half Period		active	inactive
6	CMD-6 Tap Select 1		active	inactive
7	CMD-7 Tap Select 2		active	inactive
8	CMD-8 Exclusion Window		active	inactive
9	CMD-9 Disable A0	Design Change 6-13-02	active	inactive
10	CMD-10 Disable A1	Design Change 6-13-02	active	inactive
11	CMD-11 FIPS ±5V Power Swt	Enables FIPS ±5V power	active	inactive
12	CMD-12 FIPS ±15V Power Swt	Enables FIPS ±15V power	active	inactive
13	CMD-13 EPS Event FIFO Reset	Clears EPS Event FIFO-A	active	inactive
14	CMD-14 FIPS Reset	Resets entire FIPS sensor electronics. Minimum duration should always be greater than 1 s	active	inactive
15	CMD-15 FIPS Event FIFO Reset	Clears FIPS Event FIFO-B	active	inactive

There are five hardware command registers supported by the digital processing subsystem. This table shows the command bits in command register A. In general, these commands control EPS operation. Control of FIPS is not accomplished by this bitmapped hardware register approach but rather by serial commands sent to the FIPS sensor directly from the EPU using the I^2C interface

specifications with a sensor that was thermally connected to the deck, the decision was made to isolate EPS thermally from the spacecraft with its own thermal control. A silver Teflon-coated radiator and thermostatically controlled heaters provide the basic, passive, thermal control system for EPS. The radiator points to the zenith, away from the planet, minimizing the planet heat loading onto the instrument. Blankets for multilayer insulation (MLI) cover the rest of the instrument except the aperture.

3.8 EPS Calibration

For an instrument of this type, the typical calibration procedure consists of stimulating the instrument with energetic particles, first from radiation sources, and then from accelerator beams (a fair approximation of a delta function), and recording the response of the different channels. For EPS we must scan the following variables:

- (S) Species and mass: e, H, He, O, Ar (proxy for heavier species, e.g., Fe)
- (E) Energy: 1–5 MeV (~40 energies for 10 points/decade)
- (θ) Polar angle: $\pm 10°$ ($12°$ nominal FOV)
- (ϕ) Azimuthal angle: $\pm 90°$ ($160°$ nominal FOV)

Note that even given this level of discreteness, complete characterization to the level of establishing the sensor transfer function $T_{kj}(S, E, \theta, \phi)$ for all channels requires 5 masses, 40 energies, 21 polar angles, and 180 azimuthal angles, for a total of 756,000 calibration points. Clearly, it is not possible to run this many beam values without significant infrastructure and automation. Therefore, we are dependent to a substantial degree on the separability of the transfer function. An important goal of the characterization and calibration efforts is, in fact, to establish the degree of separability. Simulations play an important role in establishing expectations for the sensor.

The amount of calibration performed prior to delivery of the flight unit was severely limited by schedule and technical constraints. However, a complete engineering model was fabricated, and calibration efforts are underway with this unit. Additionally, flight unit calibration will be performed during the long cruise phase, which involves planetary flybys for gravity assists. The calibration efforts are focused on the following areas:

1. Collimator performance and geometric factor. The size and number of collimator holes, as well as the area of the detectors, defines the geometric factor G of the sensor. The many-holes collimator design minimizes the scattering of ions and electrons at the collimator while restricting the FOV of the instrument. Side lobes exist in this collimator design. Simulations using GEANT4 software were performed to assess the collimator performance for suppressing electron scattering effects. Beam runs were performed to obtain quantitative information on the scattering properties of the collimator.
2. Ion measurements
   a. Electron optics. An important issue is the dispersion in the arrival times of the secondary electrons as a function of foil position and the angle of emittance of the secondary electrons from the foil. Such dispersion adds to the error in the measurement of time of flight.
   b. Ion energy losses. As ions traverse the TOF section of EPS through the front foil, back foil, and dead layer of the SSD (~550 Å), and eventually stop and deposit their total energy in the SSD, they will lose energy and scatter depending upon the ion's initial energy and mass and the medium through which they pass.
3. TOF measurements. When ions penetrate through the front foil, a distribution of ion velocities is created. At low energy (tens of keV), ions lose significant amounts of energy and scatter significantly when going through the front foil. However, at higher energy (hundreds of keV), the TOF spreads are mostly consequences of the uncertainties in the TOF measurement from both the electronics jitter (~1.5 ns) and secondary-electron dispersion in the TOF optics.
4. Total energy measurement. If an ion has sufficient energy left once it transits the front and back foil, it ends up in the SSD. Depending on the ion's final energy and mass when it reaches the SSD, it can penetrate through the dead layer of the SSD and produce an electronic signal to be measured.
5. TOF versus energy. Coincident measurements of a particle's energy and TOF allow the species of the particle to be determined. Calibration efforts in this area concentrate on understanding artifacts in the data due to crosstalk, high background, and other factors.

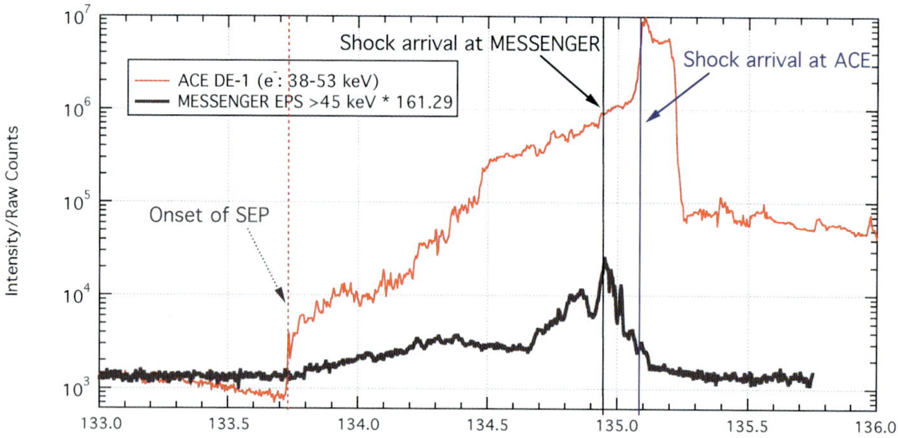

Fig. 11 EPS and ACE EPAM electron rates versus time during a solar energetic particle event that occurred during days 133 to 136 of 2005. The DE-1 channel is the lowest energy electron channel in the D sensor on EPAM

6. Efficiencies. The efficiency for detection of an ion within the SSDs is roughly ~100% (except at the very lowest energies where energy straggling can position the energy below the low-energy threshold). The efficiency for obtaining a TOF measurement is most highly dependent on the efficiency of generating secondary electrons in both the front and the rear foils.
7. Electron measurements. The EPS electron measurement strategy depends on the use of the aluminum flashing on the electron SSD. We therefore need to understand the effect of that flashing on both the ion and electron measurements within the electron SSDs.

3.9 EPS Early Flight Data

Early in MESSENGER's cruise phase the EPS sensor was on and in diagnostic mode, during which events from both the ion and electron SSDs are collected. No coincidence criteria are applied in this mode. If the particle's energy simply exceeds the detector energy threshold, the event will be collected and analyzed. By chance, a very large solar energetic particle (SEP) event occurred during this period and EPS was able to record the event. Figure 11 shows plots of the aggregate EPS electron channel rate, as well as data from the Electron, Proton, and Alpha Monitor (EPAM) instrument on the Advanced Composition Explorer (ACE) spacecraft.

4 EPPS Software

4.1 EPS and FIPS Event Data Collection

The EPPS software obtains event data through two separate FIFOs, one for EPS and one for FIPS. Event packets consist of a number of event words. The total number of words in a packet depends on the type of event. The first word within each event packet contains a synchronization pattern and an identification (ID) that specifies the event type. The synchronization pattern is used to detect the beginning of a new packet.

Depending on the event type, the EPPS hardware circuitry places event data into the appropriate FIFO as a group of words. FIPS events are placed into the FIPS FIFO as a group of seven words per packet at rates of up to 20,000 events/s. The FIPS format includes a 1-bit header that identifies the event as a proton event or a nonproton event; an 11-bit TOF value; as well as wedge, strip, and zigzag values (each 12 bits in size) from the position system.

EPS event packets are placed into the EPS FIFO as five-word groups (electron events), four-word groups (low-energy ions), seven-word groups (high-energy ions), or six-word groups (diagnostic events). The software throughput rate for FIPS charged-particle event processing as well as for EPS event processing is 5 kHz.

The software classifies events on the basis of the ID within the first word of each packet. Spectra or histograms are generated for the various event types by using lookup tables to determine the corresponding bin within each spectrum to increment. These lookup tables are accessed on the basis of some combination of the energy, time of flight, and/or mass-to-charge ratio.

4.2 FIPS Event Processing

The software maintains a mass distribution spectrum for the FIPS sensor. This spectrum consists of a collection of 256 bins (each 32-bits wide) that count the number of events corresponding to each M/Q value. In addition, the software maintains a set of five-element energy spectra for FIPS. Each FIPS spectrum corresponds to a specified M/Q range and consists of 64 32-bit bins. For events whose M/Q values fall within one of the selected ranges, an energy value is computed and used to determine which bin within the corresponding spectrum to increment. The spectra are accumulated over an integral number of voltage scans, after which they are compressed and output in telemetry. Two raw FIPS events are recorded at each scan level and reported in telemetry. Thus, up to 128 FIPS pulse-height analyzed (PHA) events may be collected per scan.

4.2.1 FIPS Integration Intervals

The basic integration intervals for FIPS science data are determined by the contents of the FIPS scan tables. The smallest integration interval corresponds to the dwell period stored in the scan table entries. Events are accumulated during each nonzero dwell period and are used to generate the various FIPS spectra. The sum of the settling times and dwell times in all of the table entries determines the duration of a single scan. Ten consecutive scans constitute one scan sequence. Some of the FIPS data items are accumulated over single scans, while others are accumulated over an entire scan sequence. Nominally, a scan and a scan sequence correspond to 65 s and 650 s, respectively.

4.2.2 FIPS Mass Distribution and Element Energy Spectra Processing

The EPPS software maintains a number of 68-element tables that are used in performing deflection system high-voltage (DSHV) scan sequences. Each of the 68 table entries has the following format:

Settling time (8 bits)	Dwell time (8 bits)	DSHV level (16 bits)	TOF threshold (16 bits)

Each table entry specifies the settling time, dwell time, DSHV voltage level, and proton threshold to be used at the corresponding step within a scan. The settling and dwell time fields are expressed in 10-ms time units. The TOF threshold deserves special mention. For plasma instruments, protons dominate the particle population. As a result they can also dominate the telemetry bandwidth of the sensor. The TOF threshold entry in the table is used to determine whether the incoming particle is a proton or heavier species. If the particle's TOF is less than the TOF threshold, the particle is identified as a proton and flagged as such. The FIPS digital logic can be commanded to send only 1 in N proton events to the event FIFO, ensuring that heavier species are not excluded from the raw event stream.

One complete scan is made up of 64 voltage steps and 4 ramp-up steps. Ten consecutive scans correspond to one "scan sequence." Table entries for which the dwell time is zero correspond to ramping the DSHV level to its initial value for the scan. Events are disabled for these ramp-up steps. Four out of the 68 table entries correspond to ramp-up steps. The remaining 64 steps correspond to DSHV settings at which events are enabled and processed.

At each step, the event FIFO is disabled and reset during the settling period for that step. After the settling period expires, event generation is enabled and events are processed for the indicated dwell period. At the conclusion of each dwell period, the EPPS software performs the following actions:

- Places the FIPS instrument in "stand-by" mode to stop the generation of event data.
- If this is the conclusion of a 10-scan sequence, signals the FIPS task to
  1. Compress and output the FIPS data in telemetry
  2. Clear the FIPS data product buffers.
- If this is not the conclusion of a 10-scan sequence, signals the FIPS task to
  1. Set the DSHV to the next voltage level as indicated by the voltage step table
  2. Set the proton event threshold as indicated by the threshold table
  3. Set the settling period to the value indicated in the corresponding table entry.
- Flushes the FIPS event FIFO and places the FIPS instrument into "Manual" mode to start the generation of event data. Sets the event collection interval (i.e., dwell time) for the next step.
- Processes events until this dwell time expires.

As events are captured during a particular step within a voltage scan, the following processing is performed for events that are extracted from the FIPS event FIFO:

1. The 9 most significant bits of the TOF value are concatenated with the 6-bit step number (0 through 63) to form a 15-bit index that identifies one of 32,000 entries in a lookup table. Each element within the table is an 8-bit value that specifies the bin that should be incremented within the mass distribution histogram.
2. The index obtained from the lookup table is used to identify which of the 256 bins to increment within a mass distribution histogram. Each of the 256 mass distribution histogram elements (bins 0–255) is maintained as a 32-bit value.
3. The bin number obtained from the lookup table is also used to determine whether the associated event falls within one of five different energy spectra ranges. If so, the current voltage step number is used to identify and increment one of the 64 elements in the corresponding energy spectra.

4.2.3 FIPS Proton and Heavy Ion Velocity Distribution Processing

The EPPS software generates an 8×8 proton velocity distribution function based on the proton events that are detected during the first 64-step voltage scan of each 10-scan sequence.

For each proton event processed, an X and a Y value are computed using the same wedge, strip, and zigzag position values.

The values $X/8$ and $Y/8$ are used as the row and column within the velocity distribution matrix that corresponds to the element or bin to be incremented. An 8×8 proton velocity distribution function is also generated for all proton events that are detected during scans 2 through 10 of each 10-scan sequence. This velocity distribution corresponds to a 9-scan accumulation interval.

The EPPS software generates separate velocity distribution functions for nonproton events that correspond to the five energy vector ranges. These five velocity distribution functions are accumulated over each 10-scan sequence. If a nonproton event maps to one of the five energy vector ranges, then the same $X/8$ and $Y/8$ values are computed for the event (as described earlier), and the corresponding bin within the selected matrix is incremented.

4.2.4 FIPS Test Scan Mode

The EPPS software system supports a command to place the FIPS instrument into a "test scan" mode in which a test scan sequence table is used in place of the normal stepping table. A test scan sequence uses the onboard pulse generation capability of the FIPS instrument. It can be used to compare the instrument response after launch to its response in ground testing and can help in identifying TOF drift, or wedge, strip, zigzag offset, or gain drift.

This Test Scan Mode is identical to the Normal Scan Mode with the following exceptions:

1. The Test Mode Stepping Table is used in place of Normal Mode Stepping Table.
2. During each step Dwell time period for data collection, a number of test events are generated using the pulse trigger command. An auxiliary 68-entry table contains the number of events (N) to trigger and the start-stop delay (D) to use in triggering the event. Each of these is an 8-bit value, so each table entry contains a pair of bytes (N, D). The count is in the high-order byte.

Scan timing and the packet generation rate are the same as for a normal scan sequence. All data products are created in the same manner as for normal scan sequences.

4.3 EPS Event Processing

4.3.1 EPS Integration Intervals

The accumulation, formatting, and reporting of EPS event and housekeeping data are synchronized to two different time intervals. These are referred to as the N1 and N2 time intervals. The N1 time interval is nominally 300 s and is an integral multiple of the N2 time interval. The N2 time interval has a nominal value of 30 s. Either or both of the intervals can be changed via command. Among the data collected at these intervals are the 35 hardware rate counter registers. To prevent these registers from overflowing, each of the intervals N1 and N2 is subdivided into 10 equal-length subintervals. Thus the basic integration period corresponds to N2/10 s. The EPS event FIFO is flushed at the conclusion of each N2/10 interval to keep the events that occur during different time intervals separate.

Hence all data products for the EPS subsystem are collected at the N2/10-s rate. Values that are reported every N2 s are produced by summing the 10 consecutive N2/10 samples. Values that are reported every N1 s are produced by summing over as many N2/10 subintervals as there are in the N1-s time period.

4.3.2 EPS Electron Event Processing

The peak energy value minus the baseline energy is used as an index into a table containing 4,096 entries. Each entry in this table is an 8-bit value that ranges from 0 to 7 and indicates which of the eight histogram bins should be incremented. If the baseline energy exceeds the peak energy, then zero is used as the index. Six histograms are maintained for electron events—one for each of the six electron channels (0 to 5) corresponding to the six look directions. The 3-bit channel number, contained in word 4 of the event packet, identifies the electron channel. Electron events are discarded under the following conditions:

1. If pile-up checking has been enabled via command and the baseline energy specified within the packet exceeds a specified threshold, which has been set via command.
2. If multiple hit checking has been enabled via command and the multiple hit flag is set within word 4 of the electron event packet.
3. If the 3-bit electron channel number in word 4 of the electron event packet does not contain a value in the range 0 to 5.

Separate counters are maintained for these three conditions and record the number of electron events that were discarded due to each condition.

The N1-s accumulation interval is subdivided into 10 equal-length subintervals. At the conclusion of each N1/10-s subinterval, the contents of the six histograms are added into the corresponding histogram buffer, which is used to accumulate the N1-s histograms. For each detector, there is also defined a pair of "super bins," each of which corresponds to some subset of the eight bins in a histogram. The contents of the bins that comprise each super bin are added together at the end of each subinterval to produce the two super bins. These two super bins are computed for each of the 10 subintervals and thus require 10 two-bin buffers for each of the six detectors.

4.3.3 EPS Low-Energy Ion Event Processing

The 11-bit TOF value is appended to the 2-bit disc-1/disc-0 identifier to produce a 13-bit number used as an index into a table containing 8,192 entries. Each contains a value that ranges from 0 to 15 and indicates which of the 16 histogram bins should be incremented. The 6-bit start segment number in word 3 of the event packet identifies the start segment.

The leftmost or most significant nonzero bit within the start segment field determines which start segment (0 to 5) is used for the event. A 16-bin histogram is maintained for each of the six start segments. This procedure yields a total of six histograms for the low-energy ion events. Each of the 16 elements within a histogram is a 32-bit integer, which is incremented for each event that maps to that particular bin.

4.3.4 EPS High-Energy Ion Processing

High-energy ion packets are discarded under two conditions:

1. Pile-up checking has been enabled via command and the baseline energy within the packet exceeds a specified threshold, which is set via command.
2. Multiple-hit checking has been enabled via command and the multiple hit flag is set within word 5 of the event packet.

A count of discarded events is maintained, but no further processing is performed for discarded events. For each event packet that is not discarded, the 12-bit peak energy value in word 1 of the event packet is used as an index into a 4,096-element array in which each element contains the 16-bit address of a row within a "matrix-rate" box definition table. Each row contains a number of "break-points" and has the following format:

N	TOF 1 low	TOF 1 high	bin 1	•••	TOF *n* low	TOF *n* high	bin *n*
16 bits	12 bits	12 bits	8 bits		12 bits	12 bits	8 bits

N specifies the number of breakpoints contained in the row. The row whose address is contained in the energy pointer table entry identified by the peak energy value within the event packet is scanned to locate the first breakpoint whose TOF-low and TOF-high values bracket the 12-bit-TOF value from the event packet. The bin number contained in the low 6 bits of the bin number field of the corresponding breakpoint identifies which one of 22 bins should be incremented in the histogram. A bin number of 0 corresponds to a background bin. If the TOF value for the event is found not to fall within any of the breakpoints, then the event is treated as a background event.

The identity of the histogram to be updated is determined by the 3-bit ion channel field from word 4 of the high-energy ion event packet. The allowed range for this channel number is 0 through 5. Thus there are six separate histograms. If the channel number falls outside of this range, the event is discarded.

4.3.5 EPS PHA Event Processing

The selection of which of the processed PHA events should be telemetered to the ground is decided by a Rotating Priority Scheme. A separate buffer is maintained for the PHA data that are output in the high-, medium-, and low-priority telemetry packets. The PHA data appearing in the high- and low-priority telemetry packets are accumulated over the N1 interval. The N2 integration interval is used for the PHA data in the medium-priority telemetry packet. The maximum number of PHA events saved per integration period is 10, 20, and 300 for the high-, medium-, and low-priority telemetry packets, respectively.

PHA events are distributed among the buffers in a round-robin fashion: the first detected event is stored in the high-priority packet buffer, the next event is stored in the medium-priority packet buffer, and the next event is stored in the low-priority packet buffer, etc. Each event allocated to a particular buffer is simply stored in the next slot within the buffer until the buffer fills up. Thereafter, a rotating priority PHA replacement scheme is used to decide which events may be displaced from the filled buffer.

All PHA events are assigned to a priority group based on the event type, 1, 2, or 3 (electron, high-energy ion, or low-energy ion), and the bin number to which the event maps. The assigned group numbers are stored in a lookup table that is accessed via the type/bin number pair. Eight such priority groups have been defined for the EPS system. An eight-element priority array is maintained for each event type. Group numbers are held in the array elements in priority order. That is, the first element contains the group number of the highest priority group, the second element contains the group number of the next-lower-priority group, etc. Hence the priority of a particular group can be determined by finding which array element contains that group number.

As each PHA event is stored in its respective buffer, the assigned priority and group number are stored along with each event. As long as each buffer contains a vacant slot, the next detected event is simply stored in the vacant slot. Once the buffer becomes full, the assigned PHA group priority for a new event will determine its relative ability to displace events that are already in the buffer. For the data in the high- and medium-priority telemetry packets, the lowest-priority event already in the buffer will be identified. This lowest-priority event will be displaced by the new event if the priority of the new event is higher.

A record is maintained of the frequency of occurrence of each group during each integration interval. The most frequently occurring group number is used at the conclusion of the interval to "rotate" the PHA priorities by pushing the most frequently occurring group to the bottom of the priority list. This rotated priority list is then used in determining the priority of PHA events in the next integration interval.

To shorten the time required to identify a candidate for possible replacement, rather than searching the 300-element PHA buffer associated with the low-priority telemetry data, a modulo 300 index is used to map directly to the candidate. If the priority of the new event is higher than the priority of this candidate, a replacement is performed and the index is incremented to point to the next candidate.

4.3.6 EPS Rate Counter Processing

4.3.6.1 Hardware Rates. The EPS system includes 35 rate counters that record in hardware the number of events of various types that have occurred. Each counter is mapped into a location within memory page 12 of the EPS processor. Three of the counters are 24-bits wide, while the remaining 32 are each 16-bits wide.

The EPS flight software system processes the rate counters at N2/10-s time intervals. At the conclusion of each N2/10-s rate counter accumulation, the software halts the operation of the counters and, after allowing the counters to settle, adds the contents of all 35 counters into the N2-rate data buffer. The counter operation is then restarted and allowed to proceed for the next N2/10 interval.

The counters are accumulated over ten consecutive N2/10 intervals to produce the N2-s rate data. Sampling the counters at the N2/10-s rate is meant to prevent the hardware rates from overflowing. The value read from each counter is added into a 32-bit word so as to obtain the cumulative count value for the entire N2-s time interval without overflowing. The content of the N2-rate buffer is summed into the N1-rate buffer to obtain the rate data that are accumulated over the N1-s time interval. N1 is a multiple of N2.

4.3.6.2 Software Rates. In addition to the 35 counters that correspond to actual hardware counters, the EPPS software maintains a group of seven "software rates." These are counters that are maintained by the software and are updated to reflect the number of events processed or rejected by the software. Table 8 identifies each EPS software rate.

The software rates are incremented on the basis of the type of the event and on whether the event is discarded due to a pileup or a multihit condition. The discarding of events due to each of these conditions can be enabled or disabled via command.

4.4 Software Status Reporting and Monitoring Functions

4.4.1 Status Data Integration Interval

Certain housekeeping and status values for both EPS as well as FIPS are collected and reported at intervals that correspond to the instrument status interval. This interval is nominally 2 s but can be changed via command. The data items contained within the instrument status packet are defined below in the status packet description

Table 8 EPS software rates

	Name of software rate counter	Description
1	Electron count	Number of events processed during the accumulation interval
2	High-energy ion	Number of events processed during the accumulation interval
3	Low-energy ion	Number of events processed during the accumulation interval
4	Electron pileup	Number of events discarded due to pileup condition
5	Electron multihits	Number of events discarded due to multiple hits
6	High-energy ion pileup	Number of events discarded due to pileup condition
7	High-energy multihit	Number of events discarded due to multiple hits

4.4.2 FIPS High-Voltage Power Supply Monitoring

The EPPS software includes the capability to monitor the MCP and post-acceleration high-voltage power supply levels. When this feature is enabled via command, the supply is shut down if its voltage level exceeds a limit previously set by command.

4.4.3 FIPS Emergency High-Temperature Scan Mode

If either one of two deflection system temperatures exceeds a limit specified by command and the FIPS high-temperature alarm monitor has been enabled via command, then the FIPS system is placed into the high-temperature mode at the conclusion of the current scan sequence. The post-acceleration level is set to a previously specified value.

While in the high-temperature scan mode, the two temperatures are checked at the end of each scan sequence to see if they both have fallen below a specified hysteresis value. If both temperatures are below this value, then the FIPS scan mode reverts to its state prior to entering the high-temperature mode and the post-acceleration level in effect prior to entering the high-temperature mode is reinstated.

4.4.4 FIPS Emergency High-Temperature Shutdown

If either one of two deflection system temperatures exceeds a limit specified by command and the FIPS high-temperature alarm monitor as well as the high-temperature emergency shutdown have both been enabled via command, then EPPS macro 5 is executed once. This macro is configured to turn off the deflection system voltage completely.

Acknowledgements The authors acknowledge the large number of scientists, engineers, and technicians at both the University of Michigan and The Johns Hopkins University Applied Physics Laboratory who contributed their technical expertise and skill to the successful development of the FIPS and EPS sensors. We particularly acknowledge John Cain, who was essential throughout the FIPS development process but passed away before the first FIPS results were obtained.

References

B.J. Anderson et al., Space Sci. Rev. (2007, this issue). doi:10.1007/s11214-007-9246-7
T.A. Bida, R.M. Killen, T.H. Morgan, Nature **404**, 159–161 (2000)
S.P. Christon, in *The Astronomy and Astrophysics Encyclopedia*, ed. by S.P. Maran (Cambridge University Press, Cambridge, 1992), pp. 427–430

S.P. Christon, S.F. Daly, J.H. Eraker, M.A. Perkins, J.A. Simpson, A.J. Tuzzolino, J. Geophys. Res. **84**, 4277–4288 (1979)
S.P. Christon, J. Feynman, J.A. Slavin, in *Magnetotail Physics*, ed. by A.T.Y. Lui (Johns Hopkins University Press, Baltimore, 1987), pp. 393–400
J.E.P. Connerney, N.F. Ness, in *Mercury*, ed. by F. Vilas, C.R. Chapman, M.S. Matthews (University of Arizona Press, Tucson, 1988), pp. 494–513
G. Gloeckler, Rev. Sci. Instrum. **61**, 3613–3620 (1990)
G. Gloeckler, K.C. Hsieh, Nucl. Instrum. Methods **165**, 537–544 (1979)
D.M. Hunten, T.H. Morgan, D.E. Shemansky, in *Mercury*, ed. by F. Vilas, C.R. Chapman, M.S. Matthews (University of Arizona Press, Tucson, 1988), pp. 562–612
W.-H. Ip, Astrophys. J. **356**, 675–681 (1990)
K. Kabin, T.I. Gombosi, D.L. DeZeeuw, K.G. Powell, Icarus **143**, 397–406 (2000)
P.L. Koehn, Ph.D. thesis, University of Michigan, Ann Arbor, 2002, 183 pp
J.C. Leary et al., Space Sci. Rev. (2007, this issue). doi:10.1007/s11214-007-9269-0
C. Martin, P. Jelinsky, M. Lampton, R.F. Malina, H.O. Anger, Rev. Sci. Instrum. **52**, 1067–1074 (1981)
B.H. Mauk, C.-I. Meng, in *Solar System Plasma Physics*, ed. by J.H. Waite, J.L. Burch, R.L. Moore. Geophysical Monograph, vol. 54 (American Geophysical Union, Washington, 1989), pp. 319–332
B.H. Mauk, D.J. Williams, R.W. McEntire, K.K. Khurana, J.G. Roederer, J. Geophys. Res. **104**, 22759–22778 (1999)
W.E. McClintock, M.R. Lankton, Space Sci. Rev. (2007, this issue). doi:10.1007/s11214-007-9264-5
A.E. Potter, T.H. Morgan, Icarus **71**, 472–477 (1987)
C.T. Russell, D.N. Baker, J.A. Slavin, in *Mercury*, ed. by F. Vilas, C.R. Chapman, M.S. Matthews (University of Arizona Press, Tucson, 1988), pp. 514–561
B. Schubert, M.N. Ross, D.J. Stevenson, T. Spohn, in *Mercury*, ed. by F. Vilas, C.R. Chapman, M.S. Matthews (University of Arizona Press, Tucson, 1988), pp. 429–460
R. Schwenn, in *Physics of the Inner Heliosphere I*, ed. by R. Schwenn, E. Marsch (Springer, Berlin, 1990), pp. 99–182
M.A. Slade, B.J. Butler, D.O. Muhleman, Science **258**, 635–640 (1992)
S.C. Solomon et al., Planet. Space Sci. **49**, 1445–1465 (2001)
A.L. Sprague, D.M. Hunten, K. Lodders, Icarus **118**, 211–215 (1995)
A.R. Vasavada, D.A. Paige, S.E. Wood, Icarus **141**, 179–193 (1999)
D.J. Williams et al., Geophys. Res. Lett. **24**, 2163–2166 (1997)
T.H. Zurbuchen, G. Gloeckler, J.C. Cain, S.E. Lasley, W. Shanks, in *Conference on Missions to the Sun II*, ed. by C.M. Korendyke. Proceedings of the Society of Photo-Optical Instrumentation Engineers, Bellingham, Wash., vol. 3442, 1998, pp. 217–224

Space Sci Rev (2007) 131: 557–571
DOI 10.1007/s11214-007-9270-7

The Radio Frequency Subsystem and Radio Science on the MESSENGER Mission

Dipak K. Srinivasan · Mark E. Perry · Karl B. Fielhauer · David E. Smith · Maria T. Zuber

Received: 22 May 2006 / Accepted: 15 August 2007 / Published online: 27 October 2007

Abstract The MErcury Surface, Space ENvironment, GEochemistry, and Ranging (MESSENGER) Radio Frequency (RF) Telecommunications Subsystem is used to send commands to the spacecraft, transmit information on the state of the spacecraft and science-related observations, and assist in navigating the spacecraft to and in orbit about Mercury by providing precise observations of the spacecraft's Doppler velocity and range in the line of sight to Earth. The RF signal is transmitted and received at X-band frequencies (7.2 GHz uplink, 8.4 GHz downlink) by the NASA Deep Space Network. The tracking data from MESSENGER will contribute significantly to achieving the mission's geophysics objectives. The RF subsystem, as the radio science instrument, will help determine Mercury's gravitational field and, in conjunction with the Mercury Laser Altimeter instrument, help determine the topography of the planet. Further analysis of the data will improve the knowledge of the planet's orbital ephemeris and rotation state. The rotational state determination includes refined measurements of the obliquity and forced physical libration, which are necessary to characterize Mercury's core state.

Keywords MESSENGER · Mercury · Telecommunications system · Radio science · Gravity science · Spacecraft tracking · Gravitational field

1 Introduction

The MErcury Surface, Space ENvironment, GEochemistry, and Ranging (MESSENGER) spacecraft (Santo et al. 2001; Leary et al. 2007), which will be the first to orbit the planet

D.K. Srinivasan (✉) · M.E. Perry · K.B. Fielhauer
Space Department, The Johns Hopkins University Applied Physics Laboratory, Laurel, MD 20723-6099, USA
e-mail: dipak.srinivasan@jhuapl.edu

D.E. Smith
Solar System Exploration Division, NASA Goddard Space Flight Center, Greenbelt, MD 20771, USA

M.T. Zuber
Department of Earth, Atmospheric and Planetary Sciences, Massachusetts Institute of Technology, Cambridge, MA 02139-4307, USA

Mercury, was launched successfully on August 3, 2004, from Cape Canaveral Air Force Base, FL. After a series of one Earth flyby, two Venus flybys, and three Mercury flybys, the spacecraft will complete its 6.6-year cruise and enter into a 12-hour, high-inclination elliptical orbit around Mercury (McAdams et al. 2007). MESSENGER's baseline mission plans are for the spacecraft to spend one Earth-year in orbit about Mercury (Solomon et al. 2001) and to collect and transmit back to Earth scientific data acquired by the spacecraft instrument suite (Gold et al. 2001).

The Radio Frequency (RF) Telecommunications Subsystem on the MESSENGER spacecraft has three major functions: (1) spacecraft command capability, (2) the highest possible quality and quantity of spacecraft telemetry and science data return, and (3) highly accurate Doppler and range tracking data to determine precisely the spacecraft velocity and position. The first two functions are essential to the operation of the spacecraft and the return of scientific data during the cruise to Mercury and orbital mission phase. The third function is integral to navigate the spacecraft and to accomplish the radio science objectives of the mission, which form a part of the MESSENGER geophysics investigation (Zuber et al. 2007).

MESSENGER carries a suite of seven science instruments, plus the spacecraft RF subsystem for the radio science (RS) experiments. While providing the metrics needed for the reconstructed spacecraft positional information for these instruments, the RF Telecommunications Subsystem is integral to two of the mission science investigations. The first investigation is the gravity mapping investigation on MESSENGER, which exclusively uses the RS instrument (the RF Telecommunications Subsystem). MESSENGER's tracking data will be used to determine the position of the spacecraft during both the cruise and orbital phases of the mission. The navigational solution of the spacecraft orbit uses a downlink signal that is Doppler shifted with respect to the signal frequency transmitted from the spacecraft telecommunications system. This Doppler shift provides the Earth line-of-sight velocity of the spacecraft, which changes due to forces acting on the spacecraft that include gravitational perturbations due to Mercury's mass distribution, spacecraft maneuvers, and radiation pressure. Gravitational perturbations signify spatial variations of density within the planet's interior, and a time-varying component in Mercury's gravity—revealed as a variable rotation rate—can quantify the amplitude of Mercury's libration. A second data type is the range from the spacecraft to the ground tracking station, which is derived from the round-trip time of propagation of a signal from the ground to the spacecraft and back.

The second science investigation that uses significantly the spacecraft telecommunications system is the topographic mapping of the surface. This investigation is performed in conjunction with the Mercury Laser Altimeter (MLA) instrument (Cavanaugh et al. 2007), which consists of a solid-state laser transmitter and receiver. By measuring the round-trip travel time of transmitted pulses, the MLA will provide precise measurements of the range of the MESSENGER spacecraft to the surface of Mercury. To convert the MLA's measured ranges to radii with respect to the planet's center of mass, it is necessary to know the position of the MESSENGER spacecraft to within 10 m. The conversion of radius to the desired measurement of topography requires subtracting the radius of the Mercury's gravitational potential ("geoid") at each point, which is also determined from the RS gravity field investigation. The RF and RS data that contribute to the topographic mapping investigation include the navigation data that provide precise spacecraft position information and orbit solutions, and occultation information from which the planetary radius at different positions on the planet can be derived by measuring the time that the RF signal is obscured by Mercury when MESSENGER passes behind the planet. These combined data will map the topography of the planetary surface and enable measurements of the planet's libration state. Although the MLA is designed to perform pulse timing measurements for ranges up to 1,800 km, signal/noise considerations limit the reliably detectable range to be ∼1,000 km and less, thus effectively restricting altimetric coverage to Mercury's northern hemisphere for the planned MESSENGER mission design (Cavanaugh et al. 2007). RS occultation information will be available in both hemispheres.

The following section describes the flight components of the radio system. Section 3 describes the plan for using RF range and Doppler data to derive the detailed orbit parameters that are used for both navigation and the gravity-mapping investigation. Finally, Sect. 4 summarizes how the RF data products are applied to the geophysics investigations at Mercury.

2 MESSENGER RF Telecommunication Subsystem

The MESSENGER RF Telecommunications Subsystem operates at X-Band: 7.2 GHz for uplink from ground stations and 8.4 GHz for downlink from the spacecraft. Communications are accomplished via the 34-m and 70-m antennas of NASA's Deep Space Network (DSN) stations in Goldstone, CA; Madrid, Spain; and Canberra, Australia.

Because of the inner-planet trajectory of MESSENGER, the Earth can be in any direction with respect to the spacecraft. This geometric constraint presents a significant RF design requirement in that high-gain coverage must be achievable in all directions. The antenna configuration shown in Fig. 1 accomplishes this requirement.

Two diametrically opposite-facing phased-array antennas (PAAs) provide the high-gain downlink signal. Each PAA is capable of electronically steering ±60° in the XY-plane of the spacecraft from the directions indicated in Fig. 1. Spacecraft rotation about the Y-axis in conjunction with the electronic scanning of the antenna beam provides the omnidirectional high-gain coverage (Srinivasan et al. 2005). Two fanbeam antennas provide medium-gain uplink and downlink capabilities, and four low-gain antennas (LGAs) complete the antenna suite of MESSENGER.

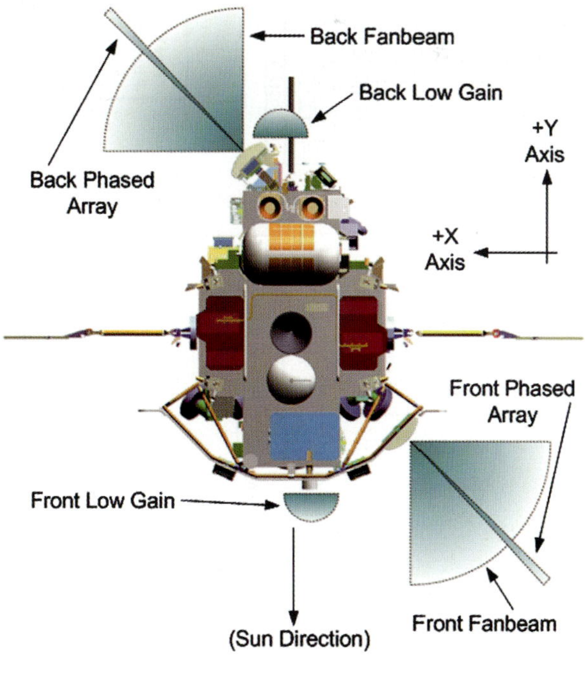

Fig. 1 The MESSENGER spacecraft antenna suite, notional patterns, and coordinate system. The $+Z$-axis is into the page. The forward and aft low-gain antennas (not pictured) point in the $-Z$ and $+Z$ axes, respectively

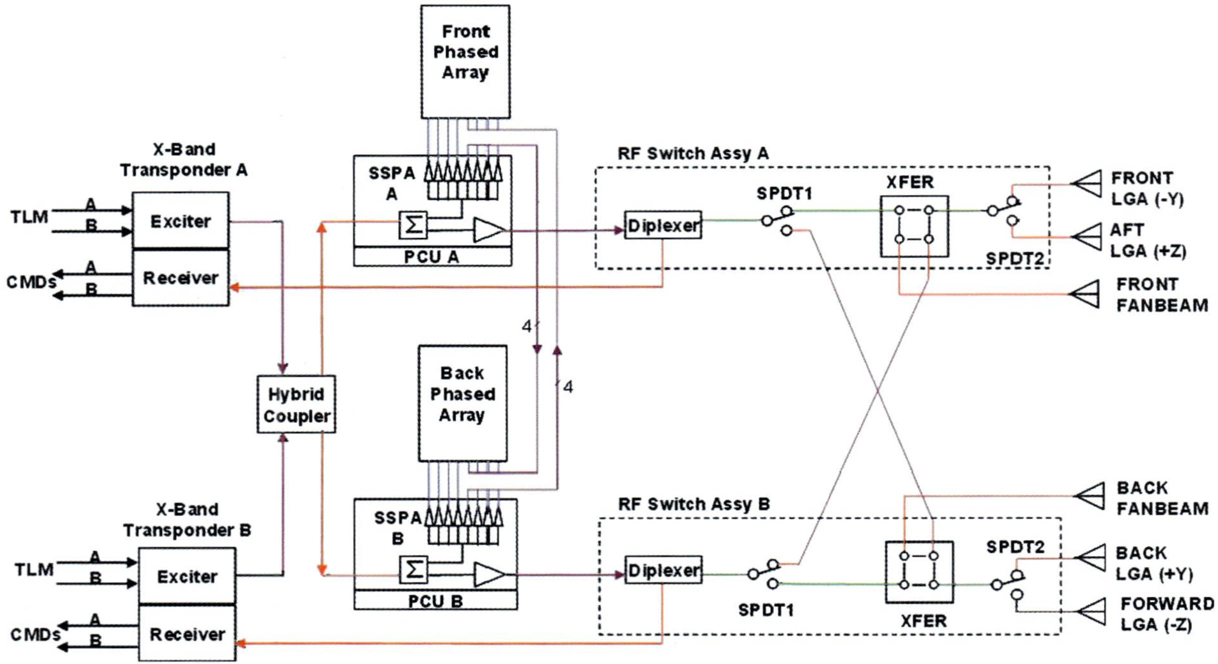

Fig. 2 The MESSENGER RF Telecommunications Subsystem block diagram. Two redundant transponders relay commands (CMDs) and accept telemetry (TLM) to and from the Command and Data Handling Subsystem of the spacecraft. The RF signal is relayed to the antennas via solid-state power amplifiers (SSPAs) and RF switch assemblies, each consisting of a diplexer, single-pole double throw (SPDT) switches, and a transfer (XFER) switch

Figure 2 shows the block diagram of the RF Telecommunications Subsystem (Bokulic et al. 2004). Redundant X-Band Small Deep Space Transponders (SDSTs), provided by General Dynamics, are responsible for receiving and demodulating the RF uplink signal, generating and modulating the RF downlink signal, and turning around uplinked ranging and Doppler components. Each SDST's downlink signal (only one is active at any given time) is routed via a passive hybrid coupler to both solid-state power amplifiers (SSPAs), shown in Fig. 3. Each SSPA can be in one of four modes: "distributed front," "distributed back," "lumped," and "off." The "distributed" modes of the SSPAs feed the RF downlink signal to either the front or the back PAA; the "lumped" mode of the SSPA feeds the RF downlink signal, via the two RF switch assemblies, to the fanbeam antenna or the LGAs. In the "distributed" mode, the RF signal is split eight ways and routed to eight "stick amplifiers." Each stick amplifier consists of a four-bit phase shifter (that controls the steering of the phased-array antenna beam), a small-signal amplifier, a driver amplifier, a power amplifier, and an isolator. The output power of each stick amplifier is approximately 34 dBm; in a distributed mode, a total of four sticks are operational per SSPA, yielding an output power of approximately 40 dBm. The "lumped" mode of the amplifier offers a 40-dBm power output for the fanbeam and low-gain antennas.

As shown in Fig. 4, each PAA is composed of eight center-fed slotted waveguides, each with 26 radiating elements (slots). The fanbeam antenna along the lower length of the antenna structure is also shown in this figure. The slotted-waveguide antennas are inherently linearly polarized; a novel combination of parasitic monopoles attached next to the radiating slots and single dummy waveguides mounted on both sides of the eight radiating waveguides produces right-hand circular polarization. This innovation yields a 3-dB improvement in link performance because the DSN antennas are circularly polarized (Stilwell et al. 2003).

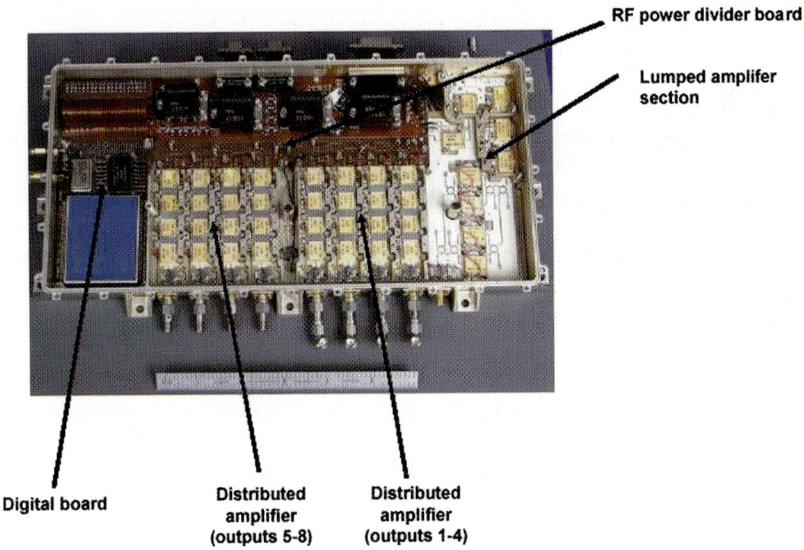

Fig. 3 The MESSENGER solid-state power amplifier. The lumped amplifier section feeds a 40-dBm RF signal to the fanbeam and low-gain antennas. The RF power divider board splits the RF signal to feed the various stick amplifiers in the distributed amplifier sections, which feed RF power to the phased-array antennas. The digital board serves as the command and telemetry interface to the spacecraft. A 6-in (15.2-cm) ruler indicates scale

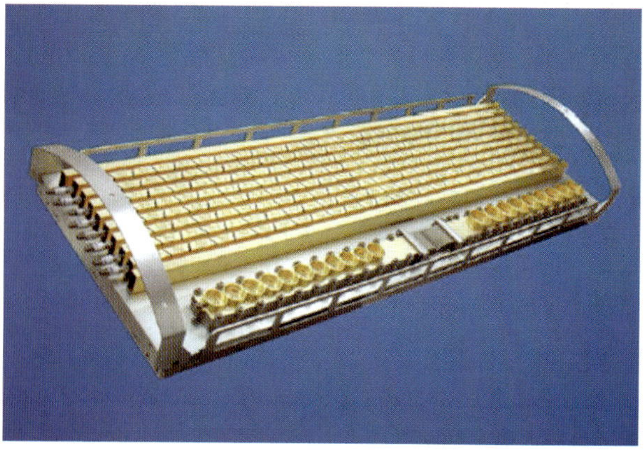

Fig. 4 A MESSENGER phased-array antenna and fanbeam antenna. Eight slotted waveguide antennas comprise the phased-array antenna. The fanbeam antenna is comprised of two helix arrays (uplink and downlink). The overall dimensions of the assembly are 28 cm by 81 cm

With one SSPA in "distributed" mode, one PAA can be illuminated in its "half-array" mode with four waveguides illuminated. With both SSPAs in either the "distributed front" or "distributed back" mode, the front PAA or the back PAA will operate in "full-array" mode (Wallis et al. 2004). This mode offers a 6-dB increase in effective isotropic radiated power (EIRP; 3-dB gain from radiating twice the power, and 3-dB gain in the antenna gain of using eight antenna elements instead of four); this mode corresponds to an increase in the allowable data rate by a factor of four. The distributed SSPA modes will be used during the high-rate ground contacts once MESSENGER is at Mercury.

3 MESSENGER Tracking Operations

3.1 Precision Orbit Determination

MESSENGER's X-band tracking provides the line-of-sight velocity of the spacecraft and the distance of the spacecraft from Earth. The specified accuracy of the velocity measurement for the RF subsystem is to within ± 0.1 mm/s over a 60-s integration period. The Doppler (Earth line-of-sight velocity) measurements are derived from the change in the carrier frequency, taking into account the frequency translation in the onboard transponder. Figure 5 shows that through the first 400 days of cruise, the average typical 1-σ Doppler residual is on the order of 0.03 mm/s. This measurement, provided by the navigation team, includes the contributions from all the non-RF sources (e.g., Earth atmospheric effects) and confirms that the RF subsystem is performing well within its specifications.

The ranging measurements are made via standard sequential tone ranging. These data assist in fitting the Doppler data to a precise orbit. The DSN transmits a series of ranging tones; the onboard transponder receives these tones and retransmits them back to Earth. The difference between the time of transmission of the ranging tones and the time of reception of the tones, along with knowledge of the transmission delay within the spacecraft, yields the spacecraft range. Measurements of the internal spacecraft delay taken during ground testing ranged from 1,356.89 ns to 1,383.74 ns depending on the RF configuration (transponder, SSPA, and antenna configuration). Thus far in flight, a set value of 1,371 ns has been used as the spacecraft delay. A more precise delay time will be used for science operations at Mercury; however, even a 20-ns error in this estimate would lead to a range error of only 3 m. Calibration operations to be performed during the cruise phase of the mission will provide

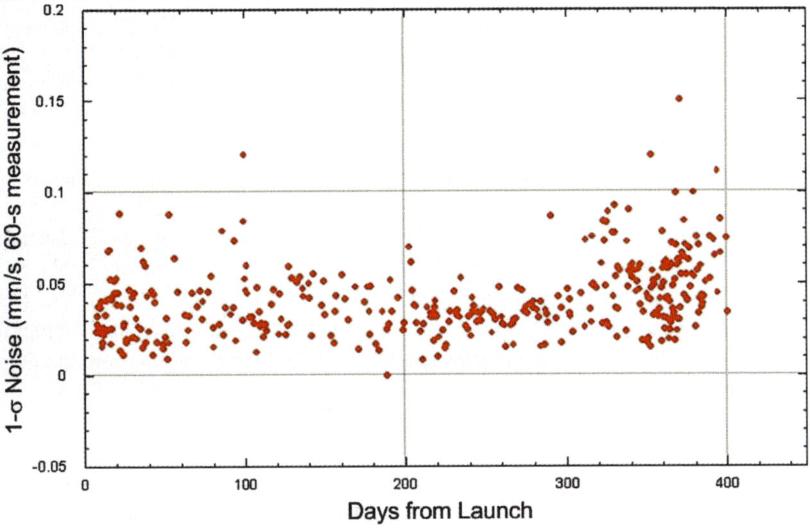

Fig. 5 Two-way Doppler residual noise per DSN ground contact. Each point represents the 1-σ Doppler noise during one DSN pass during the first 400 days of cruise. The average error is 0.03 mm/s over a 60-s integration time

the precise delay time. Thus far in flight, the typical 1-σ ranging residual is 45.25 ns, which translates to a range accuracy of 6.8 m. These values fall within the required accuracy of 10 m.

In the Mercury orbit phase, MESSENGER completes two revolutions per Earth day, and it is planned to acquire tracking data during each DSN contact pass. Currently the baseline is to have 12 eight-hour tracks per week. From these data the spacecraft position and velocity will be derived (orbit determination) and used to predict the future position of MESSENGER to enable the targeting of the instruments. In addition, these same data will be used in the analysis of the instrument data by providing precise positions at which the previous observations were made.

3.2 RF Tracking Operations

Once in Mercury orbit, the RF configuration will route the uplink through the most favorable fanbeam antenna and the downlink via a phased array for high-rate data-transmission passes. During these passes, the RF signal levels will be sufficiently high to acquire good tracking (Doppler and range) data. However, during the remainder of the time, the spacecraft will be oriented such that the main suite of instruments will be pointed at the planet. This pointing constraint, coupled with the sunshade-to-Sun pointing constraint, forces the spacecraft-to-Earth vector to lie outside of the fanbeam and phased-array antenna patterns. Therefore, during these periods, any tracking and RS data must be gathered through the use of the more omni-directional low-gain antennas. This configuration significantly lowers the signal-to-noise ratio for tracking measurements.

Two large sources of error on the Doppler tracking measurements include the solar plasma effect and the thermal noise effect. In order for the Doppler tracking data to be useful for Mercury gravitational field mapping, the instrumental accuracy of the Doppler measurements should be better than 0.1 mm/s over the integration time of the measurement, which will effectively vary from 10 s to several minutes. The raw Doppler measurements are recorded at the DSN's Central Data Recorder (CDR) at 10 samples per second. During post-processing, the integration time is selected. The longer the integration time, the more accurate the measurement; however, with longer integration times, the measurement has less spatial precision (e.g., during measurements taken at Mercury periapsis, the spacecraft can move approximately 200 km during a 60-s integration). For operations at Mercury, it is desirable to sustain integration times of no more than 10 s.

The solar plasma effect occurs when the RF signal originates in or passes through the solar plasma with a Sun-Earth-probe (SEP) angle of less than 30° (Kinman 2002). During orbital operations, this condition occurs when Mercury is on the far side of the Sun. The solar plasma causes phase scintillations in the RF signal, thereby increasing the Doppler measurement error. At the X-band frequencies used on the MESSENGER mission, the Doppler measurement error can be approximated during the mission when MESSENGER is on the far side of the Sun as

$$\sigma_{V,\mathrm{Solar}} = \frac{0.73c\sqrt{5.5 \times 10^{-6}} \cdot [\sin(\theta_{\mathrm{SEP}})]^{-1.225}}{f_c T^{0.175}}, \tag{1}$$

where $\sigma_{V,\mathrm{Solar}}$ is the Doppler measurement error (mm/s) due to the solar plasma, c is the speed of light in a vacuum (mm/s), θ_{SEP} is the SEP angle, f_c is the downlink carrier frequency (Hz), and T is the integration time (s). The actual error will also be a function of the current solar activity level, which will be highly variable in the 2011 timeframe (the peak of the solar cycle).

As shown in Fig. 6, the error caused by the solar plasma is nearly always greater than 0.1 mm/s for integration times of less than 60 s when MESSENGER is on the far side

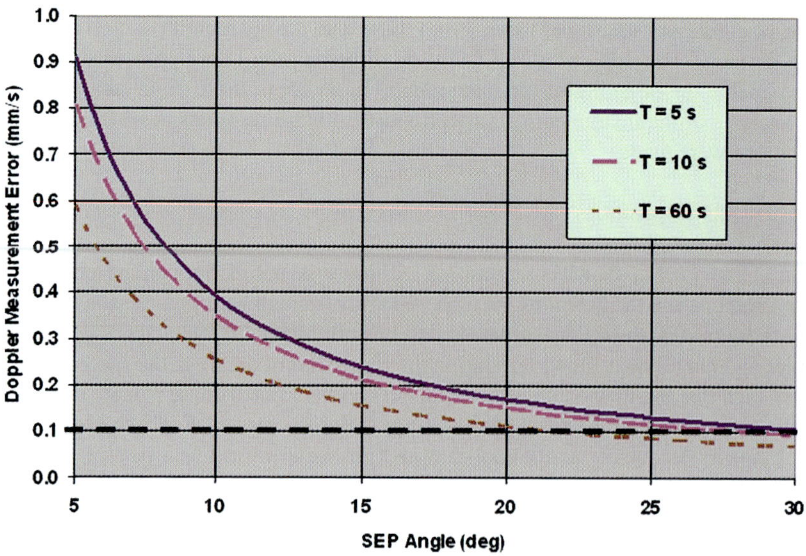

Fig. 6 Doppler measurement error induced by the RF signal passing through the solar plasma at SEP angles of less than 30° using integration times of 5, 10, and 60 s. At virtually all times during the orbital phase of the MESSENGER mission when Mercury is on the far side of the Sun from the Earth, this additional error amounts to several tenths of a mm/s

of the Sun during the orbital phase of the mission. This result indicates that the highest-resolution gravity data will be obtained when the spacecraft is on the near side of the Sun. Lower-resolution gravity will be obtainable at all times using integration times equivalent to several minutes, and these observations will be used in the estimation of low-degree terms in the gravity field and for the libration estimation.

The thermal noise effect is a function of the RF uplink and downlink signal-to-noise ratios (Kinman 2002) and can be modeled as

$$\sigma_{V,\mathrm{Thermal}} = \frac{c}{2\pi\sqrt{2}f_c T}\sqrt{\frac{B_L}{P_c/N_0|_{D/L}} + \frac{G^2 B_L}{P_c/N_0|_{U/L}}}, \tag{2}$$

where $\sigma_{V,\mathrm{Thermal}}$ is the Doppler measurement error due to the thermal noise (mm/s), c is the speed of light in a vacuum (mm/s), f_c is the downlink carrier frequency (Hz), T is the integration time (s), B_L is the one-sided downlink carrier loop bandwidth (Hz), P_C is the downlink or uplink received carrier power (W), N_0 is the downlink or uplink noise power spectral density (W/Hz), and G is the transponder turnaround ratio (880/749).

Doppler observations on the near side of the Sun will therefore be possible on the LGAs (see Fig. 7) for significant portions of time. This result is important because the gravity field may be mapped with the spacecraft in the orientation needed for other observations, which renders the PAAs nonusable for those observation times. The main source of the signal level change will be the orientation of the spacecraft; depending on where the Earth lies in the LGA pattern, the signal power level will change. At times, however, it may be necessary to orient the spacecraft to the downlink attitude so that the PAAs may be pointed at Earth for a stronger signal; this arrangement, however, may prevent the spacecraft from gathering

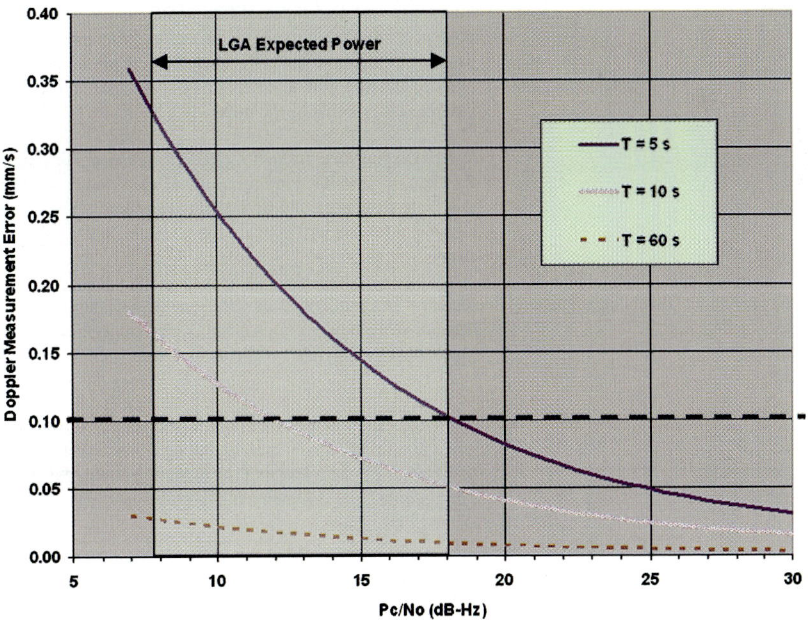

Fig. 7 Error induced by the thermal noise as a function of downlink P_C/N_0 using various integration times. When MESSENGER is on the near side of the Sun, data with less than 0.1 mm/s error can be obtained from the LGAs with a 10-s integration time for a significant portion of time

science data for the remaining science instruments. The frequency at which this action will be taken will be a balance between gathering the information for the orbit and gravity field investigations and gathering data for the other science investigations.

Because of these constraints, low-gain tracking data in the region of the periapsis may be available only when Mercury is on the near side of the Sun. Currently, the orbital operations plan calls for DSN passes to be extended or added to include coverage of every periapsis while on the near side of the Sun, and one periapsis on the high-gain phased-array antenna every four days (eight MESSENGER orbits about Mercury) while on the far side of the Sun.

3.3 Cruise-Phase Calibration Operations

To assist in the precision orbit determination needed when the spacecraft is in Mercury orbit, the RS cruise operations include calibrations and verifications for the Doppler and range measurements, including in-flight measurement and modeling of the error sources. These operations, described in Table 1, will help refine the assumed noise models and ultimately increase the accuracy of the Doppler data during the orbital mission phase.

Table 1 Cruise-phase radio science calibration operations

Test	When performed	Goals, comments
RS01: Noise calibration	2006, 2007, 2010	Measure Doppler and ranging accuracy as functions of power level and solar plasma. By changing the RF antenna and the ranging modulation index, different RF carrier and ranging powers are available to characterize the noise sources as a function of RF power. Performing this test on the near and far sides of the Sun will help separate the thermal and solar plasma components of the noise.
RS02: Ranging calibration	2007	Verify ranging delays due to spacecraft electronics and cables. By changing the RF configuration, all the various signal paths will be used to perform ranging measurements. The different ranging residuals will confirm the nominal spacecraft ranging delays.
RS03: Solar conjunction calibration	Each conjunction	Measure Doppler residuals over varying SEP angles. Due to the numerous conjunctions during cruise, data from different levels of solar activity will be available.
RS04: Lunar occultation measurement	May 19, 2007	Use DSN radio science receivers to record the time of loss and reacquisition of signal (ingress and egress occultations) when the spacecraft passes behind the Moon. When in Mercury orbit, similar measurements behind Mercury will constrain the shape of the planet.
RS05: Orbit geometry characterization	Aug.–Sep. 2007	Evaluate LGA tracking performance over a similar geometry as expected during mission phase. During this period, the trajectory arc of the spacecraft closely follows the orbital path of Mercury on the near side of the Sun. This will be an excellent opportunity to gauge RF tracking performance using the LGA.

4 Radio Science Investigations

4.1 Gravitational Field

The accuracy with which the past and future position of the spacecraft can be determined is a function of the tracking data and the orbital perturbations (which influence the motion of the spacecraft), of which the Mercury gravitational field is the largest source of error. Thus, in addition to solving for the position of the spacecraft around Mercury, the parameters that describe the gravity field of Mercury will be estimated from the same tracking data.

The X-band transponder will provide the range-rate data between the spacecraft and a DSN ground station and will be used in a matrix of observation equations for the spacecraft state that can be solved (Smith et al. 1993). The normal representation of the gravitational potential field derives from a solution to Laplace's equation in spherical geometry, which consists of a finite series of spherical harmonics of the form

$$U(r,\phi,\lambda) = \frac{GM}{r} + \frac{GM}{r}\sum_{l=2}^{N}\sum_{m=0}^{l}\left(\frac{a}{r}\right)^{l}\overline{P}_{lm}(\sin\phi)(\overline{C}_{lm}\cos m\lambda + \overline{S}_{lm}\sin m\lambda), \tag{3}$$

where U is the potential at the point (r,ϕ,λ), r is radial distance from the planetary center of mass to the spacecraft, ϕ is latitude, λ is longitude, GM is the product of the gravitational constant G and the mass M of Mercury, a is the mean radius of Mercury, $\overline{P}_{lm}(\sin\phi)$ are normalized associated Legendre functions of degree l and azimuthal order m, N is the degree and order of the solution, and $\overline{C}_{lm}$ and $\overline{S}_{lm}$ are the normalized spherical harmonic coefficients that are estimated from the tracking observations. The perturbing acceleration is derived from the above equation, integrated as part of the equations of motion and normal equations developed relating the observations to variables of the orbit and the gravity coefficients $\overline{C}_{lm}$ and $\overline{S}_{lm}$.

Many forces affect the motion of the spacecraft. To extract the accelerations due to Mercury's gravity, these other forces must be modeled. The quality of the resulting higher-order gravity-field coefficients depends directly on the accuracy of the models of these other forces. For Mercury, with little atmosphere and strong solar effects, the most important forces are the solar pressure, the planet reflectance pressure, and the planet's thermal pressure. All these forces depend on the accuracy of the spacecraft surface model, and the planet-based forces further depend on accurate knowledge of Mercury's surface properties. The spacecraft is modeled as a series of panels, each with area, angle, emission, and absorption properties that must be accurate to better than a few percent. The final spacecraft surface model will be calibrated during the gravity-solution process by extensive iterations with the gravity solution.

The resolution of the gravity field, which is determined by the degree l of the field, is generally limited by the altitude of the spacecraft above the surface. For MESSENGER, in a 12-hour eccentric orbit with an inclination of approximately 80° and minimum periapsis altitude of 200 km at 60°N latitude, the resolution varies between about 100 km to more than 1,000 km. Figure 8 shows how that resolution varies with latitude for MESSENGER. The horizontal axis is the Mercury latitude, and the vertical axis is the resolution in degree and order (l,m). The approximate block size represented by the degree and order is shown for $l = 75$, 60, and 30. The closed-loop appearance of the chart, indicating two resolutions for a given latitude, is due to the eccentricity of the orbit and results from the different altitudes of the spacecraft on the ascending and descending parts of the orbit. However, as the planet rotates on its axis, the effective resolution at all longitudes is anticipated to be

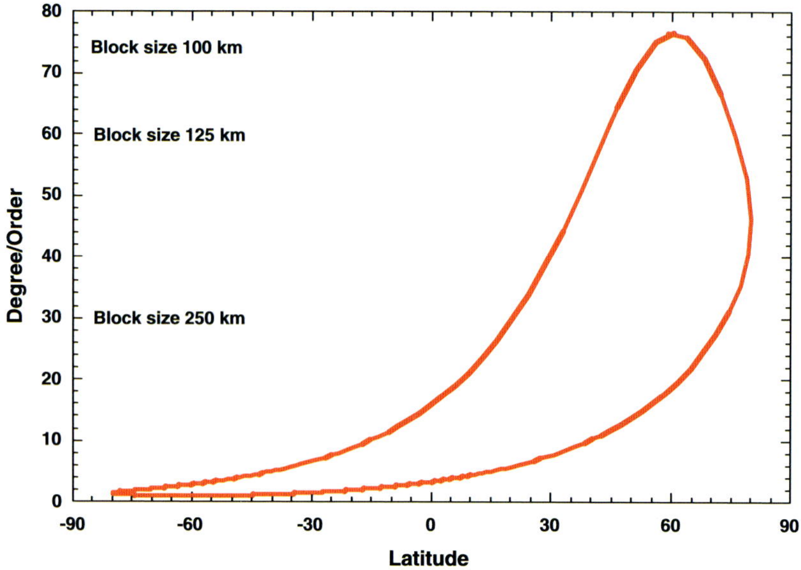

Fig. 8 Approximate resolution of the Mercury gravity field expected from the MESSENGER mission as a function of latitude. The periapsis is in the northern hemisphere at approximately latitude 60°N

Table 2 Mercury gravitational parameters

Parameter	Value
GM, km^3 s^{-2}	$22{,}032.09 \pm 0.91$
C_{20}	$-(6.0 \pm 2.0) \times 10^{-5}$
C_{22}	$(1.0 \pm 0.5) \times 10^{-5}$

From Anderson et al. (1987). Coefficients are not normalized

closer to the higher resolutions. Further, since the closest approach to the planet will be in the northern hemisphere, the resolution in the north is at an average block size of about 200 km, significantly better than that of the south, where the resolution will be on the order of 1,000 km. Unfortunately, this limitation will restrict the ability to interpret some of the geological features in the south compared with the north.

The mass of Mercury has been measured previously from radio tracking of the Mariner 10 spacecraft. The current knowledge of these parameters, summarized in Table 2, is based on historical observations as well as more recent reanalysis of combined data sets (Anderson et al. 1987, 1996). The first observations of the Mercury gravity field by MESSENGER will be during the three Mercury flybys that are expected to occur in January 2008, October 2008, and September 2009 prior to orbit insertion at Mercury in March 2011. These three flybys will provide the first gravity observations of Mercury since Mariner 10 flew by Mercury in 1974–1975 (Anderson et al. 1987). Although they will provide only brief and low-resolution observations compared with the orbital phase, these data will be able to confirm and possibly extend the earlier observations of the very lowest degree coefficients in the gravity model.

4.2 RS Topographic Mapping

The spacecraft will be occulted from Earth twice on the majority of orbits. If the spacecraft is tracked into occultation or acquired as it emerges from occultation, the time of the oc-

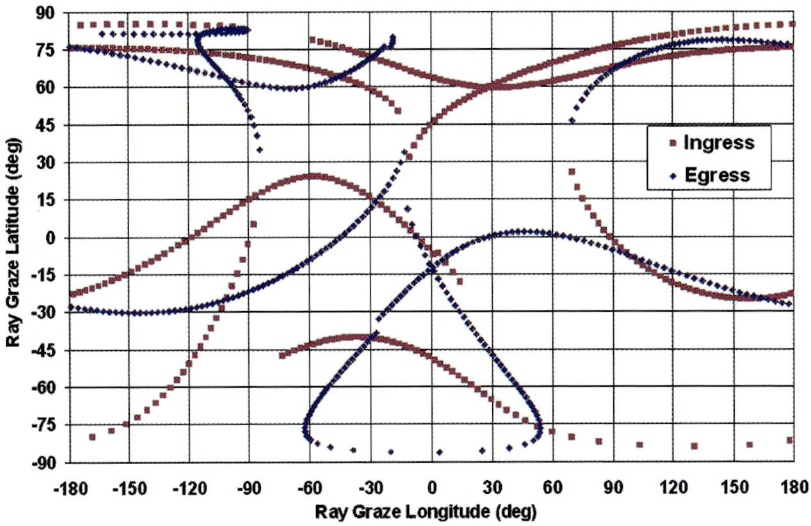

Fig. 9 Baseline distribution of Mercury occultations during the MESSENGER orbital mission phase. The final distribution will depend on the precise time of arrival of the spacecraft at Mercury and the entry into the mapping orbit

cultation can be used to estimate the radius of the planet at the grazing ray location (Kliore et al. 1973; Lindal et al. 1979). DSN open-loop radio science assets will be necessary to acquire the RF signal on the egress occultations, as locking to the signal with the normal block-V receiver is not possible for this measurement since there is not an ultra-stable oscillator onboard the spacecraft. Since the spacecraft position can be determined to within 10 m through the precision orbit determination process, it is possible to derive occultation radii to a similar level. Particularly important are occultations that will occur in the southern hemisphere where altimeter coverage will be lacking. These periods will occur several times during the mission when the spacecraft orbit and Earth are nearly coplanar. For these orbits the acquisition of at least two occultations per day will be attempted. These observations will be important in constraining the global shape of Mercury (Smith and Zuber 1996) and will significantly improve knowledge of the planet's offset between center of figure and center of mass, now estimated to be 1.030 ± 0.65 km (Anderson et al. 1996). Figure 9 shows the locations of all occultations during the nominal MESSENGER mission. The DSN tracking assets will be chosen carefully to maximize occultation coverage, particularly in the southern hemisphere.

4.3 Gravity and Topography

Combining gravity and topography measurements will allow an improved estimate of the planet's bulk density. Removing the gravitational attraction of surface topography will yield a Bouguer gravity map that shows subsurface density anomalies that can interpreted in terms of crustal thickness, which is indicative of extent of past melting and impact redistribution of crust. Also to be determined are transfer function relationships between gravity and topography, which yield information on crustal thickness that is indicative of thermal state at the time of surface and subsurface loading. Thus the determination of crustal and lithospheric thickness will permit insight into the thermal and geological evolution. Mapping of crustal and lithosphere thickness variations will be possible only in the northern hemisphere where high-resolution gravity and topography will be obtained (see Zuber et al. 2007).

4.4 Rotational State

An orbiting spacecraft is sensitive to the long-wavelength power in the gravitational field, so the low-degree terms in the spherical harmonic expansion of the gravitational potential are well constrained (Kaula 1966). Of particular interest are degree-two terms, the gravitational flattening C_{20} (unnormalized) and the equatorial ellipticity C_{22}. The coefficient C_{22} defines the long-wavelength gravitational shape of the equator and provides a strong constraint on the planetary rotation rate and obliquity (Smith et al. 1993). The other phenomenon of interest is the libration, which is a variation in the planetary rotation rate and is manifest as an oscillation of the equatorial shape. The combination of the gravitational flattening, equatorial ellipticity, obliquity, and libration magnitude provide a constraint on Mercury's core state (Peale, 1976, 1981, 1988) from the expression

$$\left(\frac{C_m}{B-A}\right)\left(\frac{B-A}{MR^2}\right)\left(\frac{MR^2}{C}\right) = \frac{C_m}{C} \leq 1, \tag{4}$$

where the parameters in (4) are determined from

$$\begin{aligned} \phi_0 &= \frac{3}{2}\frac{B-A}{C_m}\left(1 - 11e^2 + \frac{959}{48}e^4 + \cdots\right), \\ \frac{C}{MR^2} &= \frac{\left[\frac{J_2}{(1-e^2)^{3/2}} + 2C_{22}\left(\frac{7}{2}e - \frac{123}{16}e^3\right)\right]\frac{n}{\mu}}{(\sin I)/i_c - \cos I}, \\ \frac{B-A}{MR^2} &= 4C_{22}, \end{aligned} \tag{5}$$

where ϕ_0 is the amplitude of the physical libration, i.e., the maximum deviation of the axis of minimum moment of inertia from the position it would have had if the rotation were uniform at $1.5n$ where n is the orbital mean motion; i_c is the obliquity of the Cassini state, the state that Mercury's spin axis is expected to occupy; e is the orbital eccentricity; J_2 is $-C_{20}$; and I is the inclination of the orbit plane to the Laplacian plane on which the orbit precesses at the uniform rate $-\mu$. The conditions for which (4) is applicable for determination of Mercury's core state were discussed by Peale et al. (2002).

A simulation of the MESSENGER mission scenario (Zuber et al. 2007) indicated that the conditions required for application of this method are achievable (Zuber and Smith 1997; Peale et al. 2002). The combination of the libration of the gravitational field obtained from planetary tracking with the libration of the topographic field obtained from the MESSENGER altimetry (Cavanaugh et al. 2007) may permit estimation of the viscous coupling of Mercury's core and mantle (Zuber and Smith 1997).

5 Conclusions

While providing the mission-critical function of communication between the spacecraft and the operations center on the Earth, the MESSENGER RF system provides for the collection of Doppler and range data that are essential for precision navigation and the mission geophysics investigations. Reduction of cruise data and analyses of expected orbit operations indicate that the Doppler accuracy should meet the specification of 0.1 mm/s. This accuracy should resolve the spherical harmonics of Mercury's gravitational field up to order 70 in the vicinity of the periapsis and will reveal the libration of the gravity field. The RF system further supports the geophysics investigation by supplying planet-radii data acquired through the timing of the RF occultations. These data will be used with the MLA data to map the topography of Mercury, data which are then combined with the gravity field to achieve the science goals related to the state and history of Mercury's internal structure, including the presence and size of a liquid core, its coupling with the solid core, and the crustal and lithospheric thickness where data resolution permits.

References

J.D. Anderson, G. Colombo, P.B. Esposito, E.L. Lau, G.B. Trager, Icarus **71**, 337–349 (1987)
J.D. Anderson, R.F. Jurgens, E.L. Lau, M.A. Slade III, G. Schubert, Icarus **124**, 690–697 (1996)
R.S. Bokulic et al., IEEE Aerospace Conference, Big Sky, Mont., IEEAC paper 1370, 2004, 10 pp., CD-ROM 4-1503
J.F. Cavanaugh et al., Space Sci. Rev. (2007, this issue). doi:10.1007/s11214-007-9273-4
R.E. Gold et al., Planet. Space Sci. **49**, 1467–1479 (2001)
W.M. Kaula, *Theory of Satellite Geodesy* (Blaisdell, Waltham, 1966), 124 pp
P.W. Kinman, JPL DSMS Document 810-005, Rev. E (2002), 124 pp
A.J. Kliore, F.J. Fjeldbo, B.L. Seidel, M.J. Sykes, P.M. Woiceshyn, J. Geophys. Res. **78**, 4331–4351 (1973)
J. Leary et al., Space Sci. Rev. (2007, this issue). doi:10.1007/s11214-007-9269-0
G.F. Lindal et al., J. Geophys. Res. **84**, 8443–8456 (1979)
J.V. McAdams, R.W. Farquhar, A.H. Taylor, B.G. Williams, Space Sci. Rev. (2007, this issue). doi:10.1007/s11214-007-9162-x
S.J. Peale, Nature **262**, 765–766 (1976)
S.J. Peale, Icarus **48**, 143–145 (1981)
S.J. Peale, in *Mercury*, ed. by F. Vilas, C.R. Chapman, M.S. Matthews (Univ. Arizona Press, Tucson, 1988), pp. 461–493
S.J. Peale, R.J. Phillips, S.C. Solomon, D.E. Smith, M.T. Zuber, Meteorit. Planet. Sci. **37**, 1269–1283 (2002)
A.G. Santo et al., Planet. Space Sci. **49**, 1481–1500 (2001)
D.E. Smith, M.T. Zuber, Science **271**, 184–188 (1996)
D.E. Smith et al., J. Geophys. Res. **98**, 20,871–20,889 (1993)
S.C. Solomon et al., Planet. Space Sci. **49**, 1445–1465 (2001)
D.K. Srinivasan et al., IEEE Aerospace Conference, Big Sky, Mont., IEEEAC paper 1067, 2005, 11 pp
R.K. Stilwell, R.E. Wallis, M.L. Edwards, Proceedings of the IEEE International Symposium on Antennas and Propagation and United States National Committee, Canadian National Committee, International Union of Radio Science North American Radio Science Meeting, vol. 3, Columbus, OH, 2003, pp. 1030–1033
R.E. Wallis, J.R. Bruzzi, P.M. Malouf, in Antenna Measurements Techniques Association 26th Annual Meeting and Symposium, Stone Mountain, GA (2004), pp. 331–336
M.T. Zuber, D.E. Smith, Lunar Planet. Sci. **27**, 1637–1638 (1997)
M.T. Zuber et al., Space Sci. Rev. (2007, this issue). doi:10.1007/s11214-007-9265-4

Space Sci Rev (2007) 131: 573–600
DOI 10.1007/s11214-007-9261-8

Launch and Early Operation of the MESSENGER Mission

Mark E. Holdridge · Andrew B. Calloway

Received: 24 July 2006 / Accepted: 10 August 2007 / Published online: 8 November 2007

Abstract On August 3, 2004, at 2:15 a.m. EST, the MESSENGER mission to Mercury began with liftoff of the Delta II 7925H launch vehicle and 1,107-kg spacecraft including seven instruments. MESSENGER is the seventh in the series of NASA Discovery missions, the third to be built and operated by The Johns Hopkins University Applied Physics Laboratory (JHU/APL) following the Near Earth Asteroid Rendezvous (NEAR) Shoemaker and Comet Nucleus Tour (CONTOUR) missions. The MESSENGER team at JHU/APL is using efficient operations approaches developed in support of the low-cost NEAR and CONTOUR operations while incorporating improved approaches for reducing total mission risk. This paper provides an overview of the designs and operational practices implemented to conduct the MESSENGER mission safely and effectively. These practices include proven approaches used on past JHU/APL operations and new improvements implemented to reduce risk, including adherence to time-proven standards of conduct in the planning and implementation of the mission. This paper also discusses the unique challenges of operating in orbit around Mercury, the closest planet to the Sun, and what specific measures are being taken to address those challenges.

Keywords Mercury · MESSENGER · Mission operations

1 Introduction

The scientific objectives of the MErcury Surface, Space ENvironment, GEochemistry, and Ranging (MESSENGER) mission (Solomon et al. 2001, 2007) include characterizing:

a. the chemical composition of Mercury's surface,
b. the planet's geological history,
c. the nature of Mercury's magnetic field,
d. the size and shape of Mercury's core,

M.E. Holdridge (✉) · A.B. Calloway
The Johns Hopkins University Applied Physics Laboratory, Laurel, MD 20723, USA
e-mail: mark.holdridge@jhuapl.edu

e. volatiles at Mercury's poles, and
f. the nature of Mercury's exosphere and magnetosphere.

To accomplish these objectives, the MESSENGER spacecraft (Leary et al. 2007) must be safely placed into orbit around Mercury as per the mission's trajectory design (McAdams et al. 2007). The first major step in this road to Mercury was successfully carried out in August 2005 with an Earth gravity-assist maneuver and concurrent instrument operations using the Moon and the Earth as calibration sources during numerous instrument operations. This Earth flyby not only sent MESSENGER onto the next phase in its journey, it also served as a training exercise for developing and integrating complex instrument and spacecraft housekeeping activities into one sequence of activities. This Earth flyby is followed by two Venus flybys and three Mercury gravity-assist maneuvers prior to Mercury orbit insertion in 2011. In addition to these gravity-assist maneuvers, there are five required firings of MESSENGER's main engine, the first of which was accomplished on December 12, 2005, to keep the MESSENGER spacecraft on the proper path to Mercury (McAdams et al. 2007). On MESSENGER's fourth return to Mercury, the spacecraft will fire its main engine during a pair of orbit insertion maneuvers, making MESSENGER the first spacecraft and instrument suite to orbit the planet Mercury.

2 Overview of MESSENGER Mission Operations

A top-level view of data flow and division of responsibilities between organizations actively involved in MESSENGER flight operations is shown in Fig. 1. The operation is conducted at The Johns Hopkins University Applied Physics Laboratory (JHU/APL) in Laurel, MD, as the lead spacecraft development organization. MESSENGER operations are conducted within the Space Science Mission Operations Center (SSMOC) facilities at JHU/APL, used previously by Near Earth Asteroid Rendezvous (NEAR) Shoemaker and Comet Nucleus Tour (CONTOUR) missions and currently shared with the New Horizons Mission to Pluto and the Kuiper belt. The general operational architecture and working models for MESSENGER postlaunch operations have been adapted using lessons learned from these missions (Holdridge 2001, 2003).

The SSMOC is the hub of the physical data flow and overall coordination of mission operations. It serves as the home to the mission operations team that is responsible for

a. mission planning,
b. activity development and command sequencing,
c. activity review and testing,
d. real-time flight operations, and
e. postevent assessment.

The operations team consists of full-time team members who concentrate on the MESSENGER spacecraft systems as well as part-time team members who provide more generalized support and backup support in certain key areas. Operations team size was greatest during launch and early operations in order to support the large number of unique spacecraft and instrument commissioning activities. During this time the team size peaked at 15 (not including supporting members of the spacecraft development team), a number that will not be required again until the orbital operations commence at the planet Mercury in 2011. The team was downsized to the current nine members following early commissioning operations. This core team conducts the core operations functions listed above and is augmented

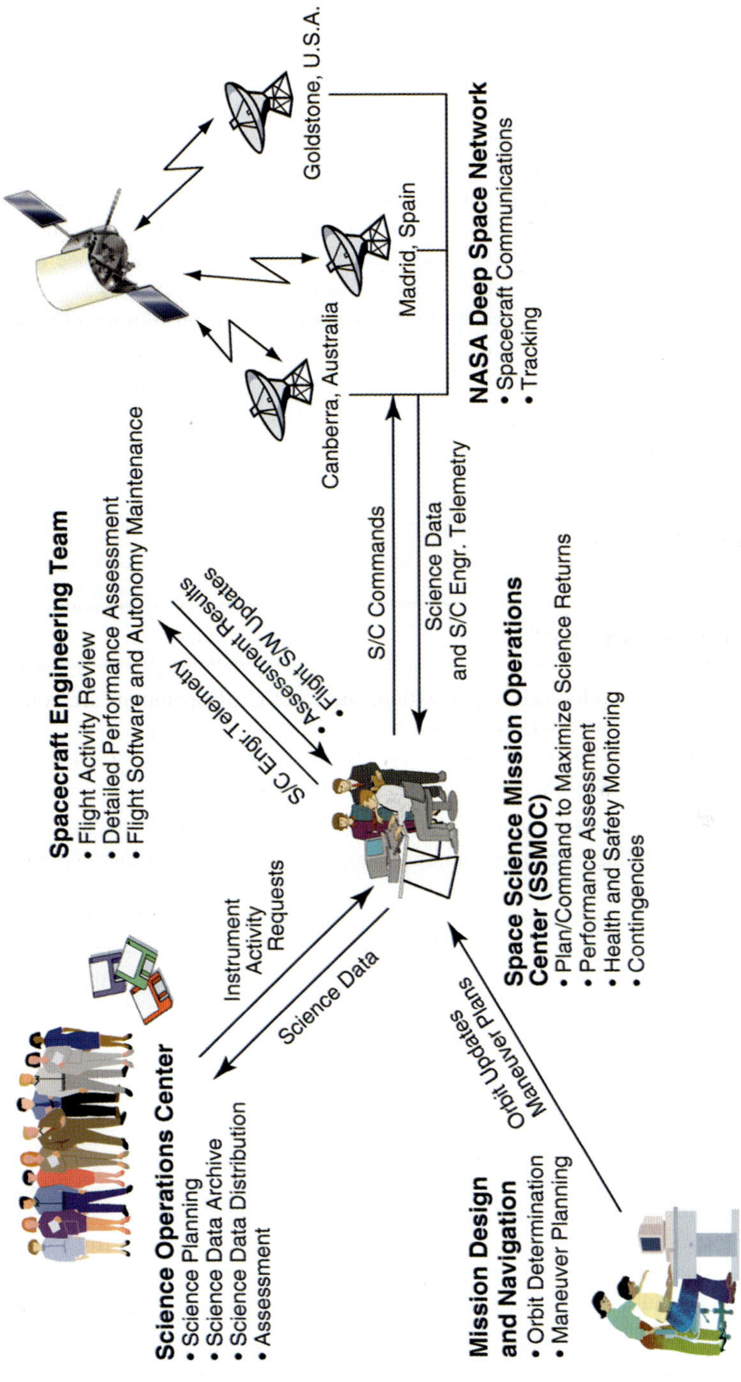

Fig. 1 Mission operations team interfaces and responsibilities

with several part-time team members who also support the New Horizons mission, a mission to Pluto that launched in January 2006. Missions to both the inner and outer planets are therefore operated side-by-side in the same mission operations center, the SSMOC, a unique opportunity for all involved.

In addition to the core team, spacecraft and instrument design engineers, also located at JHU/APL, support the operation using detailed knowledge and understanding of specific systems. These engineers are not involved in the day-to-day operations of the mission but are concurrently designing JHU/APL's next-generation flight systems.

Day-to-day science operations are managed by the Payload Operations Manager and the Science Coordinator at JHU/APL. The Science Operations Center (SOC) (Winters et al. 2007) is located at JHU/APL and serves as the operational hub of the science operation and the primary operational interface to the SSMOC. Instrument team members work directly from within the SOC or work remotely from their respective home institutions. The science operations team, consisting of instrument scientists, engineers, and command sequencers, is responsible for the opportunity analysis and planning of instrument operations. The science operations team provides the instrument-operation command loads to the SSMOC. Payload operations are determined by the instrument leads and are provided to the SSMOC for execution after testing and validation to ensure that they do not violate operational constraints. The science and mission operations teams work closely together on the development and validation of instrument command sequences.

MESSENGER navigation operations are performed by KinetX, Inc., a commercial organization that offers a range of systems engineering and software (S/W) services, including space navigation. The KinetX navigation team is responsible for the following operations functions:

a. processing of radiometric Deep Space Network (DSN) tracking data,
b. orbit determination and orbit predictions,
c. DSN antenna pointing predictions, and
d. maneuver planning and reconstruction.

The mission design team, also located at JHU/APL, works in close coordination with the navigation team, providing maneuver planning, orbital-related predictions, and mission contingency planning.

Spacecraft tracking is performed by the DSN, managed by the Jet Propulsion Laboratory (JPL) in Pasadena, CA. MESSENGER primarily uses the DSN 34-m-aperture antenna subnet for normal operations, augmented by the 70-m subnet as required to improve data return during special operations, including trajectory-correction maneuvers (TCMs). The DSN supports all telemetry acquisition, commanding, and tracking requirements throughout the MESSENGER mission.

3 Application of Space Mission Operations Standards

The MESSENGER operations team follows on the success of the teams that previously operated the NEAR and CONTOUR spacecraft. As a result, lessons learned from those past missions have been directly incorporated into standard operating practices. These lessons have also been formally incorporated into JHU/APL's Space Mission Operations Standards (SMOS), a guiding document for the proper conduct of mission operations at JHL/APL.

SMOS formalizes industry-standard approaches for the conduct of space mission operations regardless of funding levels and, as such, establishes a "floor" for the minimum acceptable operations practices. Specific operational areas SMOS addresses include:

a. configuration management,
b. assessment and monitoring,
c. simulation and testing,
d. flight constraints,
e. training and certification,
f. real-time operations,
g. contingency planning,
h. formal review practices,
i. operations schedules, and
j. user documentation.

SMOS stipulates requirements to be met by all JHU/APL mission operations. However, flexibility exists for how the Mission Operations Manager (MOM) implements each standard. Additional successful practices are also documented for consideration. SMOS is a living document and will be modified to include future operational practices proven successful on MESSENGER and deemed appropriate for other missions. The MESSENGER operation strikes a balance by minimizing both cost and risk through the adoption of SMOS standards and the use of highly experienced mission operations team members.

4 Launch and Early Operations

The MESSENGER launch was highly automated and included the key events listed in Table 1. MESSENGER was launched at night with all critical systems on battery power. The spacecraft exited eclipse after more than 20 minutes with its solar panels in their stowed positions. The solar panels were deployed 37 minutes later after the spacecraft separated from the third stage. Shortly thereafter, the first telemetry was received from the DSN Canberra (CAN), Australia, complex and then later the Goldstone (GDS), CA, complex. The spacecraft was spun down and separated from the upper stage. Once separated, the spacecraft detumbled and acquired the Sun. It did so while being monitored closely in real time from the SSMOC. Both of the spacecraft low-gain antennas were programmed to transmit for four minutes surrounding the propulsive detumble event to ensure ground monitoring of these critical activities during spacecraft tumble. The spacecraft detumble was monitored with brief outages as the spacecraft's elevation above the horizon remained near the operating limits of the station. The spacecraft climbed above 6° elevation during the subsequent GDS contact. The spacecraft successfully stabilized its attitude with the solar panels rotating from their stowed positions to the Sun direction. After discharging 17%, the single 23-A h battery entered a sustained period of recharge.

The above sequence relied on the proper configuration of the spacecraft prior to launch via command scripts and procedures developed by the integration and test team. Autonomous execution of a hard-coded firmware sequence and onboard command macros, developed by the mission operations team, implemented the above activities after onboard autonomy sensed the separation event.

Two unexpected handovers from primary to redundant units occurred during this early period, including a switch from power distribution unit (PDU) A to B, and a switch from the primary star tracker to the backup unit. These units were determined to be healthy during tests conducted on day 2 following launch. Both units were later restored to primary status.

Initial uplink (command capability) was established during the first GDS contact, 1 hour and 23 minutes after launch. During this initial period of commanding, mission operations carried out several planned activities including the reconfiguration of onboard processor

Table 1 Launch event timeline

UTC	Launch relative times	Spacecraft events
06:15:56	L+00:00:00	*August 3, 2004, MESSENGER launch*
06:36:43	L+00:20:37	*Eclipse exit*
07:12:42	L+00:56:46	S/C separation
07:12:45	L+00:56:49	*PDU separation sequence starts*
07:12:45	L+00:56:49	SSPA A on
07:12:56	L+00:57:00	Panels #1 & #2, hinge 1 deployed
07:15:57	L+01:00:01	Panels #1 & #2, hinge 2 deployed
07:17:31	L+01:01:35	30-s thruster burp completed
07:18:11	L+01:02:15	Actuate "separation complete" relay
07:18:15	L+01:02:19	*CAN reaches* 6° *elevation*
07:19:11	L+01:03:18	*MP separation sequence starts*
07:21:53	L+01:05:57	*Hawaii rise* −6° *elevation*
07:24:45	L+01:08:49	All wheels on, first chance at detumble
07:24:57	L+01:09:01	SSPA B on
07:27:57	L+01:12:01	Propulsive detumble start
07:28:11	L+01:12:15	FPP mission phase from launch to separation
07:28:58	L+01:13:02	SSPA B off
07:29:59	L+01:14:03	Nominal propulsion system safing
07:30:04	L+01:14:08	Start safing of propulsion system
07:30:07	L+01:14:11	*CAN elevation drops below* 6°
07:36:38	L+01:20:42	*GDS 6°*
07:39:34	L+01:23:38	*GDS 10°–begin uplink sweep*
08:28:11	L+02:12:15	Autonomous transition to first contact rules if not already commanded

Notes: S/C = spacecraft; PDU = power distribution unit; SSPA = solid-state power amplifier; MP = main processor; FPP = fault protection processor. Most spacecraft events were the result of executed commands; entries in italics were related operational events

memory parameters and timers for postlaunch operations. The playback of the solid-state recorder (SSR), including all engineering data recorded just prior to liftoff, rounded out the first day's active operations. Commanding was concluded after approximately eight hours, followed by a planned period of team rest during which flight controllers continued to monitor the flight systems closely.

Commissioning exercises began on day 3 of the operation starting with the guidance and control (G&C) and radio frequency (RF) communications systems. Commissioning operations, including detailed checkouts of both primary and secondary systems, continued on a daily basis through the first nine days.

The events that took place during the first five months of operations are shown in Fig. 2, together with the corresponding DSN support levels. DSN coverage was continuous during the first six days. However, commanding operations each day generally began mid-morning local time after a daily planning and coordination meeting. DSN coverage was reduced to 16 hours per day on day 7, as planned, to minimize total mission DSN costs and reduce impact to other flight projects. On mission day 21, coverage was further reduced to eight hours per day.

After day 9, instrument checkout activities began to be interspersed with the spacecraft subsystem checkouts. Instrument digital processing unit (DPU) A was first powered on August 12, 2004, along with the Magnetometer (MAG) and Gamma-Ray and Neutron

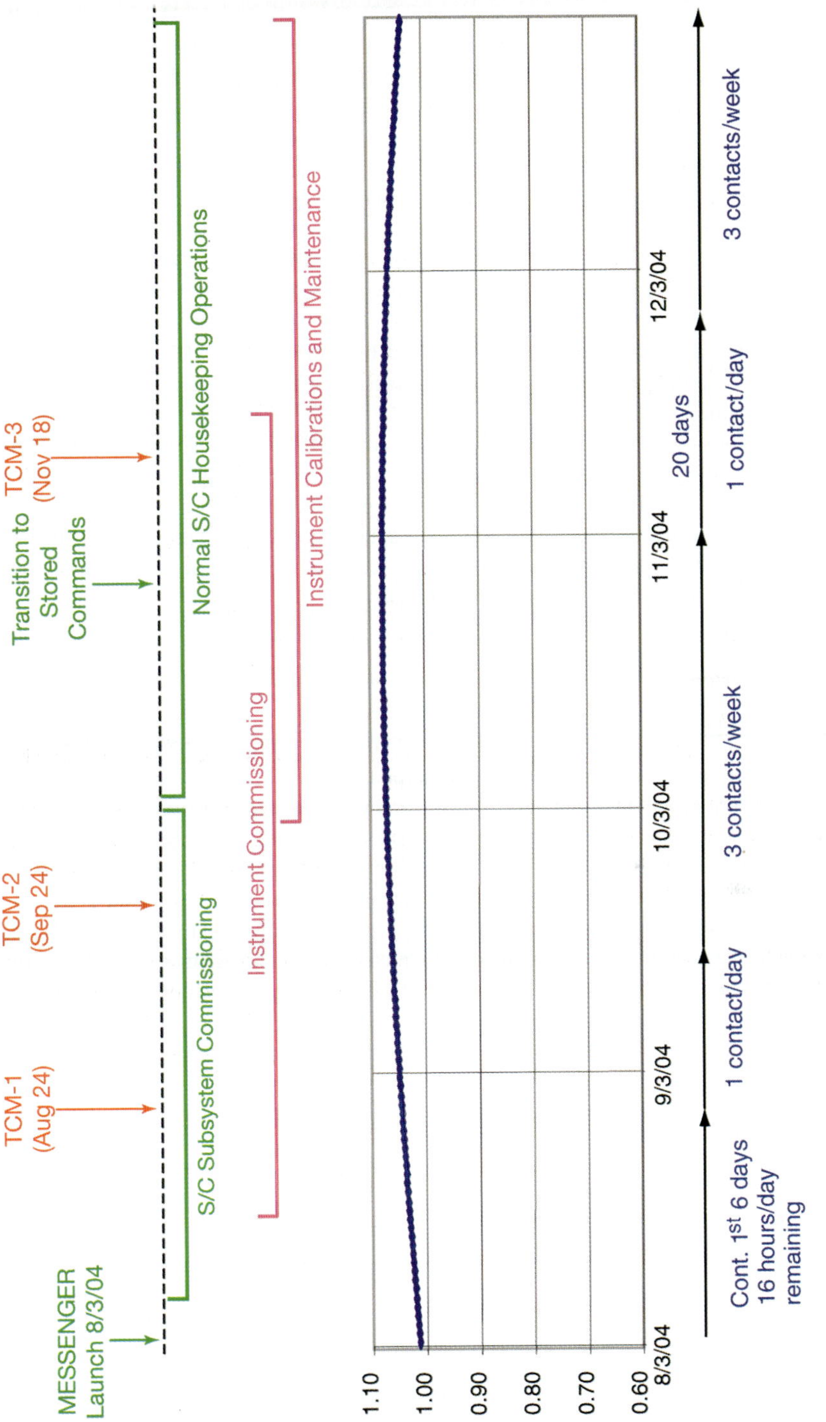

Fig. 2 Initial five-month timeline of MESSENGER flight operations

Spectrometer (GRNS) instruments. By September 27, all instruments had been successfully checked out through both the primary and secondary DPU interfaces. One remaining GRNS checkout, which included a full cool-down using an internal cooler and high-voltage power supply of the Gamma-Ray Spectrometer (GRS) sensor, was performed during a period of increased DSN support and real-time monitoring in November 2004.

The first trajectory-correction maneuver, TCM-1, was performed on August 24, 2004. This maneuver was followed by TCM-2 and TCM-3 on September 24 and November 18, respectively. Each of these maneuvers involved the mono-propellant portions of the propulsion system and the four associated 26-N thrusters (McAdams et al. 2007). These highly successful maneuvers, each monitored in real time via the DSN, placed MESSENGER on its proper trajectory for an Earth flyby on August 2, 2005. TCM-2 was also used to reduce onboard momentum build-up, normally controlled by more passive techniques including adding small tilts to the spacecraft's attitude and offsetting the angle that each solar panel presents to the Sun.

Early operations commissioning exercises were concluded by mid-November 2004 with the completion of the GRS checkout and the locking of the solar array pins after full deployment was confirmed in power system data. Additional checkout of the anti-Sun-facing Sun sensors and anti-Earth-facing communications antennas was performed around the time of a flip of the spacecraft to a sunshade-to-Sun attitude (Leary et al. 2007) in March 2005 when the geometries favored operating those components.

5 Cruise Operations

Following the successful spacecraft and instrument commissioning, MESSENGER mission operations transitioned into the planetary "cruise" phase with initiation of regular spacecraft maintenance activities and instrument calibration and maintenance activities. The cruise operation will include a total five large velocity adjust (LVA) thruster firings and six planetary flybys (one of Earth, two of Venus, three of Mercury) to provide the proper setup for Mercury orbit insertion (McAdams et al. 2007). This combination makes for an extremely challenging cruise operation. Each planetary flyby will be accompanied by a series of TCMs for targeting and trajectory cleanup, some of which can also be classified as critical. A large Mercury orbit-insertion maneuver in 2011 will mark the transition from cruise operations to the orbital phase of the mission.

Routine maintenance activities during the cruise phase include power system voltage updates, ephemeris updates, memory verification, time drift corrections, instrument calibrations, and flight recorder playbacks. The first year of the cruise operation included the following key operations events as shown in Fig. 3:

a. spacecraft "flip/flop" operations, alternating the side of the spacecraft that faced the Sun,
b. trajectory correction maneuvers (TCMs),
c. Earth flyby with full instrument operations,
d. Deep Space Maneuver 1, and
e. a flight software update.

Cruise operations beyond the first year include two flybys of Venus and three of Mercury. The first flyby of Venus occurred during a solar conjunction of the spacecraft, so payload operations were suspended. The second flyby of Venus, in addition to providing the trajectory correction and gravity assist required for reaching Mercury, will provide an opportunity to practice the science sequences for Mercury flyby 1 with the instrument payload. The Mer-

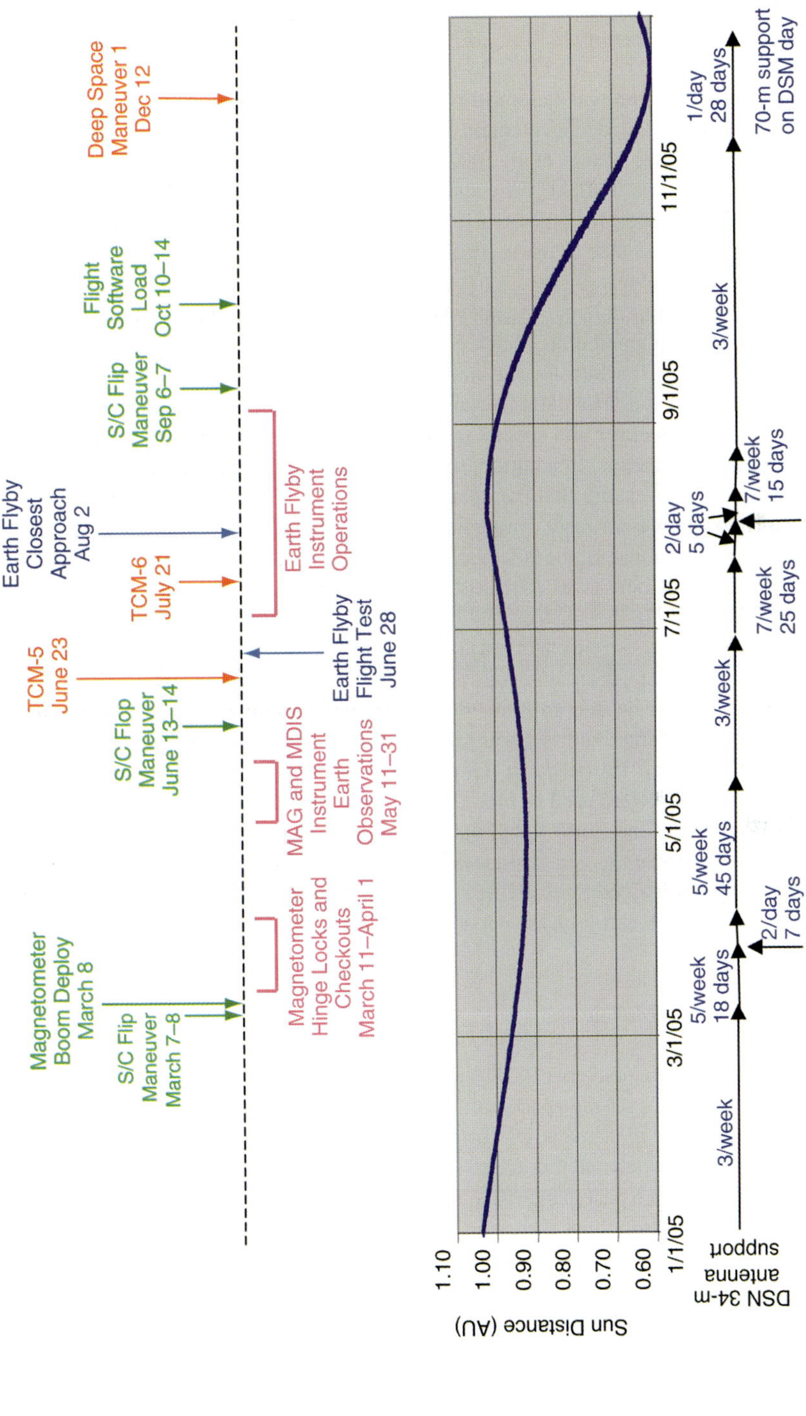

Fig. 3 Early cruise operations

cury flybys will provide science observations at near-equatorial latitudes at lower altitudes than will be possible during orbital operations.

5.1 Spacecraft Flip/Flop Operations

The spacecraft thermal design was tailored for Mercury orbital operations in relative close proximity to the Sun. One key design element for protecting the spacecraft from the intense solar energy is the sunshade (Santo et al. 2002; Leary et al. 2007). For inner cruise operations (Sun distance less than 0.95 AU), the sunshade was oriented toward the Sun with the spacecraft behind it. However, for outer cruise operations (greater than 0.95 AU from Sun) the spacecraft's unprotected side is oriented directly towards the Sun to provide adequate warmth at reduced power levels.

MESSENGER operations at launch began with the backside of the spacecraft oriented toward the Sun, where it remained until March 8, 2005, when the spacecraft and its sunshade were rotated 180° to face the Sun for the first time in the mission. This "flip" operation involved coordination of the articulating solar arrays to ensure adequate battery states of charge throughout the maneuver and special tailoring of the onboard autonomy system and guidance and control parameters. In addition, reprogramming of the spacecraft's nonvolatile memory was necessary to ensure that the spacecraft would return to this "safe state" following an anomaly or processor reset. Immediately following the March flip maneuver, the Magnetometer boom (mounted on the side opposite from the sunshade) was pyrotechnically deployed while the deployment arms were still warm from pointing toward the Sun.

In the weeks surrounding the spacecraft flip operation, changing Sun and Earth geometries permitted additional first-time checkouts of spacecraft components including the front-side RF phased-array and fan-beam antennas and shade-mounted Sun sensors.

This flip operation was reversed on June 14, 2005 ("flop" maneuver), to place the spacecraft with its sunshade oriented anti-sunward as Sun distances during the summer of 2005 were climbing on approach to the Earth. During the Earth flyby, short-duration (several hours) flip/flop maneuvers were used to place the sunshade toward the Sun for selected instrument operations. After the Earth flyby, on September 7, 2005, the sunshade was once again oriented toward the Sun as Sun distances were reduced to 0.6 AU for the first time in the mission. The shade remained Sun-oriented until March 2006 when a final flop maneuver was performed as the spacecraft traveled outside of 0.95 AU from the Sun for the final time in the mission.

5.2 Trajectory Correction Maneuvers

Five trajectory correction maneuvers (TCMs) were performed in the first year of MESSENGER flight operations to target the spacecraft (McAdams et al. 2007) after launch and on Earth return. Each of these maneuvers was implemented with generic command sequences developed and tested prelaunch and tailored to satisfy specific ΔV (velocity adjustment) targets and lessons learned from past maneuvers. Each maneuver during that period used either the 4.4-N or 26-N mono-propellant thrusters. Preparations for each TCM involved incorporating numerous maneuver parameters into command sequences to control the G&C and propulsion systems (including thruster selection, burn duration, burn attitude, and other parameters) for each maneuver. The resulting maneuver sequence was then combined with other command blocks that configured the affected subsystems including avionics and RF systems. The resulting sequences were then subjected to a number of tests using a high-fidelity spacecraft simulator and dynamic models to identify any problems with the sequence. As each TCM had specific differences, contingency simulations were conducted for every TCM to test the system response to faults and the return to a safe state using the "as flown" command sequence. The final command sequence and test results were then reviewed per a strict review process for critical events prior to uplink and execution. All maneuvers to date have been successful.

5.3 Earth Flyby

The successful execution of all TCMs provided the accurate targeting necessary for the Earth gravity-assist maneuver, or flyby. The Earth flyby altered MESSENGER's trajectory toward the inner solar system while serving as a valuable opportunity for instrument teams to calibrate instruments with the well-characterized Earth–Moon system. As the first of several planned planetary flybys, the Earth flyby served as a training exercise for coordinating the housekeeping and instrument operations. Lessons learned from this exercise are being applied to future flybys, including the three Mercury flybys planned for 2008 and 2009.

Regular optical navigation (OpNav) measurements of the Earth were conducted on approach to the Earth as end-to-end tests of the operational systems and to characterize OpNav system performance. Although not necessary for Earth flyby navigation, the optical navigation exercises will prove out the systems and processes necessary for the eventual Mercury flyby operations planned to begin in January 2008.

Nested between the TCMs and OpNav measurements, periodic instrument calibrations were performed during Earth approach as well on the outbound trajectory. Instrument operations included:

a. two-way active and passive Mercury Laser Altimeter (MLA) operations,
b. Mercury Dual Imaging System (MDIS) imaging of Earth–Moon system, including an Earth departure movie, and
c. lunar and Earth hydrogen corona observations with the Mercury Atmospheric and Surface Composition Spectrometer (MASCS).

The planning, integration, review, and testing of critical housekeeping and instrument operations necessary for successful Earth flyby operations served as an operational training exercise for the entire team. Operational processes are expected to be refined and later reapplied during subsequent flybys when the scientific rewards will be much greater. A total of six planetary flybys are scheduled for the MESSENGER mission (McAdams et al. 2007) prior to Mercury orbit insertion, making this mission trajectory, and the resulting operation, one of the most challenging of any planetary mission.

5.4 Flight Software Update

Assessment of flight systems after launch led to several recommended changes to flight software incorporated into MESSENGER's main processors (MPs), which include both G&C and command and data handling (C&DH) software on one main processor card. This assessment culminated in the development of an updated version of flight software necessary to correct "bugs" primarily in the G&C portion of the software. Updating MESSENGER's MP flight software involved several steps for the mission operations team following delivery from the flight software team. These steps included:

a. loading and verification on the spacecraft hardware simulator in SSMOC using actual flight scripts,
b. recertification of the simulator with new software and supporting testbed modifications,

c. preparation and test of flight scripts to be used during associated operations including fault protection processor (FPP) and MP state changes, and
d. development and test of flight sequences necessary for the in-flight test of the "bug fixes" to ensure proper G&C operation prior to deep-space maneuver 1 (DSM-1).

The upload and necessary processor reboot took place in October 2005, in preparation for DSM-1 in December 2005.

5.5 Deep-Space Maneuver 1

The final operation of great significance to the mission during the first phase of cruise operations was a large ΔV using MESSENGER's largest thruster, the 672-N LVA bi-propellant rocket motor. This maneuver was the first of five such LVA motor firings required during the cruise to Mercury. The maneuver, DSM-1, was successfully completed on December 12, 2005. The 325 m/s ΔV for DSM-1 is the largest of LVA motor firings planned until Mercury orbit insertion (McAdams et al. 2007). The preparations for DSM-1 followed the thorough approach used for the development, test, and review of all critical operations with steps added for external peer review and additional contingency scenario testing. Building blocks used to implement this highly successful maneuver have been placed under configuration control and will used as a basis for developing the next LVA motor firing, DSM-2, planned for execution in October 2007.

6 Mission Planning

The MESSENGER MOM is responsible for coordination with each mission operational element in the development of the mission timeline and application of lessons learned in timeline development. Detailed engineering and planning of flight activities are conducted at weekly planning meetings chaired by the MOM. Recommendations for resulting additions or changes to the mission timeline are reviewed at weekly mission management meetings, chaired by the Mission Director, for final approval. Representatives from each operational interface shown in Fig. 1 participate in these planning meetings.

The mission timeline is implemented by the mission planning and scheduling team, responsible for constructing each operation utilizing a reusable set of parameter-driven command building blocks, referred to as fragments and canned activity sequences (CASs). The architecture for fragments and CASs makes use of two levels of command hierarchy built into time-proven sequencing software referred to as SeqGen and developed by JPL. Members of the MESSENGER mission operations team have designed CASs and fragments using the SeqAdapt software, also developed at JPL. These reusable command blocks were designed to support planned subsystem and instrument operations. Each parameter-driven CAS and associated fragments are first tested individually on the MESSENGER high-fidelity hardware simulator (many were also tested on the spacecraft prelaunch), and they are then made available for reuse in the SeqGen command sequencing system, used by mission planners.

Housekeeping operations, including SSR management, RF communications, and TCMs, are assembled by the mission operations team using parameter-driven calls to existing CAS and fragment structures and then provided to the instrument teams at the start of the sequencing cycle for a given time period. Instrument inputs to mission operations are in the form of SeqGen input files that call out the appropriate CAS and include the input parameters required and timing information for each requested activity.

More than 100,000 commands were executed on the MESSENGER spacecraft during the first 13 months of postlaunch operations, including many first-time events. Examples include commissioning activities and numerous complex instrument maintenance activities, five TCMs, and calibration operations that involved special pointing of the spacecraft. During that time, only one anomaly of any significance occurred. A spacecraft safing incident occurred in September 2004 following an oversubscription of operating system memory and a command that failed as a result. Flight operations constraints were modified to prevent similar occurrences. Much of the reason for otherwise highly reliable results to date are the incorporation of a time-proven, multi-step, test-and-review process applied to every special operation performed.

All flight operations are conducted using prebuilt onboard sequences composed of absolute time-tagged commands that trigger relative time macros. All such activities are first tested with a faster-than-real-time software model of the spacecraft, known as StateSim, followed by high-fidelity, hardware-in-the-loop simulations with the spacecraft hardware simulator. Rigorous formal reviews of each first-time or critical spacecraft event are performed using onsite spacecraft system engineering support and subsystem design engineer review, with the intent of using the collective MESSENGER expertise available at JHU/APL. Activity reviews are normally conducted in two parts, with each Preliminary Design Review (PDR) intended to cover the high-level design of the activity prior to starting detailed sequencing, and a Critical Design Review (CDR) that includes review of the final sequence and its test results. Elements included in these reviews, conducted episodically since launch, include:

a. statement of goal for event and general success criteria,
b. detailed timeline of events,
c. spacecraft resources required,
d. DSN support requirements,
e. operations staffing requirements,
f. anticipated changes to autonomy rules required,
g. initial plan for dividing activities into commands (e.g., real-time vs. stored, time-tagged versus macros),
h. plan for postevent downlink of data and analysis and DSN support required,
i. identification of verification criteria and plans for simulation and test,
j. review of related simulator fidelity limitations,
k. review of any and all planned deviations in testing from the nominal case,
l. verification of activity inputs/outputs such as TCM parameters,
m. line-by-line command sequence review,
n. test and simulation results review,
o. verification that activity objectives are met,
p. verification that no flight rules or constraints are violated,
q. documentation of all problems or issues that must be resolved prior to approval of command sequence uplink,
r. review of participant readiness to support activity execution, and
s. verification that all ground systems and interfaces required can support the event.

For some particularly critical operations, additional reviews may be held prior to or between the PDR and CDR. In addition, some future critical sequences will be flight tested in advance of the actual event to ensure success. Activities that will include this added layer of testing include one-time science exercises such as planetary flybys. The Earth flyby was the first such activity.

7 Mission Control

Real-time mission control is conducted from the JHU/APL SSMOC with a real-time flight control team. To enhance training, all flight controllers responsible for conducting launch and early operations were brought on and trained prelaunch, at the start of DSN readiness testing mission simulations. The flight control team is composed of two teams, two persons each, and a lead flight controller who manages both teams and supports periodic operations on both shifts.

Since the majority of flight operations are now sequenced, the primary activities performed in real time include:

a. uplink of command sequences,
b. DSN interfacing and coordination,
c. monitoring of spacecraft health and status, and
d. management of SSR operations.

All real-time operations are procedure driven with text procedures that guide general conduct and computer scripts that control the real-time systems. Command uplinks must first be approved by the MESSENGER MOM after verifying, by means of independent review, that all test and review procedures have been followed. The only commands not requiring MOM approval are those routine commands included in the preapproved commanding procedure. These commands are necessary for clearing "sticky" status bits, resetting counters, and performing other routine operations that are not directly coupled to the onboard sequence.

8 Mission Assessment

Mission assessment is conducted both in real time by the flight control team and offline with in-depth analysis of spacecraft behavior via the telemetry archive and related analysis tools. Real-time assessment is partially automated with real-time red and yellow limit checking incorporated into the ground system software. Limit and alarm databases are maintained to minimize repeated limit violations not deemed critical to the operation. All violations are reported to members of the offline engineering team for evaluation and resolution via a real-time call list. Members of the real-time flight control team are kept up to date on all facets of the ongoing operation and are considered an extension of the engineering team as its first line of defense. The real-time team manually compares expected states as reported in the StateSim output reports (e.g., predicted command history, autonomy rule firings, Sun angles) to the actual spacecraft state, and they look for any deviations. All discrepancies are reported to the engineering team and management in real time.

Additional offline assessment and trending are performed on a regular basis by the mission operations team and extended subsystem and instrument engineering teams. Off-nominal behavior is documented in an anomaly database as an Anomaly Report (AR) where it remains open until all corrective action is completed. Corrective actions typically take the form of a flight or ground software fix, procedural change, or ground database change.

9 Recorder Operations

The MESSENGER spacecraft was designed to record and store housekeeping and science data on one of two 8-Gbit synchronous dynamic random access memory (SDRAM) SSRs using Disk Operating System (DOS)-style formatted files. MESSENGER is also one of the first missions to use the Consultative Committee for Space Data Systems (CCSDS) File Delivery Protocol (CFDP) (Krupiarz et al. 2003) for downlinking files from the SSR. This protocol uses two-way handshaking and tunable timers to ensure that all data are captured on the ground for each file in an automated fashion. Files are downlinked in a predefined priority order whenever the auto-playback mode is enabled, which is the normal method of operation. Files may also be downlinked directly by command. This design was chosen in order to fulfill science objectives during Mercury orbit, such that the highest priority science data remaining to meet mission requirements are always downlinked first, with enough flexibility to accommodate change and observations of opportunity. One of the early operational challenges has turned out to be file size management. The operations team worked closely with the engineering and instrument teams during checkout and special activities to ensure that files are closed sufficiently frequently to prevent them from growing excessively large. In general, the mission operations team prefers to keep files under 10 Mbyte in size except under special circumstances.

Fifteen directories have been defined for the SSRs. These include the Record directory, the Downlink directory, the Trash directory, the Optical Navigation directory, the Image directory, and the 10 data file priority directories shown in Fig. 4.

The SSR directories are named according to their function:

REC contains open files of actively recording packets,
TRASH contains files that are candidates for deletion,
IMG contains raw science images unprocessed by the MP,
OPNAV contains unprocessed raw images used for navigation support, and
DNL contains the current file that has been selected for downlink.

Successfully downlinked files can be configured for automatic deletion or for placement into the Trash directory. During the cruise phase, files are placed in the trash and are removed via command load after a selectable duration, nominally two weeks. During orbit phase, successfully downlinked files will be automatically deleted to preserve recorder space. Most nonhousekeeping data are compressed prior to downlink with the type of MP compression selected via command. The mission planning members of the operations team use the software simulator and its various models (discussed in Sect. 4) to produce specialized output reports. One of those reports is the predicted set of files for downlink from each track, used by the real-time team as a guideline of downlink progress. The real-time team in turn periodically performs American Standard Code for Information Interchange (ASCII) directory listings of the recorder for the planning team to ensure that the simulator output prediction

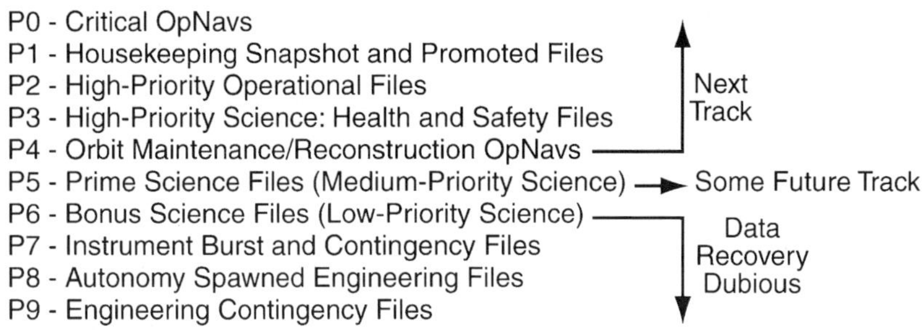

Fig. 4 SSR data directories and data priority recovery scheme

reports have not diverged due to nondeterministic activities such as operator intervention and network outages. The simulator is updated as necessary when differences are found between the model and reality. The predicted downlink reports are provided to the SOC as a reference for science planning and data analysis coordination. In addition, the SOC has access to the actual downlinked file list, which the real-time team currently produces at the conclusion of each track.

Data are collected in packets, each with its own unique application identification (APID), and each packet is then routed to a file in one of the 10 data directories. The operations team maintains an onboard file filter table, which defines the files into which a packet is placed. The filter table contains two entries for routing each packet to a file, thereby creating a flexible scheme for establishing priorities. The baseline filter table is typically changed for special activities and for maneuvers. The driver for this filter design is the downlink bit rate. The downlink bit rate fluctuates considerably during the orbit phase as shown in Fig. 5. Therefore, there will be periods when the recorder will be storing more data than can be accommodated by the bandwidth for downlink and other times when the opposite will be true. This situation also means that during the low bit-rate periods, medium-priority data will not be downlinked for extended periods of time, occasionally weeks after the original observation. The low-rate data may be deleted before they are downlinked. It is for these reasons that the science team worked closely with mission operations to define what packets and files would be routed to the higher and lower priority directories to ensure maximum payout for primary science data without saturating the recorder.

At the beginning of each real-time DSN tracking support, the onboard command load enables auto-playback and allocates a percentage of the downlink packets for recorder data (interleave rate), effectively resuming the CFDP protocol for that track. The flight software polls the 10 data directories, starting at P0, looking for the first directory that contains at least one file. A downlink list is then created in the order oldest to newest for that directory,

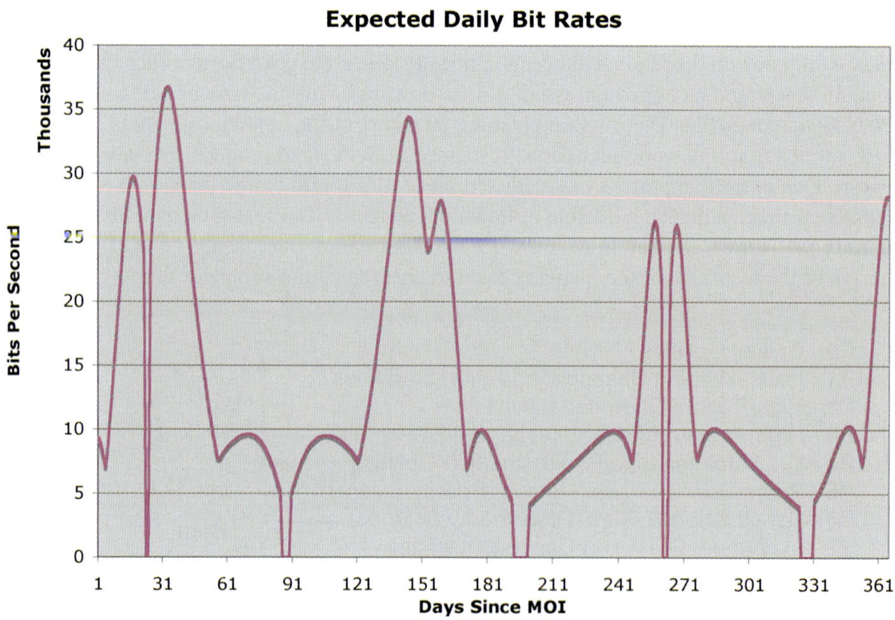

Fig. 5 Orbit-phase downlink bit-rate capability

and each file is placed in the DNL directory for downlink. Once all the files in that list are downlinked, a new directory poll is conducted for the next lower directory. This automated functionality requires very little in the way of operator intervention. The real-time team observes the progress of all the files and the protocol handshaking parameters. The team also temporarily disables the ground-originated requests for additional file downloads during command loads and other critical activities to avoid any possible command interference. The team also performs directory listings as needed. The file names contain a 10-digit ASCII representation of the mission elapsed time (MET) at which the file was opened, along with the defined priority and instance, which define the recorder state and contents. At the end of each track, auto playback is disabled and the interleave rate is set back to zero via the onboard command load, thereby pausing the CFDP flight timers until the next track. There have been a few ground system outages in the early mission, but the flight protocol has handled them all as designed by eventually retransmitting the data once the handshaking timers expired. In addition, several files were discarded by the ground system in the early operations as timer durations were being tuned and file size limitations were better defined. These files were later directly downlinked without any problems.

10 Instrument Science Planning Operations

The complexity of science operations and the logistics of large round-trip light times require that science operations be conducted via an onboard command load. This situation requires a strictly scheduled planning and command load process. Fig. 6 shows the command load

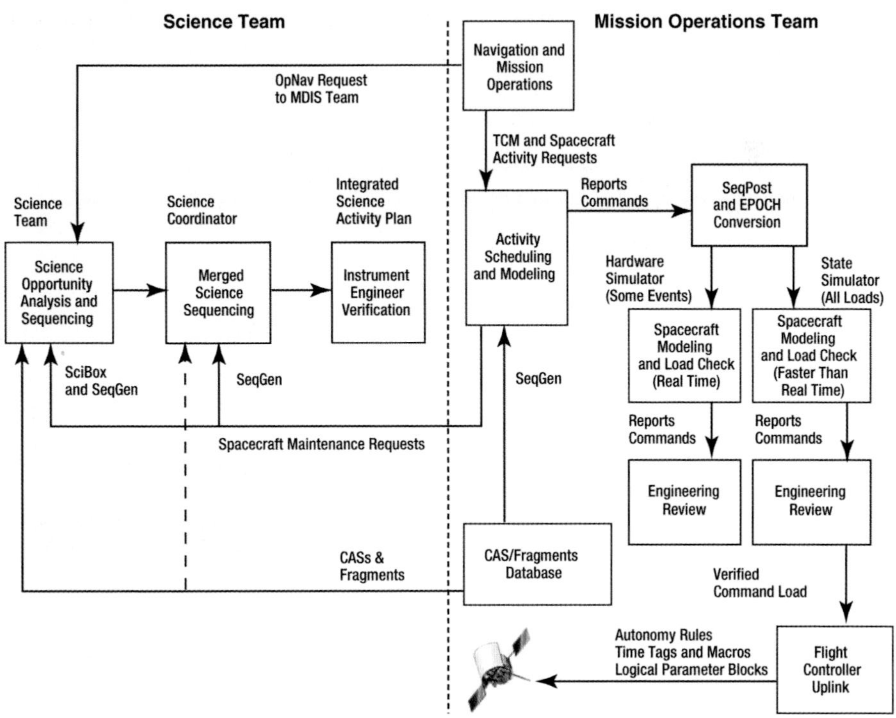

Fig. 6 Stored command process

cycle and the responsibilities of both the science and mission operations teams to produce a safe and effective command load.

The science planning team responsibilities are shown to the left of the dashed line. The team has a library of tested CASs that were produced, reviewed, and tested in conjunction with mission operations. The operations team maintains this library throughout the mission. The opportunity analysis software known as SciBox uses this library and takes as input the spacecraft maintenance schedule from the mission operations team, including planned maneuvers and DSN track schedules, along with optical navigation requests from the navigation team. A merged sequence of commands will ultimately be produced and then verified by the instrument engineers before being sent back to operations. The right side of the dashed line in Fig. 6 shows the operations team responsibilities. In addition to producing and sending the maintenance schedule, the operations team takes the integrated activity plan and performs the necessary conversions using the EPOCH real-time command software and the SeqPost command load generation software. The load is produced and run through the appropriate simulator, and the necessary load and activity reviews are scheduled and conducted. Once the final command load is validated and approved, it is then ready for uplink to the spacecraft. Two load opportunities will always be available prior to the beginning of the first time tag. Each load will be dumped from the spacecraft and compared with the ground image via load/dump comparison tools to verify that the uplinked commands were loaded error free.

To implement the science priority scheme effectively, the operations team has worked with the Science Planning Group (SPG) to come up with an eight-week planning/delivery cycle, as shown in Table 2. The mission operations initial file provides the science team with the maintenance schedule during week $N-6$. The science team uses this input to ensure that they operate the instruments so as not to interfere with SSR playbacks and TCM or momentum maintenance events, and to avoid collecting more high-priority science data

Table 2 Eight-week science operations planning cycle

	Monday	Tuesday	Wednesday	Thursday	Friday
Week $N-8$	Instrument teams begin design period for planned activities during weeks "N" and "$N+1$"				
Week $N-6$	Operations team delivers weeks "N" and "$N+1$" spacecraft housekeeping files	Science Planning Group (SPG) meeting			
Week $N-3$		SPG meeting		Initial instrument request files for weeks "N" and "$N+1$" completed	Initial instrument request files for weeks "N" and "$N+1$" reviewed and delivered to operations team
Week $N-2$	Operations team delivers StateSim review file	SPG meeting		Final instrument request files for weeks "N" and "$N+1$" completed	Final instrument request files for weeks "N" and "$N+1$" reviewed and delivered to operations team
Week $N-1$	Operations team builds and uploads week "N" load Operations team delivers final command report for week "N"				

than can be downlinked during the high-gain contacts. The SeqGen software contains flight rules and models that validate the science requests.

During week $N-3$, the science request files from each instrument team are passed on to the Science Coordinator. The Science Coordinator takes the mission operations initial request file, the OpNav request file, and each science request file and merges them in the SeqGen system. The coordinator checks for conflicts between the mission operations requests and the science requests and conflicts among the science requests and iterates internally as needed. When all conflicts have been resolved and the merged requests have been validated using SeqGen flight rules and models, the merged requests are reviewed by designated instrument engineers for instrument health and safety. If there are any problems the instrument requests are sent back to the appropriate teams for correction. The process repeats until all instrument engineers and the Science Coordinator are satisfied. The approved requests are then broken out by instrument again, and they are then sent to mission operations in the SeqGen output format.

Mission operations will receive the approved science request files during week $N-2$ and merge them with an updated version of the mission operations initial requests in SeqGen. The update includes any changes to the DSN track schedule and burn requests from the mission design and guidance and control teams. The merged requests are validated using SeqGen flight rules and models. Mission operations will adjust requests as needed, in conjunction with the Science Coordinator, to conduct the activities within available spacecraft resources (e.g., memory, power, attitude). Mission operations will perform any change (time permitting, the science teams will make changes) to science requests that is required to meet spacecraft and ground system operational limits. These changes are performed while consulting with the instrument teams. When requests pass the SeqGen level of verification, mission operations will build the command load and perform a second level of load verification using the faster-than-real-time StateSim software simulator (see above).

The command load generation process translates the output files into onboard stored commands and macros that are uploaded to the spacecraft prior to execution. SeqGen creates the ASCII spacecraft sequence file that is the translation of the CASs back to individual commands in time order. Next, the JHU/APL SeqPost program takes the file and groups the commands into reusable or temporary macros and stored time-tag commands. SeqPost builds macros based on the commands listed in time order for both the spacecraft and the instruments and eliminates duplicate macros to make efficient use of onboard memory. A Spacecraft Test and Operations Language (STOL) command load file is created defining the macros and time tag commands. A STOL command load script is created for use with the binary version of the command load created later in this process. If the command load contains a special activity that has not been performed before, that portion of the command load will be run on the hardware simulator for validation and a results review will be conducted prior to spacecraft uplink.

Following conversion for uplink, the command macros and time-tagged commands will be loaded into the MESSENGER software simulator (StateSim). This process permits a timely checkout of the merged command sequence as it will be loaded on the spacecraft. StateSim will model the loading of command macros and their calling by time-tagged commands. Any problems in the command processing will be caught and reported via an error report. An autonomy history report will show which autonomy rules were tripped. StateSim models spacecraft attitude based on the attitude commands, and it computes attitude relative to the Sun line and targets of interest. This information is useful to assess power and thermal conditions the spacecraft will experience in executing the command load. Appropriate orbits must be loaded into StateSim for proper Sun and planet position computations. The input/output files for SeqGen and StateSim are used by a second person in a manual verification procedure. After the second-person check, there is a larger engineering review of the outputs of SeqGen and StateSim to ensure spacecraft health and safety.

11 Mercury Orbital Operations

Mercury orbit-phase operations will begin after the Mercury orbit insertion (MOI) operations are completed in March 2011 (McAdams et al. 2007). Nearly 70% of the spacecraft propellant will be used during a pair of LVA bi-propellant maneuvers. MOI will be the most critical operation of the mission beyond launch, so enhanced measures will be taken to reduce any risk of aborting the maneuver. One example that will be considered during the MOI review period is to conduct the operation with a greatly reduced subset of active autonomy rules.

During the Mercury orbit phase, science data collection and capture will be the primary operational function. All instruments will be operational during each orbit and will be acquiring the data necessary for meeting the mission success criteria (Solomon et al. 2007). This activity will be performed in a store-and-forward manner. Typically each 24-hour period (two orbits) will be composed of a 16-hour data collection period followed by an 8-hour data downlink period. Both solid-state power amplifiers (SSPAs) will be used for maximum data return using the phased-array antennas. The injection design orbit parameters enable fulfillment of full science mission success criteria with a 200-km minimum altitude at 60°N periapsis latitude, a 12-hour period, and an 80° inclination as shown in Fig. 7 (McAdams et al. 2007).

Once Mercury orbit is achieved, the unique Mercury environment will require several differences from cruise operations. Differences include orbit-correction maneuvers (OCMs), more frequent solar array positional changes, increased recorder data volume with stricter management criteria, increased DSN support, and the introduction of a hot-planet keep-out (HPKO) zone, described in the following. Spacecraft Sun-keep-in (SKI) constraints will continue to be enforced as they are for cruise operations inside 0.95 AU Sun distances. SKI constraints require the spacecraft's sunshade be oriented to within 12° of the Sun–spacecraft line (Leary et al. 2007).

Conditions within the year of Mercury orbit operations will also change. For example, Mercury orbital motion will result in changing spacecraft orbits, e.g., the noon–midnight and dawn–dusk orbits shown in Fig. 8. Both thermal and instrument viewing conditions differ between these orbit extremes, so instrument operations and spacecraft orientations will vary as the type of orbit changes. Figure 8 also depicts these different viewing conditions by illustrating the prime science collection regions in red, the medium-priority science collection regions in yellow, and the lower-priority science collection regions in green. There are orbit segments, shown in white, during the noon-midnight seasons in which the instruments cannot view Mercury at all within the attitude constraints.

In contrast to the cruise phase, when TCMs and larger DSMs are performed individually to maintain the course toward Mercury, during the orbit phase three planned pairs of maneuvers will be carried out every 88 Earth days to maintain the ideal orbital parameters as the periapsis altitude naturally increases. The initial maneuver planning and execution process remain the same, but OCMs will be performed as pairs with the possibility of less than a day between burns. This close spacing between maneuvers will require use of the burn update process, which is a quick-turnaround update to a maneuver load that has already been uplinked to the spacecraft. Routine staffing will also change since orbital phase

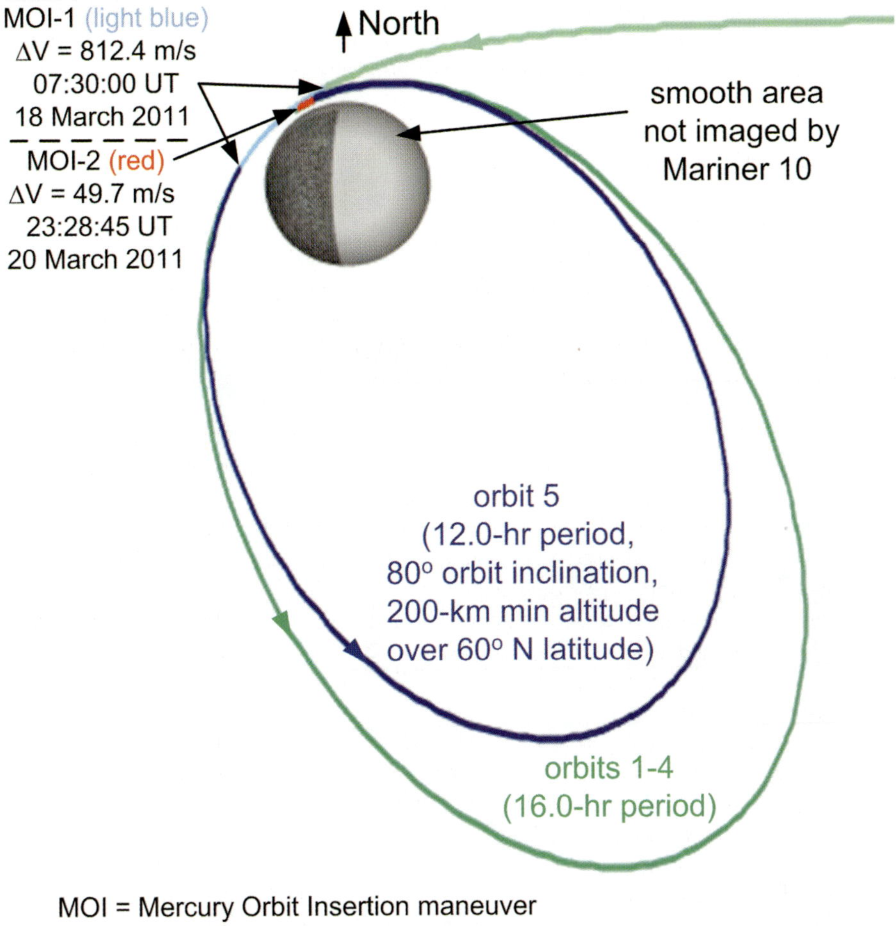

Fig. 7 Mercury orbit insertion and orbit parameters

spacecraft contacts will occur at least daily rather than the nominal three times per week during cruise. The track schedule will generally coincide with day-shift operations, however, so that full operations team and engineering support will coincide with the majority of contact time.

Recorder management and solar array management will also change once Mercury orbit is achieved. Recorder volume during the orbit phase will be greatly increased, with all instruments continuously powered and operating together. In order to avoid recorder saturation, files will be automatically deleted once they are successfully downlinked to the ground, and low-priority data will be automatically deleted without being downlinked. The solar array panel positions do not change often during the cruise phase, except for discrete activities. However, the power engineers will be providing the operations team with requests for panel position changes often during the orbit phase, as often as every two days during peak periods. These parameter block adjustments will define new solar array positions for each operational and safe mode and are designed to balance thermal and power conditions for the solar arrays over the course of the orbital phase. Because the spacecraft is operat-

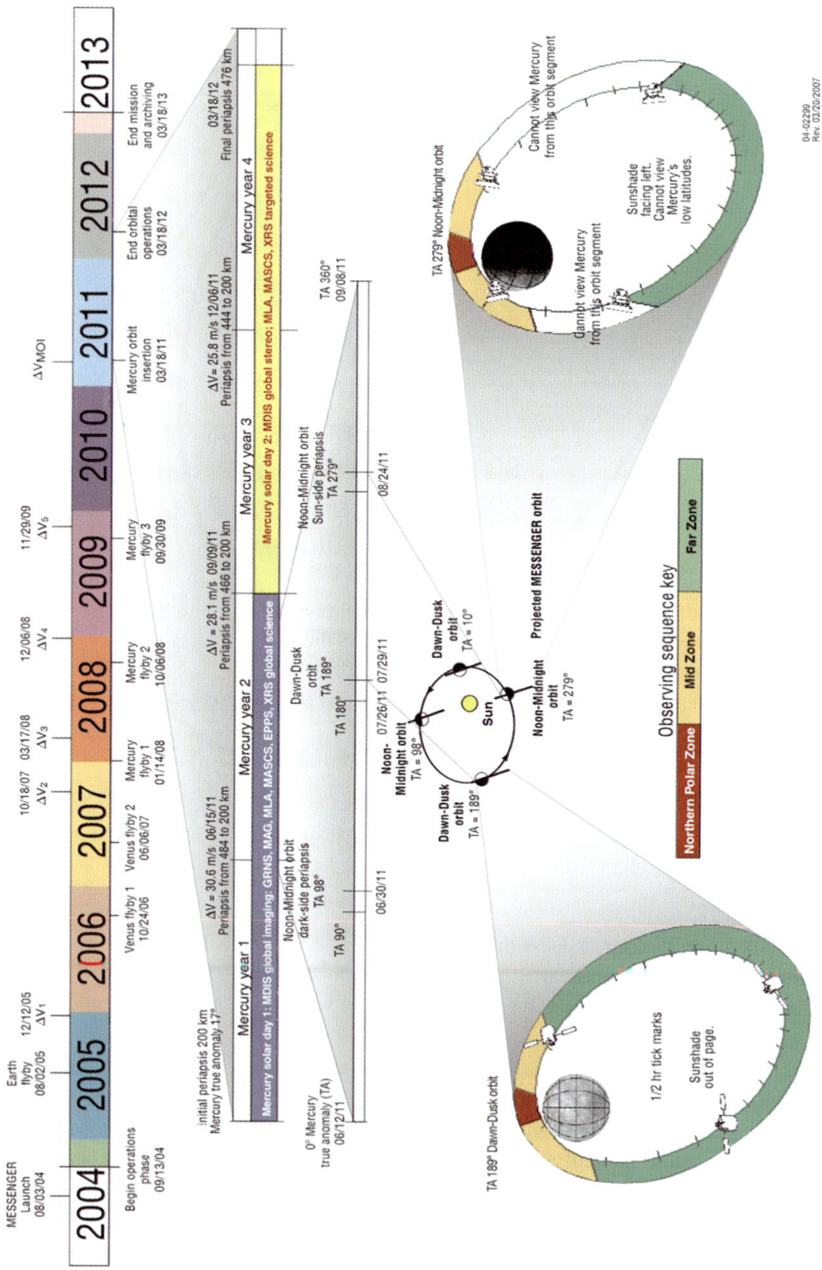

Fig. 8 Dawn–dusk and noon–midnight orbit differences

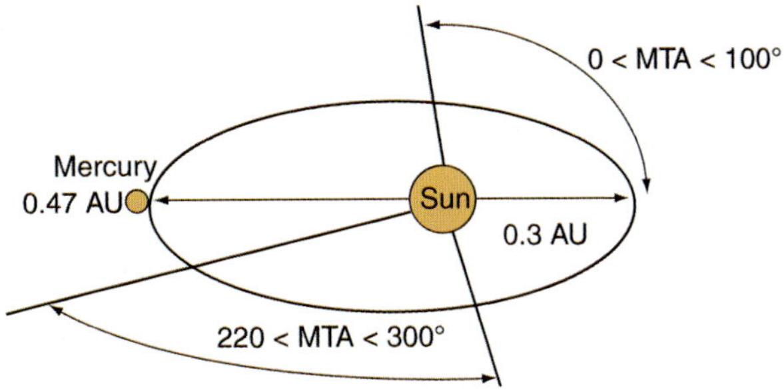

Fig. 9 Mercury True Anomaly hot planet seasons

ing so near to the Sun, the solar arrays are tilted away from the Sun line to reduce thermal effects, and the amount of off-pointing is driven by power availability.

Perhaps the biggest change from cruise operations is the introduction of a new constraint, the HPKO zone. Infrared radiation from the planet surface during a Mercury True Anomaly (MTA) of 0–110° and 220–300°, shown in Fig. 9, causes a thermal spike of sufficient magnitude that pointing restrictions and additional solar array control measures are necessary.

The hot planet orbits comprise about half the Mercury orbits and occur twice per Earth day during the MTA ranges described. The Mercury HPKO zone specifies a range of spacecraft body-fixed directions in order to keep the top deck of the spacecraft pointed away from the planet. There are five conditions that must be satisfied simultaneously for the onboard HPKO zone logic to be active:

1. Mercury is in one of the two defined true anomaly regions in its orbit about the Sun;
2. the spacecraft is over the sunlit hemisphere as it orbits the planet;
3. the spacecraft is over the northern hemisphere as it orbits the planet;
4. the top deck of the spacecraft is pointing within a range of nadir (the direction from the spacecraft to the target planet center) defined as the $-Z$ body-frame axis pointing less than 90° from nadir, plus a user-defined angular offset, currently set to 0.5°; and
5. the HPKO functionality must be in the enabled state via command.

The G&C subsystem can detect a violation of the zone, even during safe-mode transitions, and take action. For example, if the downlink geometry would require a hot planet violation to achieve safehold attitude, the G&C subsystem will rotate the spacecraft such that the hot planet constraint is maintained at the sacrifice of the downlink until the spacecraft exits the zone boundary again. The HPKO season will also require the operations team to sequence the panels edge-on to the Sun for the worst 20 minutes of each orbit as shown in Fig. 10. This action will reduce the thermal effects on the arrays throughout the Mercury orbit phase.

12 Mission Operations Center and Ground System

MESSENGER mission operations are conducted from the SSMOC. The SSMOC is a shared facility for JHU/APL space science missions. MESSENGER and New Horizons have operated concurrently from the facility following the New Horizons launch in January 2006, and other missions use a portion of the SSMOC for data and voice routing. Members of both teams meet to discuss ground system issues and activities that could impact one or both operations, and they also attend weekly mission management meetings to remain informed on upcoming spacecraft activities for each mission. This coordination minimizes impact and ensures adequate personnel coverage for both missions. The SSMOC facility, shown in

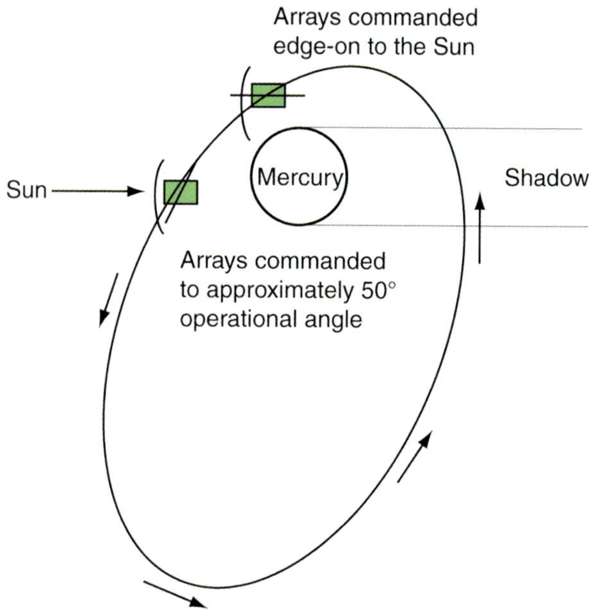

Fig. 10 Solar array control over hot planet crossings

Fig. 11, consists of personnel offices, a hardware simulator room, a situation meeting room, an equipment room, the real-time command and monitoring area (known as the Pit), and a payload monitoring area, along with a small kitchen and a copy room.

A mission T1 line contains the data and voice channels to and from the SSMOC. Voice boxes are strategically placed throughout the control center for effective communications during critical activities. Four overhead projectors and display screens are also frequently used during critical activities for telemetry data viewing and three-dimensional attitude displays. The MESSENGER ground system uses a distributed architecture consisting of UNIX-based Sun workstations and personal computers, printers, and X terminals for both real-time and offline operations, as shown in Fig. 12.

The MESSENGER ground system is supported by three networks: the NASA Restricted IONet network, the Operations DMZ (De-Militarized Zone) Network, and the JHU/APL Internal Network. The IONet is the primary command and control network, the DMZ is primarily used for offline planning and analysis capabilities, and the JHU/APL Internal network supports all of the personnel PCs and e-mail. The two command workstations reside on the restricted IONet. Since it is a NASA network, the IONet is governed by the Goddard Space Flight Center (GSFC) security audits, risk assessment plan, and security plan. The IONet is also in a secure keycard access area of the SSMOC. The EPOCH 2000 software provides the core functionality for commanding and telemetry monitoring of the MESSENGER spacecraft combined with JHU/APL applications such as a telemetry chooser filtering tool. All communications with the spacecraft are performed using the NASA Deep Space Mission Services at JPL. Telemetry received by a Deep Space Communications Complex (DSCC) is sent as Standard Formatted Data Unit (SFDU)-encapsulated telemetry transfer frames to the SSMOC. Commands originated from the SSMOC are transmitted to the DSCC using the CCSDS Space Link Extension (SLE) Forward Command Link Transmission Unit (CLTU) service. The EPOCH software operates in conjunction with Oracle-based ASCII flat files that the operations team maintains for proper command formatting and telemetry decommutation.

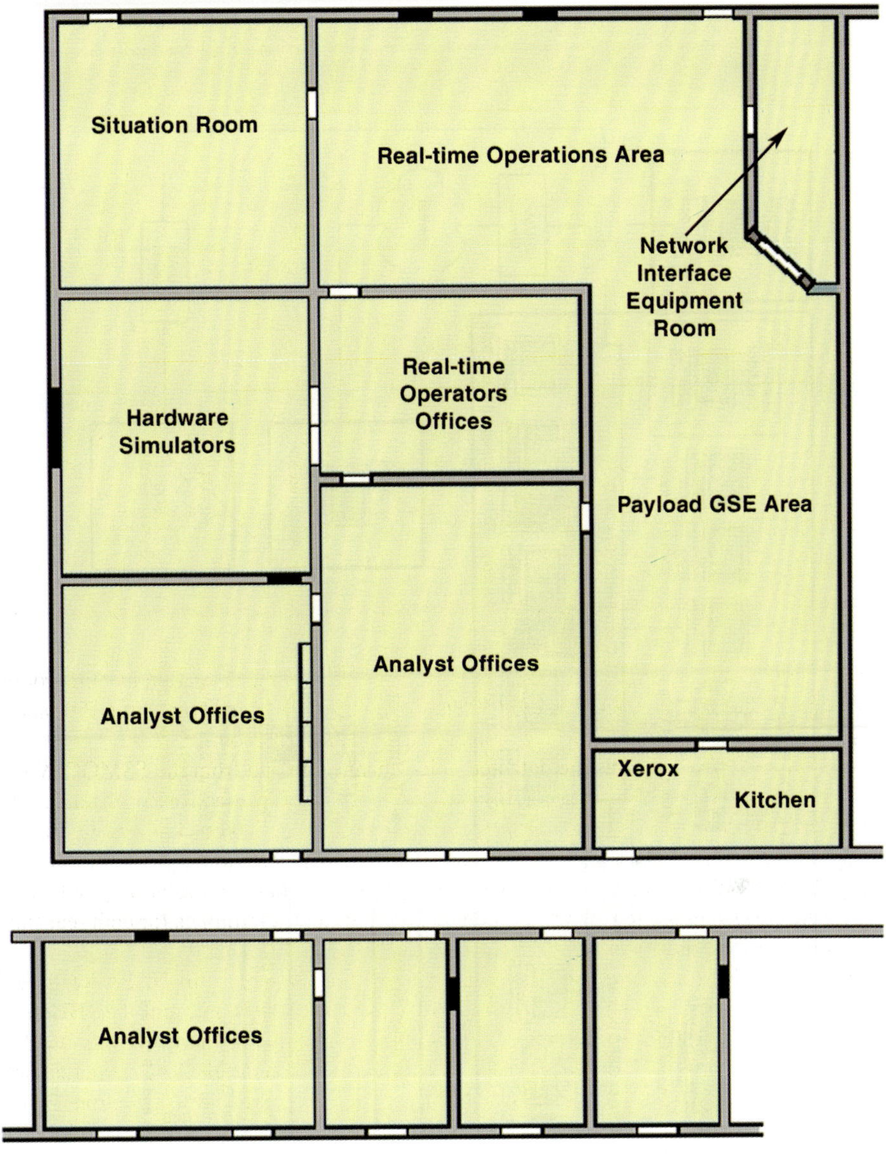

Fig. 11 Space Science Mission Operations Center layout

During real-time MESSENGER contacts, the DSN forwards ground station monitor data to the SSMOC including receiver status information, command queue, automatic gain control (AGC), and other information useful for link diagnosis. File delivery from the SSR is handled by the JPL-core CFDP ground software combined with JHU/APL support software. The load, dump, and compare (LDC) suite consists of several tools that generate STOL command scripts for loading products to the spacecraft. Examples of LDC applications on the DMZ include the autonomy rule compiler, computed telemetry compiler, storage variable compiler, parameter load tool, processor load tool, macro load tool, time parameter load

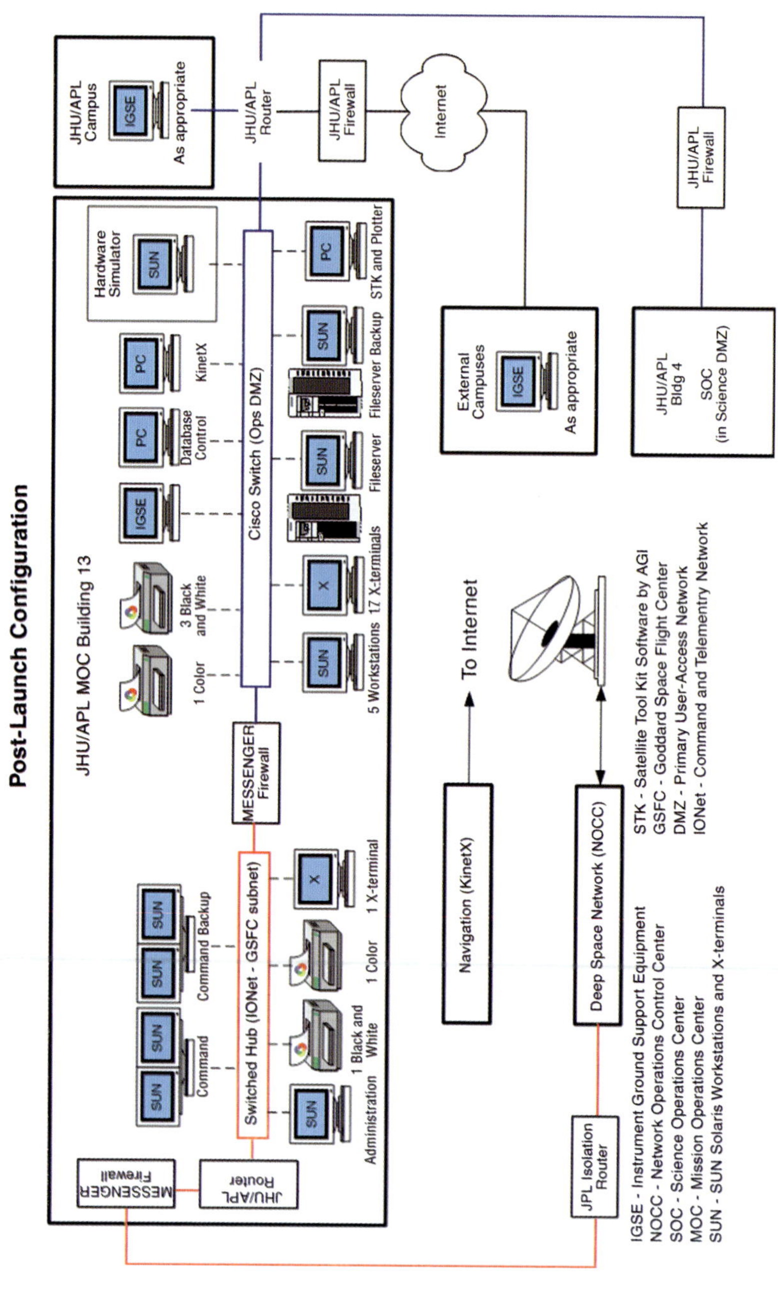

Fig. 12 JHU/APL MESSENGER ground system architecture

tool, and ephemeris load tool. Offline functions are primarily performed on the DMZ, but the necessary real-time software that supports memory object dumps and comparisons, either cyclic redundancy check (CRC) or byte-for-byte, resides on the IONet.

The DMZ network also supports planning, assessment, monitoring, and data archival functions and the hardware simulator, with dedicated machines supporting these functions. The primary components of the planning and scheduling system include the JPL SeqAdapt and SeqGen sequence generation software, the StateSim software simulator developed by an outside contract, the JHU/APL SeqPost software, and mission operations keyword generation software for DSN support. Various pieces of "glueware" have also been developed to ensure that all of these packages work together. The assessment suite consists of an engineering dump package for text-formatted archive data retrieval, plotting software and automation Perl scripts, timekeeping flight correlation reports, an alarms tool for reporting limit violations within archived data, and data gap analysis and reporting tools. The DMZ network is also used by external operational support such as the KinetX navigation team, the mission design team, Science Operations Center personnel, and engineering support personnel A software application known as Big Brother is used to monitor processes, central processing unit usage, and disk space on all of the SSMOC machines, and text alerts are dispatched to appropriate ground support personnel whenever any of the red alarm settings are surpassed. Telemetry monitoring as well as archived data playback can be performed from any of the DMZ machines licensed to run the EPOCH software.

The hardware simulator consists of an Avtec Portable Telemetry Processor (PTP) front-end command interface, testbed emulation software, G&C dynamics and power models, a command and telemetry workstation supported by EPOCH 2000, a dual-string Integrated Electronics Module, and engineering models for each of the instruments and the data processing units. This configuration provides the operations team with considerable fidelity for testing spacecraft sequences and command scripts. Since the simulator includes real engineering model MPs and FPPs, it is especially useful for testing command and data handling command sequences and autonomy responses. The PDU and G&C sensors are examples of primary spacecraft components that were emulated for the testbed.

13 Summary

The MESSENGER spacecraft is off to an excellent start on a six-and-one-half year journey through the inner solar system prior to Mercury orbit insertion. All planned objectives through this early cruise phase of the mission, including spacecraft and instrument commissioning, have been met. The early operations have gone very smoothly, a result that can be attributed largely to the application of lessons learned from past missions including adherence to time-proven standards of mission conduct. The operation will continue to build on these successes as it moves into the inner cruise phase of the mission. There are many challenges ahead, including flyby operations at Venus and Mercury, conjunctions, solar eclipses, and Mercury orbit insertion. The proven processes and standards of conduct that are in place will help ensure that all of these objectives are as successful as launch and early operations have been.

Acknowledgements The authors acknowledge the contributions of JHU/APL personnel for providing several of the figures and tables used in this paper. Gabrielle Griffith provided the authors with the ground network configuration material. In addition, the mission design and orbit information was provided by Jim McAdams. Rick Shelton defined the mission planning and command sequence diagrams, and Bob Nelson provided the MOC layout material and the communications bit rate information. The orbital operations figures were developed by Andy Santo.

References

M.E. Holdridge, American Astronautical Society/American Institute of Aeronautics and Astronautics Astrodynamics Specialists Conference, Paper AAS 01-375, Quebec City, Canada, 2001, 20 pp

M.E. Holdridge, in R.W. Farquhar (ed.), Acta Astronautica **52**, 343–352 (2003)

C.J. Krupiarz et al., in *Proceedings of the 5th International Academy of Astronautics International Conference on Low Cost Planetary Missions*. Special Publication SP-542 (European Space Agency, Noordwijk, the Netherlands, 2003), pp. 435–442

J.C. Leary et al., Space Sci. Rev. (2007, this issue) doi:10.1007/s11214-007-9269-0

J.V. McAdams, R.W. Farquhar, A.H. Taylor, B.G. Williams, Space Sci. Rev. (2007, this issue) doi:10.1007/s11214-007-9162-x

A.G. Santo et al., in *53rd International Astronautical Congress*. The World Space Congress, Paper IAC-02-U.4.04, Houston, TX, 2002, 11 pp

S.C. Solomon et al., Planet. Space Sci. **49** 1445–1465 (2001)

S.C. Solomon et al., Space Sci. Rev. (2007, this issue). doi:10.1007/s11214-007-9247-6

H.L. Winters et al., Space Sci. Rev. (2007, this issue). doi:10.1007/s11214-007-9257-4

Space Sci Rev (2007) 131: 601–623
DOI 10.1007/s11214-007-9257-4

The MESSENGER Science Operations Center

Helene L. Winters · Deborah L. Domingue · Teck H. Choo · Raymond Espiritu · Christopher Hash · Erick Malaret · Alan A. Mick · Joseph P. Skura · Joshua Steele

Received: 24 July 2006 / Accepted: 7 August 2007 / Published online: 25 October 2007

Abstract The MESSENGER Science Operations Center (SOC) is an integrated set of subsystems and personnel whose purpose is to obtain, provide, and preserve the scientific measurements and analysis that fulfill the objectives of the MErcury Surface, Space ENvironment, GEochemistry, and Ranging (MESSENGER) mission. The SOC has two main functional areas. The first is to facilitate science instrument planning and operational activities, including related spacecraft guidance and control operations, and to work closely with the Mission Operations Center to implement those plans. The second functional area, data management and analysis, involves the receipt of science-related telemetry, reformatting and cataloging this telemetry and related ancillary information, retaining the science data for use by the MESSENGER Science Team, and preparing data archives for delivery to the Planetary Data System; and the provision of operational assistance to the instrument and science teams in executing their algorithms and generating higher-level data products.

Keywords MESSENGER · Mercury · Science operations · Science planning · Data products · Data archiving · Planetary Data System

1 Introduction

The MErcury Surface, Space ENvironment, GEochemistry, and Ranging (MESSENGER) spacecraft will be the first to orbit the planet Mercury. The mission has been designed to address the following six broad scientific questions:

- What planetary formational processes led to Mercury's high ratio of metal to silicate?
- What is the geological history of Mercury?
- What are the nature and origin of Mercury's magnetic field?

H.L. Winters (✉) · D.L. Domingue · T.H. Choo · A.A. Mick · J.P. Skura · J. Steele
The Johns Hopkins University Applied Physics Laboratory, 11100 Johns Hopkins Road, Laurel, MD 20723-6099, USA
e-mail: helene.winters@jhuapl.edu

R. Espiritu · C. Hash · E. Malaret
Applied Coherent Technology, 112 Elden St., Suite K, Herndon, VA 22070, USA

Table 1 MESSENGER instrument payload

Instrument	Description
Mercury Dual Imaging System (MDIS)	Narrow-angle imager and wide-angle multispectral imager. Pointing is assisted with a pivot platform. Maps landforms, surface spectral variations, and topographic relief from stereo imaging.
Gamma-Ray and Neutron Spectrometer (GRNS)	Gamma-Ray Spectrometer (GRS) measures the emissions from radioactive elements and gamma-ray emission stimulated by cosmic rays. Maps elemental abundances in crustal materials. Neutron Spectrometer (NS) provides sensitivity to hydrogen in ices at the poles and average sub-spacecraft atomic weight of surface material.
X-Ray Spectrometer (XRS)	Measures the fluorescence in low-energy X-rays stimulated by solar gamma rays and high-energy X-rays. Maps elemental abundances of surface materials.
Magnetometer (MAG)	Measures the detailed structure and dynamics of Mercury's magnetic field and searches for regions of magnetized crustal rocks.
Mercury Laser Altimeter (MLA)	An infrared laser transmitter coupled with a receiver that measures the round-trip time of a burst of laser light reflected off Mercury's surface, yielding a distance measurement. Produces highly accurate measurements of topography and measures Mercury's physical libration.
Mercury Atmospheric and Surface Composition Spectrometer (MASCS)	Ultraviolet-Visible Spectrometer (UVVS) measures composition and spatial and temporal variations of exospheric species. Visible-Infrared Spectrograph (VIRS) maps surface reflectance to determine mineral composition.
Energetic Particle and Plasma Spectrometer (EPPS)	Measures the composition, spatial distribution, energy, and time variability of charged particles within and surrounding Mercury's magnetosphere. Plasma is measured by the Fast Imaging Plasma Spectrometer (FIPS) and higher-energy particles by the Energetic Particle Spectrometer (EPS).
Radio Science (RS)	Uses the Doppler effect to measure Mercury's mass distribution, including spatial differences in crustal thickness and corresponding gravitational field variation. Determines Mercury's radius versus position from radio signal occultations.

- What are the structure and state of Mercury's core?
- What are the radar-reflective materials at Mercury's poles?
- What are the important volatile species and their sources and sinks near Mercury?

These questions, described and discussed in greater detail by Solomon et al. (2001, 2007), led to the selection of the instrument payload (Gold et al. 2001), listed in Table 1. The mission science objectives are met by acquiring data sets that will achieve the following project-level requirements:

- Provide major-element maps of Mercury to 10% relative uncertainty on the 1,000-km scale and determine local composition and mineralogy at the ~20-km scale.
- Provide a global map with >90% coverage (monochrome) at 250-m average resolution and >80% of the planet imaged stereoscopically, provide a global multi-spectral map at 2 km/pixel average resolution, and sample half of the northern hemisphere for topography at 1.5-m average height resolution.

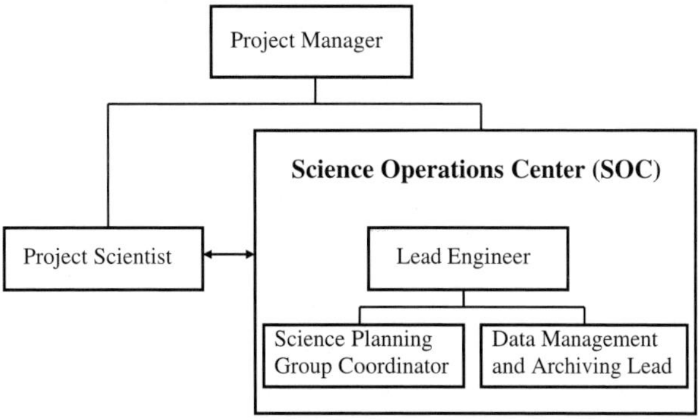

Fig. 1 SOC management structure within the MESSENGER organization. The Project Manager reports directly to the Principal Investigator, who is responsible for all aspects of the MESSENGER mission

- Provide a multipole magnetic-field model resolved through quadrupole terms with an uncertainty of less than ~20% in the dipole magnitude and direction.
- Provide a global gravity field to degree and order 16 and determine the ratio of the solid-planet moment of inertia to the total moment of inertia to ~20% or better.
- Identify the principal component of the radar-reflective material at Mercury's north pole.
- Provide altitude profiles at 25-km resolution of the major neutral exospheric species and characterize the major ion-species energy distributions as functions of local time, Mercury heliocentric distance, and solar activity.

Science operations within the MESSENGER project are the responsibility of the Project Scientist and are directed by the Science Steering Committee (chaired by the project's Principal Investigator). The day-to-day tasks of ensuring that the project meets its science goals reside within the Science Operations Center (SOC) and its Science Planning Group (SPG).

The MESSENGER Science Team is divided into four discipline groups (Geology, Geophysics, Geochemistry, and Atmosphere and Magnetosphere), each with a designated discipline group leader who represents the group on the Science Steering Committee. The Project Scientist and the SPG Coordinator are also members of the Science Steering Committee, thus flowing Science Team needs and insight into MESSENGER's daily science operations. Additional members of the Science Steering Committee include the Payload System Engineer and the Project Manager.

The overall responsibility for the daily science operation tasks resides within the MESSENGER SOC, headed by the SOC Lead Engineer. The SOC is an integrated set of subsystems and personnel who are responsible for acquiring, distributing, and archiving the science observations and data products that fulfill the science objectives of the MESSENGER mission. SOC management structure is depicted in Fig. 1. The SOC supports all instrument operations associated with the mission timeline (McAdams et al. 2007). The SOC is being developed by The Johns Hopkins University Applied Physics Laboratory (APL) in partnership with Applied Coherent Technology Corporation (ACT). A physical SOC facility is located on the APL campus, but the SOC can also be accessed remotely.

2 SOC Overview

2.1 SOC Responsibilities

The SOC has two main areas of responsibility: (1) science acquisition and (2) data management and archiving. Science acquisition includes the planning and commanding of instrument activities and related spacecraft guidance and control (G&C) activities, in close coordination with the Mission Operations Center (MOC), as well as data validation and progress assessment in meeting the mission science goals. Science acquisition is the responsibility of the SPG, a team composed of scientists and instrument engineers charged with planning and implementing instrument operations. To meet these responsibilities, the SPG uses the Science Planning, Framing, and Toolbox (SciBox) utility, a science planning and operation analysis software tool.

The second area of responsibility, data management and archiving (DMA), includes processing science-related telemetry, providing access to all science-related data, generating science data products, monitoring data coverage, and generating and delivering to NASA's Planetary Data System (PDS) an archive of the MESSENGER science observations. DMA is handled by ACT and is accomplished primarily through the use of their Planetary Information Processing Environment (PIPE) utility, a software tool that processes science-related telemetry and generates relevant data products.

2.2 SOC Software Tools

The two main software tools of the SOC are SciBox and PIPE, used to accomplish the science acquisition and DMA tasks, respectively.

2.2.1 SciBox

SciBox is a Java-based toolset offering science planners a suite of utilities to aid in the planning of instrument activities. It provides an end-to-end data acquisition plan for obtaining observations of Mercury that will meet MESSENGER mission objectives. This end-to-end, full-mission plan is based on instrument goals and constraints, as well as G&C and spacecraft operational constraints. The SciBox toolset breaks down the mission-long data acquisition plan into weekly increments that it then translates into sets of instrument command sequences and guidance and control command sequences that can be loaded to the spacecraft for execution. The weekly instrument command requests can be modified using the SciBox toolset, which then automatically reoptimizes the remainder of the full-mission plan, thus ensuring that mission objectives can be met. The full mission plan, also referred to as the baseline operation plan, is reexamined and reoptimized by the SPG in eight-week cycles, in order to accommodate changes in operation and science acquisition strategies. The Science Steering Committee establishes priorities for instrument activities, which are incorporated into SciBox as instrument constraints. All command sequence requests generated by SciBox are reviewed by the SPG and the MOC before being sent to the spacecraft.

2.2.2 PIPE

PIPE is a proprietary software utility of ACT, customized for use by the MESSENGER project. It is used to process all science-related telemetry packets and images, decommutate and decompress them, and generate raw data and image files known as Experiment Data Records (EDRs) for access by the MESSENGER Science Team.

PIPE is also used in the generation and ingestion of higher-level science data. Data products are produced according to predefined formats agreed upon by the MESSENGER project and the PDS and communicated through individual instrument Software Interface Specification (SIS) documents that will be archived with the PDS. All data are cataloged in a Relational Database Management System (RDBMS) for access by the Science Team and to facilitate the generation and delivery of the MESSENGER data archive to the PDS.

2.2.3 *Utility Interface*

The science acquisition and DMA utilities, SciBox and PIPE, respectively, do not function in isolation. Both software utilities are used to monitor the acquired data set; thus they must share information.

The PIPE utility is used to view and analyze data taken during instrument observations. It provides the instrument teams with an interface for assigning data quality flags based on the usefulness of the observation for meeting the MESSENGER science objectives. Any observation resulting in data of insufficient quality to meet the mission's science objectives must be replanned and reflected appropriately in the data coverage maps. Therefore, the PIPE software transfers both the quality flag values and the coverage maps to the SciBox software utility.

3 Science Acquisition

The science acquisition responsibilities of the MESSENGER SOC are fulfilled through the MESSENGER SPG, a component of the SOC. The SPG consists of each instrument's core team plus the Payload System Engineer and is led by the SPG Coordinator. The instrument core teams are composed of the Instrument Scientist or Science Sequencer (the SPG's main point of contact for each instrument), Instrument Engineer, Cognizant Co-Investigator, and MOC instrument lead.

The science acquisition responsibilities are divided into four categories: (1) observation planning, (2) instrument commanding, (3) data validation, and (4) science objective assessment. Observation planning includes the planning and generation of an integrated observing strategy for all instruments. Instrument commanding is the development of instrument commands and the translation of the observation plan into command sequences for upload to the spacecraft through the MOC. Validation includes observation validation and observation data quality assessment. Science objective assessment includes the monitoring of data coverage and mission status. These responsibilities and the tasks contained therein (detailed in the following sections) are the mechanism through which the MESSENGER project operates its instrument payload, providing the Science Team with the data required to meet the MESSENGER science objectives and mission goals.

3.1 Observation Planning

Observation planning is the development and generation of an integrated instrument observation strategy, which must meet the scientific mission success criteria, as set forth in MESSENGER's Program Level Requirements document.

The science observation strategy starts with an end-to-end mission plan that integrates all instrument activities over the full mission time span. This science observation strategy demonstrates the means to acquire the data set needed to meet the mission success criteria, while remaining flexible. The end-to-end mission plan allows for modifications due to unforeseen events, such as Deep Space Network (DSN) outages, poor data quality, or the discovery of interesting scientific properties that warrant more focused observations. This plan is generated during the mission cruise phase, prior to the first Mercury flyby, and kept up to date throughout the mission by the SPG Coordinator. This end-to-end mission planning capability is provided by the SPG's science planning tool, SciBox.

The end-to-end plan is then subdivided into time increments commensurate with the instrument commanding schedule (i.e., weekly). The science planning tool, SciBox, provides a suggested weekly instrument operation plan and command load that are commensurate with fulfilling the science observation strategy. These weekly operation plans can be modified and reingested into SciBox for reoptimization of the end-to-end mission plan. This capability ensures that modifications do not impact the ability of the mission to meet its success criteria. All command loads are reviewed by the SPG Coordinator and MOC. The MOC tests and validates each load prior to uplink to the spacecraft for execution.

3.2 Instrument Commanding

Instrument commanding includes the development of instrument commands and the translation of the science observation strategy into executable command sequences. The responsibilities under observation commanding include command library generation, command sequence generation, and weekly command sequence evolution.

Command library generation includes the design, development, and testing of instrument commands. The commanding of the instrument payload is conducted using Canned Activity Sequences (CASs) containing a set of commands designed to perform a specific function, each command tagged with a relative time of execution. The CASs are designed and reviewed by the Science Sequencers and Instrument Engineers, and constructed and tested by the MOC, after which they are placed under configuration management by the MOC.

Command sequences are generated through the conversion of the science observation strategy into executable commands. SciBox takes weekly increments of the end-to-end mission plan and converts these into weekly instrument requests. Requests are one or more CASs grouped to perform instrument operations. All CASs within a request are given a relative time of execution, based on the science observation strategy. Each instrument's requests are available from SciBox after all operational conflicts have been resolved on the basis of instrument prioritizations supplied by the Science Steering Committee. The Science Sequencers can modify these requests, using SciBox to examine the impact on the full-mission plan to ensure that the mission success criteria can still be met for all instruments.

Command sequences evolve through an iterative delivery to the MOC team of a viable set of instrument commands, on a weekly time scale. Table 2 shows the schedule and information exchange between the SPG and the MOC for each week of instrument operations while in Mercury orbit. Multiple command loads, each in a different phase of their construction, are being developed and addressed each week. They are all reviewed at the weekly SPG meeting to monitor their status and address any issues. Each command load request is built by the same set of instrument team leads within the SPG. Six weeks prior to the execution of the command load on board the spacecraft, the SPG receives the spacecraft housekeeping files from the MOC. These files indicate time periods for such activities as Deep Space Maneuvers (DSMs), tracks, burns, and spacecraft maintenance tasks, during which instrument operations cannot be scheduled. At this time the imaging team receives its inputs from the Navigation Team for any requested optical navigation images. The spacecraft housekeeping and navigation requests are incorporated into the SciBox planning tool, and the tool delivers a suggested command load (baseline operation plan) for the activities that will execute during week N. During this week, any change requests to the command load provided by SciBox are submitted to the Payload Operations Manager (POM) and SPG Coordinator and are resolved prior to the end of the week. The suggested command load is optimized with the approved changes to ensure that the mission science objectives are met. Five weeks prior to the execution of the command load, the SPG Science Sequencers build their initial instrument requests based on the SciBox-suggested command load and any changes that have been approved. Each set of instrument requests is reviewed and approved by the Instrument

Table 2 Instrument command load delivery schedule when in Mercury orbit

Week	Monday	Tuesday	Wednesday	Thursday	Friday
$N-6$	MOC delivers Initials; SciBox ingests MOC Initials	All baseline operation plan change requests submitted to POM	All baseline operation plan change requests are resolved		
$N-5$	Instrument Sequencers complete Initial Command Requests		Instrument Engineer and Scientist review Initial Command Requests	POM merges and checks Science Initials	POM delivers Science Initials to MOC
$N-4$	MOC builds Initial Load, runs "mid-scheds"				
$N-3$	MOC delivers review of Initial Load	Instrument Sequencers complete Final Command Requests	Instrument Engineers review Final Command Requests	POM merges, reviews, and delivers Science Finals to MOC	
$N-2$	MOC builds final load				
$N-1$	MOC reviews and uplinks load				
N	Spacecraft executes load				

Engineer and the Payload System Engineer. After review and approval they are delivered to the POM, who ensures there are no conflicts between individual instrument requests or with the spacecraft housekeeping file prior to delivering the instrument command requests to the MOC. During week $N-4$ (four weeks prior to the load's execution on the spacecraft), the MOC simulates and tests the command load and works with the SPG to resolve any instrument conflicts with the SPG. Three weeks prior to the execution of the load, the MOC provides the SPG with the simulator results, and any outstanding issues are resolved prior to the final delivery of the instrument request files. During this week the SPG Science Sequencers build their final instrument requests. Each set of instrument requests is reviewed and approved by the Instrument Engineer and the Payload System Engineer. After review and approval they are delivered to the POM for final verification that there are no conflicts between the individual instrument requests or with the spacecraft housekeeping file. Two weeks prior to the execution of the command load, the MOC builds the full command load. One week prior to execution on the spacecraft, the MOC reviews and uplinks the commands to the spacecraft.

3.3 Data Validation

Data are validated by examination of the instrument measurements obtained from the spacecraft. Observations must first be validated to determine that the data downlinked from the spacecraft correspond to the commanded observations executed on the spacecraft. This validation is accomplished through data examination, downlink reports, and solid-state recorder (SSR) resource management and modeling, which is performed by both the MOC and SPG independently.

The quality of those observations must also be assessed to verify the usability of the measurement for achieving the science objectives. Because of the large volume of data being acquired by each instrument, the data quality validation and assessment process is automated, and warning flags are issued where manual examination may be warranted. The assessment criteria for this automated process are currently being defined by the MESSENGER Science Team.

When data are lost due to command execution failures, lost because of spacecraft or Deep Space Mission System (DSMS) downlink errors, or determined to be of poor quality, the observations are rescheduled. The results of both validation tasks are fed back into the science observation strategy to ensure that the data coverage required to meet the mission success criteria is achieved.

3.4 Science Objective Assessment

Data coverage maps are produced by the SOC PIPE and SciBox tools and are monitored to determine whether science objectives are being met. Mission status is monitored by evaluation of this data coverage in terms of successful achievement of the mission science objectives and success criteria.

3.5 Science Planning, Framing, and Toolset (SciBox) Utility

A challenge for the SOC is to schedule and operate the MESSENGER instruments to their full capability and to acquire all data needed to fulfill the mission objectives. The primary task in meeting this challenge is to determine the available observation opportunities. These opportunities are restricted by the limited range of safe spacecraft pointing, available power, limited SSR size, conflicting pointing geometries among instruments, and available downlink bandwidth.

The approach to meeting this challenge is to use science planning and operation simulation tools to develop a mission-long, integrated schedule so that resources may be fully and optimally used. SciBox is the science planning and operation simulation tool that will allow the SPG to identify observation opportunities efficiently, develop a conflict-free schedule, and convert automatically the scheduled opportunities into instrument command sequences for uplink to the spacecraft and its instruments. The following sections describe the science tool's functionality in addressing the challenge of acquiring all needed science data.

3.5.1 SciBox Capabilities

The SciBox planning tool:

- Presents each instrument scientist with a list of all opportunities to acquire data to meet the instrument's science objectives.
- Allows the Science Team to choose which opportunities are to be used.
- Allows the Science Team to take advantage of opportunities to take measurements when a feature of interest is discovered on Mercury.
- Helps the scientists and engineers visualize what the spacecraft may be able to do at any point in the mission. The visualization tools include display of the spacecraft position and orientation relative to Mercury, display of the field-of-view of each applicable instrument, and display of plots to track different aspects of the spacecraft.

- Models the power requirements of each instrument on the spacecraft, allowing the planners to see the potential impact of power usage by each instrument for comparison with the total power available. Using this feature provides assurance that the power required to perform an activity does not exceed the projected power availability.
- Tracks the state of all mission objectives by interfacing with PIPE to obtain records of data that have been downlinked and considering what data have been taken but remain to be downlinked, for comparison with what data have yet to be taken to complete mission objectives. This tracking of objectives affords the insight needed to select subsequent observation opportunities.
- Monitors the use of the SSR and models each instrument's data storage usage.
- Allows planners to determine quickly if there are any scheduling conflicts between instruments. These conflicts must be discovered as soon as possible to allow instrument scientists to replan observations appropriately.
- Takes all scheduled opportunities and creates an instrument command load request to be sent to the spacecraft via the MOC. This sequence contains all instrument commands as well as related G&C spacecraft pointing commands.
- The SciBox tool offers comprehensive planning features. It can simulate the operations of the MESSENGER spacecraft, plan instrument activities, and generate the commands to control spacecraft pointing and instrument operations.

3.5.2 SciBox Components

SciBox is composed of several software components that follow the flow of the science-acquisition process. The end-user interfaces primarily through a Visualization Tool, the base component of the SciBox utility. The remaining components include the opportunity analyzer, schedule editor, state predictor, command generator, status monitor, and mission planner. Inputs, or operational guidelines, to SciBox and its components include instrument and spacecraft operational constraints (such as instrument pointing requirements and spacecraft restrictions on pointing relative to the Sun), observation requirements (such as range to planet values and viewing geometry angles), and coverage requirements.

The Visualization Tool (Fig. 2) provides an interface for each of the planetary encounters during the mission cruise phase (Earth flyby, two Venus flybys, three Mercury flybys) and in Mercury orbit. It is based on a framework that has two display areas: the frame area, which extends along the top and the left-hand side of the window, and the workspace area, which contains the instrument application. Along the top of the frame is the time panel, which allows the user to update dynamically all views in time using the buttons provided. The upper-left area contains a three-dimensional view of the spacecraft and the target body, allowing the user to visualize how the spacecraft is oriented in relation to the planet, in this case Earth. Below this display is an area that allows the user to specify the pointing orientation of the spacecraft. To the right of these two displays is the workspace area. Displayed in Fig. 2 are the Earth flyby analyzer and Schedule Editor for the MDIS instrument. The Schedule Editor window has five display areas. At the top is a panel that allows the user to select the magnitude range and spectral types of the stars to be shown in the displays. Directly below that panel is a display that shows the Field of View (FOV) of the instrument suite against the star field and planets. The display offers the option to visualize in one of multiple planes, breaking down a complex problem by reducing the number of dimensions presented at one time. Below that display is a panel containing a Zoomed Fixed Space View, a Zoomed Instrument View, and an Instrument Controller panel. The Zoomed Fixed Space View shows a view from the spacecraft at a user-selected Right Ascension and Declination

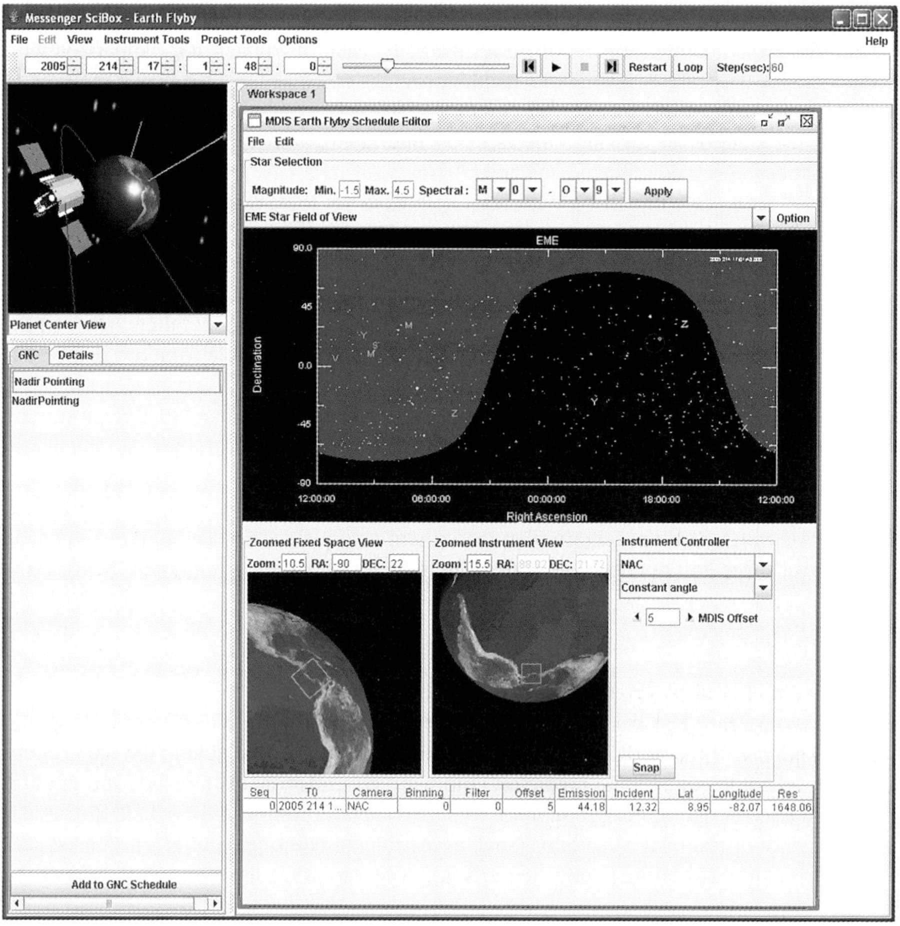

Fig. 2 SciBox provides visualizations to aid in science acquisition planning

with the instrument's FOV shown in the view. The Zoomed Instrument View shows a view from the perspective of the instrument with the FOV shown. The Instrument Controller Panel allows the user to select which of the instrument's cameras to use, what mode the camera will be in, as well as additional controls for the offset pointing of the camera. Finally, at the bottom of the window, is a table containing the task schedule for that instrument.

The instruments have multiple observation modes, such as nadir-pointing, raster scanning, and limb scanning. For each mode of every instrument SciBox provides an Opportunity Analyzer that takes into account such factors as spacecraft ephemeris data, instrument operational constraints, spacecraft operational constraints, instrument resource allocations, and scientific goals, and then provides the user with a list of observation opportunities within the planned mission lifetime. Parameters that determine opportunities can be manipulated for optimal scientific return, depending on the mode and instrument. Activities can then be selected from the list of opportunities and added to the schedule, which is displayed in the Schedule Editor.

For each instrument SciBox provides a Schedule Editor that integrates the schedules from that instrument's modes, scans for conflicts, and allows the user to select from the chosen activities. The G&C pointing schedule is also displayed at this level so the user can be aware of the spacecraft pointing, for example, when determining whether an instrument is pointing nadir to the planet. This interface provides the user with the integrated information needed to resolve conflicts and select an optimal set of activities.

Information from the Schedule Editor, in addition to spacecraft resource information (such as SSR status and available downlink bandwidth), is provided to SciBox's State Predictor component. The State Predictor gives an anticipated spacecraft resource status report (presented via the Visualization Tool interface) in areas such as available SSR space and power usage. From this predicted state, the user can then adjust observation selections, if needed, to keep the payload within its resource allocations.

Once a set of activities has been selected that fall within the instrument's resource allocations, the Command Generator component of the SciBox utility is used to convert the chosen activities into a suggested command load. Previously defined observation selections, as well as definitions for spacecraft CASs, are input into the Command Generator. The Command Generator takes the scheduled science observations or activities and generates spacecraft commands that will accomplish those observations.

The resulting command sequence is then passed into the Status Monitor, which provides a global coverage field displaying the status of all science observations, past, present, and future. The monitor displays areas on Mercury that have been analyzed, based on input from PIPE (as well as whether the data associated with the observation have been downlinked or not), areas whose observations remain to be taken, and those observations that suffered some sort of failure during data acquisition or downlink. (Observations of these areas can then be rescheduled to occur later in the mission.)

The final component of SciBox automatically optimizes the scheduled science observations. Changes, if any, are suggested in the form of updated science opportunities. The entire process is executed iteratively until no further changes are suggested; commands and schedules are then delivered as output.

3.5.3 SciBox Delivery

MESSENGER SciBox is being developed in a series of builds scheduled for completion prior to the first Mercury flyby. This development schedule affords an opportunity to perform additional testing, including verification of would-be command sequences and visualization versus actual trajectory and orientation. Several maintenance phases are scheduled beyond the development phase.

4 Data Management and Archiving

The SOC's DMA area of responsibility includes two broad tasks: (1) operations support and (2) PDS-compliant archive generation. Operations support includes providing the instrument teams with both pre-flight and in-flight support as needed, processing telemetry as received from the spacecraft, generating predefined science data products, and monitoring the data coverage to ensure that science objectives are being met. Operations support provides access to the data and data products by the Science Team and other project personnel. PDS-compliant archive generation includes the generation, validation, and delivery of the MESSENGER archive data set to the PDS.

4.1 Instrument Team Support

Instrument team support by the SOC for pre-flight operations and in-flight operations provides data ingestion capabilities, product generation capabilities, and data access. In-flight support also includes data monitoring (such as coverage, data quality evaluation, and sensor performance).

4.1.1 Pre-flight Support

Pre-flight support from the SOC was requested by and provided to the MDIS instrument team in the form of assistance during instrument calibration and instrument integration and testing. Using the PIPE software utility MDIS calibration data were automatically copied from the instrument's Optical Calibration Facility (OCF) to the remotely located MESSENGER SOC within approximately 30 seconds of acquisition. The SOC provided a Web interface to the data organized by test that included capabilities for searching the image database for particular configurations, dark correction, and de-smearing of data. PIPE also provided a variety of plotting and profiling options used for data analysis in near-real time to search for any instrument anomalies and to validate test results.

4.1.2 In-flight Support

In-flight support is being provided for all MESSENGER instruments as well as the Radio Science experiment. Data ingestion capabilities include telemetry processing and receipt of higher-level data products from the Science Team. A key component provided by in-flight support is access to the raw, formatted data from the spacecraft.

4.2 Telemetry Processing

Data are telemetered to the ground via the DSN using Consultative Committee for Space Data Systems (CCSDS) File Delivery Protocol (CFDP). Instrument and relevant spacecraft data are transferred to the MOC as frames of packetized telemetry, according to the CCSDS standard, and supplemented with metadata, such as ground receipt time. The packetized telemetry is extracted from the frames, and the image data taken by the MDIS cameras are extracted from the packets and stored by the MOC in a telemetry archive file system (Holdridge and Calloway 2007). The MOC is responsible for merging newly created packets with previously archived data. The MOC then forwards these files to the SOC for processing into EDR products, which the Science Team and other MESSENGER project members can access for calibration and data analysis. The MOC retains a copy of its telemetry so that any telemetry can be re-requested by the SOC.

Image files from the MDIS cameras include science observations and Optical Navigation (OpNav) images. The MOC sends the image files to the SOC where they are converted into PDS-compliant binary image format and Flexible Image Transport Standard (FITS) format for access by both the Science and Navigation Teams. OpNav images designated as "critical" by the Navigation Team are converted to FITS format directly by the MOC, where they are made accessible, for use by Navigation Team members. These critical OpNav images will also be made available to the Science Team through the SOC.

4.3 Data Product Generation

The data products generated for the MESSENGER mission include EDRs and Reduced Data Records (RDRs). The EDRs, formatted raw data, are produced by the SOC directly from the spacecraft telemetry. The RDRs are produced primarily by the MESSENGER Science Team and delivered to the SOC. Calibrated Data Records (CDRs), a type of RDR, are produced by the SOC using algorithms provided by the Science Team. In addition, the SOC provides ancillary data (such as formatted engineering data) that are used by the Science Team for data reduction and product generation, such as are also archived to the PDS.

4.3.1 Science Product Generation

The SOC is responsible for producing all EDRs from the spacecraft telemetry (received from the MOC) and providing access to all EDRs by the Science Team in a timely fashion. EDRs consist of unprocessed instrument-count data and may include (depending on the instrument) a description of the observation geometry (such as boresight, spacecraft, and target). In those instances where onboard compression has been applied, the EDRs contain decompressed instrument data.

The RDRs are subdivided into three groups: CDRs, derived data products (DDPs), and derived analysis products (DAPs). CDRs are processed data, derived from the EDRs, consisting of decompressed and calibrated data, where the data values are expressed in physical units. Tables 3 and 4 (taken from the MESSENGER Data Management and Science Analysis Plan) list the instruments and their associated DDPs and DAPs that the MESSENGER mission will produce and deliver to the PDS. Detailed descriptions of the content and format of these products are contained in the RDR SIS documents for each instrument, which will be archived with the PDS along with the data products.

4.3.2 SPICE Kernels

The MESSENGER project uses the Spacecraft Planet Instrument Camera-matrix Events (SPICE), NASA's Navigation and Ancillary Information Facility (NAIF) information system, to generate ancillary data needed to interpret the MESSENGER data products. SPICE consists of a standardized set of files, known as kernels, and a toolkit of applications useful in producing and manipulating the ancillary data. Table 5 summarizes the SPICE information that will be kept by the MESSENGER project.

Some of the SPICE kernels are also used by the SOC software utilities. SciBox uses the SPK files in producing its trajectory model. It also uses the content of, or information similar to that found in, the Instrument, Frames, and Planetary Constant Kernels to determine what a given instrument can "see" at any given time. PIPE uses the C kernels to create the coverage maps. The remaining kernels are made available to the Science Team for data analysis and product generation. SPICE files are also archived to the PDS as part of the project's ancillary data set.

4.4 Data Monitoring

Data monitoring is needed for the SPG to perform data validation and science objective assessment. Data monitoring includes maintaining coverage maps, evaluating data quality, and monitoring instrument performance.

As data are telemetered to the ground and received by the SOC, the PIPE software creates coverage maps on an instrument-by-instrument basis. These maps outline the current

Table 3 Derived data products

Instrument	Data product	Description
MDIS	Cataloged images	A viewable catalog for browsing through the image data base
GRNS	γ-ray spectra, neutron flux	A library of γ-ray spectra and neutron flux measurements correlated to the planet surface
XRS	X-ray spectra	A library of X-ray spectra correlated to the planet surface
MAG	**B**-field vectors	A library of magnetic field vectors
MLA	Range profiles, radiometry	A library of range profiles and associated radiometry information
MASCS/UVVS	Limb tangent height spectra	A library of emission spectra associated with position above the planet surface
MASCS/VIRS	Surface reflectance spectra	A library of reflectance spectra associated with position on the planet surface
EPPS	Particle energy vs. composition and angle distribution	A tabular or graphical library of measured particle energy as a function of composition and angular distribution
RS	Doppler data, ranging data, occultation times	A library of Doppler, ranging, and occultation information as a function of date of acquisition

observation coverage in relationship to the surface of Mercury. The SciBox software then uses these maps to overlay locations on the surface that have been observed and for which the associated data remain to be downlinked from the SSR. These maps do not include data that have been received but were determined to be of insufficient quality to meet the science goals.

The MESSENGER Science Team, in conjunction with the Science Steering Committee, provides guidelines to the instrument teams for evaluating the quality of data and determining whether it meets the mission's science objectives. Data quality flags are then applied (via PIPE) to the observation data by the instrument teams. This information is passed to the SciBox planning utility to indicate to the planning tool which observations need to be taken again. These quality flags will also be included as part of the archive to the PDS. Because of the large data volume, data quality assessment is performed by an automated process that examines basic statistical information and provides an indicator for items that may need more detailed examination by instrument team personnel.

In addition to science observations, the instrument teams include in their command loads health and safety operations, along with calibration operations. The data from these operations are used to monitor instrument performance. The SOC is responsible for processing these data sets in the same manner as the science data.

Table 4 Derived analysis products

Instrument	Analysis product
GRNS, XRS	Global element map
MDIS, MASCS/VIRS	Spectral unit map
MDIS	Global monochrome map
MDIS	Stereo maps
MDIS	Multispectral image catalogue
MLA	Northern hemisphere topography map
MLA	Altimetric profiles
RS	Northern hemisphere gravity model
MAG	Multipole internal magnetic field model
MAG, EPPS	Time-dependent magnetosphere model
MLA, RS	Libration amplitude
MLA, RS	Right ascension and declination of Mercury's rotational pole
RS	Spherical harmonic gravity field
RS	Low-degree global shape
MASCS/UVVS	Exosphere model
MASCS/UVVS, EPPS, GRNS, XRS	Volatile species and sources

4.5 Data Archiving

All NASA missions are required to archive their scientific observations and associated data with the PDS. The PDS archives the data in self-described data volumes in a nonproprietary format, to provide open access to all observations for use beyond the life of the mission. Supporting documentation and ancillary data (such as SPICE and engineering data) are, therefore, submitted as part of the PDS delivery. Table 6 lists the PDS nodes to which archive deliveries will be made. Table 7 lists an estimate of total data volume to be delivered to each of the PDS nodes.

The MESSENGER project, working in partnership with the PDS, has created a Data Archive Working Group (DAWG) to facilitate the archival process. EDR and RDR data formats are pre-defined by the Science Team as described in each instrument's SIS document, which is reviewed and approved by both the MESSENGER project and the PDS. This agreement allows the archive generation to be automated and constructed in parallel with receipt of the data within the SOC for Science Team usage. Table 8 shows the data release schedule. Following the first and second Mercury flybys the MESSENGER project will deliver all information required to convert EDRs into CDRs so that the PDS archive could support a data analysis program. Should resources permit, following these flybys actual CDRs will also be delivered in conjunction with the EDRs and calibration information.

4.6 PIPE Architecture

The Planetary Information Processing Environment, or PIPE, is an integrated suite of software tools and applications, most of which are built upon the core MSHELL engine. MSHELL is the name for the proprietary scripting language and software engine developed by ACT for use in image and data processing. PIPE also includes Hypertext Markup

Table 5 SPICE summary

Data File/Kernel type	Source	Description	Generation schedule
Spacecraft and Planetary Kernels (SPKs): Ephemeris			
Planetary SPK	Navigation Team with NAIF support	Full-mission file that contains positions for Sun, Mercury, Venus, Earth, Moon, planetary barycenters	As needed
Merged Spacecraft Reference Trajectory SPK	Navigation Team with NAIF support	Latest official full-mission merged reference trajectory that contains only the spacecraft ephemeris; created by merging a series of reconstructions and short-term predictions onto the modeled whole-tour SPK	2–3 days after each Navigation Team release
Reference SPK	Navigation Team	Spacecraft trajectory information	Part of each Navigation Team release
C-Kernels: Attitude			
Spacecraft High Rate Daily CKs	Mission Operations	Daily spacecraft attitude kernels generated by mission operations at variable sampling resolutions depending on downlink availability and spacecraft configuration (can be up to 1 Hz)	Weekly
MDIS Attitude History CK	SOC	1-Hz spacecraft attitude data bracketing all MDIS observations. This kernel is generated from APID 0x191 (or 0x1B1) data packets. It represents the same data as the Spacecraft High Rate Daily CKs, but at a different sampling frequency, and it falls under MDIS SSR and downlink control	As needed
MDIS Pivot Platform CK	SOC	MDIS pivot platform CK produced from each image header	As needed
Frame Kernels: Reference frames			
Reference Frame FKs	SOC	Reference Frame Specifications	As needed

Table 5 (*Continued*)

Data File/Kernel type	Source	Description	Generation schedule
Orbiter and Instruments FK	SOC	MESSENGER Orbiter and Instruments' frames	Upon instrument team's review or the addition of new frames
Dynamics FK: Ephemeris-based frames	SOC	Ephemeris-based frames	When new dynamics frames are requested by instrument teams
Instrument Kernels: Instrument Field-of-Views (FOVs)			
EPPS IK	SOC	EPPS instrument field of view size, shape, and orientation	As needed
GRNS IK	SOC	GRNS instrument field of view size, shape, and orientation	As needed
MAG IK	SOC	MAG instrument field of view size, shape, and orientation	As needed
MASCS IK	SOC	MASCS instrument field of view size, shape, and orientation	As needed
MDIS IK	SOC	MDIS instrument field of view size, shape, and orientation	As needed
MLA IK	SOC	MLA instrument field of view size, shape, and orientation	As needed
Spacecraft Clock Kernel: Spacecraft Clock			
Spacecraft Clock SCLK		Spacecraft clock coefficients	As needed
Planetary Kernels: Planetary Constants			
Planetary Constants PcK		Target body size, shape, and orientation	As needed
Leapseconds Kernel: Leapseconds			
Leapseconds LSK		Leapsecond tabulations	As needed

Language (HTML) query pages and Java applications, which provide the front-end user interface to the PIPE software through which data can be requested, viewed, and analyzed. PIPE uses a Commercial Off-the-Shelf (COTS) RDBMS to store configuration profiles for generated data products as well as storing metadata, information that can be queried by the Science Team to locate products of interest quickly.

Table 6 Destination PDS nodes

Data set	Destination PDS node
MDIS	Imaging
GRNS	Geosciences
XRS	Geosciences
MAG	Planetary Plasma Interactions (PPI)
MLA	Geosciences
MASCS	Atmospheres
EPPS	PPI
RS	Geosciences
SPICE Ancillary Data	Navigation and Ancillary Information Facility (NAIF)

Table 7 Data volumes by PDS node

PDS node	Estimated data volume (GB)
Atmospheres	41
Geosciences	79
Imaging	391
NAIF	6
PPI	47

PIPE has been configured to fulfill the data management and archiving requirements for the SOC. These requirements include the processing of all science-related data products, access to these data products, data coverage monitoring, and delivery of PDS-compliant data archives of MESSENGER science observations.

4.6.1 PIPE Capabilities

Before mission customization, PIPE provides baseline capabilities, including automatic assimilation and processing of spatial data; cataloging, integration, and distribution of data; flexible configuration from a single machine to a cluster of machines; easy incorporation of new data formats for ingestion and delivery; Web-based graphical user interfaces and image-processing functionality; and extendibility via user-provided algorithms and applications, with only a standard Web browser required by the user.

In support of the MESSENGER mission, PIPE has been customized to meet the following requirements:

- Ingest and catalog data in the SOC, including telemetry packets from the MOC (used to generate EDRs) as well as the RDR data products produced by the instrument teams.
- Generate data products in a timely manner. For PIPE this is a data-driven process, where data products are created upon receipt of the source data for a given data product (for example, telemetry packets which are used to create EDRs). PIPE is also responsible for the generation of data product archives for submission to the PDS at several stages of the mission (Table 6).
- Provide interfaces for local and remote access to data products, inclusion of Science Team-provided algorithms, and monitoring of data product generation.

Table 8 PDS delivery schedule

Mission phase	Dates	Deliverables	Release date
Launch	August 3, 2004	On-ground calibration data	February 2006
Earth–Moon flyby	August 2, 2005	EDRs of all flight data	December 5, 2007 (VF2 + 6 months)
Venus flyby 1 (VF1)	October 24, 2006	EDRs of all flight data	December 5, 2007 (VF2 + 6 months)
Venus flyby 2 (VF2)	June 5, 2007	EDRs of all flight data	December 5, 2007 (VF2 + 6 months)
Mercury flyby 1 (M1)	January 14, 2008	EDRs of all flight data	July 14, 2008 (M1 + 6 months)
		CDRs for MAG and calibration documentation for MDIS and MASCS	July 14, 2008 (M1 + 6 months)
Mercury flyby 2 (M2)	October 6, 2008	EDRs of all flight data	April 6, 2009 (M2 + 6 months)
		CDRs for MAG and calibration documentation for MDIS and MASCS	April 6, 2009 (M2 + 6 months)
Mercury flyby 3 (M3)	September 29, 2009	EDRs of all flight data	March 29, 2010 (M3 + 6 months)
		CDRs from all investigations	March 29, 2010 (M3 + 6 months)
Mercury orbit insertion (MOI)	March 18, 2011		
Mapping	March 18, 2011 to March 18, 2012	EDRs of all flight data	September 18, 2011 and March 18, 2012 (Every 6 months after MOI)
		CDRs from all investigations	September 18, 2011 and March 18, 2012 (Every 6 months after MOI)
End of orbital operations (EOO)	March 18, 2012	EDRs of all flight data	September 18, 2012 (EOO + 6 months)
		CDRs from all investigations	September 18, 2012 (EOO + 6 months)
		DDPs and DAPs from all investigations	March 18, 2013 (EOO + 12 months)

Examining the individual components that compose the PIPE system offers insight into how these requirements are fulfilled.

4.6.2 PIPE Components

The three main components of PIPE can be categorized as Ingestion and Product Generation, Data Storage, and Data Access. Each component relies on and provides information to the other two components. Ingestion and Product Generation registers data in Data Storage, which then permits the data to be viewed via Data Access.

Ingestion and Product Generation uses ingestion engines that ingest source data and generate products. An ingestion engine consists of the PIPE software scripts, applications, and database tables that are used to create a given data product. PIPE software scripts parse incoming data and determine if they satisfy pre-defined criteria for ingestion and product generation. Criteria include current PIPE server load and the source data requirements, such as waiting for the completion of large data file transfers or multiple file transfers. Software applications may be called by the script to decompress any compressed data and repackage binary data into a given EDR or higher-level data product format. Multiple database tables record metadata that have been extracted from source data as well as provide the configuration profiles for the ingestion engine to generate a given data product.

The ingestion engine consists of both ingestion and product generation because the PIPE process is data driven, in that it attempts to create data products upon receipt of source data by the SOC. This occurs even in the case of data products that span long periods of observation. In the case of such long-term data products, the ingestion process is still triggered by incoming data, but queries are made to the database tables to determine if sufficient data have been collected before generating a data product.

The Data Storage PIPE component consists of an RDBMS and a file system architecture working in tandem to store the data. The database stores the metadata information and tabular data, registers data products, and monitors data processing at different stages. Tabular data products, such as those from spectrometer-based instruments, are stored in separate databases; the nature of such tabular data allows them to be more fully described. Scripts are then written to access these databases and map parameters back into the default PIPE software architecture.

The file system architecture is used to store data files and recognize an incoming data stream as part of an event-driven process. For example, there are designated file paths that PIPE periodically checks to determine whether data have arrived and need to be ingested. In addition, the PIPE file system tree allows local and remote users to browse through data products at the file system level, sorting by instrument, product type, or day of year.

PIPE Data Access can occur through a variety of interfaces depending on the level and type of data access desired. The two main user interface access categories are called Coverage-Level and Data Browsing-Level Access. In addition, there is a third category of access that includes the interface between SciBox and PIPE.

Coverage-Level Access is typically used to determine the data received to date. It consists of both HTML form-based and Java-based interfaces. The HTML form-based interface allows the user to filter the list of desired products by multiple criteria; see, for example, Fig. 3. This capability is especially useful for spectrometer-based instruments where the desired data may be filtered by geometry, time, and instrument parameters. The Java-based interface allows the user to "rubber-band" a region of interest on the interface itself as well as the integration of multiple data products into a coverage image mosaic. One example of its use is a query for multiple instrument observations (such as from MDIS and MASCS) over the same location on Mercury.

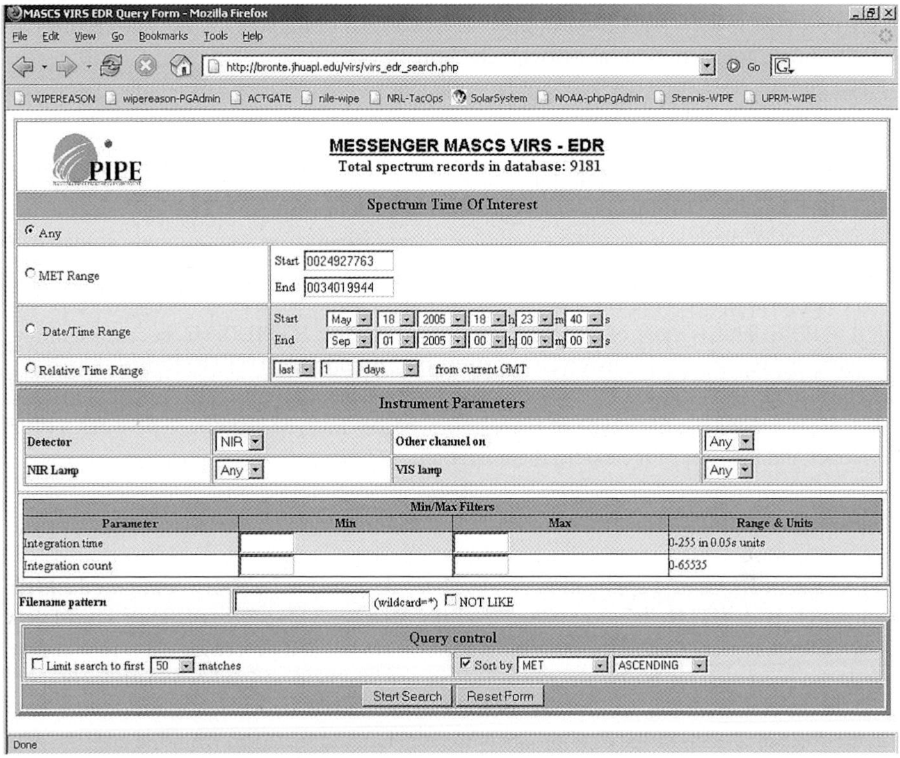

Fig. 3 An example of a MESSENGER Data Coverage HTML Web page

Both types of Coverage-Level access will allow a user to request a list of query-matched data products, thumbnail images of the data products, various download and packaging options, and additional options that may be added per Science Team requests. For example, the Science Team may request the integration of analysis tools or applications of their algorithms as one of the results of the Coverage-Level Access interface. The results of a Coverage-Level Access interaction may also be sent by the user to a Data-Level Access interface to allow analysis directly on the data itself.

Data Browsing-Level access is also subdivided into an HTML form-based or Java-based interface. This access type is at the level where the user can manipulate the data product, perform analysis, or request a download of the data product for analysis on his or her local machine. The default PIPE Data Browsing Java interface contains menu options that allow the user to perform basic plots or determine statistical information, such as mean and standard deviation. The Science Team may request that additional analysis algorithms be applied to the data product and be made available via custom menu options. Working Data Browser interface examples may be found at http://solarsystem.wipecentral.net.

External applications may access PIPE via two methods. The first is via an HTML and Extensible Markup Language (XML) interface and the second is through the use of custom HTML forms. The HTML/XML interface requires a predefined XML schema that the external application must follow in order to request the appropriate level of data products. An XML schema already exists for external applications to request data product information at the coverage level or data browsing level. A working example of XML queries may be found at http://goes.wipecentral.net/wipexml.htm.

The current XML schema may not be sufficient to filter requests of tabular-based data products however, such as those generated from the MESSENGER spectrometer instruments. Additional XML schema to handle this and other external application queries will be defined over the course of the mission. The results of the XML query will be an XML document that contains information in various XML tags. One or more XML tags may contain pointers to Web directories that the external application can parse and use to download data products.

HTML form access requires that the external application build an HTML form and submit it to PIPE. This is more of a custom level of access than HTML/XML because it allows the query and result to be in any given hypertext transfer protocol (HTTP) format. For example, the result could be an HTTP data stream that allows the external application to download a data file immediately. Each type of HTML form access must be negotiated between a PIPE developer and the user of the external application.

4.6.3 PIPE Customization

PIPE is a network-centric application that has been designed to be easily configured and reconfigured via an HTML interface called the PIPE DataBase Manager (PDBM). The PDBM allows administrators to control how PIPE locates, stores, and processes data. In addition, a database is created for every MESSENGER instrument, and within each instrument database are the configuration profiles for each data product. These configuration profiles control the format of each of the data products to be generated by PIPE, from the specific data parameters to the data type (e.g., integer versus floating point) to the byte size of each data parameter. This flexibility allows PIPE to incorporate changes to the predefined data formats that may be needed during the course of the mission.

PIPE has also been designed to be a modular system. Additional computers may be added to an existing PIPE system in order to increase processing power or increase file storage space. In addition, the database server, Web server, and main PIPE processing duties may be distributed among multiple computers.

5 Facility Configuration and Standards

The SOC facility is physically located on the APL campus in Laurel, MD. It is comprised of servers, workstations, and supporting peripheral devices. The facility is available to all project members.

The PIPE software, which supports the SOC's DMA tasks, is hosted on a Windows server running a Microsoft SQLServer RDBMS. Data products are stored on a Linux-based server. The Science Team and their associates may access PIPE software either locally or remotely.

The science planning and commanding software, SciBox, is served from a Windows workstation and may be accessed by the Science Team and their associates either via a Web interface or from a compact disk in a stand-alone mode. SciBox is a Java-based application that can be operated on any system supporting Java 1.5 or higher, with Java3D.

The long duration of the MESSENGER mission will necessitate system upgrades, so the current configuration will undoubtedly change throughout operations.

6 Summary

Science acquisition, data management, and data archive responsibilities for the MESSENGER mission reside with the MESSENGER Science Operations Center. The SOC is an integrated set of personnel and subsystems responsible for planning and commanding instrument operations; for processing, distributing and providing access to the acquired data set; for validating, monitoring, and assessing the data set in terms of meeting the mission's science objectives; and for archiving the mission's science products with the PDS. The SOC accomplishes these tasks in partnership with ACT and through the functions of its SPG by the use of carefully designed and versatile software utilities (SciBox and PIPE) with a heritage of space mission applications.

References

R.E. Gold et al., Planet. Space Sci. **49**, 1467–1479 (2001)

M.E. Holdridge, A.B. Calloway, Space Sci. Rev. (2007, this issue). doi:10.1007/s11214-007-9261-8

J.V. McAdams, R.W. Farquhar, A.H. Taylor, B.G. Williams, Space Sci. Rev. (2007, this issue). doi:10.1007/s11214-007-9162-x

S.C. Solomon et al., Planet. Space Sci. **49**, 1445–1465 (2001)

S.C. Solomon, R.L. McNutt Jr., R.E. Gold, D.L. Domingue, Space Sci. Rev. (2007, this issue). doi:10.1007/s11214-007-9247-6